AF809769

Surveying Instruments and their Operational Principles

1 Fialovszky

Authors

Lajos Fialovszky

Formerly at the Geodetical Institute of the Budapest University of Technology
H–1111 Budapest, Műegyetem rkp. 3/9.

1., 2., 5.2.3., 6.1.2.3. and 6.1.2.4.

Gyula Holéczy

Geodetical Institute of the Budapest University of Technology
H–1111 Budapest, Műegyetem rkp. 3/9.

4., 5., 6. (with exception of 5.2.3., 6.1.2.3. and 6.1.2.4.)

András Krauter

Geodetical Institute of the Budapest University of Technology
H–1111 Budapest, Műegyetem rkp. 3/9.

8., 9., 10.

Tibor Lukács

Institute of Geodesy, Cartography and Remote Sensing
H–1051 Budapest, Guszev u. 19.

12.1., 13.1., 14.

Péter Májay

Geodetical Institute of the Budapest University of Technology
H–1111 Budapest, Műegyetem rkp. 3/9.

3.

L. Tamás Sándor

H–1021 Budapest, Kuruclesi út 45/a.

13.2., 13.3.

Mrs. Ilona Staudinger

Institute of Geodesy and Cartography
H–1149 Budapest, Bosnyák tér 5.

11., 12.2.

Péter Valkó

Formerly at the Department of Electronic Instruments of the Budapest University of Technology
H–1521 Budapest, Goldmann Gy. tér 3.

7.

Reviewers

L. Fialovszky (4—6., 8—12., 14.)
Gy. Holéczy (5.2.3., 6.1.2.3., 6.1.2.4.)
I. Joó (2.)
I. Kardos (13.)
A. Krauter (7.)
A. Lisziewicz (1.)
L. Miskolczy (3.)

Developments in Geotechnical Engineering, 62

Surveying Instruments and their Operational Principles

Editor-in-Chief

Lajos Fialovszky

Associate Editor

Gyula Holéczy

Elsevier
Amsterdam–Oxford–New York–Tokyo 1991

This book is the revised and enlarged version of
GEODÉZIAI MŰSZEREK, edited by L. Fialovszky
Műszaki Könyvkiadó, Budapest

Translation by
Gy. Deme (1.)
J. Verő (2.3., 2.4., 2.5., 11., 12., 13., 14.2.)
Á. Wallner (2.1., 2.2., 3., 4., 5., 6., 7., 8., 9., 10., 14.1.)

General revision of the English translation by
Professor *Alwyn Robbins,* Oxford

Joint edition published by
Elsevier Science Publishers B.V., Amsterdam, The Netherlands
and
Akadémiai Kiadó és Nyomda Vállalat, Budapest, Hungary

The distribution of this book is being handled by the following publishers

for the U.S.A. and Canada
Elsevier Science Publishing Co., Inc.
655 Avenue of the Americas,
New York, New York 10010, U.S.A.

for the European countries, Democratic People's Republic of Korea, Republic of Cuba, Socialist Republic of Vietnam and People's Republic of Mongolia
Kultúra Hungarian Foreign Trading Co., P.O.Box 149, H-1389
Budapest, Hungary

for all remaining areas
Elsevier Science Publishers
Sara Burgerhartstraat 25,
P.O.Box 211, 1000 AE Amsterdam, The Netherlands

Library of Congress Cataloging in Publication Data

Geodéziai műszerek. English.
Surveying instruments and their operational principles / edited by Lajos Fialovszky; [translation by Gy. Deme, J. Verő, Á. Wallner; general revision of the English translation by Alwyn Robbins].
p. cm.
Rev. and enlarged translation of: Geodéziai műszerek.
Includes bibliographical references.
ISBN 0-444-98829-7 (U.S.)
1.Geodesy–Instruments. 2. Surveying–Instruments. 3. Optical instruments. I. Fialovszky, Lajos. II. Title.
QB328.A1G4613 1990
526'. 1'028–dc20 89-17035
CIP

ISBN 0-444-98829-7 (Vol. 62)

Printed in Hungary by Akadémiai Nyomda, Budapest
© L. Fialovszky–Gy. Holéczy, 1991
© English translation – Gy. Deme–J. Verő–Á. Wallner, 1991

All rights reserved. No part of this publication may be reproduced by any means, or transmitted or translated into machine language without the written permission of the copyright owner.

Further titles in this series

1. G. Sanglerat — The Penetrometer and Soil Exploration
2. Q. Záruba and V. Mencl — Landslides and their Control
3. E. E. Wahlstrom — Tunneling in Rock
4. R. Silvester — Coastal Engineering, 1 and 2
5. R. N. Yong and B. P. Warkentin — Soil Properties and Behaviour
6. E. E. Wahlstrom — Dams, Dam Foundations, and Reservoir Sites
7. W. F. Chen — Limit Analysis and Soil Plasticity
8. L. N. Person — Rock Dynamics and Geophysical Exploration
   Introduction to Stress Waves in Rocks
9. M. D. Gidigasu — Laterite Soil Engineering
10. Q. Záruba and V. Mencl — Engineering Geology
11. H. K. Gupta and B. K. Rastogi — Dams and Earthquakes
12. F. H. Chen — Foundations on Expansive Soils
13. L. Hobst and J. Zajíc — Anchoring in Rock
14. B. Voight (Editor) — Rockslides and Avalanches, 1 and 2
15. C. Lomnitz and E. Rosenblueth (Editor) — Seismic Risk and Engineering Decisions
16. C. A Baar — Applied Salt-Rock Mechanics, 1
    The In-Situ Behavior of Salt Rocks
17. A. P. S. Selvadurai — Elastic Analysis of Soil-Foundation Interaction
18. J. Feda — Stress in Subsoil and Methods of Final Settlement Calculation
19. A. Kézdi — Stabilized Earth Roads
20. E. W. Brand and R. P. Brenner (Editors) — Soft-Clay Engineering
21. A. Myslivec and Z. Kysela — The Bearing Capacity of Building Foundations
22. R. N. Chowdhury — Slope Analysis
23. P. Bruun — Stability of Tidal Inlets
    Theory and Engineering
24. Z. Bažant — Methods of Foundation Engineering
25. A. Kézdi — Soil Physics
    Selected Topics
26. H. L. Jessberger (Editor) — Ground Freezing
27. D. Stephenson — Rockfill in Hydraulic Engineering
28. P. E. Frivik, N. Janbu, R. Saetersdal and L. I. Finborud (Editors) — Ground Freezing 1980
29. P. Peter — Canals and River Levées
30. J. Feda — Mechanics of Particulate Materials
    The Principles
31. Q. Záruba and V. Mencl — Landslides and their Control
    Second, completely revised edition
32. I. W. Farmer (Editor) — Strata Mechanics
33. L. Hobst and J. Zajíc — Anchoring in Rock and Soil
    Second, completely revised edition
34. G. Sanglerat, G. Olivari and B. Cambou — Practical Problems in Soil Mechanics and Foundation Engineering, 1 and 2
35. L. Rétháti — Groundwater in Civil Engineering
36. S. S. Vyalov — Rheological Fundamentals of Soil Mechanics
37. P. Bruun (Editor) — Design and Construction of Mounds for Breakwaters and Coastal Protection
38. W. K. Chen and G. Y. Baladi — Soil Plasticity
    Theory and Implementation
39. E. T. Hanrahan — The Geotechnics of Real Materials
    The ε_g, ε_k Method

40. J. Aldorf and K. Exner — Mine Openings
Stability and Support
41. J. E. Gillott — Clay in Engineering Geology
42. A. S. Cakmak (Editor) — Soil Dynamics and Liquefaction
43. A. S. Cakmak (Editor) — Soil-Structure Interaction
44. A. S. Cakmak (Editor) — Ground Motion and Engineering Seismology
45. A. S. Cakmak (Editor) — Structures, Underground Structures, Dams and Stochastic Methods
46. L. Réthâti — Probabilistic Solutions in Geotechnics
47. B. M. Das — Theoretical Foundation Engineering
48. W. Derski, R. Izbicki, I. Kisiel and Z. Mroz — Rock and Soil Mechanics
49. T. Ariman, H. Hamada, A. C. Singhal, M. A. Haroun and A. S. Cakmak (Editors) — Recent Advances in Lifeline Earthquake Engineering
50. B. M. Das — Earth Anchors
51. K. Thiel — Rock Mechanics in Hydroengineering
52. W. F. Chen and X. L. Liu — Limit Analysis in Soil Mechanics
Second, completely revised edition
53. W. F. Chen and E. Mizuno — Nonlinear Analysis in Soil Mechanics
54. F. H. Chen — Foundations on Expansive Soils
Second, completely revised edition
55. J. Verfel — Rock Grouting and Diaphragm Wall Construction
56. B. N. Whittaker and D. J. Reddish — Subsidence
Occurrence, Prediction and Control
57. E. Nonveiller — Grouting
Theory and Practice
58. V. Kolár and I. Nemec — Modelling of Soil-Structure Interaction
59A. R. S. Sinha — Underground Structures
Design and Instrumentation
59B. R. S. Sinha and L. Ozdemir — Underground Structures
Instrumentation and Constructions
60. R. L. Harlan, K. E. Kolm and E. D. Gutentag — Water-Well Design and Construction
61. I. Kazda — Finite Element Techniques in Groundwater Flow Studies
with Applications in Hydraulic and Geotechnical Engineering
62. L. Fialovszky — Surveying Instruments and their Operational Principles

Professor Lajos Fialovszky
1898–1987

Professor Fialovszky died two weeks after the final version of this manuscript was received by the publishers.

Professor Lajos Fialovszky died in Budapest on 17th March 1987, in his 89th year. He was born on 18th November 1898 in Nagybecskerek, Hungary (now part of Yugoslavia), the fourth of six children. His father was a surgeon and specialist in internal medicine. He took part in the First World War as an officer in the Hungarian Army, and he was awarded several honours for his exemplary service. After the war he was employed as an official in charge of ballistics in the Institute of Military Engineering, Budapest, and in 1932 earned a diploma in engineering at the Budapest University of Technology.

In the late thirties he served as a field officer in warfare technology. First he was a department manager in the Ministry of Industrial Affairs, later becoming director of the Precision Machine Tool Factory. After Hungary entered the war allied with Germany, his opposition to the official pro-Nazi policy resulted in him spending 11 months in a military prison for political offenders.

In 1945 his service with the newly formed Hungarian People's Army began, and after his official rehabilitation he was promoted to colonel, and soon appointed deputy commander in the Institute of Military Engineering. While this represented the climax of a military career spanning 30 years, after which many others may have chosen an honorable retirement, he began at the age of 50 his second career, this time in the sphere of civil life.

From 1949 he directed a research group as a constructive engineer, first at the Gamma Optical Works, and then at the Hungarian Optical Works (MOM, Budapest) where he successfully initiated the re-start and modernization of design and manufacture of geodetic instruments in Hungary. Several geodetic instruments of his design were awarded patent rights in both Hungary and abroad. His work and life was interrupted in 1957 after the failure of the People's Uprising, when he spent five months in prison. Except for this, for over 20 years, until his retirement, he played an important role in the planning and production of Hungarian surveying instruments, in addition to which he was responsible for the introduction of studies in geodetic instrument science in university departments of Geometer Engineering.

In 1951 Lajos Fialovszky received the degree of Technical Doctor in physics, geodesy and mathematics at the Budapest University of Technology for his work on "Mean errors of point connections, sensitivity and light-loss of optical systems".

In 1955 he obtained the degree of "Candidate of Technical Sciences" in surveying instruments theory from the Hungarian Academy of Sciences. The subject of his dissertation was "Optimal planning of anallactical telescopes". His rapid progress in science resulted in a successful defence of his thesis "Automatic planning of optical systems" for the academic degree of Doctor of Technical Sciences (the highest scientific degree in Hungary) in geodesy and optics in 1961. In 1968, his scientific — and third — career of some 20 years was crowned with the award of the title of university professor from the Budapest University of Technology.

At the age of 72, despite being formally retired, he continued his active scientific work in the Geodetical Institute of the Budapest University of Technology and at the Department of Astronomy of the Eötvös Loránd University. He contributed to the international research effort in determining more precisely the gravity constant*, and his collaboration with Hungarian and Soviet scientists was widely acknowledged. His broad intellectual ability was demonstrated by the fact that, at the age of 85, he began work in the field of space research. His last essay, "Relative motion of a free masspoint in a spacecraft revolving around the Earth", was published in Hungarian and English**, as well as in Russian. In appreciation of his outstanding life's work he was awarded a gold medal by the Government — on the recommendation of the Hungarian Academy of Sciences — on the occasion of his 85th birthday.

Professor Lajos Fialovszky was a member of the Presidium of the Hungarian Society for Geodesy, the Hungarian Society for Optics, Precision Engineering and Acoustics, the Hungarian "Eötvös Loránd" Society of Physics, and associate member of the Polish Geodetic Society.

His intellectual legacy is in all 73 scientific works — among these books — in optics, precision engineering, geodesy, geophysics and astronomy, published in Hungary, in Germany, in England, in the USA, in Poland and in the Soviet Union. His surviving family includes his wife, two children and five grandchildren.

The preparation of this English edition of his book was his very last project.

* See, e.g., L. Fialovszky: Nonlinear effects in the Eötvös method of the determination of the gravity constant. In: *The constant of gravitation,* Akadémiai Kiadó, Budapest, 1979, pp. 221–245.

** L. Fialovszky, Acta Geod., Geophys. et Mont. (Akadémiai Kiadó, Budapest), 18/4, 1983.

Contents

Preface

The development of surveying and geodesy as one of the most ancient branches of Earth sciences is closely connected with the development of surveying instruments. It is correct to state — especially in recent decades — that the appearance of new surveying instruments and the development of earlier types influenced and continues to influence the ever increasing tasks of surveying and the development of methods and procedures used.

The reciprocal statement is also true. The rapid development of the natural sciences, especially of physics and of technology, continuously gives rise to new demands, tasks and requirements for surveying. In order to solve these problems, new instruments utilizing recent developments in the natural sciences are needed, as well as a general updating of existing instruments. This situation can be summarized by stating that the theory and practice of the science of surveying on the one hand, and surveying instruments on the other, are in continuous interaction with each other.

The consequence of the ever increasing range of surveying instruments has been that the widening field of knowledge in this area has been separated from the main body of surveying, in the strictest sense, at a great number of universities and high schools, and surveying instruments has been introduced as an independent discipline.

The first edition of this book in Hungarian in 1979 met with great interest among Hungarian surveyors, going out of print within a few weeks. Interest in the publication of the book in a world-wide language was reported from other countries. This demand should be met by the present (revised) edition which takes into account both valuable comments and proposals from Hungarian readers and the most recent developments in surveying. These developments have taken place both in traditional opto-mechanical instruments and — perhaps even more importantly — in the field of electronic instruments.

Due to the wide variety of surveying instruments this book discusses only the constructional principles of instruments necessary for the most general tasks of surveying, and describes some representative types. Even within this more restricted field some focuses were found. Older types of instrument already out of use, or used only exceptionally, are described in broad outline, mentioning basic principles, methods of testing and possibilities for their modernization. New categories of instruments (gyroscopes, compensation levelling instruments, electronic theodolites, distance meters and tacheometers) are dealt with in more detail. Care has been taken to include the automatization which is rapidly spreading in all fields of instrument manufacture, being necessary to reduce manual operations. This automatization appears in field work, in recording and interpreting measurement results and in map construction.

The book has three parts. Part I (Chapters 1 to 6) deals with the principles and constructional details of the different types of opto-mechanical field surveying instruments (angle measuring instruments, height measuring instruments, precise projection instruments, distance meters and tacheometers), and it also gives methods for testing them. The base for this part is Chapter 1, dealing with principles of optics and optical elements and including also elements of lasers. Part II (Chapters 7 to 10) describes electronic surveying instruments which are in rapid development. This part is based on Chapter 7, which deals with basic principles and the elements of electronics. Part III (Chapters 11 to 14) describes instruments and tools for the processing of field measurements.

The authors of the book are specialists in different fields of surveying instruments. Their aim was to ensure the most correct treatment possible of the wide range of knowledge presented.

The aim of the present volume is to describe and discuss instruments for surveying, their constructional and operational characteristics and their evaluation and testing, firstly for builders, architects and technicians, but further for teachers of surveying in order to help them to solve surveying tasks, to choose the correct types of instrument and to understand their proper use. In addition to the above, however, as a result of long experience in the use of surveying instruments, the authors endeavour to express ideas about the further development and trends to be followed in this area, thus to assist the development and construction of different kinds of instruments. We believe that these ideas will also be useful as a guide to the professional manufacturers.

We have to thank the Publishing House and Printing Office of the Hungarian Academy of Sciences which took over the task of publication on realizing the need to have the present volume in English.

Last but not least, we would like to thank to those firms who kindly sent us valuable photos and other materials for the book.

L. Fialovszky

Part I

Opto-mechanical Surveying Instruments

1 Fundamentals and Elements of Optics; Levels

1.1 Fundamentals of Physical Optics

Optical radiation is a part of the electromagnetic spectrum with wavelengths from 1 pm $=10^{-12}$ m to 0.3 mm. *Natural light* is that part of the optical radiation which stimulates certain cells in our eyes (with wavelength from 380 to 770 nm). The optical radiation in the wavelength below 380 nm is called *ultraviolet* radiation while that above 770 nm is called *infrared* radiation. Optical radiation has dual characteristics: it is comprised of discrete units called photons and at the same time it has a wave-like behaviour. Some phenomena, e.g. light emission, light absorption and photoelectricity, can be explained as evidence for particle theory; others, e.g. reflection, refraction interference, diffraction and polarization, illustrate the wave theory of light.

1.1.1 Light Emission

Within the atoms of a light source, when the electrons have been stimulated by external energy, they fall back to their previous levels and radiate their free energy in the form of particles or photons. The loss of energy can occur spontaneously (*spontaneous emission*) or due to conditions established artificially (*stimulated emission*). Natural light sources as well as incandescent lamps and arc lights radiate in the form of spontaneous emission. The spectrum of their light corresponds to their composition and their light intensity varies according to the wavelength. Fluorescent lamps, however, radiate light in discrete wavelengths corresponding to the transitions of the stimulated gas atoms and there are lines in its spectrum. Each spectrum line is coherent to a certain approximation (coherent light has a definite wavelength and phase). Stimulated emission provides coherence of the highest degree. If nearly all the atoms of an appropriate optical body are stimulated simultaneously (*inverse population*) and regain their energy state in coincidence, a *laser beam* is generated. The laser beam can be focussed easily, therefore may have great intensity.

In vacuo the velocity of propagation of light is: $c_0 = 2.9979\,245\,87 \times 10^8 \pm 3 \times 10^{-1}$ m/s. Transparent materials (called *optical media*) have various *optical densities* which are characterized by the refractive index n of the medium. In a medium represented by the refractive index n the velocity of propagation of light is $c_n = c_0/n$. The refractive index of a medium depends on the wavelength of the light, but to a lesser degree it varies with temperature as well as with pressure. This is especially true for gases and fluids. The dependence of the refractive index on wavelength is represented by the colour *dispersion* of a medium. The rate of colour dispersion is expressed by the *Abbe number* of the medium:

$$\nu = \frac{n_d - 1}{n_F - n_c}, \tag{1.1}$$

where n_d, n_F and n_c are the refractive indices of the medium which refer to the *Fraunhofer lines* of the respective wavelength in the solar spectrum: d is the line of yellow helium with wavelength 587.3 nm; F is the line of bluish-green hydrogen with wavelength 286.1 nm; c is the line of red hydrogen with wavelength 656.3 nm.

1.1.2 Light Absorption

The intensity A of the light passing through an optical medium decreases according to the formula:

$$A = A_0 \exp(-\alpha s), \tag{1.2}$$

where A_0 is the intensity of the incident light, α is the attenuation factor, s is the path covered.

The light absorption of optical media (e.g. air, glass, etc.) depends on the wavelength of the light, varying as a function of the wavelength. Light absorption harms length measurement because visible and optically measurable distances are shortened by it.

There is a special case of light absorption, when an absorbing medium emits electrons of quantity proportional to the light energy absorbed. By means of this phenomenon light can be transformed into electrical signals (see Chapter 7).

1.1.3 Light Interference

The interaction of oscillations (waves) of equal wavelength is called interference. Two interfering waves amplify each other if they meet with equal phase and extinguish each other if they meet with a phase difference of 180°. In other words, if the difference between two light paths is $k \cdot \lambda$, where k is an integer, λ is the wavelength of the light, the two waves amplify each other and, if the difference is $(2k+1)\lambda/2$, they extinguish each other. Light interference occurs only if coherent waves meet.

The phenomenon of interference is used for several practical purposes, e.g., for distance measurement (see Chapter 8), but it is also used for comparing two optical surfaces which have nearly equal curvatures. If light rays meet two continuous spherical surfaces A and B whose radii of curvature differ slightly (*Fig. 1.1*), the interaction of the light rays reflected from the two surfaces causes the phenomenon of interference. This is manifested by interference lines, also called *Newton's rings* (i_1, i_2, . . ., i_x), which can be observed from the direction of the reflected rays with the unaided eye, telescope or microscope. Points on an interference line always correspond to point pairs on the two surfaces where the surfaces are equidistant. At points on two adjacent interference lines the light path between the two surfaces differs with the wavelength λ of the light. Since, however, the light falling on surface B, and reflected, travels the distance between the surfaces twice, the distances between the surfaces along two adjacent interference lines differ by $\lambda/2$.

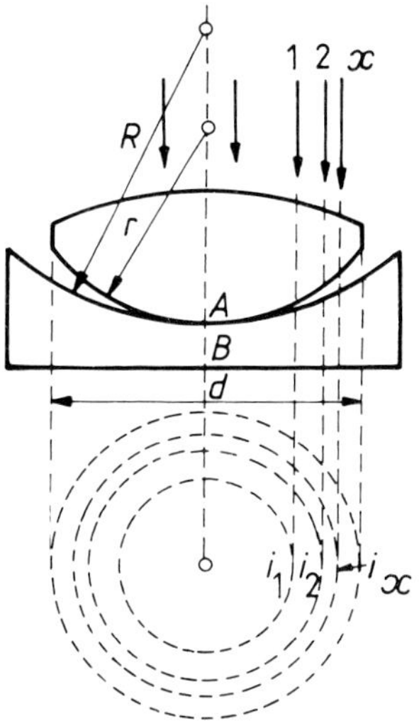

Figure 1.1. Control of the curvature of a lens surface by a test lens

In optical workshops this phenomenon is used as a measuring method for checking lens surfaces, whereby the radius of curvature r of the lens to be checked is compared with the master lens B with surface curvature of known radius R. The deviation between r and R is:

$$\Delta r = \frac{R}{1 + \dfrac{d^2}{4\varkappa\lambda R}}, \tag{1.3}$$

where d is the diameter of the lens to be tested, $\varkappa$ is the number of interference lines (Newton's rings). In order to create a specification for Δr, i.e. a tolerable deviation between the radii of curvature r and R, the optical designer prescribes the maximum number of the Newton's rings on the basis of Eq. (1.3) by:

$$\varkappa = \frac{d^2}{4\lambda r(1+\Delta r)} \Delta r. \tag{1.4}$$

In modern optical instruments, including geodetic instruments, optical surfaces are coated with a thin, reflectivity-reducing film or films. The thickness of the films lies within the order of magnitude of the wavelength of the light. At the boundary surfaces of two media, with reflectivity indices n_1 and n_2, the reflectivity number of the resultant reflection in the case of a perpendicular incidence angle is:

$$r_{1,2} = \frac{n_1 - n_2}{n_1 + n_2}. \tag{1.5}$$

The intensity of the reflected light is $r_{1,2}^2$ times the intensity of the incident light. The curve a in *Fig. 1.2* (nearly a straight line) illustrates the percentage of reflection from a glass surface with a refractive index of $n_2 = 1.52$ in the visible wavelength

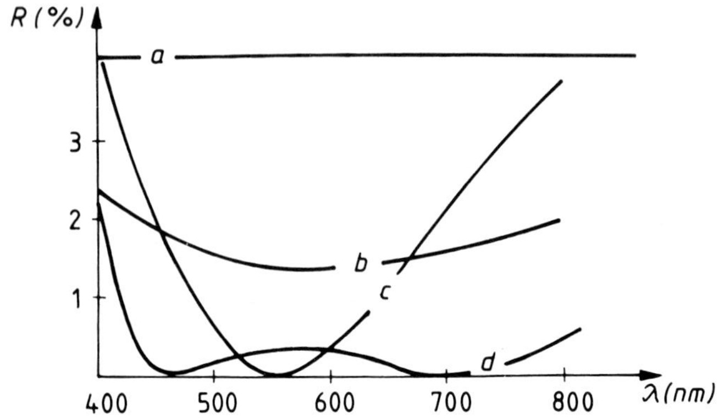

Figure 1.2. Reflection-reducing effect of a film coating

region. Curve *b* — plotted against the wavelength λ — illustrates the reflection-reducing effect of a film coating made of a material (MgF_2) of refractive index $n = 1.38$ and applied with thickness a quarter of the wavelength $\lambda = 550$ nm. A film composition providing zero reflection for a given wavelength, e.g. for $\lambda = 550$ nm can be produced by applying two different components (curve *c*). Curve *d* shows the reflection of a reflection-reducing, multi-layer system for a wide band of the spectrum, which meets the most rigorous requirements: the reflection is less than 0.4 percent in the region 420 nm $\leq \lambda \leq$ 800 nm.

Multi-ray interference is corrected by the interference filter which acts as a generator of monochromatic light: its maximum transparency lies between 20 and 70 percent.

Reflectivity-increasing film combinations are made of absorption-free compounds. These combinations consist of films with refractive indices alternately high and low, each film having the thickness of a quarter of the light wavelength. The attainable maximum reflectivity can surpass even 99.85 percent. Mirrors with such high reflectivity are used in laser techniques.

1.1.4 Diffraction

If a light beam meets the boundary between a transparent and an opaque medium, part of the light changes its propagation direction and enters the shadow area (diffraction). When passing through a narrow slit (diaphragm), parts of the light beam diffract at both edges of the aperture and different wave fronts interfere, thus generating an interference pattern consisting of several fringes. Light passing through a small circular aperture forms a ring-shaped diffraction pattern. The phenomenon of diffraction appears in all optical systems with finite diameters, it disturbs the imaging process and restricts the resolving power of the optical system (see Subsection 1.5.3). Transparent and opaque fringes, with width similar to the wavelength of the light, form the amplitude grid of diffraction. An optical diffraction grating has a wave-shaped surface of similar dimension.

1.1.5 Polarization

Light, considered as an electromagnetic wave, can be polarized. The electric and magnetic wave vectors of natural light can take any direction in the plane perpendicular to the direction of the propagation of light. If the oscillation direction of the wave vectors shows regularity, the light is polarized. Polarization can be generated by reflection or by passing through an anisotropic medium, becoming plane, elliptical or circular polarization.

Optical anisotropy (double refraction) may occur in natural materials, but it can also be brought about by electric or magnetic fields (*electro-optical* and *magneto-optical effects*). The phenomenon of polarization is utilized in the modulators of electro-optical distance meters (see Chapters 7 and 10).

1.2 Fundamental Rules of Geometrical Optics

The straight line perpendicular to the wave front of a beam of light is called the *light ray* and the rays within a spatial angle are called beam or bundle of light rays. Light travels along a straight line in homogeneous optical media. According to the concepts of geometrical optics, each light ray propagates independently without any mutual interaction.

The incident ray, the normal to the reflecting surface at the point of incidence, and the reflected ray all lie in the same plane. The incidence angle ε between the incoming ray and the normal, and the reflection angle ε' between the reflected ray and the normal, are equal in absolute value but have opposite signs (*Fig. 1.3, a*).

At the boundary surface of two media of different optical densities, having refractive indices n and n', a light ray is refracted by an amount depending on the ratio of the refractive indices (*Fig. 1.3, b*). The incident ray, the refracted ray, and the normal to the boundary surface at the point of incidence all lie in the same plane. For an

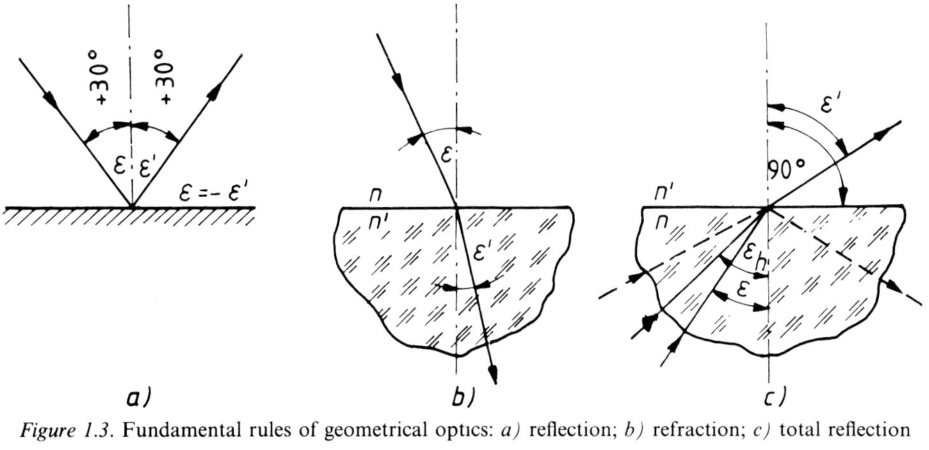

Figure 1.3. Fundamental rules of geometrical optics: *a)* reflection; *b)* refraction; *c)* total reflection

incidence angle ε and refraction angle $\varepsilon'$the *Snell–Descartes rule* is valid:

$$\frac{\sin \varepsilon}{\sin \varepsilon'} = \frac{n'}{n}. \tag{1.6}$$

The ratio between the propagation velocities of light in the two media is:

$$\frac{c_n}{c_{n'}} = \frac{n'}{n}.$$

A light ray orthogonal to a boundary surface between two media does not change its direction.

If a light ray in air ($n \cong 1$) enters glass with a refractive index $n' > 1$, then Eq. (1.6) gives:

$$\sin \varepsilon' = \frac{\sin \varepsilon}{n'}. \tag{1.7}$$

When a light ray from an optically more dense medium enters a less dense one ($n > n'$), its angle of incidence is less than its refraction angle ($\varepsilon < \varepsilon'$ *Fig. 1.3*, *c*). The incidence angle cannot be increased unrestrictedly because the refraction angle cannot exceed 90°. If the incidence angle ε exceeds the limiting angle ε_n, then total reflection occurs, and the boundary surface between the two media acts as a mirror.

In the case of total reflection, the limiting angle of incidence is:

$$\sin \varepsilon_n = \frac{n'}{n}. \tag{1.8}$$

If the light ray enters air ($n' = 1$) from glass ($n > 1$), the limiting angle is:

$$\sin \varepsilon_n = \frac{1}{n}. \tag{1.8a}$$

1.3 Components of Optical Instruments

1.3.1 Plane Mirror

A plane mirror is usually a thin metal layer which forms the coating of the very accurately machined plane surface of a glass plate. Such a metal coating has far from perfect reflecting properties hence, in order to improve its reflectivity, reflectivity-increasing coatings are used (see Subsection 1.1.3).

On meeting a mirror, a conical beam of light from a point source retains its conical feature with respect to the mirror image C' of the centre C *(Fig. 1.4)*. Each point of an object is regarded as the point source of a conical light beam. If one rotates a plane mirror by an angle α, the reflected ray is deflected by an angle 2α *(Fig. 1.5)*. Two plane mirrors inclined to each other by an angle τ deflect the incident beam by an angle $2\tau = \delta$ *(Fig. 1.6)*, which is independent of the incidence angle of the beam *(Fig. 1.7)*. If a mirror system, consisting of an odd number of mirrors perpendicular to a plane, is rotated by an angle α, the reflected ray is deflected by an angle 2α. If

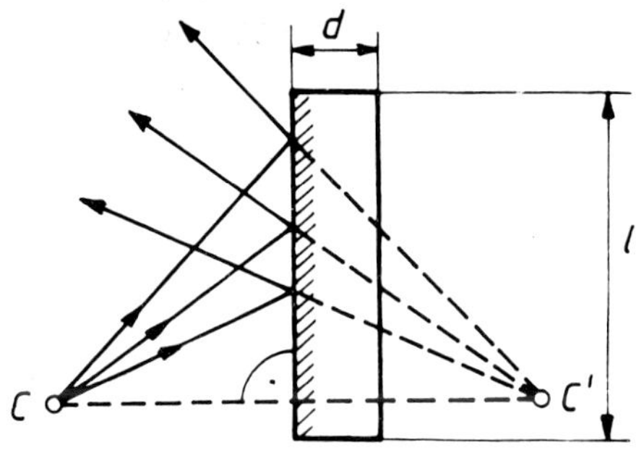

Figure 1.4. Image formed by a plane mirror

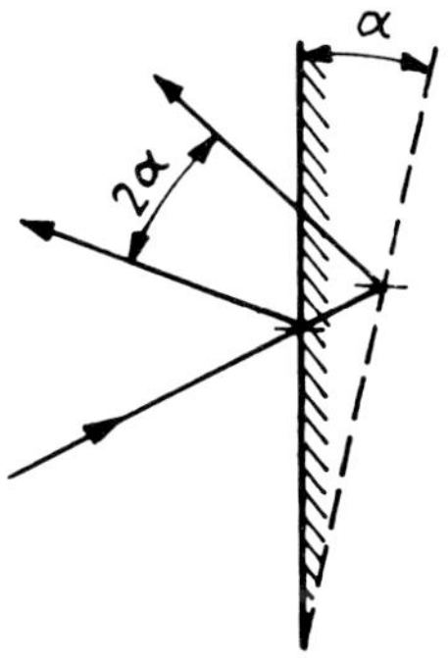

Figure 1.5. Effect of rotating a plane mirror

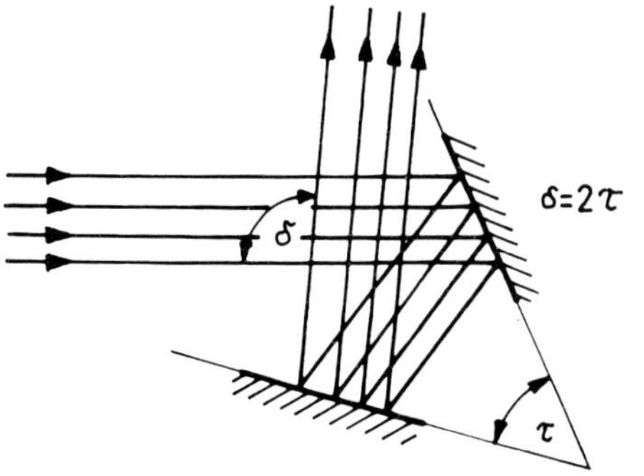

Figure 1.6. Propagation of parallel rays reflected by two mirrors

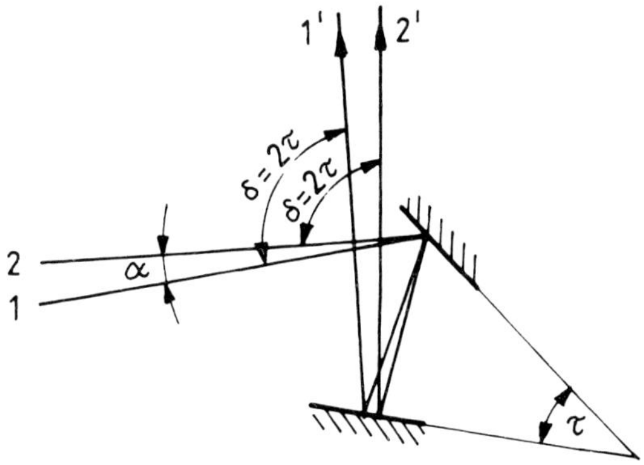

Figure 1.7. Rotation of two mirrors

the system consists of an even number of mirrors, the direction of the reflected beam remains unchanged.

The reflecting surface of a plane mirror usually differs slightly from a plane, being either a concave or a convex surface due to defects in manufacture. This defect can be characterized by the focal length of the finished mirror surface or by the number of interference lines (Newton's rings) generated between the mirror and a master plane surface imposed on the mirror (see Subsection 1.1.3). In order to assure a satisfactory plane surface, an optically flat mirror must be made from a glass body of a required thickness. This necessary thickness d is 1/5 to 1/7 of the largest dimensional l of the mirror surface *(see Fig. 1.4)*.

Plane mirrors coated on their rear are not used in image projecting systems due to their unfavourable characteristics. The use of plane mirrors at all is only suitable if it gives a reasonable saving in mass and dimension compared to prisms having the same function.

Mirrors which both reflect and transmit light are called semi-transparent or half-silvered mirrors *(Fig. 1.8)*. Mirrors which reflect only a certain part of the light spectrum, while transmitting the remainder of the light rays, are called filter mirrors *(Fig. 1.9)*. The ideal plane mirror is free of image forming defects.

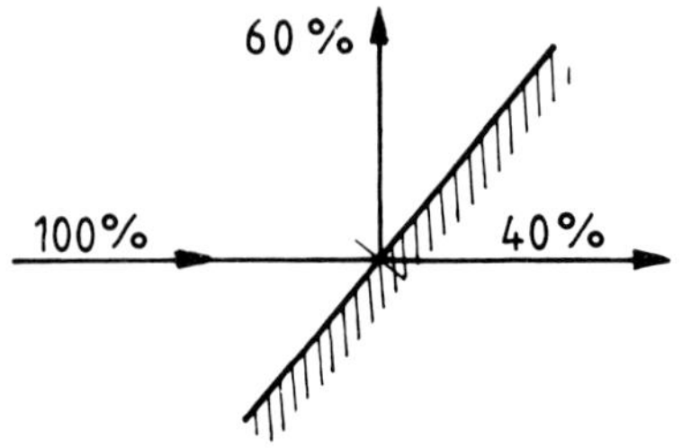

Figure 1.8. Semi-transparent mirror

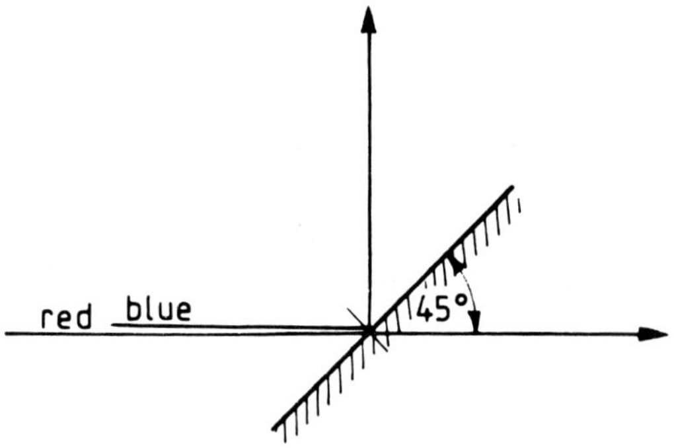

Figure 1.9. Filter mirror

1.3.2 Plane Parallel Optical Plate

A plane parallel optical plate is an optical medium between two parallel surfaces with a certain refractive index n and a certain thickness d; the medium is usually glass but it may be crystal or any other material. When a light beam is transmitted through a plane parallel plate, its centre C is shifted along its axis to the point C' *(Fig. 1.10)*. The shift of the centre of a light beam with a narrow aperture is:

$$\Delta s = \frac{n-1}{n} d. \tag{1.9}$$

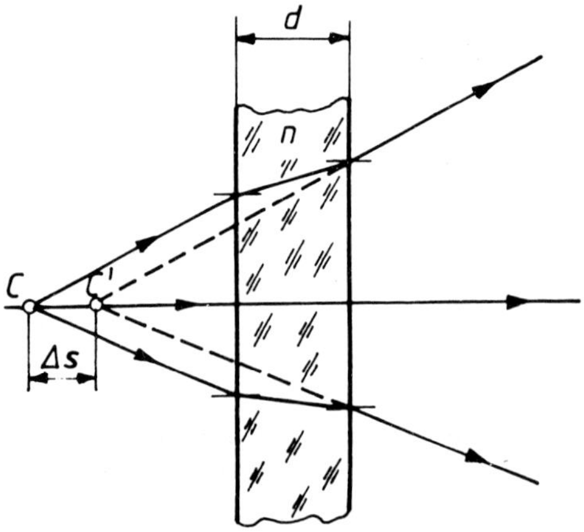

Figure 1.10. Image formed by a plane parallel plate

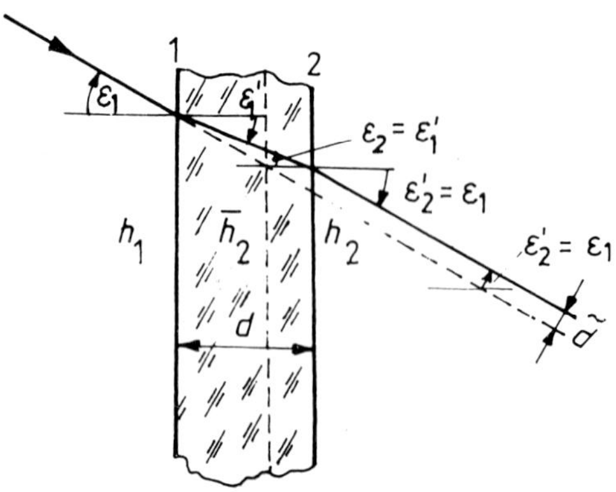

Figure 1.11. Ray path in a plane parallel plate

A light beam meeting a plane parallel glass plate of thickness d and refractive index n, with an incidence angle ε_1, is shifted parallel to itself, after transmission through the plate, by the distance:

$$\tilde{d} = d \sin \varepsilon_1 \left(1 - \sqrt{\frac{1 - \sin^2 \varepsilon_1}{n^2 - \sin^2 \varepsilon_1}} \right) \tag{1.10}$$

(see *Fig. 1.11*).

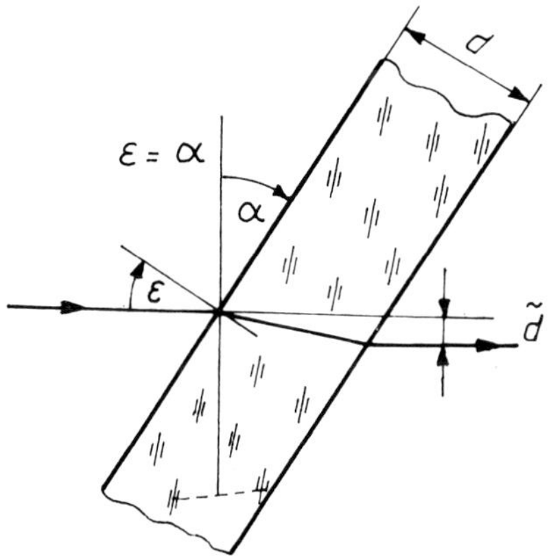

Figure 1.12. Rotation of a plane parallel plate

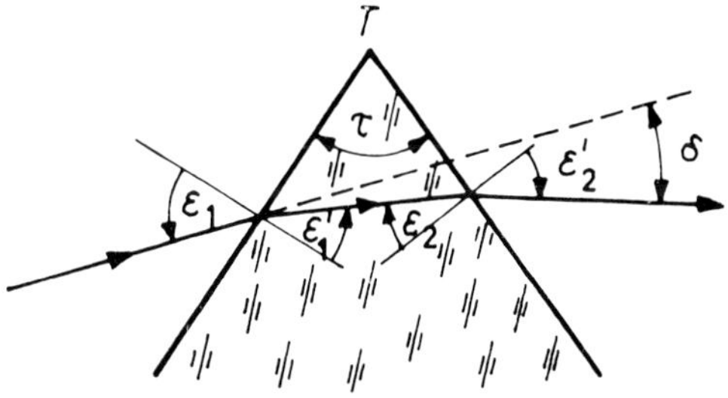

Figure 1.13. Ray path in a deflecting prism

If ε_1 is small, the following formula gives a good approximation:

$$\tilde{d} \approx \frac{n-1}{n} d \cdot \varepsilon_1 \text{ (rad)} \tag{1.11}$$

which means that the shift of the beam is proportional to the incidence angle (or to the rotation angle of the plane parallel plate). By rotating a plane parallel plate through an angle α, the beam can be parallel shifted by the small distance $\tilde{d}$ *(Fig. 1.12)*. This is utilized in optical micrometers (see Point 2.2.5.3). Defects of the boundary surfaces occurring during manufacture can be characterized similarly as for plane mirrors. A plane parallel plate is free of image forming defects only if a beam of parallel light passes through it with incidence angle $\varepsilon \approx 0$.

1.3.3 Prisms

According to their main operating use, prisms are distinguished as deflecting prisms and reflecting prisms.

Deflecting prisms. In the primary section (i.e. in the plane perpendicular to the deflecting edge T) of a simple prism of prism angle τ and of refractive index n, the light ray entering with an incidence angle ε_1 is deflected in the primary section by the angle δ with respect to the direction at incidence *(Fig. 1.13)*. This angle can be calculated from Eq. (1.6) with the substitutions $n'=n$, $n=1$, $\varepsilon=\varepsilon_1$ and $\varepsilon'=\varepsilon'_1$, giving:

$$\varepsilon'_1=\arcsin\left(\frac{\sin\varepsilon_1}{n}\right)$$

and

$$\varepsilon'_2=\arcsin\{n\sin(\tau-\varepsilon'_1)\}.$$

The deflecting angle is then:

$$\delta=\varepsilon_1+\varepsilon'_2-\tau. \qquad (1.12)$$

If the prism angle $\tau\leqq 6°$, the prism should be regarded as an optical wedge. If the incidence angle $\varepsilon\leqq 1°$, the ray deflection of the optical wedge is:

$$\delta=(n-1)\tau. \qquad (1.13)$$

If an optical wedge is shifted perpendicular to its plane of symmetry in the direction of the light ray by the distance d *(Fig. 1.14)*, the light ray is also parallel shifted by the amount:

$$\Delta y=(n-1)d\tau \text{ (rad)}. \qquad (1.14)$$

The most important image forming error of an optical wedge is chromatic aberration (colour error). However this can be eliminated to a high degree if two wedges with different prism angles τ_1 and τ_2, made of glass of different chromatic characteristics (i.e. having different Abbe numbers v_1 and v_2), are chosen so that the following equations are satisfied *(Fig. 1.15)*:

$$\delta=(n_1-1)\tau_1+(n_2-1)\tau_2$$

and

$$\frac{\tau_1}{\tau_2}=\frac{v_1}{v_2}. \qquad (1.15)$$

Two parallel rays entering with a separation y remain parallel after deflection, but

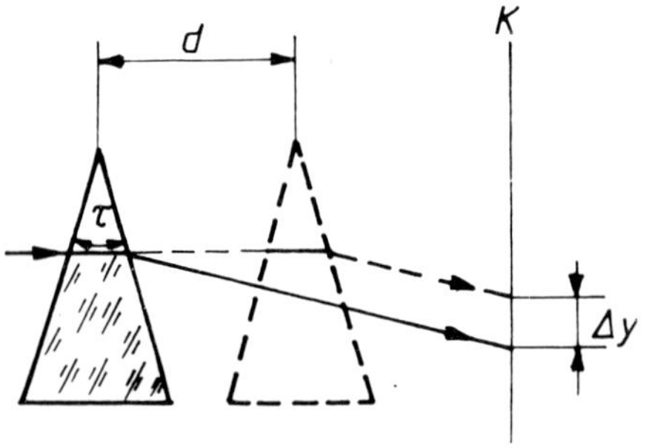

Figure 1.14. Effect of the shift of an optical wedge

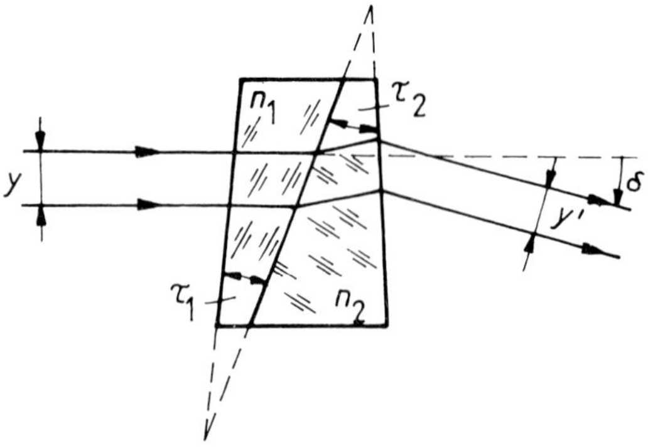

Figure 1.15. Achromatic optical wedge-pair

their separation changes to y'. The linear optical magnification of a pair of optical wedges is:

$$\beta = \frac{y'}{y}. \tag{1.16}$$

Reflecting prisms and prism systems. The output of a rectangular prism can be considered as the combined effect of a plane mirror and a plane parallel plate *(Fig. 1.16)*. If a prism is turned in its primary section (i.e. in the plane perpendicular to the reflecting surface and to the two deflecting surfaces of the prism) through an angle α, then due to the effect of reflection the incident ray turns through an angle 2α, and due to the effect of the plane parallel plate, it is parallel shifted through the distance $\bar{d}$ defined by Eq. (1.10).

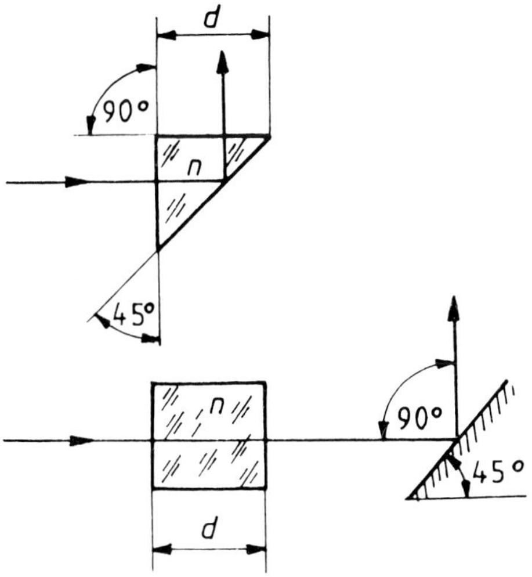

Figure 1.16. Rectangular relfecting prism

The possible optical trains through a rectangular prism using one reflecting surface are shown in *Fig. 1.17*, *a* and *b*, while those using two reflecting surfaces are shown in *Fig. 1.17*, *c* and *d*.

The biprism shown in *Fig. 1.18* is used in theodolites and levels, as well as in certain distance meters (see Point 6.1.2.3), as an image cutting or image uniting prism or as an image splitting edge.

Figure 1.19 shows prisms with two reflecting surfaces. These include the very frequently used type *b* (pentagonal prism) as well as the triangular prism *c* which do not act on the principle of total reflection, the prism sides marked in the figure being therefore mirror coated. A triple mirror prism or corner prism *(Fig. 1.20, a)* deflects the incident rays through 180°, a roof prism *(Fig. 1.20, b)* reverses the image. A light splitting prism is shown in *Fig. 1.21, a*, an image uniting prism in *Fig. 1.21, b*, both types being mainly used in autocollimation systems. The optical trains of some combined prism systems are illustrated in *Fig. 1.22, a, b, c* and *d*.

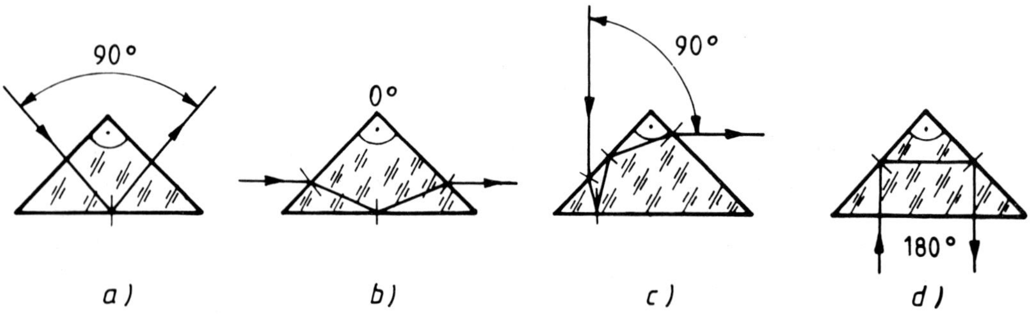

Figure 1.17. Ray paths in rectangular prisms

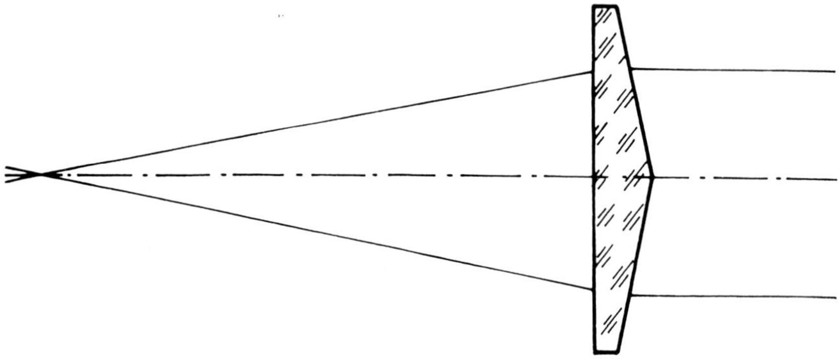

Figure 1.18. Dual wedge (biprism)

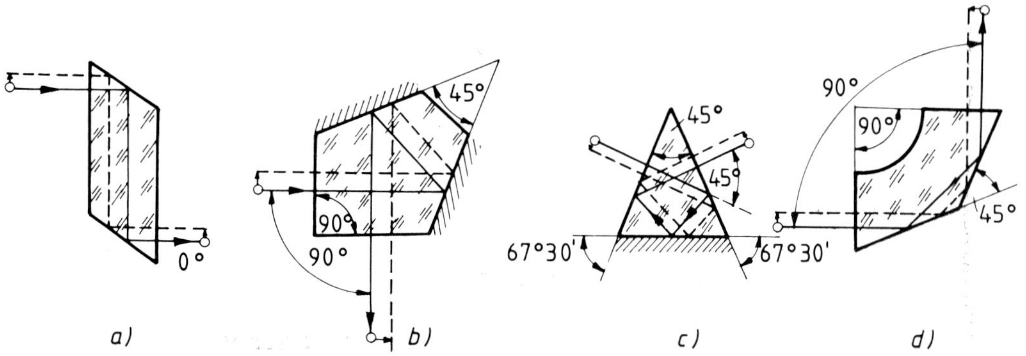

Figure 1.19. Prisms with two reflecting surfaces, deflecting the rays by a constant angle: *a)* rhombic prism; *b)* pentagonal prism (pentaprism); *c)* Schmidt prism; *d)* Wollaston prism

The effect of errors in the boundary angles of reflecting prisms. Due to imperfect manufacture the nominal values of the boundary angles between prism surfaces are inevitably subjected to slight errors. The main effect is that the light beam entering the prism is not deflected to the prescribed direction but slightly deviates from it.

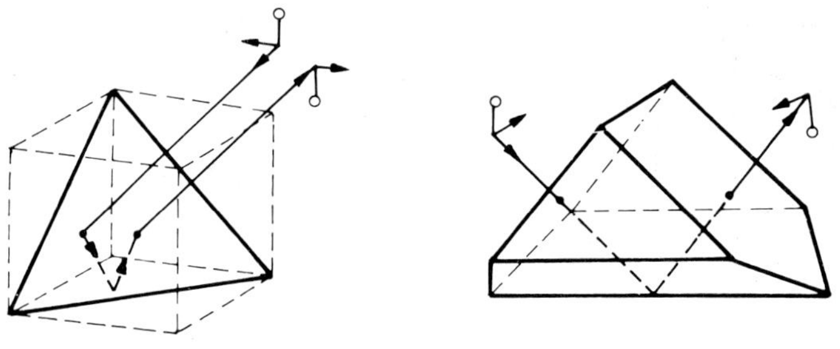

Figure 1.20. Prisms reversing the total image: *a)* corner prism (triple mirror prism); *b)* Amici prism (roof prism)

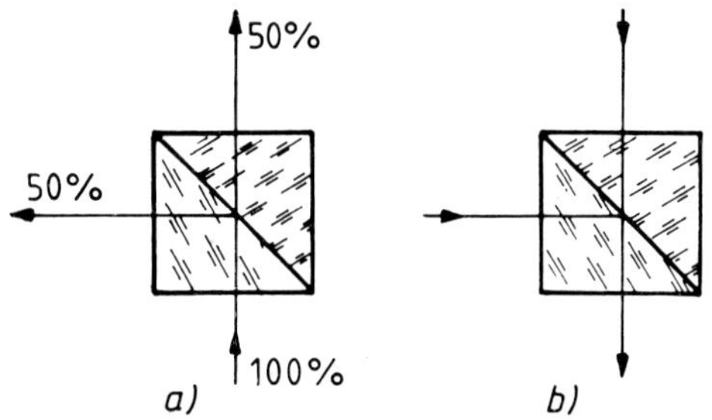

Figure 1.21. a) light splitting prism; *b)* image uniting prism

If the rectangular prism *ABC* has the following erroneous boundary angles *(Fig. 1.23)*:

$$A=90^\circ+\varepsilon_1; \quad B=45^\circ+\varepsilon_2; \quad C=45^\circ-\varepsilon_1-\varepsilon_2$$

then the deviation of the exit beam from the nominal direction is, in the case of a small incidence angle α_1:

$$\delta=-2n\varepsilon_1. \tag{1.17}$$

In other words the deviation angle depends on the refractive index n of the glass and on the error ε_1 of the right angle of the prism, but it is independent of the error of the 45° boundary angles. Consequently the right angle of a prism should be machined exactly.

If the intersection edge of the boundary surfaces at the right angle of a rectangular prism is inclined by a small angle ε *(pyramidal error)* to the hypotenuse surface of the prism *(Fig. 1.24)*, then the ray, which enters the body of the prism at an incidence angle α_1 in the incidence plane parallel to the edge A, exists at the angle α_2 in a plane parallel to the incidence plane. In the case of small α_1, the direction of the exit ray deviates from its nominal direction by an angle whose value may be calculated from Eq. (1.17).

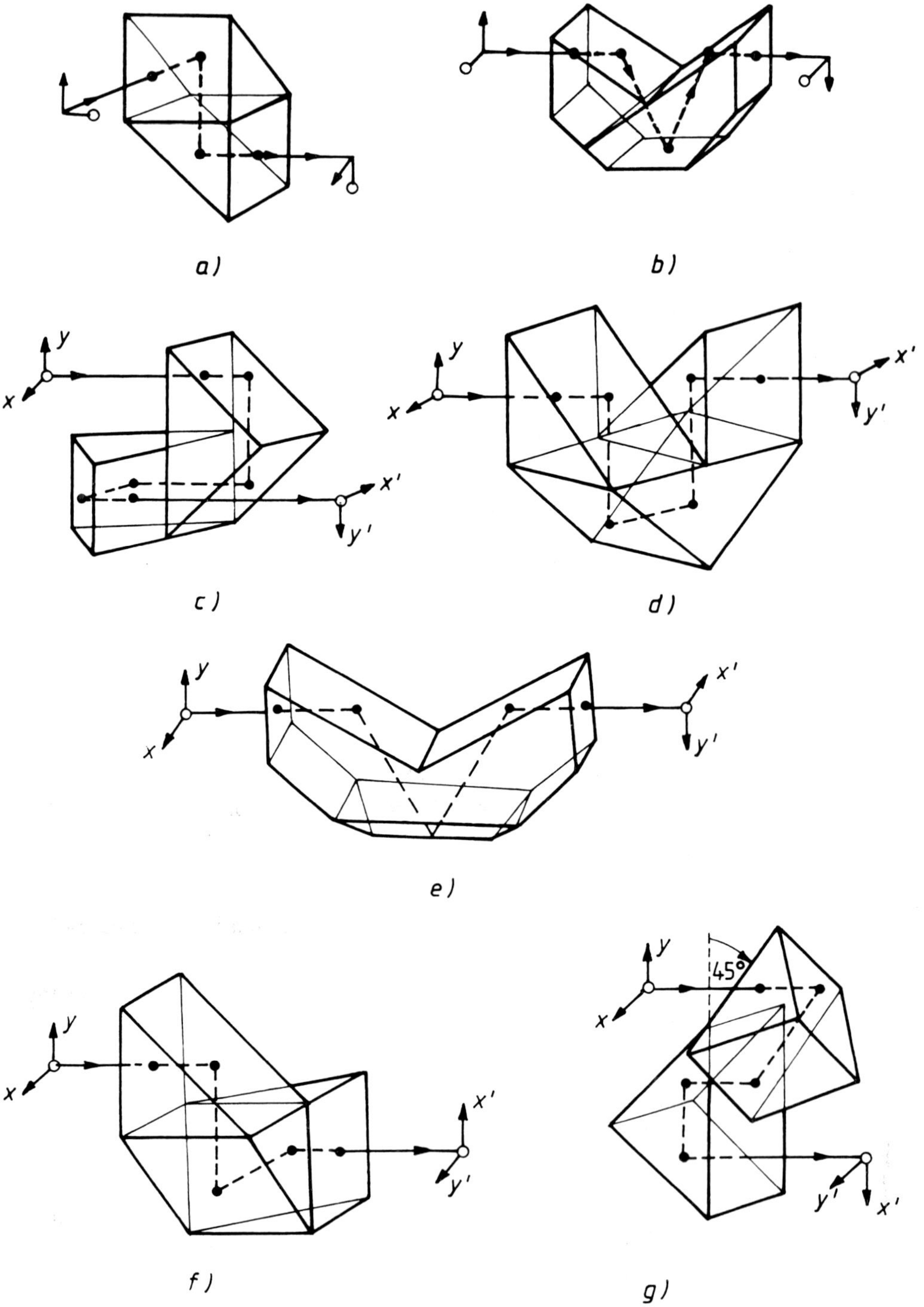

Figure 1.22. Prism systems: *a)* sphenoidal, deflection of rays by 90° with image reversal by 90°; *b)* Abbe prism — forming erect images without side reversal; *c)* Porro first prism — deflecting the image by 180°; *d)* Porro second prism — deflecting the image by 180°; *e)* Abbe roof prism; *f)* triple prism — deflecting the image by 90° with side reversal; *g)* Porro first prism of 45° — deflecting the image by 90° (without side reversal)

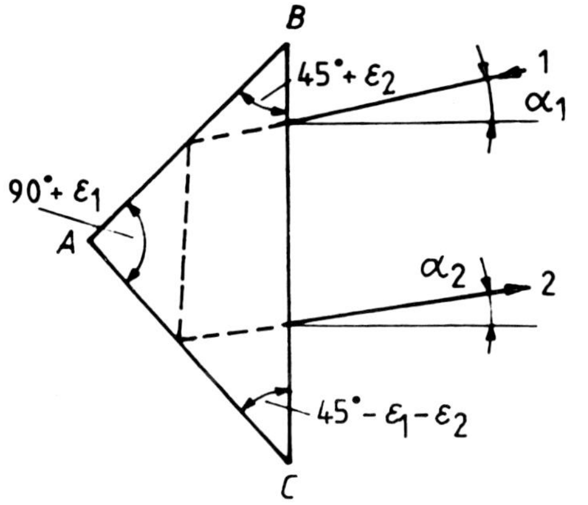

Figure 1.23. Effect of boundary angle errors of a rectangular prism

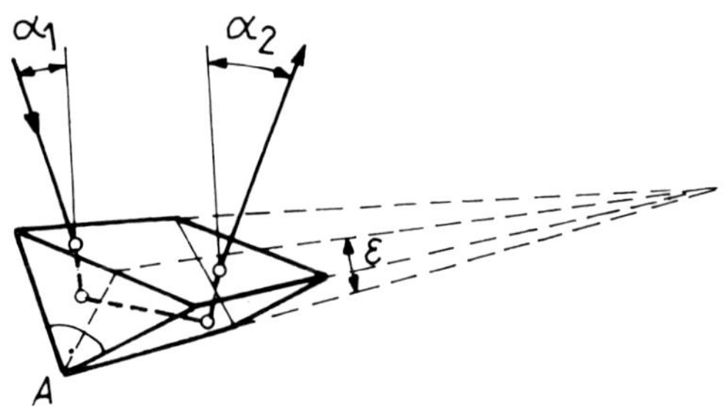

Figure 1.24. Effect of the pyramidal angle error of a rectangular prism

If the boundary surfaces of a pentagonal prism *(Fig. 1.25)* form a right angle with a small deviation ε_1 (thus $A = 90^\circ + \varepsilon_1$), and the two mirror surfaces form the prism angle with a small error ε_2 ($C = 45^\circ + \varepsilon_2$), then a light ray entering the prism in the primary section with an incidence angle α_1 deviates from the correct exit direction, in the case of small α_1, at an angle:

$$\delta = -n(\varepsilon_1 - 2\varepsilon_2). \tag{1.18}$$

It is most unsatisfactory if the roof sides of a roof prism (the surfaces *ABFG* and *BCEF* in *Fig. 1.26*) form a right angle with an error ε, because this results in double imaging. The reason for this is that the parallel light rays 1 and 2, entering the prism on the right and left sides of the plane of symmetry *FBH* intersecting the roof edge of the prism, are deflected on leaving the prism by the same amount but in opposite directions. The angle between the directions of the rays 1′ and 2′ is:

$$\delta = \sqrt{8}n\varepsilon \approx 4.2\,\varepsilon, \tag{1.19}$$

where $n \cong 1.5$ is the refractive index of the material of the roof prism.

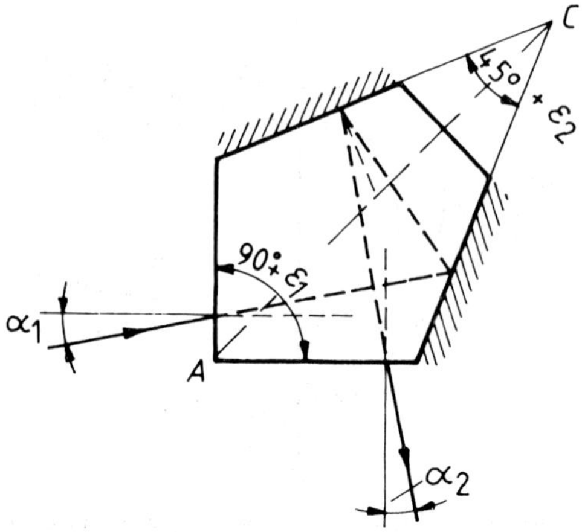

Figure 1.25. Effect of boundary angle errors of a pentagonal prism

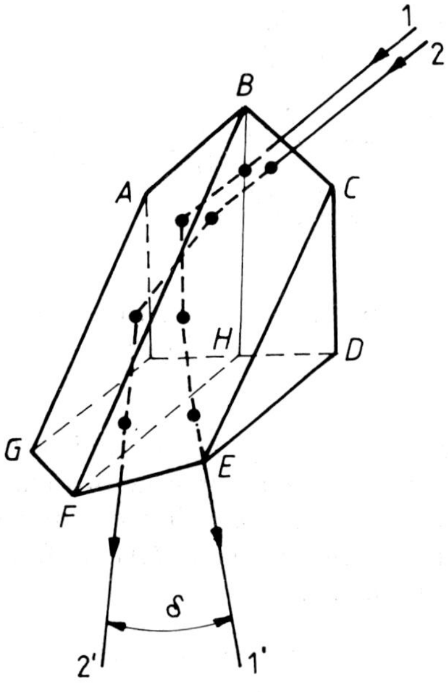

Figure 1.26. Effect of the roof angle error of a roof prism

It is obvious from Eq. (1.19) that the deviation is more than four times greater than the error in the angle between the roof surfaces. The right angle made by the roof surfaces should therefore be machined very precisely.

The triplet mirror prism *(Fig. 1.27, a)* is used as a reflector for electronic distance measurement instruments (see Chapters 8, 9 and 10). If the right angles made by the three total reflecting surfaces 1, 2 and 3 have the small angular errors $\varepsilon_{2,3}$; $\varepsilon_{1,3}$ and

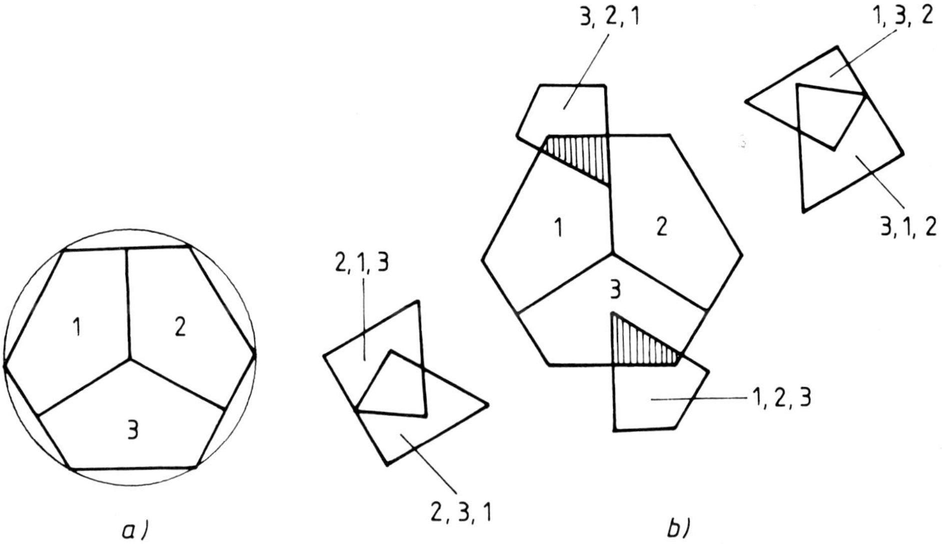

Figure 1.27. Beam divergence due to small angular errors of a triplet mirror prism

$\varepsilon_{1,2}$, then they cause a parallel light beam meeting the front of the prism approximately orthogonally to be reflected, not as one single beam, but — corresponding to the following six possible sequences of the reflecting surfaces (1, 2, 3; 1, 3, 2; 2, 1, 3; 2, 3, 1; 3, 1, 2 and 3, 2, 1) — as six individual beams having nearly triangular cross sections and diverging in different directions. If the errors of the orthogonal boundary surfaces are known (measured), the directions of the individual beams can be determined by vector analysis in the following way. The vectorial form of the law of reflection

$$\vec{\mathbf{S}}_1' = \vec{\mathbf{S}}_1 - 2(\vec{\mathbf{S}}_1\vec{\mathbf{P}}_1)\vec{\mathbf{P}}_1 \tag{1.20}$$

is applied in turn to the three reflecting surfaces. One obtains:

$$\vec{\mathbf{S}}_3' = \vec{\mathbf{S}}_1 + 2(\vec{\mathbf{S}}_1\vec{\mathbf{P}}_1)\{-\vec{\mathbf{P}}_1 + 2(\vec{\mathbf{P}}_1\vec{\mathbf{P}}_2)\vec{\mathbf{P}}_2 + 2(\vec{\mathbf{P}}_1\vec{\mathbf{P}}_3)\vec{\mathbf{P}}_3 - 4(\vec{\mathbf{P}}_1\vec{\mathbf{P}}_2)(\vec{\mathbf{P}}_2\vec{\mathbf{P}}_3)\vec{\mathbf{P}}_3\} + \\ + 2(\vec{\mathbf{S}}_1\vec{\mathbf{P}}_2)\{-\vec{\mathbf{P}}_2 + 2(\vec{\mathbf{P}}_2\vec{\mathbf{P}}_3)\vec{\mathbf{P}}_3\} - 2(\vec{\mathbf{S}}_1\vec{\mathbf{P}}_2)\vec{\mathbf{P}}_3, \tag{1.21}$$

where $\vec{\mathbf{S}}_i$ = the unit vector of the rays meeting the surfaces $i = 1, 2, 3$ of the triplet mirror prism,
$\vec{\mathbf{S}}_i'$ = the unit vector of the rays reflected from the three surfaces, and
$\vec{\mathbf{P}}_i$ = the unit vector of the normals of the three (near-orthogonal) surfaces.*

If Eq. (1.21) is applied to the six variants of the order of sequence of the reflecting surfaces 1, 2 and 3, the diverging directions of the six partial reflections of the incident

* *Lajos Fialovszky*: Einfluss der Winkelfehler des Tripelprismas auf die reflektierten Strahlen beim elektrooptischen Entfernungsmessen, 10th Intern. Wiss. Koll. TH Ilmenau (1965) and *L. Fialovszky*: Effect of angular errors of trihedron mirrors and trihedron prisms used as reflectors of electro-optical distance measurements to the reflected light-beam (in Hungarian). MTA VI. Oszt. Közl., Budapest, Vol. 37 (1966).

light beam are obtained. If e.g. the three angular errors of the three orthogonal sides are $\varepsilon_{2,3} = -1$ sec of arc, $\varepsilon_{1,3} = 0.5$ sec of arc and $\varepsilon_{1,2} = 0.75$ sec of arc, calculation based on Eq. (1.21) gives the cross sections of the six partial light beams reflected by a triplet mirror prism to be of diameter 10 cm. *Figure 1.27, b* shows the cross sections of the diverging partial reflections at a distance of 4000 m from the reflecting prism, together with the cross section of the parallel light beam entering a regular hexagonal prism. It can be seen from the figure that only the hatched parts of two of the six partial reflections coincide with the cross section of the incident light beam, i.e. only the hatched parts return into the aperture of the instrument emitting light or laser beams, assuming that the *aperture* and the reflecting prism are of the same size. Thus, for the angles of the triplet prisms used as reflectors for electro-optical distance meters, the tolerable errors of manufacture are even less than in the above example. For the measurement of distances of several km or several 10 km, the reflector should be composed of a large number *k* of triplet prisms because, at a greater distance from the reflector, the reflected 6*k* light paths, diverging by different amounts, form a more or less homogeneous pattern due to the random distribution of the angular errors. Although this pattern has an unnecessarily large diameter, an adequate amount of light enters the receiving aperture of the electro-optical distance meter. Losses of the reflected energy received in an electro-optical distance measuring instrument can also be caused by other factors; they are treated in Chapter 8.

1.3.4 Fibre Optics

Light rays entering the polished front end of a glass fibre are totally reflected many times by its cylindrical jacket before they emerge at the opposite end *(Fig. 1.28)*. The internal reflection loss of light rays entering obliquely is reduced by coating the glass fibre with a glass layer of a lower refractive index. Loss is also caused by absorption. The light-conducting fibres used usually have a diameter of 10–40 μm. In small fibre diameters the major part of the light is diffused and the spatial saturation factor is also diminished. The light permeability of a fibre is little decreased by bending as long as the radius of curvature is about 20 times greater than the fibre diameter.

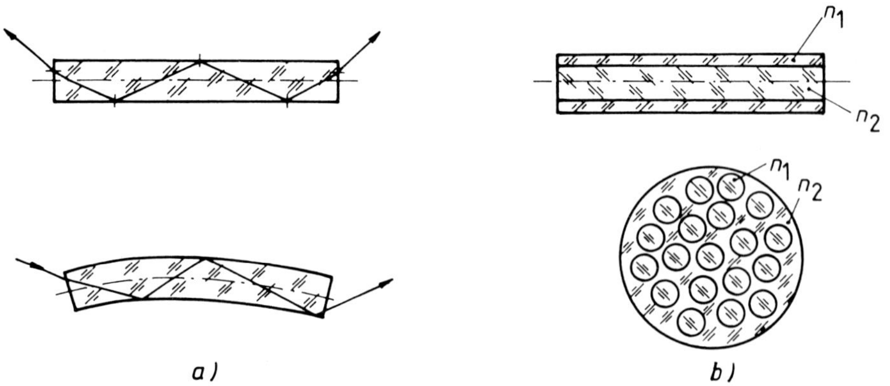

Figure 1.28. Fibre optics: *a)* optical train in a glass fibre; *b)* bundle of glass fibres

The fibre core, usually made of flint glass of a greater refractive index ($n_1 > 1.6$), is coaxially coated by a sheath usually made of crown glass of a lower refractive index ($n_2 < 1.52$) *(*see *Fig. 1.28, b).* The absorption loss of a fibre made of especially clear optical glass is 10–30 percent per metre.

By bundling fibres into a coil a great amount of light energy or image information can be transmitted. For image transmission an identical arrangement of the fibre ends at both ends (entrance and exit) is required *(fibre matrix).* The resolving power of a fibre coil nearly corresponds to that of twice the diameter of the fibres used. A concentric and coherent light beam loses its concentricity and coherency when passing through a fibre coil of cores having the same refractive index. This disadvantage can be eliminated by using fibres of varying refractive indices, the fibres being arranged in the cross section of the coil with a parabolic distribution of their refractivities. When using especially clear materials, this solution has the further advantage that the light permeability of the fibre coil is increased (i.e. the damping factor decreased) and this enables information to be transferred by light over a longer distance.

1.3.5 Optical Lenses and Lens Systems

The optical axis of an optical system (i.e. the straight line connecting the centres of curvature of the centred optical surfaces) is taken as the z axis *(Fig. 1.29)*; the meridional plane (i.e. the plane containing the optical axis) lying in the plane of the figure is taken as the y–z plane in a right-handed coordinate system. The x axis is perpendicular to the plane of the figure. The positive direction of the z axis is determined by the direction of propagation of the light rays. In the plane of the figure this direction is from left to right, the positive direction of the y axis is upwards and the positive direction of the x axis is inwards with respect to the plane of the figure. The distances (s, z, f) along the z axis are negative for points to the left of the reference point (in the figure, the centre T of the optical surface). The radius r of the refractive surface is positive if its centre of curvature C lies to the right of the centre of the surface T, and negative if C lies to the left. The aperture angle σ between the light ray and the optical axis is positive if the light ray is inclined downwards to the right.

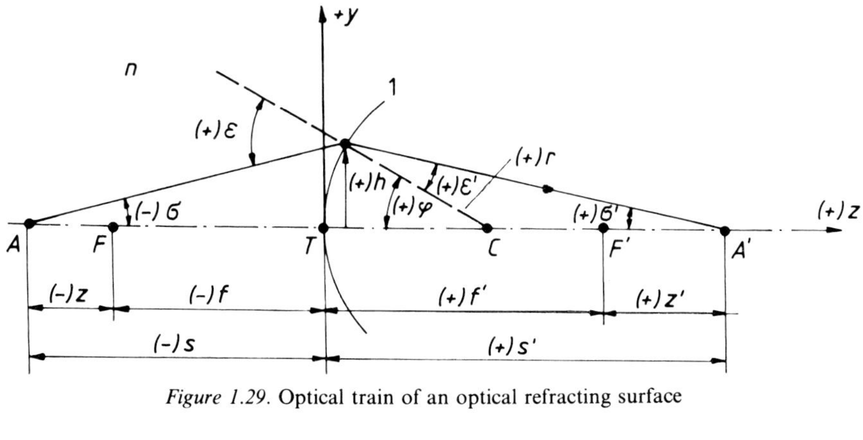

Figure 1.29. Optical train of an optical refracting surface

The sign of the central angle φ between the axis and a ray through a point on the surface at height h is determined by the relation $\sin \varphi = h/r$. The sign of the incidence angle ε of a light ray meeting the refracting surface can be obtained from the expression $\varepsilon = \varphi - \sigma$. The space to the left and to the right of the optical surface is called object space and image space respectively. Details in the image space are denoted with the same notation as their conjugate counterparts in the object space but with a superscript. Thus, if the refractive index of a medium in the object space is n, then that of the medium in the image space is n'. The sign rule is equally valid for details in the object space and in the image space. In the figure the signs in parentheses before the notations both in the object and image spaces show the actual signs of data on the basis of their positions in the figure.

From the path length s of a light ray in the object space, which originates at an object point A with the aperture angle σ on the optical axis and which meets the optical surface at an incidence angle ε, one can compute the path length s' of the light ray in the image space to the image point A' on the optical axis, as well as the aperture angle σ' in the image space, by means of the following formulae:

$$\sin \varepsilon = \frac{s-r}{r} \sin \sigma,$$

$$\sin \varepsilon' = \frac{n}{n'} \sin \varepsilon,$$

$$\sigma' = \sigma + \varepsilon - \varepsilon',$$

$$s' = r + r \frac{\sin \varepsilon'}{\sin \sigma'}. \tag{1.22}$$

Paraxial or near-axis zones are the zones of the optical system in which all the angles (σ, σ', ε, ε' and φ) are small, and thus, in the expanded series of the sine functions of these angles, all terms of order higher than the first can be neglected. Optical systems are designed as a first approximation for the paraxial zones. For these zones Eq. (1.22) can be reduced to the following simple linear formula:

$$n\left(\frac{1}{r} - \frac{1}{s}\right) = n'\left(\frac{1}{r} - \frac{1}{s'}\right). \tag{1.23}$$

Equation (1.23), the so-called *Abbe invariant*, shows that the products of multiplying the refractive index by the difference between the reciprocal of the radius and the reciprocal of the path length are equal before and after refraction. The same holds for all consecutive surfaces within an optical system. One obtains from Eq. (1.23):

$$s' = \frac{n'}{\dfrac{n'-n}{r} + \dfrac{n}{s}} \tag{1.24}$$

by means of which the path length of a light ray emerging from the refracting surface can be computed in the paraxial zone. Since no angles are present in Eq. (1.24), at the same s within the paraxial zone s' remains practically unchanged if the aperture angle σ of the incident ray is less than 1°. This means that, in the paraxial zone, light rays originating at any point (object point) on the optical axis intersect at one point (image point) on the axis.

If a light ray arriving from infinity ($s = -\infty$) and parallel to the axis enters at height h, then the corresponding s' is the focal length f' on the image side; it can be obtained from Eq. (1.24) as:

$$f' = \frac{n'r}{n'-n}. \tag{1.25}$$

Further, the path length s on the object side corresponding to an infinite image distance ($s' = \infty$) is the focal length of the optical surface on the object side:

$$f = -\frac{nr}{n'-n}. \tag{1.26}$$

The ratio of the two focal lengths is equal to the ratio of the two refractive indices:

$$\frac{f}{f'} = -\frac{n}{n'}. \tag{1.27}$$

A point lying in the optical axis at distance $f'(f)$ from the optical surface is the focus of the surface $F'(F)$ on the image side (and on the object side, respectively).

From Eqs (1.24), (1.26) and (1.27) one obtains:

$$\frac{f}{s} + \frac{f'}{s'} = 1. \tag{1.28}$$

Equation (1.28) is the *Gaussian equation* of image forming, expressing both the relation between the focal length on the object side and that on the image side and the relation between the path lengths on the two sides.

If the positions of object point A and image point A' are referred to the positions of the foci F and F', respectively, the substitutions $s = z + f$ and $s' = z' + f'$ give the *Newtonian relationship* between the product of the distances, object point–respective focus and image point–respective focus, and the product of the two focal lengths:

$$zz' = ff'. \tag{1.29}$$

An optical system contains a certain number (k) of centred optical surfaces (i.e. their centres lie on the common optical axis) *(Fig. 1.30)*. Equations (1.22) and (1.24)

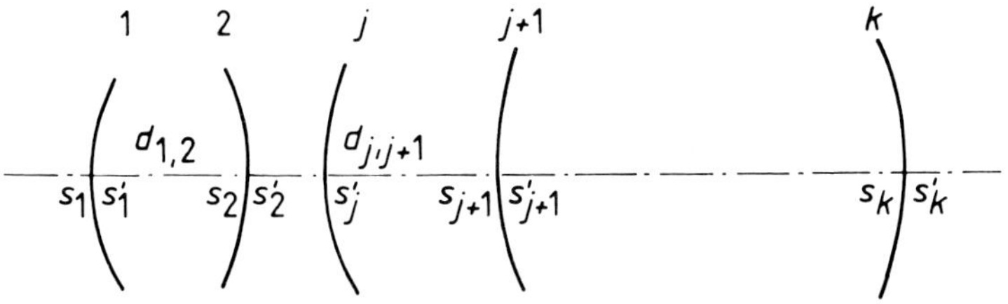

Figure 1.30. Optical system consisting of several refracting surfaces

can also be used to calculate the path lengths s'_{j+1} measured from the $(j+1)$-th surface for light rays passing through any $(j+1)$-th surface of the intermediate optical surfaces. The path length in the object space can be computed from the path length s'_j in the image space of the previous surface j, as

$$s_{j+1} = s'_j - d_{j,j+1}, \tag{1.30}$$

where $d_{j,j+1}$ is the distance of surface $j+1$ from surface j.

In order to determine the focal length in the image space of a system consisting of k optical surfaces, consider an axis-parallel ray ($s_1 = -\infty$) near the optical axis. If in front of the first optical surface 1 and beyond the last surface k the medium is air ($n_1 = n'_k = 1$), and the path lenghts are obtained from Eqs (1.24) and (1.30), then the focal length of the optical system is:

$$f' = \frac{s'_1 s'_2 \ . \ . \ . \ s'_k}{s_2 s_3 \ . \ . \ . \ s_k}. \tag{1.31}$$

The focal length of the object side is the same but has opposite sign ($f = -f'$). According to the rule of signs in Eq. (1.31), the path lengths s_j and s'_j have positive and negative sign if they lie on the right and left side, respectively, of the surface j.

For a system having two surfaces with radii r_1 and r_2, a surface separation $d_{1,2}$ and consisting of a material of refractive index n, in other words for a single lens *(Fig. 1.31)*, the reciprocal of the image-side focal length of a lens surrounded on both sides by air can be obtained from Eq. (1.31) to be:

$$\frac{1}{f'} = (n-1)\left\{\frac{1}{r_1} - \frac{1}{r_2} + \frac{n-1}{n}\frac{d_{1,2}}{r_1 r_2}\right\}. \tag{1.32}$$

The object side focal length is the same but has opposite sign. According to the rule of signs, radius r_1 of lens 1 should be taken as positive in Eq. (1.32) if the centre of curvature lies behind the surface when looking at it in *Fig. 1.31* from the direction of the light ray arriving from the left, i.e. if the lens surface is *convex*. Radius r_2 of surface 2 is negative if surface 2 is convex when looking at it from beyond (from the right). Such a lens is called *biconvex*.

The image-side principal point H' (the object-side principal point H) of the lens is the point on the optical axis distant $f'(f)$ from the image-side focus F' (the object-side focus F) *(see Fig. 1.31)*. The distance with correct sign between the principal point H' and the axial point T_2 of the rear surface 2 (i.e. from the *reference point*) is:

$$S_{H'} = \overline{T_2 H'} = \frac{r_2 d_{1,2}}{n(r_1 - r_2) - (n-1)d_{1,2}}. \tag{1.33}$$

Analogously, the distance with correct sign between the principal point H and the axial point T_1 of the front surface 1 is:

$$S_H = \overline{T_1 H} = \frac{r_1 d_{1,2}}{n(r_1 - r_2) - (n-1)d_{1,2}}. \tag{1.34}$$

The plane perpendicular to the optical axis through the principal point H is called the object-side *principal plane* and, analogously, the plane perpendicular to the axis and containing H' is called the image-side principal plane.

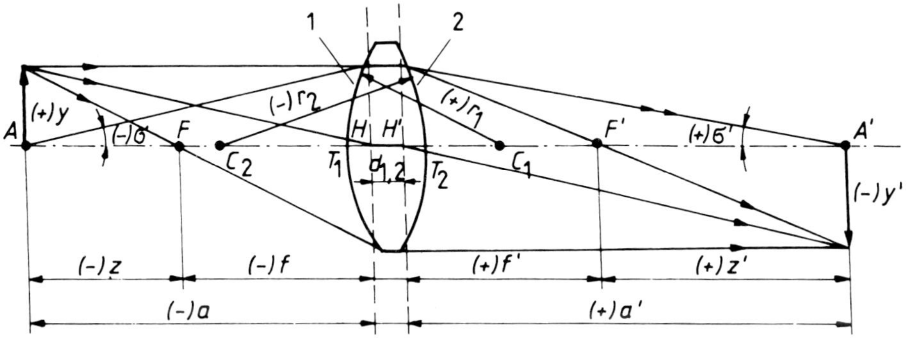

Figure 1.31. Ray path in a thick lens

The distance between the image-side and object-side principal planes is:

$$\overline{HH'} = \overline{T_1T_2} - \overline{T_1H} + \overline{T_2H'} =$$

$$= d_{1,2}\left\{1 - \frac{r_1 - r_2}{n(r_1 - r_2) - (n-1)d_{1,2}}\right\}. \tag{1.35}$$

In order to simplify calculation the thickness of the lens can often be neglected (if $d \ll r$); then Eq. (1.32) gives:

$$\frac{1}{f'} = (n-1)\left(\frac{1}{r_1} - \frac{1}{r_2}\right). \tag{1.36}$$

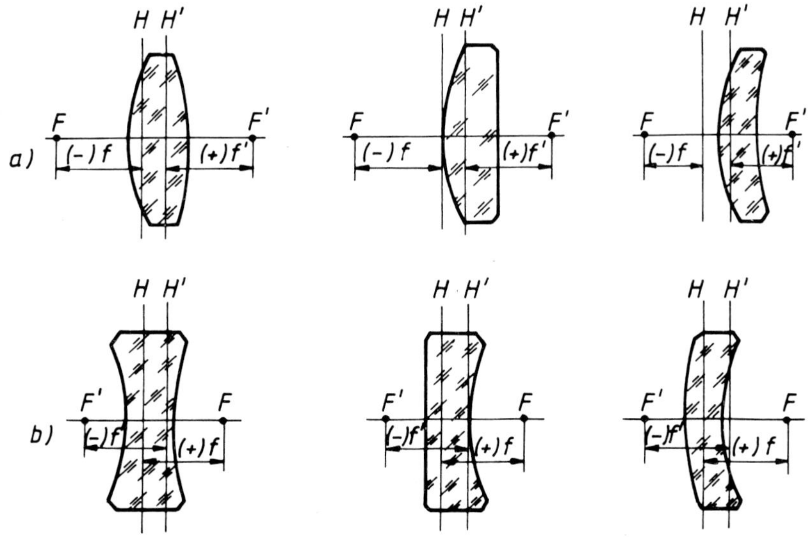

Figure 1.32. Different lens types: *a)* collecting lenses (from left to right: biconvex, plane-convex, convex-concave); *b)* dispersing lenses (from left to right: biconcave, plane-concave, convex-concave)

A lens in air forms the image of an object point A, lying on the optical axis at distance $a=\overline{HA}$ from the object-side principal plane, at the point A', which lies on the optical axis at distance $a'=\overline{H'A'}$ from the image-side principal plane. The fundamental relation between object distance and image distance is:

$$-\frac{1}{a}+\frac{1}{a'}=\frac{1}{f'}. \tag{1.37}$$

For thin single lenses $a=s$ and $a'=s'$, thus

$$-\frac{1}{s}+\frac{1}{s'}=\frac{1}{f'}. \tag{1.38}$$

A lens with a positive image-side focal distance f' is called a *collecting lens* (or condenser), because parallel light rays passing through it are *converged.* A lens with a negative image-side focal distance f' is called a *dispersing* lens because parallel light rays passing through it are *diverged.* The former is thickest on its optical axis, while the latter is thinnest there *(Fig. 1.32)*. In the figures the approximate positions of the two focal points and principal planes of the lenses are also shown.

The *lateral* imaging *scale* of an optical lens *(*see *Fig. 1.31)* is the ratio between the corresponding axis-perpendicular dimension of image and object:

$$\beta=\frac{y'}{y}=\frac{f}{z}=\frac{z'}{f'}=\frac{a'}{a}. \tag{1.39}$$

The ratio between the corresponding axial dimensions of image and object along the optical axis, $\Delta z'$ and Δz, is called the axis-parallel scale, or *axial scale* of a lens:

$$\alpha=\frac{\Delta z'}{\Delta z}=-\frac{z'}{z}=\beta^2. \tag{1.40}$$

The *angular scale* of an optical lens, in other words the *ratio of convergence*, is the ratio between the angles made with the axis by any light ray from the object point on the axis and the corresponding *conjugated* ray to the image point on the axis:

$$\gamma=\frac{\sigma'}{\sigma}=\frac{a}{a'}=\frac{1}{\beta}. \tag{1.41}$$

The following connection thus holds between the three scales:

$$\alpha\gamma=\beta. \tag{1.42}$$

An optical system consisting of several centred surfaces can be substituted by a single "equivalent" lens. Its data are obtained from an axis-parallel ray ($s_1=-\infty$) passing through the optical system, using Eqs (1.24) and (1.30). The resultant focal length of the equivalent lens can be obtained from Eq. (1.31). The distance between the image-side principal plane H' of the equivalent lens and the axial point T_k of the last optical surface k is:

$$\overline{T_kH'}=s'_k-f'. \tag{1.42a}$$

The resultant focal length on the object side should also be determined from an axis-parallel ray, entering the optical system from the object space ($s_k^* = \infty$ where the asterisk refers to path lengths for the optical surfaces with reverse sequence) by means of Eqs (1.24) and (1.30), but here the optical surfaces have the reverse sequence: $k, k-1, \ldots, 1$. The resultant focal length f on the object side can be obtained by using Eq. (1.31) as above, since $f=-f'$ if there is air on both sides of the optical system. The distance between the object-side principal plane H of the equivalent lens and the axial point T_1 of surface 1 (i.e. the first refracting surface) is:

$$\overline{T_1 H} = s_1 - f. \tag{1.42b}$$

If the resultant focal points and principal planes of an equivalent lens are known, both image-forming formulae, namely Eq. (1.38) deduced from the Gaussian equation and Eq. (1.29) (Newton's image-forming equation), can be used for the optical system. The image point A' of an object point A lying anywhere on the axis can be determined by means of these equations.

The resultant focal length for a lens system consisting of two lenses can be computed by

$$\frac{1}{f'} = \frac{1}{f'_1} + \frac{1}{f'_2} - \frac{e_{1,2}}{f'_1 f'_2}, \tag{1.43}$$

or by

$$f' = \frac{f'_1 f'_2}{f'_1 + f'_2 - e_{1,2}}, \tag{1.44}$$

where $e_{1,2}$ is the algebraic distance (i.e. the distance with sign according to the sign rule) between the object-side principal point H_2 of the second lens and the image-side principal point H'_1 of the first lens. (In case of thin lenses $e_{1,2}$ is the distance between the first and second lens.)

The reciprocal of the resultant focal length of a compound lens system consisting of three lenses is:

$$\frac{1}{f'} = \frac{1}{f'_1} + \frac{1}{f'_2} + \frac{1}{f'_3} - \frac{e_{1,2}}{f'_1}\left(\frac{1}{f'_2} + \frac{1}{f'_3}\right) - \frac{e_{2,3}}{f'_3}\left(\frac{1}{f'_1} + \frac{1}{f'_2}\right) + \frac{e_{1,2}e_{2,3}}{f'_1 f'_2 f'_3}. \tag{1.45}$$

The equation can be extended for compound lens systems consisting of more than three lenses.

1.3.6 Spherical Mirrors

The Abbe invariant referring to optical lens surfaces as given in Eq. (1.23) also holds for spherical mirrors if $n=1$ and $n'=-1$ are used as refractive indices. The imaging equation of a spherical mirror can be deduced from Eq. (1.24) as:

$$\frac{1}{s'} + \frac{1}{s} = \frac{2}{r}. \tag{1.46}$$

If an object is at infinity ($s=-\infty$), its image distance is equal to the focal length ($s'=f'$), hence from Eq. (1.46) one obtains $f=r/2$; the focal length of a spherical mirror is thus equal to half the radius of curvature.

1.3.7 Aberration in Lenses

The correct image-forming of optical lenses is limited by errors caused either by manufacturing technology or by the laws of optics themselves. The errors due to manufacturing technology are the consequences partly of variations in the quality of optical glass (see Subsection 1.3.8) and partly by geometric defects occurring in the machining of the surfaces (see Subsection 1.3.9). Image-forming errors of ideally machined lenses of ideal glass due only to reasons of optics are called aberrations. The image-forming errors of an optical system are jointly caused by the lenses composing it and by other optical elements.

From the point of view of geometrical optics, optical image-forming can be regarded as ideal if the image formed by a lens (or lens system) is created by *collinear transformation* of the object, namely if the concentric bundle of light rays leaving each point of the object is reunited also as a concentric bundle of light rays in the corresponding point of the geometrically similar image in the image space. A concentric bundle of rays entering an optical system, however, never emerges concentrically, due to aberration, hence the rays form a *dispersed light spot* of finite extent instead of a point.

In geometrical optics image-forming errors are generally characterized on the basis of the comparison between an image generated by selected rays of different colours and the ideal image. In the course of a geometrical-optical analysis all rays are taken as having the same intensity, thus such an analysis can give no information about image-forming errors of physical origin caused by inequalities in the illumination of an object *(contrast)*. For a comprehensive characterization of the quality of optical image-forming the transfer function of the optical system can be used (see Subsection 1.3.8).

One of the main goals of optical calculation is to minimize image-forming error by optimizing the internal parameters of the system. Geometrical image-forming errors are divided into *monochromatic* and *chromatic* aberrations.

Two types of monochromatic aberrations are distinguished: image-forming error caused by light rays entering from object points lying along the axis (spherical aberration), and image-forming errors caused by light rays entering from object points not lying on the optical axis *(coma, astigmatism, image-field curvature* and *distortion)*.

The chromatic aberrations can also be divided into two groups, namely *axial* or *magnifying* chromatic aberrations.

Spherical aberration. The rays of a monochromatic light bundle, from an object lying at a finite or infinite distance, intersect the optical axis at different points depending on their aperture angles with the optical axis and their distance from the optical axis *(Fig. 1.33, a)*. The image forming rays close to the optical axis (paraxial rays) can be regarded as ideal. The deviation between the axial intersection in the

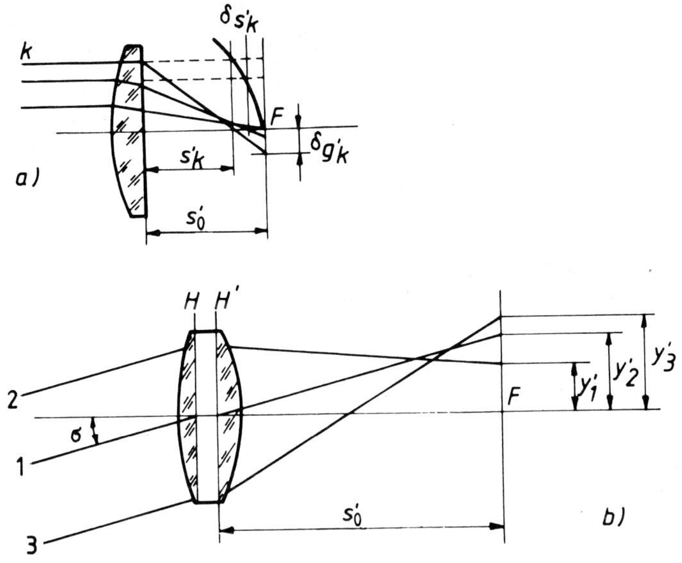

Figure 1.33. a) spherical aberration; *b)* coma

image-side space of a ray k, more distant from the axis, and of a paraxial ray 0, is called spherical aberration. It can be expressed as:

$$\delta s'_k = s'_k - s'_0 . \tag{1.47}$$

The rays between 0 and k cause a dispersion (in other words, transversal spherical aberration) with radius $\delta g'_k$. Spherical aberration can be reduced by using at least two lenses, namely one convex and one concave, because the spherical aberrations of these two lens types are of opposite sign.

Coma. A bundle of incident rays coming from an object point lying off the optical axis produces an irregular and lengthened dispersion ring *(Fig. 1.33, b)*. With increasing inclination of the bundle of rays (i.e. increasing angle σ between principal ray 1 and the optical axis), the effect of coma will also be increased, and this leads finally to a complete fading of the image. Thus the angle of view of a lens cannot be increased without limit. Coma cannot be completely eliminated even by using more lenses, therefore the optical angle of the inclined bundles of rays should be restricted by applying *diaphragm apertures* within the system. Coma is important mainly in lenses (lens systems) of greater power and wide angle of view.

Astigmatism and Curvature of the Image-field. Astigmatism is a defect caused by the fact that the image-side intersection points P'_m of the different rays which come from an object point P lying off the optical axis and which propagate in the meridional plane (in which the optical axis also lies) are not at the same distance from the lens as the intersection points P'_s of the rays propagating in the saggital (or equatorial) plane, which is perpendicular to the meridional plane *(Fig. 1.34)*. Image points lying

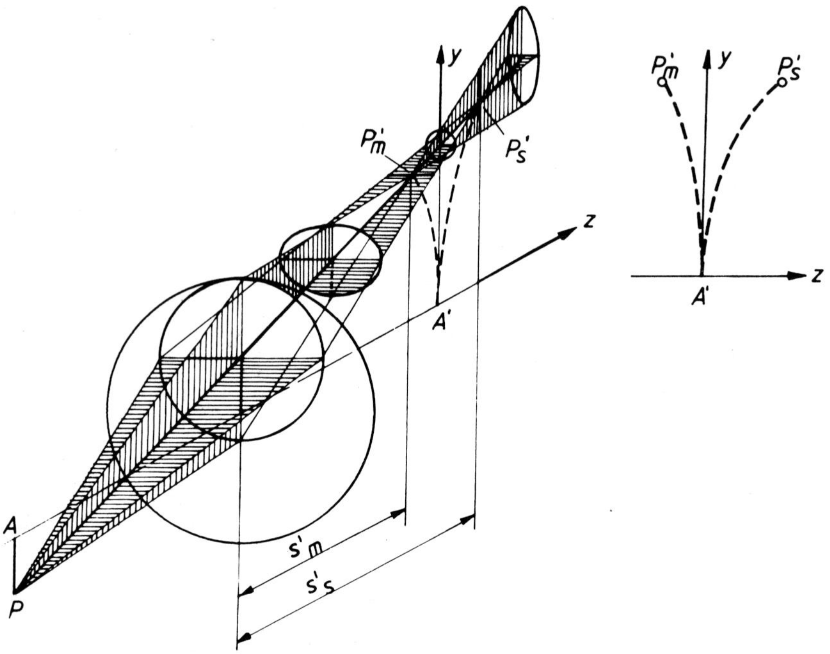

Figure 1.34. Astigmatism and curvature of image-field

on the optical axis have no astigmatism, but with increasing distance of the point from the optical axis, the meridional and saggital image points are split along two lines, thus producing an astigmatism of $s'_s - s'_m$.

The curvature of the image-field is the average image curvature between the meridional and the saggital planes. In practice astigmatism can be observed in the image of a cross-shaped linear object. The meridional rays form the correct image of the horizontal line *(Fig. 1.35, b)* of the cross *(Fig. 1.35, a)*, while the saggital rays form the correct image of the vertical line *(Fig. 1.35, d)*, but at different image distances. In the ideal image plane both line images are slightly blurred *(Fig. 1.35, c)*. Astigmatism and image-field curvature are primarily image-forming errors of wide angle lenses (lens systems).

Distortion. A uniform lateral scale is not maintained for each point of the optical image generated by a lens (or a lens system). If this scale is β_0 close to the optical

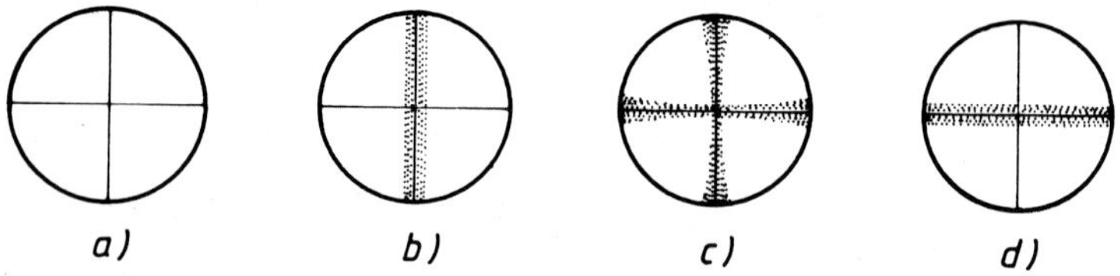

Figure 1.35. Astigmatic imaging: *a)* object; *b)* image generated by meridional rays; *c)* imaging in the correct image plane; *d)* image generated by saggital rays

axis in the image-side space, and β_i somewhat more off the axis, then the relative distortion is:

$$\delta_i = \frac{\beta_i - \beta_0}{\beta_0} = \frac{\Delta\beta}{\beta_0}. \tag{1.48}$$

The distortion of *telescopic* optical systems (see Subsection 1.5.3) can be characterized by the angular magnifications related to the respective image points:

$$\delta_i = \frac{\Gamma_i - \Gamma_0}{\Gamma_0} = \frac{\Delta\Gamma}{\Gamma_0}. \tag{1.49}$$

The influence of distortion can be illustrated by means of the image of a square grid which is distorted into a pillow-like or barrel-like form, depending on the sign of distortion. Image sharpness is not influenced by distortion.

Chromatic aberrations: *chromatic variation* and *chromatic magnification.* An axis-parallel light beam of compound colour is decomposed into its colours when passing through a lens, and the rays of different colours (i.e. of different wavelengths) converge at different points on the optical axis in the image-side space. The chromatic variation of a lens is usually defined as the separation of the focal points of the blue ray F, of wavelength $\lambda_F = 486.1$ nm, and of the red ray C, of wavelength $\lambda_C = 656.3$ nm *(Fig. 1.36, a)*:

$$\Delta s' = s'_F - s'_C. \tag{1.50}$$

Chromatic variation changes with the incident height h of the light rays. This is the chromatic variation of spherical aberration. The chromatic variation of light rays propagating infinitely close to the optical axis ($h \to 0$) is called paraxial chromatic aberration. The focal point F' of a lens in the image-side space is defined as the image formed by the axis-parallel and paraxial yellow light rays of wavelength $\lambda_d = 578$ nm (denoted by d).

An inclined light ray of compound colour of an object point not lying on the optical axis generates an elementary spectrum due to colour dispersion *(Fig. 1.36, b)*. The rays F form the image y'_F, the rays C form the image y'_C, of the object y. The size difference between the two images is called chromatic magnifications:

$$\frac{y'_F - y'_C}{y'_d} = \frac{\Delta y'}{y'}. \tag{1.51}$$

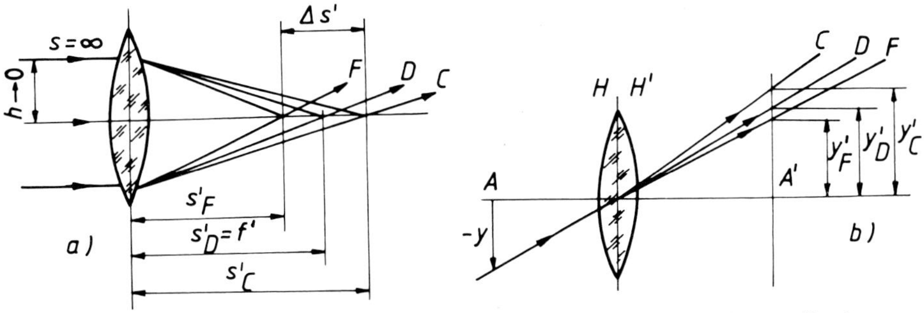

Figure 1.36. Chromatic aberrations: *a)* chromatic variation; *b)* chromatic magnification

Chromatic aberrations can be reduced by using in the optical system a convex and a concave lens made of glass of different refractive indices and dispersion values (*Abbe* numbers).

1.3.8 Transfer Function of Optical Systems

The image-forming errors yield special information about the quality of an optical system. Knowledge of them, however, is not always enough for the unambiguous definition of the quality of image forming. Therefore the *contrast transfer function* has recently come into use for the characterization of the image-forming quality of an optical system. It has the advantage that it can be calculated in the design before an optical system is constructed and, furthermore, it can be controlled by measurement in existing optical systems.

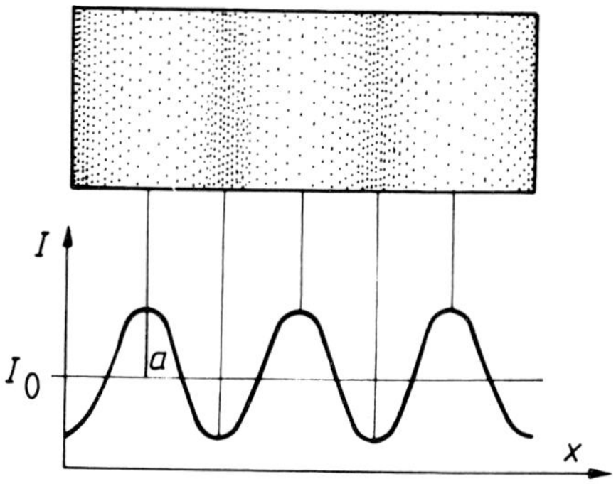

Figure 1.37. Sinusoidal optical grid

Optical grids of different density, consisting of parallel dark and light strips, are attached in front of the optical system as objects on which the intensity distribution I of the dark and light strips is sinusoidal *(Fig. 1.37)*:

$$I = I_0 + a \sin px\,, \tag{1.52a}$$

where

I_0 = the mean intensity,
p = the cyclical frequency,
ν = $p/2\pi$, the space frequency given in units of line pairs/mm if x is given in mm,
a = the amplitude.

In practice, however, optical grids with transparent and opaque strips (generating *square waves*) are used. Analogously the intensity distribution of the image of an object, formed by the optical system, is:

$$I' = I'_0 + a' \sin px\,. \tag{1.52b}$$

The contrast level is expressed by:

$$K = \frac{I_{max} - I_{min}}{I_{max} + I_{min}}\,. \tag{1.52c}$$

For the object:

$$I_{max} = I_0 + a$$

and

$$I_{min} = I_0 - a\,,$$

by means of which the contrast level of the object is:

$$K = \frac{a}{I_0}. \tag{1.52d}$$

Similarly the contrast level of the image is:

$$K' = \frac{a'}{I'_0}. \tag{1.52e}$$

The contrast transfer function is the ratio of the contrast levels of the image and of the object:

$$\frac{K'}{K} = \frac{I_0 a'}{a I'_0}. \tag{1.52f}$$

The function (1.52f) is usually modified by means of a constant multiplier so that its maximum is 1 at the space frequency $\nu = 0$ *(Fig. 1.38)*.

The contrast level of an image continuously decays as more and more dense object grids are used. By increasing ν the contrast transfer function decreases; this is caused partly by diffraction due to the limited aperture of the optical system, partly by the image-forming errors and the defects of manufacture.

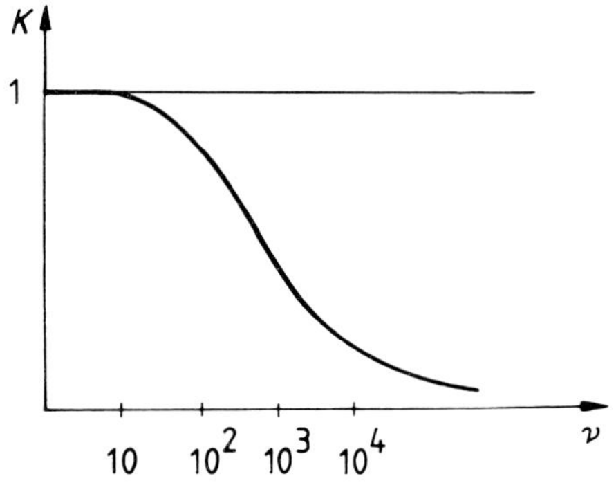

Figure 1.38. Constrast transfer function

For telescopes which form an image at infinity from an object at infinity, the space frequency is usually expressed in units of line pairs per second of arc, instead of line pairs per mm.

1.3.9 Manufacture of Optical Elements

The material of the optical elements used in geodetic instruments is in general optical glass, seldom quartz glass or optical crystal, and sometimes a liquid or a gas. The optical characteristics of glass, primarily its refractive index n_D and dispersion

$(n_F - n_C)$, are determined by the quantity and the quality of the basic material (SiO_2) as well as that of the materials (e.g. BaO and PbO) used in the alloy. The Abbe number ν characteristic for crown glass is $\nu > 50$, that for flint glass is $\nu < 50$.

Deficiencies in glass manufacturing technology can cause the following defects: occurrence of bubbles, veiling, wire-shaped defects, inclusion of impurities, crystallization, incorrect vitrification, internal tensions and discolouration. Glass defects may considerably deteriorate the image-forming quality of an optical system otherwise very carefully constructed and corrected. The requirements of optical glass vary according to the place and function in the optical system of the element in question. For example, fully bubble- and impurity-free glass must be required for optical elements located in the image plane or near to it, thus primarily for the cross-hair plate of a geodetic telescope or a reading microscope. The physical characteristics of optical glass are published in the glass catalogues of the glass factories.

Sometimes it is reasonable to use the more expensive and labour intensive quartz glass or optical crystal instead of optical glass. Quartz glass is much more transparent to ultraviolet and infrared rays than optical glass. Some optical crystals are *anisotropic* (birefracting), thus good for light polarization. From the point of view of electro-optical tacheometers, ADP (ammonium dihydrogen phosphate), KDP (potassium dihydrogen phosphate) and nitrobenzene are of importance, due to their electro-optical characteristics (see Chapter 10).

For the production of the elements of an optical system, design calculations are made and carefully corrected on the basis of aberration analyses, then detailed technological specifications are prepared. These contain the machining processes, the type codes of machine tools to be used, including the type and quality of holding, machining and cutting devices, measuring and auxiliary instruments, grinding and polishing materials, cleaning agents and adhesives. The number of the common work pieces to be machined and last, but not least, the mechanical dimensions as well as the tolerances of the optical data are also included.

The machining processes can be divided into two groups as follows:

1. shaping of mechanical dimensions (diameter, thickness, spherical radius, inclination angle of planes, etc.);
2. finishing, cleaning and controlling of the surfaces (fine-optical processing).

1.4 The Human Eye

The majority of geodetic and surveying instruments can be used visually, therefore the physiological and optical characteristics of the human eye must be taken into consideration at their design stage. The optical system of a human eye *(Fig. 1.39)* consists of the cornea *1*, the aqueous humour *2*, the lens *3* which changes curvature in order to change focus when viewing objects at different distances, the vitreous body *4*, and the pupil *5* which accomodates to the light intensity entering the eye.

The average human eye as an optical system has a focal length of 16.7 mm; this corresponds to 59.7 dioptres (1 dioptre = 1000 mm divided by the focal length in mm). The focal length of the eye changes according to the distance of the observed object

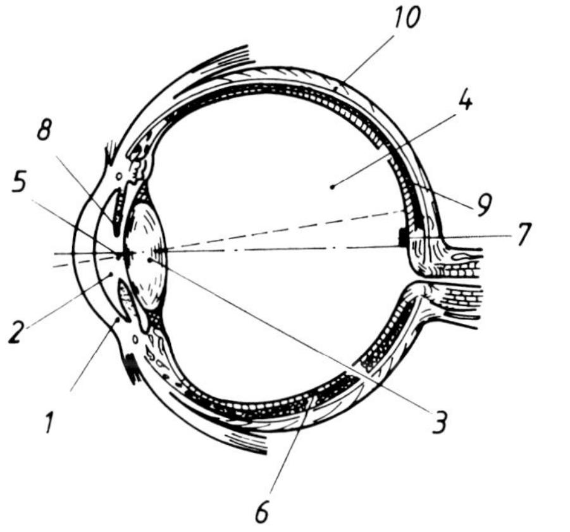

Figure 1.39. The human eye: *1* cornea; *2* aqueous humour; *3* eye-lens; *4* vitreous body; *5* pupil; *6* retina; 7 fovea; *8* iris; *9* veiny skin; *10* horny skin

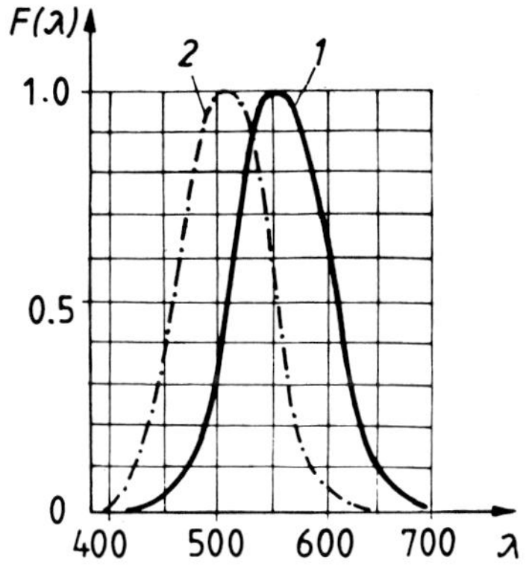

Figure 1.40. Sensitivity of the eye as a function of the light wavelength: *1* sensitivity in day-time; *2* sensitivity at twilight

(accomodation). The efficiency of accomodation depends on age: at age 10 the minimum distance of clear sight is about 100 mm (10 dioptres), but at age 60 it is about 1000 mm (1 dioptre).

The sharpest image formed on the retina *6* of the average human eye is that of an object at a distance of $\delta = 250$ mm (clear sight distance). Depending on light intensity the diameter d of the pupil changes between 2 and 8 mm *(adaptation)*; in daylight of medium intensity it is $d = 4$ mm. The retina contains blood vessels, branches and ends of the visual nerve, light sensitive cones and rods. The centre of clear sight on the retina is a spot of 0.3 mm diameter called the fovea *7* which contains only cones, the rods being situated somewhat farther from it. The number of small cones decreases with increasing distance from the fovea and sight transmittance is more and more taken over by the rods. The number of cones is about 6 million, of the rods about 120 million, while the number of visual nerve ends amounts to about 1 million. The size (diameter) of the cones increases from 1 to 2.5 μm from the centre of the fovea towards the outer areas of the retina. The diameter of the rods is about 2 μm. The field of view of the eye is elliptical in shape: the angle of view in the vertical is about 130°, in the horizontal about 200°.

The human eye perceives light rays of various wavelengths differently. In day-time the strongest excitation is generated by rays of wavelength $\lambda = 555$ nm; the sensitivity $F(\lambda)$ of the eye decreases at both increasing and decreasing wavelengths *(Fig. 1.40)*. At twilight the greatest sensitivity shifts towards shorter wavelengths.

The *resolving power* of the normal human eye is 1 minute of arc. The angle of view of a still readable capital letter on the test-chart used in ophthalmology (e.g. the letter *E*) is 5 min of arc, while that of some individual elements (e.g. the three horizontal lines of the letter *E* or the two separations between them) is 1 min of arc. The reciprocal value of the resolution of one individual element, expressed in minutes of arc, is called visual acuity. For some exceptional people visual acuity may even reach the value 2 or 3 which corresponds to 30 or 20 sec of arc. Visual acuity relates only to the fovea and its value decreases rapidly with increase of the angle of view. *Figure 1.41* shows how visual acuity decreases as the direction of the light ray departs from the eye axis relative to the fovea. If the illumination of the object becomes weaker, the pupil size increases and the rods gradually take over the role of the cones. With poor illumination the human eye is colour-blind and visual acuity also decreases with decrease of illumination. The visual acuity of 1 min of arc of the normal eye decreases to 5 min of arc in moonlight. Visual acuity also depends both on the contrast of the object observed and on the colour of the light, being a maximum in yellow and green. The eye as an optical system has a chromatic aberration of 1.5 dioptres.

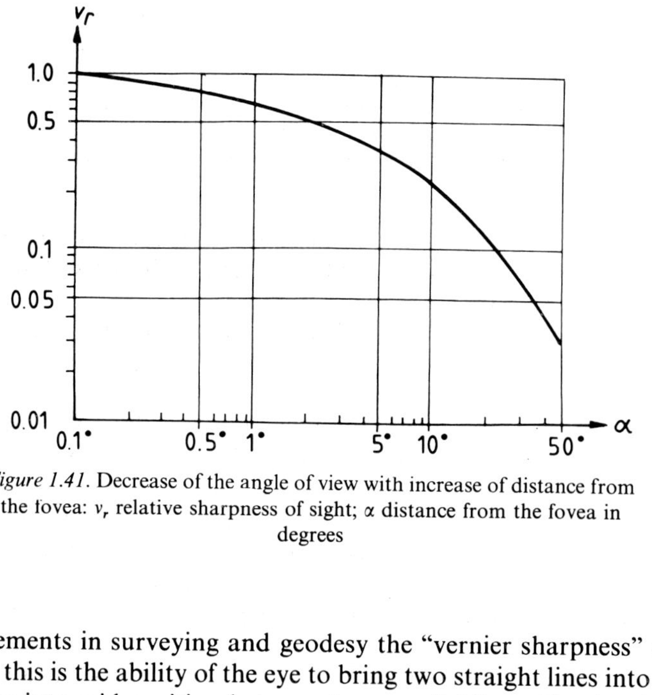

Figure 1.41. Decrease of the angle of view with increase of distance from the fovea: v_r relative sharpness of sight; α distance from the fovea in degrees

For measurements in surveying and geodesy the "vernier sharpness" of the eye is of importance: this is the ability of the eye to bring two straight lines into coincidence and to set a line into mid-position between two parallel lines. Vernier sharpness is 5 to 10 times better than visual acuity as the eye can perceive a dark line of thickness 1 sec of arc against a light background. Against a dark background the thickness of a white line or the radius of a white spot has less significance, only its illumination is of importance.

For optical instruments the estimating ability of the human eye is frequently utilized, by means of which a fraction of an interval (e.g. the number of decimal parts) can be estimated. Within the range of clear sight the size of interval most appropriate for estimation can vary between 0.7 and 2.0 mm. The accuracy of estimation depends on the light contrast between the object and its background.

1.5 Optical Equipments Used in Surveying Instruments

For the observation of small objects or of those at a great distance, when the resolving power of the eye is insufficient, optical magnifying equipment is used. In addition to their observation the possibility of accurate sighting is an important requirement; for this purpose optical equipment with a line of sight is suitable.

1.5.1 Magnifying Glass

The minimum size observable for a normal eye under favourable visibility conditions has an angle of view of 1 min of arc. If one intends to view objects of a smaller angle, a magnifying glass must be used. The magnifying glass is a collecting lens with a short focal length. It should be held so that the object is within the focal length on the object side, i.e. the absolute value $|s|$ of the object distance should be smaller than the focal length f' of the lens. In this position the convex lens forms a virtual, erect and magnified image of the object *(Fig. 1.42)*. The eye should be in the position where the image is formed at the distance δ of clear sight (i.e. comfortable sight). The magnification may be found as follows.

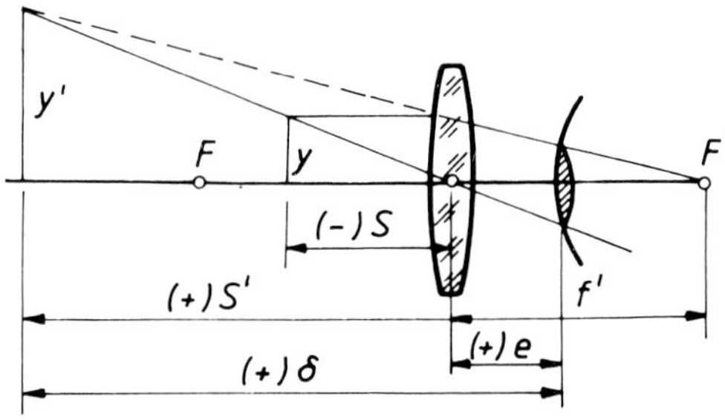

Figure 1.42. Ray path in a magnifying glass

If the object size is y and the image size y', then the lateral (or linear) scale is:

$$\beta = \frac{s'}{s} = \frac{-(\delta - e)}{s}.$$

The ratio between the angles subtended by the object and image at the distance δ, called angular magnification, is the same, namely:

$$\Gamma = \frac{-(\delta - e)}{s}. \tag{1.53}$$

From Eq. (1.38), however, one obtains:

$$\frac{1}{s} = -\frac{1}{f'} + \frac{1}{s'} = \frac{-1}{f'} + \frac{1}{-(\delta - e)},$$

and substituting it into Eq. (1.53) the angular magnification of the magnifying glass is:

$$\Gamma = \frac{\delta}{f'} - \frac{e}{f'} + 1. \tag{1.54}$$

The magnification thus depends not only on the focal distance f' of the lens, but also on the distance δ of comfortable sight as well as on the distance e between the eye and the magnifying glass. Since the magnification is nearly in direct proportion to the distance of comfortable sight, the magnification is higher for a long-sighted person and lower for a short-sighted person. Maximum magnification occurs at $e=0$; then:

$$\Gamma = \frac{\delta}{f'} + 1. \tag{1.55}$$

According to this equation one should move the eye close to the lens when using a magnifying glass. Equation (1.55) also clearly shows that magnification can only occur if f' is shorter than the distance δ of comfortable sight, which is 250 mm. The higher the magnification, the shorter is the focal length f'. Minimum magnification occurs at $e=f'$, in which case one obtains from Eq. (1.54):

$$\Gamma = \frac{\delta}{f'}, \tag{1.56}$$

where $\delta=250$ mm. This is the so-called standard magnification of a lens.

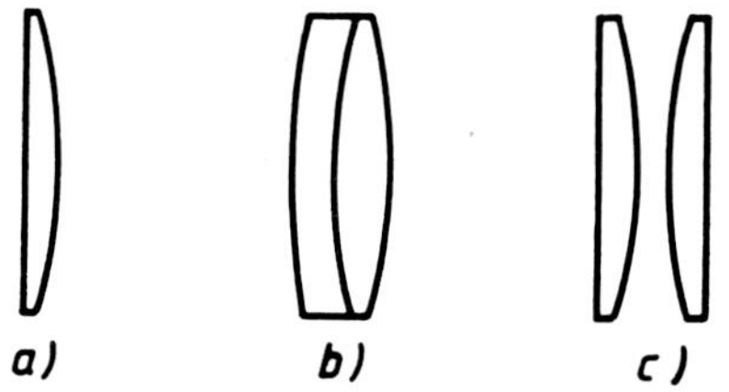

Figure 1.43. Magnifying glass types: *a)* plano-convex; *b)* cemented; *c)* two lenses with air separation

In general, any collecting lens with an appropriate focal length can be used as a magnifying glass. For low magnification (2.5 times) the simple *plano-convex* lens is very suitable, with its curved surface towards the eye *(Fig. 1.43, a)*. For higher magnification image-forming errors must be corrected, hence compound lens systems must be used. In order to reach a magnification of up to 6 times, a lens system consisting of two lenses cemented to each other, i.e. an achromat, is used *(Fig. 1.43, b)*. A magnifying glass consisting of two plano-convex lenses with their convex surfaces towards each other *(Fig. 1.43, c)* is also popular. Higher magnification and a larger field of view can be attained with more complex lens systems (consisting of 3 or 4 lenses), but such magnifying systems are not used in geodesy or surveying.

1.5.2 Microscope

Microscopes are used in the reading devices of geodetic and surveying instruments in order to magnify the angles subtended by the intervals between graduation lines. The three main parts of a microscope are: the eyepiece *1*, the objective lens system *2* and the illumination system *3 (Fig. 1.44)*. The task of the illumination system is to illuminate the object to be viewed well and uniformly, by means of daylight or an artificial light source. The objective lens system of the microscope is a combination of collecting lenses which generates, in the focal plane of the eyepiece, an enlarged image y' of the object y lying at a distance $(-f_{ob}+z)$ from the principal plane of the object lens system. From *Fig. 1.44*, the image is generated at distance $z'=d$ from the focal point on the image side of the objective lens. The distance between the image-side focal point F'_{ob} of the objective lens and the object-side focal point F_{oc} of the eyepiece is called the optical *tube-length*. Knowing this one obtains from Eq. (1.29) by substituting $z'=d$ and $f'=f_{ob}$:

$$z=\frac{-f_{ob}^2}{d}. \tag{1.57}$$

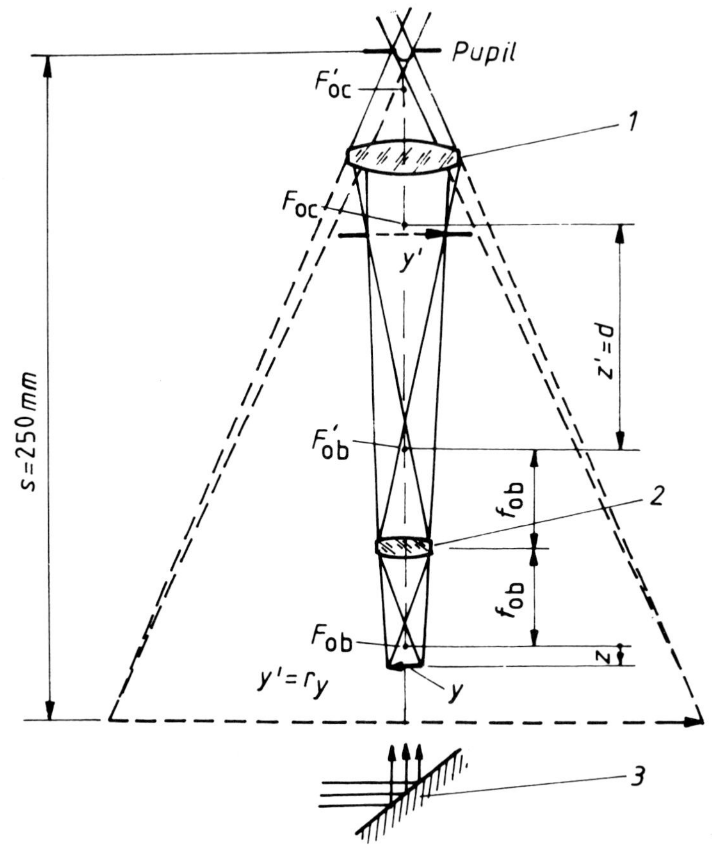

Figure 1.44. Microscope

The linear image-forming scale of the objective can be obtained from Eq. (1.39) by substituting $z'=d$:

$$\beta_{ob}=\frac{y'}{y}=-\frac{d}{f'_{ob}} \tag{1.58}$$

The image magnified by the objective lens is observed through an eyepiece having an angular magnification of Γ_{oc} and a focal length of f_{oc}. The resultant magnification of a microscope is:

$$\Gamma_{micr}=\beta_{ob}\Gamma_{oc}, \tag{1.59}$$

or by using Eqs (1.58) and (1.56):

$$\Gamma_{micr}=-\frac{d}{f_{ob}}\frac{250}{f_{oc}}=-\frac{250}{f}, \tag{1.60}$$

where f is the resultant magnification of the microscope:

$$f=-\frac{f_{oc}f_{ob}}{d}. \tag{1.61}$$

From Eq. (1.60) it can be recognized that the microscope acts as a magnifying glass which yields a fully reversed image.

In addition to the higher magnification, the microscope has the significant advantage with respect to the magnifying glass that the focal plane of the eyepiece acts as an image plane where a cross-hair plate (or reticule plate) can be placed. The reticule plate of the reading microscope can contain a single line (measuring mark or *index*), by means of which the interval between two adjacent graduations can be subdivided by estimation. Furthermore, the reticule plate can contain the subgraduation of a graduation interval of the observed primary graduation, by means of which reading accuracy can be increased. The former type is called a *line microscope* (or estimation microscope), and the latter a *scale microscope.*

The magnification of the optical equipment needed for viewing the graduations is chosen so that the optically magnified length of the smallest interval should be 0.7 to 2 mm. If a magnification higher than $\Gamma=14\ldots20$ is required, it is suitable to use a microscope. The magnifying power of reading microscopes used in geodesy amounts to $\Gamma=50\ldots80$ times.

1.5.3 Telescope

The aim of a telescope is to enlarge the angle of view of a distant object, thus to make recognizable more details than is possible by observing it with an unaided eye. Object and image are both at infinity but their angles of view are different. Such an optical system consisting of several refractive surfaces or lenses, which satisfies the condition that both the object and image distances are infinite, is called an *afocal* or *telescopic* system. The resultant focal length of such a system is infinitely long.

The telescopic system is effected by the combination of two lenses or lens systems, in which the image-side focal point of the first lens or lens system coincides with the object-side focal point of the second lens or lens system. The distance between the principal plane H'_1 of the first lens (system) on the object side and the principal plane H_2 of the second lens (system) on the object-side is:

$$e_{1,2} = f'_1 - f_2 = f'_1 + f'_2, \tag{1.62}$$

from which the focal distance of the total system is, from Eq. (1.44):

$$f'_0 = \infty. \tag{1.63}$$

One of the two lenses (lens systems) must be a collecting lens, because at least one lens must have a real focal point. Hence a telescopic system may consist of two collecting lenses (lens systems) or of a collecting lens (lens systems) and a dispersing lens (system).

The optical train of the so-called *Kepler telescope* or *astronomical telescope*, consisting of two collecting lenses and yielding a reversed image, is shown in *Fig. 1.45*.

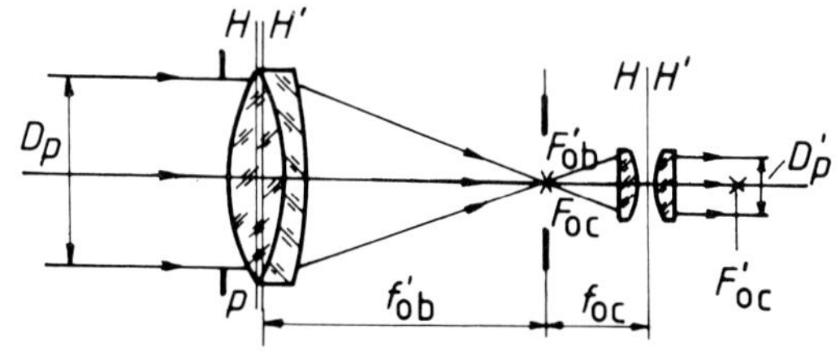

Figure 1.45. Optical train in a Kepler telescope

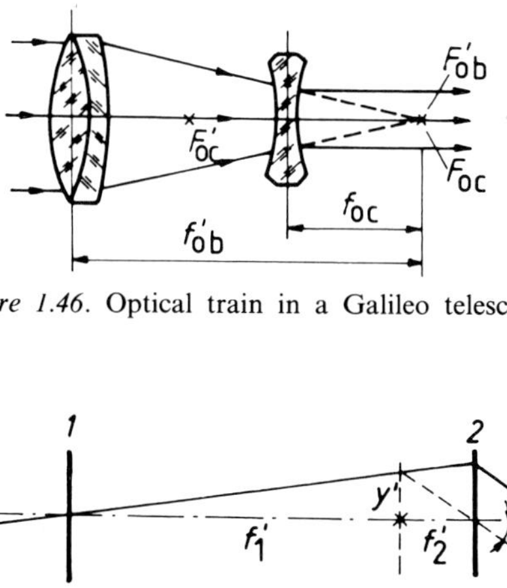

Figure 1.46. Optical train in a Galileo telescope

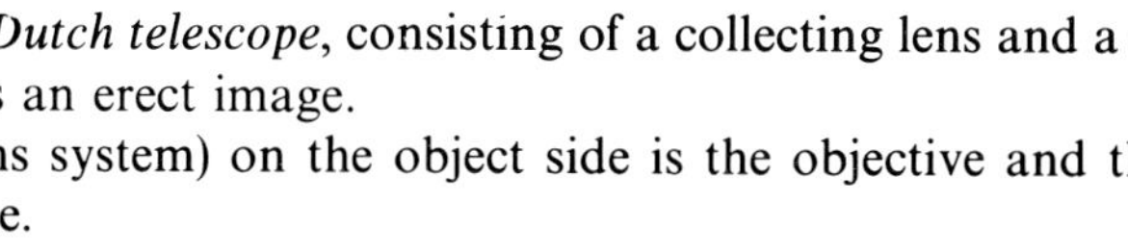

Figure 1.47. Angular magnification of a Kepler telescope

The *Galileo* or *Dutch telescope*, consisting of a collecting lens and a dispersing lens *(Fig. 1.46)*, yields an erect image.

The lens (or lens system) on the object side is the objective and that on the eye side is the eyepiece.

In the Kepler telescope an object of length y in the object plane, perpendicular to the axis, subtends an angle σ *(Fig. 1.47)*, given by:

$$\tan \sigma = \frac{y'}{f'_1}. \tag{1.64}$$

The image y' generated by the objective lens *1* is seen through the eyepiece *2* to subtend an angle σ' given by:

$$\tan \sigma' = \frac{y'}{f_2}. \tag{1.65}$$

The angular magnification of a telescope is the quotient of these two values:

$$\Gamma = \frac{\tan \sigma'}{\tan \sigma}. \tag{1.66}$$

In the paraxial zone the angular magnification is:

$$\Gamma = \frac{\sigma'}{\sigma} = \gamma, \tag{1.67}$$

or taking Eqs (1.64) and (1.65) into consideration:

$$\Gamma = \frac{f'_1}{f_2} = -\frac{f'_1}{f'_2}. \tag{1.68}$$

From Eq. (1.68) it is obvious that the magnification of the Kepler telescope has a negative sign (reversed image) and that of the Galileo telescope has a positive sign (erect image). If the optical train of an outer axis-parallel ray is taken into consideration *(Fig. 1.48)*, then one can write:

$$\frac{h_1}{h_2} = \frac{f'_1}{f_2} = -\frac{f'_1}{f'_2}, \tag{1.69}$$

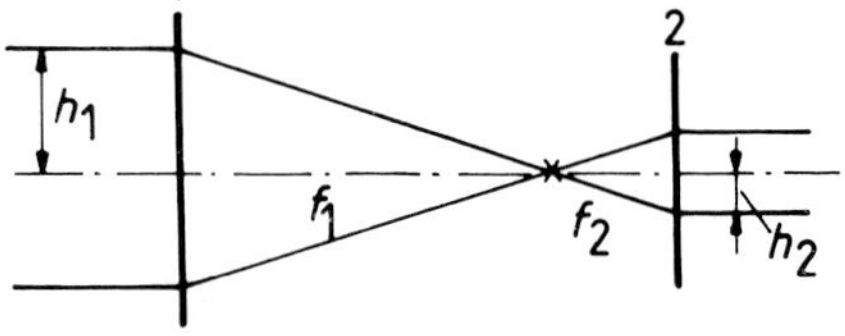

Figure 1.48. The diameter of the entrance pupil as a function of telescope magnification

where h_1 is the semi-diameter of the entrance pupil D_p, h_2 is the semi-diameter of the exit pupil D'_p. The angular magnification of a telescope can thus also be expressed as the ratio of the entrance and exit pupil diameters:

$$\Gamma = \frac{D_p}{D'_p}. \tag{1.70}$$

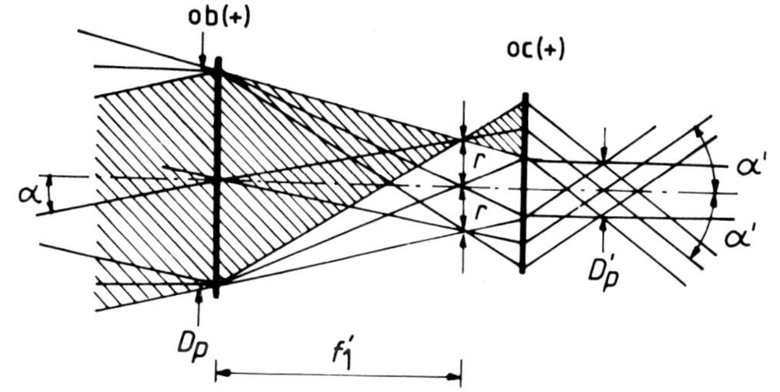

Figure 1.49. Angular magnification of a Kepler telescope

Though Eqs (1.67), (1.68) and (1.70) are mathematically identical, nevertheless the magnifications determined from the measurement of the angles, of the focal length and of the pupils show slight differences due to the different effects of the image forming errors. In practice magnification is determined by means of Eq. (1.70) since, under manufacturing conditions, the simplest measurement is that of the pupils.

The entrance pupil D_p of the Kepler telescope is determined by the free aperture of the objective *(Fig. 1.49)*. The image generated from the entrance pupil by the part of the optical system behind the objective, i.e. by the eyepiece, is the exit pupil D'_p which is formed in the space behind the eyepiece as a real image (apparent on a screen). The observer's eye has to be positioned at the exit pupil in order to see the total field of view of the telescope.

The field of view of the Kepler telescope is determined by the diaphragm R surrounding the field of view, which is realized by a mechanical diaphragm (field stop) with a diameter $2r$, placed in the plane of the intermediate image. The image of the field stop generated by the objective in the object space is an entrance window with diameter D_a, which bounds the extension of the object apparent through the telescope. The complete viewing angle of the visible object is the angle 2α between two outside principal rays:

$$\tan \alpha = \frac{r}{2f'_1}. \tag{1.71}$$

The field of the image (subjective field of view) is bounded by the image on the edge of the field stop R, formed by the eyepiece, which is an exit window with diameter D'_a. The angle of view $2\alpha'$ of the image field is determined by:

$$\tan \alpha' = \frac{r}{2f'_2}. \tag{1.72}$$

The image-forming quality of a telescope is determined primarily by the quality of the objective. A light ray originating from a distant object, e.g. from a star, is diffracted by the entrance pupil of the objective (i.e. by the internal edge of the objective lens seating). This phenomenon is called *Fraunhofer diffraction* and a diffraction image corresponding to the size of the entrance pupil is generated in the focal plane, the *Airy disc (Fig. 1.50)*. The images of two close object points can still be visually distinguished if the maximum points of their diffraction images are at least far enough apart for a zone darker by 3 . . . 5 percent to separate them. In *Fig. 1.51, a* the maximum points of the Airy discs of each of two object points coincide with the minimum points of the other: this phenomenon is called *Rayleigh resolution.* In this case the distance between the two maximum points in the focal plane of the objective is $t_0 = 2r_0$, where r_0 is the radius of the Airy disc:

$$r_0 = \frac{0.61}{D_p} \lambda f'_{\text{ob}} \text{ (mm).} \tag{1.73}$$

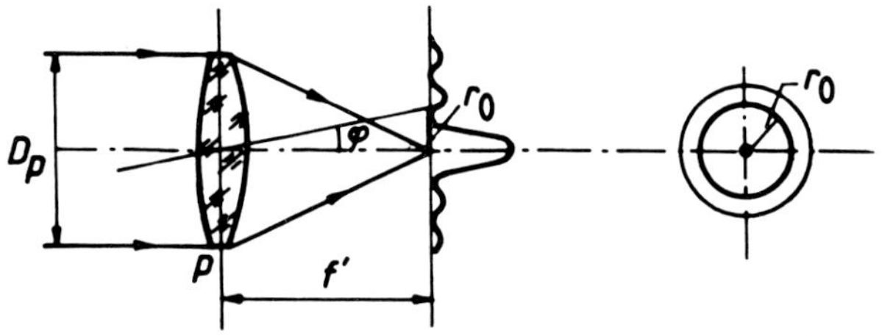

Figure 1.50. Resolving power of the objective

The contrast of the resolution is 26 percent. The distance between the maximum points can be reduced, however, to $t_{\min} = 0.85 \times 2r_0$ *(Fig. 1.51, b)* to give the desired minimum contrast of 3–5 percent:

$$t_{\min} = \frac{2 \times 0.61}{D_p} \lambda f'_{\text{ob}} 0.85. \tag{1.74}$$

The two nearest points which can still be resolved are separated by the angle:

$$\varphi \cong \tan \varphi = \frac{t_{\min}}{f'_{\text{ob}}} = \frac{1.22 \lambda 0.85}{D_p}. \tag{1.75}$$

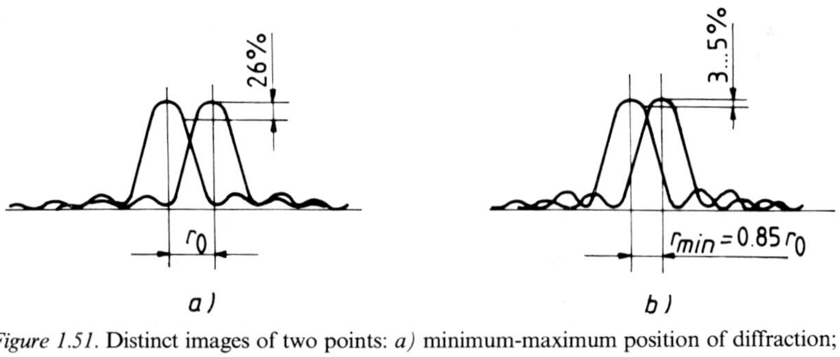

Figure 1.51. Distinct images of two points: *a)* minimum-maximum position of diffraction; *b)* minimum contrast position

If the wave length of the transmitting light is $\lambda = 555$ nm, then the value φ in seconds is:

$$\varphi'' = \frac{120''}{D_p}, \tag{1.76}$$

with the entrance pupil D_0 expressed in nm.

This is the resolving power of an ideal objective, but it is impaired in practice by the aberrations of the objective, by glass defects and by defects in machining. The resolving power of an actual objective is about 20 percent less than that of an ideal one. Knowing the entrance pupil diameter D_p necessary to achieve the required resolving power, the focal length of the objective should be chosen so that, by correcting the imaging errors, the parameters remain within tolerance interval.

The eyepiece used for viewing the image, generated on the reticule, should have a magnification at which the angle subtended by the smallest object resolvable by the objective should exceed the resolving power of the human eye.

The brightness of a point-type object increases proportionally to the square of the telescope magnification. The brightness of the image of a larger object, covering part or the whole of the field of view, remains unchanged apart from the light loss within the telescope. Consequently, viewing through a telescope gives improved brightness for a point-type object, and thus enhances the difference in contrast between the object and its background.

Furthermore, as a result of undesirable reflections from the optical surfaces due to imperfect mirror reflection and light absorption of optical glass, only part of the incident light can pass through the telescope. The ratio of the amounts of light passing through the entrance pupil and the exit pupil is called *damping*. In order to reduce damping, a *reflection-decreasing* film is applied to light-transmitting surfaces and a *reflection-increasing* interference coating to light-reflecting surfaces (see Subsection 1.1.3). Their use is also reasonable because undesired reflections give diffuse light rays which impair imaging.

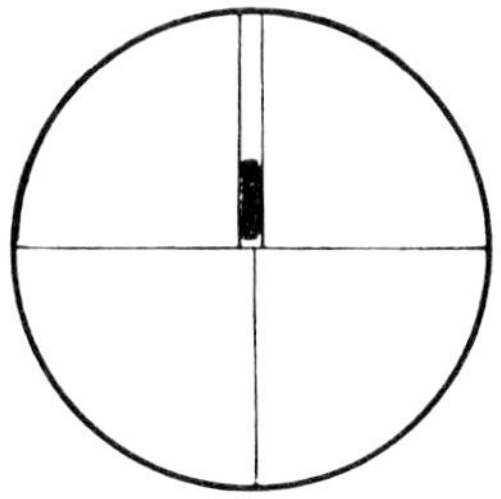

Figure 1.52. Sighting with dual vertical cross-hair lines

In general a signal at a geodetic or surveying control point is imaged on the cross-hair plate as a very narrow line which must be brought into coincidence with the axis of symmetry of the double vertical cross-hairs *(Fig. 1.52)*. Since the vernier sharpness of the human eye can reach the value $\varphi_n = 10''$ (see Subsection 1.1.4), the theoretical standard deviation of pointing with a telescope of magnification Γ is:

$$\mu_i = \frac{\varphi_n}{\Gamma} = \frac{10''}{\Gamma}. \tag{1.77}$$

The accuracy of pointing can be enhanced by the use of a *double image eyepiece* in which the image of the object is split into two halves which are made to coincide *(Fig. 1.53)*. An image splitting system is arranged in front of the eyepiece which consists of a trapezoidal prism *1* and of a roof (trapezoidal) prism *2*. The reflecting surface *3* is applied to those parts of the surfaces at which the two prisms are cemented together. The light beam entering the prism system is split into two parts at the front edge *a* of the reflecting surface *3*. One of the beams enters prism *1*, the other the roof prism *2*. The beams are reflected from the prism surfaces opposite to the reflecting surface *3*; their optical paths are thus of equal length and they meet again at the rear edge *b* of the reflecting surface *3* which serves as the splitting edge. The prism system should be in such a position in the telescope that the rear edge *b* of the reflecting surface *3* lies in the image plane of the telescope. In this case each of the two partial light beams generates a partial image, and they can be made to coincide along the edge. One of the half images is side correct, while the other half image generated by

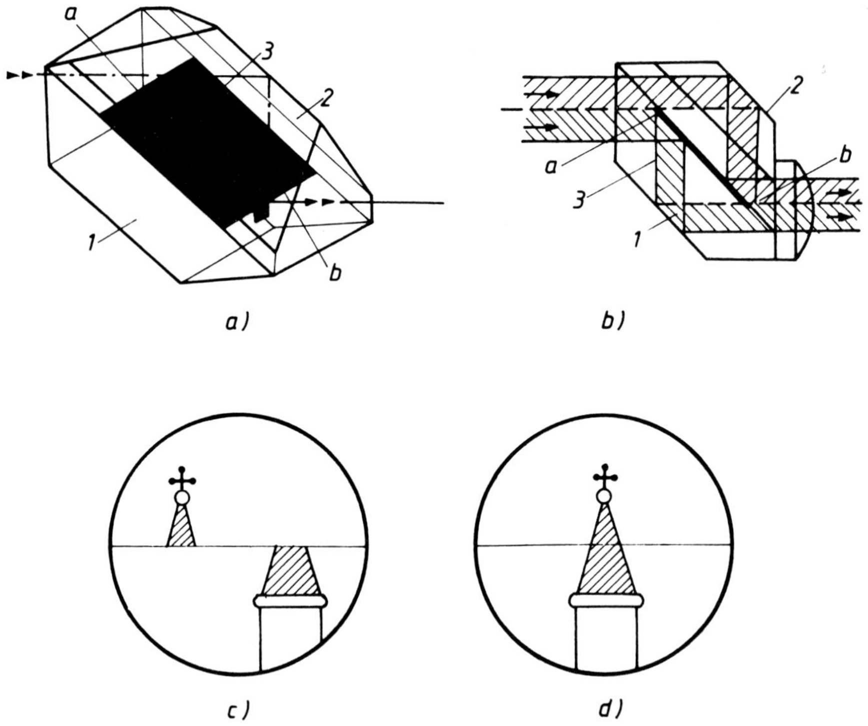

Figure 1.53. Sighting with double image eyepiece: *a)* image splitting prism system (Geodetic and Geophysical Research Institute of the Hungarian Academy of Sciences, Sopron); *b)* side view of the image splitting prism system; *c)* non-coincident half-images; *d)* coincident half-images

the light beam passing through the roof prism *2* is side-reversed. Thus, by changing the sighting direction, the two half images together cover twice the distance that the vertical hair covers when pointing with a conventional cross-hair. In this way pointing accuracy is increased.

The resolving power of the human eye is utilized optimally if the exit pupil of a telescope is $D'_p = 1.0 \ldots 1.2$ mm. If the acceptable pointing error is μ_i'', then from Eq. (1.77) the telescope magnification required is:

$$\Gamma_{\min} = \frac{10''}{\mu_i''}. \tag{1.78}$$

Knowing the magnification and the exit pupil, the diameter of the objective can be obtained from Eq. (1.70) as:

$$D_p = \Gamma D'_p. \tag{1.79}$$

Commensurate with the diameter D_p of the entrance pupil, the shortest objective focal length acceptable from the point of view of aberration is selected. On the basis of the data obtained in this way the optical system of a geodetic telescope can be designed.

Telescope objective. The simplest objective consists of two lenses, the simple astronomical telescope having just this. The image of an object at a finite distance appears in an image plane which differs from the focal plane *(Fig. 1.54)*. In order to bring the image of an object at distance $(f + z)$ to the cross-hair plane, the latter should be moved by the distance z' which can be calculated by Eq. (1.29).

The shortest observing distances of geodetic telescopes used in practice lie between 0.6 and 5 m. When moving the tube supporting the reticule plate the line of sight of the telescope wobbles, mainly due to defects in the moving mechanism of the reticule tube (Point 2.2.5.2). A further disadvantage of the simple telescope is the open mechanism needed for moving the reticule. These disadvantages are greatly reduced in a telescope with internal focussing *(Fig. 1.55)*. The collecting lens L_1 and the dispersing lens L_2, being at a variable separation e, give together a *teleobjective* with resultant focal length [from Eq. (1.43)]:

$$f = \frac{f_1 f_2}{f_1 + f_2 - e}. \tag{1.80}$$

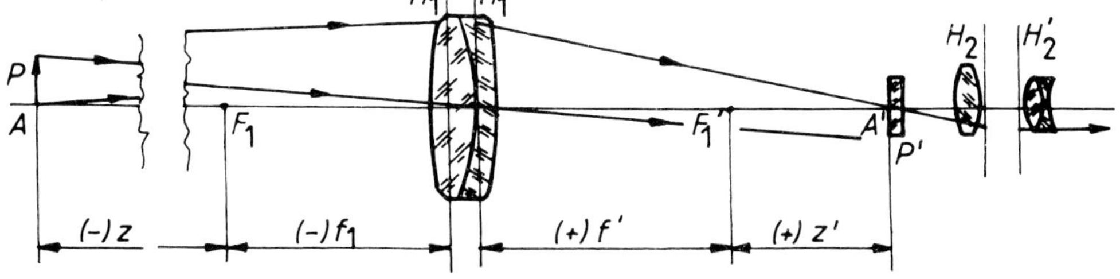

Figure 1.54. Image of a close object generated in the movable image plane

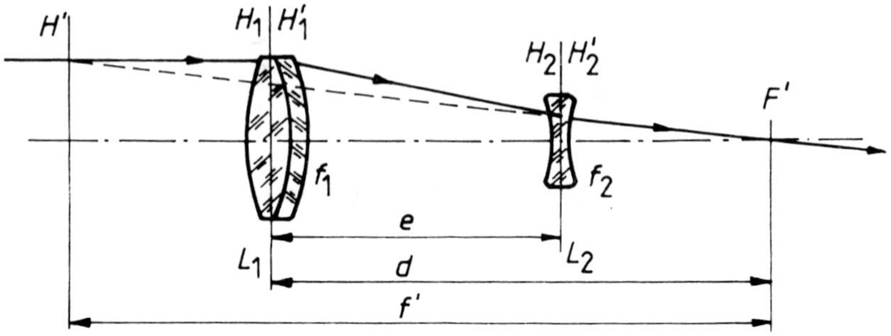

Figure 1.55. Teleobjective

If the object distance changes between $s=-\infty$ and $|s|=|s|_{min}$ (s may have negative sign), by moving the dispersing lens along the axis (i.e. by increasing the distance e) the image can always be adjusted into the fixed cross-hair plane. The distance between the lens L_1 and the rear focal point F' of the equivalent lens, which is in the cross-hair plane, is:

$$d=f+e\left(1-\frac{f}{f_1}\right). \tag{1.81}$$

In the case of a given (constant) d, the distance between tne two lenses, which depends on the object distance, can be obtained from:

$$e=\frac{d+s_1'}{2}-\sqrt{\left(\frac{d+s_1'}{2}\right)^2-f_2(d-s_1')}, \tag{1.81a}$$

where

$$s_1'=\frac{s_1-f_1}{s_1+f_1},$$

and s_1 is the object distance (with negative sign) measured from the objective L_1. For infinite object distances $s_1'=f_1$, hence

$$e=e_0=\frac{d+f_1}{2}-\sqrt{\left(\frac{d-f_1}{2}\right)^2-f_2(d-f_1)}. \tag{1.81b}$$

The mechanism controlling the motion of the focussing negative lens can be manufactured more precisely, thus the *wobble of the line of sight* caused by manufacturing deficiences is smaller (Subsection 2.2.5.2).

In order to improve the imaging ability of the teleobjective and at the same time to decrease the length of the telescope, teleobjectives with *mirrors* or with *mirror lenses* are also used. The use of mirrors primarily decreases chromatic aberration. The optical train of a mirror telescope is shown in *Fig. 1.56, a*, that of a mirror lens telescope in *Fig. 1.56, b*. Teleobjectives with mirror lenses are, however, more sensitive to changes of temperature.

Panfocal telescopes are seldom used as geodetic telescopes. A panfocal telescope consists, besides the eyepiece, of at least three lenses or lens systems, two of them

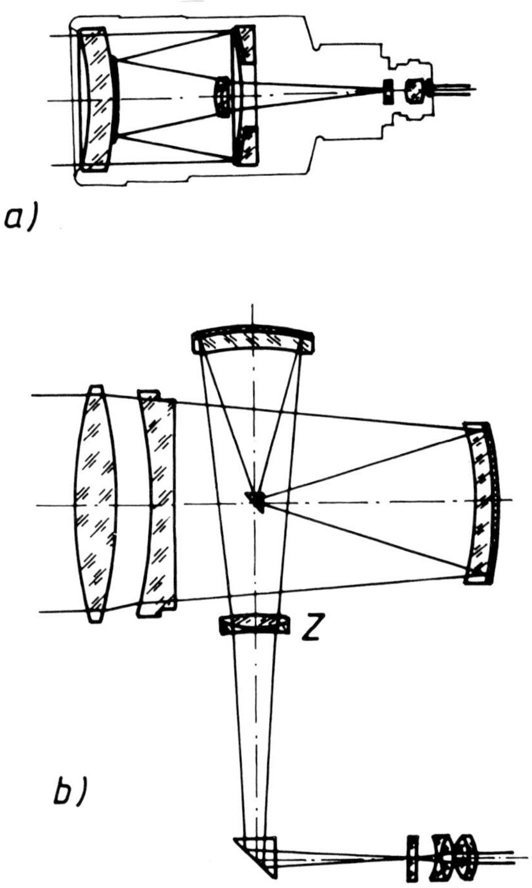

Figure 1.56. Mirror telescope: *a)* the telescope of the theodolite *Theo 010* (VEB Jenoptik, Jena); *b)* the mirror lens telescope of the theodolite *DKM3* (Kern, Aarau)

being moved by focussing. The objective lens and the cross-hair plate (reticule) are fixed. The magnification of such telescopes varies, as does the angle of view. The telescope magnification is a maximum, and the angle of view a minimum, when the observing distance is over 100 m. The telescope magnification decreases and the angle of view increases with decrease of the observing distance; the decrease of magnification can even be 1/6, and the increase of the angle of view 6, depending on the telescope construction. For example, in the case of the precision level *N3* (*Wild,* Heerbrugg) manufactured with a panfocal telescope, the decrease of the observing distance from 100 m to 0.28 m brings about a decrease of telescope magnification from 45 × to 12 ×, as well as an increase of the angle of view from 1° 21′ to 4° 3′. It should be noticed, however, that the focal length of a panfocal telescope changes to such a great extent that its cross-hair plate cannot be provided with stadia hairs.

Cross-hair plate. A hair system (i.e. line system) of proper form and thickness is applied onto the carefully machined surface of the plane parallel glass plate (reticule) which is placed in the image plane of the objective. The form and thickness of the cross-hairs are devised according to the requirements of optical pointing and reading, respectively. The thickness of the lines, as well as their contrast, depends on the manufacturing method used. In modern instruments cross-hair lines with a thickness of 2 to 20 μm are either engraved with a diamond or applied by etching or by micro-photography. In order to protect the cross-hair plate from impurities, it is covered with another glass plate. The black cross-hair lines are only visible if there is sufficient contrast against a light background. If the background is dark, the cross-hair plate is illuminated artificially.

In order to adjust the line of sight of the telescope, the cross-hair plate is mounted in a casing which can be moved in the plane normal to the line of sight (Point 2.2.5.2). The free aperture of the casing ring usually determines the field of view, hence it also acts as a *diaphragm.*

Telescope-eyepiece. For viewing the optical image generated in the reticule plane, a compound lens system is used. Its functional principle is the same as that of a simple magnifying glass (Subsection 1.5.1), but it has a higher performance than the latter. Due to the great variety of types of eyepiece, only a few basic ones are shown in *Fig. 1.57.*

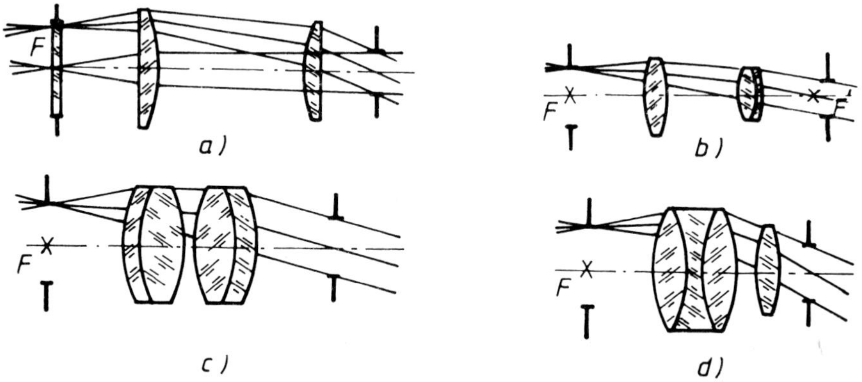

Figure 1.57. Eyepieces: *a)* Ramsden eyepiece; *b)* Kellner eyepiece; *c)* symmetrical eyepiece; *d)* orthoscopic eyepiece

The normal human eye sees the image of the cross-hair lines sharply only if the focal plane of the eyepiece coincides with the cross-hair plane. For short sighted eyes the eyepiece should be moved towards the cross-hair plate, and for far sighted eyes away from it. In order to compensate a visual defect of an observer of up to ± 5 dioptres, eyepieces of geodetic telescope are slightly movable.

To observe near-zenith directions, an eyepiece with a *bent ray path (Fig. 1.58)* is used, or the eyepiece can be substituted by a microscope of equivalent magnification *(Fig. 1.59).*

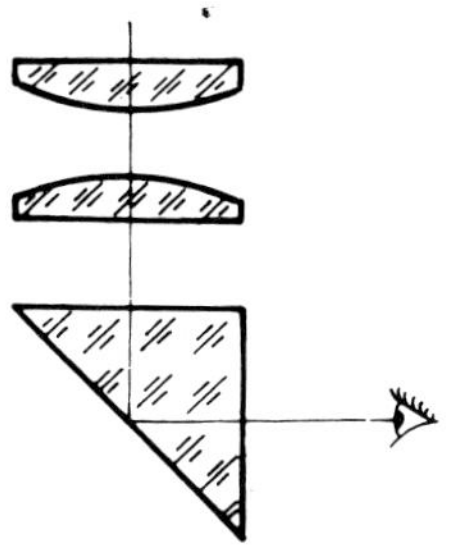

Figure 1.58. Eyepiece with a bent ray path

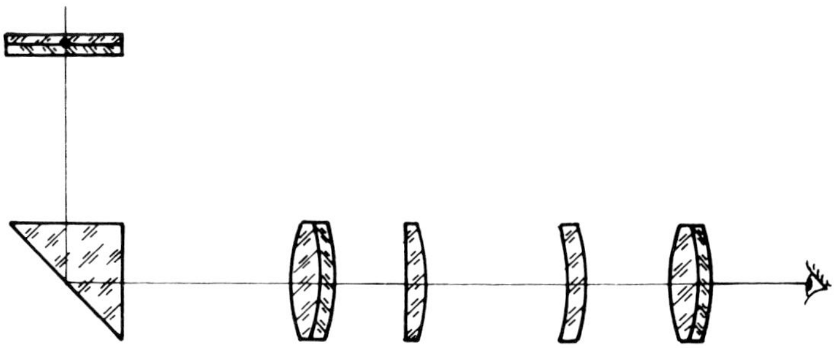

Figure 1.59. Eyepiece microscope

Image-reversing optical systems. From the point of view of pointing it is unimportant whether the image of the viewed object is *erect* or *reversed*. Recently, however, telescope giving an erect image have been preferred in geodetic and surveying instruments. The scheme of the optics of such a telescope is shown in *Fig. 1.60.*

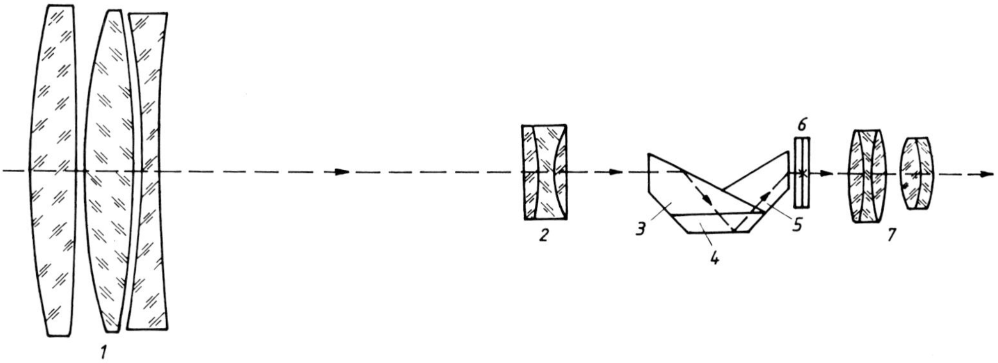

Figure 1.60. Telescope providing an erect image: objective *1,* focussing lens *2,* image reversing system consisting of trapezoid prisms *3* and *5,* roof prism *4,* cross-hair plate *6,* eyepiece *7*

1.5.4 Collimator and Autocollimator

For testing and adjusting geodetic telescopes a distant target is usually necessary (Subsection 2.2.8). In the laboratory a collimator is used instead of a target at an infinite distance *(Fig. 1.61)*. In the focal plane *2* of the objective lens *1* of the collimator, there is a properly illuminated *test plate (Fig. 1.62)* or a graduated *scale for testing the resolving power*, its image substituting for the infinitely distant target.

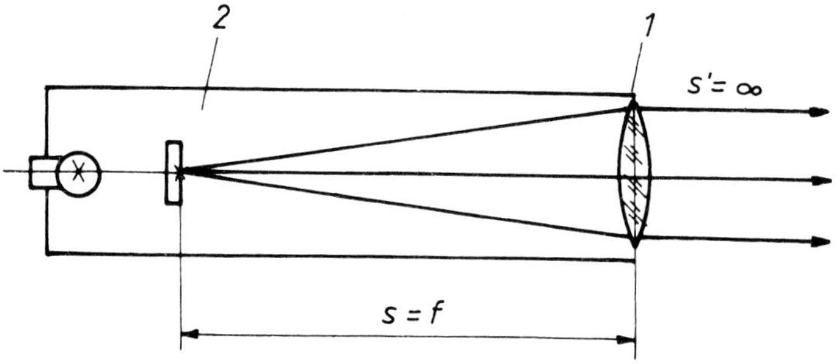

Figure 1.61. Collimator

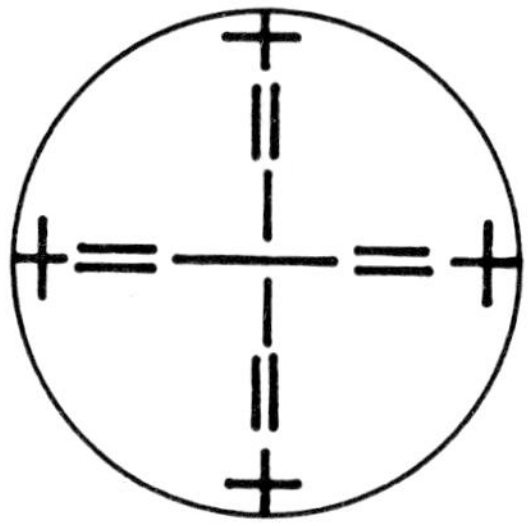

Figure 1.62. Test plate of a collimator

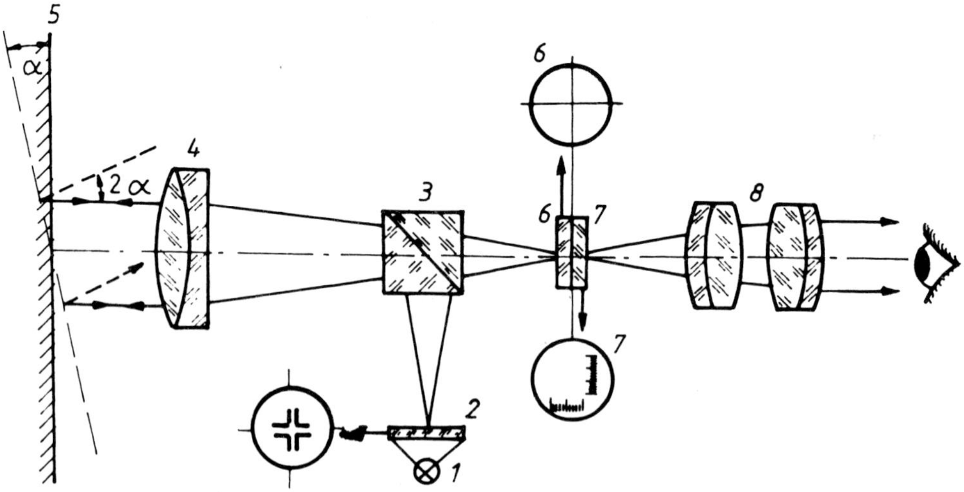

Figure 1.63. Autocollimator

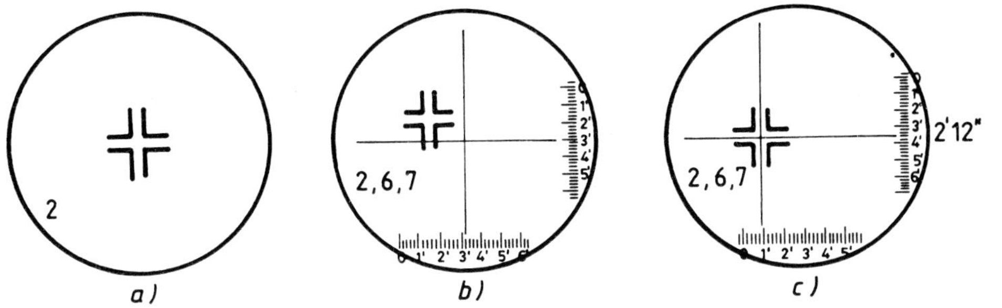

Figure 1.64. The reading of an autocollimator; *a)* illuminated test plate 2; *b)* sighting position; *c)* reading position ($H=0'\ 56''$; $V=3'\ 28''$)

The telescope of a geodetic or surveying instrument focussed to infinity can also serve as a collimator if its cross-hair plate is evenly illuminated from the side of the eyepiece. The precondition of the reliable use of a collimator is that its free aperture and focal length should be greater than those of the telescope to be tested.

A suitable instrument for the accurate measurement of small angles and angle deflections is the autocollimator used as an accessory of the telescope, a variant of which is shown in *Fig. 1.63*. The beam of rays, starting from the light source *1* and passing through the reticule *2*, enters the splitting prism *3* (see *Fig. 1.21*, *a*), passes through the lens *4* and is then reflected from the reflecting surface *5* turned away with angle α, being deflected through the deflecting angle 2α. The beam now passes again through the lens *4*, the prism *3*, the movable cross-hair plate *6* and the graduated reticule *7* and is observed in the eyepiece *8*. *Figure 1.64*, *a* shows the reticule *2*, *Fig. 1.64*, *b* the united image of the three reticules visible in the autocollimator, and *Fig. 1.64*, *c* the reading position after adjustment of the movable reticule *6*.

1.5.5 Laser

Laser (Light Amplification by Stimulated Emission of Radiation) means a monochromatic and coherent light source with small beam divergence. The activated medium in lasers can be: a mixture of different gases (He–Ne, CO_2, N, Ar, etc.), a solid crystal (ruby, neodymium, YAG, etc.), liquid (alcoholic solutions of salts or fluorescent organic combinations of rare earths, e.g. of Eu salts). They can be stimulated by electric current, by external light, by a flash lamp or sometimes by another laser. Lasers can operate in a continuous mode or in a pulse mode.

It was one of *Einstein*'s hypotheses that, in atomic systems exposed to irradiation, a stimulated emission must also appear in addition to absorption and spontaneous emission. Stimulated emission can occur if an atom in a stimulated condition is exposed to an electromagnetic radiation having a frequency (wavelength) which corresponds to the transient (irradiated) frequency of the atom. In this state the stimulated atom emits a coherent radiation which has the same direction and wavelength as the incident radiation. Besides the stimulated emission a precondition of the laser operation is that the saturation of certain energy levels *(population)* must be higher than that of a certain lower energy level *(inverse population)*.

In practice *He–Ne lasers* are those most frequently used and their production costs are the lowest. The output of these lasers lies between 0.1 and 50 mW depending on their size. In them helium atoms are stimulated by electric current and by collision with neon atoms, the latter becoming stimulated also. In the case of neon the laser transitions occur at about wavelengths 632.8, 1150 and 3309 nm. The denser population of the energy level is maintained by the *metastable* state (relatively long life) of the helium atoms. A typical He–Ne laser is shown in *Fig. 1.65*. Tube *1* is filled with a *helium–neon* mixture of 1:6 mixing ratio and the gas pressure is 3×10^{-3} bar. The two windows *2* of the tube make a *Brewster-angle*, so that the light travelling along the longitudinal axis of the tube is *linearly polarized* (oscillating in the plane normal to its incidence plane).

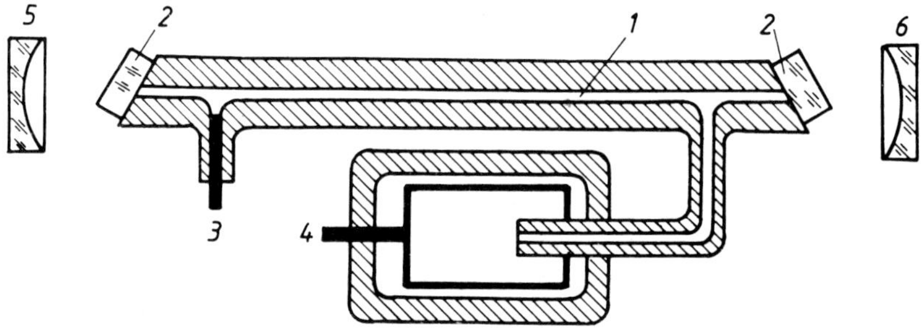

Figure 1.65. He–Ne gas laser

The active medium of the laser is stimulated by direct current coupled to anode *3* and cathode *4* (4–15 mA, 1200–4000 V). In order to create stimulated emission, the tube holds mirrors *5* and *6* of high reflectivity. The end mirror *5* has a reflectivity of at least 99.8 percent, the first mirror one of about 98 percent. The two mirrors constitute a *Fabry–Perot resonator* of length equal to the distance *L* between them, and whose axis coincides with that of the active medium in the tube. The waves emitted from each point of the active medium are partially reflected from the mirrors and return to their starting point after every circuit. The condition for amplification of these elements is $2L = m\lambda$ where λ is the wavelength of the laser, and *m* an integer. Standing waves or modes with wavelengths of the *m* different values are created in the resonator. TEM_{00m} means the *base mode* of the laser, i.e. the *axial mode* corresponding to *m*, and the intensity distribution $\exp(-r^2)$ in the plane normal to the axis of the laser.

Due to its stable construction the laser with *internal mirrors (Fig. 1.66)* is suitable for field use, the tube and the resonator (the two mirrors) forming a closed-mechanical unit *(a directional laser)*.

An important characteristic of the laser is its *divergence*: the divergence of a laser beam is the aperture angle of the cone determined by circles with radii corresponding to $1/e^2$, the reciprocal square of the maximum intensity *e* in the planes perpendicular to the propagation direction of the laser. In terms of its construction data the laser divergence is:

$$\Theta = 2\left(\frac{\lambda}{\pi R_2}\sqrt{\frac{R_2 - L_2}{L_2}}\right)^{\frac{1}{2}}, \tag{1.82}$$

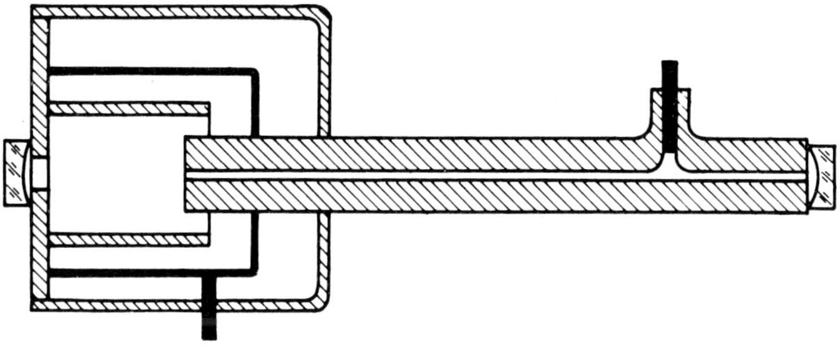

Figure 1.66. He–Ne laser with internal mirrors

where

$$L_2 = L\frac{u_1}{u_1+u_2},$$

$$u_i = 1 - \frac{R_i}{L} \qquad (i=1,2).$$

R_2 is the radius of the end mirror, R_1 the radius of the first one and L is the distance between them.

The beam emerging from the laser propagates in a cone, whose aperture angle corresponds to the divergence, as a spherical wave whose centre lies behind the first mirror at a distance:

$$\frac{R_2 L_2}{R_2 - L_2}.$$

In this virtual position the diameter of the laser beam (the "virtual beam neck") corresponds to the reciprocal square of the maximum intensity e:

$$d_0 = \frac{2R_2}{R_2 - L_2}\left(\frac{\lambda}{\pi}\sqrt{L_2(R_2-L_2)}\right)^{\frac{1}{2}}. \qquad (1.83)$$

The laser *collimator* decreases the laser divergence by using two optical systems: a positive or negative lens system with a short focal length and an objective lens with a long focal length *(Fig. 1.67)*. The convergence relation [Eq. (1.41)] of this optical

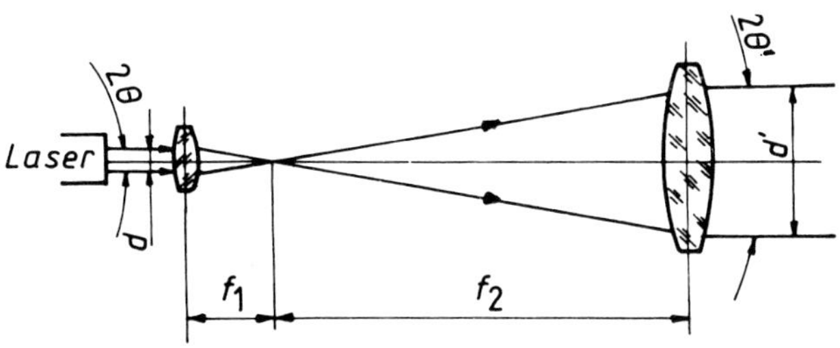

Figure 1.67. Divergence decreasing laser collimator

system is: $\gamma < 1$; its lateral imaging scale (Eq. 1.39) is $\beta > 1$. Hence, corresponding to scale β, the diameter d of the entrance pupil increases to the diameter d' of the exit pupil and the divergence of the laser beam decreases by the ratio d/d' from 2Θ to $2\Theta'$.

1.5.6 Mounting of Lenses

The mounting of optical lenses should be strain-free. Thus, in the case of the very frequently used solution shown in *Fig. 1.68*, sufficient air separation is left between the mount and the lenses to prevent the mount causing any strain in the lenses, even in cold weather. The spherical surfaces *2* are in contact with the spring *1* using cornered intermediate rings so that, in spite of the air separation, the lenses cannot move sideways. Changes in length caused by thermal expansion are also compensated by the spring *1*. The objective can be fixed at the proper distance from the reticule by means of the *intermediate ring 3* of variable thickness, fixed externally to the objective mount. Optical elements of smaller diameter, being less exposed to strain caused by thermal expansion, are fixed by threaded rings or by *flanges*, but fixing with flanges without causing any strain requires great skill.

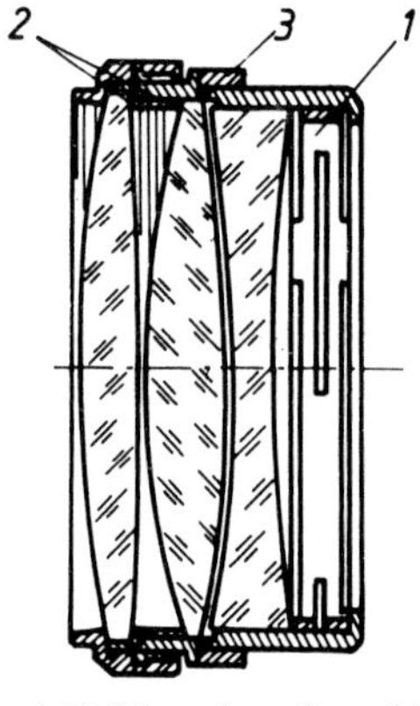

Figure 1.68. Mounting of an objective

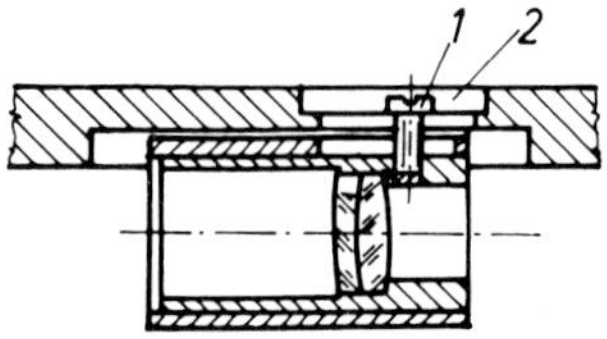

Figure 1.69. Cemented lenses fixed in the mount with cement

Today small optical elements are fixed by cement as shown in *Fig. 1.69*. For cementing the following rules should be observed:

no more than three drops of cementing material should be applied with angular separations of 120°;

the cement should be in contact only with one lens;

while setting, the lens should be held in place by a *spring*.

Cements which harden without any change in volume, and which are slightly elastic after setting, are the most advantageous. *Figure 1.69* shows a solution for the simultaneous adjusting and fixing of a small objective lens. The cylindrical screw *1* can be moved in the notch *2*. After loosening the screw the lens can be set to its correct position by moving the cylindrical screw, and then tightening it.

1.6 Spirit Levels

A spirit level is a sealed glass tube nearly filled with liquid, the internal surface of which is partly or wholly a ground and polished surface of revolution, with convex curvature when looking from without. The liquid does not completely fill the glass tube, so that there remains a small space filled with gas, the so-called *bubble*. When in operation the bubble touches the surface of revolution. If the liquid when still is influenced only by the force of gravity, the centre C of the bubble occupies the highest point of the surface of revolution *(Fig. 1.70)*. The local plumb line is defined by that perpendicular *(normal)* to the surface of revolution which passes through the centre of the bubble or, in other words, by the direction of the line joining the bubble centre C and the curvature centre K.

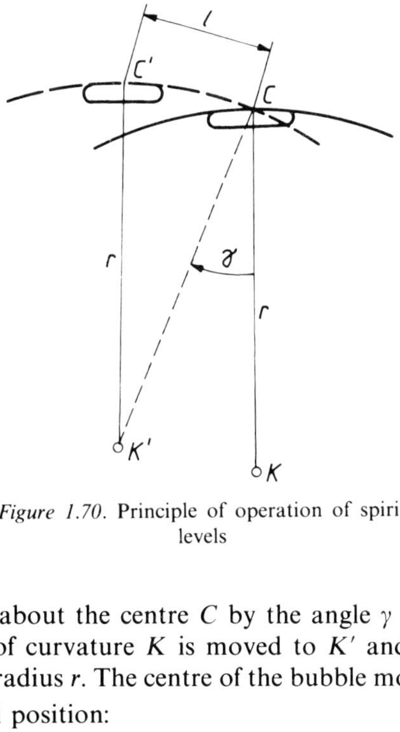

Figure 1.70. Principle of operation of spirit levels

If the level is turned about the centre C by the angle γ with respect to the local plumb line, the centre of curvature K is moved to K' and the centre point of the surface of revolution of radius r. The centre of the bubble moves through the distance $l=\overline{CC'}$ from its original position:

$$l=r\frac{\gamma}{\rho}. \tag{1.84}$$

If γ is measured in sec of arc, then ρ above is one radian expressed in sec of arc: $\rho''=206\,265''$.

Spirit levels can be used in geodetic or surveying instruments to set axes into some desired position with respect to the local vertical or horizontal, as well as to measure small deviations from this position. According to their shape distinction is made between *tubular* and *circular* levels (bull's eye levels). According to their method of attachment to an instrument, spirit levels can be *fixed* or *free*.

1.6.1 Tubular Levels

A tubular level consists of a cylindrical glass tube whose inner surface is fashioned as a barrel shaped *(toroidal)* surface of revolution by grinding and polishing *(Fig. 1.71)*. For observing the bubble position graduations are applied to the outside of the glass tube; the graduation interval is 2 mm or 1 Pars *(Parisian line)* =2.2558 mm. Position C of the bubble centre is determined by reading both ends of the bubble and taking the arithmetic mean. The graduation of a tubular level can be *geodetic*, either *continuous (Fig. 1.72, a)* or *truncated (Fig. 1.72, b)*, or *astronomical (Fig. 1.72, c)*. The central angle ε corresponding to a graduation unit p of the curved inner surface of the spirit level is the level *constant*, which can be expressed in sec of arc as:

$$\varepsilon'' = \frac{p}{r}\rho''. \tag{1.85}$$

The reciprocal of the level constant is called the level *sensitivity*:

$$\frac{1}{\varepsilon''} = \frac{r}{p\rho''}. \tag{1.86}$$

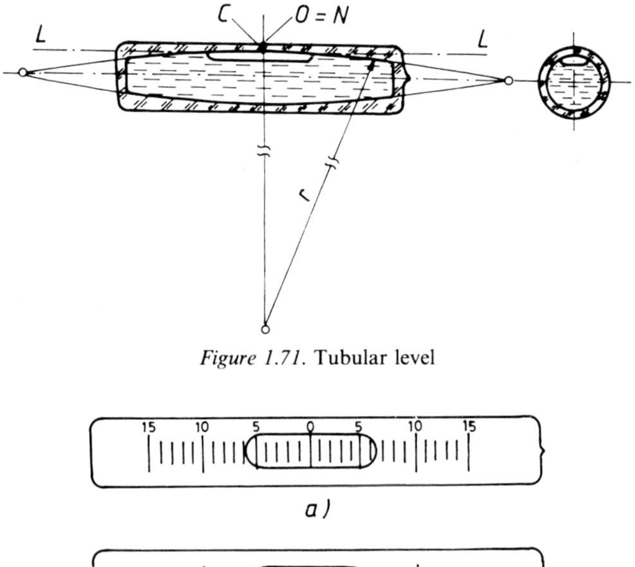

Figure 1.71. Tubular level

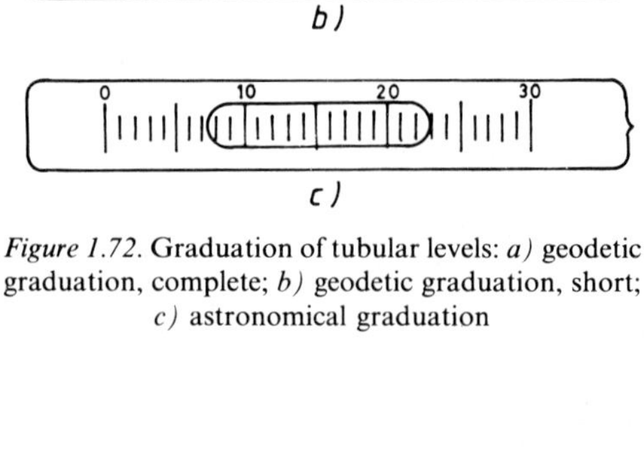

Figure 1.72. Graduation of tubular levels: *a)* geodetic graduation, complete; *b)* geodetic graduation, short; *c)* astronomical graduation

The line tangent to the curved inner surface of the level tube at the centre O of the graduations, i.e. at the *normal point* N, is the level *axis* L (see *Fig. 1.71*). If the point corresponding to the mean value of the readings of both ends of the bubble coincides with the graduation centre O, the axis of the spirit level is horizontal. From the point of view of measurements in geodesy and surveying the aim is always to set a main element (e.g. the vertical axis, or the horizontal axis) of the instrument into the vertical or horizontal. Therefore, tubular levels are attached permanently to main elements of the instrument (*plate* level, *index* level or *telescope* level), or they are removable (*striding* level, *plane table* level). It is a very important requirement in both cases that the axis of the tubular level should be either parallel to the instrument axis to be adjusted, or perpendicular to it, and that this relationship should be maintained for a long time.

The axis of a tubular level attached to the vertical axis should be perpendicular to the vertical axis. If there is a small error, the tangent point B of that line tangent to the curved inner surface of the level tube which is perpendicular to the vertical axis, does not coincide with the normal point N. In such cases the tubular level should be *adjusted*, i.e. the points B and N should be brought into coincidence: $B \equiv N$. With a well adjusted tubular level the bubble centre C, the tangent point B and the normal point N all coincide ($B \equiv N$) when the bubble is set.

In *reversible* levels the inside is ground along both the top and bottom, reversible levels thus having two bubble axes which must be parallel to each other. Reversible levels can be used in both positions.

The quality of a tubular level is determined by the precision of manufacture and by the properties of the materials used. The adhesion between the ground surface and the liquid, the internal friction of the liquid, tube deformations and changes of bubble length due to temperature, all have a detrimental effect on the accuracy of tubular levels. Nevertheless, operational precision and accuracy of measurement depend primarily on the level sensitivity and the reading accuracy of the ends of the bubble. Graduated tubular levels are dimensioned so that the required accuracy can be achieved by estimating readings to half a graduation unit.

The setting accuracy of a tubular level is significantly increased by a bubble reading system which compares the positions of the ends of the bubble *(coincidence level)* *(Fig. 1.73, a)*. The level bubble is illuminated from below by a rotatable mirror; a

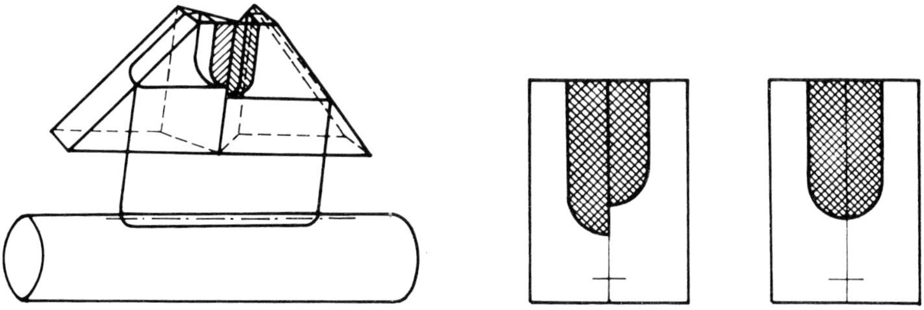

a)   b)   c)

Figure 1.73. Coincidence level

prism system, consisting of two sphenoids (wedges) placed above the bubble, projects the images of the ends of the bubble alongside each other. If the tubular level is slightly inclined, the half images of the bubble ends move in opposite directions, and show twice the amount of the bubble deviation *(Fig. 1.73, b)*. If both bubble ends coincide, the bubble is levelled *(Fig. 1.73, c)*. Bubble ends can be set by an unaided eye to an accuracy of about ± 0.1 mm; this corresponds to one twentieth of a graduation unit. The accuracy of setting can be further refined by using a magnifying glass of magnification Γ. The setting error of a tubular level can thus be reduced to $0.05\ \varepsilon/\Gamma$.

1.6.2 Mounting of a Tubular Level

The mounting of a tubular level must allow for the thermal expansion of the tube, maintaining the level in adjustment. With the mount shown in *Fig. 1.74, a* the glass tube of the level, embedded in wax or plaster, is placed in the metal mount, which can be moved relative to its support by the adjusting screws *1* and *2*. *Figure 1.74, b* shows a construction where the glass tube of the level is fixed at both ends to three points *1*, *2* and *3*. At one end of the level the elastic fixing allows sufficient play to accomodate thermal expansion. The adjustment of the level is made by the spring-actuated slow-motion (adjusting) screw *4* at one end of the level tube.

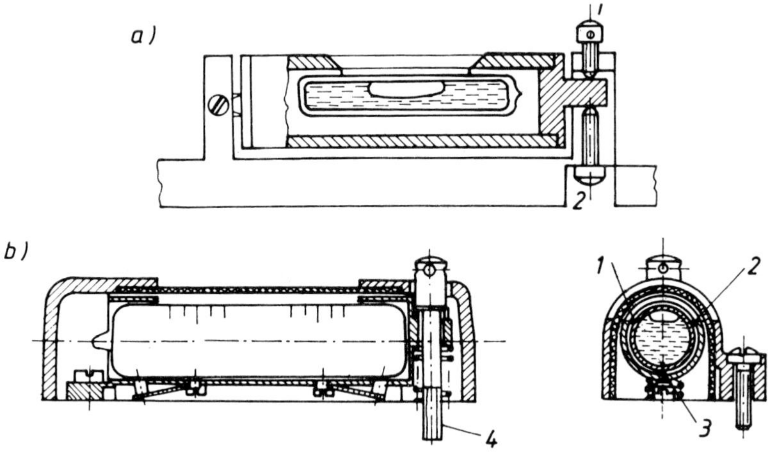

Figure 1.74. Mount of a tubular level and its adjusting device: variant *a*, variant *b*

1.6.3 Determination of the Sensitivity of Tubular Levels

The sensitivity of free tubular levels can be determined by a level trier, the simplest design of which is shown in *Fig. 1.75*. Its main part is a T-shaped *balance beam* resting on three supporting pins. Two of the pins are rigid and the line connecting them is perpendicular to the longitudinal axis of the *T-beam*, while the third pin lies in the

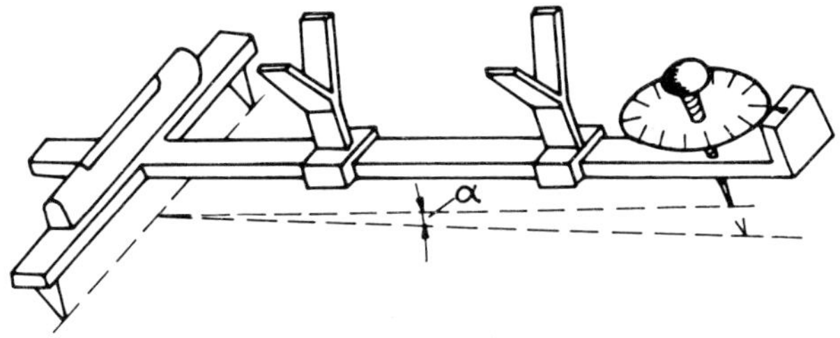

Figure 1.75. Simple level trier

longitudinal axis and can be adjusted vertically by means of a fine adjusting screw. A full (360°) turn of this screw gives an inclination of the T-beam in the vertical plane of angle:

$$\alpha'' = \frac{m}{a}\rho'', \tag{1.87}$$

where m = the pitch of the fine adjusting screw,
a = the effective length of the T-beam of the trier, i.e. the distance between the centre of the precision screw and the line connecting the two rigid pins (rotation axis of the T-beam),
α'' = the sensitivity of the level trier.

The position of the fine adjusting screw can be read on the scale of the drum fixed to it, the number of revolutions on a stable scale.

In order to test a tubular level with a level trier, the level should be placed onto the T-beam with its axis parallel to the axis of the T-beam. By means of the fine adjusting screw the bubble is moved to the end of its run, with both ends within the graduations. Now the positions a_1 and b_1 of the ends of the bubble and the position m_1 of the fine adjusting screw are read. Next the bubble is moved to the opposite end of its run and its ends (a_2, b_2) and the position m_2 of the fine adjusting screw are again read. The mean sensitivity of the level is obtained from:

$$\varepsilon = \frac{2(m_2 - m_1)}{(a_2 + b_2) - (a_1 + b_1)}. \tag{1.88}$$

By means of the level trier the precise curvature of the inside surface of a level tube can also be checked. If equal bubble deviations are obtained by turning the fine adjusting screw in equal steps, the curvature of the inside surface is exact. Otherwise the level sensitivity is unequal at different bubble positions.

The construction of level triers is now more refined. The *Freiberg* level trier *(Fig. 1.76, a)* is based on the principle of *Höncke*'s Hannover level trier. The table *2* can be tilted by the measuring screw *1 (Fig. 1.76, b)*; the angular levers *3*, *4* are connected to each other and to the housing *8* by the leaf-springs *5*, *6* and *7*. The construction ensures a tilting free of obstruction and a high angular sensitivity. This type of level trier is used for testing tubular levels with or without a mount as well as for testing levels in instruments (up to a mass of 15 kg). A special mass compensator ensures that the load acting on the measuring screw is always 100 p. Thus the effect of different loads on measuring accuracy is eliminated. The measuring range of the

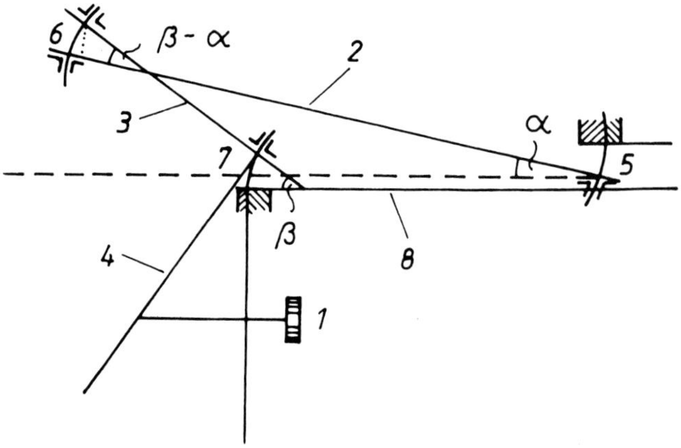

b)

Figure 1.76. Freiberg level trier (Freiberger Präzisionsmechanik, Freiberg, GDR); *a)* photo; *b)* principle of operation

instrument is 300 sec of arc, one graduation unit of the measuring screw is 25 sec of arc and the average measuring error is ± 2 sec of arc. The levels are protected against undesirable heat effects by a removable plastic box. The length of the angular lever *4* can be slightly altered, thus the nominal angular value of the measuring screw can be exactly set. The tip of the measuring screw rests in an eccentric position on the *agate plate* of the tilting lever: this plate can be turned and, if damaged, replaced.

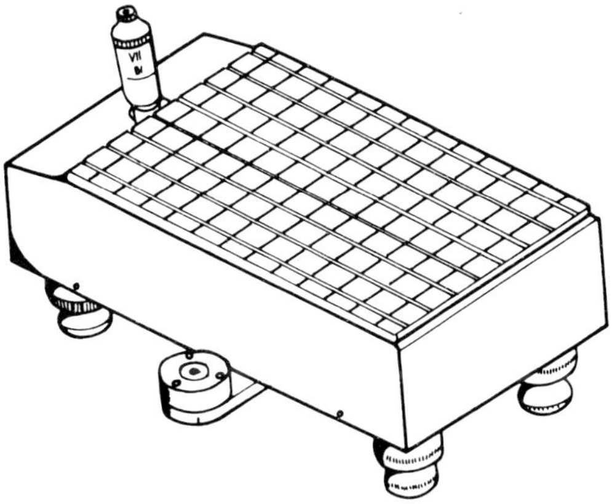

Figure 1.77. LIKON level trier (Breithaupt, Kassel)

The base plate of the *LIKON* level trier *(Fig. 1.77)* rests on three foot screws. The tilting table (measuring surface) made of *invar* is connected to the base plate by a mechanical lever. The measuring surface can be tilted by means of a micrometer screw. One side of the table (the inclination axis) is connected by leaf-springs to the base plate. One end of the two-armed lever, connected to the base plate also by a leaf-spring, is joined to the spindle of the micrometer screw, its other end to the measuring surface. One interval on the measuring drum of the micrometer corresponds to an inclination of 1 sec of arc of the tilting table. Levels in mounts with a plane base surface can be placed directly onto the measuring surface. Levels without a mount or with an cylindrical mount can be placed onto the measuring surface in V-shaped magnetic supports.

Conventional level testing methods are characterized by the gradual change of the inclination angle of the level axis. Thus one can test those points on the ground surface of a level, where the bubble comes to rest. The inclination axis of a level trier can be a prism edge, a cylindrical axis resting in bearings, a ball bearing or a joint with a leaf-spring. Its measuring device can be a measuring screw, an electrical sensor or an interference sensor; it can be operated manually or automatically, in the latter case the observation being made by photo-recording. Recently, level triers with a continuously changing inclination angle have also been constructed. Thus the internal surface of a level can be tested from point to point. The semi-automatic and automatic level trier of *Alpár* and *Bummer* (Geodetic and Geophysical Research Institute, Sopron, Hungary) for detecting uneven bubble motion is based on this principle. The characteristic motion curve of the bubble ends is drawn or photo-recorded from which the defects of the internal surface and the curvature of the level tube can be determined, as well as the level sensitivity.

The sensitivity of spirit levels mounted on geodetic and surveying instruments can be determined using the instrument itself.

The sensitivity of graduated telescope levels and levels of levelling instruments can be simply determined by using a levelling staff. The instrument (theodolite, levelling instrument) is set up at a known (measured) distance $t = 30$ to 50 m from the vertical levelling staff, with one of the foot screws pointing in the direction of the staff, and the staff is sighted. By means of this foot screw one end of the bubble should be set exactly at the zero graduation line of the tubular level, and the position l_1 of the horizontal cross-hair of the telescope is read on the staff for this telescope position *1*. Then the same end of the bubble is set by means of the same foot screw to the graduation line n, and the position l_2 of the horizontal cross-hair line is read on the staff for position *2*. The sensitivity of the spirit level is:

$$\varepsilon'' = \frac{l_2 - l_1}{t} \frac{\rho''}{n}. \tag{1.89}$$

In order to eliminate the disturbing effect of liquid adhesion (stickiness) and of thermal change in the length of the bubble, it is advisable to read both ends (a and b) of the bubble at positions *1* and *2* (a_1, b_1 and a_2, b_2) and to determine the level sensitivity by the following equation:

$$\varepsilon'' = \frac{l_2 - l_1}{t} \frac{2\rho''}{(a_2 + b_2) - (a_1 + b_1)}. \tag{1.90}$$

1.6.4 Adjustment of a Tubular Level

A tubular level needs to be adjusted, by means of its adjusting screws, in order to bring about a defined relative position between its axis and a straight line which is to be constructed. This straight line is usually the *vertical* or the *horizontal*. The level is adjusted if its axis is respectively normal to the vertical and parallel to the horizontal. Accordingly there are two basic types of adjustment of a level.

1. The adjustment of a tubular level for setting the vertical consists of the following steps:

a) setting the vertical axis vertical, and

b) setting the bubble centre C to the centre of the level graduations, i.e. to the normal point N, by means of the adjusting screws of the level.

It should be mentioned that the vertical axis can also be set vertical if the adjustment of the level is slightly erroneous, provided that the error does not exceed the operational range of the level, i.e. the tangential point B lies within the operational range. To use this method the vertical axis is set approximately vertical in two directions perpendicular to each other, i.e. in both directions the bubble is centred by tilting the vertical axis to a certain extent. The tangential point B is now determined; to do this, position C_1 of the bubble centre is determined at any position of the level, then the level is turned through 180° around the approximately plumbed vertical axis and position C_2 of the bubble centre is determined. The arithmetical mean of positions C_1 and C_2 yields the tangential point B. If the bubble centre C of an erroneously adjusted level is set to the tangential point B in two positions approximately perpendicular to each other, the vertical axis is vertical.

In practice the setting of the bubble centre is difficult, therefore, only one end of the bubble is read in both the direct and the reversed level positions, and this end is then set to the arithmetic mean of the two readings. Now step b) is to be carried out.

2. The adjustment of a tubular level for setting the horizontal consists of the following steps:

a) levelling its horizontal axis,

b) elimination of the occasionally existing cross divergence error by means of the horizontal adjusting screws of the level,

c) setting the bubble centre *C* to the level graduation centre, i.e. to the normal point *N*, by means of the vertical adjusting screws of the level.

The horizontal axis can also be levelled by means of a somewhat erroneously adjusted level, by determining the tangential point *B* according to the method of level adjustment described above. It should be mentioned that, in the case of a free tubular level, the reversal of the level around the approximately plumbed vertical axis by 180° is made by reversing the free level.

The axis of a level, erroneously adjusted in relation to the horizontal axis, and the horizontal axis itself are usually diverging straight lines. The angle made by their horizontal projections is the *cross divergence error*, or the horizontal adjustment error of the level. The angle made by the projections in the vertical plane of the horizontal axis is the vertical adjustment error of the level. The presence of a cross divergence error can be recognized in the following manner. The bubble is centered and the level rotated back and forth around its horizontal axis by approximately equal amounts. If the bubble centre moves in opposite directions with opposite rotations, the level has a cross divergence error. Its practical effect is usually small but it cannot be neglected in all cases. It can be eliminated by means of the horizontal adjusting screws and the adjustment of the level is then continued with step 2c).

The adjustment of spirit levels on levelling instruments is treated in detail in Point 3.1.2.6.

1.6.5 Circular Levels

The circular level or box level is a cylindrical sealed glass pot, the inside of the top of which is a *concave spherical cap (meniscus) (Fig. 1.78)*. The radius of curvature of the cap determines the level constant or sensitivity. The graduations of the level consist of concentric circles, the centre of which is the *principal point* of the level. The *tangential plane* at the principal point is the axial plane of the level.

The sensitivity of a circular level lies between 3 and 60 min of arc, i.e. it is usually much lower than that of tubular levels. By means of a circular level an axis tilted in any direction can be set vertical, though with lower accuracy. Circular levels are mainly used for the approximate plumbing of the vertical axis of an instrument, for plumbing the collimation axis of an optical plummet, further for levelling plane tables as well as for setting levelling staves into the vertical.

For the adjustment of a circular level the vertical axis (vertical line) should be plumbed independently from the circular level.

Circular levels are fixed by plaster in an adjustable metal casing *(Fig. 1.78, a)*. Recently instead of plaster, cement is used for fixing box levels *(Fig. 1.78, b)*. Circular levels can be adjusted by three adjusting screws and they must always be in adjustment, before use.

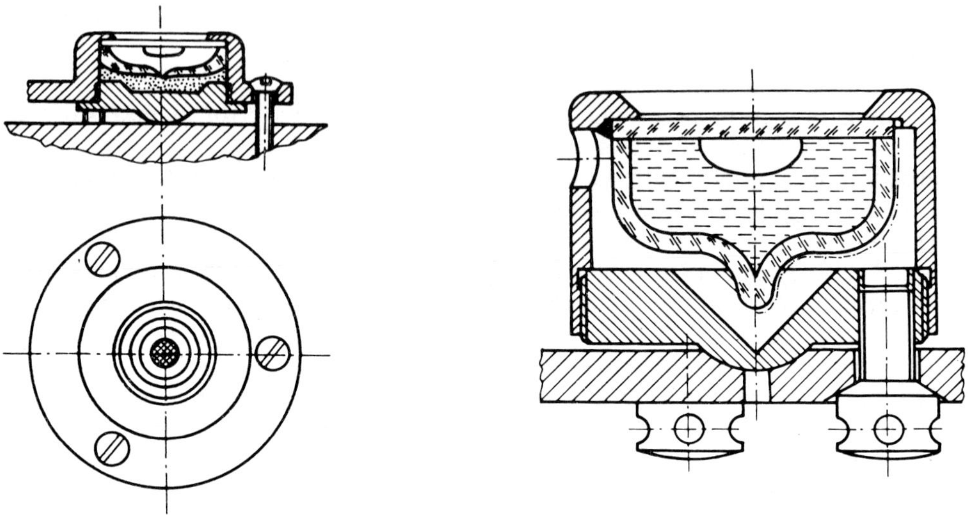

Figure 1.78. Circular level: *a)* embedded in plaster; *b)* cemented in the mount

1.6.6 Electronic and Electrolytic Levels

The operating principle of electronic levels is as follows: two *inductive resistors* connected into a *bridge* are changed by the movement of a *metal pendulum* and the resultant induced electric voltage is measured *(Fig. 1.79).*

The two *resistances* R_1 and R_2 are constant (R_1 is adjusted only when adjusting the bridge for the zero position of the metal pendulum). The reactances L_1 and L_2 change in accordance with the position of the pendulum swinging between them. The current flowing in the zero branch of the bridge, proportional to the displacement of the pendulum, is indicated by the *galvanometer G.*

The electronic level *Talyvel (Fig. 1.80, a)* consists of two units: the level unit weighing 1.1 kg, and the indicator unit weighing 3.8 kg. The oscillator *1 (Fig. 1.80, b)*

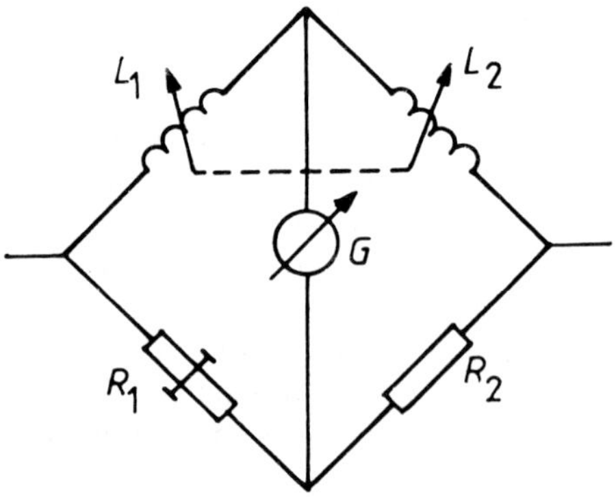

Figure 1.79. Principle of the electronic level

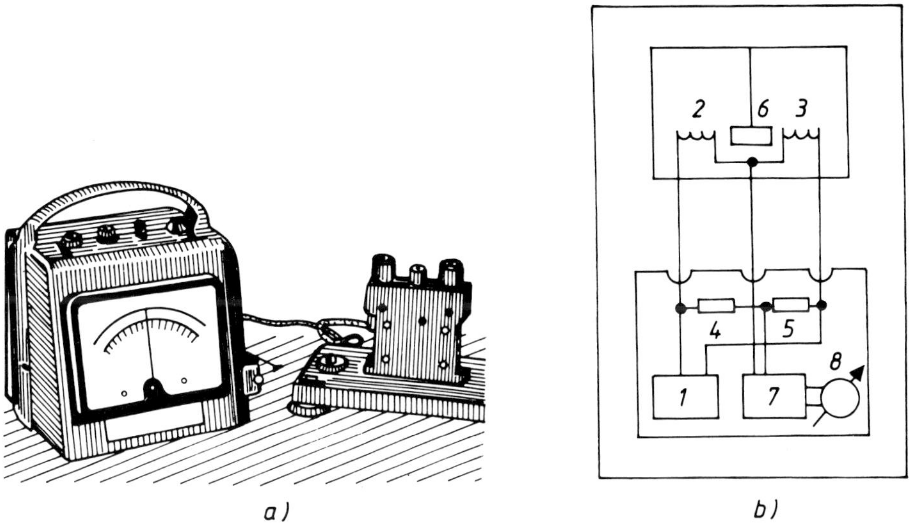

Figure 1.80. Electronic level *Talyvel* (Rank Precision Industries Ltd, England): *a)* sketch; *b)* block diagram of circuitry

is fed from 9 V batteries and supplies the bridge, consisting of resistances *2*, *3*, *4* and *5*, with low frequency a.c. The suspended pendulum *6* of high permeability moves between the inductive resistors *2* and *3*. If the bridge is in equilibrium, no current flows. When the level unit is inclined, the position of pendulum *6* changes with respect to the inductive resistors *2* and *3*, and a.c. is induced by the changed reactances. The induced a.c. is amplified by the amplifier *7* and measured by the galvanometer *8*. The deflections of the galvanometer are proportional to the changes of inclination which can be measured in three measuring ranges $\pm 50''$, $\pm 100''$ and $8'$, with a reading accuracy of $2''$, $4''$ and $20''$, respectively.

The electronic level manufactured by the firm *Tese S.A.* (Remens, Switzerland) has two measuring ranges ($\pm 150''$ and $\pm 30''$), the average error is $\pm 1''$ for one measurement, the time needed for one measurement is 3s and the mass of the instrument is 4 kg.

The advantages of electronic levels are that they can be read immediately (because there is practically no inertia) and that small inclination angles can be measured very accurately.

The operating principle of an *electrolytic* level is as follows: a number of parallel connected *electrodes*, such as four, are placed at the bottom of a spirit level in circular arrangement. The common opposite electrode is placed above the bubble. Current flows through the bottom electrodes, through the liquid which is an electrolyte, and through the upper electrode. If the inclination of the level changes and the bubble moves, the resistance of the electrolyte and thus the current flowing through it also change. This change is measured. Such an electrolytic level was produced by the firm *Regelungs- und Messtechnik Schmidt* (Reinbeck, FRG). The level operates with a supply voltage of 2 V, in the measuring range of $\pm 8''$, its average error is $\pm 1''$ and its mass is 16 g.

2 Instruments for Measuring Angles

In surveying the horizontal angle $\varphi_{3,2}$ *(Fig. 2.1)* is measured from the point P'_1, lying above (or below) a surface point P_1, and the legs of the *horizontal angle* $\varphi_{3,2}$ are the horizontal projections $P'_1P''_2$ and $P'_1P''_3$ of the spatial directions $P'_1P'_2$ and $P'_1P'_3$ determined by P'_2 and P'_3 situated above or below the points P_2 and P_3 (horizontal angular measurement). Further, the angle between the spatial direction $P'_1P'_2$ and its horizontal projection $P'_1P''_2$, i.e. the *elevation angle* $\alpha_{2,1}$ of the point P'_2 with respect to P'_1, as well as the elevation angle $\alpha_{3,1}$ of point P'_3 with respect to P'_1 (defined similarly) are also measured (vertical angular measurement). Instead of the elevation angle, the angle made by the spatial angle between $P'_1P'_2$ and the local vertical, passing through the points P_1 and P'_1, is often measured; this is the *zenith angle* $\xi_{2,1}$ of the point P'_2 referred to the point P'_1; the zenith angle $\xi_{3,1}$ of the point P'_3 is similarly measured. The angle between the spatial directions $P'_1P'_2$ and $P'_1P'_3$, i.e. the *position angle* $\gamma_{3,2}$ of the points P'_2 and P'_3 referring to point P'_1 does not belong to this system and is generally not measured directly in surveying.

The instrument for horizontal angular measurement and setting out is the *theodolite*. Horizontal angles can also be measured or set out, however, by *tacheometers* (Chap-

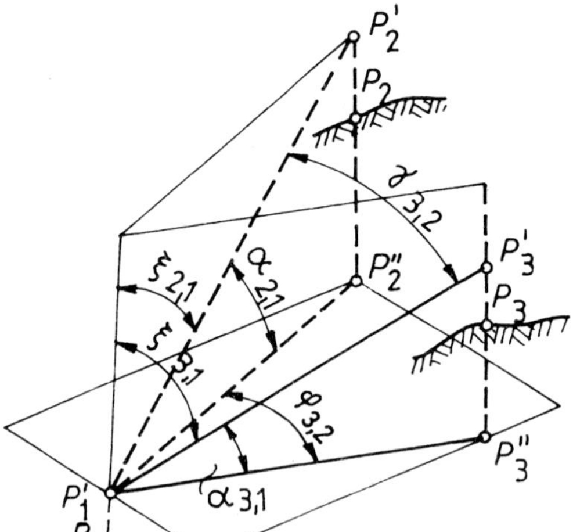

Figure 2.1. Angles measured in surveying

ter 6) or by a *levelling instrument* supplied with a *graduated horizontal circle* (Chapter 3). Certain defined horizontal angles (90° and 180°, sometimes 45°) can be set out with lower accuracy by means of simple devices for setting out angles, such as a cross-staff, prism or mirror. Elevation angles or zenith angles are also measured with the theodolite or the tacheometer, since these are now always made with a graduated vertical circle. If position angles are required, they can only be measured by means of a hand-held angle measuring device, e.g. a *sextant*. The horizontal projection of the points is determined by a projecting instrument or device.

Several systems of measures are still used for angular measurements. Thus, in geodesy and surveying, both the old sexadecimal system of 360 degrees with subdivisions of minutes of arc and seconds of arc and the new centesimal system of 400 gons (new degrees) with subdivisions of centesimal min of arc (^{c}) and centesimal sec of arc (^{cc}) are used. In practice a combined system seems to be promising, where the total circle is divided into 360 old degrees, but a decimal subdivision of the degree is adopted. This system is already used in school education and in pocket calculators.

2.1 Devices for Setting out Directions and Angles

By means of certain devices, angles of 90°, 180° or 45° can be set out, and positions can be set onto a given straight line. Cross-staffs, drums and mirrors to set out angles are already obsolete, therefore they are not dealt with here.

2.1.1 Survey Rod

The most simple device to set out a direction is the survey rod or ranging pole. This can be used, not only to mark the point to be observed, but also to continue a straight section already set out, by hand-ranging or by aligning one's position onto the straight line. The survey rod is a straight rod of 3 to 6 cm diameter, 2 to 3 m length, made of wood or metal; one of its ends is made with a hard tip, the other is often made so that another survey rod can be attached to it in order to obtain a higher mark *(Fig. 2.2)*. It is generally painted red and white or black and white in alternating 20 cm lengths. There are survey rods of equilateral triangle cross section, the tip here being a continuation of one of the edges; six of them can be bundled to make a thick survey stake.

2.1.2 Prisms for Setting out Angles

Prisms for setting out angles are devices by means of which angles of 90° and 180° (sometimes of 45°) can be set out. The hypotenuse of the simple prism square *(Fig. 2.3, a)* is silver coated. The frame of the prism has a handle into which a string or a projecting rod can be hung in order to set it above the point. If the incident ray enters the prism through its perpendicular side AB, near the edge A *(Fig. 2.3, b)* with angle of incidence ε_1, then $\varepsilon_4 = \varepsilon_1'$ and $\varepsilon_4' = \varepsilon_1$ due to the double reflection on the planes AC and BC. Since one-one legs each of the angles ε_4' and ε_1 (the normals to

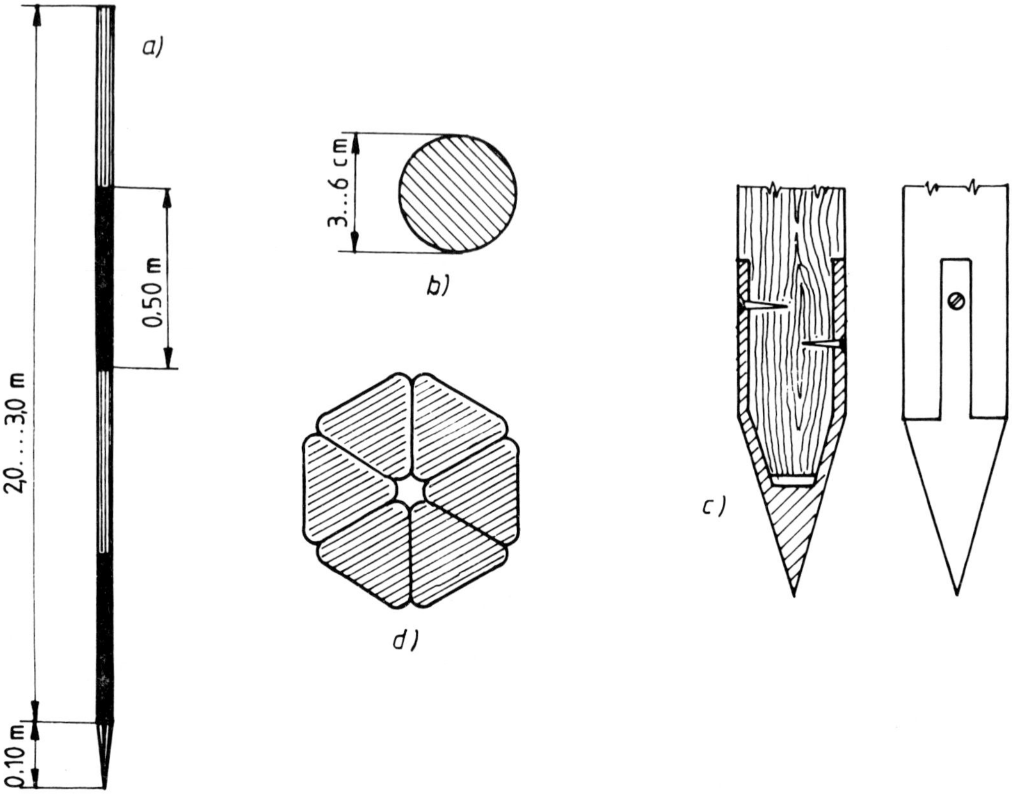

Figure 2.2. a) survey rod; *b)* cross section of a survey rod; *c)* hard tip of a survey rod; *d)* survey rods with triangular cross section

the planes *AC* and *AB*) are always perpendicular to each other, the two other legs, i.e. the entrance and exit rays, are also perpendicular to each other. Thus the deflection angle of the prism is $\delta = 90°$ and this is independent of the incidence angle ε_1 of the entering ray. By looking above or below the prism the perpendicular direction can be set out.

The path of a ray which also enters the prism through the surface *AB* at incidence angle ε_1, but near the edge *B*, is shown in *Fig. 2.3, c*. The sum of the angles in the quadrangle *EAFD* is:

$$(90° - \varepsilon_1) + 90° + (90° - \varepsilon_2') + (180° - \delta) = 360°$$

from which $\delta = 90° - (\varepsilon_1 + \varepsilon_2')$. Since, however, $\varepsilon_1' = \varepsilon_2$, and therefore $\varepsilon_1 = \varepsilon_2'$, consequently $\delta = 90° - 2\varepsilon_1$. Accordingly the deviation angle δ depends on the incidence angle ε_1.

If the prism is rotated around an axis perpendicular to its principal section plane, the image moves. This ray path cannot be used for setting-out. (In this ray path there is only one reflection, while in the former ray path, illustrated in *Fig. 2.3, b*, there are two reflections.)

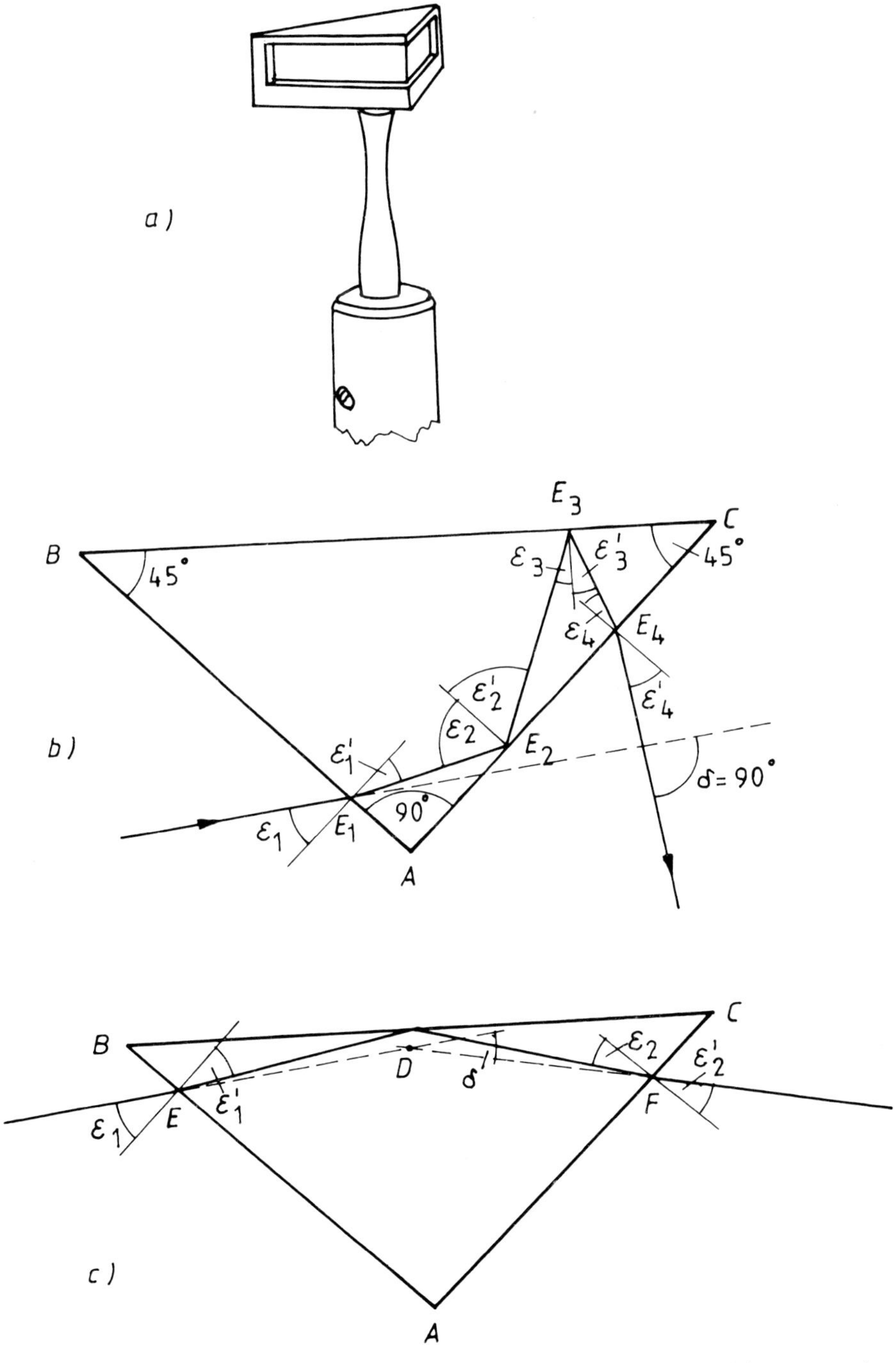

Figure 2.3. a) simple rectangular prism for setting out; *b)* first variant of the ray-path in a rectangular prism; *c)* second variant of the ray-path in a rectangular prism

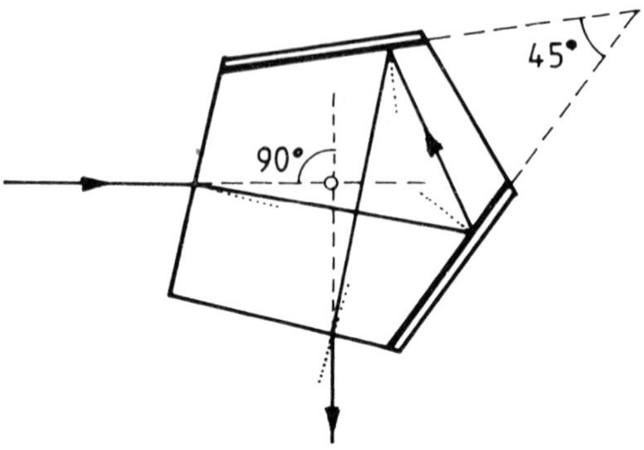

Figure 2.4. Simple pentaprism for setting out

The simple *pentaprism (Fig. 2.4)* as a device for setting-out has the advantage that it yields a brighter image, the vertex of the rectangle to be set out lies at the centre of the prism and the effective (image-forming) beam of rays is wider. If the top and bottom surfaces of the prism have reflecting coatings, the prism can also be used on steeper lines.

The device with a double prism for setting out angles consists of two identical rectangular or pentaprisms *(Fig. 2.5)*, sometimes *Wollaston prisms* (see *Fig. 1.19, d*), placed above each other in a common mount. Each of the two prisms can be used to set out a rectangle, while their common use enables alignment into a straight line or the setting-out of a direction perpendicular to the straight line. The accuracy of the setting-out is ± 1 min of arc; the precondition is that the error of the angle between

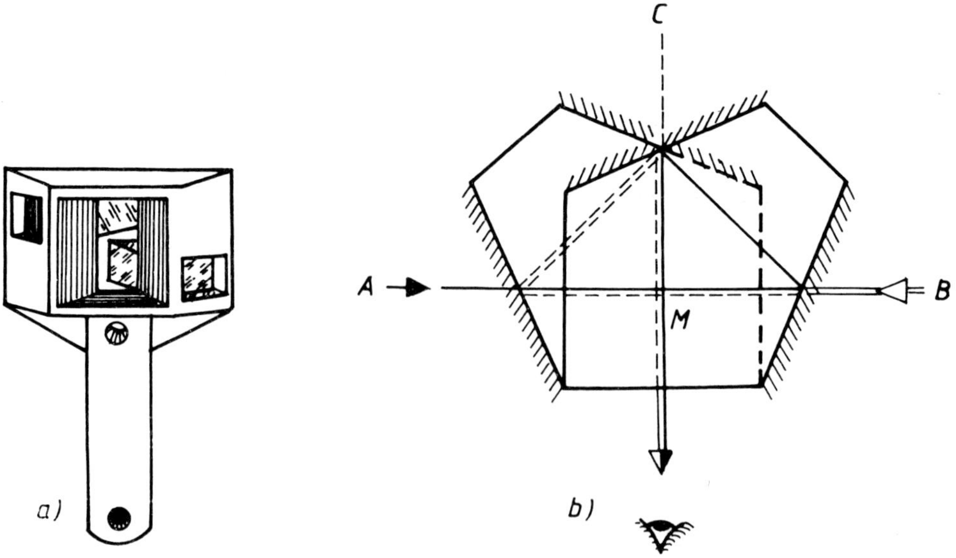

Figure 2.5. a) setting-out device with two rectangular prisms; *b)* the ray-path in a double pentaprism

the sides of the prism must be less than 20 sec of arc (Subsection 1.3.3). A telescopic *projecting rod* or a *plummet with string* is part of the prism system for setting out angles. When using the prism one sees two different marks at once, namely the mark already set out and the mark to be set out in the direction required. If the latter is ranged by hand, so that both marks appear in the same vertical line, the setting out is completed.

2.2 Theodolites

The theodolite is the most important and most accurate instrument of angle measurement, by means of which both horizontal and elevation angles can be measured. The geodetic telescope of the theodolite used for observing can be rotated around the vertical axis of the theodolite and tilted around its horizontal axis. The direction of the observed point is read on the graduated horizontal circle (horizontal limb), while the vertical and zenith angles, respectively, are read on the graduated vertical circle. The other parts of the theodolite are necessary for setting up above the station point, adjusting the vertical axis into the plumb line, rotating and tilting the telescope and clamping it in any rotated or tilted position, and for other supplementary operations. The tripod also belongs to the theodolite which is placed on the tripod before measurement.

Constructionally the main parts of the theodolite are: the base plate, the tribrach (they are often united into one unit), the central part and the top part of the instrument. Before measurement the base plate is attached to the tripod, the tribrach to the base plate (if they are separated) and the central part to the tribrach (or lower part). The top part can be rotated around the vertical (or standing) axis with respect to the central part and the tribrach fixed to each other or the top and central parts can be rotated together with respect to the tribrach, or — with newer types — the central part can be rotated with respect to the lower and top part fixed together.

With the older construction system of theodolites only two main parts were distinguished: the unrotatable lower part, which is identical with the tribrach, and the alidade rotatable on it (there was no central part); later the central part was regarded without this notation as part either of the alidade or of the lower part.

2.2.1 The Tripod

The instrument tripod to which the theodolite is fixed, must be stable, quickly settable, and made rigid without any stress. It consists of the *tripod head* and of three *legs (Fig. 2.6)*; the legs can be solid or telescopic, and they are joined to the tripod head by bolts. The usually wooden or light metal legs consist mainly of two stems with quadrilateral or circular cross section, stiffened by two H-shaped gussets. The bottom ends of the legs are fitted with pointed *metal shoes* with treads on them. In the case of simple theodolites alone, the instrument is placed on the conical pivot of the tripod head.

For most types of theodolites a tripod with a plate head is used, on which the upper surface of the tripod head is a flat plane *(Fig. 2.7)*. The upper ends of the legs are fixed to the bottom extensions of the tripod head. The legs can be rotated in the

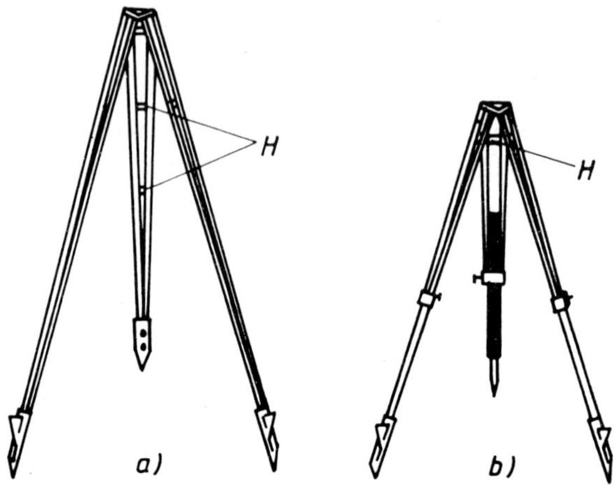

Figure 2.6. Tripod *a)* with solid legs; *b)* with telescopic legs

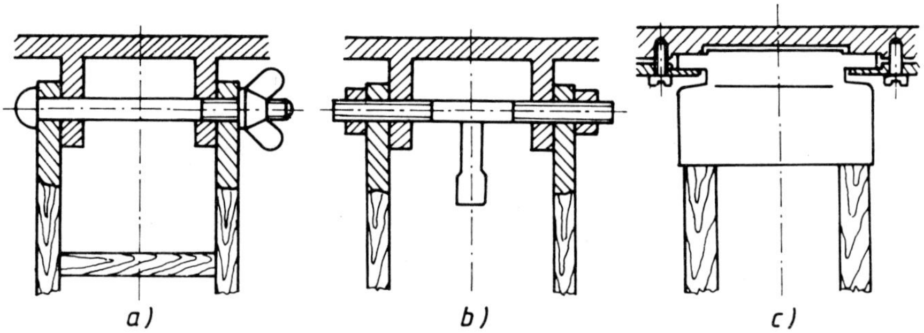

Figure 2.7. Tripod head *a)* with wing nut; *b)* with clamping lever; *c)* with cylindrical pivot

latter perpendicularly to the surface plane of the tripod head, and clamped in any position either by a *wing nut (Fig. 2.7, a)* or by a *clamping lever (Fig. 2.7, b)*. In the case of *friction clamping (Fig. 2.7, c)* clamping by hand is superfluous. Having set up the tripod, firstly the legs must be trodden into the ground and only after that are the legs to be clamped to the tripod head, in order to avoid harmful stresses. When removing the instrument, however, firstly the legs must be loosened, in order to avoid damage to the legs and to spare the joint screws. The instrument is placed on the tripod head *1* and fastened to it by tightening the joining screw *7* which is screwed into the nut of the spring plate *2* or base plate *3* of the instrument base *(Fig. 2.8)*. The joining screw can be shifted in the hole *6* of the clamping bridge *5* and the latter can be turned around the pin *4*. This enables the instrument to be moved in any direction by some cm for an exact setting over the mark. The hole in the joining screw ensures a free field of view for the *optical plumet* (see *Fig. 2.65*). Into this hole a bush with a hook can be put, from which the string of the plummet is hung. The vertical axis of the theodolite, placed on the tripod head and set up over the mark, is adjusted into the vertical by the three foot screws *7*.

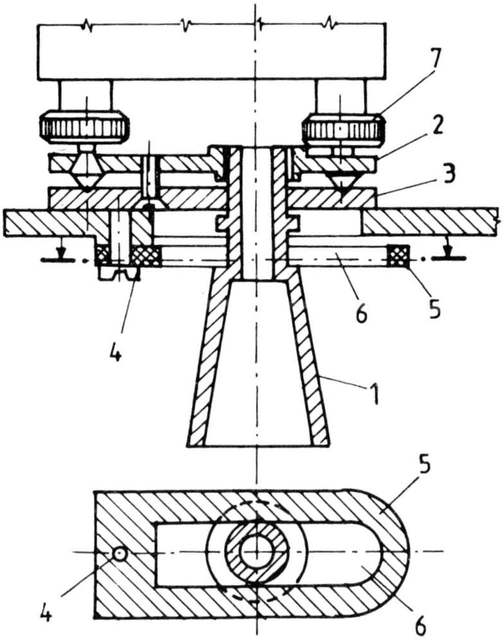

Figure 2.8. Fixing the instrument to the tripod by spring plate and fastening screw

In Great Britain, Japan and the US, tripods are also in use, which have an 89 mm diameter head on which there is a thread for screwing on the instrument. The centring of the instrument is ensured by moving this head on a plate, and also moving the plate itself on the tripod head in a perpendicular direction, without changing the direction of the vertical axis of the instrument *(Fig. 2.9)*.

When setting up these tripods, the surface plane of the tripod head must be brought approximately into the horizontal by eye. The firm *Kern* produces a centring tripod *(Fig. 2.10)* for its theodolites by means of which a preliminary vertical adjustment

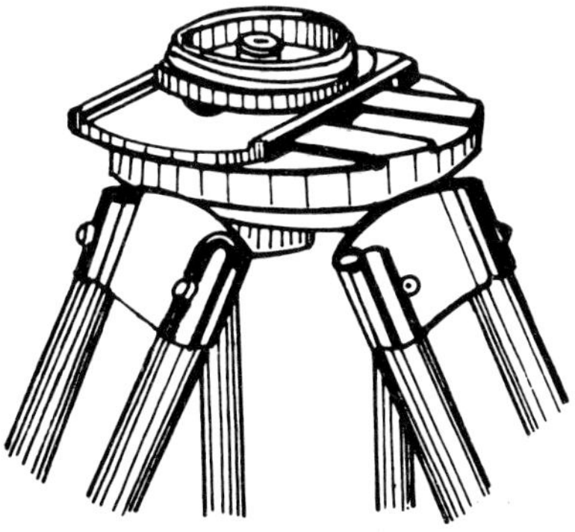

Figure 2.9. Tripod head with movable threaded plate

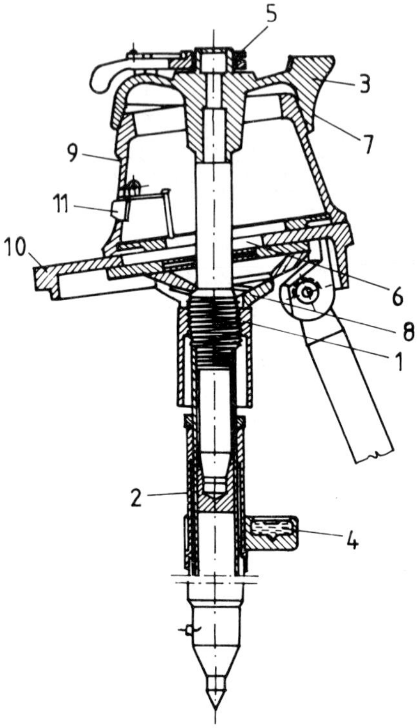

Figure 2.10. Centring tripod head (Kern, Aarau)

can be made more quickly and more exactly, and the set up over the point can be completed before attaching the instrument. After loosening the clamping screws *1* the tripod is set up in the usual way, but the legs are so trod into the ground that — by putting the point of the telescopic projecting rod *2* on the ground mark — the upper surface of the plate *3* (the surface on which the instrument stands and which is perpendicular to the rod) is made nearly horizontal, and the centring bush *5* at the centre of the plate is nearly vertically above the ground mark. The plate *3* is rigidly connected to the rod and is made horizontal by centring the bubble of the box level *4*. Thus free tilting of the projecting rod is possible, because a larger hole *6* is made in the tripod head, and also because the bottom rim *7* of the plate *3* and the upper rim *8* of the joining screw of the tripod *1* are shaped as spherical caps. Thus the rod can be turned freely in any direction on the two corresponding spherical surfaces of the movable conical part *9* of the tripod head and, together with the conical part, it can also be shifted in any direction on the surface plate *10* of the tripod head. When setting up the tripod its head is gently adjusted by the press-button *11*, ensuring that the bubble of the box level remains centred. Now the joining screw *1* of the tripod is tightened; this clamps the plate *3* to the tripod head and at the same time the conical head *9* to the plate *10* of the tripod head. If the constant of the box level is 10 min of arc, the upper surface of the plate *3*, on which the instrument stands, can be brought into the horizontal with an accuracy of ± 2 min of arc.

2.2.2 Tribrach and Lower Part of the Theodolite

The tribrach or instrument base enables the vertical axis (axis system) to be adjusted into the vertical and it maintains the stability of the axis system during measurement. The spring plate *2* mounted on the base plate *3* serves as the seat of the foot screws and clamps them to the base plate or to the tripod head (see *Fig. 2.8*). In the case of some theodolites of low or medium accuracy, the base plate is at the same time the base of the metal instrument box.

By means of the three foot screws, 120° apart, the vertical axis is adjusted into the exact vertical with the help of the alidade spirit level which is on the lower or upper part of the instrument. The foot screws should turn smoothly, not too easily and not too stiffly. For adjusting the motion a wear-adjusting screw is used. The foot screws are protected against the harmful effect of dust and moisture by a protecting cover.

On the uniform tribrach of the theodolite *Te-B23* (MOM, Budapest) *1* is the wear adjustment screw *(Fig. 2.11/a)*. The spring plate *2* presses the bottom of the foot screw *3* to the base plate *4* with an adjustable pressure. Nut *5* is attached to the spring plate 2 and clamps the lower part (tribrach) of the theodolite to the tripod. The pressure can be adjusted by changing the space *6*; this is independent of the clamping of the instrument to the tripod.

In the case of the forced centring instrument base standardized in several countries (see *Fig. 2.66*, *a* and *b*) the removable cylindrical pivot *2* of 34 mm diameter fits into the bush *4*. The pivot is rigidly mounted to the lower part of the instrument which carries the bearing of the vertical axis system. The pivot is clamped to the bush by the clamping screw *7*; this clamping can be loosened. The bottom tips of the foot screws seat in the grooves of the foot plate.

On the instrument bases of *Kern* theodolites the role of the foot screws is taken over by knobs *1* with eccentric discs with vertical axes, as shown in *Fig. 2.11/b*. The

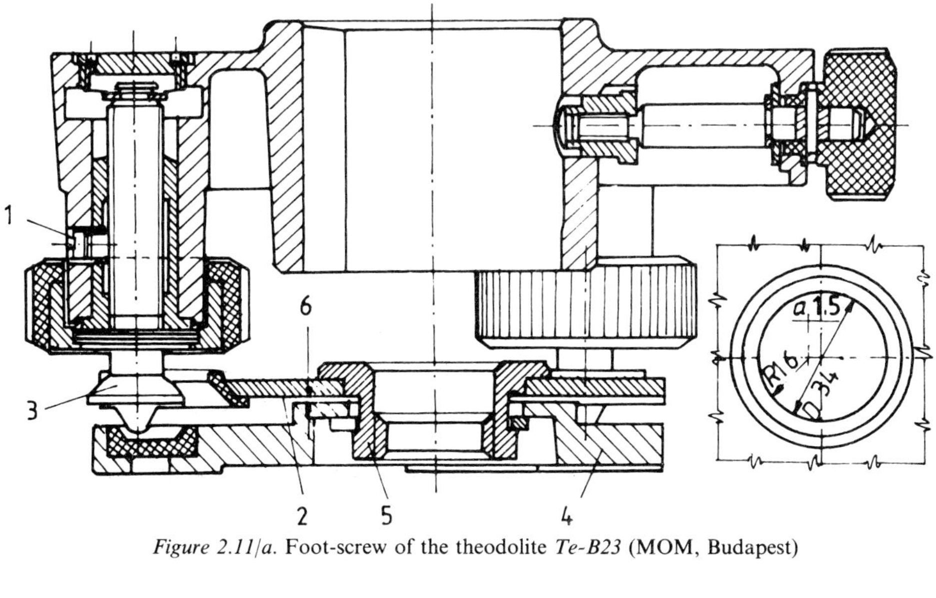

Figure 2.11/a. Foot-screw of the theodolite *Te-B23* (MOM, Budapest)

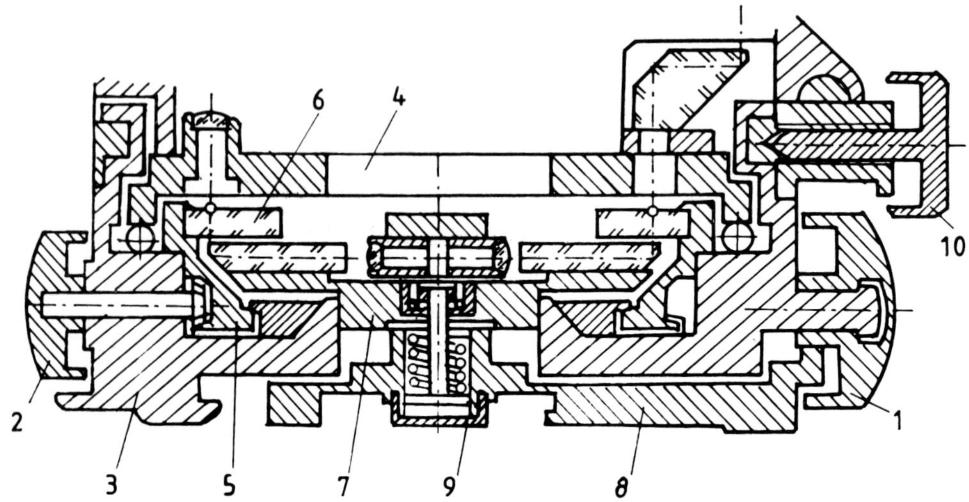

Figure 2.11/b. Instrument base of the theodolite *DKM 1* (Kern, Aarau): *1* foot-screw with horizontal axis; *2* knob rotating the horizontal circle; *3* bottom part of the instrument; *4* rotating ring of the top part of the instrument; *5* the central part of the instrument (bearing ring of the horizontal circle); *6* graduated horizontal glass circle; *7* support of the optics projecting one reading over the other; *8* base plate; *9* centring pivot; *10* horizontal centring screw

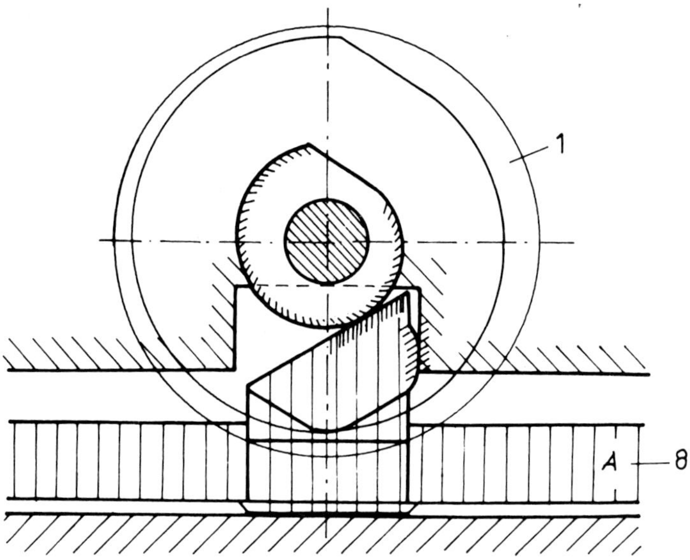

Figure 2.11/c. Kern foot-screw with horizontal axis

principle of the operation of the eccentric discs *1* is shown in *Fig. 2.11/c.* The theodolite is brought into its measuring position by turning the three eccentric discs, which are arranged similarly to foot screws; they contact the inclined plane surface. The theodolite rests on the foot plate *8* under its own weight. This base construction,

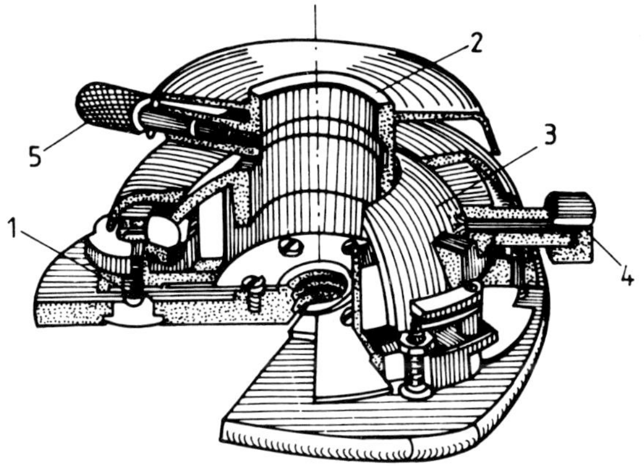

Figure 2.12. Tribrach of the theodolite *Th 2* (Opton, Oberkochen)

however, allows the vertical axis to be adjusted only within a small range: therefore a Kern centring tripod is necessary (see *Fig. 2.10*) in order that the upper surface plane of the tripod head may be set closely to the horizontal.

The instrument base of the theodolite *Th 2* (Opton, Oberkochen) is a combination of the fast operating ball-and-socket system and the slow-motion foot screw system (*Fig. 2.12*). Here, in two directions perpendicular to each other, fine-tilting foot screws

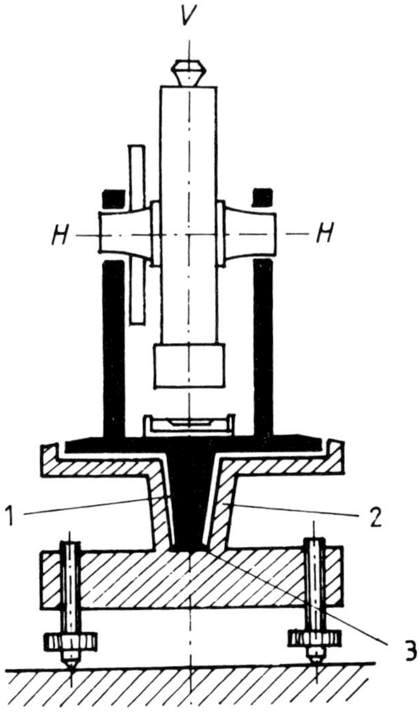

Figure 2.13. Simple conical vertical axis

1 are used, by means of which the approximate adjustment carried out by the ball-and-socket *3* of the socket *2* of the pivot-socket forced-centring system (Point 2.2.6.2) can be further refined. The connecting ball can be clamped by the clamping screw *4*. Clamping screw *5* is the clamping element of the forced centring.

2.2.3 Vertical Axis of the Theodolite

In theodolites of older types a vertical axis of conical shape *(Fig. 2.13)* is used; it can be rotated in the conical bush of the lower part which forms one unit with the instrument base and the central part. The axis *1* and the bush *2*, made of different

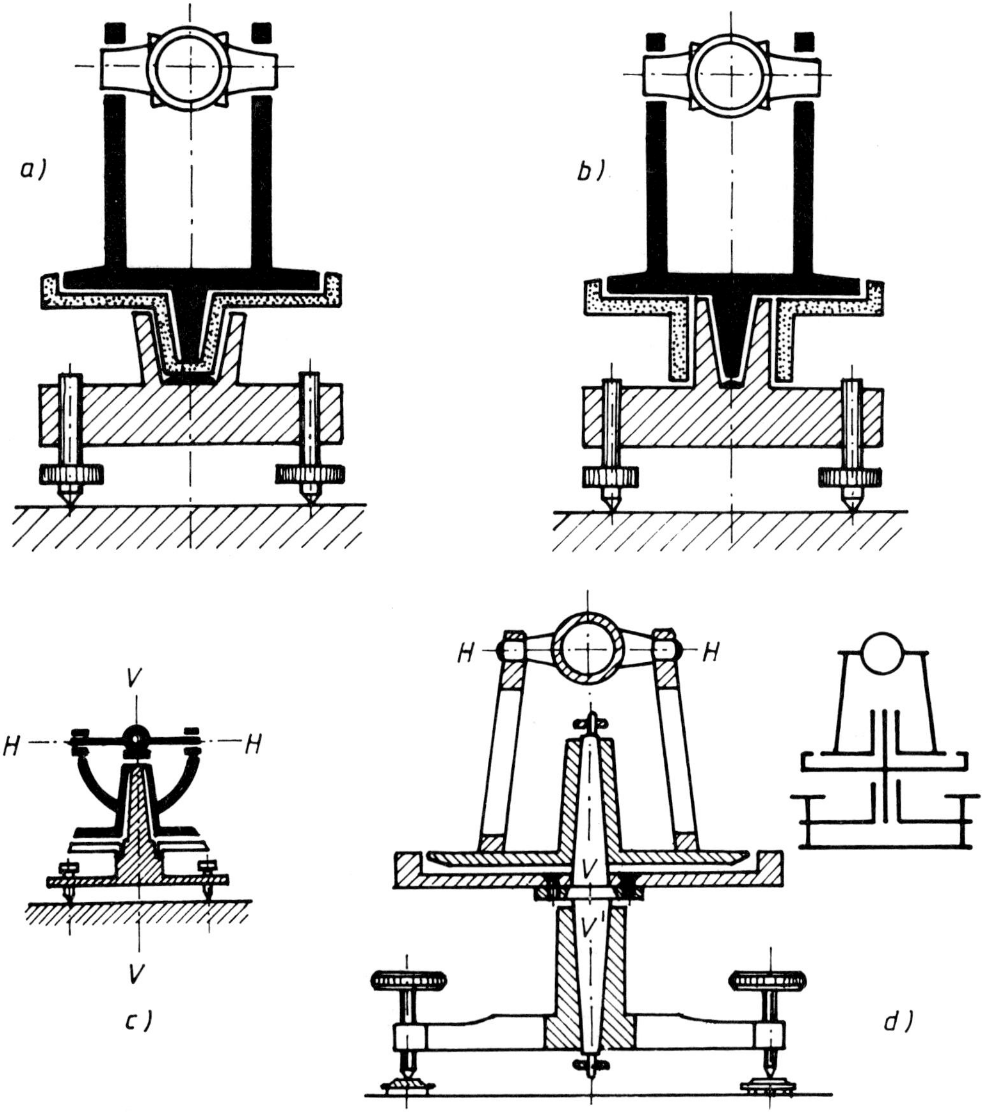

Figure 2.14. Obsolete conical axis systems: *a)* Reichenbach; *b)* Borda; *c)* Repsold; *d)* Ramsden

material, can be easily worked and lapped. The weight of the upper part is transmitted to the lower part partly on the conical surface, partly on the end-face of the axis or on the lower support *3* holding the lower part. Consequently, the increasing clearance and surface unevenness due to wear must be eliminated by lapping from time to time. In order to decrease friction and wear between the conical surfaces, an oil film of some μm thickness is applied.

Later a double conical axis system was used in theodolites. For instance, in the *Reichenbach axis system (Fig. 2.14, a)*, the top part of the instrument can be rotated around one axis with respect to the central part, and the central part can be rotated around the other axis with respect to the lower part. The disadvantage of this system is that the rotation of the top part can cause a small motion of the central part which bears the horizontal graduated circle, thus degrading measuring accuracy. *Borda*'s *(Fig. 2.14, b)*, *Repsold*'s *(Fig. 2.14, c)* and *Ramsden*'s *(Fig. 2.14, d)* vertical axis

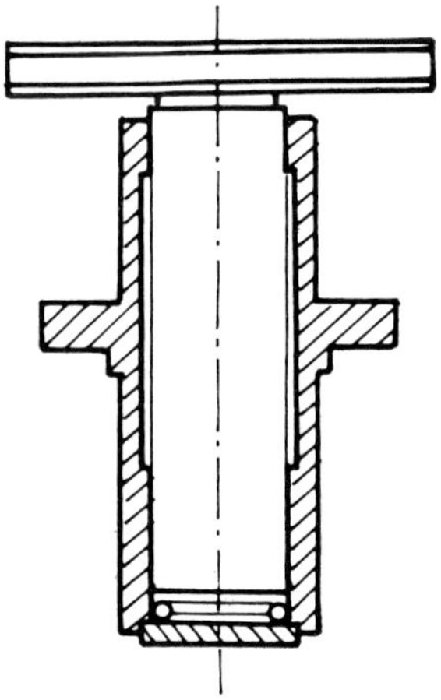

Figure 2.15. Cylindrical axis supported by a ball-bearing (VEB Jenoptik, Jena)

systems enable the top part and the central part to be rotated separately with respect to the lower part. A conical axis system is no longer used.

The *cylindrical axis*, recently in general use, has the advantage over the conical that there is no weight transmission on the bearing surfaces; the weight of the top part is totally transmitted to the lower either on the upper end-face or on a ball-bearing or on a thrust-ball bearing at the bottom of the axis *(Fig. 2.15)*. Thus no axis wear occurs; axis and socket need not be made from different materials as modern tempered steel can be used for both. The exactness of the rotation of the axis is further increased by decreasing the diameter of the greater part of the cylindrical pivot, in the middle, by 0.5 to 1.0 mm so that the pivot contacts the socket only at its upper and lower ends. These upper and lower contacting surfaces only guide the axis, assuming that the mass of the upper part is well balanced with respect to the axis. The ratio between the 0.5 μm clearance, or the $\Delta = 0.4$ μm interspace due to the not completely compressible oil film, and the axis length $h=80$ mm determines the wobble of the axis; correspondingly the wobble is about $2\Delta\rho''/h \approx 2''$.

The cylindrical pivot of the semi-kinematic standing axes in *Wild* theodolites and in the A-series of *Jenoptik* theodolites are guided by centring ball-bearings *(Fig. 2.16/a)*, instead of by contact sliding surfaces; the balls roll on the conical surface with a half vertex angle α at the upper edge of the socket, and the weight is also transmitted by them. The height of the lower cylindrical guiding surfaces is reduced at the same time to 5 mm. Due to the fact that the weight is transmitted on a ball-bearing, the rotation of the axis is very fine. The random wobble of the axis is smaller than with a cylindrical axis, since it is only the consequence of the clearance Δ between the lower sliding surfaces of the axis. If the lower part of the axis pivot is shifted by Δ in the guiding ring of the socket, then the axis tilts around the common intersection point M of the normals at the contact point A_1 of the upper conical guiding surface *(Fig. 2.16/b)*. Thus the random axis wobble is less than Δ/h' (where h' is the distance between the bottom edge of the lower cylindrical sliding surface and the point M, in other words the fictitious length of the axis).

Regular axis wobble has two components. One is the consequence of the deviation from a precise shape of the five guiding surfaces (the two lower sliding surfaces and the three upper ball-guiding surfaces). The period of this component is 360°. The other is the effect of the inequality of the ball diameters, its period being defined by the rotation angle of the axis at which the balls again reach their original position. The half vertex angle α of the guiding conical surface is such that the balls should roll without sliding on the circle at the point A_1 with radius

$$R_1 = R + r(1+\cos\alpha), \tag{2.1}$$

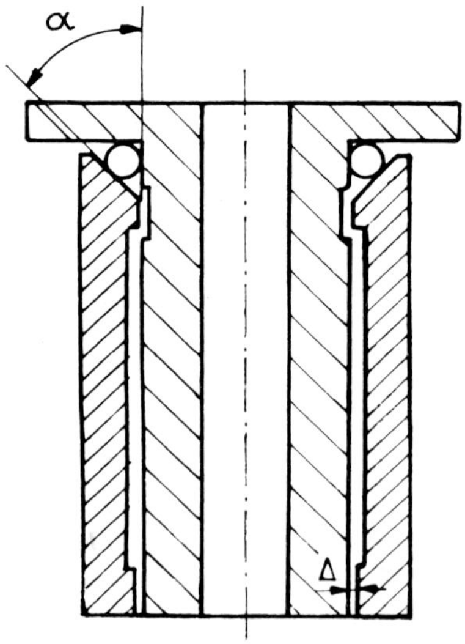

Figure 2.16/a. Semi-kinematic vertical axis used in Wild-theodolites and in the series *A* VEB Jenoptik theodolites

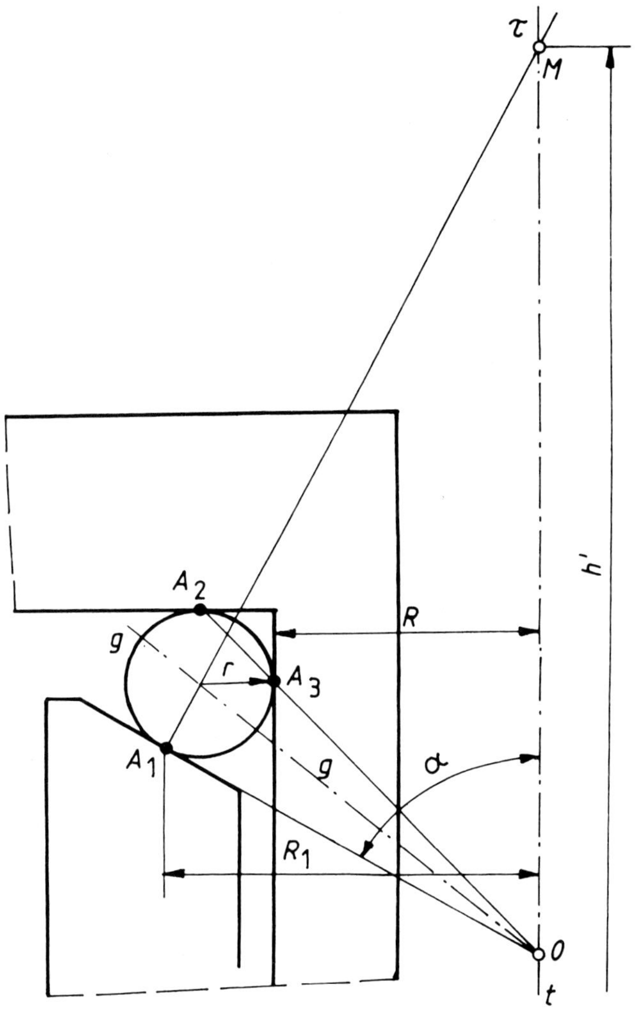

Figure 2.16/b. Part of the semi-kinematic vertical axis

also on the circle at point A_2 with radius $(R+r)$ and on the circle at point A_3 with radius R. The condition for this is that the straight line drawn through the points A_2 and A_3, and the tangent drawn from the intersection point O of the rotation axis $t-t$ to the ball with radius r, should be the generatrix of the conical surface. This gives the following relation between the radii R and r and the half vertex angle α:

$$\frac{R}{r} = \frac{1+\cos\alpha}{\sin\alpha - \cos\alpha}. \tag{2.2}$$

In this case the rolling on the three contact points is truly free from sliding if the surface is regarded as absolutely rigid and non-distorting.* Using the notation $d=2r$ and $D=2R$, one obtains from Eq. (2.2):

$$\tan\alpha=\frac{D+d\sqrt{\dfrac{2D}{D+d}}}{D-d}. \tag{2.3}$$

If, e.g., $d=4$ mm and $D=20$ mm then $\alpha=57°\ 30'$. The rolling of the balls along the circle with radius R_1 is the resultant of two rotations: the rotation of the ball around its own axis $g-g$ and the rotation around the axis $t-t$. The ratio between the angular rotation φ_t and the motion of the balls around the axis $t-t$ (angular rotation φ_{A_1}) is:

$$\frac{\varphi_t}{\varphi_{A_1}}=\frac{R+r}{R}(1+\cos\alpha). \tag{2.4}$$

In order to return to their original position, the balls must make a revolution of 360° around the vertical axis $t-t$. Substituting $\varphi_{A_1}=360°$, the second period of axis wobble, i.e. the period of the component arising due to the inequality of the ball diameters, is:

$$\varphi_t=\frac{R+r}{R}(1+\cos\alpha)360°=\frac{D+d}{D}(1+\cos\alpha)360°. \tag{2.5}$$

In the example ($D=20$ mm; $d=4$ mm) the period of the second component is:

$$\varphi_t=\frac{20+4}{20}(1+0.537)360°\simeq 663°.$$

The lower part of the semi-kinematic vertical axis of the theodolite *Te-B23* (MOM, Budapest) *(Fig. 2.17, a)* has a spherical shape and is guided by the balls on the spherical surface of radius R of the lower part of the instrument along a circle of large diameter *(Fig. 2.17, b)*, in addition to the upper cylindrical sliding guidance. The axis wobble caused by the unavoidable small inequalities in the diameter $2r$ of the balls is decreased by the large diameter of the ball-guided circle. Axis wobble is further decreased by the circumstance that it is determined by the fictitious axis length h' between the common centre M of the two ball-guiding spherical surfaces lying on the axis and the level of the upper centring and guiding slide-surface of the axis. Thus, if the axis is shifted within the clearance Δ between the guiding slide-surfaces, the axis point M remains in position without any play.

The ball-guiding surfaces of the vertical axis of *Kern* theodolites *(Fig. 2.18)* are horizontal planes. The fictitious axis length h' as defined according to *Fig. 2.17, b*, which is important with respect to axis wobble, is infinite (point M lies at an infinite distance in the straight line of the axis). Such a vertical axis can be manufactured with a very small construction height and the possible play in the clearance between the guiding slide-surfaces does not cause any axis wobble.

* *G. Kisfalusi*: Finommechanikai golyóscsapágyak kinematikai vizsgálata. Finommechanika—Mikrotechnika, Budapest, 16 (1977), 305–316.

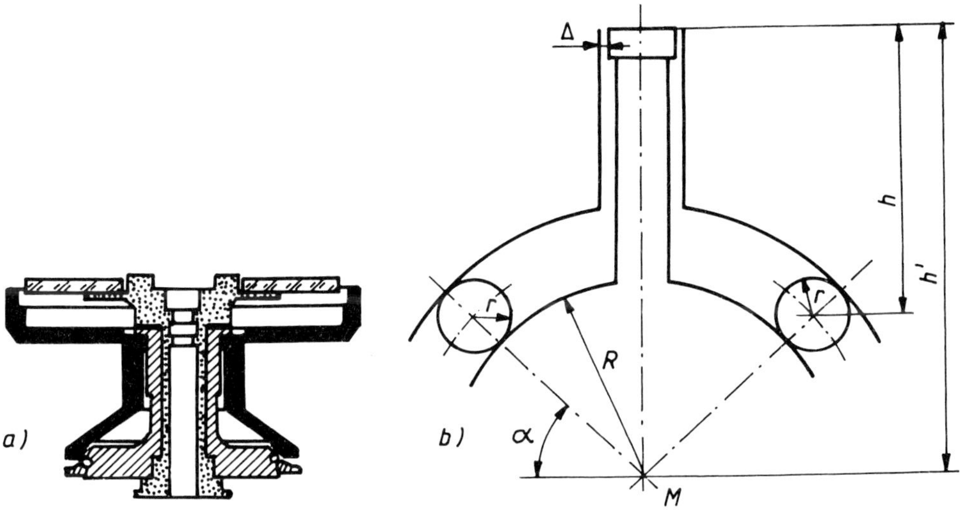

Figure 2.17. Vertical axis of the theodolite *Te-B23* (MOM, Budapest): *a)* cross section; *b)* clearance of the axis

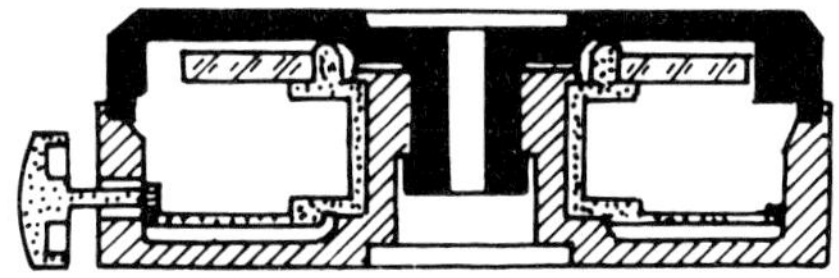

Figure 2.18. The vertical axis of Kern theodolites

Theodolites with a double vertical axis are called repetition theodolites if the limb-holding central part of the instrument can be rotated only by hand — against friction — with respect to the top part of the instrument; further, only the top part can be clamped after coarse rotation and finely adjusted after clamping by a slow motion screw. By means of a repetition theodolite horizontal angles can be measured at different parts of the graduated circle.

The name angle-multiplication theodolite means that both the central and the top parts of the theodolite can be set and adjusted with separate clamping and slow motion screws independently of each other. By means of a multiplication theodolite a multiple of the angle to be measured can be determined and the corresponding proportion yields the angle. Its disadvantage is that, during measurement, the two screws of similar function can easily be confused, and the construction is more complicated. In some newer theodolites an interchanging axis-coupling device is used (see *Fig. 2.63*), by means of which the central part of the instrument can be directly clamped either to the lower or to the top part. This enables both a desired circle reading to be set on a point and multiplying angle measurement to be made with the same clamping and fine-adjustment (slow motion) screw. It is not easy, however, to manufacture an axis-coupling which completely eliminates a small jerky rotation of the limb circle when clamping and unclamping. Hence recently on some theodolites made by *Kern* and *Opton* the repetition axis is again supplied with a slow motion screw.

2.2.4 Central Part of the Theodolite

The central part of the theodolite consists of the horizontal graduated circle (limb) and its supporting elements, also the construction elements which rotate the graduated circle around the vertical axis. Formerly the graduated circle was made of metal and the graduations were engraved either on its own flat conical surface or on a silver or China-silver band embedded into this surface. Now only glass disks, of 3 to 6 mm thickness with two parallel surface planes, are used as graduated circles which are rim-mounted. Finely marked thinner (some μm thick) division lines can be made on a glass ring and its diameter can therefore be smaller. The graduation interval can depend on the accuracy class 1°, $1^\circ/3$, $1^\circ/6$ or $1^\circ/12$ and 1^g, $1^g/2$, $1^g/5$ or $1^g/10$, respectively. Each 12th, 6th or every degree line and each 10th, 5th or every graduation line, respectively, is numbered in such a way that the numbers increase in the reading device from left to right (or sometimes downwards), and they also increase if the instrument is turned clockwise. The graduation lines are 2 to 8 μm thick depending on the reading device.

Graduation lines can be applied by one of the following three methods:

— the graduation lines are directly engraved onto the glass circle by an automatic graduating machine;

— the graduation lines are cut into a wax coat with which the glass circle has been previously coated and then the glass at the graduation lines is etched with acid and painted with black paint;

— the graduation is performed photographically.

The *automatic graduating machine*, set up in an air-conditioned vibration-proof room, has a base circle which is a gear rim of 50 to 100 cm diameter containing 1440 or 2160, seldom 720, gear teeth; it is protected with a cover ring and driven by a mechanism in steps corresponding to the required division intervals. In a generally used variant *(Fig. 2.19, a)*, the driving machine rotates the double gear arc *2* by the transmission of disc *1* *(Fig. 2.19, b)* which at each half revolution engages the gear *4* on the axis of the toroidal worm screw *3*. The threads of the worm screw engage the gear rim *5* of the graduator which is periodically turned, together with the mounted glass circle to be graduated, as long as the teeth of the gear arc *2* engage the toothed wheel *4*. During the pauses in rotation the engraving device is activated *(Fig. 2.19, e)* and engraves a graduation line onto the glass circle. The toroidal performance of the worm screw *3*, as seen in *Fig. 2.19, b*, has the advantage with respect to the earlier cylindrical worm screw *(Fig. 2.19, c)* that each thread of the toroidal screw equally engages the same number of teeth of the gear rim *5* *(Fig. 2.19, d)*, and therefore the effect of errors of the synchronously operating teeth of the gear rim is eliminated. If the numbers are applied by acid etching they are engraved with an *etching pantograph* *(Fig. 2.19, f)* onto the glass circle. Such a pantograph can also be used for engraving lines and signs onto a glass plate, thus, e.g., to engrave a cross-hair.

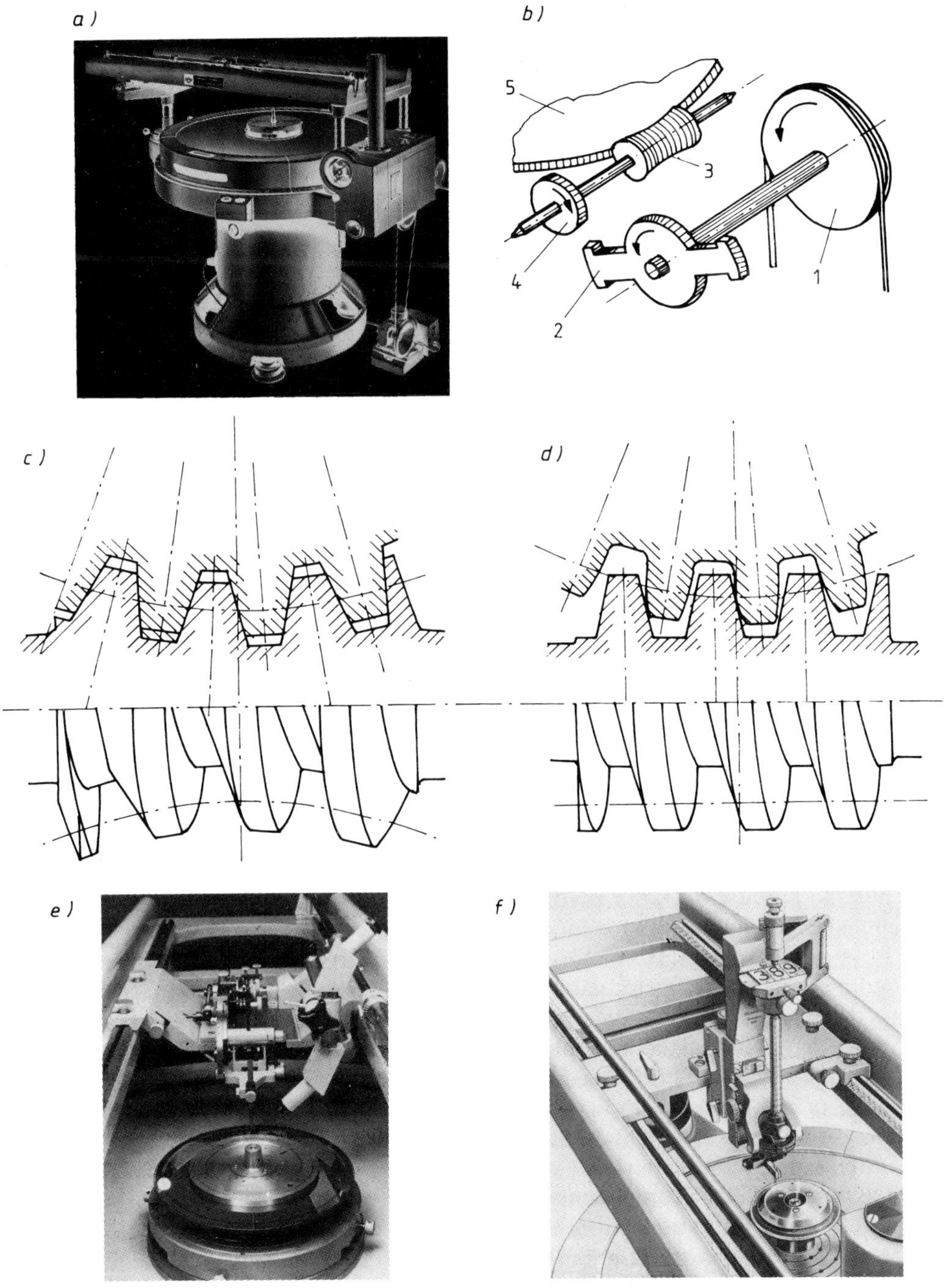

Figure 2.19. a) Automatic graduator *TKF 1000/1* (VEB Feinmess, Dresden); *b)* Worm and gear segment of a graduator; *c)* Toroidal worm of the graduator *TKF 1000/1; d)* Tangential worm; *e)* Engraving device of *TKF 1000/1; f)* Pantograph *PZG*

Recently computer controlled electronics is also used in high precision graduating machines. For the sake of economical production certain selected and exactly graduated glass-circles are often copied by precision photographic methods. Circle graduation of high accuracy can be produced by using thin-sheet techniques. This is used in the *Diadur method* developed by *Heidenhain*, by means of which 1 μm thin division lines and numbering can be applied to hard glass surfaces with good contrast and covering (light-tight) properties.

Errors in the positions of the graduation lines (graduation error) can occur for different reasons:

1. Errors in the connection between the circle to be divided and the graduator, which can be as follows:

a) An eccentricity between the gear rim and the glass circle clamped to it. Since the graduation centre of the glass circle is centred at the centre of the rotation axis of the central part of the instrument, when mounted in the theodolite, a centring error of the graduation centre of the horizontal circle is an actual source of graduation error.

b) Error of the graduation plane. This occurs if the graduation plane of the glass circle is not perpendicular to the rotation axis of the gear rim.

c) Stresses and deformations in the glass arising when the glass circle is clamped on the plane surface of the gear rim.

2. Deficiences of the graduating machine: manufacturing deficiences of the gear rim and the toroidal worm screw.

3. Wobbling of the rotation axis and deficiences of the engraving device.

4. External effects (temperature change, wind and ground movement).

The effects listed under item 1 have an effect on the positions of the graduation lines over the whole circle, with periodic distribution following graphically a more or less sinusoidal curve (regular graduation error with long periodicity). Effects under 2 cause partially random, partially short-periodic regular errors, generally of greater amplitudes. The periodicity corresponds to one or more revolutions of the toroidal screw. Effects under 3 and 4 usually produce graduation errors of random character.

Thus the graduation error is composed of systematic components (described by functions) and random components. The systematic components are usually given in the form of trigonometric series expansion and illustrated graphically as ordinates plotted against the circle graduation as abscissa. In *Fig. 2.20* the top diagram $\Delta\varphi_{reg}$ shows the resultant total systematic error corresponding to the abscissa φ (graduation position), and below it the systematic error components of periods 2π, π, $2\pi/3$, $\pi/2$ and $2\pi/5$ are shown. In *Fig. 2.21, a* the ordinates show the random graduation errors. The sum of the random graduation error and the resultant systematic error at a certain abscissa φ is the position error of the graduation line in question *(Fig. 2.21, b)*. With readings at diametrically opposite points on the graduated circle, the effect of the systematic error components of odd order is cancelled in the mean.

The effect of graduation errors on direction measurements can be decreased, or partly eliminated, by a special arrangement of the graduation system, by the respective arrangement of the reading system and by using special measuring methods. Thus, e.g., *Kern* uses on the graduated circles of its theodolites two graduation systems of different diameters which are marked on the circle in two independent working phases

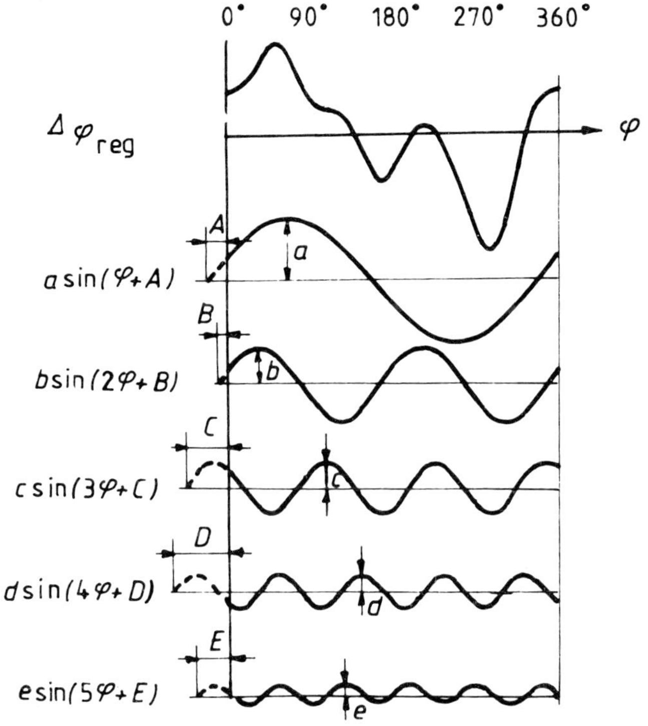

Figure 2.20. Systematic errors of graduation lines

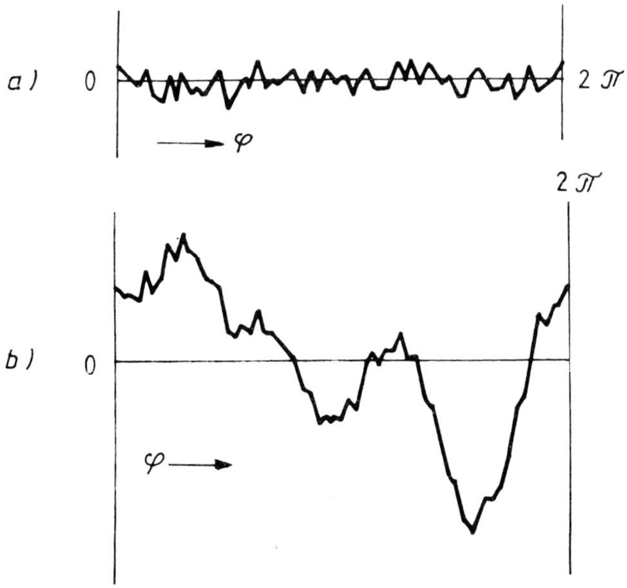

Figure 2.21. a) Random errors of the graduation lines of a circle; *b)* Position errors of graduation lines

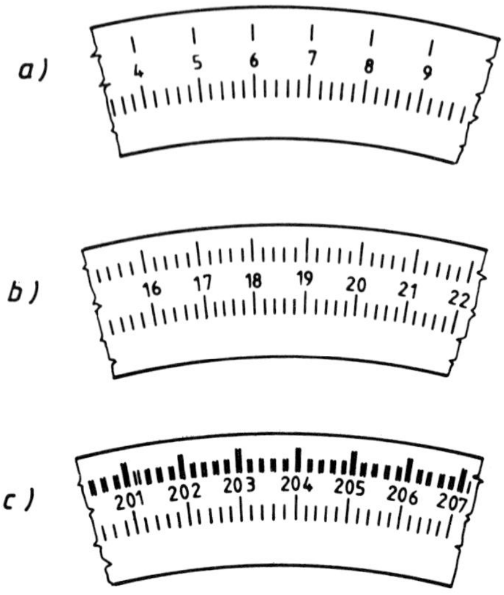

Figure 2.22. Double graduation lines (Kern, Aarau)

(Fig. 2.22). One of them is made with a large division interval, e.g. 1^g, the other with a small one, e.g. $1^g/5$ *(Fig. 2.22, a)*, or both are made with the same graduation interval *(Fig. 2.22, b)*; sometimes one is made with double division lines and the other with single lines *(Fig. 2.22, c)*, but only one of them is numbered. The two independent graduation systems are projected against each other in the reading device.

If the graduated surface of the glass circle is a reflecting mirror-surface in the ray path of the reading device, it is coated with silver and then with a protecting sheet. If the light ray meets the graduated circle in the reading device, then the graduated surface is protected against dust by a thin glass ring cemented over it.

The numbering of the horizontal graduated circle increases clockwise. On the horizontal graduated circles of some theodolites double numbering is applied in both directions in order to facilitate the setting out of angles left-wards.

2.2.5 Upper Part of the Theodolite

The upper part of the theodolite consists of the following parts: the casing with the telescope support, the horizontal axis, the geodetic telescope, the vertical circle, the reading device, the spirit level or levels of the vertical axis (alidade-level), the index level of the vertical circle or the inclination compensator and the clamping and fine adjustment devices of the horizontal and vertical circles.

2.2.5.1 Casing of the Upper Part

The pivot or socket of the vertical axis is rigidly fixed to the bottom of the casing; it rotates in the socket or around the pivot rigidly mounted on the lower part of the instrument (sometimes to its central part). Most of the optical reading device is placed inside the casing and the telescope supporting columns; the spirit level(s) of the vertical axis is mounted on this casing and the index-spirit level of the vertical circle or the tilt compensator is in the left-hand column. In the two columns of the telescope support there are the beds of the two pivots of the horizontal axis which have equal diameter; they are rigidly mounted on the telescope and the cylindrical surfaces of the pivots are accurately machined. In modern theodolites the vertical axis is hollow and the path of rays of the reading device is led through it, as well as the light rays incident on the mirror illuminating the cross-hair reticule. The mirror is at the intersection of the vertical and horizontal axes. The telescope-supporting columns are made so that the telescope can be reversed objective downwards and sometimes also eyepiece downwards, but at least the former in order to be able to measure with the telescope in two positions.

The bearing of the pivots of the horizontal axis in the telescope-supporting columns can be cylindrical or V-shaped; in the latter case the seating on four points determines the axis position unambiguously. Since the bearings of the horizontal axis must bear the weights of the telescope, the horizontal axis and the parts rigidly mounted to them (e.g. the graduated vertical circle), and thus are under unchanged pressure, the thickness of the oil film between the sliding surfaces changes less than on the vertical axis. In order to make the revolution of the horizontal axis easier, part of the weight of the larger telescopes is born by load-lightening devices.

Larger instruments were constructed in early days in such a way that the telescope, together with the pivots of the horizontal axis, could be removed from the bearings on the support columns and replaced after rotating the telescope by 180° around its axis into a reversed position, thus interchanging the pivots.

The horizontal axis must be perpendicular to the vertical axis. In older instruments it was possible to correct the perpendicularity error of the horizontal axis (in short: horizontal axis error): the bearing of the telescope was separated from the column by a small slit *(Fig. 2.23)* which could be narrowed or widened by the pull-screw *1* and the push-screw *2*. In newer theodolites the cylindrical bearings of the horizontal axis can be adjusted to the required height by turning an eccentric ring in the support, and clamped by tightening a screw. In modern instruments no device is applied for the correction of the horizontal axis error, because the simple and stable construction of the vertical and horizontal axes ensures that the perpendicularity of the two axes is long-lasting.

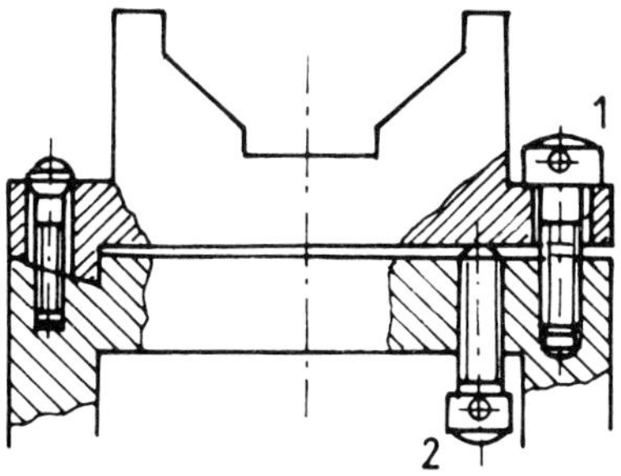

Figure 2.23. Bearing of a horizontal axis with adjusting screws

2.2.5.2 The Geodetic Telescope

The telescope described in Subsection 1.5.3 can be used as a sighting device (i.e. as a geodetic telescope) if a cross-hair is put in its image plane, and if the intersection of the hairs (its formal centre) is one point of the line of sight used for pointing.

The geodetic line of sight of a geodetic telescope is, according to *Helmert's definition*, the locus of points at different distances of which the objective lens system forms an image on the centre of the cross-hair with an undisturbed telescope.

The astronomical line of sight of a geodetic telescope is the direction of the principle ray entering the objective, which — after passing through the objective system — meets the image plane of the telescope, focussed to infinity, at the centre of the cross-hair.

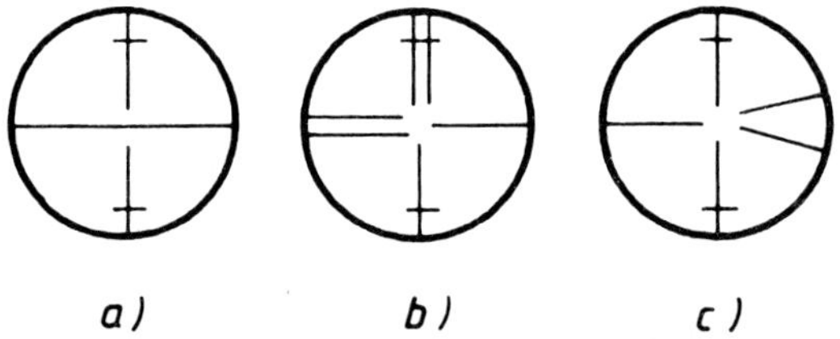

Figure 2.24. Forms of cross-hairs

Figure 2.25. Mount of a cross-hair; front view (left); lateral section (right)

Different kinds of cross-hairs are used *(Fig. 2.24)*. The mounting ring of the reticule which contains the cross-hair can be moved in the tube of the telescope by a small amount in both the vertical and the horizontal direction by means of adjusting screws, and it can also be rotated *(Fig. 2.25)*. This is needed for adjustment (Point 2.2.8.1). The cross-hair marked on a plane parallel glass plate with a very fine surface, is usually covered by a glass to protect it against dust. The internal circumference of the mounting ring is at the same time the boundary diaphragm of the field of view of the geodetic telescope. The total aperture angle of the field of view is d_b/f_0, where d_b is the diameter of the aperture and f_0 is the focal length of the objective system.

A geodetic telescope is pointed to a certain point in the field (or pointing target) if the real image of the point (pointed mark) formed by the objective lens system lies on the centre of the cross-hair.

The pointing accuracy (mean square error of pointing) of a geodetic telescope is defined as the mean square error of the directions determined by pointing in a certain target with the telescope several times (*n*-times), the error being given in angular measure. Its practical determination is discussed in Point 2.2.8.6.

Pointing ability depends on the optical resolving power of the telescope (Subsection 1.5.3), and on the thickness and sharpness of the cross-hairs. It is considerably influenced by such external factors as: the optical resolving power of the observer' eye, the illumination of the point observed (target), its background, shape, contrast and distance, and finally the weather factors such as water vapour content, wind and air vibration. Among constant external factors the mean square error of pointing of a well-constructed and well-manufactured geodetic telescope increases by 0.5 to 1 power of the telescope magnification within certain magnification limits, if the pointing is made by day. For pointing at points on the Earth's surface, telescopes with a magnification higher than 60 × are very seldom used, even on theodolites used for angle measurements in geodesy. For observing stars a higher magnification can well be used, e.g. the theodolites *Wild T4* and *Siemens Tpr* can be used with 80 × telescope magnification, achieved by changing the eyepiece.

The optical resolving power of the telescope depends first of all on the resolving power of the objective lens or lens system (Subsection 1.5.3), because the eyepiece magnifies not only the real image formed by the objective in the plane of the reticule, but also the image-forming errors of the objective. Thus, if an eyepiece of considerably higher magnification is used with a given objective as the optimal eyepiece for it, the surplus magnification gained by the telescope is only a so-called empty magnification surplus which does not increase the resolving power and pointing accuracy.

The surfaces of the optical elements in the optical systems of modern geodetic telescopes which have contact with air, are given one or two coatings of reflection reducing film (Subsection 1.1.3). The effect of such a coating is not only that the light efficiency (i.e. the ratio between the light leaving the eyepiece and that entering the objective lens), and with it the light power of the telescope, is increased by 30 to 60 percent, but mainly that the contrast-reducing effect of the scattered rays reflected from the different optical surfaces and then re-reflected forwards, is held to a minimum. The image does not become grey, but remains sharp and contrasty.

The diameter of the entrance pupil, i.e. the diameter of the free aperture of the outside objective lens, is chosen so that the exit pupil has a diameter of 1.3 to 2.0 mm. As such exit pupils are smaller than the average pupil of the eye, the spherical image-forming error of the observer has less influence and the image formed on the retina is sharper. For telescopes with mirror-lenses a somewhat greater entrance pupil must be used, because here a small part of the internal effective part of both the entrance and exit pupils is covered.

In the *simple geodetic telescope (Fig. 2.26)* the image plane of the objective changes its position depending on the distance to the point observed, therefore the reticule casing *1* must be mounted in the separate cross-hair tube *2* which can be moved backwards and forwards inside the main tube of the telescope *4* by the cross-hair

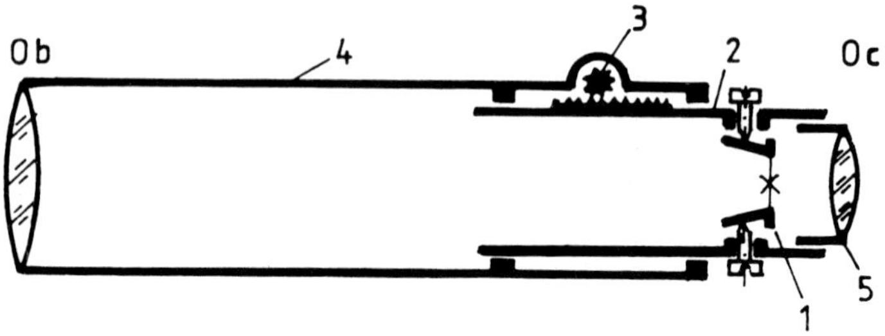

Figure 2.26. Simple geodetic telescope

adjusting screw *3*. The nearest observing distance is obtained at maximum extension of the reticule tube. The eyepiece casing *5* can be moved independently in the reticule tube in order to enable the observer to observe the virtual image, formed (by the eyepiece) of the real image of the observed point, at the optimum visibility distance of his own eye.

The stability of the line of sight of the telescope is very important. If the reticule tube of a simple geodetic telescope has a sideways clearance p_1 in the main tube which holds the objective L with centre O and focal distance f, then the centre K of the cross-hair can be shifted sideways by the same amount p_1 to the position K' *(Fig. 2.27/a)*. The straight line of sight passing through the lens centre O consequently has a sideways wobble which is, in the case of observing a point P at an infinite distance:

$$\alpha_{1,\infty} = \frac{p_1}{f} \rho''. \tag{2.6}$$

In the case of a finite observing distance, f is substituted in Eq. (2.6) by the image distance between the objective and the reticule:

$$s' = \frac{sf}{s-f}$$

with positive sign according to the sign rule given in Subsection 1.3.5, where s is the distance of the sighted point P from the objective with negative sign (object distance). In this case the wobble of the line of sight due to p_1 is:

$$\alpha_{1,t} = p_1 \frac{s-f}{sf} \rho'', \tag{2.7}$$

which is less than in Eq. (2.6), in other words the wobble decreases with decreasing observing distance.

If the centre K of the cross-hair of a simple geodetic telescope is shifted sideways into the position K' by a small distance q_1 from the straight line *i-i* passing through the optical centre O of the objective lens L and parallel to the direction of straight motion of the reticule tube *(Fig. 2.27, b)*, then the line of sight is the straight line defined by the point O' of the objective lens L which is at distance q_1 from the optical centre O, and the object-side focal point F of the lens (independently of the observing distance $t = |s| + a$, measured from the horizontal axis *H-H*). The observed point P,

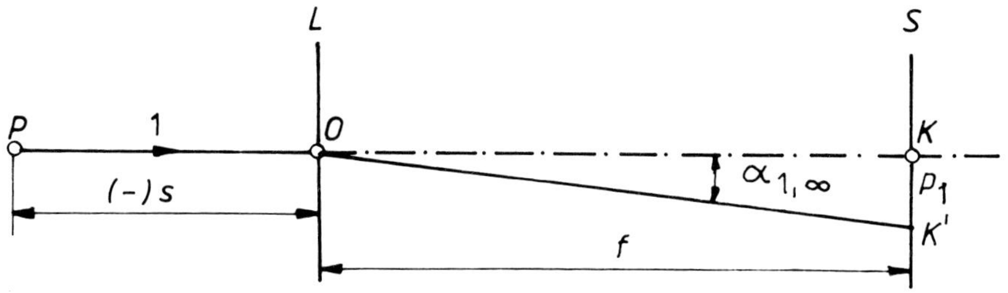

Figure 2.27/a. Wobble of the line of sight of a simple geodetic telescope due to the play of the cross-hair tube

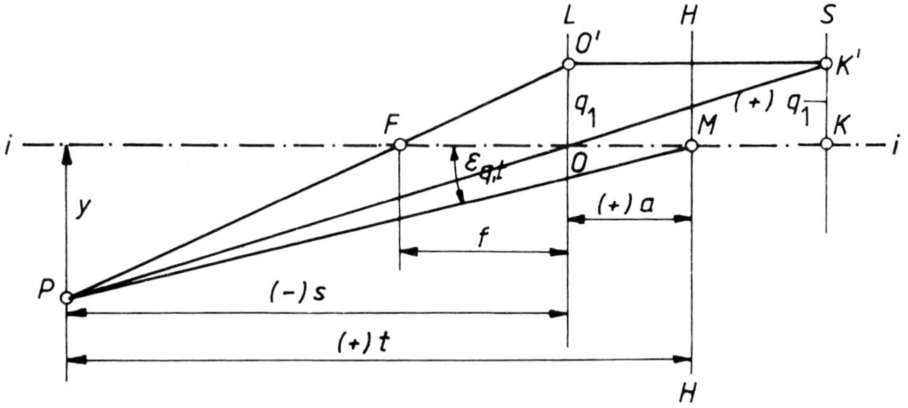

Figure 2.27/b. Distance-dependent deflection of the line of sight of a simple geodetic telescope

at distance $t=|s|+a$ from the intersection M (the instrument centre) of the straight line *i-i* and the horizontal axis *H-H*, lies at a lateral distance from the straight line *i-i*:

$$y=\frac{|s|-f}{f}q_1 \tag{2.8}$$

as can be seen in the figure, and the direction MP forms with the straight line *i-i* the angle:

$$\varepsilon''_{q,t}=\frac{y}{|s|+a}\rho''=\frac{y}{t}\rho''. \tag{2.9}$$

Substituting Eq. (2.8) into Eq. (2.9) gives:

$$\varepsilon''_{q,t}=\frac{|s|-f}{t}\frac{q_1}{f}\rho''. \tag{2.10}$$

In the case of an infinite observing distance ($t=\infty$) this deviation is:

$$\varepsilon''_{q,\infty}=\lim_{\substack{(t\to\infty)\\(|s|\to\infty)}}\frac{|s|-f}{t}\frac{q_1}{f}\rho''=\frac{q_1}{f}\rho''. \tag{2.11}$$

The difference between the deviations at finite and infinite distances, due to the shift q_1 of the cross-hair centre, is from Eqs (2.10) and (2.11):

$$\varepsilon_i = \varepsilon_{q_1,t} - \varepsilon_{q_1,\infty} = -\frac{f+a}{t}\frac{q_1}{f}\rho'' . \tag{2.12}$$

This is added to the collimation error as the first additional collimation error (Point 2.2.8.1) which is cancelled in the arithmetic means of the two measurements made in the two telescope positions. For instance, if $f=200$ mm, $a=100$ mm and $q_1=0.1$ mm, then the additional collimation error ε_i of the measured direction is at distances $t=2$ m, 10 m, 100 m and $t>1000$ m, from Eq. (2.12): $-15.4''$, $-3.1''$, $-0.3''$ and $0''$, respectively.

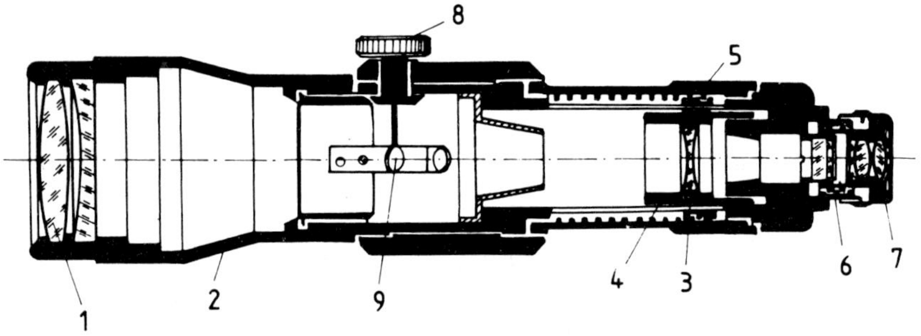

Figure 2.28. Geodetic telescope with internal focussing lens

In telescopes with internal focussing *(Fig. 2.28)* now used in surveying instruments, the lens *1* (with positive focal length of the teleobjective system and having a resultant focal length which changes with change of object distance (see Subsection 1.5.3) is rigidly mounted in the main telescope tube *2*. The second lens *3* with negative focal length is mounted in the internal tube *4* which can be moved along the main tube by the focussing screw *5*. This ensures that the image plane of the teleobjective is

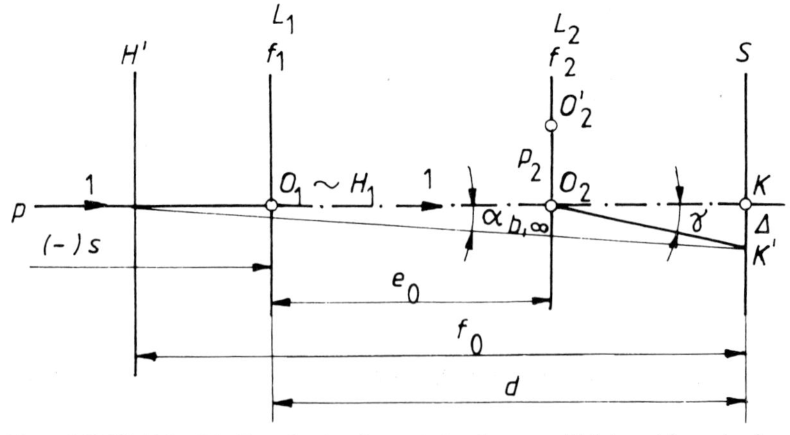

Figure 2.29. Wobble of the line of sight of a geodetic telescope with internal focussing lens due to the clearance of the mount of the focussing lens

always in the same plane *6* of the tube, independently of the distance of the observed point. The plane *6* is the plane of the reticule plate which is fixed in the main tube. The eyepiece mount *7* can be separately moved in the main tube, thus enabling the observer to observe both the virtual image formed of the cross-hair and the real image of the observed point, at the optimum visibility distance of his eye. The knob *8* at the side of the telescope is used to adjust the mirror *9* which illuminates the cross-hair.

If, in an internal focussing telescope, the mount of the movable negative lens L_2 with focal length f_2 has a sideways clearance p_2, then the centre O_2 of the lens L_2 can also be moved sideways by the same amount to O'_2 (*Fig. 2.29*). Due to this the principal ray *1*, coming from the distant point observed and incident on the centre O_1 of the fixed objective L_1, can be deflected, when passing through the concave lens L_2, by the angle

$$\gamma = \frac{p_2}{f_2}\rho'' , \tag{2.13}$$

and the image of the observed point can be moved sideways on the reticule plate by the small amount:

$$\Delta = \frac{1}{\rho}\gamma(d-e_0) = \frac{p_2}{f_2}(d-e_0) . \tag{2.14}$$

In Eq. (2.14) d is the constant optical structural length between the rear principal plane of the lens L_1 and the reticule plate S. The length f_2 is the negative focal length of the image-focussing lens L_2 which can be computed in the course of the construction of the telescope from the known (previously determined) resultant focal length f_0 of the teleobjective system adjusted to infinity, and from the focal length f_1 of the fixed objective lens, also previously determined, by means of the following formula:

$$f_2 = \frac{-f_1 f_0 (d-f_1)}{(f_0-f_1)^2} . \tag{2.15}$$

The quantity e_0 in Eq. (2.14) is the distance between the image-side principal plane of the lens L_1 and the object-side principal plane of the lens L_2 when adjusted to infinity; it can be calculated from the expression:

$$e_0 = \frac{f_1+d}{2} - \sqrt{\left(\frac{d-f_1}{2}\right)^2 - f_2(d-f_1)} . \tag{2.16}$$

In *Fig. 2.29* the plane H' is the image-side principal plane of the objective lens system consisting of the lenses L_1 and L_2, when focussed to infinity. The principle plane H' lies at the following distance from the lens L_1 (more exactly from its object-side principal plane H_1):

$$\overline{H'H_1} \approx \overline{H'O_1} = -(f_0-d) .$$

When observing the centre K of the cross-hair, S is set to the point K' moved sideways by Δ; hence the angular deflection of the line of sight of the telescope is:

$$\alpha_{2,\infty} = \frac{\Delta}{f_0}\rho'' ,$$

or by using Eq. (2.13):

$$\alpha_{2,\infty} = \frac{p_2}{f_0 f_2}(d - e_0)\rho'' . \tag{2.17}$$

In the case of finite (short) observing distances, instead of e_0, the value

$$e = \frac{s_1' + d}{2} - \sqrt{\left(\frac{d - s_1'}{2}\right)^2 - f_2(d - s_1')} , \tag{2.16a}$$

valid for finite distances, is to be used in Eq. (2.17), where

$$s_1' = \frac{s_1 f_1}{s_1 + f_1} ,$$

and instead of f_0, the resulting focal length

$$f = \frac{f_1 f_2}{f_1 + f_2 - e} \tag{2.18}$$

must be used. In the case of a geodetic telescope with internal focussing, the sideways clearance p_2 of the mount of the concave lens also causes a sideways wobble of the line of sight, which decreases with decreasing observing distance.

The ratio of values (2.17) and (2.16):

$$\frac{\alpha_{2,\infty}}{\alpha_{1,\infty}} = \frac{p_2}{p_1} \frac{d - e_0}{f_2} \frac{f}{f_0} \tag{2.19}$$

is the ratio between the wobbles of the line of sight of a simple and an internal focussing telescope with equal objective focal length, when observing at infinity. This

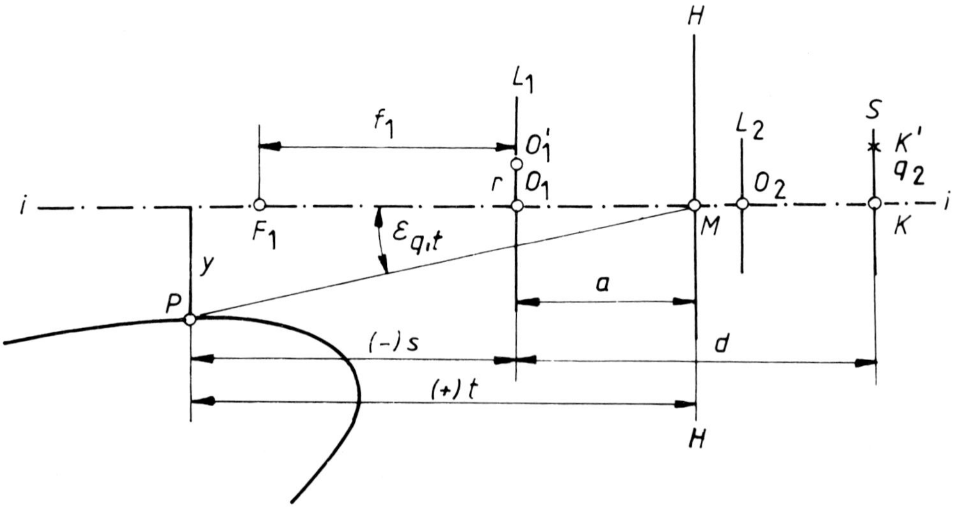

Figure 2.30. Distance-dependent deflection of the line of sight of a geodetic telescope with internal focussing lens

ratio is less than 1 even if $p_2 = p_1$. However, the moving mechanism of the concave lens of an internal focussing telescope is totally enclosed inside the telescope tube (and thus free of external mechanical influences as well as of the effects of weather and dirt), while the cross-hair tube of a simple telescope is open (it is only partly inside the telescope tube). Consequently $p_2 < p_1$ and so the wobble of the line of sight of the internal focussing telescope is considerably less than that of the simple telescope.

The line of sight of a geodetic telescope with internal focussing is a straight line — by definition — only if the centre K of the cross-hair lies on the straight line determined by the centre O_1 of the fixed objective lens L_1 and the centre O_2 of the image focussing lens L_2, but generally this is not true. Let the centre K of the cross-hairs of a geodetic telescope, with optical structural length d and resultant focal length f_0 at infinite focus, be moved into position K' through the small distance q_2, the movement being sideways with respect to the straight line *i-i* of the motion of the centre O_2 of the concave lens L_2, which passes through the centre O_1 of the fixed objective lens L_1 *(Fig. 2.30)*. Then the line of sight of the telescope by definition is the arm of the very flat hyperbola which is near to the *i-i* axis, lies in the horizontal and is defined by the equation*:

$$s_1^2 q_2^2 f_0 + s_1 q_2 y (f_1^2 + f_0^2) + y^2 f_1^2 f_0 + 2 s_1 q_2^2 f_1 f_0 + \\ + y q_2 f_1 \frac{d(f_1^2 + f_0^2) - 2 f_1^2 f_0}{d - f_1} + q_2^2 f_1^2 f_0 = 0 \,. \tag{2.20}$$

From Eq. (2.20) the lateral distance from the straight line *i-i* of the observed point P, at the finite distance $t = |s_1| + a$, is:

$$y = \frac{-B - \sqrt{B^2 - 4AC}}{2A}, \tag{2.21}$$

where

$$A = f_1^2 f_0 \,,$$

$$B = q_2 \left\{ f_1 \frac{d(f_1^2 + f_0^2) - 2 f_1^2 f_0}{d - f_1} + s_1 (f_1^2 + f_0^2) \right\},$$

$$C = q_2^2 f_0 (f_1^2 + 2 s_1 f_1 + s_1^2) \,.$$

Using Eq. (2.9), the direction MP from the instrument centre M to the observed point P at the distance t deviates from the straight line *i-i* by the angle

$$\varepsilon''_{q_2 t} = \frac{y}{|s_1| + a} \rho'' \,, \tag{2.22}$$

for which y, depending on the distance $(-)s = a - t$, can be computed from Eq. (2.21).

* The deduction of (2.20) can be found in the papers of L. Fialovszky: Konstruktion des fadendistanzmessenden Fernrohres mit Innenfokussierung (Feingerätetechnik, 1957). Focal-length-adjusting device for constructing anallactic distance-measuring telescopes (in Hungarian). (MTA Műsz. Tud. Közl., Budapest, XIII, No. 1—4, 1954) and Konstrukcja lunety dalmierczej z wewnętrznym ogniskowaniem (Geod. i Kart., Warszawa, Tom VI, Zeszyt 2, 1955).

The deflection angle due to q_2 for a telescope focussed to infinity is, analogously to Eq. (2.11):

$$\varepsilon_{q2\infty} = \lim_{(|s_1| \to \infty)} \frac{y}{|s_1| + a} \rho'' = \frac{q_2}{f_0} \rho''$$

and subtracting this from Eq. (2.22) one obtains the 1st additional collimation error

$$\varepsilon_i = \varepsilon_{q2t} - \varepsilon_{q2\infty} \, . \tag{2.23}$$

Taking e.g. the data of the theodolite *Te-B21* (MOM, Budapest) ($f_1 = 97.3$ mm; $f_2 = -33.28$ mm; $f_0 = 209.8$ mm; $d = 111.5$ mm; $a = d/2$; $e_0 = 76.38$ mm), Eq. (2.23) gives the deflection of the line of sight from the axis of the telescope, focussed to infinity, at distances $t = 2$ m, 10 m, 50 m and 100 m as $\varepsilon_i = 14.02''$, $2.01''$, $0.38''$ and $0.19''$, respectively, if the centre of the cross-hair is moved sideways by $q = 0.2$ mm from the line of motion of the centre of the image-focussing lens. This deflection also has the character of a collimation error (2nd additional collimation error); its effect on direction and angle measurements is also cancelled by the mean of measurements carried out in two telescope positions (Point 2.2.8.1).

If — due to imprecise centring — the centre O_1 of the fixed objective lens L_1 is also moved sideways from the straight line *i-i* of the motion of the centre O_2 of the image focussing concave lens L_2 by the small amount r, then the equation of the hyperbolic line of sight [Eq. (2.20)] is also somewhat changed. In this case the linear term:

$$\Delta y = \frac{r}{f_1} s_1$$

is added to the y coordinate of the point corresponding to the object distance $(-)s_1$ on the line of sight. Thus the modified equation of the hyperbolic line of sight is:

$$\begin{aligned} &s_1^2 \left[q_2^2 f_0 + r q_2 \frac{(f_0^2 + f_1^2)}{f_1} + r^2 f_0 \right] + \\ &+ s_1 y [q_2 (f_1^2 + f_0^2) - 2 r f_1 f_0] + y^2 f_1^2 f_0 + \\ &+ s_1 \left[2 q_2^2 f_1 f_0 - r q_2 \frac{d(f_1^2 + f_0^2) - 2 f_1^2 f_2}{d - f_1} \right] + \\ &+ y q_2 f_1 \frac{d(f_1^2 + f_0^2) - 2 f_1^2 f_0}{d - f_1} + q_2^2 f_1^2 f_0 = 0 \, . \end{aligned} \tag{2.20a}$$

If $r = 0$ but $q_2 \neq 0$, in other words if the fixed objective lens L_1 is centred but the centre K of the cross-hairs is not centred, then the general hyperbola of the line of sight characterized by Eq. (2.20a) is modified into a special hyperbola [Eq. (2.20)], differing by a very small amount.

If $r \neq 0$ and $q_2 = 0$, in other words if the lens L_1 is not centred, but the cross-hair centre K is centred, then the hyperbolic line of sight of Eq. (2.20a) is a straight line which starts from the point where the straight line *i-i* of motion of the centre O_2 of the image focussing lens L_2 meets the fixed objective lens L_1 (i.e. its object-side principal plane H_1); the line deflects sideways from the straight line *i-i* by the angle r/f_1 rad and passes through the object-side focal point F_1 of the lens L_1.

If $r=0$ and $q_2=0$, in other words if both the centre O_1 of the lens L_1 and the cross-hair centre K are centred (both points lie on the straight line of motion of the centre O_2 of the lens L_2), then both hyperbolic lines of sight, described by Eqs (2.20) and (2.20a), are straight lines and the straight line of sight is identical with the optical axis.

The internal focussing telescope is shorter than the simple telescope as is shown by the graph in *Fig. 2.31*, where the values of $f_{1,\mathrm{red}}=\lambda f_1$, $f_{2,\mathrm{red}}=\lambda f_2$ and $e_{0,\mathrm{red}}=\lambda e_0$ corresponding to the reduced structural length $d_{\mathrm{red}}=\lambda d=100$ mm — all reduced by the factor $\lambda=100/d$ — are plotted against the resultant focal length $f_{0,\mathrm{red}}=\lambda f_0$. For constructing *Fig. 2.31* the anallactic point (i.e. the point where, when $q_2 \neq 0$, the relevant asymptote of the hyperbolic line of sight characterized by Eq. (2.20) intersects the optical axis of the telescope) has been taken behind the fixed objective lens L_1 at half the distance d. The ordinates of the curve $f_{1,\mathrm{red}}$ are:

$$f_{1,\mathrm{red}}=\frac{-B_1-\sqrt{B_1^2-4A_1C_1}}{2A_1}, \tag{2.24}$$

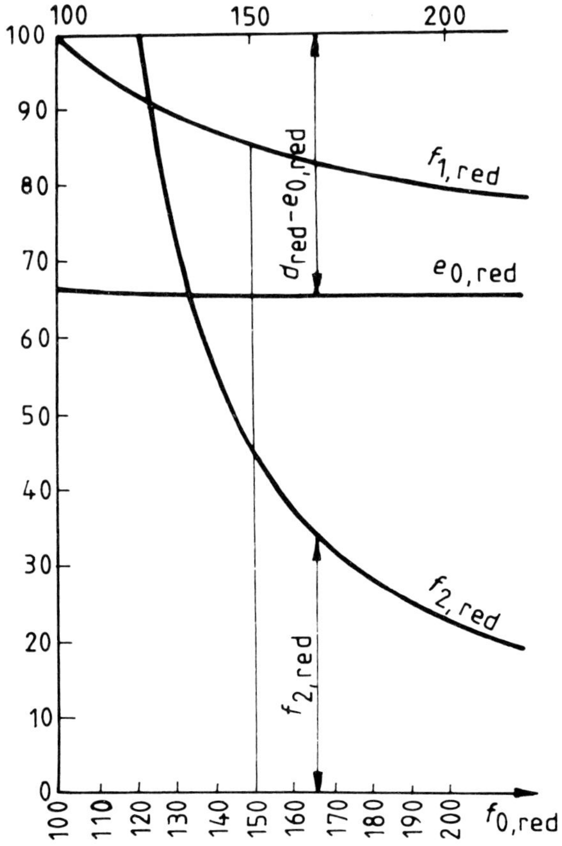

Figure 2.31. Connections between the basic characteristics of a telescope with internal focussing

where

$$A_1 = 2f_{0,\mathrm{red}} - \frac{d_{\mathrm{red}}}{2},$$

$$B_1 = -\frac{d_{\mathrm{red}}}{2}(d_{\mathrm{red}} + f_{0,\mathrm{red}}),$$

$$C_1 = \frac{d_{\mathrm{red}}}{2} f_{0,\mathrm{red}}.$$

The ordinates of the curve $f_{2,\mathrm{red}}$ are determined by the formula:

$$f_{2,\mathrm{red}} = \frac{-f_{1,\mathrm{red}} f_{0,\mathrm{red}}(100 - f_{1,\mathrm{red}})}{(f_{0,\mathrm{red}} - f_{1,\mathrm{red}})^2} \tag{2.25}$$

and, using Eq. (2.16), the ordinates of the curve $e_{0,\mathrm{red}}$ are determined by the formula:

$$e_{0,\mathrm{red}} = \frac{100 + f_{1,\mathrm{red}}}{2} - \sqrt{f_{1,\mathrm{red}}\left(f_{2,\mathrm{red}} - 50 + \frac{f_{1,\mathrm{red}}}{4}\right) + 100(25 - f_{2,\mathrm{red}})}. \tag{2.26}$$

The shortening of the telescope with respect to that of the simple telescope can be expressed by the shortening factor $\lambda = d/f_0$.

If e.g. an internal focussing telescope of structural length $d = 160$ mm, and of resultant focal distance $f_0 = 240$ mm, is to be produced (i.e. a telescope shortening of $\lambda = 160/240 = 0.66$ is required), then one finds on the graphs for $f_{0,\mathrm{red}} = \lambda f_0 = 150$ mm the corresponding values $f_{1,\mathrm{red}} = 85.2$ mm; $f_{2,\mathrm{red}} = -45.0$ mm and $e_{0,\mathrm{red}} = 65.6$ mm. By dividing these by the reducing factor $\lambda = 0.66$, one obtains the basic optical data for the telescope: $f_1 = 127.8$ mm, $f_2 = -67.5$ mm and $e_0 = 98.4$ mm.

The shortening of the telescope, i.e. the increase of $f_{0,\mathrm{red}}$, is limited by the fact — as seen on the graph — that by increasing $f_{0,\mathrm{red}}$ the reduced focal length $f_{1,\mathrm{red}}$ of the fixed objective lens L_1 decreases. It is well known that it is very difficult, if not impossible, to plan and construct a lens (or lens system consisting of several components) with the entrance pupil diameter required for the light power of the telescope without the occurrence of disturbing image-forming errors. If $f_{0,\mathrm{red}}$ is too long, i.e. the telescope is too short, the effect of the so-called secondary spectrum remaining, in spite of making the lens achromatic, is primarily harmful to image quality. For this reason also, the shortening must be kept within bounds and it is advisable not to go beyond the value 166 mm for $f_{0,\mathrm{red}}$.

The anallactic point of a telescope is generally understood to be that point on its optical axis at which a fixed angle makes an intercept, on a straight line through an observed point and perpendicular to the line of sight, proportional to the distance of the observed point from it. The anallatic point of a simple geodetic telescope is the objective-side focal point of the objective lens, in other words the anallactic point lies at a distance f_1 in front of the objective.

A telescope with internal focussing has no constant anallactic point: more precisely, at short distances the anallactic point moves within a small interval with changing distance. Therefore, one refers to the "quasi-anallactic point" of an internal focussing telescope; this is understood to be the anallactic point corresponding to infinity or to a suitably chosen finite distance. The former is nothing more than the intersection point of the optical axis and the relevant asymptote of the hyperbolic line of sight. The distance b of this point from the fixed lens L_1 (more precisely, from the image-side principle plane H'_1) is:

$$b = f_1 - \frac{f_1^2(f_0 - f_1)}{(d - f_1)(f_0 + f_1)}. \tag{2.27}$$

The distance b can also be determined from the formula

$$b = f_1 - \frac{f_1^2}{\sqrt{(d - f_1)(d - f_1 - 4f_2)}} \tag{2.28}$$

if the resultant focal length f_0 of the objective system is unknown, but the focal length f_2 of the focussing lens is known.

If the anallactic point corresponding to a certain finite distance is chosen as the quasi-anallactic point, then this point is the intersection of the optical axis and the suitably chosen chord of the hyperbolic line of sight. The anallactic point, and in connection with it the anallactic telescope, play an important role in geometric (optical) distance meters (Section 5.2).

Figure 1.60 shows a widely used example of telescopes yielding an erect image which have recently come into use. Here the reversed image generated by the objective lens system is made erect by the image-reversing prism system placed between the objective and the cross-hair reticule.

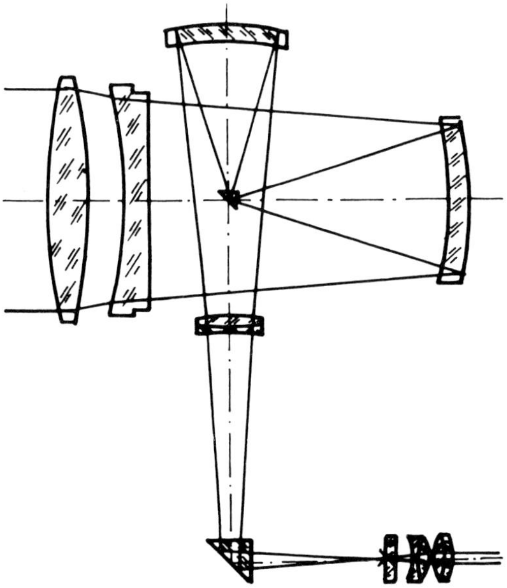

Figure 2.32. Scheme of the telescope with a mirror-lens used in the theodolite *DKM 3* (Kern, Aarau)

Certain theodolites of high or extremely high precision are manufactured with a combined mirror-lens telescope *(Fig. 1.56)* or with a bent mirror-lens telescope *(Fig. 2.32)*. The mirror-lens objective is easier to make achromatic, but its manufacture and mounting is more difficult.

A geodetic telescope is supplied with a coarse sighting device for the quick finding of the point to be observed. This can be a simple foresight and sighting slot or a finding collimator.

Telescopes are usually fitted with a centring mark for a centred set up below a roof point.

2.2.5.3 Reading Devices

Several methods have been developed for the reading of graduated circles. Some of them are outdated, as e.g. the graduation image of a circle to be read with a 30 s vernier *(Fig. 2.33)* is so long that it can be read only with a movable magnifier. This was one of the reasons why more comfortable, faster and more accurate reading devices have been developed.

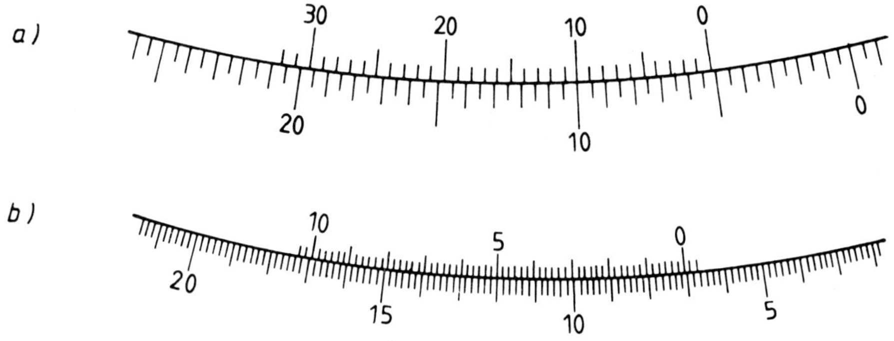

Figure 2.33. Circle reading with vernier

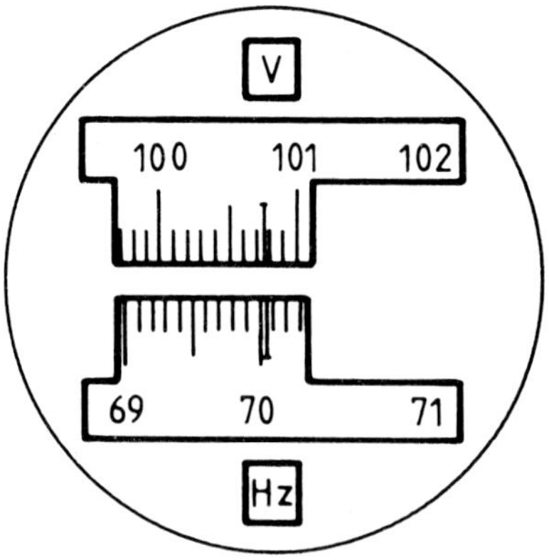

Figure 2.34/a. Image of the reading in the estimating microscope of the minute-theodolite *Th 51* (Opton, Oberkochen)

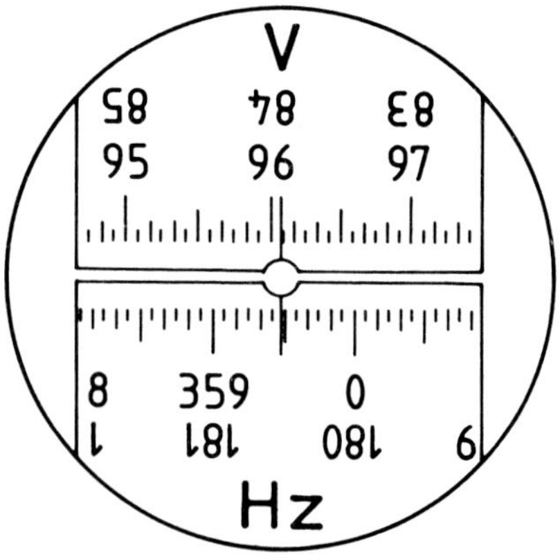

Figure 2.34/b. Reading in the estimating microscope of the small theodolite *Theo 080* (VEB Jenoptik, Jena)

The direct reading of circle graduations, either by magnifier or by estimating microscope, vernier microscope, microscope with a scale or a moving hair, is circumstantial due to the fact that the observer has to move to the reading device after pointing. Therefore in modern theodolites the reading, or the two diametrical readings, are optically projected within the casing and columns of the top part of the theodolite into the reading microscope mounted beside the telescope. In the field of view of the estimating microscope, shown in *Fig. 2.34/a*, the 10^c graduations of the horizontal and vertical circles can be read by estimation with an accuracy of 1^c. A similar microscope reading is shown in *Fig. 2.34/b*, where the 5 min graduation can be read with an accuracy of 30 s. On the horizontal and vertical circles of both theodolites numbers run in both directions, i.e. from right to left and from left to right; this allows the reading or the setting of the supplemental angle without calculation. The theodolite *4149 A* (Filotecnica Salmoiraghi, Milano) has a similar estimating microscope with a reading hair, in the field of view of which a very densely graduated circle is observed *(Fig. 2.83)*.

The scale microscope beside the telescope is also a simple, but more accurate circle reading device. The readings on the horizontal and vertical circles are visible in the microscope together with the corresponding microscope graduation projected onto them *(Fig. 2.35)*. The graduated circle can be easily read with an accuracy of 0.5^c on the simple and clear microscope scale, which consists of lines with thickness corresponding to that of the circle graduation lines. These lines are about 0.01 mm thick and are easily visible in the microscope which has a magnification of 25 ×. The lines of the microscope graduations, placed in the image plane of the objective lens, should have a thickness which, when magnified by the eyepiece, is equal to the thickness of the graduation lines magnified by the microscope. The reading microscope of the theodolite *K1-S* (Kern, Aarau) is similar but with the difference that numbers in both directions are applied to the horizontal circle *(Fig. 2.36)*. Thus the calculation of supplemental angles, necessary for some setting out of works is superfluous.

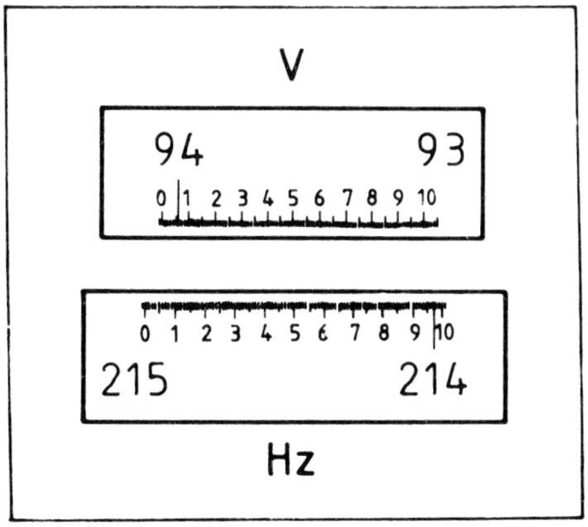

Figure 2.35. Field of view of the scale microscope of the theodolite *T16* (Wild, Heerbrugg). The reading shown is: $V = 94^g06.5^c$; $H = 214^g96.5^c$

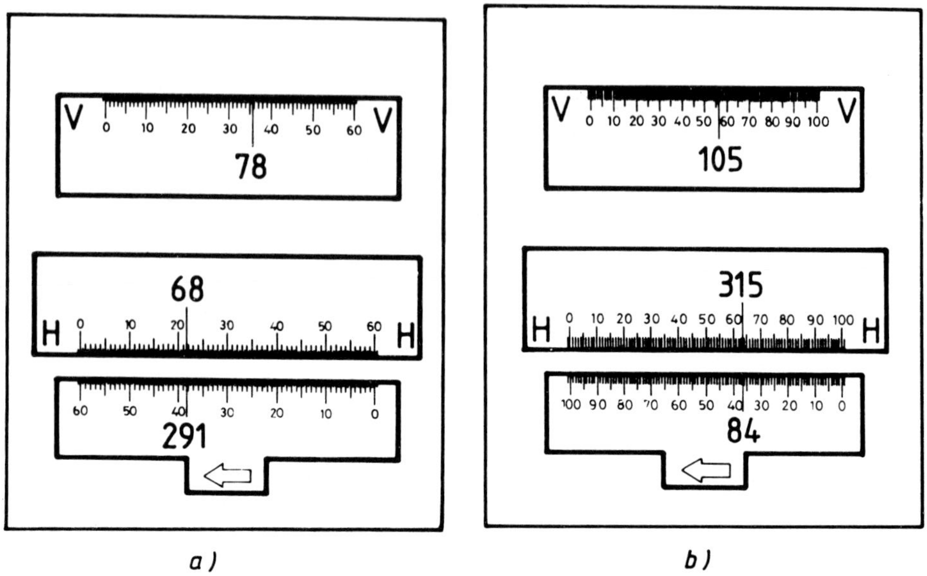

Figure 2.36. Field of view of the scale microscope of the engineer's theodolite *Kl-S* (Kern, Aarau); *a)* reading of *H* on the circle graduation of 360° increasing to the right: 68° 21.8′; increasing to the left: 291° 38.2′; reading of *V*: 78° 35.7′; *b)* reading if *H* on the circle graduation of 400^g: 315.635^g (increasing to the right) and 84.365^g (increasing to the left); $V = 105.565^g$

The requirement for high accuracy justifies the application in circle-reading devices of the optical micrometer, which has ousted the microscope with a movable hair in surveying instruments. The general principle of the operation of optical micrometers *(Fig. 2.37)* is that lens *1* forms an image of the circle graduation in the image plane *3*, which is supplied with a suitable diaphragm, after passing through the plane parallel glass plate *2* which is positioned in the ray-path. The micrometer graduation *5*, rotatable around the axis *4* and rigidly connected to the plane parallel glass plate, lies almost in the image plane *3*. From that and from the circle graduation, lens *6* forms another image in the image plane *7*, and this is observed in the microscope. The plane parallel glass plate, being itself the optical micrometer, and the synchronously movable micrometer scale can be rotated by turning the micrometer knob *8*. The plane parallel plate moves the image of the graduated circle with respect to the index mark *9* in the image plane *3* by an amount which corresponds to the amount of knob rotation. The movement can be read on the micrometer scale. The movement of the image due to the rotation of the micrometer, and the magnification of the lens *1*, must be synchronized; consequently the latter is mounted in such a way that it is adjustable. The error remaining after adjustment is the so-called error of run of the optical micrometer.

This type of micrometer is also used in the theodolite *Te-D13 (Fig. 2.38)*. Its advantage is that the operation of the micrometer is unaffected by play due to the rigid connection between the plane parallel glass plate and the micrometer scale.

The diametrically opposite reading devices were developed as a consequence of a connection between the readings of a circle graduation and the precision of the

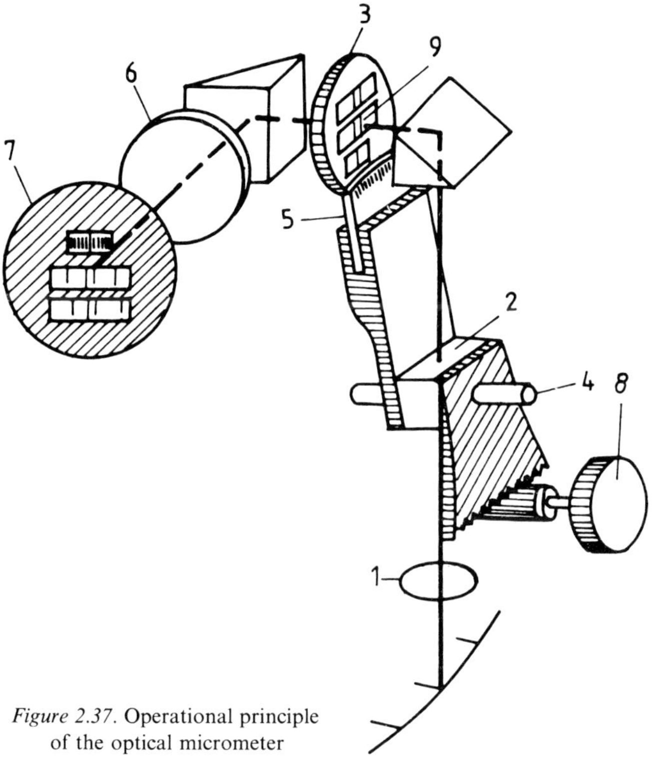

Figure 2.37. Operational principle of the optical micrometer

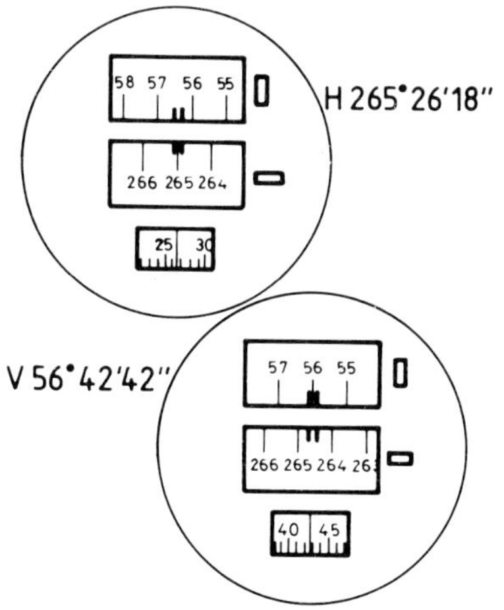

Figure 2.38. Field of view of the reading microscope of the theodolite *Te-D13* (MOM, Budapest)

theodolite axes. Reading error due to the eccentricity of the graduated circle with respect to the vertical axis (Point 2.2.8.3) does not influence the mean value of the diametrically opposite readings I and II connected to the telescope and, further, random errors of graduation and reading also have smaller effects on this mean. Instead of using two nearly diametrically opposite reading devices, the two readings are projected above or beside each other by means of an optical projecting system. The mean value of the two readings of the graduated circle is read directly in the field of view of the reading microscope. The two readings can be united into one image in the field of view in the following ways:

— the two images touch each other, the directions of the numberings being opposite to each other;

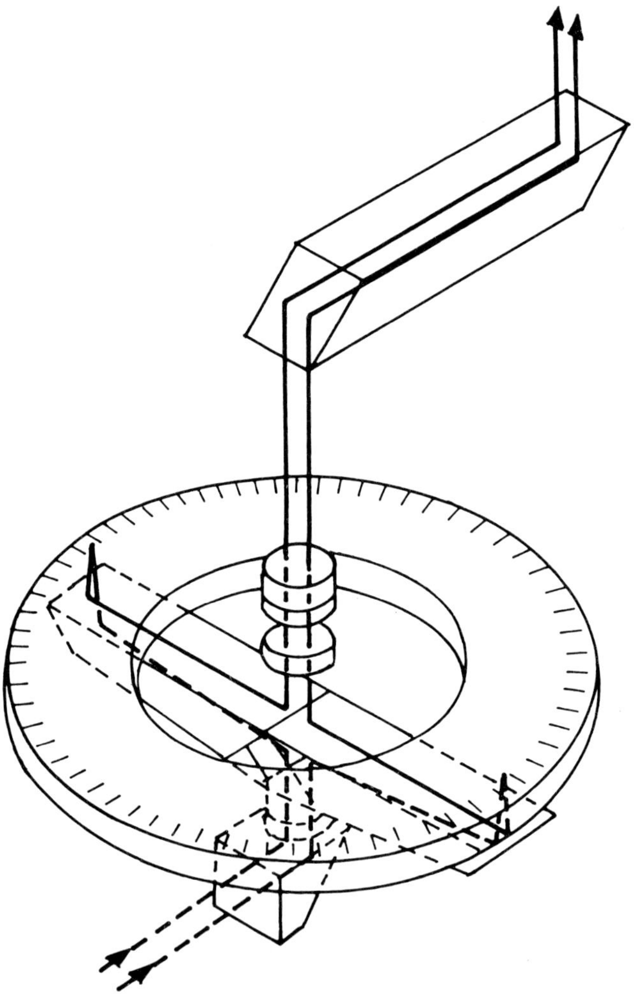

Figure 2.39. Optical train of the horizontal reading in the theodolite *T2* (Wild, Heerbrugg)

— the images are projected over each other with opposite numbering directions;
— the images are projected over each other with the same numbering directions.

The two images can be projected symmetrically beside each other, e.g. through the hollow vertical axis and through one telescope support, as in the theodolite *T2* *(Fig. 2.39)*, or one reading is projected over the other — as is more frequent — by, e.g., the various methods shown in *Fig. 2.40*. The two images of the circle readings are visible in the microscope as one image *(Fig. 2.41)*. In the first case the rays are usually reflected from the graduated surface of the circle, in the latter the rays pass through the glass circles.

In order to get opposite numbering directions in the image of the two readings, one prism of the optical system projecting the first reading over the second must be a roof-edge prism *(Fig. 2.40, a)*. If the aim is to unite the graduation lines of the readings into line-pairs, the projecting system must contain besides the lens (or lens system) two simple deflecting prisms *(Fig. 2.40, b)*. The diametrical projection can also be realized by going round the vertical axis *(Fig. 2.40, c)*.

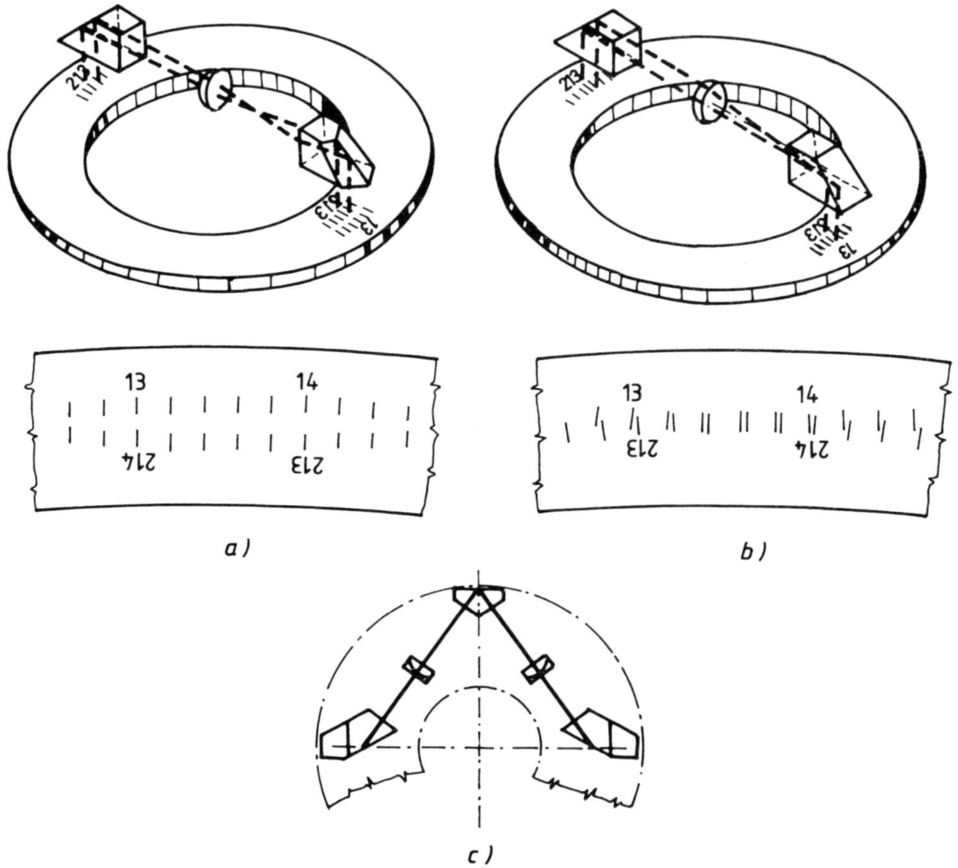

Figure 2.40. Projection of the first reading over the second: *a)* with a deflecting prism, a lens and a roof-edge prism; *b)* with a lens and two deflecting prisms; *c)* by going round the vertical axis

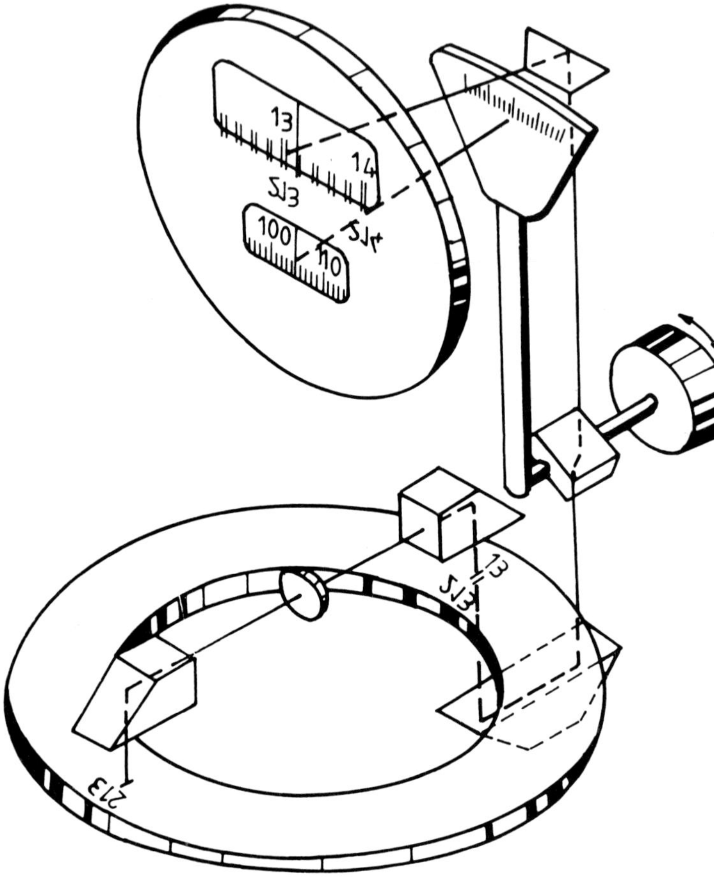

Figure 2.41. Projection of two diametral graduation lines into a double line

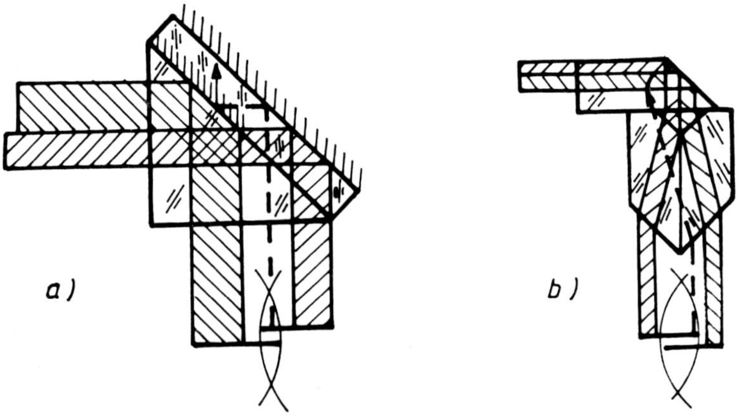

Figure 2.42. Image-splitting system *a)* in the theodolite *The 010* (VEB Jenoptik, Jena); *b)* in the theodolites *T2* and *T3* (Wild, Heerbrugg)

The images of the two diametrically opposite readings with opposite numbering should be divided from each other by a clear division line. For this reason an image-splitting system is installed into the reading ray path *(Fig. 2.42)*; this projects the parts of the two partially overlapping images besides the division line and eliminates the disturbing overlap *(Fig. 2.43)*. The fictitious reading index of the reading image is the point where the difference between the two readings is exactly 200^g and $180°$, respectively, i.e., where the fictitious indices I_2 and I'_1 coincide. When rotating the graduated circle or the top part of the instrument, the images of both reading positions move in opposite directions, but I_2 and I'_1 remain unmoved. If separate readings were made at the two circle positions, then the differences a_1 and a_2 between the preceding graduation line and the reading index should be determined, and then the mean value of the two readings calculated. Here, however, the mean value is directly read if the numerical value of one graduation line is verified, and its distance from the diametrically opposite graduation line $(a_1 + a_2)$ is determined. This distance is determined by the optical micrometer.

The rays forming the images of the reading are usually deflected by two optical elements in the micrometer *(Fig. 2.44)*; these move the images of the two graduation

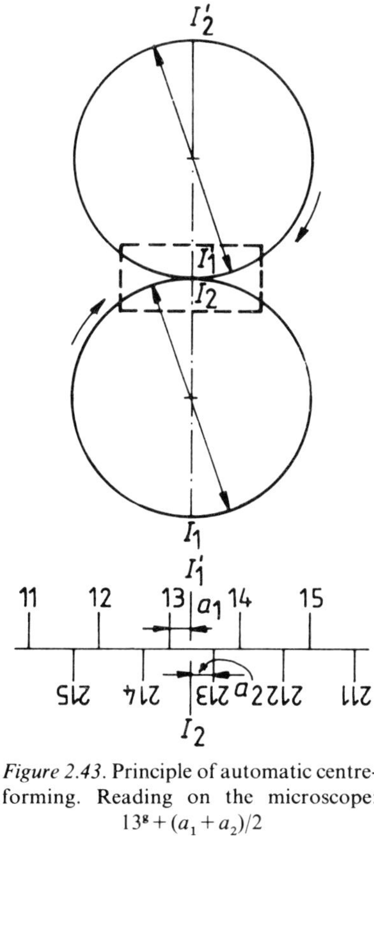

Figure 2.43. Principle of automatic centre-forming. Reading on the microscope: $13^g + (a_1 + a_2)/2$

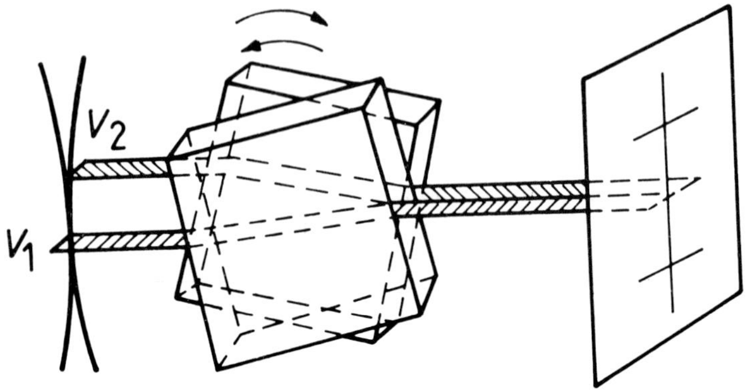

Figure 2.44. Effect of the two plane parallel glass plates on the coincidence microscope

lines always by half the distance between them in order to bring them into coincidence, and the accuracy of their setting is thus doubled. The rays arriving from reading I pass through one of the two plane parallel glass plates, which are rotatable around a common axis by equal amounts but in opposite directions, and the rays arriving from reading II pass through the other. By rotating the two plane parallel plates with the screw of the optical micrometer, the two diametral lines V_1 and V_2 can be brought into coincidence. The original distance between the two diametrically opposite graduation lines can be read directly in angular measure on the micrometer scale which moves synchronously with the plane parallel plates. The moving of these plates is often controlled by a spiral groove *(Fig. 2.45)*. Each lever, rigidly connected to a plate rotatable around an axis, is coupled to this groove which bears the micrometer scale. The two levers are rotated more or less proportionally by the rotation of the micrometer screw.

The optical micrometer with a sliding wedge *(Fig. 2.46)* is easier to manufacture, furthermore it is free of several errors occurring with micrometers with a plane parallel plate. The principal section plane of both pairs of optical wedges, which have the same deflection angle, is perpendicular to the images of the graduation lines, and the refracting edges of each wedge pair are on opposite sides. The bottom wedge pair is fixed, the top one bearing the micrometer scale is joined to the micrometer screw by a gear rack. By rotating the micrometer screw the top wedge pair is moved together with the micrometer graduation; thus the images of the two readings can be made to coincide. The disadvantage of a micrometer with a moving wedge compared with a micrometer with a plane parallel plate is that it is larger and therefore needs more space.

When the micrometer is set to coincidence, the reading is made either at the index in the centre of the field of view of the microscope and on the micrometer scale *(Fig. 2.47, a)*, or by counting the graduation intervals between the complementarily numbered graduation lines and on the micrometer scale *(Fig. 2.47, b)*.

The diametrically opposite readings in theodolites which have micrometers with symmetry adjustment are projected onto each other. A second graduation with a

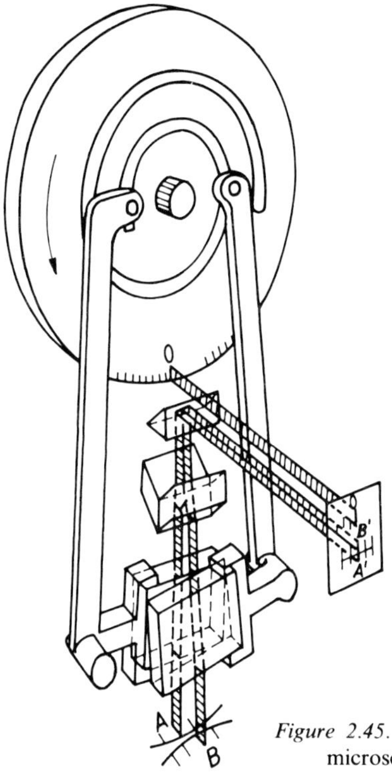

Figure 2.45. Control of the coincidence microscope by a spiral groove

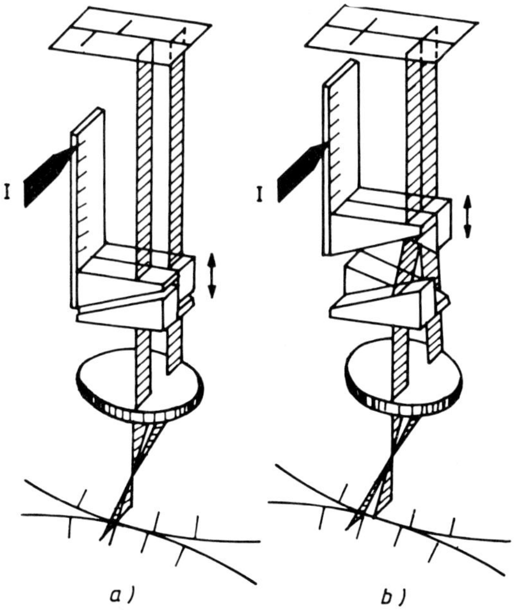

Figure 2.46. Scheme of the coincidence micrometer bringing the readings into coincidence by movable wedges *a)* in the zero position; *b)* in the reading position

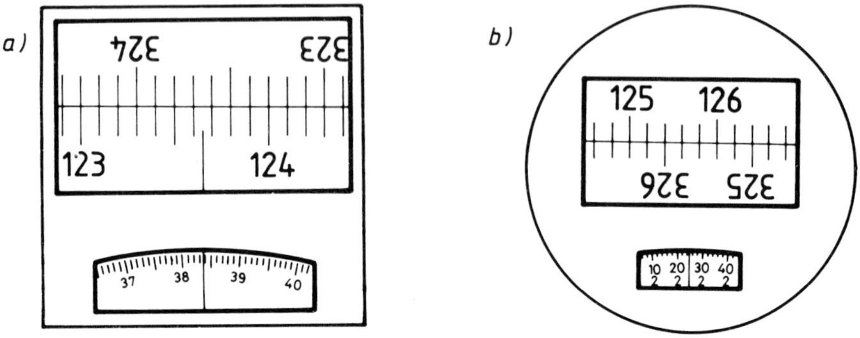

Figure 2.47. Circle reading with micrometer *a)* on the index plus the micrometer scale (theodolite *Wild T3*):Hz = 123.68838^g; *b)* by counting the graduation lines plus the micrometer scale (theodolite MOM *Te-B23*):Hz = 125.7225^g

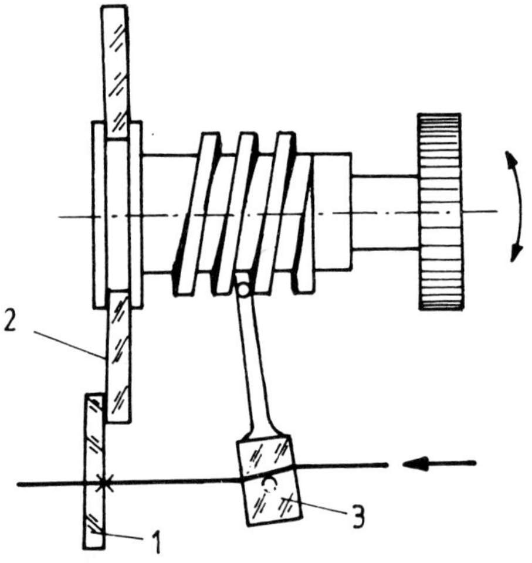

Figure 2.48. Control of the micrometer with a worm screw (on the theodolite Kern *DKM* 2): *1* reticule of the microscope; *2* microscope scale; *3* glass plate of the micrometer

different diameter is applied to the circles of such theodolites *(*see *Fig. 2.22).* The readings of the two circle graduations *(Fig. 2.22, b)* must be projected into images of double lines. For the measurement of the distance between the reading index, which is at the centre of the image plane (field of view) of the microscope, and the next pair of graduation lines, only one micrometer element needs to be put into the reading optical train, but it must be placed where the rays have already passed through both reading positions, e.g. as shown in *Fig. 2.41.* The adjustment of the optical micrometer can also be made by a worm screw *(Fig. 2.48).*

If one of the two circle graduations is double-lined *(*see *Fig. 2.22, c),* then again the image of the first reading must be moved with respect to the second by means

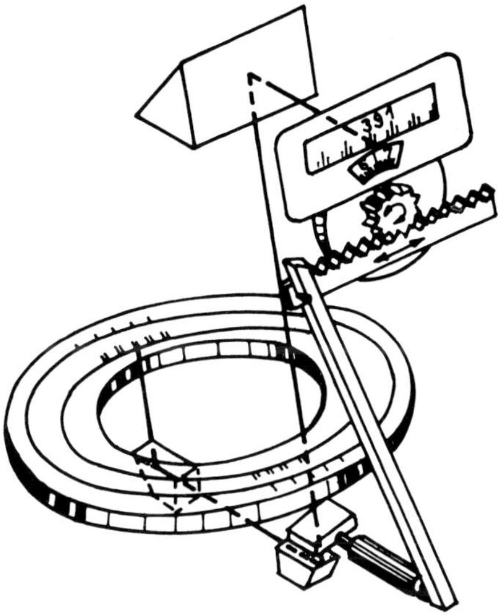

Figure 2.49. Scheme of the optical micrometer used for both readings (in the theodolite *Kern DKM 3*)

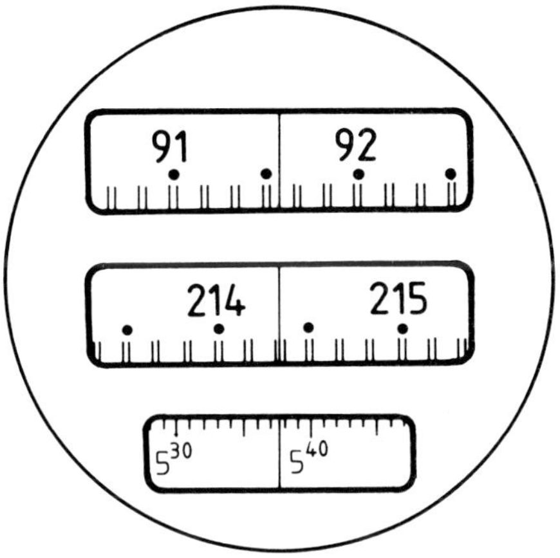

Figure 2.50. Field of view of the microscope of the theodolite *DKM 2* (Kern, Aarau): Reading: 214°20′ + 5′37.6″

of an optical micrometer consisting of one single element placed into the optical train between the two readings *(Fig. 2.49)* in order to straddle the single lines symmetrically with the double ones.

The firm *Kern* (Aarau) always uses two circle graduations in their theodolites. In one case two single-lined concentric graduations with different radii are applied to the circle *(Fig. 2.22, b)* and projected onto each other so that they are shifted sideways by a small amount, double graduation lines being thus visible in the reading microscope. The difference between the two graduation intervals of the two graduations with different radii is adjusted by a small magnification difference between the two readings. The double lines visible in the reading microscope must be brought to be symmetrical about the index line on the reticule of the microscope *(Fig. 2.50)*. In another case the circle of greater radius is double-lined *(Fig. 2.22, c)*. The diametrically opposite positions of the readings are projected onto each other by a roof-edge prism *(Fig. 2.40, a)* in such a way that, by rotating the top part of the theodolite, the images of the two readings projected onto each other move in opposite directions. The double graduation lines are projected onto the single graduation lines, and they can be brought into symmetry by the micrometer placed between the two readings *(Fig. 2.51)*. The inclusion of the three circle graduations, manufactured independently of each other, decreases the effect of graduation error. The reading system with two circle graduations has the further advantage that no dividing line, and hence no image-splitting prism system, is needed.

In modern theodolites other methods of circle reading are also used. The theodolite *Th 3* (Opton, Oberkochen), for example, has a combined microscope reading device

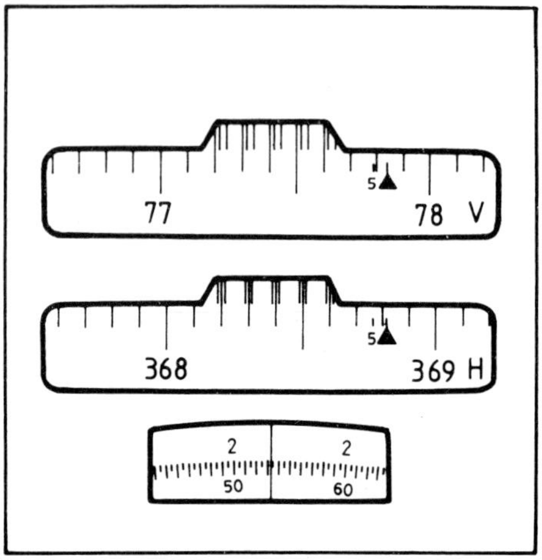

Figure 2.51. Field of view in the reading microscope of the theodolite *DKM 3* (Kern, Aarau): $H = 368.82534^{g}$

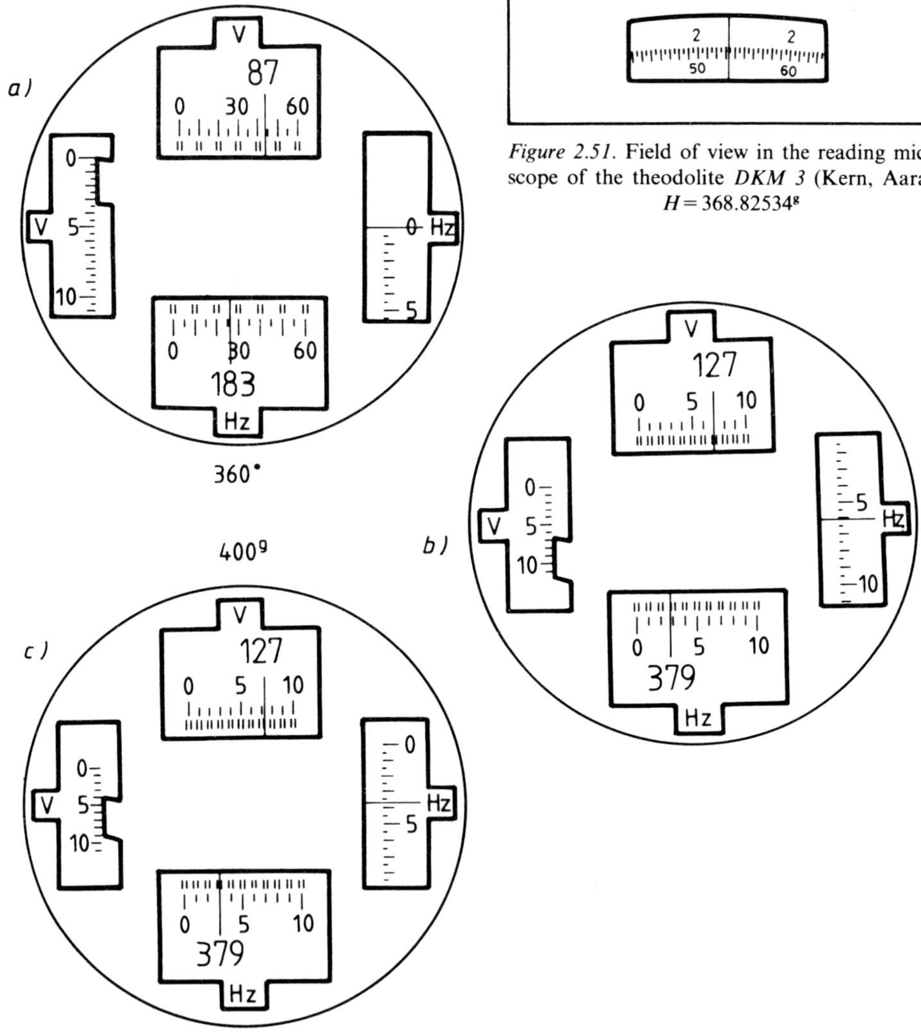

Figure 2.52. Reading images of the microscope of the theodolite *Th 3* (Opton, Oberkochen). The readings are: *a)* without micrometer: Hz $= 183°26'$; $V = 87°44'$; *b)* with micrometer: $V = 127.7^{g} + 6.4^{c} = 127.764^{g}$; *c)* with micrometer: Hz $= 379.3^{g} + 3.7 = 379.337^{g}$

(Fig. 2.52) by means of which (1) for lower accuracy, direct readings with the microscope stationary can be made on the circle graduation with an accuracy of 1 min of arc or 0.01^g by estimation; (2) for higher accuracy, readings can be made on the circle graduation and on the micrometer graduation by adjusting the circle graduation line to the next double line with the micrometer screw, obtaining an accuracy of 3 sec or 0.1^c.

Recently reading methods are sought which require less attention from the observer. Thus partly or totally digitized circle reading systems have been developed from the diametrically opposite circle readings previously described. The reading is made by the numbers displayed excluding the last digit.

On the digital display of the theodolite *T1* (Wild, Heerbrugg) *(Fig. 2.53)* direct

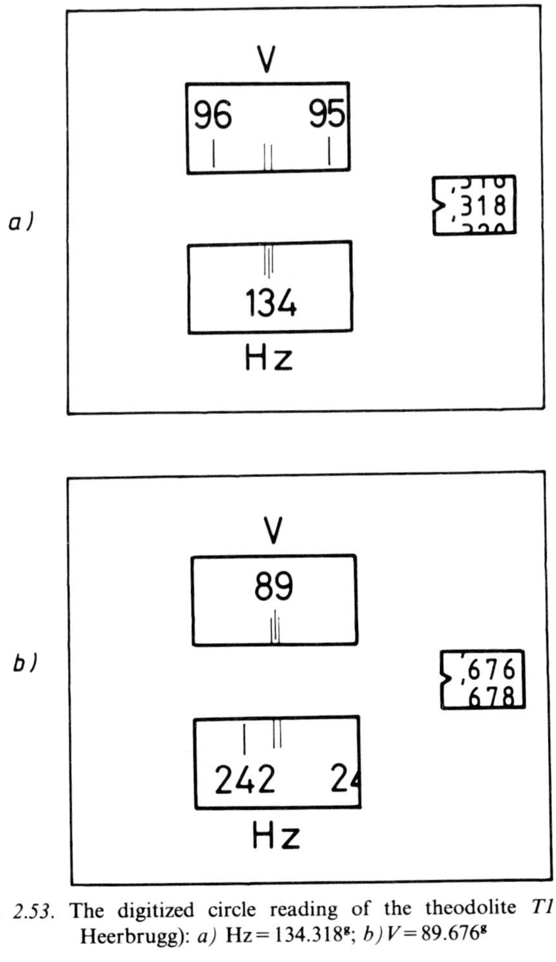

Figure 2.53. The digitized circle reading of the theodolite *T1* (Wild, Heerbrugg): *a)* Hz = 134.318^g; *b)* V = 89.676^g

reading can be made to 20^{cc} and 10^{cc} can be estimated. In the microscope of the theodolite *DKM 2A* (Kern, Aarau) *(Fig. 2.54)* small marks additional to the numbers

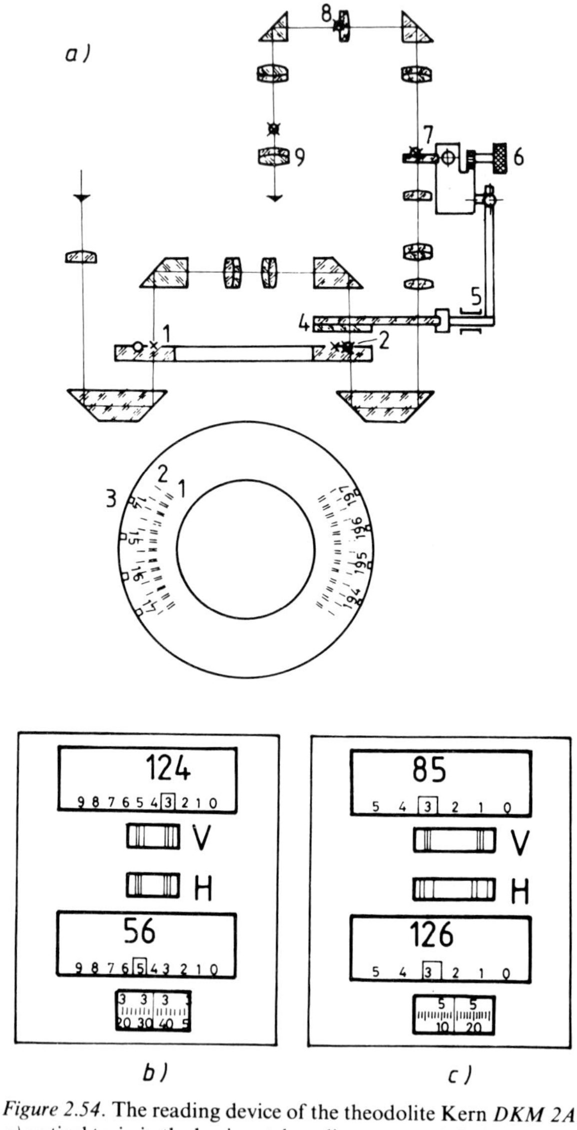

Figure 2.54. The reading device of the theodolite Kern *DKM 2A*; *a)* optical train in the horizontal reading system: *1* the inner circle graduation with double lines; *2* the outer circle graduation with single lines; *3* identification marks; *4* micrometer plate; *5* embedding of the micrometer-plate support; *6* micrometer screw; *7* micrometer scale; *8* field of view diaphragm; *9* eyepiece; *b)* field of view of the microscope. Reading on the 400^g horizontal graduation: $H = 56^g53^c34^{cc}$; *c)* on the vertical circle graduation: $V = 85°35'14''$

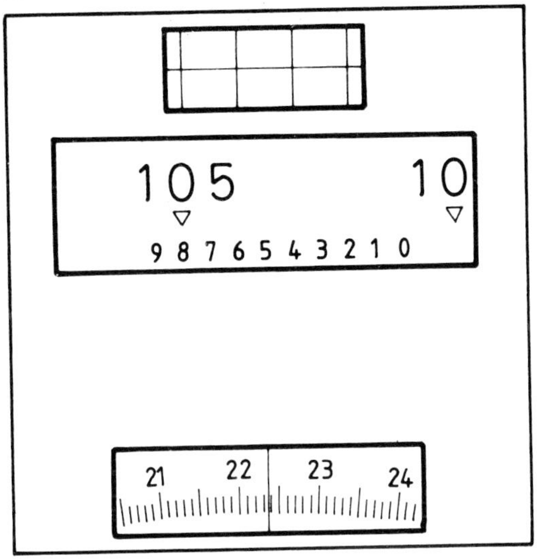

Figure 2.55. Partly digitized reading in the new model of the theodolite *Wild T2*: 105.82237^g

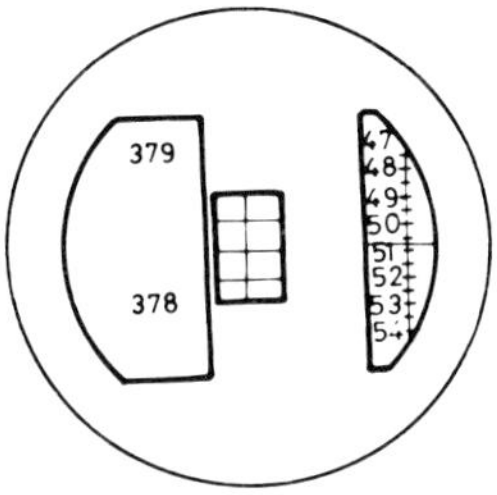

Figure 2.56. Field of view of the microscope of the theodolite *Th 2* (Opton, Oberkochen (400^g): $H = 378.8506^g$

of the graduation lines frame the digits of 10^c or 10 min to be read. There is a small window below the main scale in which an auxiliary scale, sub-dividing the 10^c or 10 min divisions of the main scale, may be read after setting the micrometer. This scale is numbered every 10^{cc} or 10 sec. The new variant of the theodolite *T2* (Wild, Heerbrugg) has a similarly digitized reading *(Fig. 2.55)*. The theodolite *Th 2* (Opton, Oberkochen) is also made with a partially digitized circle reading *(Fig. 2.56)*. Its micrometer is an optical moving wedge; the numbers in the field of view of the microscope must be read from left to right. The digits 0.01, 0.001 and 0.0001 grad (1^c, 10^{cc} and 1^{cc}) are found on the erect graduation of the micrometer with downward numbering.

The integer 10^c values of the partially digitized reading of the precision theodolite *Theo 010A* (Jenoptik, Jena) can be read in the two slits between the micrometer-scale windows: even numbers in the upper, odd numbers in the lower slit; the second digit of the values c and the values cc are visible in the right-hand window *(Fig. 2.57)*.

Examples of some totally digitized readings are shown in *Fig. 2.58*. The angle can be read here on the graduations of the glass circle and of the micrometer directly down to 10″, 6″, 5″ and 1″, respectively.

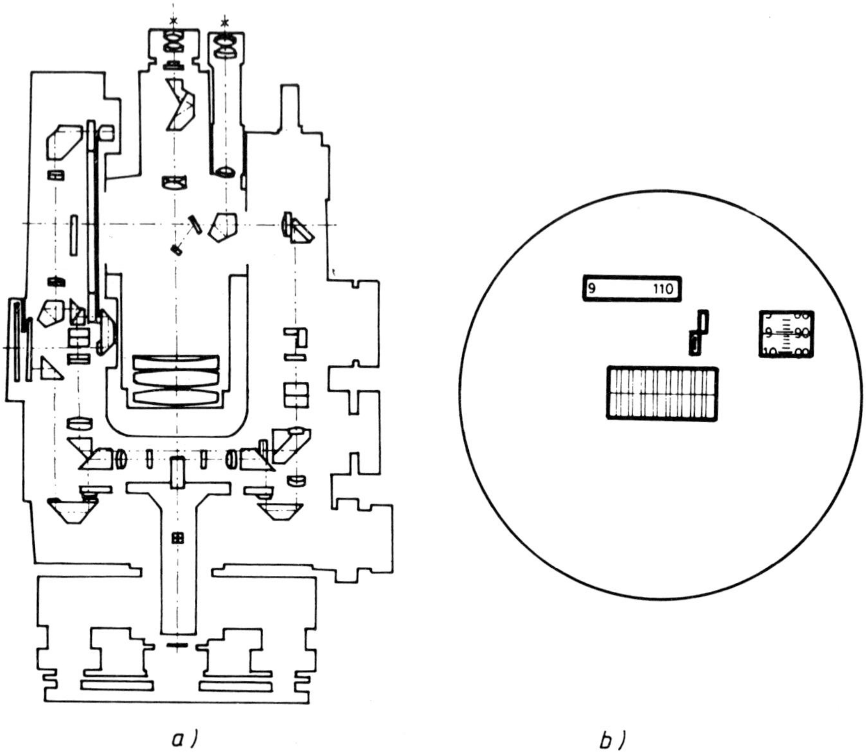

Figure 2.57. Reeding device of the theodolite *Theo 010A* (VEB Jenoptik, Jena): *a)* optical train; *b)* reading: $110^{g}19^{c}91^{cc}$

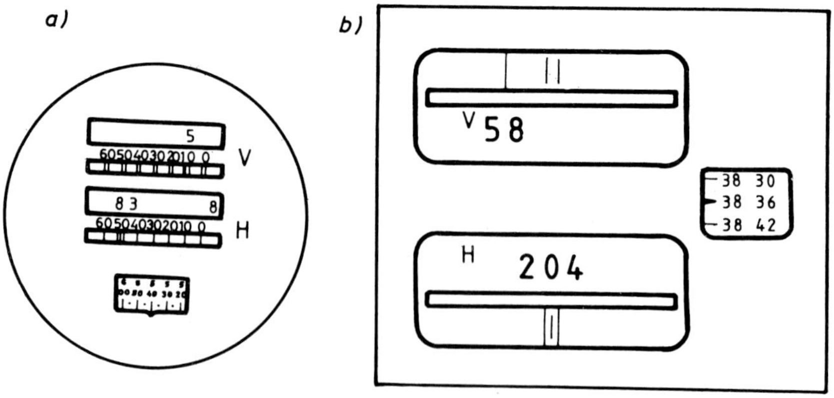

Figure 2.58. Examples of digital readings: *a)* theodolite *Nikon NT-3A* (Nippon Kogaku, Japan): $H=83°55'40''$; *b)* theodolite *Pentax TH-06D* (Asahi, Japan): $H=204°38'36''$

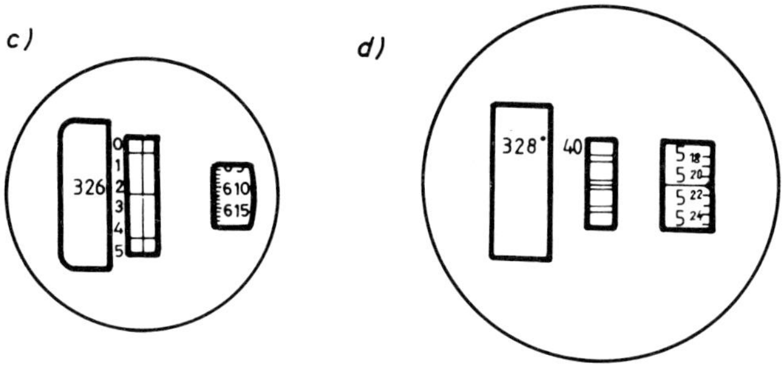

c) theodolite *Pentax TH-01* (Asahi, Japan): $H = 326°26'12''$; *d)* theodolite *Nikon NT-5* (Nippon Kogaku, Japan): $H = 328°45'21''$

2.2.5.4 Devices for Clamping and Fine Adjustment

For the proper use of the theodolite axes they must be clamped (fixed) during the measurement, in such a way that the two clamped parts of the instrument must be rotatable with respect to each other by a small amount. For pointing with proper accuracy the sensitivity of the unaided hand is insufficient. After a rough pointing the axes are clamped to the fine adjustment device, usually by means of a clamp fastened by a screw. The fine adjustment device is nothing more than a slow motion gear.

Figure 2.59, a shows the simple axial clamping and fine adjustment (slow motion screw) system used in theodolites of lower accuracy. By tightening the clamping screw

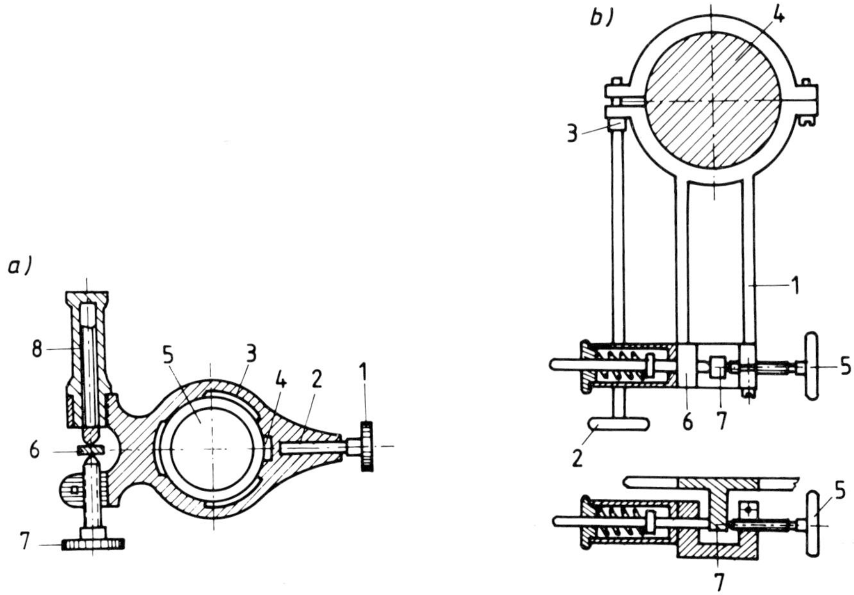

Figure 2.59. Simple axial clamping and fine adjustment systems: variant *a)*, variant *b)*

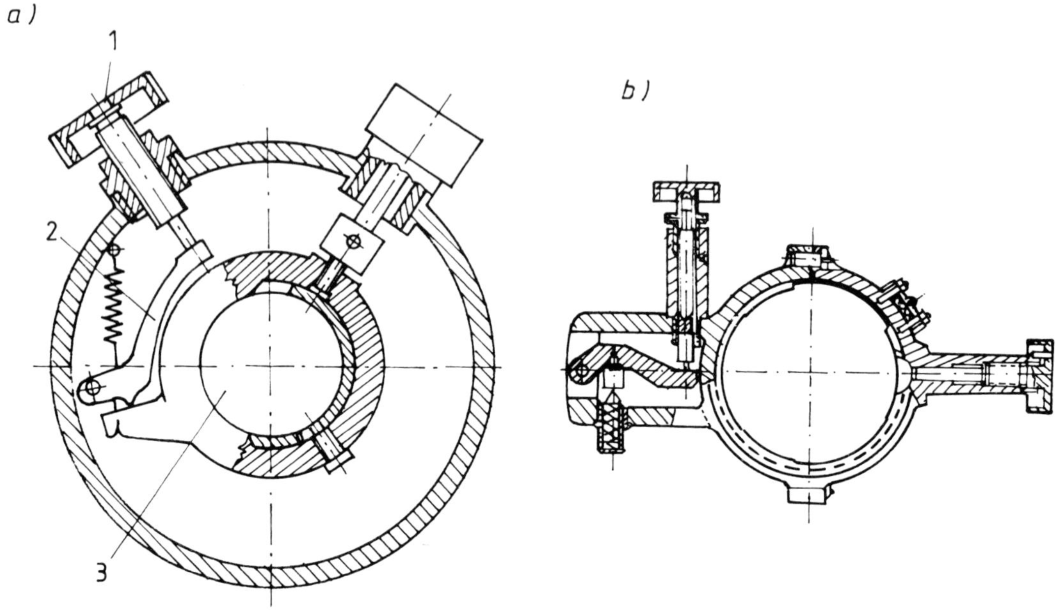

Figure 2.60. Fine adjustment screw with transmission:
a) variant used in the theodolite *Theo 010* (VEB Jenoptik, Jena);
b) variant used in the theodolite *T3* (Wild, Heerbrugg)

1 the ring *3*, supplied with the nose *2*, firmly clamps the shaft *5* of the lower part of the instrument by pressure on element *4*. Since the nose *6* of the top part of the instrument lies between the fine adjustment screw *7* and the counter-spring *8*, the top part is clamped, but it can be turned minutely by the fine adjustment screw *7*.

A similar simple method is shown in *Fig. 2.59, b*, where the frame *1* is clamped to the axis *4* by means of the clamping screw *2* through the block *3*. The fine adjustment screw *5* and the spring-loaded counter-acting pin *6* are fixed to the frame; the nose *7* fixed to the rotating base is straddled by them. The frame clamped to the axis is connected by *7* to the bearing of the axis. The angle fixed in this way can be changed within a narrow range by the fine adjustment screw. The rate of motion of the fine adjustment depends on the pitch of the screw *5* and on the distance of the nose *7* from the rotation axis.

In theodolites of higher precision the rate of motion of the fine adjustment is lessened, for example, by connecting the fine adjustment screw *1* to the nose *3* through the interposed lever *2*, as in *Fig. 2.60, a*, or by means of the arrangement shown in *Fig. 2.60, b*.

In order to simplify the handling of the theodolite a coaxial clamping and fine adjustment screw system has often been used recently. Such an instrument is outlined in *Fig. 2.61*. The axis *1* is connected by the brake blocks *2* to the lever *3* which can be rotated by a small amount through the connecting rod *4* by rotating the fine adjustment knob *5* which operates the fine screw *6*. The brake blocks are activated

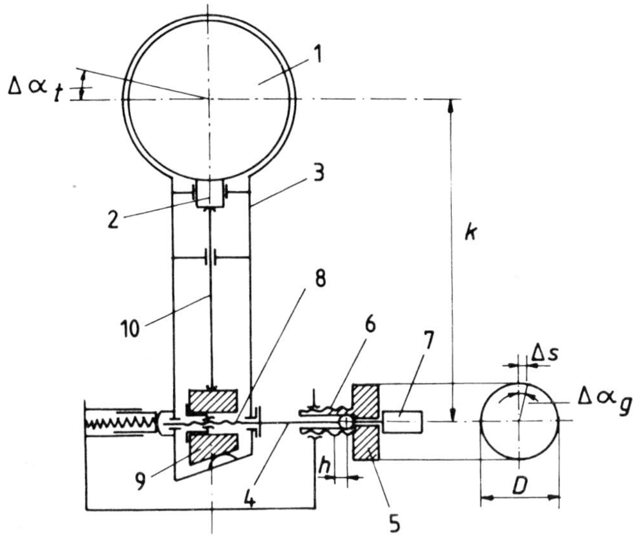

Figure 2.61. Variant of the coaxial clamping and fine adjustment system (MOM, Budapest)

by turning the clamping knob *7* which operates the screw *8* and the interposed rod *10* which can be moved by the wedge-transmission *9*.

Jenoptik uses in its new theodolites (*GEOmatt System*) a coaxial horizontal and vertical clamping screw and coaxial horizontal and vertical fine adjustment screws *(Fig. 2.62, a)*. The vertical and horizontal axes can be simultaneously unclamped by pressing together the ends of the horizontal clamping lever *1* and of the vertical clamping lever *2* which are opposite the fine adjustment screw systems *4* and *5*. Both axes can be simultaneously clamped by pressing together the ends of the two clamping levers near the fine adjustment screws. The vertical and horizontal axes can be clamped or unclamped separately, however, by pressing only the horizontal clamping lever *1* or the vertical clamping lever *2*. For routine measurement of horizontal angles the image of the vertical circle reading (i.e. the upper part of field of view) in the field of view of the reading microscope can be covered by turning the knob *3 (Fig. 2.62, b)*. The horizontal fine adjustment screw *4 (Fig. 2.62, a)* and the vertical fine adjustment screw *5* are also coaxial.

On certain theodolites, in addition to the clamping and fine adjustment device on the top part, a similar device is used on the central part containing the horizontal graduated circle. This permits multiplying angle measurements to be made by temporarily clamping the central part (i.e. the horizontal circle) to the top part, e.g. by means of clamping with leaf springs *(Fig. 2.63)*. In its strained position the leaf spring *1* presses the leaf *2*, which is fixed to the central part of the instrument, to the top part *3*, so clamping the latter.

A separate rotating system is used for the central part (i.e. for the horizontal graduated circle) in theodolites of medium and high precision, instead of the multiplying clamping system by means of which the horizontal glass circle can be set

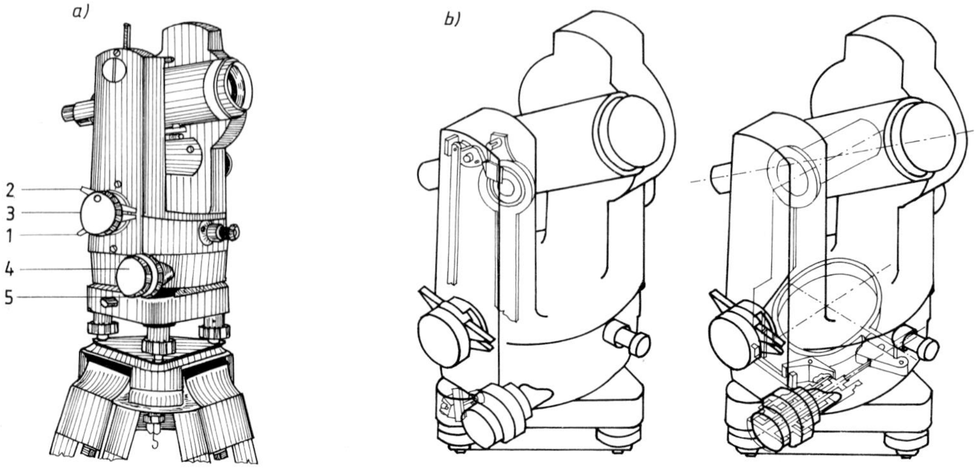

Figure 2.62. Coaxial clamping system of the horizontal and vertical axes, coaxial fine-adjustment screw (VEB Jenoptik, Jena)

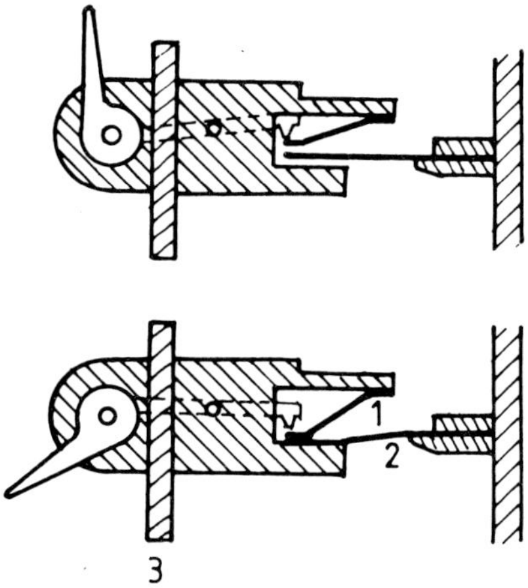

Figure 2.63. Pliable strip clamping of the central part (horizontal circle) of the instrument to the part in the theodolite *Wild T16*

to a required reading with the bottom and top parts clamped to each other. Its construction is usually a geared disc, rigidly fixed to the central part, which can be rotated by the clutched rotating knob by an arbitrary amount, but this knob is disengaged from the disc when in its rest position. In the theodolites made by *Kern*, as well as in the *Opton* theodolite *Th 2*, the rotating mechanism of the horizontal circle has both a fast and a slow motion.

Friction clamping and endless fine adjustment is seldom used in theodolites. The adjustable clamping device clamps the axis to the bearing by friction, with a fine adjustment device interposed. The friction is sufficiently strong for rigid clamping to be ensured during the operation of the fine adjustment device, which usually consists of an endless, continuous worm gearing (between a worm-wheel and a worm-spindle). The axis can be rotated by hand against the friction clamping.

Figure 2.64 shows a construction where the rotating part *1* is connected to the fixed part *5* through the worm gear connection *2–3* and through the bearing *4*. There is a friction connection *6* between the worm wheel *3* and the fixed part *5*. The friction is induced by the spring *7*. The rotatable part can be rotated through the worm gear connection against the friction by means of an appropriate force (for

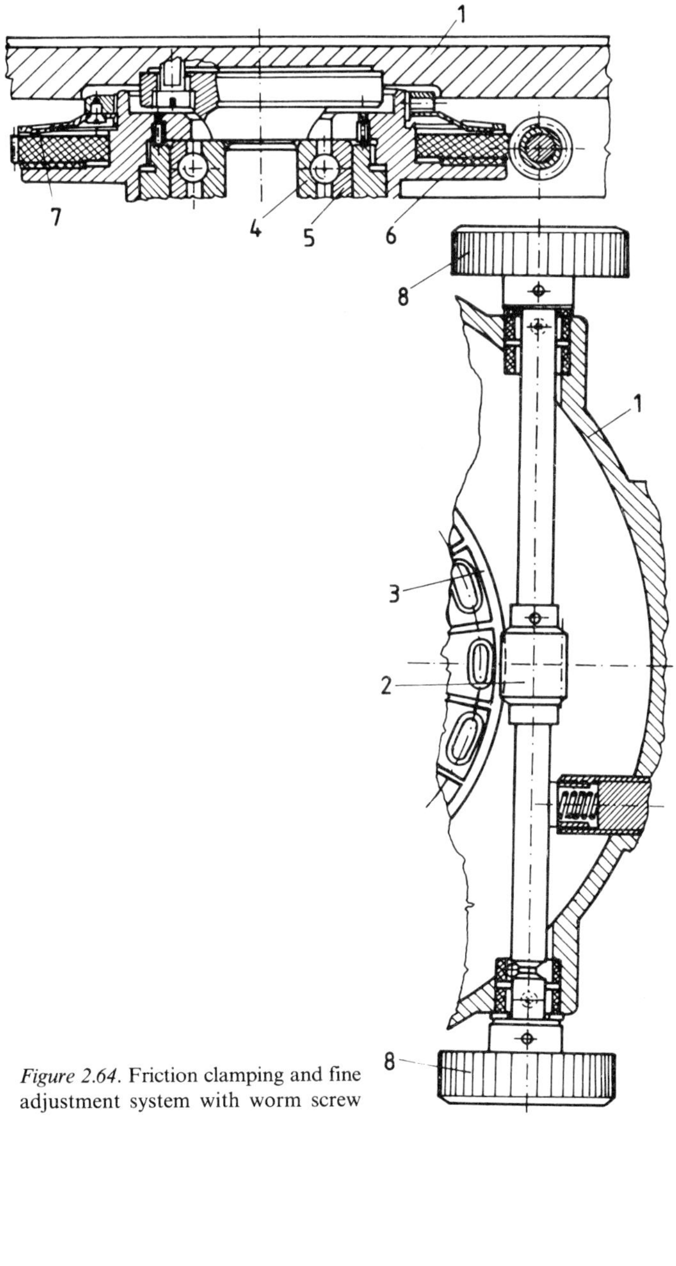

Figure 2.64. Friction clamping and fine adjustment system with worm screw

rough pointing). For fine adjustment one of the knobs *8* or *8'* should be rotated, when the worm moves with respect to the standing (fixed) worm wheel and hence the rotatable part also moves. The moving rate is equal to the gear number of the worm wheel if it has a single thread.

2.2.6 Supplementary Devices and Accessories

2.2.6.1 Centring Devices

The theodolite is set up above the field point by means of a centring device. For simple measurements a plummet, or plumb-bob is sufficient but, for more exact measurements or in wind, a centring rod or optical centring is used.

The plummet consists of a thin cord and of a weight of 0.3 to 0.5 kg attached to it. The weight is made of lead, brass or iron, in the shape of a solid of revolution, and it ends in a sharp point. The cord length can be changed in different ways (by knotting, by clamp, by a friction leaf and counterweight, by a spring-roll-up).

The centring rod is a telescopic steel tube, with a circular level, which can be extended and hung on the hook of the fixing screw of the tripod. On the rod there is usually a length graduation, by means of which the height above the ground point of the tripod head plate or of the theodolite horizontal axis can be measured. The theodolite is first set up roughly by eye above the ground mark, then the end point of the rod is set to the ground point, the clamping screw is unclamped and the instrument is moved on the tripod head plate, keeping the bubble of the circular level of the centring rod central. When using the *Kern* centring tripod (see *Fig. 2.10*), adjustment of the centring rod simultaneously roughly adjusts the vertical axis into the plumb line.

The optical projector or optical plummet can be a separate device *(Fig. 2.65, a* and *b)* or it can be installed in the theodolite. The former are manufactured so that centring by them is possible either above a ground point or below a roof point. It has a broken telescope for horizontal viewing, supplied with a cross-hair; it can be adjusted into the vertical by means of the two tubular spirit levels on the tribrach, by tilting it around the standing axis. The two positions of the rotatable prism in front of the objective lens permits centring above a ground mark or below a roof mark. Other optical plummets have a broken telescope with two objectives, viewing downwards and upwards, and the rotatable prism is between them.

The integrated optical plummet is placed in the top part and in the hollow standing axis, in the *Wild* theodolites *T16* and *T1-A* for example, or in the bottom part of the instrument, e.g. in the theodolite *Wild T2*. The theodolite is centred by adjusting its vertical axis to the vertical and then moving it on the tripod head until the image of the station point is brought to the centre of the cross-hair of the optical plummet. Due to a possible small deviation of the line of sight of the optical plummet from the vertical axis, a small difference can occur when the theodolite is rotated through 180°. This error can be eliminated by moving the theodolite through half this difference. If the optical plummet is adjusted, its centre does not move when rotating the theodolite. If it does move, the reticule of the plummet must be adjusted. If the plummet is built into the tribrach of the instrument, the above test is made by changing the position of the tribrach on the tripod by 120°.

a)

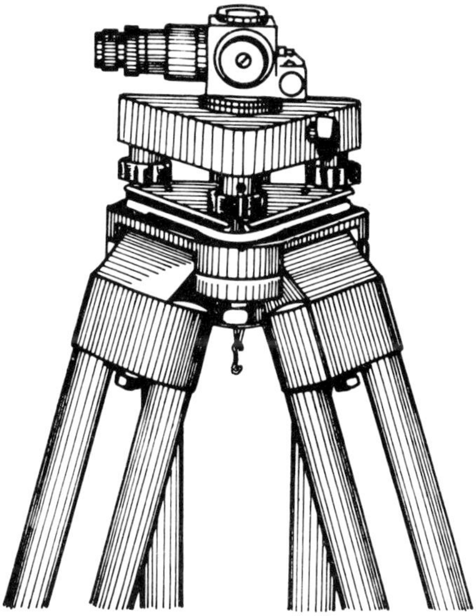

b)

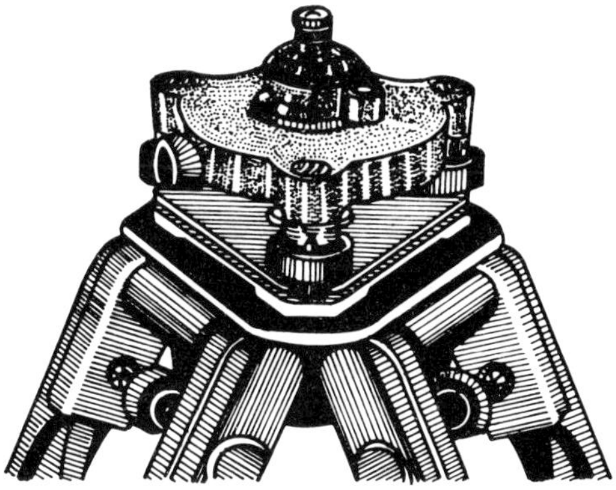

Figure 2.65. a) optical plummet (VEB Jenoptik, Jena); *b)* automatic optical plummet (MOM, Budapest)

Setting up above a ground point is facilitated and faster with an automatic optical plummet *(Fig. 2.65, b)*; here it is sufficient to adjust the tribrach into the vertical within 1°, because within this range the built in pendulum automatically sets the axis of the plummet vertical.

The centring error is ±3 to ±5 mm with a plumb bob, ±1 to ±2 mm with a centring rod and ±0.5 mm with an optical plummet. The firm *Wild* also produces a roof projector for the theodolites *T1A*, *T16* and *T2*. It can be put on the telescope, after the latter has been made horizontal using the vertical circle and its index, and it can then be used as a separate optical plummet.

A special accessory is the objective prism (pentagonal prism) attachable to the tube of the objective lens, by means of which elevated sights can be made. A zenith sight can be made by setting the telescope vertical and using an eyepiece-prism.

2.2.6.2 Devices for Forced Centring

A device for forced centring enables the units to be centred to be interchanged on the tribrach, i.e. a theodolite, target, subtense bar, or optical plummet, in such a way that the vertical axes of each can be set to the same position with an accuracy of some hundredths of mm. The instrument reference point is in practice the vertical constraint centralizer axis.

There are two groups of constructional solutions. In the case of the first group, if the plumbing has been already made with one unit, then the vertical axis of the unit to be centred is not only centred but also plumbed when fitted to the centring elements.

In case of the second group, the forced centring units do not set the standing axes of the units vertical, hence the vertical axis of each unit to be centred must be plumbed separately. The three-grooved forced centring device, the Kern system of forced centring (see *Fig. 2.10*) and the forced centring device with a Freiberg sphere belong to this group.

The forced centring system with cylindrical pivot and socket *(Fig. 2.66, a)* belongs to the first group; it consists of the cylindrical stud *2* at the bottom of the top part *1 (Fig. 2.66, b)* which can be removed, together with the central part of the theodolite,

Figure 2.66. Pivot-socket forced centring on the theodolite *Theo 030* (VEB Jenoptik, Jena): *a)* view; *b)* scheme

and of the cylindrical socket *4* attached to the tribrach *3*, which also holds the foot-screws *(Fig. 2.66, a* and *b)*. The cylinder jacket is lengthened on one side: the stud of the upper part is pushed at point *5* against the two edges *(6–6)* of this lengthened portion by the clamping screw (the two edges are 90° apart); this ensures an exact clamping.

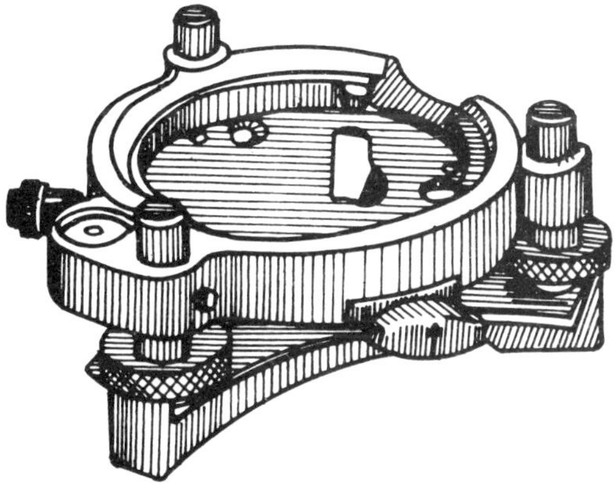

Figure 2.67. Forced centring plate on the theodolite *Wild T2*

In the case of the forced centring device with a dish *(Fig. 2.67)*, which also belongs to the first group, the plane surface of the base of the unit to be centred sits on three points on the tribrach dish. The forced centring is realized by fitting the cylindrical surface of the flange, at the bottom of the unit to be force-centred, to the inner cylindrical surface of the ring in the dish; the accuracy is about 0.1 mm.

In the case of the three-grooved forced centring belonging to the second group, the unit to be centred should be set together with its tribrach onto the centring plate of the tripod head or of a pillar base-plate, so that the points of the three foot-screws fit into the three radial grooves lying 120° apart. Now the vertical axis of the instrument must be made vertical by the foot-screws.

The instrument set on a Kern system centring base (see *Fig. 2.10)* is already plumbed from the outset, but the vertical axis is only plumbed approximately; it must therefore be plumbed precisely by means of the three eccentric foot-screws with horizontal axes (see *Fig. 2.11)*.

Different constructions are also used for forced centring by different firms. The Freiberg centring sphere, of 10 to 20 mm diameter mounted at the bottom of the instrument to be centred, is used in mine surveying; the sphere is fitted into a tube rigidly fixed to the tribrach which can be moved on the tripod head, which must have already been centred. After attaching the unit to be centred, it must be levelled by its foot-screws. The Freiberg sphere is a simple and suitable device for forced centring, especially in the case of setting up the instrument on a pillar. It is thus used on the special precision theodolites *T3* (Wild, Heerbrugg) and *Theo 010* (Jenoptik, Jena), as well as on the dam-site surveying instruments made by *Freiberger Präzisionsmechanik* and *Fennel,* Kassel.

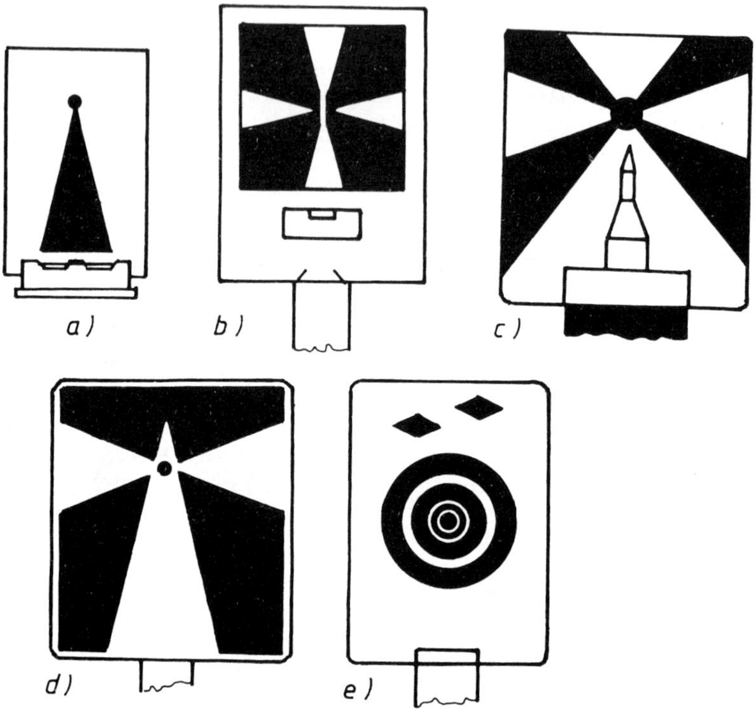

Figure 2.68. Targets: *a)* Fennel; *b)* Kern; *c)* MOM; *d)* VEB Jenoptik; *e)* Opton

Different types of targets are marketed (e.g. those shown in *Fig. 2.68*). For plumbing they are usually supplied with a box level or sometimes with tubular levels. For underground or night use the targets are supplied with electrical illuminating devices on their reverse side.

2.2.6.3 Accessories of Theodolites

In order to be able to use theodolites for many purposes, different types of accessories are manufactured. They are: optical plummets, target plates, subtense bars, gyro-attachments, spirit levels for the horizontal axis and for levelling, tubular and box compasses, overhead plummets, distance-measuring attachments, objective prisms, eyepiece prisms, autocollimation eyepieces and eyepiece lights, etc.; they are dealt with in the corresponding sections of the book.

In poor illumination conditions and for night measurement, illuminating devices are used for illuminating the reticule and the circle readings. The power supply (battery) is usually fixed to the tripod, the reflecting mirrors are replaced by electric bulbs.

Observing the Sun is made possible by the *Roelofs solar prism* which can be attached in front of the telescope *(Fig. 2.69)*. It consists of three glass wedges in succession: the first deflects half the received beam of rays from the Sun in the direction of the horizontal hair of the cross-hair; the second wedge, also covering half the objective, deflects the two partial beams of rays in the direction of the vertical hair, and the

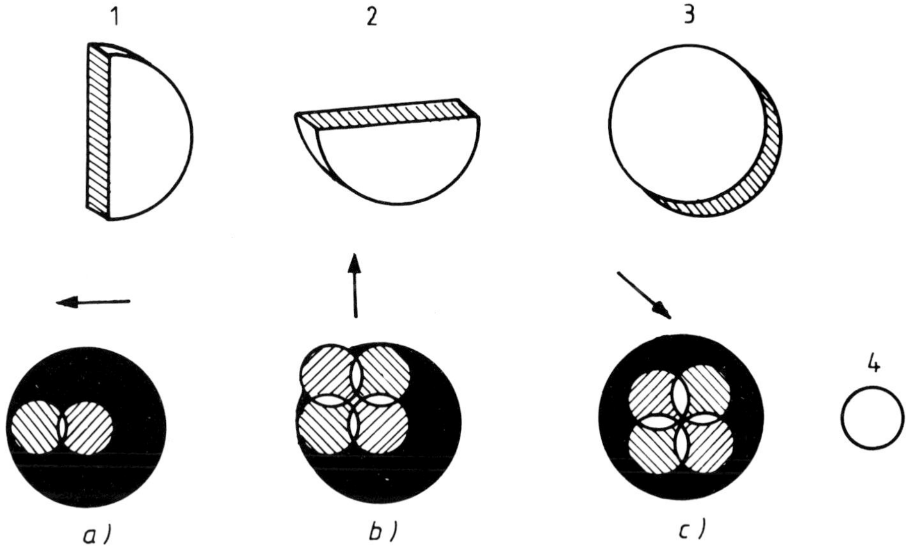

Figure 2.69. Operating principle of the Roelofs solar prism: *a)* double image due to the effect of wedge *1*; *b)* four-fold image due to the effect of wedges *1* and *2*; *c)* wedge *3* moves the four-fold image to the centre of the field of view of the telescope

third wedge, totally covering the objective lens, deflects the beam, now separated into four parts, in the direction 45° by halving the direction between the two hairs by a properly chosen deflecting angle. When pointing to the Sun, four images of the Sun symmetrically shifted with respect to each other are visible through the three wedges in the telescope. Having approximately pointed to the Sun, the cross-hair is exactly adjusted to the cross composed by the four strips of Sun images visible in the field of view by means of the vertical and horizontal fine adjustment screws of the theodolite. The line of sight then points to the centre of the Sun. Such a solar prism is manufactured by *Wild* for the theodolite *T2*.

By means of the Wild meridian finder *(Fig. 2.70/a)* also attached to the objective of the telescope of the theodolite *T2* (Wild, Heerbrugg), a quick geographical position determination of approximate accuracy can be carried out. One of the two prisms of the meridian finder deflects the line of sight of the telescope by the angle corresponding to the polar distance of the star α *Ursae Minoris* (Polaris), while the other deflects it by the angle corresponding to the polar distance of the star β *Ursae Minoris*. The planes of the ray deflections caused by the two prisms define an angle which is equal to the difference between the right ascensions of the two stars. The line of sight of the telescope points to the north pole if the images of both stars are visible approximately at the centre of the cross-hair. Observing procedure is firstly to point to Polaris. Then attach the meridian finder and rotate it until star β comes near to star α. Now wait until the distance between the two stars is a minimum, maintaining star α at the centre of the cross-hair. The mean of the readings of this position and of the initial pointing to Polaris gives the direction of North, but this must be corrected for the change of the star coordinates. The readings on the vertical

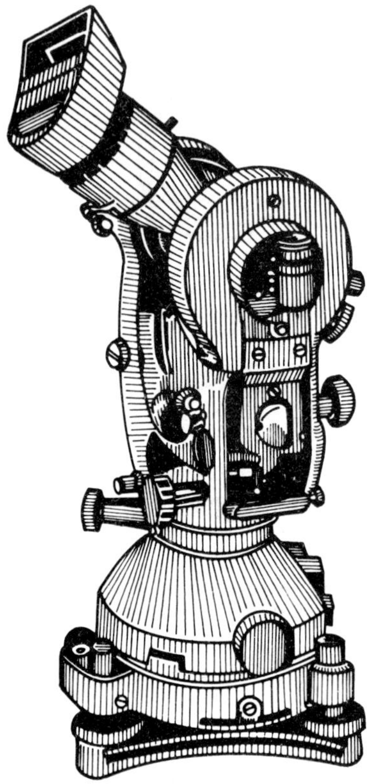

Figure 2.70/a. Theodolite *T2* (Wild, Heerbrugg) with the meridian finder attached

Figure 2.70/b. Meridian finder attached to the theodolite *Theo 020* (VEB Jenoptik, Jena) (manufacture ceased in 1974)

circle, corrected for refraction, give the approximate value of the geographical latitude φ.

By means of the meridian finder made for the theodolite *Theo 020* (Jenoptik, Jena), the meridian can be determined not only by pointing to the star, but also by pointing to the Sun *(Fig. 2.70/b)*. The most important element of the attachment put onto the objective is the mirror, which can be rotated around the axis of the telescope as well as around an axis perpendicular to it. Having pointed to the terrestrial target and read the horizontal circle, the meridian finder is attached to the telescope in front of the objective and the bubble of the diagonal spirit level of the meridian finder is adjusted to centre. For setting the declination δ of the Sun (or star), the zenith angle $\zeta_\delta = (45° - \delta/2)$ is set on the vertical circle and the bubble of the spirit level is adjusted to centre. Then the zenith angle $\zeta_\varphi = (90° - \varphi)$ is set on the vertical circle. By rotating the mirror around the axis of the telescope, and the telescope around the vertical axis, the star or the Sun is pointed. The reading on the horizontal circle now gives the direction of North. The measurement is repeated in the second telescope position, and the mean of the two readings must be corrected by a correction that can be found in a table delivered with the instrument. The standard deviation of the azimuth measurement is ± 1 min of arc.

A 60° prism, attachable in front of the objective of a telescope set horizontal, and mercury in a small dish can be used together for the determination of local time and latitude; as a unit it is called a prism astrolable and stars at a zenith angle of 30° can be observed with it. Incident rays perpendicular to the upper plane surface of the 60° prism, have a zenith distance of 30° *(Fig. 2.71)* and enter the telescope parallel to the telescope axis. The rays incident at an angle of 60° to the mercury surface T, which acts as a mirror, are reflected from it, are reflected again on the upper surface of the prism and enter the telescope also parallel to its axis. If a star defined by the zenith angle of 30° approaches the almucantar, the two images visible in the field of view of the telescope approach each other and coincide at the instant of transit on the almucantar of the zenith angle $\zeta = 30°$. This instant is read on a clock. From this the latitude of the observing point can be calculated (or if the latter is known, one can calculate the correction of the time read). The exact position of the prism with respect to the telescope can be determined by means of an autocollimation eyepiece.

The *Fialovszky meridian-finder reticule plate* can be mounted in telescopes of small theodolites, with a field of view of a minimum of 2°, by means of which the meridian direction can be determined at night after having pointed to the star α *Ursae Minoris* (Polaris) *(Fig. 2.72)*. On the reticule the radii of the meridian-finder circles marked by 80, 90 and 00 (meaning the years 1980, 1990 and 2000) correspond to the polar distance of the star Polaris in the respective years. If the zenith angle $\zeta = (90° - \varphi - \rho)$ corresponding to the latitude φ and increased by the astronomical refraction ρ, is set on the vertical circle of the theodolite, and the image of the star Polaris is set on the meridian-finder circle corresponding to the year of the measurement (or to an interpolated position), by rotating the theodolite around its vertical axis, then the line of sight of the telescope is pointed to the north pole. When using a small theodolite having $18\times$ magnification and a reading accuracy of 20 sec of arc, the meridian can be determined by this method with an accuracy of ± 1 min of arc if Polaris is in a position which differs from its upper or lower culmination by at least 2^h of hour angle. In practice this means if, when looking at the sky with an unaided eye, the

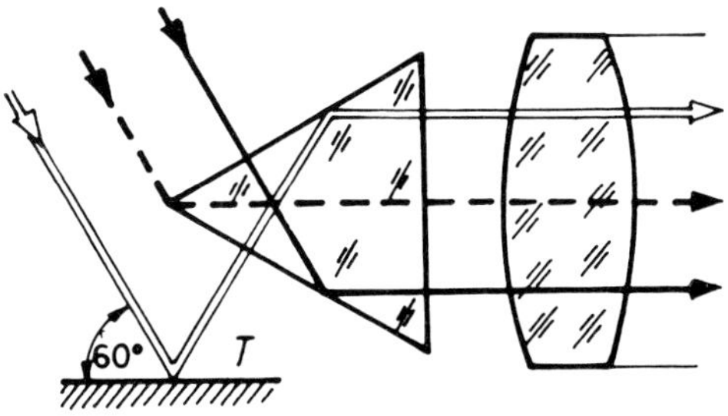

Figure 2.71. Principle of the prismatic astrolabe

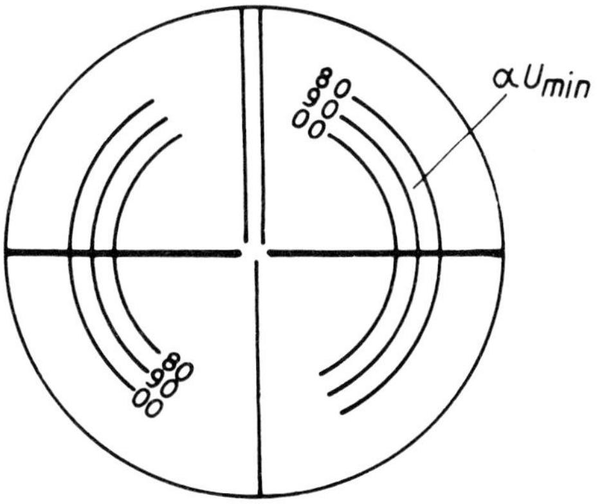

Figure 2.72. Reticule of the meridian finder

direction determined by the line joining Polaris and η *Ursae Majoris* (the furthermost outside star of the head of the Great Bear) makes an angle with the vertical through Polaris equal to or greater than about 30°. If the star η *Ursae Majoris* is visible to the right (to the left) of Polaris, the image of Polaris should be set on the left (right) side of the meridian-finder circle corresponding to the year (or interpolated).

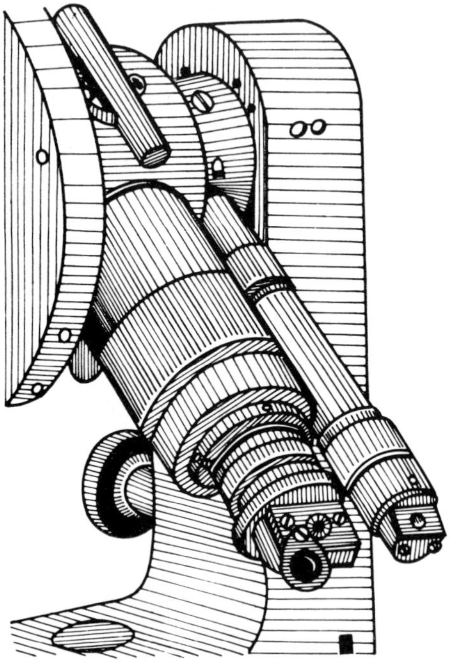

Figure 2.73. Eyepiece prism

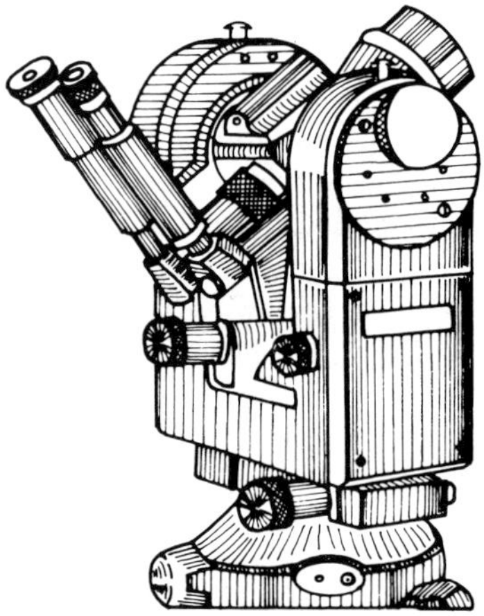

Figure 2.74. Bent eyepiece on the theodolite *T2* (Wild, Heerbrugg)

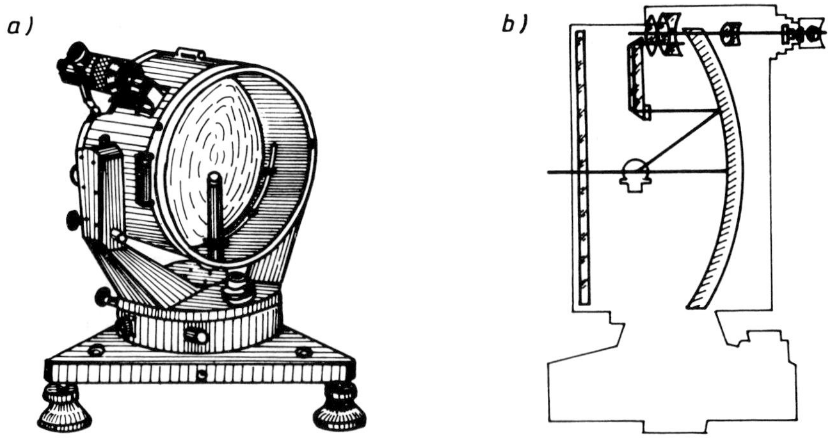

Figure 2.75. Geodetic signal lamp *TSG 200* (VEB Jenoptik, Jena). Effective distance: 100 km, dimensions: 29 × 34 × 38 cm; *a)* view; *b)* optical scheme

By means of a glass plate micrometer attachable to the telescope objective, small lateral deflections from the line of sight can be determined, e.g. for setting up machines, with an accuracy of ± 0.05 mm up to a distance of 15 m.

A lens (convergent) attachable to the telescope tube in front of the objective, makes it possible to observe at distances shorter than the normal minimum for the theodolite.

Steep sighting is facilitated by an eyepiece prism *(Fig. 2.73)* or a bent eyepiece *(Fig. 2.74)* attachable in front of both the telescope eyepiece and the eyepiece of the reading microscope.

For observing the Sun a sun-filter is attached in front of the eyepiece to protect the eye of the observer. The harmful effect of the Sun shining into the telescope can be eliminated by a sun shade attachable to the objective.

For observing on pillars a cast iron base plate is used. In the case of first-order triangulation the points to be observed are often marked by a geodetic signal lamp *(Fig. 2.75)*.

2.2.7 Classification of Theodolites

The designations of different types of theodolites differ between the manufacturing firms and are often based on tradition. Theodolites can be named and classified according to their different characteristics. Thus they can be distinguished on the basis of the reading device, namely vernier, microscope (with estimating, scale, coincidence microscopes, etc.) and recording theodolites (see Section 2.3). According to the connection between the central and bottom parts of the theodolite one speaks of simple, repetition, angle-multiplying theodolites and of theodolites with a rotatable horizontal circle. According to the accuracy of the circle reading in a rather generalized way, the names one minute theodolite and one second theodolite are also used. Compass theodolites (Section 2.5), theodolites for the observation of rapidly moving objects (Subsection 2.4.3), gyro-theodolites (Subsection 2.5.2) and balloon theodolites (Subsection 2.4.3), as well as electronic theodolites (Chapter 10) form special categories. Classification according to intended use is: architect's theodolites or small theodolites which are the most simple ones; engineer's (or tacheometer) theodolites are mostly used for traversing, for measuring detail and for setting out; precision theodolites which are instruments for third and fourth order triangulation as well as for precision setting-out, but with special manufacture they can also be used for first- and second-order triangulation and high precision setting-out. In this classification the mean square error of a direction measured in two telescope positions is usually used, rather than any other instrumental characteristic. Therefore it is suitable to group theodolites into the following categories:

I. the mean square error of measurement with theodolites of highest precision is $\leqq 0.7''$

II. the mean square error of measurement with high precision theodolites is $\leqq 2''$

III. the mean square error of measurement with theodolites of medium accuracy is $\leqq 7''$

IV. the mean square error of measurement with theodolites of low accuracy is $\leqq 30''$

Table 2.1. Some models of low precisi

Type (figure)	Manufacturing firm	Theodolite				Graduated glass circles				
		magni-fication Γ	diameter of objective (mm)	length (mm)	shortest pointing distance (m)	diameter		reading device		
								graduation unit		system
						H (mm)	V (mm)	H	V	
1	2	3	4	5	6	7	8	9	10	11
TEKAT	Breithaupt, Kassel	18	30	126	1.1	65	47	5′ 5^c	5′ 5^c	scale microscop
KT 1	Ertel, München	18	25	160	1.5	95	70	10′ 20^c	10′ 20^c	reading index
BTU (2.76a)		24	30	—	1.5	—	—	5′ 10^c	5′ 10^c	estimating microscop
FTS	Fennel, Kassel	19	25	130	1.3	80	60	1′ 2^c	2′ 5^c	scale microscop
TH 11	Filotecnica Salmoiraghi, Milano	22	27	128	1.0	80	57	5′ 5^c	10′ 10^c	estimating microscop
TG 4br	Galileo, Firenze	22	28	115	1.8	49	49	2′ 5^c	2′ 5^c	vernier- -microsco
KOS (2.76b)	Kern, Aarau	19	24	—	0.75	79	69	5′ 10^c	5′ 10^c	estimating microscop
Te–Ni 1	MOM, Budapest	17.5	25		1.5	102	84	1° 1^g	1° 1^g	direct reading
Te–N2		17.5	25		1.5	102	84	1° 1^g	1° 1^g	direct reading
GT–60	Sokkisha, Tokyo	16	21	115	1.0	70	70	20′ 50^c	30′ 50^c	reading index
T 30	Soviet Union	20	29	137	1.0	70	70	10′	10′	estimating microscop
T 05 (2.76c)	Wild, Heerbrugg	19	25	—	0.8	67	50	2′ 10^c	2′ 10^c	estimating microscop
Theo 080 A (2.76d)	VEB Jenoptik, Jena	18	32	125	0.9	61	61	5′ 10^c	5′ 10^c	estimating microscop

eodolites with a glass circle

ccuracy of the ompen-sator (″)	Spirit-level constants				Mass			Other characteristics
	level of vertical axis (″)	index level of the vertical circle (″)	tele-scope level (″)	tubular level (′)	instru-ment (kg)	box (kg)	tripod (kg)	
12	13	14	15	16	17	18	19	20
—	—	—	30	4	3.0	2.5	3.5	usable as hanging theodolite
—	40	—	40	—	2.3	1.9	3.9	telescope supported only on one side
—	60	—	40	10	2.7	1.8	4.2	circles illuminated by built in LEDs
—	60	60	40	8	3.0	3.0	5.2	
—	—	—	30	10	2.4	0.9	5.7	bottom of the box built together with the instrument
—	50	—	30	30	2.7	2.7	4.7	bottom of the box built together with the instrument
—	—	—	—	2	4.3	1.8	—	additional percent graduation on the vertical graduated circle; telescope level can be attached
—	20			10	2.6	—	—	also usable for levelling, range of operation of compensator ± 10 min of arc
—	20			10	2.6	—	—	also usable for levelling, sensitivity of levelling bubble 20 sec of arc
—	137	—	60	—	3.0	2.5	—	
—	45	—	20	—	2.0	1.0	3.8	bottom of the box built together with the instrument
—	60	—	40	10	3.1	2.2	5.0	spirit level for slope measuring (± 10%); circle illumination by built-in LEDs electro-optical distance meter Citation CI 450 attachable
—	120	—	—	—	2.0	2.2	5.0	

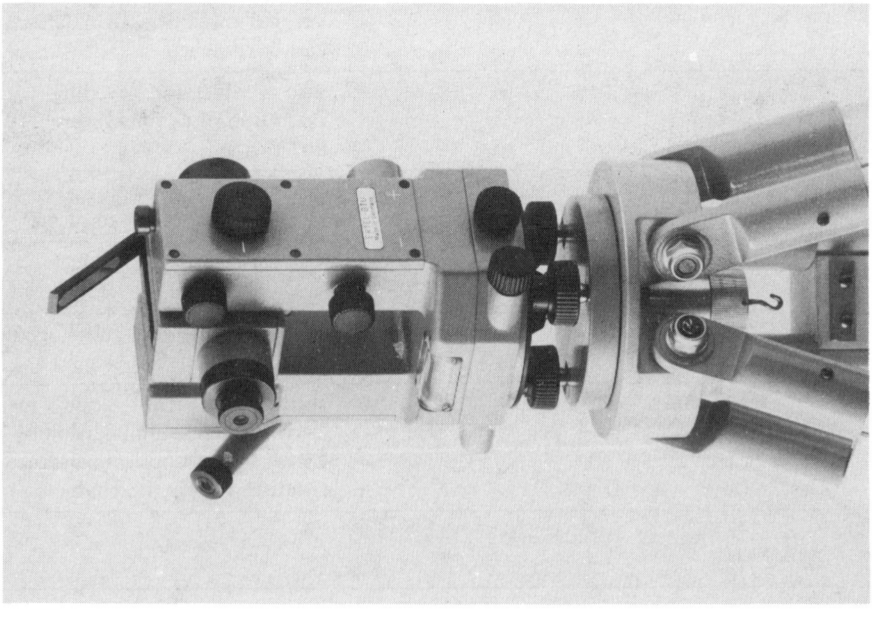

Figure 2.76. a) theodolite *BTU* (Ertel, München)

b) theodolite *KO-S* (Kern, Aarau)

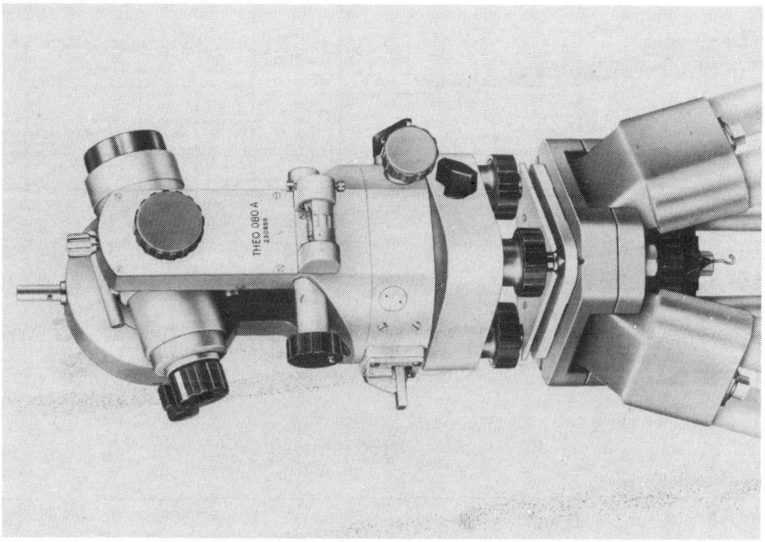

c) theodolite *T 05* (Wild, Heerbrugg)

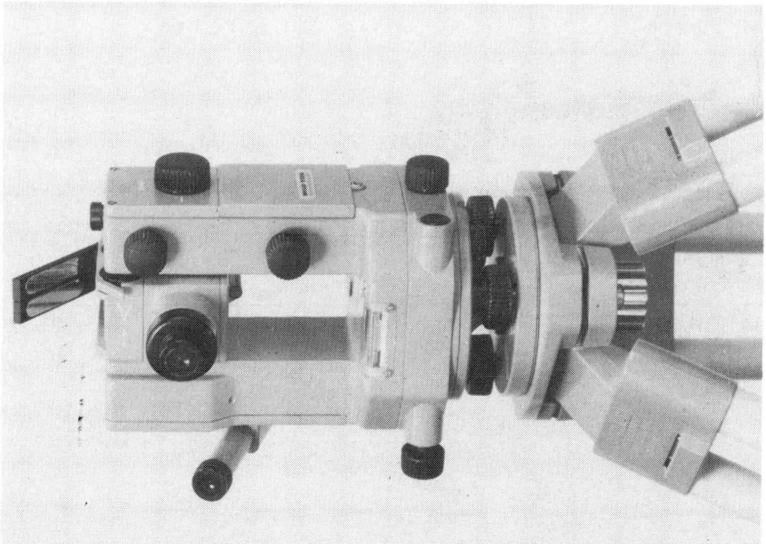

d) theodolite *Theo 080 A* (VEB Jenoptik, Jena)

Table 2.2. Some models of medium

Type (figure)	Manufacturing firm	Theodolite				Graduated glass circles				
		magnification Γ	diameter of objective (mm)	length (mm)	shortest pointing distance (m)	diameter		reading device		
								graduation unit		system
						H (mm)	V (mm)	H	V	
1	2	3	4	5	6	7	8	9	10	11
FT1 A	Fennel, Kassel	30	40	175	1.25	90	70	1′ 1^c	1′ 1^c	scale microscope
FT 20A		30	40	175	1.25	90	70	20″	20″	scale microscope
FTH 1		30	40	175	1.25	90	—	1′ 1^c	—	scale microscope
K1–S (2.77a)	Kern, Aarau	30	45	—	1.7	89	70	30″ 1^c	30″ 1^c	scale microscope
K1–M (2.77b)		30	45	—	1.5	89	70	6″ 20^{cc}	6″ 20^{cc}	optical micrometer
Te–D33	MOM, Budapest	24	40	175	2.0	77	68	1′ 1^c	1′ 1^c	optical micrometer
Te–D43		25	40	175	2.0	77	68	1′ 2^c	1′ 2^c	scale microscope
Te C13		30	40	175	2.5	77	65	10″ 20^{cc}	10″ 20^{cc}	symmetry adjusting optical micrometer
NT–2A	Nikon, Tokyo	30	45	165	1.3	84	84	20″ 50^{cc}	20″ 50^{cc}	optical micrometer
NT–3A		30	45	165	1.3	84	84	10″ 20^{cc}	10″ 20^{cc}	optical micrometer
NT–2S		30	45	165	1.3	84	84	30″ 1^c	30″ 1^c	scale microscope
Th 42 (2.77c)	Opton, Oberkochen	30	40	155	1.6	98	85	20″ 1^c	20″ 1^c	scale microscope
FG–1B	Pentax, Tokyo	28	40	—	1.6	110	70	20″ 1^c	1′ 2^c	vernier- -microscope

precision theodolites

Accuracy of the compensator (″)	Spirit-level constants				Mass			Other characteristics
	level of vertical axis (″)	index level of the vertical circle (″)	telescope level (″)	tubular level (′)	instrument (kg)	box (kg)	tripod (kg)	
12	13	14	15	16	17	18	19	20
*	40	—	—	8	4.7	3.8	6.7	automatic vertical index; electro-optical distance meter FEN 2000 attachable
*	40	—	—	8	4.7	3.8	6.7	automatic vertical index; electro-optical distance meter FEN 2000 attachable
—	40	—	—	8	4.3	3.8	6.7	
±6	30	—	—	—	4.6	2.4		electro-optical distance meter DM 502 or DM 503 attachable
±3	30	—	—	—	4.6	2.4		electro-optical distance meter DM 502 or DM 503 attachable
±1	30	—	—	6	4.8	2.5	5.1	
±1	30	—	—	6	4.8	2.5	5.1	type Li Te–D43 with built-in laser
—	20	30	—	6	5.4	2.8	5.1	
±2	30	—	—	10	4.7	2.6	4.0	
±1	30	—	—	10	4.7	2.6	4.0	
±1	30	—	—	10	4.7	2.6	4.0	
±1	30	—	—	10	4.7	4.2	6.3	electro-optical distance meter Eldi 2S attachable
—	90	—	40	—	4.9	2.0		

Table 2.2. (continued)

Type (figure)	Manufacturing firm	Theodolite: magnification Γ	Theodolite: diameter of objective (mm)	Theodolite: length (mm)	Theodolite: shortest pointing distance (m)	Graduated glass circles: diameter H (mm)	Graduated glass circles: diameter V (mm)	Reading device: graduation unit H	Reading device: graduation unit V	Reading device: system
1	2	3	4	5	6	7	8	9	10	11
TH–20D		30	42	—	1.3	78	78	50^{cc}	50^{cc}	optical micrometer
TH–10DC (2.77d)		30	42	—	1.3	78	78	20^{cc}	20^{cc}	optical micrometer
TH–06D		30	42	—	1.3	78	78	10^{cc}	10^{cc}	optical micrometer
TH–60S		30	42	—	1.3	78	78	1^{c}	1^{c}	scale microscope
TM 20H	Sokkisha, Tokyo	30	45	160	1.3	76	76	20″ 50^{c}	20″ 50^{c}	optical micrometer
TS 20A		30	45	160	1.3	82	76	1^{c}	1^{c}	scale microscope
TM 6		30	45	170	1.3	80	70	6″ 20^{cc}	6″ 20^{cc}	symmetry adjusting optical micrometer
TS 6		30	45	170	1.3	95	85	1′ 1^{c}	1′ 1^{c}	scale microscope
T5	USSR	25	36	173	2.0	—	—	1′	1′	scale microscope
TL–6DE	Topcon, Tokyo	30	42	152	1.5	70	70	6″	6″	symmetry adjusting optical micrometer
T1 (2.77e)	Wild, Heerbrugg	30	42	—	1.7	79	79	6″ 20^{cc}	6″ 20^{cc}	optical micrometer

Table 2.2. (continued)

ccuracy of the ompen-sator (″)	Spirit-level constants				Mass			Other characteristics
	level of vertical axis (″)	index level of the vertical circle (″)	tele-scope level (″)	tubular level (′)	instru-ment (kg)	box (kg)	tripod (kg)	
12	13	14	15	16	17	18	19	20
*	30	—	—	8	4.0	3.5		circle illumination by LED, with automatic switch-off (30 s); automatic vertical index
*	30	—	—	8	4.0	3.5		circle illumination by LED, with automatic switch-off (30 s); automatic vertical index; electro-optical distance meter PM 81 attachable
*	30	—	—	8	4.0	3.5		circle illumination by LED with automatic switch-off (30 s); automatic vertical index
*	30	—	—	8	4.0	3.5		circle illumination by LED with automatic switch-off (30 s); automatic vertical circle
±0.5	30	—	—	10	4.6			electro-optical distance meters RED 2, RED 3 or REDmini attachable
	40		40	10	3.9			electro-optical distance meters RED 2 or REDmini attachable
±0.5	30	—	—	10	5.6	3.3		electro-optical distance meters RED 2 attachable
±0.5	30	—	—	10	5.6	3.3		electro-optical distance meters RED 2, RED 3 or REDmini attachable
—	30	15	—	—	3.5	7.0	6.0	
±0.25	30	—	—	10	4.0	3.3		
±1	30	—	—	8	5.8	2.8	6.5	attachable: gyroscope GAK 1, electro-optical distance meters Di 4, Di 20, Ci 410, Ci 450; Roelofs' prism, laser-eyepiece; objective prism can be mounted

Table 2.2. (continued)

Type (figure)	Manufacturing firm	Theodolite: magnification Γ	Theodolite: diameter of objective (mm)	Theodolite: length (mm)	Theodolite: shortest pointing distance (m)	Graduated glass circles: diameter H (mm)	Graduated glass circles: diameter V (mm)	Reading device: graduation unit H	Reading device: graduation unit V	Reading device: system
1	2	3	4	5	6	7	8	9	10	11
T16 (2.77f)	Wild, Heerbrugg	30	42	—	1.7	94	79	1′ 1^c	1′ 1^c	scale microscope
T1A		28	40	150	1.5	73	65	20″ 1^c	20″ 1^c	optical micrometer
Theo 020B (2.77g)	VEB Jenoptik, Jena	30	40	—	1.5	86	86	1′ 1^c	1′ 1^c	scale microscope
Theo 015B (2.77h)		30	40	—	1.5	86	86	6″ 20^{cc}	6″ 20^{cc}	optical micrometer

Table 2.3. Some models of hig

Type (figure)	Manufacturing firm	Theodolite: magnification Γ	Theodolite: diameter of objective (mm)	Theodolite: length (mm)	Theodolite: shortest pointing distance (m)	Graduated glass circles: diameter H (mm)	Graduated glass circles: diameter V (mm)	Reading device: graduation unit H	Reading device: graduation unit V	Reading device: system
1	2	3	4	5	6	7	8	9	10	11
T4200 (2.78a)	Filotecnica Salmoiraghi, Milano	30	40	172	2.5	90	90	1″ 2^{cc}	1″ 2^{cc}	optical micromete
DKM 2	Kern, Aarau	32	45		1.5	80	70	1″ 1^{cc}	1″ 1^{cc}	optical micromete
DKM 2–A (2.78b)		32	45		1.5	80	70	1″ 1^{cc}	1″ 1^{cc}	symmetry adjusting optical micromete

Table 2.2. (continued)

ccuracy of the mpen- sator (″)	Spirit-level constants: level of vertical axis (″)	index level of the vertical circle (″)	tele- scope level (″)	tubular level (′)	Mass: instru- ment (kg)	box (kg)	tripod (kg)	Other characteristics
12	13	14	15	16	17	18	19	20
±1	30	—	—	8	5.3	2.8	6.5	photo-camera P32, further electro-optical distance meters Di 4, Di 20, Ci 410, Ci 450 attachable
±1	30	—	—	8	5.0	2.0	6.5	attachable: gyroscope GAK 1, electro-optical distance meters Di 4, Di 20, Ci 410, Ci 450; photo-camera P 32
±1	30	—	—	8	4.5	4.4	6.5	telescope spirit level, objective prism, autocollimating eyepiece, direction mark are attachable; coaxial clamping and slow-motion screws
±1	30	—	—	8	4.8	4.4	6.5	telescope spirit level, objective prism, autocollimating eyepiece, direction mark are attachable; coaxial clamping and slow-motion screws

recision theodolites

Accuracy of the compen- sator (″)	Spirit-level constants: level of vertical axis (″)	index level of the vertical circle (″)	tele- scope level (″)	tubular level (′)	Mass: instru- ment (kg)	box (kg)	tripod (kg)	Other characteristics
12	13	14	15	16	17	18	19	20
±0.3	20	—	—	10	6.1	5.2	7.2	
—	20	20	—	—	3.6	1.8	5.3	
±0.3	20	—	—	—	6.8	1.8	5.3	horizontal axis micrometer in model DKM–AM; electro-optical distance meter DM 503 attachable to model DKM–AE

Table 2.3. (continued)

Type (figure)	Manufacturing firm	Theodolite				Graduated glass circles				
		magnification Γ	diameter of objective (mm)	length (mm)	shortest pointing distance (m)	diameter		reading device		
								graduation unit		system
						H (mm)	V (mm)	H	V	
1	2	3	4	5	6	7	8	9	10	11
Te–B23 (2.78c)	MOM, Budapest	30	45	174	2.0	93	60	1″	1″	coincidence adjusting optical micrometer
T2	USSR	25	35	182	1.5	90	65	1″	1″	coincidence adjusting optical micrometer
T2 (2.78d)	Wild, Heerbrugg	30	40		2.2	90	70	1″ 1^{cc}	2″ 1^{cc}	coincidence adjusting optical micrometer
Theo 010B (2.78e)	VEB Jenoptik, Jena	30	40		1.5	86	86	1″ 2^{cc}	1″ 2^{cc}	coincidence adjusting optical micrometer

Table 2.4. Some models of highes

Type (figure)	Manufacturing firm	Theodolite				Graduated glass circles				
		magnification Γ	diameter of objective (mm)	length (mm)	shortest pointing distance (m)	diameter		reading device		
								graduation unit		system
						H (mm)	V (mm)	H	V	
1	2	3	4	5	6	7	8	9	10	11
Tpr	Siemens, AG, W-Berlin	32 60 80	63	—	10	200	140	0.2″	0.2″	coincidence adjusting optical micrometer
DKM 3	Kern, Aarau	45	68	—	5.0	104	104	0.5″ 0.5^{cc}	0.5″ 0.5^{cc}	symmetry adjusting optical micrometer

Table 2.3. (continued)

Accuracy of the compen-sator (″)	Spirit-level constants				Mass			Other characteristics
	level of vertical axis (″)	index level of the vertical circle (″)	tele-scope level (″)	tubular level (′)	instru-ment (kg)	box (kg)	tripod (kg)	
12	13	14	15	16	17	18	19	20
±1.0	20	—	—	6	5.5	3.7	6.2	coaxial clamping and fine-adjusting screws
—	15	20	—	—	5.2	4.5	5.3	coaxial clamping and fine-adjusting screws
±0.3	20	—	—	8	6.0	2.2	6.5	attachable: Gyroscope GAK 1, electro-optical distance meter Di 4, Ci 415, Ci 450, photo camera P 32
±0.3	20	—	—	8	4.7	4.4	6.5	telescope level, objective prism, auto-collimation eyepiece, direction mark attachable; coaxial clamping and fine-adjusting screws

precision theodolites

Accuracy of the compen-sator (″)	Spirit-level constants				Mass			Other characteristics
	level of vertical axis (″)	index level of the vertical circle (″)	tele-scope level (″)	tubular level (′)	instru-ment (kg)	box (kg)	tripod (kg)	
12	13	14	15	16	17	18	19	20
—	2	—	—	5	31	15	—	automatic vertical index; readings recorded on film
—	10	10	—	—	12.2	3.1	5.3	autocollimation eyepiece

Table 2.4. (continued)

Type (figure)	Manufacturing firm	Theodolite				Graduated glass circles				
		magni-fication Γ	diameter of objective (mm)	length (mm)	shortest pointing distance (m)	diameter		reading device		
						H (mm)	V (mm)	graduation unit H	graduation unit V	system
1	2	3	4	5	6	7	8	9	10	11
DKM 3A (2.79a)	Kern, Aarau	45	68	—	5.0	104	104	0.5″ 0.5^{cc}	0.5″ 0.5^{cc}	symmetry adjusting optical micrometer
Th 2 (2.79b)	Opton, Oberkochen	30	40	—	1.6	100	85	1″ 1^{cc}	1″ 1^{cc}	coincidence adjusting optical micrometer
Microptic 3	Rank Precision Ind. Ltd, UK	40	50	170	1.8	98	76	0.2″ 1^{cc}	0.2″ 1^{cc}	symmetry adjusting optical micrometer
T1	USSR	30 40	62	255	5.0	135	90	1″	1″	coincidence adjusting optical micrometer
OT 02		24 30 40	60	265	5.0	135	90	0.2″	0.2″	coincidence adjusting optical micrometer
T3 (2.79c)	Wild, Heerbrugg	24 30 40	60	260	4.6	140	95	0.2″ 1^{cc}	0.2″ 1^{cc}	coincidence adjusting optical micrometer
T4		60 80	60	—	100	250	145	0.1″	0.2″	coincidence adjusting optical micrometer
Theo 002 (2.79d)	VEB Jenoptik, Jena	H: 38 59 75 V: 32 50 65	65	—	4000	250	200	0.2″	0.2″	coincidence adjusting optical micrometer

Table 2.4. (continued)

Accuracy of the compensator (″)	Spirit-level constants				Mass			Other characteristics
	level of vertical axis (″)	index level of the vertical circle (″)	tele-scope level (″)	tubular level (′)	instru-ment (kg)	box (kg)	tripod (kg)	
12	13	14	15	16	17	18	19	20
—	10	10	—	—	14.2	3.9	5.3	astronomical variant; Horrebow level with 1 sec constant, autocollimating eyepiece
±0.2	20	—	—	10	5.2	4.8	6.3	electro-optical distance meter Eldi 2 attachable
—	10	20	—	—	8.0	6.5	6.2	
—	7	12	—	—	11.8	5.8	8	Horrebow-level with 4 sec constant
—	7	12	—	—	11	5	8	
—	6.5	12	—	—	11.2	3.7	5.7	manufactured with autocollimation eyepiece T 3A; Roelofs's prism, laser-eyepiece attachable
—	—	2	—	4	50	41		bent telescope; $f_0 = 578$ mm; telescope-level 1 sec; Horrebow-level 1 sec
±0.1	10	—	—	30	60	50	18	bent telescope, $f_0 = 938$ mm; readings recorded on film; compensator-stabilized line of sight with operation range ±0.03′, aligning time ±0.05 sec; target finding telescope

Figure 2.77. a) theodolite K1-S (Kern, Aarau)

b) theodolite Kl-M (Kern, Aarau)

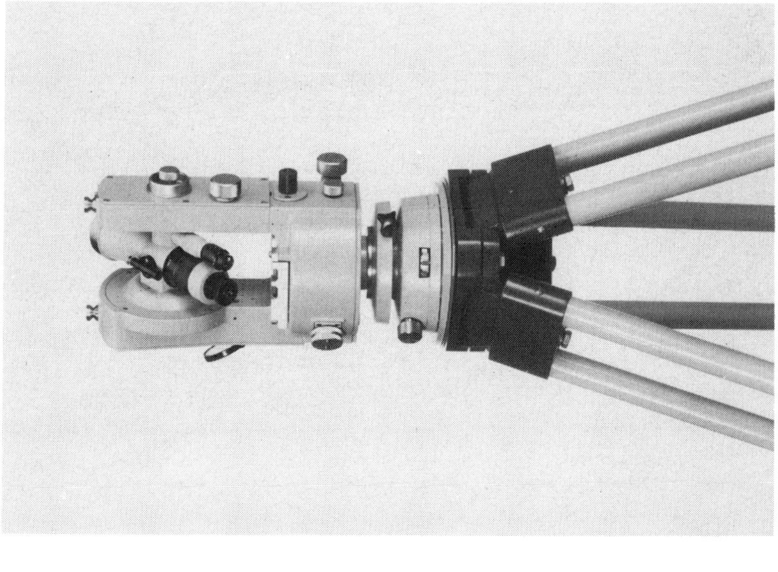

c) theodolite *Th 42* (Opton, Oberkochen)

d) theodolite *Th-20 DC* (Pentax, Tokyo)

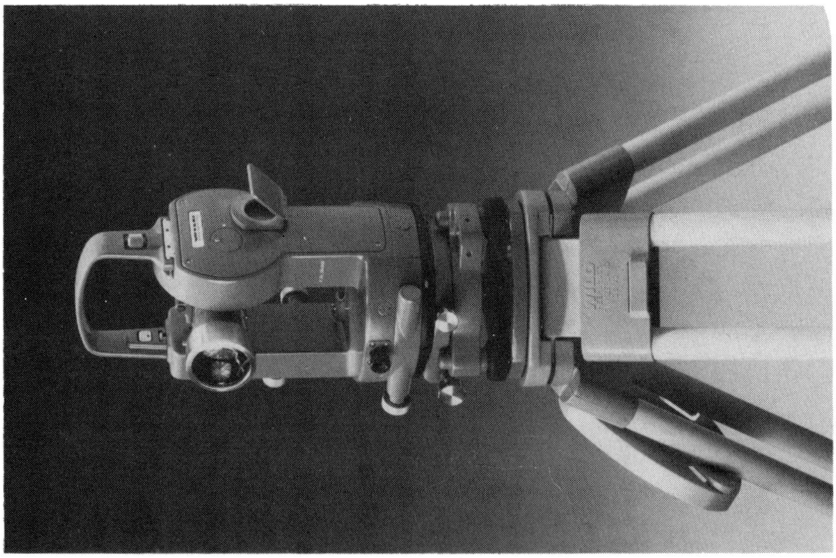

Figure 2.77. e) theodolite *T1* (Wild, Heerbrugg)

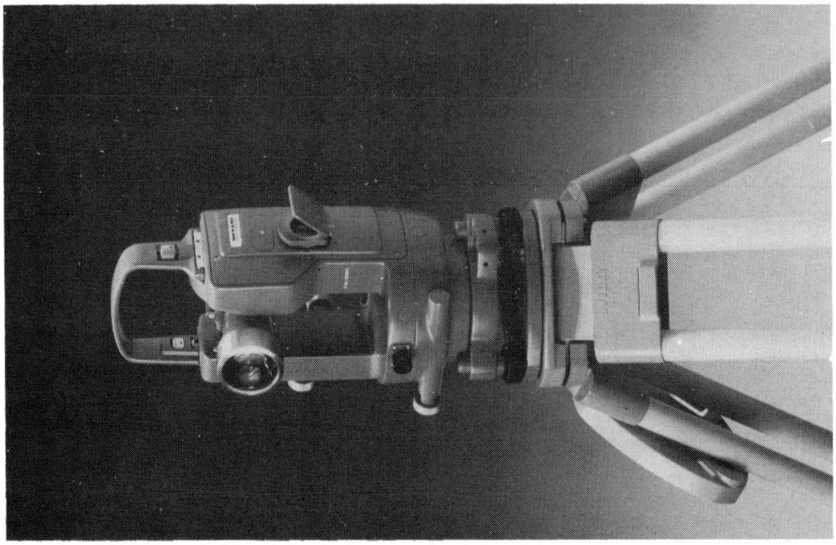

f) theodolite *T16* (Wild, Heerbrugg)

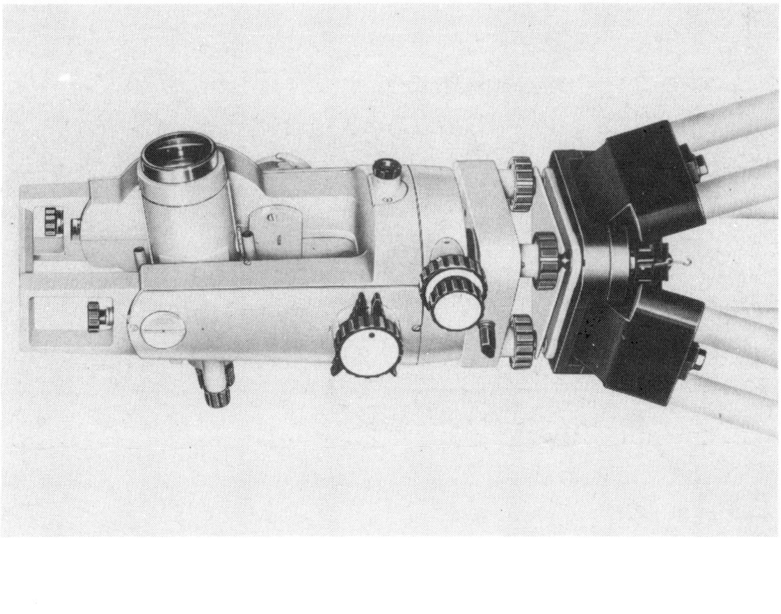

g) theodolite *Theo 020B* (VEB Jenoptik, Jena)

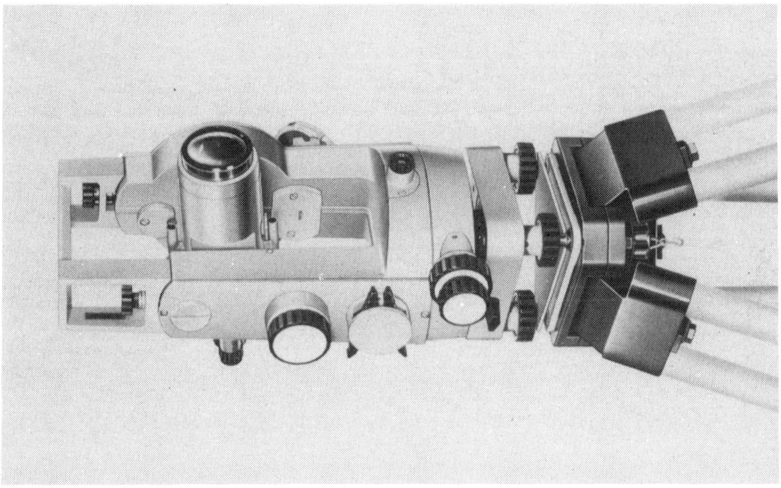

h) theodolite *Theo 015B* (VEB Jenoptik, Jena)

Figure 2.78. a) theodolite *T 4200* (Filotecnica Salmoiraghi, Milano)

b) theodolite *DKM 2A* (Kern, Aarau)

c) theodolite *Te-B23* (MOM, Budapest)

d) theodolite *T2* (Wild, Heerbrugg)

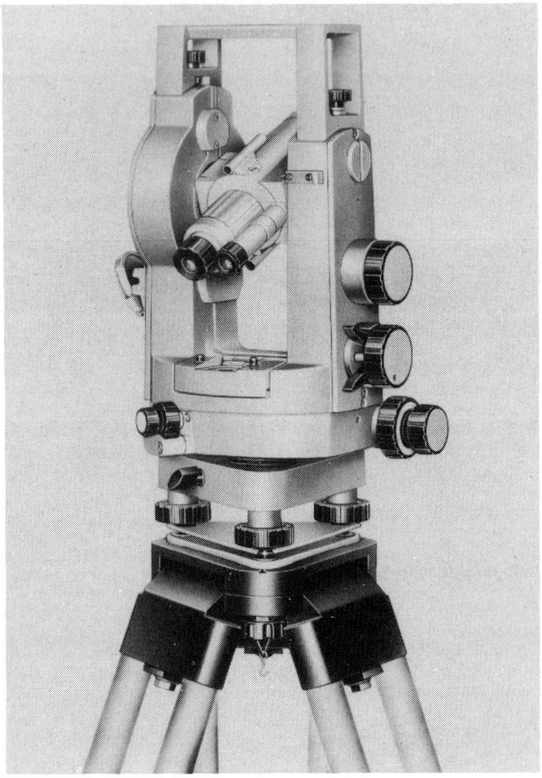

Figure 2.78. e) theodolite *Theo 010B* (VEB Jenoptik, Jena)

In *Tables 2.1, 2.2, 2.3* and *2.4* theodolites with glass circles are listed grouped into the categories I, II, III and IV, which are marketed by the best known manufacturers of geodetic instruments. Their most important characteristics are also given in the tables.

Theodolites of *low accuracy (Table 2.1; Figs 2.76, a* to *2.76, d)* are usable for simple measurements and setting out in architecture, and for measurement of detail. Thus they are small, light and easy to handle. The diameter of their horizontal glass circle is 50 to 80 mm, that of their vertical circle 40 to 70 mm. They usually have telescopes with a magnification of 17 to 20 ×. In consequence of their limited accuracy they have a simple reading device, such as an estimation microscope or a graduated microscope, reading directly to 10 min or 10^c with minutes estimated. For this purpose one reading is sufficient. The bottom part of the instrument box is at the same time the base-plate for some such theodolites. Some earlier types have no alidade tubular levels, because for the achievable measuring accuracy it is sufficient to plumb the vertical axis by means of a box level.

Theodolites of *medium accuracy (Table 2.2; Figs 2.77, a* to *2.77, h)* are used mainly for triangulation, small triangulation and simple tacheometry. Due to their broad field of use they are the most wide-spread of all theodolites. The diameter of their horizontal glass circle is 73 to 90 mm, that of their vertical circle 53 to 80 mm. Reading is usually made only in one position, because the relatively small eccentricity of the graduated circle has no harmful effect. They have in general a scale microscope or a reading microscope with a simple optical glass plate micrometer as a reading device; the direct reading is between 1 min of arc (1^c) and 10 sec of arc (30^{cc}), and their tenths can be estimated. In the case of an optical micrometer a magnification of 20 × of the microscope is sufficient. The scale microscope requires a magnification of up to 70 ×; consequently the lines of the circle graduations and of the microscope scale must be fine (2 to 3 μm). Present technology enables the production of such thin lines that scale microscopes have recently come into wide use due to their simple and comfortable reading. The magnification of the theodolite telescopes is 20 × to 25 ×. The instruments usually have a repetition axis system and are supplied with the stadia hairs needed for optical distance measurements. The clamping and fine adjustment systems are often mounted on a common axis. Theodolites with friction clamping and fine adjustment with a worm-screw are also manufactured.

The main application fields of *high precision theodolites (Table 2.3*; *Figs 2.78, a* to *2.78, e)* are second and third order triangulation, precision traversing, precision tacheometry, high precision engineering survey and optical distance measurement with a subtense bar. The diameter of the horizontal glass circles of the instruments belonging to this category is 80 to 100 mm, that of their vertical circles 70 to 90 mm. In order to eliminate the effect of the eccentricity of the graduated circle, modern instruments have a reading microscope with an optical micrometer yielding automatically the mean of the two diametrically opposite readings. The magnification of their microscopes is 40 × to 50 ×, that of their telescopes about 30 ×, and of their free objective diameter about 40 ×. They have no repetition device, but an arbitrary reading can be set on the horizontal circle.

Theodolites of *highest precision (Table 2.4*; *Figs 2.79, a* to *2.79, d)* are used for first order triangulation, determination of geographical position, setting out of tunnels. The diameter of their horizontal circle is 100 to 250 mm, that of their vertical circle 90 to 150 mm. Readings are made by reading microscopes yielding automatically the mean and by optical micrometers. In certain theodolite types the telescope is broken or supplied with a mirror lens. The telescope magnification is 30 × to 65 ×, the diameter of the free aperture of the objective lens about 60 mm. In most of the types, different telescope magnifications can be obtained by means of interchangeable eyepieces. These instruments are very stable and have a large mass; they are fitted with supplementary devices for increasing the measuring accuracy, such as e.g. an eyepiece microscope (for star-tracking), a striding level (in order to take into account the tilt of the horizontal axis), sometimes a checking telescope fixed to the tripod (in order to take into account the deformation of the tripod).

a)

c)

Figure 2.79. a) astronomical theodolite *DKM 3A* (Kern, Aarau); *b)* theodolite *Th 2* (Opton, Oberkochen); *c)* theodolite *T3* (Wild, Heerbrugg); *d)* astronomical theodolite *Theo 002* (VEB Jenoptik, Jena)

2.2.8 Test and Adjustment of Theodolites

The result of direction measurements and angle measurements with theodolites are influenced by:

1. manufacturing and assembling errors of the instrument, further by small displacements of different elements;
2. the errors in setting up the instrument;
3. the external conditions of the measurement (i.e. the physical state of the air between the instrument point and the observed point);
4. the visual acuity and the routine of the observer.

From the point of view of instrument theory the errors in (1) and (2) come into account. Instrumental reasons cause partly random, partly systematic measuring errors. Random errors are caused among others by the clearance of moving elements, by random differences in the positions of the graduation lines, by varying stresses occurring at the clamps and by elastic deformations. These error sources must be reduced to a minimum, partly by the manufacturers by means of adequate construction, proper technology, careful assembly and close control, partly by the users by means of proper handling and by the choice of appropriate measuring methods. In the following sections are discussed the error sources in the instrument and those due to its improper adjustment, also the systematic errors caused by the inaccurate setting up of the instrument.

2.2.8.1 Errors of the Axes and Lines of Sight

In the axes and in the line of sight of a theodolite the following errors may occur:

1. A deviation by an angle ε_k of the astronomical line of sight of a geodetic telescope from the plane perpendicular to the vertical axis, i.e. the so-called collimation error which is taken as positive if the line of sight is deflected from this plane in the first telescope position to the right as observed from the eyepiece. One has always to take account of a *collimation error*, because due to reasons of manufacture the mechanical axis of the telescope is not exactly perpendicular to the horizontal axis, and because the optical centres of the lenses of the optical system of the telescope can be centred only to a limited accuracy.
2. The deviation ε_i of the direction of the geodetic line of sight to a point of finite distance (i.e. the observed point) — determined by Eqs (2.12) and (2.23), respectively — from the astronomical line of sight. This error is zero at long observing distances and increases with decreasing distance (first additional collimation error). The deflection ε_i is positive if the geodetic line of sight deflects in the first telescope position to the right when observed from the eyepiece, i.e. if the geodetic line of sight is convex when observed from the left hand side.
3. The eccentricity e of the astronomical line of sight with respect to the vertical axis. In general this is also always present, partly due to imperfect machining, partly in consequence of the necessary adjustment of the reticule. Its error effect on the measurement of direction is reciprocally proportional to the observing distance d:

$$\varepsilon_e'' = \frac{e}{d}\rho''. \tag{2.29}$$

The numbers e and ε_e are positive if the astronomical line of sight lies in the first telescope position to the right when observed from the eyepiece (second additional collimation error); it approaches zero with increasing observing distance.

4. The deviation ε_h of the horizontal axis from the plane normal to the vertical axis (perpendicularity error of the horizontal axis). This error is kept very small by modern manufacturing technology, nevertheless one has to account for it in the case of precise measurements. The deviation ε_h is positive if the horizontal axis is deflected in the first telescope position to the right downwards from the plane normal to the vertical axis, when observed from the eyepiece.

The effects of the above-mentioned errors of the axes and of the line of sight on direction or angle measurement are usually treated in the literature independently from each other. Since, however, these errors occur together in practice, the error of direction measurement due to the jointly existing errors are studied here without restriction to the case of small axis errors.* In *Fig. 2.80, a* O is the instrument centre (the intersection point of the vertical and the horizontal axis), OC the plumbed vertical axis, H'—H' the actual position of the horizontal axis which deflects from the true horizontal position H—H by the angle ε_h, P is the observed point at a finite distance d, ε_k is the collimation error, ε_i is the direction deviation from the direction adjusted to an infinite distance (first additional collimation error), ε_e is the eccentricity error of the line of sight due to the eccentricity e, expressed in angular measure by Eq. (2.29) (second additional collimation error), and the direction error δ_{I} occurring due to the joint effect of the four axes- and line-of-sight-errors in the first telescope position. From the spherical triangle PCB *(Fig. 2.80, b)* one gets from the cosine theorem applied to the side PB:

$$\sin(\varepsilon_k+\varepsilon_i+\varepsilon_e) = -\sin\alpha\sin\varepsilon_h+\cos\alpha\cos\varepsilon_h\sin\delta_{\text{I}},$$

whence the direction error in the first telescope position, due to the joint effect of the four errors of the axes and the line of sight respectively, is:

$$\delta_{\text{I}} = \arc\sin\left\{\frac{\sin(\varepsilon_k+\varepsilon_i+\varepsilon_e)+\sin\alpha\sin\varepsilon_h}{\cos\alpha\cos\varepsilon_h}\right\}. \tag{2.30a}$$

Transiting the telescope around its horizontal axis and turning it around its vertical axis, point P is observed in the second telescope position; now the absolute values of errors ε_k, ε_i, ε_e and ε_h remain the same, but their signs become opposite, thus the error of the measurement in the second telescope position is, by changing the signs of ε_h, ε_i, ε_e and ε_h in Eq. (2.30a):

$$\delta_{\text{II}} = \arc\sin\left\{\frac{\sin(-\varepsilon_k-\varepsilon_i-\varepsilon_e)+\sin(-\varepsilon_h)\sin\alpha}{\cos\alpha\cos(-\varepsilon_h)}\right\} =$$

$$= -\arc\sin\left\{\frac{\sin(\varepsilon_k+\varepsilon_i+\varepsilon_e)+\sin\alpha\sin\varepsilon_h}{\cos\alpha\cos\varepsilon_h}\right\} = -\delta_{\text{I}}. \tag{2.30b}$$

* *L. Fialovszky:* Common effects of axis errors and eccentricity in the line of the theodolite on the measured horizontal directions and the vertical angles, MOM Review 4 (Budapest, 1973); A teodolit tengelyhibáinak együttes hatása, MTA Műsz. Oszt. Közl. VIII (1952); Wplyw bledu okladu osiowegu teodolitu na pomiar katoc poziomich, Geodezja i Kartografia, Warszawa, Tom III (1954).

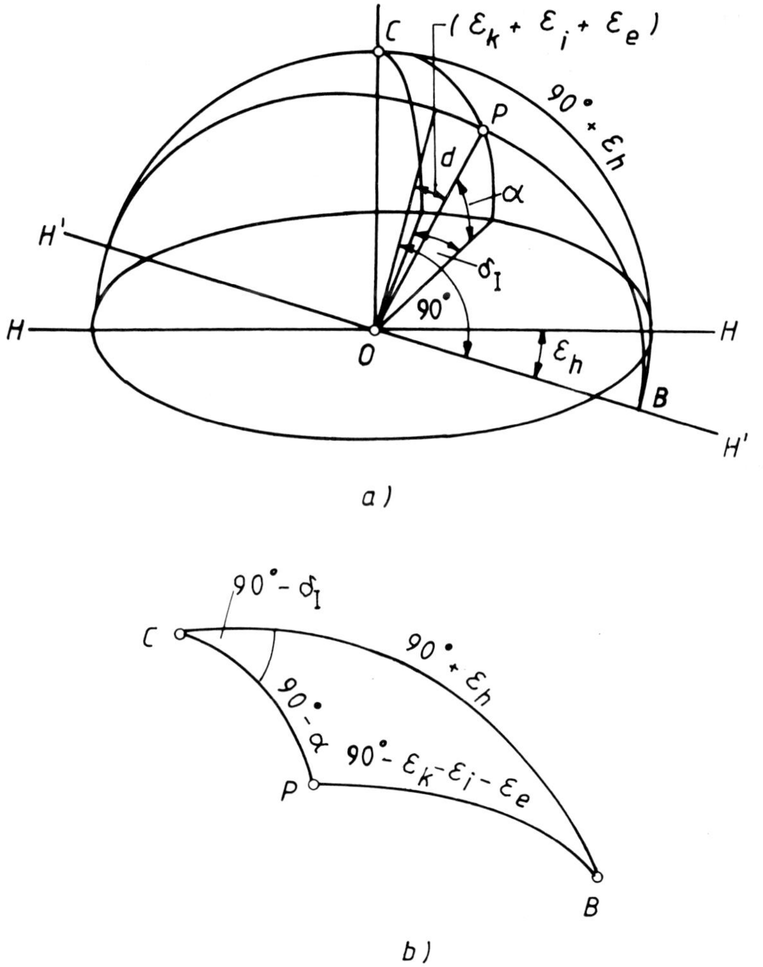

Figure 2.80. a) common effect on direction measurement of the collimation error ε_k, eccentricity error ε_e, direction deflection ε_i, and perpendicularity error of the horizontal axis ε_h; *b)* partial view

As may be seen, Eq. (2.30b) is an odd function, therefore the error δ_{II} of the measured direction in the second telescope position is the same as the error δ_I in the first telescope position, but with opposite sign. Hence the mean value of the directions measured in the two telescope positions is free of the effects of all the four errors of axes and line of sight, respectively, independently of the size of the errors (assuming that the vertical axis is vertical). If it is further taken into account that in practice all the four errors of axes and line of sight, respectively, are small, then one obtains from Eq. (2.30a):

$$\delta_I = \frac{\varepsilon_k}{\cos\alpha} + \frac{\varepsilon_i}{\cos\alpha} + \frac{\varepsilon_e}{\cos\alpha} + \varepsilon_h \tan\alpha. \tag{2.31}$$

The first term on the right-hand side of Eq. (2.31), i.e.:

$$\delta_{\mathrm{I},k} = \varepsilon_k \sec \alpha \tag{2.32}$$

is the effect of the collimation error ε_k, the second term:

$$\delta_{\mathrm{I},i} = \varepsilon_i \sec \alpha \tag{2.33}$$

is the effect of the direction deflection ε_i, the third term:

$$\delta_{\mathrm{I},e} = \varepsilon_e \sec \alpha \tag{2.34}$$

is the effect of the eccentricity e of the line of sight, and finally the fourth term:

$$\delta_{\mathrm{I},h} = \varepsilon_h \tan \alpha \tag{2.35}$$

is the effect of the perpendicularity error ε_h of the horizontal axis. For instance, with measurement in one telescope position only, the errors of direction measurements caused separately by the errors of the axes and of the line of sight — if they are taken as equal, as $\varepsilon_k = \varepsilon_i = \varepsilon_e = \varepsilon_h = 10''$ in the case of the elevation angles: $\alpha = 0°$; 10°; 20°; 30°; 40°; 50°; 60°; 70° and 80° — are $\delta_{\mathrm{I},k} = \delta_{\mathrm{I},i} = \delta_{\mathrm{I},e} = 10''$; 10.2″; 10.6″; 11.5″; 12.6″; 15.0″; 20.0″; 29.4″ and 57.05″, respectively, and: $\delta_{\mathrm{I},h} = 0$; 1.76″; 3.64″; 5.77″; 8.39″; 11.9″; 17.3″; 25.5″ and 56.7″, respectively.

It should be noted that when $\alpha \cong 90°$ the approximation formulae cannot be used and the direction errors cannot be calculated.

The collimation error can be determined and eliminated in the following way:

1. If the theodolite to be tested has two diametrically opposite readings projected beside each other, a distant point P, at about the same height as the theodolite, is observed in the first telescope position and the horizontal circle (l_{I}) is read. The point P is now observed in the second telescope position and the horizontal circle (l_{II}) is read. Reversing the difference $\Delta l = (l_{\mathrm{II}} - l_{\mathrm{I}})$ by 180°, twice the collimation error is obtained, i.e. $2\varepsilon_k = (\Delta l \pm 180°)$. To eliminate this error the reading corrected by the collimation error, i.e. $l'_{\mathrm{II}} = (l_{\mathrm{II}} - \varepsilon_k)$, is set on the horizontal circle, then the cross-hair is moved sideways by means of the horizontal adjustment screws until the telescope again points to P in its second position. As a check the reading $l'_{\mathrm{I}} = (l'_{\mathrm{II}} \pm 180°)$ is set on the horizontal circle, and now the line of sight must point to P in the first telescope position.

2. If the theodolite has two independent diametrically opposite reading indices, the former method is modified inasmuch as in both telescope positions both indices (1 and 2) should be read. From the mean l_{II} of the two readings $l_{\mathrm{II},1}$ and $l_{\mathrm{II},2}$ obtained in position II, and from the mean l_{I} of the readings $l_{\mathrm{I},1}$ and $l_{\mathrm{I},2}$ obtained in the position I, the collimation error is:

$$\varepsilon_k = \frac{1}{2}(l_{\mathrm{II}} - l_{\mathrm{I}}) \pm 180°.$$

For its elimination the reading $(l_{\mathrm{II}} - \varepsilon_k)$ is set on the horizontal circle and the cross-hair is adjusted so that it points to point P in the second telescope position.

3. The tests and adjustments described in 1 and 2 can also be carried out in a closed room. In front of the theodolite to be tested another theodolite of at least the

same accuracy is set up at such a height that the objective centres of the two telescopes, when set horizontal, are at the same height within 1 ... 2 mm. The telescopes of both theodolites are focussed to infinity and pointed at each other. The telescope of the second theodolite as a collimator plays the role of the distant point P. From here on the method described in 1 and 2, respectively, should be followed.

4. If the theodolite has only one reading index, the test made by one of the methods described above does not yield the collimation error alone, but the sum of the collimation error and the direction error caused by the eccentricity of the centre of the circle graduations, which varies as a function of the circle position. Therefore the collimation error of such a theodolite must be determined in the manner discussed in Point 2.2.8.3.

The sum of the additional collimation errors $\varepsilon_i(d)$ and $\varepsilon_e(d)$ which depend on the distance d can be determined as follows: the total collimation $\varepsilon_{k,\mathrm{tot}}(d)=\varepsilon_k+\varepsilon_i(d)+\varepsilon_e(d)$ is determined for different distances, as e.g. for $d=2$; 10; 50 and 100 m, and the net collimation $\varepsilon_{k,\mathrm{tot}}(d=\infty)=\varepsilon_k$ is determined for a very long distance.

From this the values

$$\varepsilon_i(d)+\varepsilon_e(d)=\varepsilon_{k,\mathrm{tot}}(d)-\varepsilon_k$$

can be calculated. The direction deviation $\varepsilon_i(d)$ and the eccentricity error $\varepsilon_e(d)$ — and thence the linear eccentricity e — cannot be determined separately, but this is not necessary.

In theodolites with an adjustable horizontal axis the perpendicularity error ε_h of the horizontal axis can be determined and eliminated in the following way:

1. At first the collimation error is eliminated very carefully in the manner described. Next an elevated point P is observed in the first telescope position, and the horizontal circle (l_I) and the zenith angle ζ on the vertical circle are read after centring the bubble of the index-level. Now P is observed in the second telescope position and the horizontal circle is read (l_II). The effect of the error ε_h of the horizontal axis is:

$$\delta_h=\frac{l_\mathrm{II}-l_\mathrm{I}\pm 180^\circ}{2},$$

from which, by using Eq. (2.35) (and by taking into account that $\tan\alpha=\cot\zeta$) one obtains:

$$\varepsilon_h=\frac{l_\mathrm{II}-l_\mathrm{I}\pm 180^\circ}{2\cot\zeta}. \tag{2.36}$$

To eliminate this error the reading corrected by δ_h, i.e. $l'_\mathrm{I}=l_\mathrm{I}+\delta_h$ is set on the horizontal circle, and then the seat of the horizontal axis is raised or lowered until the line of sight in the first position points again to P. For confirmation, it should be checked that, by setting the reading $l'_\mathrm{II}=l'_\mathrm{I}\pm 180^\circ$ on the horizontal circle, the line of sight in the second telescope position points to P.

2. The previous method can be carried out more reliably in a closed room. The telescope of the theodolite to be tested is made free of collimation error and focussed to infinity; in front of it another theodolite of at least the same accuracy is set up as a collimator at an elevation angle of about 40° to 45°, and its telescope is also focussed

to infinity. The telescope of the theodolite to be tested in position I is precisely pointed to the cross-hair centre of the second theodolite playing the role of a collimator; the direction l_I is read on the horizontal circle and, after centring the bubble of the index level, the zenith angle ζ is read on the vertical circle. The theodolite to be tested is now pointed in position II to the cross-hair centre of the second theodolite and the direction l_{II} is read. Henceforth the method described in 1 is followed.

3. The perpendicularity error of the horizontal axis can also be determined with a striding level, if there is one available for the theodolite. The striding level, with known constant ε and unknown adjustment error Δ, is placed on the two cylindrical surfaces specially machined for this purpose at each end of the axis of the theodolite, which must be levelled very carefully. The striding level is placed on the levelled theodolite in position I, as shown in *Fig. 2.81, a*. In the figure, point C is the centre of curvature of the arc of the longitudinal section of the inner polished surface of the striding level; with a continuously graduated level, l_A is the distance of the nominal centre A of the graduations from the first graduation line, and l_I is the distance of the centre of the bubble from this line; *1* is the horizontal axis and *2* the horizontal direction. From the figure one obtains in level position I:

$$l_I\varepsilon = l_A\varepsilon - (\varepsilon_h + \Delta),$$

and in level position II obtained by reversing the striding level *(Fig. 2.81, b)*:

$$l_{II}\varepsilon = l_A\varepsilon - (\Delta - \varepsilon_h) = l_A\varepsilon + \varepsilon_h - \Delta,$$

and by subtracting the latter from the former:

$$\varepsilon(l_{II} - l_I) = 2\varepsilon_h,$$

whence:

$$\varepsilon_h = \varepsilon\frac{l_{II} - l_I}{2}. \tag{2.37}$$

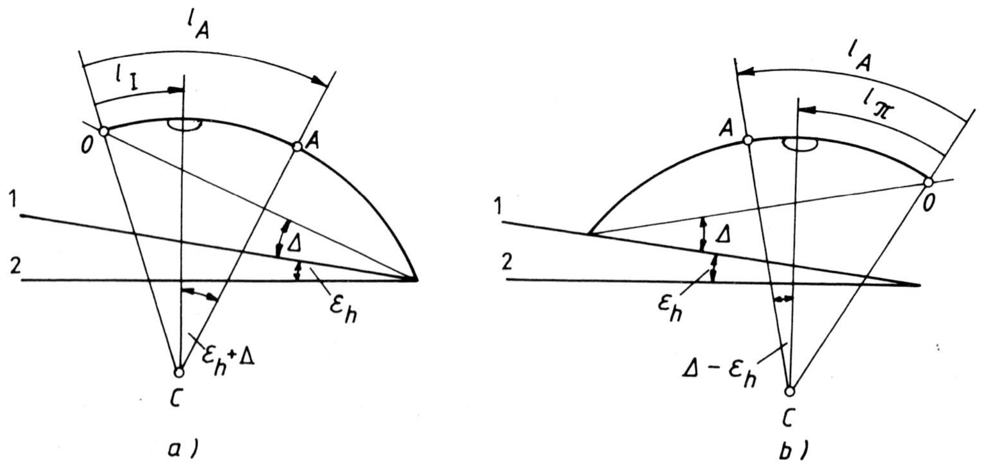

Figure 2.81. Determination of the perpendicularity error of the horizontal axis with a striding level: *a)* first level position; *b)* second level position

Denoting the readings at the ends of the bubble in position I by l'_{I} and l''_{I}, and in position II by l'_{II} and l''_{II}, the error of the horizontal axis is obtained, by using Eq. (2.37) with correct sign, as:

$$\varepsilon_h = \varepsilon \frac{(l'_{\mathrm{II}} + l''_{\mathrm{II}}) - (l'_{\mathrm{I}} + l''_{\mathrm{I}})}{4} . \tag{2.38}$$

The perpendicularity error of the horizontal axis can be determined in this way for high precision measurements. Its effect on the measurement of a direction can be calculated from Eq. (2.35) and then eliminated.

The perpendicularity error of modern theodolites with covered horizontal axes can be corrected only in a workshop. The error of the horizontal axis in such instruments is generally less than 10 secs of arc.

The error $\delta_{\mathrm{I}(2,1)}$ (due to the error of the horizontal axis and of the line of sight) of the measurement of the horizontal angle $\varphi_{2,1}$, made by the directions 1 and 2 when observing the points P_2 and P_1 from the stand point O, is from Eq. (2.31):

$$\delta_{\mathrm{I}(2,1)} = \varepsilon_k \left(\frac{1}{\cos \alpha_2} - \frac{1}{\cos \alpha_1} \right) + \frac{\varepsilon_i(d_2) + \varepsilon_e(d_2)}{\cos \alpha_2} - \frac{\varepsilon_i(d_1) + \varepsilon_e(d_1)}{\cos \alpha_1} + \varepsilon_h (\tan \alpha_2 - \tan \alpha_1) , \tag{2.39}$$

if both directions were measured only in the first telescope position. In the formula $d_2 = OP_2$ and $d_1 = OP_1$ are the distances of points P_2 and P_1 from the instrument stand point, and α_2 and α_1 are the elevation angles of the directions to the points P_2 and P_1. From Eq. (2.39) it is clear that the effect of the perpendicularity error of the horizontal axis, and of the error of the line of sight, on an angle measurement carried out in only one telescope position is zero only if $\alpha_2 \approx \alpha_1$, and at the same time $d_2 \approx d_1$, or both d_2 and d_1 are long, as the additional collimation errors ε_i and ε_e are then zero.

2.2.8.2 Errors in Setting up Theodolites

Eccentric set-up near a fixed point. A direction measurement in an eccentric set-up must be corrected by the centring correction:

$$\eta'' = \frac{r}{d} \rho'' \sin \varepsilon . \tag{2.40}$$

In Eq. (2.40) r is the distance between the vertical line passing through the instrument centre and the fixed point (i.e. the linear measure of the eccentricity), d is the horizontal distance to the observed point, ε is the angle between the direction from the eccentric instrument centre to the fixed point and the direction to the observed point (i.e. the orientation angle of the eccentricity) in the sense that the left leg of the angle points to the fixed point. If $\varepsilon < 180°$, then η is positive, in the opposite case it is negative.

The plumbing error of the vertical axis. If the vertical axis V—V of a theodolite *(Fig. 2.82, a)* is tilted in the vertical plane ACB by the angle ε_v from the plumb line OC passing through the centre O of the theodolite, then instead of the correct

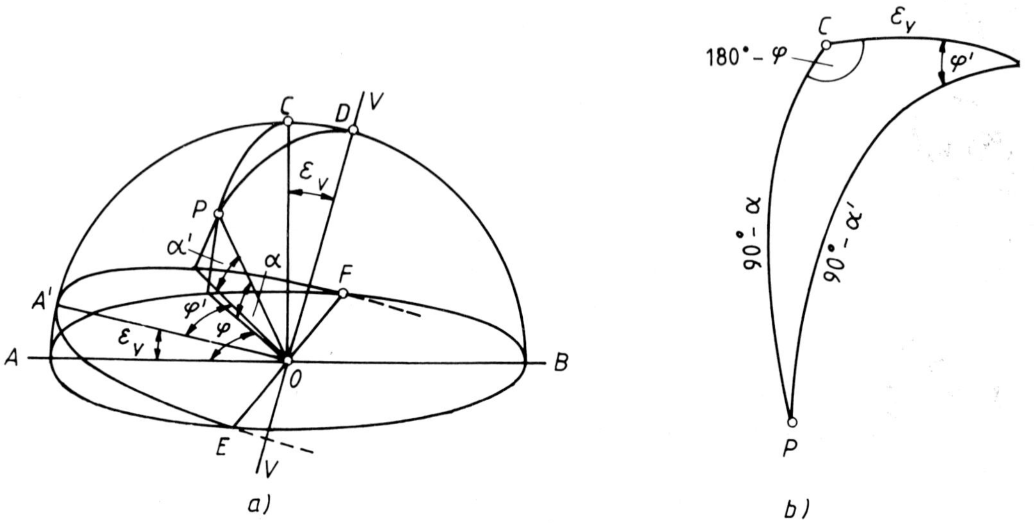

Figure 2.82. Effect on direction measurement of vertical axis levelling error

horizontal angle φ of the direction pointing to a point P at elevation angle α and referred to slope-plane ACB, the erroneous angle φ' is measured on the horizontal circle $EA'F$, tilted in the slope-plane by the angle ε_v, and instead of α we read the erroneous angle α'. The direction error $\delta_v = \varphi - \varphi'$ can be determined as follows. Since δ_v is small, one can write:

$$\sin\varphi - \sin\varphi' = 2\sin\frac{\varphi-\varphi'}{2}\cos\frac{\varphi+\varphi'}{2} \approx \delta_v \cos\varphi',$$

whence:

$$\delta_v = \frac{\sin\varphi - \sin\varphi'}{\cos\varphi'}. \tag{2.41}$$

From the spherical triangle PCD in *Fig. 2.82, b* one obtains from the cosine theorem applied to the side PC:

$$\sin\alpha = \cos\varepsilon_v \sin\alpha' + \sin\varepsilon_v \cos\alpha' \cos\varphi',$$

or by neglecting the second and higher order terms which are very small:

$$\sin\alpha \approx \sin\alpha' + \varepsilon_v \cos\alpha' \cos\varphi'.$$

Whence:

$$\cos\alpha = \sqrt{1-(\sin\alpha' + \varepsilon_v \cos\alpha' \cos\varphi')^2},$$

or again neglecting the small terms of second and higher order:

$$\cos\alpha = \cos\alpha'(1-\varepsilon_v \tan\alpha' \cos\varphi'). \tag{2.42}$$

One also obtains from the spherical triangle PCD:

$$\sin\varphi = \sin\varphi' \frac{\cos\alpha'}{\cos\alpha}$$

and by substituting Eq. (2.42):

$$\sin \varphi = \frac{\sin \varphi'}{1 - \varepsilon_v \tan \alpha' \cos \varphi'} \approx$$

$$\approx \sin \varphi' + \varepsilon_v \tan \alpha' \cos \varphi' \sin \varphi' ,$$

and by using this and Eq. (2.41):

$$\delta_v = \varepsilon_v \tan \alpha' \sin \varphi' \approx \varepsilon_v \tan \alpha \sin \varphi . \tag{2.43}$$

In this formula

$$\varepsilon_v \sin \varphi = \varepsilon_v' \tag{2.44}$$

is the projection of the tilt ε_v of the vertical axis in the vertical plane of the horizontal axis.

Hence:

$$\delta_v = \varepsilon_v' \tan \alpha , \tag{2.45}$$

i.e. it is equal to the direction error of Eq. (2.35), occurring due to the error ε_v' of the horizontal axis. In the second telescope position neither the absolute value nor the sign of the tilt ε_v of the horizontal axis changes, thus the error δ_v is also present in the mean value of the data measured in the two telescope positions. If the vertical axis is not plumbed correctly, the measurements in the two telescope positions do not totally eliminate the effects of the error of the horizontal axis and of the line of sight. Therefore, the vertical axis must always be set vertical very carefully before measurement, and its setting must be checked several times if several measurements are carried out at the same stand point.

The tilt of the vertical axis causes no error if $\alpha = 0$, and if $\alpha \neq 0$ but $\varphi = 0$, i.e. if the vertical axis is tilted in a plane perpendicular to the vertical plane containing the (correctly levelled) horizontal axis. The effect of the error is maximum if the two planes coincide.

The effect of the error due to the tilt of the vertical axis can be eliminated by correcting the reading on the horizontal circle by the correction calculated from Eq. (2.45). For this purpose ε_v can be determined by the striding level placed on the cylindrical surfaces of the pivots of the horizontal axis.

Using Eq. (2.43) the error of the horizontal angle $\varphi_{2,1}$, made by the directions *2* and *1* to the points P_2 and P_1, due to the tilt ε_v of the vertical axis is:

$$\delta_{2,1} = \varepsilon_v (\tan \alpha_2 \sin \varphi_2 - \tan \alpha_1 \sin \varphi_1) , \tag{2.46}$$

either measured in one or in both telescope positions. The tilt of the vertical axis does not cause any error in the measurement of the angle if the right hand side of Eq. (2.46) is zero, i.e. if:

$$\tan \alpha_2 \sin \varphi_2 = \tan \alpha_1 \sin \alpha_1 . \tag{2.47}$$

This is true

1. in the trivial case if $\alpha_1 = \alpha_2 = 0$, or
2. if Eq. (2.46) is satisfied by the angles φ_2 and φ_1, made between the direction of tilt and the directions *2* and *1*, pointing to P_2 and P_1, respectively. Since the vertical axis has a small wobble when the top part of the instrument is rotated, due to defects

in manufacture (see Subsection 2.2.3), only the projection of the vertical axis in the vertical plane of a defined direction can be precisely set vertical. Case 2 comes about if the projection of the vertical axis on the vertical plane, deflected from direction *1* by the angle

$$\bar{\varphi}= \text{arc tan}\frac{C_{1,2}-\cos\varphi_{2,1}}{\sin\varphi_{2,1}} \tag{2.48}$$

is precisely vertical*. In Eq. (2.48):

$$C_{1,2}=\frac{\tan\alpha_1}{\tan\alpha_2}. \tag{2.49}$$

According to this a single angle measurement can be improved in the following way. The instrument is set up approximately over the point and the vertical axis made approximately vertical; the approximate values of the elevation angles $\bar{\alpha}_1$ and $\bar{\alpha}_2$ of the points P_1 and P_2, and the approximate value $\bar{\varphi}_{2,1}$ of the horizontal angle $\varphi_{2,1}$ to be measured, are then determined to an accuracy of 1° to 2°, and the approximate $\bar{\varphi}$ is calculated from Eq. (2.48), e.g. by means of a pocket calculator. For the precise set up over the point one foot-screw is positioned to the right of the left hand direction *1* by an angle of about $\bar{\varphi}$. The vertical axis is precisely levelled in the vertical plane through this foot-screw, i.e. — in the case of theodolites which have a tubular spirit level for adjusting the vertical axis situated on the top part of the instrument parallel to the telescope axis — the levelling is made in the two positions in which the telescope is parallel to the other two foot-screws. The angle is now measured exactly in this position of the vertical axis.

A closer analysis of Eq. (2.48) reveals the following. If $\alpha_2 \neq 0$, but $\alpha_1 \approx 0$, the vertical axis must be levelled precisely in direction *2*. If both α_1 and α_2 are large, but of the same (opposite) sign, the optimum direction of the precise levelling lies between (outside) the angle legs *1* and *2*. If the absolute values of the large α_1 and α_2 are nearly equal and their signs are the same (opposite), then the optimum direction of the precise levelling is the direction bisecting (perpendicular to) the angle $\varphi_{1,2}$ to be measured.

The above method can be used for precise traversing, especially in the case of slope printings.

In certain theodolites a compensator compensating for the tilt of the vertical axis is used, e.g. in the theodolite *4149-A* (Filotecnica Salmoiraghi, Milano), where the automatic vertical index (Point 3.2.2.2) is implemented with a so-called optical level. It allows on the one hand the checking of the correct operation of the compensator, on the other it shows the component of the tilt of the vertical axis in the vertical plane of the horizontal axis. The main feature of the device *(Fig. 2.83)* is that a circular mark on a glass plate is placed into the optical train of the rays illuminating the reading of the vertical circle, and its image is projected onto the reading area of

* *L. Fialovszky*: Zur Reduktion des Stehachsenfehlers bei Horizontalwinkelmessungen, Zeitschrift für Vermessungswesen 103 (1978) and Decreasing of the effect of the verticality error of the vertical axis to the horizontal angle measurement (in Hungarian), Geod. és Kart. 30, Budapest, 1978/2.

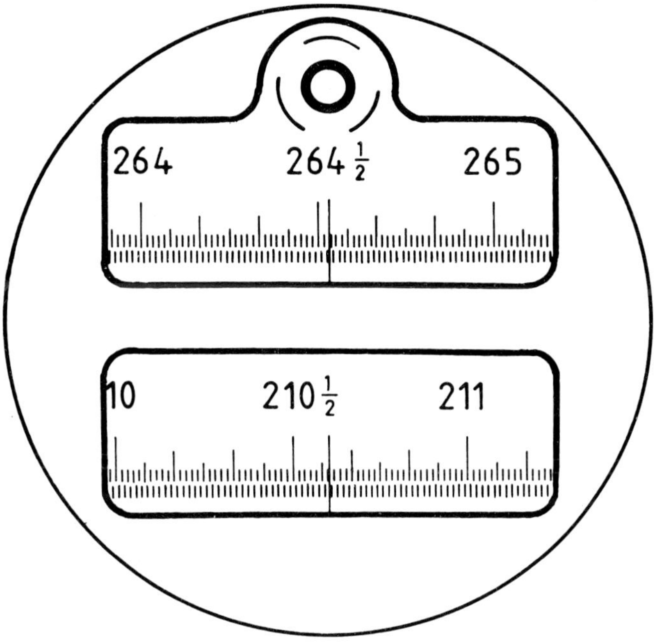

Figure 2.83. Field of view of the estimating microscope, with the circle mark of the optical level, of the theodolite *4149 A* (Filotecnica Salmoiraghi). Reading: $V=264°31.8'$, $H=210°36.2'$

the vertical circle by means of a lens and two reversing prisms. In this way not only can the correct operation of the automatic vertical index be checked (the circular mark must be within the limiting circle visible at the top of the field of view of the microscope), but the difference in height between the circle and the centre of the boundary circle also shows the lateral tilt of the vertical axis; this can be eliminated by the foot-screws with an accuracy of ± 5 sec of arc (semi-automatic elimination).

Kern (Switzerland) improved the liquid compensator of the theodolite *DKM 2A* (Point 3.2.2.2) to include a micrometer for the horizontal axis and built it into the theodolite *DKM 2-AE*, thus enabling the projection $\varepsilon_{v,h}$ of the tilt angle ε_v of the vertical axis (i.e. the tilt of the vertical axis) to be measured in the vertical plane containing the horizontal axis simultaneously with the direction measurement. The horizontal axis micrometer *(Fig. 2.84, a)* utilizes the fact that the sections of the liquid surface of the liquid compensator *1*, not only in the observing direction but also in the horizontal axis direction, remain horizontal when the vertical axis is tilted. The projecting optics *(Fig. 2.84, a)*, consisting of the lenses *2* and *2'*, produce the image *6* of the image *5* of the line-index on reticule *3*, projected on the altitude circle *4*; this is due to the effect of the projection of the tilt of the vertical axis in the direction of the horizontal axis (in the first telescope position) which is rotated with respect to the imaginary circle which passes through the image *5* and is concentric with the graduated circle. Its image *7* appears in the microscope in a position moved by the distance *10* from the double index line *9* visible in the image plane of the microscope

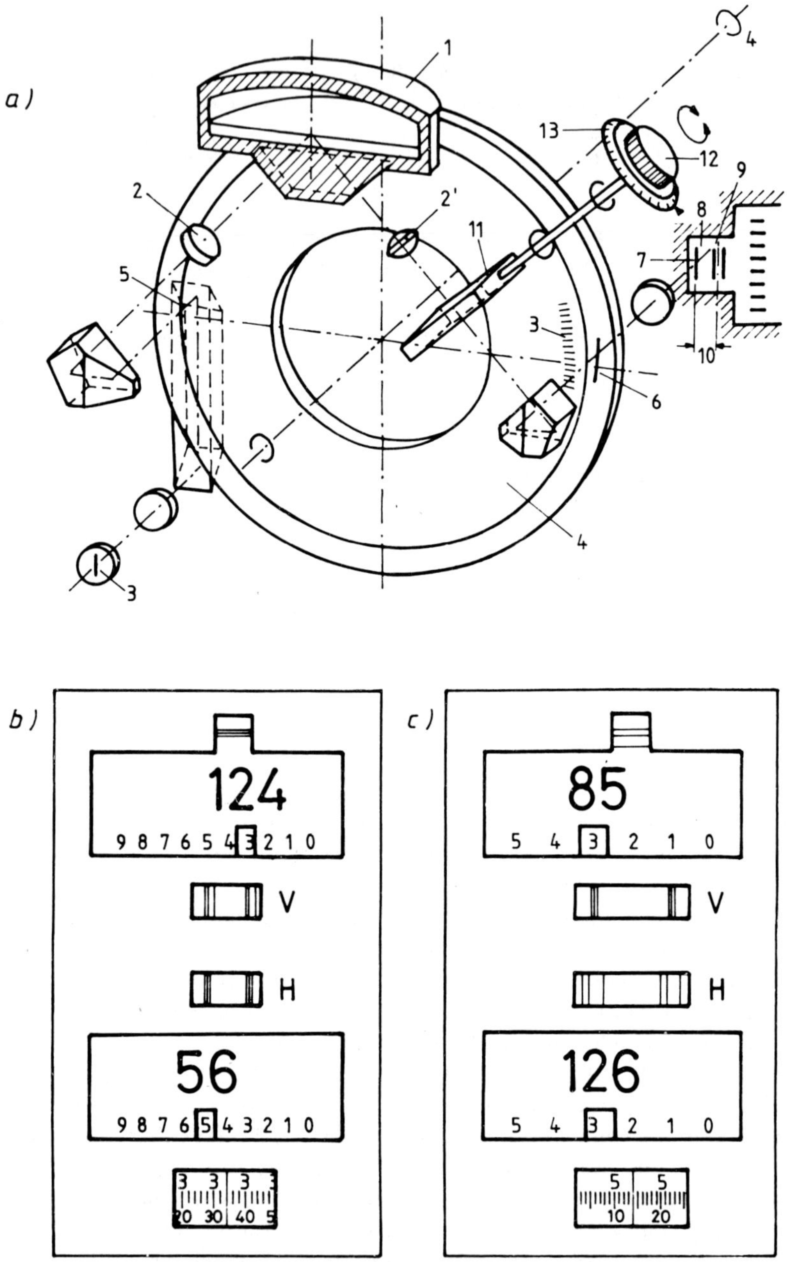

Figure 2.84. Micrometer of the horizontal axis of the theodolite *DKM 2-AE* (Kern, Aarau): *a)* theoretical scheme; *b)* the line-image *7* is not set within the double index *9* in the *V* (vertical) window in the field of view of the reading microscope; *c)* the line-image *7* is set within the double index *9*

(in the diaphragm aperture *8*). The shift of the image is proportional to the tilt angle ε_I of the graduation plane of the vertical circle, and to the same tilt of the horizontal axis. The plane parallel glass plate *11*, placed in the ray path of the transmitting optics *2-2'*, can be rotated by means of the micrometer screw *12*, and in this way the image *7* can be moved until it coincides with the centre of the double index line *9* (*Fig. 2.84, b* and *c*). Thus after measurement in the first telescope position,

$$\varepsilon'_I = \beta + \varepsilon_I = \beta + \varepsilon_{v,h} + \varepsilon_h$$

can be read on the micrometer scale *13* (*Fig. 2.84, a*), where $\varepsilon_{v,h}$ is the projection of the tilt onto the vertical plane containing the horizontal axis; ε_h is the perpendicularity error of the horizontal axis and β is a constant. Following the same method in the second telescope position, one reads on the micrometer scale *13*:

$$\varepsilon'_{II} = \beta - \varepsilon_{II} = \beta - \varepsilon_{v,h} + \varepsilon_h \, .$$

Half the difference between the two micrometer readings, i.e.:

$$\frac{\varepsilon'_I - \varepsilon'_{II}}{2} = \varepsilon_{v,h}$$

yields the projection $\varepsilon_{v,h}$ of the tilt ε_v of the vertical axis in the vertical plane containing the horizontal axis. Using this, the directions obtained as the mean of the horizontal circle readings of the pointings in both telescope positions can be corrected, using Eq. (2.45).

A tilt compensator with a pendulum mirror is mounted into the optical train of the optical system of the theodolite *Theo 002* (VEB Jenoptik, Jena). This compensator eliminates the effect of the error of the horizontal axis and of the tilt of the vertical axis completely automatically.

2.2.8.3 Eccentricity of the Centre of Graduation and of the Rotating Axis of the Horizontal Circle

The graduation centre of the horizontal graduated circle is the common intersection point of the graduation lines. It is identical with the point on the surface of the graduated circle which coincided with the vertical axis of the dividing machine when the graduation was applied. The graduated horizontal circle is very carefully centred when assembling the theodolite in the workshop, nevertheless a linear eccentricity $e_{2,1}$ of some hundredths mm or of some μm remains between the graduation centre *1* and the axis *2* of the central part of the instrument which contains the circle (*Fig. 2.85, a* and *b*). Due to defects of manufacture, however, the rotation axis *2* of the central part does not lie exactly in the vertical axis *3* but has a linear eccentricity $e_{3,2}$ with respect to it. The graduation centre *1* is eccentric with respect to the standing axis *3*; the amount and direction of eccentricity $e_{3,1} = e$ depends on the rotated position of the central part (horizontal circle). The eccentricity e, important from the point of view of direction measurement, is the resultant of the eccentricities $e_{2,1}$ and $e_{3,2}$, taken as vectors making an angle α which depends on the rotated position of the circle. Its amount (absolute value) is a maximum in that circle position where the

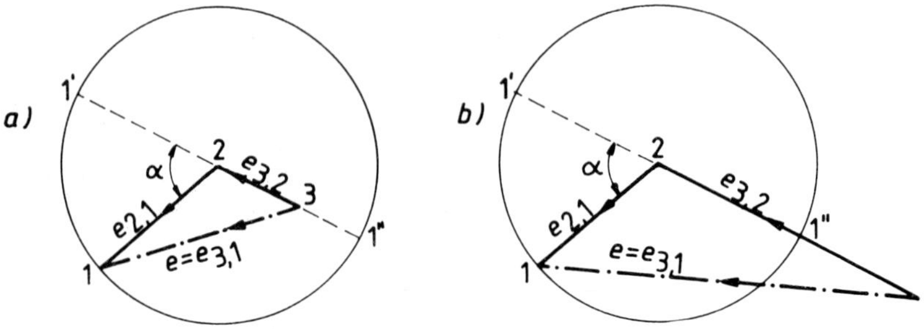

Figure 2.85. Eccentricity data of theodolites: $e_{2,1}$=eccentricity of the circle graduation centre *1* with respect to the rotation axis *2* of the central part; $e_{3,2}$=eccentricity of the axis *2* with respect to the vertical axis *3*; $e_{3,1}=e$=eccentricity of the graduation centre *1* with respect to the vertical axis *3*

eccentricity $e_{2,1}$ lies in the same direction as the eccentricity $e_{3,2}$ (direction 2, 1′ in *Fig. 2.85*, *a* or *b*), when:

$$|e_{\max}|=|e_{3,2}|+|e_{2,1}|\,.$$

The resultant e is a minimum in the circle position 180° opposite to this, where $e_{2,1}$ points in opposite direction to $e_{3,2}$, i.e. in the direction 2, 1″ shown in *Fig. 2.85*, *a* and *b* respectively, and

$$|e_{\min}|=|e_{3,2}|-|e_{2,1}|\,.$$

A) Effect of Eccentricity on Direction Measurements

(a) If a direction is measured either with a theodolite having only one reading index, or with a theodolite having two reading indices in diametrically opposite positions but only one of them is read, and if the measurement is carried out only in the first telescope position, then the effect ε of eccentricity e on the direction measurement depends on the position angle between the directions of the eccentricity e and the beam 3, I of index I *(Fig. 2.86, a)*. One obtains on index I the erroneous reading $l'_{\mathrm{I}}=l_{\mathrm{I}}-\varepsilon$, instead of the correct reading l_{I}. The error $\varepsilon=l_{\mathrm{I}}-l'_{\mathrm{I}}$ is, from the triangle 1, 3, I, and taking $\varepsilon\approx\varepsilon'$ and $r\approx r'$:

$$\varepsilon''=\frac{e}{r}\rho''\sin\omega\,. \tag{2.50}$$

The error is a maximum if $e=e_{\max}$ and at the same time $\omega=90°$ or 270°, i.e. if the eccentricity e is perpendicular to the index beam 3, I.

(b) For the study of the effect of graduation-centre eccentricity it is supposed that: (1) the theodolite has no collimation error, i.e. the astronomical line of sight is perpendicular to the horizontal axis; (2) the horizontal axis has no tilt; (3) the line of sight is not eccentric; finally, (4) the vertical axis is correctly levelled. Having observed a distant point in the first telescope position, the erroneous reading $l'_{\mathrm{I}}=(l_{\mathrm{I}}-\varepsilon)$ is obtained on the reading index I due to the eccentricity e of the centre of the graduated circle, instead of l_{I} *(Fig. 2.86, a)*. After transiting the telescope around its horizontal axis and rotating the top part of the instrument by 180° around the vertical axis —

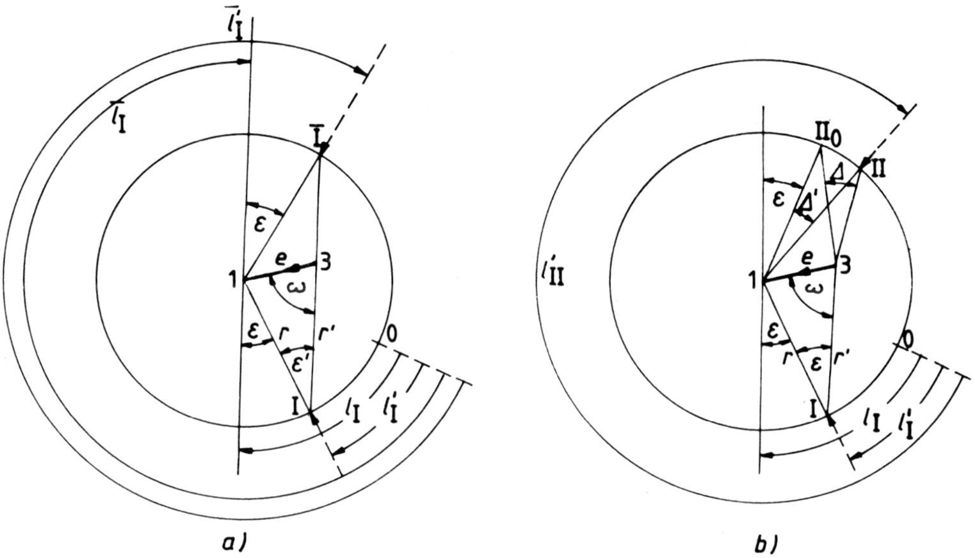

Figure 2.86. Effect of the eccentricity $e=e_{3,1}$ of the graduation centre l on direction measurements

i.e. in the second telescope position — the same distant point is observed. Now the erroneous reading $\bar{l}'_{\mathrm{I}}=\bar{l}_{\mathrm{I}}+\varepsilon=l_{\mathrm{I}}+180°+\varepsilon$ is obtained on the index I due to the eccentricity e, instead of the correct reading $\bar{l}_{\mathrm{I}}=l_{\mathrm{I}}+180°$ (index I is now in the position marked by I in *Fig. 2.86, a*). If the direction is measured in both telescope positions with a theodolite having only one reading index, or with a theodolite having two indices but only one of them is read, then the arithmetic mean of the erroneous readings l'_{I} and $\bar{l}'_{\mathrm{I}}$, changed by 180°, is:

$$\frac{l'_{\mathrm{I}}+\bar{l}'_{\mathrm{I}}\pm 180°}{2}=\frac{l_{\mathrm{I}}-\varepsilon+l_{\mathrm{I}}+\varepsilon}{2}=l_{\mathrm{I}}, \tag{2.51}$$

which is free of the effect of eccentricity e.

(c) In the case of a direction measurement in the first telescope position with readings on two diametrically opposite indices, *Fig. 2.86, b* shows that the erroneous reading $l'_{\mathrm{I}}=l_{\mathrm{I}}-\varepsilon$ is obtained on index I, instead of the correct reading l_{I}. The same figure also shows that the erroneous reading on index II is $l'_{\mathrm{II}}=l_{\mathrm{I}}\pm 180°+\varepsilon+\Delta$, where $\Delta\approx\Delta'$ is the deviation angle of the index II from the exact diametrically opposite position II_0 as a result of manufacturing defects. The arithmetic mean of the erroneous readings l'_{I}, and l'_{II} changed by 180°, is:

$$\frac{l'_{\mathrm{I}}+l'_{\mathrm{II}}\pm 180°}{2}=l_{\mathrm{I}}+\frac{\Delta}{2}, \tag{2.52}$$

which is free of the effect ε of the eccentricity e. In Eq. (2.52) half the error of index II appears, but this is a constant error independent of the direction itself which thus has no effect on the result of the direction measurement.

A direction measurement made in both telescope positions with readings on the two diametrically opposite indices is free of the effect of the eccentricity e.

In modern theodolites, where the two readings are projected into one image, the circle readings are always free of the eccentricity error independently of whether one or both telescope positions have been used.

B) Determination of Eccentricity Elements

In spite of the fact that the effect of the eccentricity of the centre of the graduated circle can be easily eliminated, it may be sometimes necessary to determine the eccentricity elements of a theodolite.

(a) The eccentricity elements of the centre of the graduated circle in theodolites with one reading index, which are even today very popular and frequently used, can be determined by the following method which yields at the same time the collimation error of the theodolite (Point 2.2.8.1). Three collimators are set up *(Fig. 2.87)* roughly horizontally at the height of the theodolite in directions differing by about 90° from each other (for a laboratory test). The collimators can be substituted by theodolites with telescopes focussed to infinity. The same test can be carried out in the field without collimators: instead of them three targets are put out in directions differing by 90° at about the same height as that of the theodolite, and the distance between target and theodolite should be at least 100 m. It should be mentioned here that if only collimation error is to be determined then only two collimators or targets are required, in directions 180° apart.

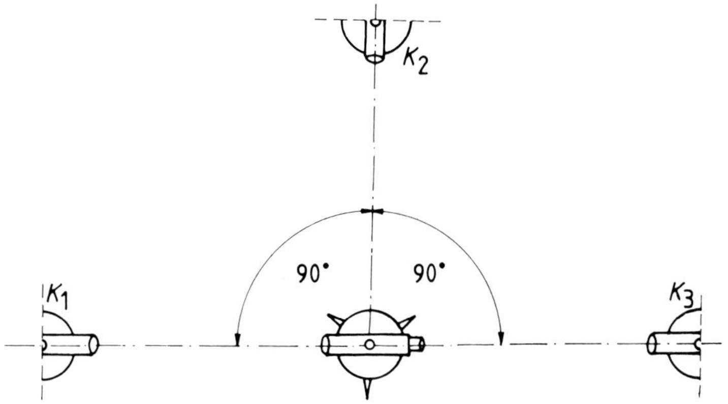

Figure 2.87. Three collimators determining the eccentricity elements and the collimation error of a theodolite with one reading index

Let the cross-hair centre K_1 of the left hand side collimator (or target) be observed with the theodolite telescope in the first position, by rotating the top part of the instrument around the vertical axis *3* *(Fig. 2.88, a)*. The position l_1 of the index *i* is read on the horizontal circle. The reading l_1 is erroneous by

$$\varepsilon_1'' = \frac{e}{r}\rho'' \sin\omega \tag{2.53}$$

due to the eccentricity of the centre *1* of the graduated circle with respect to the vertical axis *3*; the position angle of the eccentricity with respect to the beam 3, *i* is

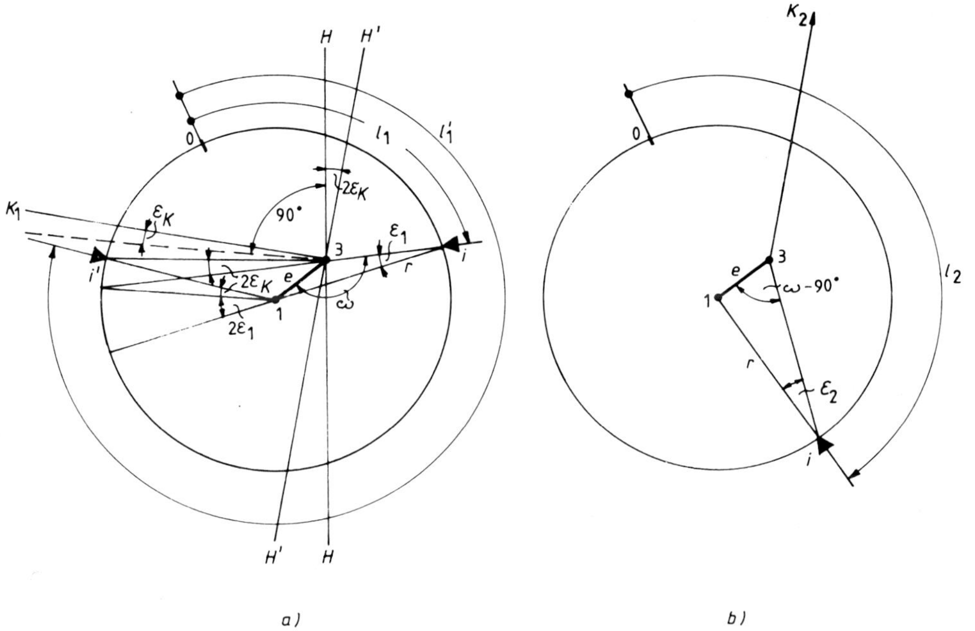

Figure 2.88. Determination of the eccentricity data of a theodolite with one reading index

denoted by ω. Now transit the telescope around the horizontal axis $H-H$ and rotate the top-part of the instrument around the vertical axis *3*. In this second telescope position the left hand side collimator K_1 (or target) is observed again. Now the horizontal axis is deflected by the double collimation error $2\varepsilon_k$ into the position $H'-H'$. Now read l_1' which is the new position i' of the index after rotation around the axis *3* by the angle $(180°+2\varepsilon_k)$. From the figure one can write:

$$l_1'-l_1=180°+2\varepsilon_1+2\varepsilon_k, \tag{2.54}$$

or by substituting Eq. (2.53):

$$l_1'-l_1=180°+\frac{2e}{r}\rho\sin\omega+2\varepsilon_k. \tag{2.55}$$

Now return to the first telescope position and observe the collimator K_2 (or the corresponding target) 90° to the right *(Fig. 2.88, b)*, reading l_2 on the circle. The position angle of the eccentricity is now $(\omega-90°)$; consequently the angular error due to the eccentricity e is:

$$\varepsilon_2=\frac{e}{r}\rho\sin(\omega-90°)=-\frac{e}{r}\rho\cos\omega. \tag{2.56}$$

Turning over into telescope position II the collimator K_2 or the right hand target is observed (this position is not shown in *Fig. 2.88*), giving the reading l_2'. Similarly to Eq. (2.54) one can write:

$$l_2'-l_2=180°+2\varepsilon_2+2\varepsilon_k,$$

or by substituting Eq. (2.56):

$$l'_2 - l_2 = 180^\circ - \frac{2e}{r}\rho\cos\omega + 2\varepsilon_k. \tag{2.57}$$

Now the same test is made to collimator K_3 (or target) at about 180° from collimator K_1 (or target). From the readings:

$$l'_3 - l_3 = 180^\circ + 2\varepsilon_3 + 2\varepsilon_k$$

or

$$\varepsilon_3 = \frac{e}{r}\rho\sin(\omega - 180^\circ) = -\frac{e}{r}\rho\sin\omega$$

and after substitution:

$$l'_3 - l_3 = 180^\circ - \frac{2e}{r}\rho\sin\omega + 2\varepsilon_k. \tag{2.58}$$

The sum of Eqs (2.55) and (2.58) is:

$$l'_1 + l'_3 - (l_1 + l_3) = 4\varepsilon_k,$$

from which twice the collimation error is:

$$2\varepsilon_k = \frac{l'_1 + l'_3 - (l_1 + l_3)}{2}. \tag{2.59}$$

From Eq. (2.55):

$$\frac{2e}{r}\rho\sin\omega = l'_1 - l_1 - 180^\circ - 2\varepsilon_k, \tag{2.60}$$

and from Eq. (2.57):

$$\frac{2e}{r}\rho\cos\omega = -(l'_2 - l_2 - 180^\circ - 2\varepsilon_k). \tag{2.61}$$

If Eq. (2.60) is divided by Eq. (2.61) the acute angle (ω) is:

$$(\omega) = \arctan\frac{l'_1 - l_1 - 180^\circ - 2\varepsilon_k}{-(l'_2 - l_2 - 180^\circ - 2\varepsilon_k)} = \arctan\frac{|(2.60)|}{|(2.61)|}. \tag{2.62}$$

From this one obtains for the position angle ω:

$$\omega = \begin{cases} (\omega) \text{ if both (2.60) and (2.61) are positive,} \\ 180^\circ - (\omega) \text{ if (2.60) is positive and (2.61) is negative,} \\ 180^\circ + (\omega) \text{ if both (2.60) and (2.61) are negative,} \\ 360^\circ - (\omega) \text{ if (2.60) is negative and (2.61) is positive.} \end{cases} \tag{2.63}$$

The eccentricity can be obtained from Eq. (2.55), as:

$$e = \frac{(l'_1 - l_1 - 180° - 2\varepsilon_k)r}{2\rho \sin \omega} \tag{2.64}$$

or from Eq. (2.57), as:

$$e = \frac{-(l'_2 - l_2 - 180° - 2\varepsilon_k)r}{2\rho \cos \omega}. \tag{2.65}$$

Eq. (2.64) is the more precise if $|\sin \omega| > |\cos \omega|$.

If the central part of the theodolite (its horizontal circle) can be rotated with respect to the top and bottom parts clamped together, the eccentricity e varies as a function of the position angle ω and therefore the linear eccentricity e calculated according to the above method refers only to the circle position used in the test. For the calculation of the maximum and minimum values of the eccentricity the former observations should also be made after rotating the circle through 180°. Therefore, the telescope must be rotated back into the first position and index i should be set to the original reading l_1, with the telescope again pointing to the collimator K_1. Let the central part of the instrument now be rotated by 180° with respect to the bottom and top parts clamped together. The previous observations should be carried out now from this position, i.e. $\bar{l}_1 = (l_1 + 180°)$. As a result the eccentricity $\bar{e}$ and the position angle $\bar{\omega}$ are obtained. From e and $\bar{e}$, as well as from the difference $\Delta\omega = \bar{\omega} - \omega$, the eccentricity of the centre of the graduated circle *1* with respect to the rotation axis *2* can be calculated from *Fig. 2.89, a* or *b* as:

$$e_{2,1} = \frac{1}{2}\sqrt{e^2 - \bar{e}^2 - 2e\bar{e}\cos \Delta\omega}, \tag{2.66}$$

and the eccentricity of the rotation axis *2* with respect to the standing axis *3* as:

$$e_{3,2} = \sqrt{e^2 + e_{2,1}^2 - 2ee_{2,1}\cos \beta}, \tag{2.67}$$

where

$$\cos \beta = \frac{e^2 + 4e_{2,1}^2 - \bar{e}^2}{4ee_{2,1}}. \tag{2.68}$$

The eccentricities $e_{2,1}$ and $e_{3,2}$ can be obtained with sufficient accuracy by measuring them in *Fig. 2.89, a* or *b*, constructed from the data e, $\bar{e}$ and $\Delta\omega$. The maximum and minimum eccentricities are:

$$e_{max} = e_{3,2} + e_{2,1}, \tag{2.69}$$

$$e_{min} = e_{3,2} - e_{2,1}. \tag{2.70}$$

It may be noted that more accurate values (practically free of the effect of graduation errors) of the eccentricities $e_{2,1}$ and $e_{3,2}$ can be obtained if the above tests are carried out not only by starting from $l_{1,1}$, but also from starting positions uniformly distributed around the circle, as e.g. $(l_1 + j30°)$ $(j = 0, 1, \ldots, n = 11)$, in which case the most reliable values of the eccentricities are obtained from the mean values of $n + 1 = 12$ of $e_{2,1,j}$ and $e_{3,2,j}$, respectively.

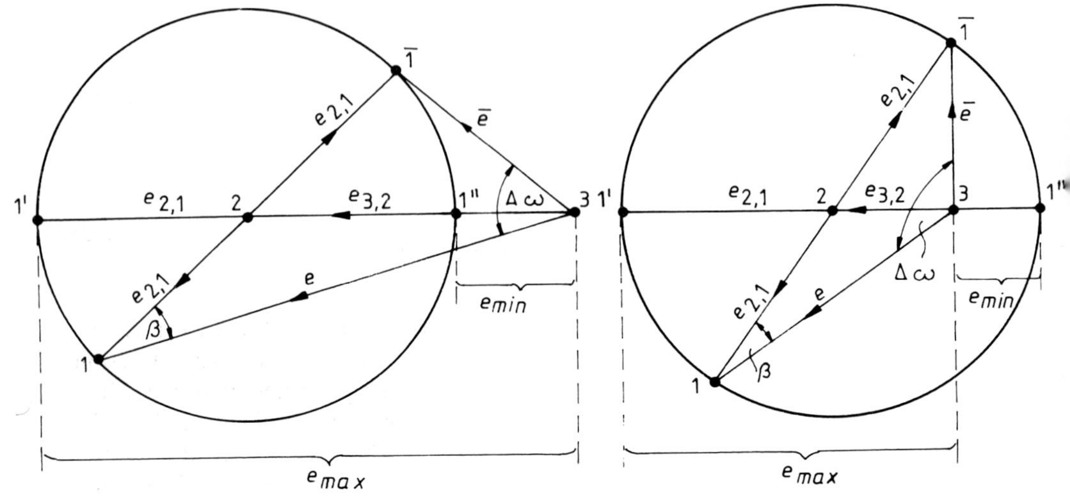

Figure 2.89. Eccentricity data of a theodolite with one reading index

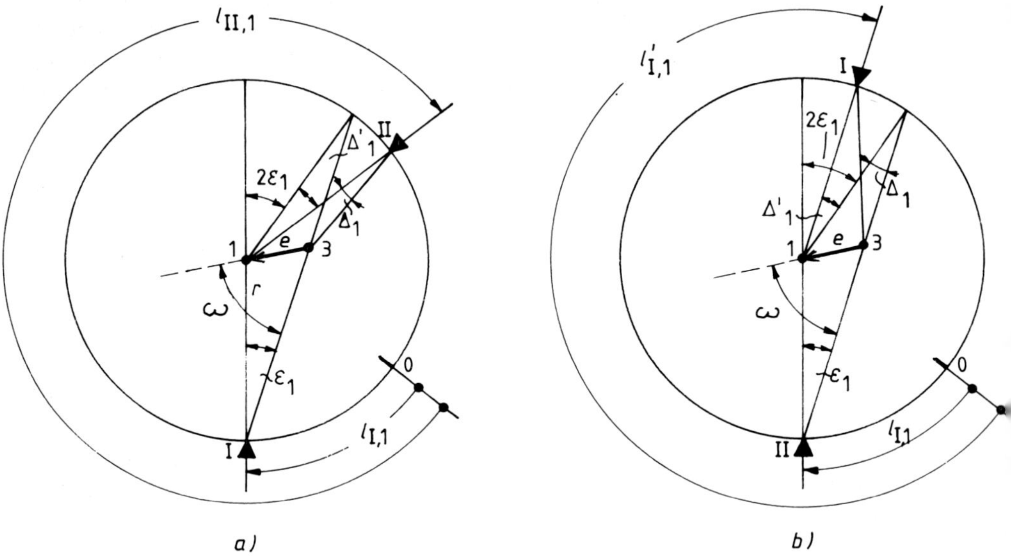

Figure 2.90. Determination of the eccentricity elements of a theodolite with two diametral reading indices

(b) The eccentricity elements of the above so-called simple theodolites, with two diametrically opposite r reading indices, can be determined most easily as follows. The positions $l_{I,1}$ of index I, and $l_{II,1}$ of index II are read on the horizontal circle (*Fig. 2.90, a*). Now, by rotating the top part of the instrument, index II is set to the reading $l_{I,1}$ (*Fig. 2.90, b*), and $l'_{I,1}$ is read on index I. From *Fig. 2.90, a* one obtains

$$l_{II,1} - l_{I,1} = 180° + 2\varepsilon_1 + \Delta_1,$$

noting that $\Delta'_1 \approx \Delta_1$, and from *Fig. 2.90, b*:

$$l'_{\mathrm{I},1} - l_{\mathrm{I},1} = 180^\circ + 2\varepsilon_1 - \Delta_1,$$

in which ε_1 is the error due to the eccentricity, and Δ_1 the angle error of index II. From them one obtains by subtraction:

$$l_{\mathrm{II},1} - l'_{\mathrm{I},1} = 2\Delta_1,$$

and by addition:

$$l_{\mathrm{II},1} + l'_{\mathrm{I},1} - 2l_{\mathrm{I},1} = 4\varepsilon_1 + 360^\circ,$$

from which

$$\Delta_1 = \frac{l_{\mathrm{II},1} - l'_{\mathrm{I},1}}{2}, \tag{2.71}$$

and

$$\varepsilon_1 = \frac{e}{r}\rho \sin\omega = \frac{l_{\mathrm{II},1} + l'_{\mathrm{I},1}}{4} - \frac{l_{\mathrm{I},1}}{2} - 90^\circ. \tag{2.71a}$$

By rotating the top part of the instrument, index I is set to the reading $l_{\mathrm{I},2} = l_{\mathrm{I},1} + 90^\circ$ and the position $l_{\mathrm{II},2}$ of index II is read. Next index II is set to reading $l_{\mathrm{I},2}$, by rotating the top part, and the position $l'_{\mathrm{I},2}$ of index I is read. Now one obtains analogously to Eq. (2.71):

$$\Delta_2 = \frac{l_{\mathrm{II},2} - l'_{\mathrm{I},2}}{2}, \tag{2.71b}$$

and analogously to Eq. (2.71a), but substituting ω by $(\omega - 90^\circ)$:

$$\varepsilon_2 = \frac{e}{r}\rho \sin(\omega - 90^0) = -\frac{e}{r}\rho\cos\omega =$$

$$= \frac{l_{\mathrm{II},2} + l'_{\mathrm{I},2}}{4} - \frac{l_{\mathrm{I},2}}{2}. \tag{2.71c}$$

From Eqs (2.71a) and (2.71c) the acute angle (ω) is:

$$(\omega) = \arctan \frac{|l_{\mathrm{II},1} + l'_{\mathrm{I},1} - 2l_{\mathrm{I},1}|}{|-(l_{\mathrm{II},2} + l'_{\mathrm{I},2} - 2l_{\mathrm{I},2})|} =$$

$$= \arctan \frac{|(2.71\mathrm{a})|}{|(2.71\mathrm{c})|}. \tag{2.72}$$

From Eq. (2.72) the position angle ω is:

$$\omega = \begin{cases} (\omega) \text{ if both (2.71a) and (2.71c) are positive,} \\ 180^\circ - (\omega) \text{ if (2.71a) is negative and (2.71c) is positive,} \\ 180^\circ + (\omega) \text{ if both (2.71a) and (2.71c) are negative,} \\ 360^\circ - (\omega) \text{ if (2.71a) is positive and (2.71c) is negative.} \end{cases} \tag{2.73}$$

Knowing ω, the eccentricity is obtained from Eq. (2.71a) in the form:

$$e = \frac{l_{\mathrm{II},1} + l'_{\mathrm{I},1} - 2l_{\mathrm{I},1}}{4\rho} \frac{r}{\sin\omega}, \tag{2.74}$$

or from Eq. (2.71c) in the form:

$$e=-\frac{l_{\mathrm{II},2}+l'_{\mathrm{I},1}-2l_{\mathrm{I},2}}{4\rho}\,\frac{r}{\cos\omega}. \tag{2.75}$$

Eq. (2.74) is the more accurate if $|\sin\omega|>|\cos\omega|$. If index I is set on reading $l_{\mathrm{I},1}$, the direction of e is deflected from the direction of the index beam 3, I by the angle ω.

The eccentricity elements ω determined by Eq. (2.73), and $e=e_{3,1}$ determined by Eqs (2.74) and (2.75) respectively, are valid for the circle position used in the test. In order to determine the constant eccentricity data $e_{3,2}$ and $e_{2,1}$, the previous observations described in (b) should be repeated with the circle rotated through 180°. For this purpose index I is set to the reading $\bar{l}_{\mathrm{I},1}=l_{\mathrm{I},1}\pm180°$. The observations described should be repeated, starting from this position. As result $\bar{\Delta}_1$, $\bar{\Delta}_2$, $\bar{\omega}$ and $\bar{e}$ are obtained. By means of the data e, $\bar{e}$ and $\Delta\omega=\bar{\omega}-\omega$, the constant eccentricities $e_{2,1}$ and $e_{3,2}$ as well as the maximum and minimum total eccentricities, i.e. $e_{\max}$ and $e_{\min}$, can be determined according to the test discussed in Point **B**(a). A more exact value of the angle error of index II is:

$$\Delta=\frac{\Delta_1+\Delta_2+\bar{\Delta}_1+\bar{\Delta}_2}{4}. \tag{2.76}$$

The remarks at the end of Point **B**(a) — after Eq. (2.70) — are also valid here.

(c) The determination of the eccentricity elements of modern theodolites with two readings projected into one image are not dealt with here, because the direction measurement is not influenced by the eccentricity in such instruments, whatever measuring method is used.

2.2.8.4 Test of the Graduation of Horizontal Circles

The error $\Delta\varphi_i$ of the i'th graduation line, $\varphi_1, \ldots, \varphi_i, \ldots, \varphi_n$, of the horizontal circle is the angular difference from its ideal position. The ideal graduation points are the fictitious graduation lines which differ from their neighbour by equal graduation intervals, the sum of whose errors is:

$$[\Delta\varphi_i]_{i=1}^{n}=0. \tag{2.77}$$

It follows that even the graduation line O is not error-free. Theoretically it is possible to determine the actual error $\Delta\varphi_i$ of each graduation line, e.g. in such a way that all intervals are measured with an optical micrometer and the results, being erroneous due to run, are corrected by equal corrections so that the sum of the corrected intervals is 360°. The difference between nominal intervals and corrected intervals yields the graduation interval errors, with the effects of eccentricity and observation error included. This method, however, is not used because

1. it is extremely lengthy,
2. it is not sufficiently accurate, and
3. the so-called diametral error is more important in practice. That is to say, circle readings on theodolites are mostly obtained as the mean of two diametrically opposite readings, therefore one usually speaks of a diametral error ΔD_φ being free of the eccentricity error of the graduated circle, which is the difference between the mean

of the two diametrically opposite readings and the error-free value. For this the relation

$$[\Delta D_{\varphi_i}]_{i=1}^{n/2} = 0 \tag{2.78}$$

is valid.

On the horizontal circle of the theodolite, fewer diameter errors are investigated than there are graduation lines within half the graduated circle. The largest part of the different methods of circle graduation testing consists of measurements of a defined test angle (or angles) repeated several times, and starting from different initial graduation lines, uniformly distributed. From the results obtained the trends of the systematic diameter errors are determined and some data characteristics for graduation accuracy are calculated.

The test angle can be set up by: (a) using two marks or collimators, e.g. with the circle graduation testing device *PRUTE* (Breithaupt, Kassel) shown in *Fig. 2.91, a,* which consists of a base plate which can be levelled, and of two collimators on it which are adjustable to a test angle between 30° and 50°; (b) with a refracting prism or mirror attached in front of the telescope objective of the theodolite, the same target or collimator being observed at first directly and then through the prism or mirror; (c) by moving the theodolite telescope between two stops.

Methods (a) and (b) are of equal accuracy while (c) is faster but less accurate. Method (a) is usually used.

The *Wild method* is a simplification of the so-called *rosette method* developed by *Bruns*. The test consists of the measurement of an angle of $\beta = 36°$ ($\beta = 40^g$ in the case of circles with 400 graduations) in four rosettes, in five circle positions for each rosette, forward in the first and reversed in the second telescope position. The starting positions of the different rosettes are: 0°, 9°, 18°, 27° (or 0^g, 10^g, 20^g, 30^g in case of 400 graduations). The different circle positions within a rosette differ by 36° (or 40^g). Thus 40 values are obtained for the angle measured: 10 from each rosette, 5 in the first and 5 in the second telescope position. Having set the telescope to the starting position 0^0 (0^g) and pointed to the left hand target in the first telescope position, the precise diametral value is read on the horizontal circle; then, by rotating the telescope around its vertical axis, the right hand target is observed and the diametral value read. The difference between the two readings is the value α_1 obtained for the test angle β. Next the top part of the instrument is clamped to its central part (i.e. to the horizontal circle), the clamp between the central and bottom parts is loosened, the telescope is rotated back together with the central part, and the left hand target is observed again and a reading made; after again clamping the central part and loosening the top part, the right hand target is observed and a reading made: the difference of the latter two readings gives another value α_2 for the test angle β, differing by a small amount from the first. These measurements are continued along half the horizontal circle and then backwards in the second telescope position. The following relationships exist between the true value α of the angle β and the values α_i obtained in the two telescope positions, e.g. in rosette 1:

between 0° and 36°: $\alpha = \alpha_1 + u_{36} - u_0 + \lambda_1$,

between 36° and 72°: $\alpha = \alpha_2 + u_{72} - u_{36} + \lambda_2$,

....

between 144° and 180°: $\alpha = \alpha_5 + u_0 - u_{144} + \lambda_5$.

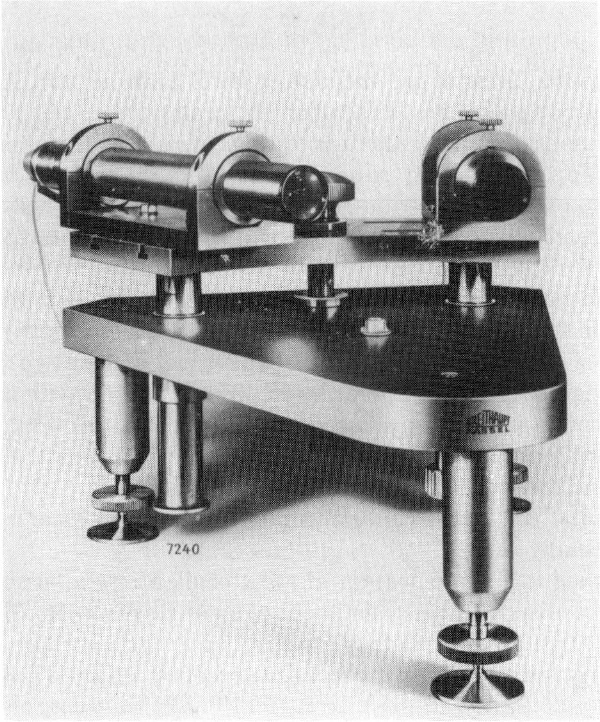

a)

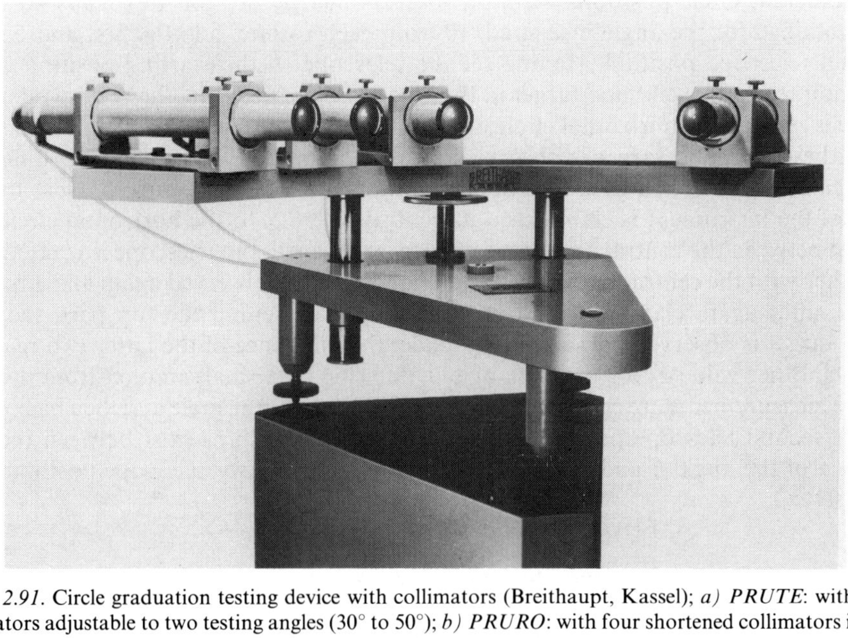

b)

Figure 2.91. Circle graduation testing device with collimators (Breithaupt, Kassel); *a) PRUTE*: with two collimators adjustable to two testing angles (30° to 50°); *b) PRURO*: with four shortened collimators in the four directions (0^g, 15^g, 37^g, 90^g)

Here the α_i $(i=1, 2, \ldots, 5)$ are the results measured in two telescope positions in the i-th circle position, λ_i are the random observation errors of the measured results α_i, u_i are the errors of the corresponding graduation lines. The most reliable value of α is:

$$\alpha = \frac{[\alpha_i]_{i=1}^{5}}{5}, \tag{2.79}$$

in which the sum of the graduation errors u and the random observation errors λ_i are eliminated.

The corrections to the measured results α_i are:

$$v_1 = \alpha - \alpha_1 = u_{36} - u_0,$$

$$v_2 = \alpha - \alpha_2 = u_{72} - u_{36},$$

....

$$v_5 = \alpha - \alpha_5 = u_0 - u_{144}, \tag{2.80}$$

from which the graduation errors can be obtained (by taking $u_0 = 0$) as:

$$u_0 = 0,$$

$$u_{36} = v_1,$$

$$u_{72} = v_1 + v_2,$$

$$u_{108} = v_1 + v_2 + v_3, \tag{2.81}$$

....

Referring them to the mean $u_k = \dfrac{[u_i]}{5}$, one obtains:

$$(u_i) = u_i - u_k. \tag{2.82}$$

Measurements and calculations are carried out for all four rosettes. Plotting the values (u_i) as ordinates against $\varphi = 0°, 9°, 18°, \ldots, 171°$ as abscissae and fitting a

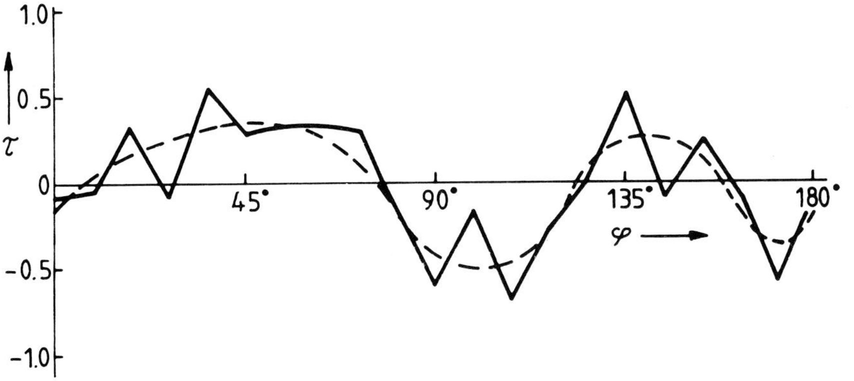

Figure 2.92. Graph of systematic diametral errors

continuous line to the broken line of their end points the graph of the systematic diametrical errors is obtained *(Fig. 2.92)*. The mean square diametral error is:

$$\tau = \pm\sqrt{\frac{[(u_i)(u_i)]}{20}}. \tag{2.83}$$

This, however, can also be determined by the formula

$$\tau = \pm\sqrt{\frac{m_1^2 - m_2^2}{2}}, \tag{2.84}$$

where

$$m_1 = \sqrt{\frac{vv}{n-1}} \tag{2.85}$$

is the angular mean square error calculated from the differences $v_i = \alpha_i - \alpha$ of all n α_i values from their mean values, and

$$m_2 = \sqrt{\frac{1}{n}\left[\left(\frac{\Delta}{2}\right)^2\right]} \tag{2.86}$$

is the angular mean square error computed from the differences Δ of the results obtained at the same circle positions in two telescope positions.

In the *Heuvelink method*, now most usually used, the test angle $\beta = 45°$ (50^g) is measured at $n = 9$ (and 10, respectively) circle positions uniformly distributed (i.e. at successive intervals of 20° or 20^g) forward and back in s series (e.g. $s = 4$). The mean α_i is calculated from the results measured in the different circle positions i, being burdened by observation error and diametral error. The mean of all α_i values is the most probable value of the test angle:

$$\alpha = \frac{[\alpha_i]_{i=1}^{n}}{n}. \tag{2.87}$$

From the distances $x_i = \alpha_i - \alpha$ the square of the total standard deviation is:

$$M^2 = \frac{[x_i x_i]}{n-s} \tag{2.88}$$

which also contains observation error and circle graduation error. The square of the total standard deviation of a direction is half of this, i.e.: $M^2/2$. The square of the standard deviation of one single direction measurement should be calculated as:

$$m^2 = \frac{1}{4}\left[\frac{[\Delta\Delta]}{n_0} - \gamma^2\right], \tag{2.89}$$

where Δ is the difference between one forward and one backward measurement of the test angle β, and

$$\gamma = \frac{[\Delta]}{n_0} \tag{2.90}$$

is the constant error of direction measurement and n_0 the number of the differences Δ.

The mean total diametral graduation error τ is obtained from the standard deviations of direction as [Eq. (2.88)]:

$$\tau=\sqrt{\frac{M^2}{2}-\frac{m^2}{4}}. \tag{2.91}$$

In this method the systematic diametral graduation error is expressed by the trigonometric series

$$\Delta D_\varphi=a\sin(2\varphi+A)+b\sin(4\varphi+B)+c\sin(6\varphi+C)+\ \ldots \tag{2.92}$$

where the odd terms are missing. The constants A, B, C can be calculated from the measured results α_i by means of the formula

$$\tan(l\beta+K)=\frac{-[x_i\sin 2l\varphi]}{[x_i\cos 2l\varphi]}, \tag{2.93}$$

$$(\mathrm{K}=A, B, C \quad \text{if} \quad l=1, 2, 3),$$

and the constants a, b, c, by means of:

$$k=\frac{-[x_i\sin 2l\varphi]}{n\sin l\beta\sin(l\beta+K)} \tag{2.94}$$

$$(k=a, b, c; \quad l=1, 2, 3; \quad K=A, B, C; \quad \beta=45°).$$

It is advisable to plot the diametral graduation errors determined from Eq. (2.92) as ordinates against the values φ from 0° to 180° as abscissae.

If the first one, two or three terms of the series (2.92) are eliminated from the total mean square error of the diametral error given by Eq. (2.91), then the remaining random diametral errors are obtained as:

$$\tau'=\sqrt{\frac{M'^2}{2}-\frac{m^2}{4}}, \qquad \tau''=\sqrt{\frac{M''^2}{2}-\frac{m^2}{4}},$$

$$\tau'''=\sqrt{\frac{M'''^2}{2}-\frac{m^2}{4}}, \tag{2.95}$$

in which:

$$M'=\sqrt{\frac{[x'x']}{n-s-2}}, \qquad M''=\sqrt{\frac{[x''x'']}{n-s-4}},$$

$$M'''=\sqrt{\frac{[x'''x''']}{n-s-6}}, \tag{2.96}$$

where

$$[x'x']=[x_ix_i]\ -\frac{2}{n}[x_i\sin 2\varphi]^2-\frac{2}{n}[x_i\cos 2\varphi]^2,$$

$$[x''x'']=[x'x']\ -\frac{2}{n}[x_i\sin 4\varphi]^2-\frac{2}{n}[x_i\cos 4\varphi]^2, \tag{2.97}$$

$$[x'''x''']=[x''x'']-\frac{2}{n}[x_i\sin 6\varphi]^2-\frac{2}{n}[x_i\cos 6\varphi]^2.$$

The values τ, τ', τ'', τ''' calculated from Eqs (2.91), (2.95) characterize the graduation accuracy of the tested circle (of the theodolite).

In connection with Heuvelink's original method it should be noted that, when using only one test angle β in any circle position, only relative graduation errors are measured between two graduation lines separated by β. Thus the method does not reveal errors with period β. This is shown by the fact that the fourth term of the series (2.92) — which should yield the systematic graduation error of period $\beta = 45°$ — cannot be determined, because $\sin 4\beta = 0$ should appear in the denominator of Eq. (2.94) when used to obtain the amplitude d of the fourth term. Moreover, the constants A, B, C, and a, b, c as well as the diametral error ΔD_{φ}, also τ, τ', τ'', τ''', depend on the test angle β. In the case of equal test angles β the reliability of the different constants is also different. Therefore the characteristics of graduation accuracy determined by one test angle β are not unambiguous.

The original Heuvelink method was later improved in several ways. One of the most important improvements is the Roelofs method which uses six test angles ($\beta = 15^g$, 22^g, 37^g, 53^g, 75^g and 90^g) and includes all 200 graduation lines of the half-circle. Its system and method of calculation is well suited to determine the first seven terms of the trigonometric series [Eq. (2.92)]. The mean square errors of the seven amplitudes are nearly equal. The practical application of Roelofs method is considerably facilitated by the circle graduation testing device *PRURO* of Breithaupt *(Fig. 2.91, b)*; this enables the 6 test angles, necessary for the use of Roelofs method, to be obtained from any two of the four collimators, which are fixed in directions which make successive angles of 15°, 22°, and 53°.

Graduation accuracy characteristics are the essential qualifying parameters of a theodolite.

The effect of the long period systematic graduation errors, discussed above, can be eliminated in practice by proper measuring methods, namely by measuring the required angle in several positions uniformly distributed around the horizontal circle.

An error of the toroidal worm-screw of the graduating machine causes systematic graduation error of short period (Point 2.2.4.2). If n is the tooth number of the geared master circle, then the short period of graduation error is usually the lowest common multiple of $360 \times 60/n$ and the graduation unit i. Thus, e.g. if $n = 1440$, then for $i = 5$ min of arc the period is 15 min of arc, and for $i = 4$ min of arc it is 60 min of arc. In order to determine these short period graduation errors, the intervals within a graduation interval, such as from 1° to 3°, are measured with an optical micrometer at different points uniformly distributed around the circle. The error of run is determined by repeated measurements (Point 2.2.8.5) and the mean value of the same intervals, measured on different parts of the circle, is corrected by it. The differences between the corrected mean intervals and their nominal values yield the short period graduation errors, which can also be represented in the form of a graph by plotting them against e.g. the graduation positions over 1° as abscissae. The effect of the short period graduation errors can be eliminated by commencing the separate measurements of the required angle at graduation lines uniformly distributed within the short period.

2.2.8.5 Errors of Reading Devices

Reading devices with a scale microscope are adjusted by the manufacturers so that the difference between the images of two adjacent lines of the circle graduation, generated in the image plane of the reading microscope, is equal to the width of the microscope scale. Some small displacements of the different elements in the optical reading system, due to excessive strain in the instrument primarily during transportation, may cause a deviation from the adjusted position by 1 to 2 per mille which is the so-called error of run. In the case of the scale microscope it is generally unnecessary to take this error into account because it is less than the reading error. In the case of a microscope with an optical micrometer, often consisting of more than 100 scale units, the error of run must be taken into account if it is greater than half the scale unit of the micrometer. In order to determine the run the scale line O of the scale of the coincidence-adjusting optical microscope is set to the index of the graduation, and the central graduation line D of the upper half image in the reading microscope is made to coincide with the graduation line $(D+180°)$ of the lower half image, using the horizontal fine adjustment screw of the theodolite *(Fig. 2.93, a)*. Next the graduation line $(D+180°)$ is brought into exact coincidence with the graduation line $(D-1)$ *(Fig. 2.93, b)* by means of the screw of the optical micrometer, and the micrometer scale is read (e_1) with an accuracy of 1/10 scale unit. In the upper half image the run is:

$$r_1 = e - e_1 \tag{2.98}$$

where e is the width of the scale of the optical micrometer in scale units. Next the graduation line $(D+180° - 1)$ is made to coincide with the graduation line D of the

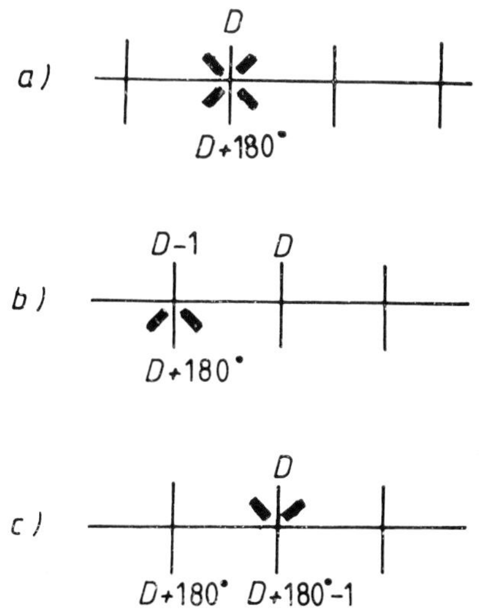

Figure 2.93. Determination of the run of an optical coincidence micrometer

upper half image *(Fig. 2.93, c)*, using the micrometer screw, and the position e_2 of the micrometer index is read. The run of the lower half image is:

$$r_2 = e - e_2, \tag{2.99}$$

which can slightly differ from the run r_1 of the upper half image, because the optical trains from the two diametrically opposite reading positions are not equal. *Figures 2.93, b* and *c* differ only in that two different graduation lines are brought into coincidence. All three coincidences should be repeated several times. It is advisable to repeat the test on different, uniformly distributed, parts of the horizontal circle and in the meantime also to vary the subgraduation lines with a uniform distribution within two adjacent graduation lines of one degree, in order to eliminate the effect of short period circle graduation errors. It is also possible, however, that several adjacent graduation lines (e.g. 10) are tested in one circle position. The correct value of the run is the mean of the average runs obtained for the upper and the lower half image:

$$r = \frac{r_1 + r_2}{2}. \tag{2.100}$$

The necessary correction to the circle reading due to the systematic error of run is

$$\Delta l = \frac{r}{e} l, \tag{2.101}$$

where l is the reading on the micrometer.

The random error of the reading device of a scale microscope is the consequence of the estimation of a graduation line, projected in the image plane of the microscope, within a scale unit. The standard deviation of an estimation is about 1/20 of the graduation unit if the scale unit of the microscope graduation is seen in the microscope as about 1 mm (i.e. it is seen to subtend about 13 min of arc) at the clear visibility distance of $\delta = 250$ mm and, in addition, the widths of the microscope scale lines are seen in the microscope to subtend about one tenth of this (about 80″). For instance, in order to read a circle graduated in degrees and with a graduation diameter of d mm, on a microscope scale of width 60 scale units, with a mean square error of $\pm 1/20$ mm, a microscope magnification Γ_{micr} must be used because in this case one graduation interval of the graduated circle is seen in the microscope as 60 mm at the clear visibility distance δ: in other words $\Gamma_{\mathrm{micr}} d\pi/360 = 60$ mm. From this one finds the ideal microscope magnification to be:

$$\Gamma_{\mathrm{micr}} = \frac{21\,600}{d\pi}, \tag{2.102}$$

the ideal thickness of the circle graduation line to be:

$$w_c = \frac{1}{10\Gamma_{\mathrm{micr}}} \tag{2.103}$$

and the thickness of the microscope scale line to be:

$$w_m = \frac{1}{10\Gamma_{\mathrm{ep}}}, \tag{2.104}$$

where Γ_{ep} is the magnification of the microscope eyepiece. As an example, with $d=80$ mm, $\Gamma_{micr}=80$, $w_e=1.25$ μm, and the reading limit of the scale microscope is 3 secs of arc.

The mean square error of a reading on a microscope with an optical micrometer depends on the accuracy of the coincidence of the two images of the graduated circle projected beside each other. Two lines with sharp edges can be made to coincide by an unaided eye with a sight-angle accuracy of about ± 10 secs of arc. In order to bring the diametral graduation lines of a circle of graduation diameter d, with an accuracy of ± 1 sec of arc, into the field of view of the microscope by means of an optical micrometer, the magnification Γ_{micr} of the microscope should be chosen so that

$$\Gamma_{micr}\frac{d\pi\rho''}{360\times 60\times 60\times 250}=10''$$

from which

$$\Gamma_{micr}=\frac{15\,700}{d\pi} \qquad (2.105)$$

(e.g. with $d=100$ mm, $\Gamma_{micr}=50$).

A 0.1 mm thick line can be brought into a symmetrical position between two other lines about 1 mm apart by an unaided eye with an accuracy of 10″ in sight angle at the clear visibility distance. Hence Eq. (2.105) is again valid to give the magnification of a reading microscope, with a symmetry-adjusting optical micrometer, necessary for the required accuracy of 1 sec of arc. It must be mentioned that the microscope magnification given by Eq. (2.105) assures a reading accuracy equal to the mean square error of adjustment of coincidence or symmetry, if the reading on the scale of the optical micrometer has a greater accuracy.

The mean square error of a coincidence or symmetry-adjusting reading microscope of a theodolite is determined by coinciding (adjusting to symmetry) n-times, by repeated displacement and readjustment with the micrometer screw and making the readings l_i on the micrometer scale. The mean square error of the reading is

$$m_l=\pm\sqrt{\frac{[\Delta l_i^2]_{i=1}^n}{n-1}}, \qquad (2.106)$$

where

$$\Delta l_i=l_i-\frac{[l_i]_{i=1}^n}{n}. \qquad (2.107)$$

2.2.8.6 Test of Theodolite Telescopes

The mean square error of pointing of a telescope can be determined theoretically on the basis of the definition given in Point 2.2.5.2 in the following way. A target set up at a distance of 100 to 150 m is observed several times (e.g. n-times) and the horizontal circle is read (l_i), the horizontal fine adjusting screw being moved after a pointing and re-adjusted for a new pointing. It should be emphasized that only the horizontal fine adjusting screw should be touched during this series of measurement and that the instrument itself should not move. The approximate mean square error of the direction can be obtained from the readings l_i, as:

$$(m_{\text{dir}}) = \pm \sqrt{\frac{[\Delta l_i^2]_{i=1}^n}{n-1}}, \tag{2.108}$$

where

$$\Delta l_i = l_i - \frac{[l_i]_{i=1}^n}{n}. \tag{2.109}$$

The approximate mean square error of the direction determined in this way still contains the mean square error of the reading on the reading device (Point 2.2.8.5). Therefore the value of Eq. (2.108) must be corrected in the case of a reading microscope with an optical micrometer by the mean square error of the reading as determined by Eq. (2.106). Having eliminated Eq. (2.106) from Eq. (2.108) the mean square error of the pointing is:

$$m_{\text{dir}} = \pm \sqrt{(m_{\text{dir}})^2 - m_l^2}. \tag{2.110}$$

The mean square error of pointing can be determined more accurately by means of the following method of testing which requires, however, a special device, namely a target which can be moved by means of a fine adjusting screw against a precisely manufactured mm scale on a metal rod or cylinder rigidly fixed (e.g. to a wall). The position of the target can be read on the micrometer scale with an accuracy of 1/100 mm. The test is made as follows: The target is fixed rigidly and precisely observed with the theodolite from a distance of about 100 to 120 m. During the test neither the theodolite nor its screws must be touched. The operator controlling the target moves it by a small amount, then the observer at the theodolite looks into the telescope and instructs the operator by waving (or by telephone or radio) to readjust the target exactly into its original position by means of the fine adjustment screw. The target operator reads and records the displacement l_i expressed in 10^{-2} mm units. The operation is repeated n-times and the mean square error solely of pointing is:

$$m_{\text{dir}} = \pm \frac{\rho''}{s} \sqrt{\frac{[\Delta l_i^2]_{i=1}^n}{n-1}} 10^{-5}, \tag{2.111}$$

where

$$\Delta l_i = l_i - \frac{[l_i]_{i=1}^n}{n}, \tag{2.112}$$

and s is the distance of the target in m.

The wobble, distance-dependent deflection and eccentricity of the line of sight of the telescope are discussed in Point 2.2.8.1.

2.2.8.7 Final Adjustment of Theodolites

After completion of the manufacture and assembly of a theodolite, or after repair, a check is made with a final adjustment collimator device *(Fig. 2.94)*. This test device consists of three collimators with reticule and cross-hair which are placed above each other at different elevation angles. The elevation angle of the central collimator can be changed within a small range with a slow motion screw, and it can be adjusted

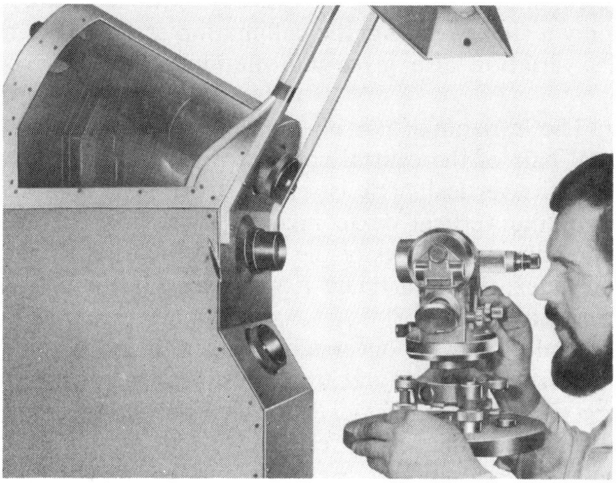

Figure 2.94. Final adjustment collimator *KN-24* (MOM, Budapest)

into the horizontal with a striding level. Its cross-hair is also supplied with stadia hairs with stadia constant 100. It is required that the elevation angle of the central collimator should be $\alpha=0$, that the parallactic angle determined by the stadia hairs should be $0°\,34'\,22.6''$, and that the cross-hairs should be vertical and horizontal respectively. The elevation angles of the other two collimators are $\alpha=+30°$ and $\alpha=-30°$, respectively. There is a graduation on both the left and right hand sides of the vertical hair of the collimator whose elevation angle is $-30°$. The theodolite to be tested is set up and fixed on a table in front of the collimators and levelled. The table can be moved vertically and it is adjusted with the theodolite on it to make the horizontal axis of the theodolite the same height as the axis of the central collimator.

By means of the final adjustment collimator the following can be tested:

a) The tilt of the horizontal axis. The telescope is pointed in its first position firstly at the centre of the cross-hairs of the upper collimator, then at the horizontal hair of the lower collimator, and the position of the vertical hair of the theodolite is read on the graduation of the collimator. The same operation is repeated in the second telescope position. The horizontal axis is adjusted so that the same reading is obtained on the scale of the lower collimator in both telescope positions. This test can also be carried out with a slightly oblique vertical axis and cross-hair, and the collimation error of the theodolite does not affect the test because the absolute values of the elevation and slope angles are equal.

b) Tilt of the cross-hair. The cross-hair of the theodolite is brought into coincidence with the cross-hair of the central collimator by turning its adjustment screws.

c) The stadia constant 100. By moving the objective lens of the theodolite along the telescope axis, the stadia hairs are made to cover the stadia hairs of the horizontal collimator.

d) Collimation error. The vertical hair of the central collimator is observed in both telescope positions. The deviation of the difference between the two readings from

180° is twice the collimation error. It should be mentioned that, in the case of a theodolite with only a single reading, the collimation error is determined in several positions, and its variation also gives information about the eccentricity of the horizontal circle.

e) Index error. The horizontal hair of the telescope is brought into coincidence with the horizontal hair of the central collimator (the telescope being horizontal), then the reading on the vertical circle is set to 90° with the micrometer set at zero. Next the bubble of the vertical circle index level is brought to centre (or into coincidence) by the adjusting screws of the level. In the case of theodolites with an automatic vertical index, the adjustment is made by adjusting the compensator or by moving the cross-hair.

When testing theodolites with one reading index, if the reading in the second telescope position deviates after the previous adjustment ($\alpha = 0°$; $z = 90°$) from $\alpha = 180°$ and $z = 270°$, respectively, or if the index error changes with change in the elevation angle of the pointings, then the vertical circle is eccentric.

f) The levelling spirit level. In the case of theodolites supplied with a spirit level for levelling, the horizontal hair of the telescope is pointed during the test to the horizontal hair of the central collimator, then the bubble of the level is brought to centre (or to coincidence) by means of the level adjustment screws.

2.3 Recording Theodolites

A most important area in the development of the complex automation of surveying is the automatic data recording of field measurements. For the sake of economic machine processing in the office and of automation, measurement data must be recorded in a form that enables immediate data processing in computers.

2.3.1 Operational Principles

From the point of view of the automation of field work, three main phases of surveying field work can be separated after the instrument has been set up: the pointing operation, the reading operation and the recording of some supplementary data (e.g. the serial number of the station and of the observed station, etc.). The third phase can be most easily automated as in this phase only some numerical data are to be recorded by an adequate device, together with notes made earlier by the booker. The most difficult problem is to automate pointing; it has not been solved yet. Readings on the circle after pointing can be automated either by analogue or by position measuring systems. In the former case change of angular value is given in the form of change of another physical quantity (e.g. electric voltage); in the latter it is based on pulse counting or on the reading of the angular values in a binary encoded form. Digitally operating positioning systems are more advantageous as their data can be more easily processed on (equally digital) computers. Orientation angles measured in the field (and possibly other quantities also) are expressed by binary encoded numerical values.

Instead of a horizontal circle with the usual graduation (or as well as it) a glass disc with special encoding is used which, on the one hand, fixes the necessary directions with sufficient accuracy and, on the other, makes possible automatic reading — mostly by photoelectric methods, or simply photographically. Thus a reading can be made either immediately, or by the use of a photograph or paper tape.

The most important demands for recording theodolites are as follows:

— The angle reading device should ensure a reading accuracy of 1 sec of arc so that the instrument may be used for triangulation and optical distance measurement (for the measurement of parallactic angles).

— The mass and dimensions of the instrument should not greatly exceed those of traditional theodolites.

— The instrument should be simple enough to be handled by a surveyor with average experience in angle measurement.

— The position of the zero graduation on the code disc should be determinable unambiguously.

— For recording, film, paper or magnetic tape should be used; it should be possible to use them both together and separately.

— The encoded data should be easily (directly) processable by electronic computers.

— Together with the observed data, the serial numbers of the stations as well as of the observed stations, together with any other necessary remarks, should be able to be recorded.

It is advantageous to record the data on magnetic tape as its small dimensions give an advantage in field work.

A pulse counting device transmits a pulse if the quantity to be measured changes by a previous set value. This value is determined on the basis of the necessary accuracy of the angle measurement.

The reading of binary encoded angular values may be carried out by a photocell, by a magnetic head or by some other method. Direct reading of angular values has the advantage with respect to pulse counting that the zero position is unambiguously determined and the possibility of false angle reading is smaller.

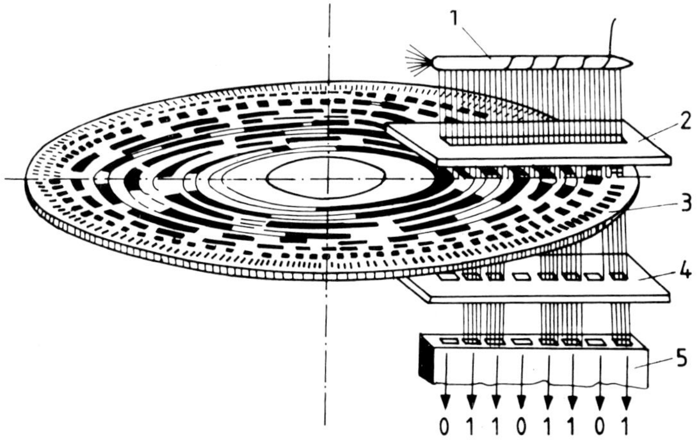

Figure 2.95. Photoelectric circle reading device

Photoelectric reading devices have the highest resolution of all digital angle detecting devices; in addition they are small and, even with small encoding grids, angular readings with an accuracy of some seconds of arc can be realized.

A photoelectric device to read binary encoded angular values *(Fig. 2.95)* consists of the light source *1* of the light beam limiter *2*, the encoding disk *3*, the reading light diaphragms *4* and the block of photocells *5*. The encoding disc is divided into concentric angular strips (each strip corresponding to a digit) which consists of transparent and non-transparent (black and white) segments with a prescribed regularity. Each angular strip has its own photocell. The light source illuminates the code disc through the beam limiter. If the light falls on a transparent part of an angular strip then its photocell produces a current which can be detected; if the light meets a non-transparent part then the photocell produces no current. The combination of signals and of lack of signal gives the rotation angle of the encoding disc encoded in a binary system according to the code used. The resolution of a transformer device operating on this principle is

$$\Delta\alpha = \frac{\alpha_{max}}{2^n}, \tag{2.113}$$

where α_{max} is the greatest angle to be measured (360°), n is the number of digits in the binary system, i.e. the number of angular strips on the disc. The higher the resolution required, the greater is the number of decimals required.

The reading of the encoded angles is indeterminate at the boundaries of the divisions; in such cases the choice of recording either of the possible values (0 or 1) is random. *Figure 2.96, a* shows a code disc with eight digits in a simple binary system. Let us suppose that the series of photocells is along the beam *OA*, where all the strips are transparent, corresponding to the binary number 11111111. By switching over to the position *OB* the number 00000000 is read. If one photocell (e.g. that corresponding to the seventh digit from the right) is moved a little in the clockwise direction, the number 10111111 is read, corresponding to the direction of the beam *OC*. This is a blunder.

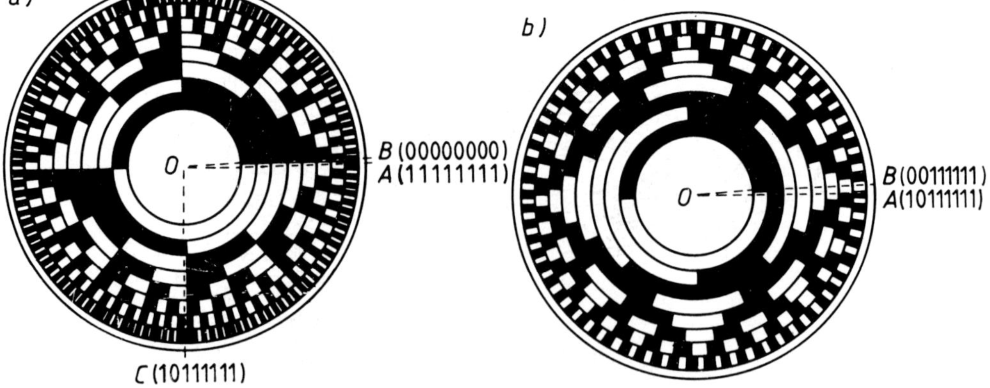

Figure 2.96. a) simple glass disc encoded in the binary system; *b)* glass disc cyclically encoded in the binary system

There are several ways to eliminate this ambiguity. Good results are obtained by the so-called logical reading where the electrical device determines the correct reading for the digit in question by taking into account the number of the previous digit. In photoelectric transformers, where the photocell and the light source are radially positioned, cyclic encoding is used. Here the signal combinations are distributed along the disk so that, if one switches over from one angular value to the next, then only one of the digits changes. *Figure 2.96, b* shows by a dotted line the position of the reading diaphragms for two adjacent numerical values. By switching over from the position *OA* (10111111) to the position *OB* (00111111) only one of the digits changes. Thus, with cyclic encoding, the ambiguity at the boundary of the division change affects only one digit. The difficulty in the application of the cyclic code lies in the necessary special connection and the need for other supplementary devices. Therefore it is not advantageous to transform the cyclic code in the theodolite itself into a simple binary code. It is better to record the data of the cyclic code on paper tape and to carry out the transformation in the computer.

In addition to the encoding discs presented in *Figs 2.95, 2.96, a* and *b*, there are many other varieties. Some encoding systems record directly the sine- and cosine-values of the angles.

2.3.2 Some Types of Recording Theodolites

The theodolite *FLT 3* (Fennel, Kassel) *(Fig. 2.97, a)* has encoded horizontal and vertical circles and is able to record the seconds of arc. The graduation unit on both circles is 0.2^g. An annular code consisting of black and white signal combinations

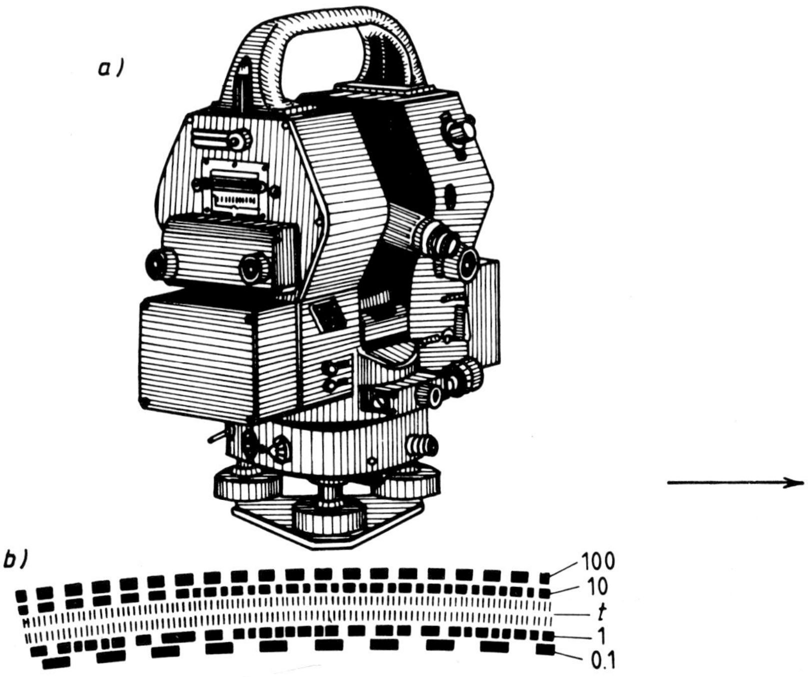

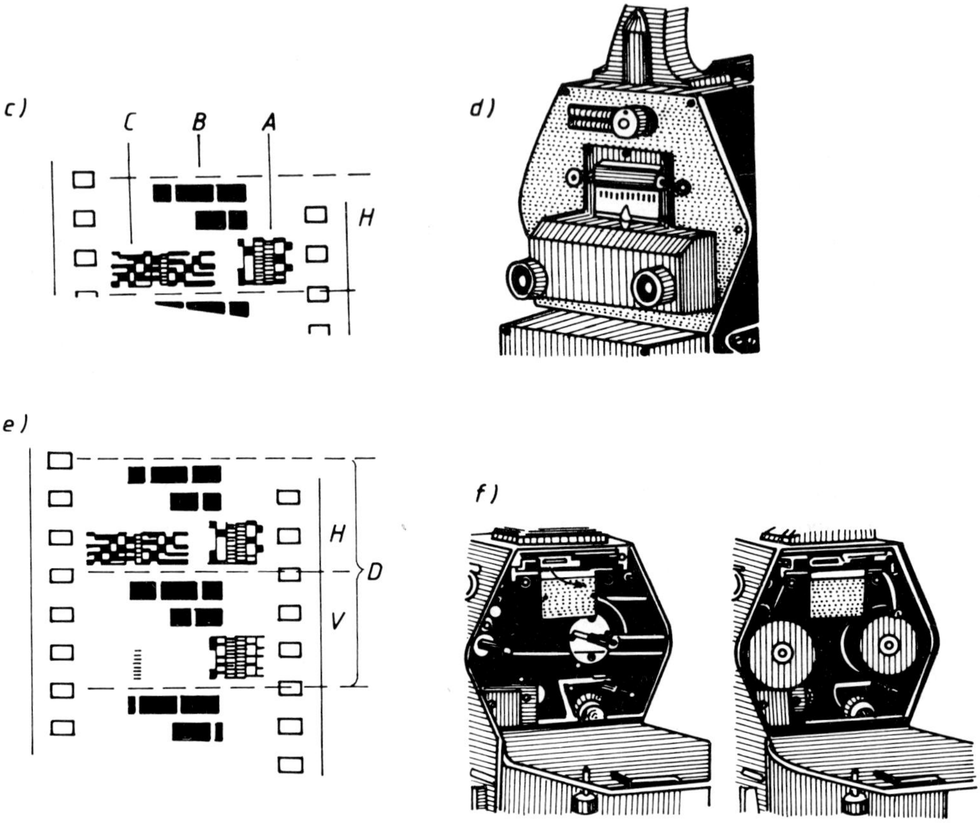

Figure 2.97. a) recording theodolite *FLT 3* (Fennel, Kassel); *b)* cross section of the annular code with code series 100^g, 10^g, 1^g, 0.1^g and with two timing signal series; *c)* reading image recorded on film; *d)* digital setting device for the recording theodolite *FLT 3*; *e)* horizontal and vertical angular readings recorded simultaneously; *f)* device photographing the readings for the recording theodolite *FLT 3*

(Zuse-type) is used on the code disc *(Fig. 2.97, b)* which contains four information channels and two timing channels. The four information channels correspond to 100, 10, 1 and 0.1 grad, respectively. The diametrically opposite images of the circular graduations are optically united. Due to the opposite motion of the graduations, the graduation interval of 0.2^g gives a code interval of 0.1^g. The encoding has a decimal system. The annular code contains 4000 different groups of codes which determine the main reading of 0.1^g unambiguously. Signal combinations were selected to give numbers between 0.0 and 399.9 on the horizontal circle, and between 400.0 and 799.9 on the vertical circle.

The circle reading, which is photographed, contains the following parts *(Fig. 2.97, c)*:

a) The digital representation *A* gives the coarse value of the circle position with an accuracy of 0.1^g.

b) The analogue representation of the graduation lines *B* shows them as white lines on a black field: the position of the lines in one of the fields with respect to the two adjacent lines in the other field yields — expressed in milles — the second, third and fourth decimal digit of the truncated reading. Together with the digitally recorded coarse reading, this gives the reading on the circle with an accuracy of 1^{cc}.

Other data in connection with a measurement (station number, sighted station, weather data, etc.) can be recorded by a 12 decimal setting device (see *Fig. 2.97, d*). At each decimal place any number between 0 and 9 can be set. These data are recorded by the instrument in a similarly encoded form as combinations of black and white signals in an oblong field *C (Fig. 2.97, c)*, which is centrally distributed by a timing channel into two equal parts containing 6 channels each.

The readings on both the horizontal and the vertical circle can be recorded in an encoded form. If only horizontal angles are measured then the complete record consists of the elements denoted by *A*, *B* and *C (Fig. 2.97, c)*. If elevation angles are also recorded, then an additional field denoted by *V* appears on the record, which contains the code of the vertical circle *(Fig. 2.97, e)*.

The film transport is automatic *(Fig. 2.97, f)*. The film roll is about 10 m in length, therefore 680 results can be recorded on one roll if only horizontal angles are measured, 340 with both horizontal and vertical angles.

The recorded information can be evaluated and transformed using the interpreting device *Zuse Z 84*. As the films used in recording theodolites contain the information

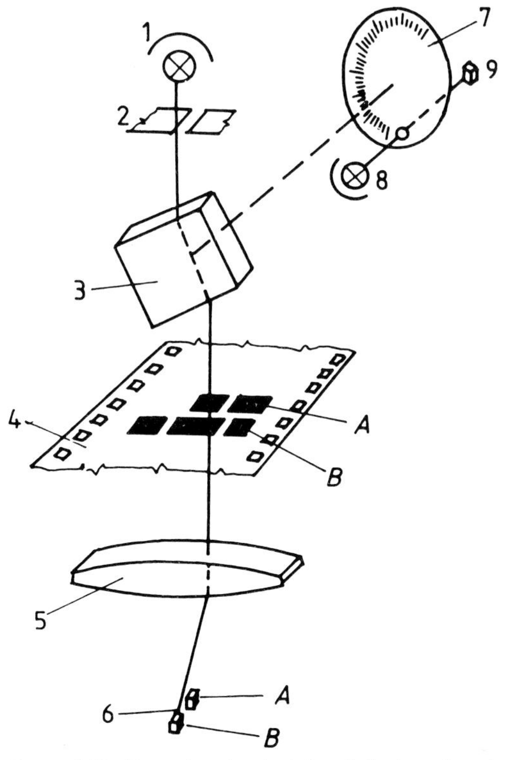

Figure 2.98. Operational principle of the transforming and evaluating device *Zuse Z 84*

partially in digital form *(A* in *Fig. 2.97, c)* and partially in analogue form *(B* in *Fig. 2.97, c)*, the analogue part must be evaluated separately. The interpreting device *(Fig. 2.98)* has a light source *1* from which light passes through the diaphragm *2* and the rotating prism *3*, and the narrow beam illuminates on the film *4* the two analogue images denoted by *B (Fig. 2.97, c)*. The transmitted light beam passes through the cylindrical lens *5* to meet the corresponding photodiodes *A* and *B 6*. The segments within the graduations are measured by the circle *7* which is rigidly mounted to the rotating prism *3* and which is divided into 18,000 graduations. The light source *8* is located on one side of the graduated circle, the photodiode *9* is on the other. An electronic counter device counts the rotation angle which is later recorded on paper tape together with the encoded values. The device *Zuse Z 84* has a START and a STOP mode of operation. The evaluation of one image together with punching and transport takes 3/4 second. The total information content of the image is transformed into 49 signals of the tape punch on 5 channels.

In addition to the electronic control there is also a mechanical device built into the evaluating instrument to check the film supply, and there is another automatic system to check that the power supply units are in continuous operation. The transformer is transistorized and is a normal desk size.

Power is supplied to the FLT3 by batteries. The main data of the instrument are: magnification 30 ×, diameter of the objective 40 mm, diameter of the horizontal circle 107 mm, basic graduation 0.2^g, encoded basic graduation 0.1^g, accuracy of readings 1^{cc}, film width 35 mm, film length 10 m. The film is transported electrically. An artificial light source is used for illumination.

Figure 2.99. Recording theodolite *Ko-B1* (MOM, Budapest)

The recording theodolite *Ko-B1* (MOM, Budapest) *(Fig. 2.99)* has two main parts: the theodolite and the electronic unit. The basic graduation of both the horizontal and vertical circles is 10 minutes of arc. The graduated circle contains the codes for the integer grades and for the ten minute graduations. Encoded signals are read in series by an electronic device. Diametrically opposite signals are projected into each other, brought into coincidence by a continuously operating optical micrometer and then measured. The electronic device yields three coincidence signals within one second of time on both the horiziontal and the vertical circle. The optical micrometer operates on the basis of pulse counting. The glass circle of the micrometer is divided into 600 units and thus the basic graduation interval of 10 minutes of arc can be read with an accuracy of 1 second of arc. As the optical micrometer has a continuous operation, the instrument is also suitable for following moving targets, and data can be recorded every 0.2 second of time. Additional data (number of the station, number of the observed station, weather data, etc.) can be recorded by eight 10-digit numbers. The electronic unit has a 7-digit display. The digital display gives either horizontal directions or elevation angles, depending on the actual connection. Data are, however, recorded independently of the digital display in a parallel form.

The reading speed is sufficient for the use of a tape punch with a speed of 150 characters/second. With a tape punch of lower speed, the speed of reading is determined by the punch.

The operation-mode switch on the electronic unit enables the following modes of the tape punch to be set:

a) recording of numbers on the number setting device only,

b) recording of the averages of 1, 2 or 5 readings, according to the accuracy requirements, together with the eight identifying numbers,

c) continuous recording of the numerical values (when following moving targets, the highest angular speed is 2 grades/second around both axes).

Principal data of the instrument are: magnification 36 ×, diameter of the objective 55 mm, smallest displayed angular value 1 second of arc, mass of the instrument 12.5 kg, of the electronic unit 13.5 kg, power supply 24 V battery.

2.4 Special Theodolites

2.4.1 Balloon Theodolites

Balloon theodolites are used for the continuous observation of meteorological balloons. Such a theodolite is e.g. the balloon theodolite *GARAT* (Breithaupt, Kassel) which measures the sidereal and elevation angles of meteorological balloons at predetermined intervals *(Fig. 2.100)*. From the height of the hydrogen-filled balloon — computable from its known ascent velocity — also from the sidereal and elevation angles read at certain intervals, the ascent path of the balloon can be determined, and from it the direction and velocity of the wind at different heights is computed. The theodolite has one horizontal and two vertical circles. Readings are made at two diametrically opposite points. The telescope is roughly pointed at the balloon with the aid of the coarse sight on the objective head by manual rotation around both axes. For fine pointing and for following thumbwheels are used for both the sidereal and the vertical motion. The magnification of the telescope is 16 ×, the free diameter of the objective 48 mm, its field of view 3.5°; the constant of the alidade level 45 secs of arc; the diameter of the horizontal metal circle 142 mm, that of the vertical metal

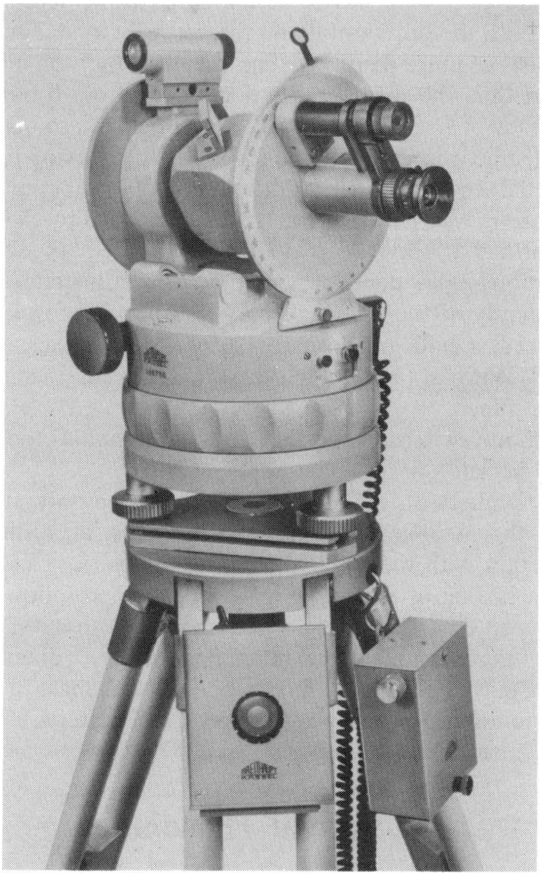

Figure 2.100. Balloon theodolite *GARAT* (Breithaupt, Kassel)

circle 130 mm; the graduation unit on the circle is 1°; the accuracy of reading on the vernier 6 minutes of arc; the mass of the instrument 4.0 kg.

The theodolite *Pilot Balloon (Fig. 2.101)* has two telescopes: a main and a search telescope. Both are bent telescopes with a common eyepiece, the switch-over being made by a rotatable mirror. A special feature of this theodolite is that both the horizontal and the vertical circle are in a horizontal position, thus only one index is needed for the two circles. The diameters of the graduated circles are 134.4 mm. The magnification of the main telescope is 20 ×, its field of view 2°. The magnification of

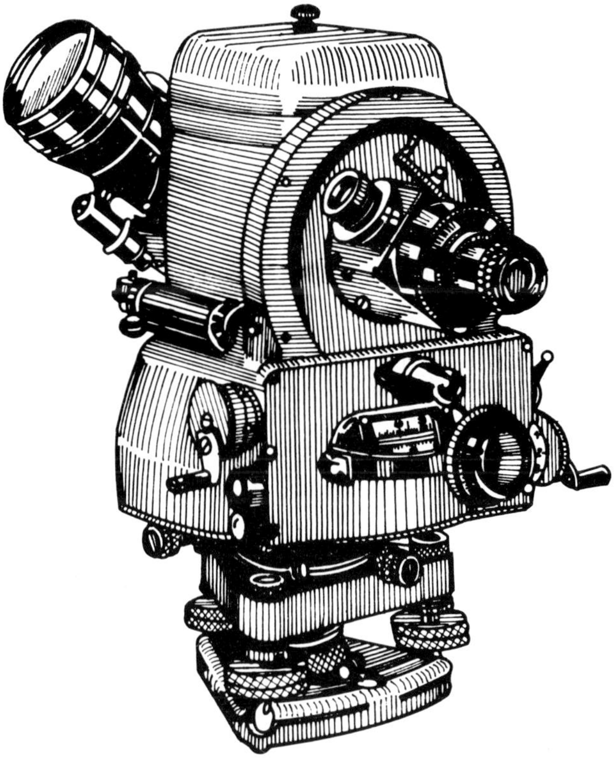

Figure 2.101. Theodolite *Pilot Balloon* (Hilger and Watts)

the auxiliary telescope is 5 ×, its field of view 8°. The graduation unit of the horizontal and vertical circles is 0.5°; the smallest angle observable by the micrometer is 0.1°. The mass of the theodolite is 28 kg, that of the tripod 25 kg.

2.4.2 Potentiometric Theodolites

Potentiometric theodolites are used to follow moving targets with a maximum angular speed of 1° to 2°/s and horizontal and elevation angles can be computed *(Fig. 2.102)*. The continuous following of targets is simplified by the use of continuous control screws. The continuous readings are displayed electronically and recorded if necessary.

This type of instrument is used mostly to monitor the movement of measuring ships and airplanes but it is also an important tool in the traffic control service of busy airports. Both the main and the auxiliary (search) telescopes are bent. The magnification of the main telescope is 16 ×, the diameter of the objective is 60 mm, its field of view 3°. The magnification of the auxiliary telescope is 3.8 ×, its field of view 16°. In the field of view of the auxiliary telescope a circle bounds the field of view of the main telescope. The angular rotation is measured continuously by a

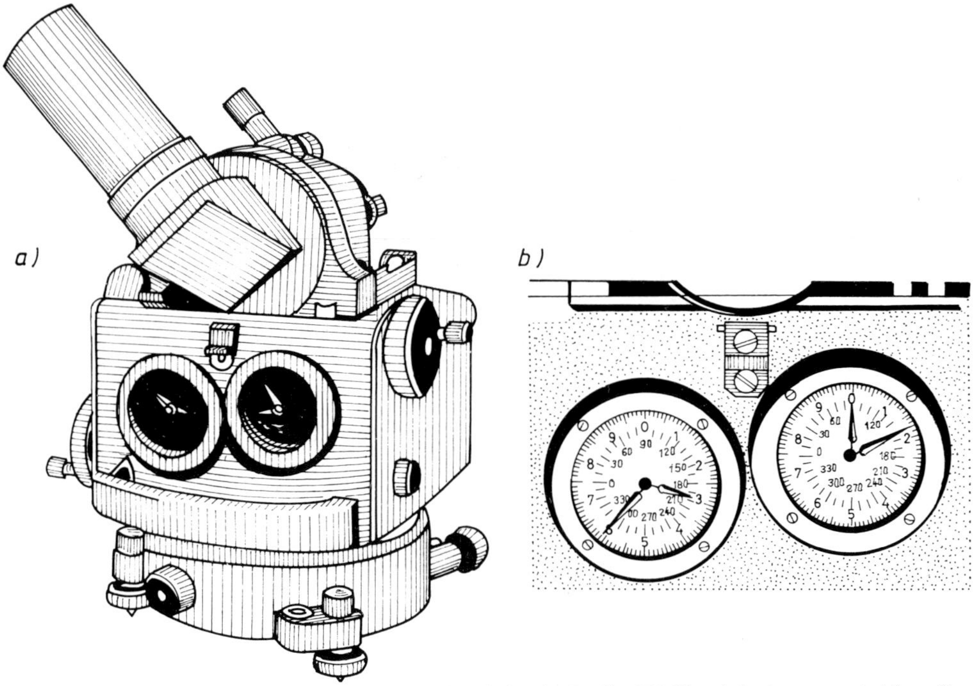

Figure 2.102. Potentiometric recording theodolite (Askania, Berlin, FRG): *a)* the instrument; *b)* reading meters

potentiometer. The shorter hand of the reading 'clock' *(Fig. 2.102, b)* shows degrees (from 0° to 360°), the longer hand sub-divisions of 0.05°; 0.01° can be read by estimation. The mass of the theodolite is 17 kg. The recording cylinders connected to the instrument record continuously in graphical form but angular values can also be recorded and displayed digitally.

2.4.3 Other Special Theodolites

Kinotheodolites serve to observe fast moving targets (airplanes, rockets, missiles, spacecraft) and their position is continuously recorded. It is possible to record quickly the horizontal and elevation angles characterizing the position of the targets. The moving target can be followed by control screws, by hand or automatically. Examples are the theodolite *Kth* (Siemens AG, Berlin) and the kinotheodolite *EOTS* (Contraves AG, Zürich).

Instruments used to follow artificial satellites contain instead of the main telescope a measuring camera which photographs at a large scale the instantaneous position of the luminous satellite and the nearby celestial bodies. The shutter device of the camera operates according to a previously compiled program. Time marks are also recorded. Most such instruments have several operational modes to carry out the observations. In a fixed (rigid) position of the camera the moving target can be photographed as a point by a proper choice of exposure interval. In another mode of operation the control device built into the instrument moves the optical axis of the camera during the exposure according to a previously prepared programme, so that it follows either the artificial satellite or the surrounding fixed stars. Several exposures are made on the same negative and they result in point-like images (of artificial satellites and fixed stars) which are measured by a precise monocomparator. Results are then processed on computers.

Such instruments are : the *Baker–Nunn* observing camera widely used in the US, and the satellite theodolite *BC-4* developed from the universal theodolite *Wild T4*. A modern instrument is the satellite observing equipment SBG. In the Soviet Union, the satellite observing cameras AFU 75, FAS and VAU are used.

2.5 Orientation Theodolites

It is often the case that the quantity to be measured is not the horizontal angle between the two directions at the station of measurement towards two other points visible from the station, but the angle (azimuth) between the direction from the station to one point and North. For this purpose orientation theodolites are used: compass theodolites and gyro-theodolites.

2.5.1 Compass Theodolites

2.5.1.1 Compass

The compass is a magnetic needle provided with a pointing device. The needle, made of tungsten or chromium steel plate and standing on its edge (edge-bar needle) *1 (Fig. 2.103)*, lies with its agate cup *2* on the vertical steel pin *3*. In order to protect the pin, the needle can be lifted by the arresting screw *4* with the help of the lever *5* and pressed against the glass cover *7* of the needle house *6*. The needle should be so lifted when not in use and prior to any transportation. The magnetic needle has high remanence (magnetism remaining after the magnetizing field has ceased), and high coercive force (an opposite field which eliminates the remanence).

Due to the geomagnetic field the compass turns into the direction which makes an angle with the direction of geographic north N_g that corresponds to the magnetic declination Δ, i.e. into the direction of magnetic north N_m *(Fig. 2.104)*. The declination Δ is positive or negative if magnetic north deviates from geographic north N_g towards the east or the west, respectively. The magnetic needle deviates from the horizontal in the vertical plane (magnetic inclination). Both horizontal and vertical deviations change with geographical position.

The angle μ between geographic (astronomical) north N_g and map (grid) north N_k, i.e. the direction of the axis X of the grid, is called the (projection) convergence of the meridian. It is positive/negative, if map north deviates from geographic North towards the east/west. The angle between magnetic north N_m and map north N_k is the magnetic orientation angle ϑ which is positive/negative if map north deviates

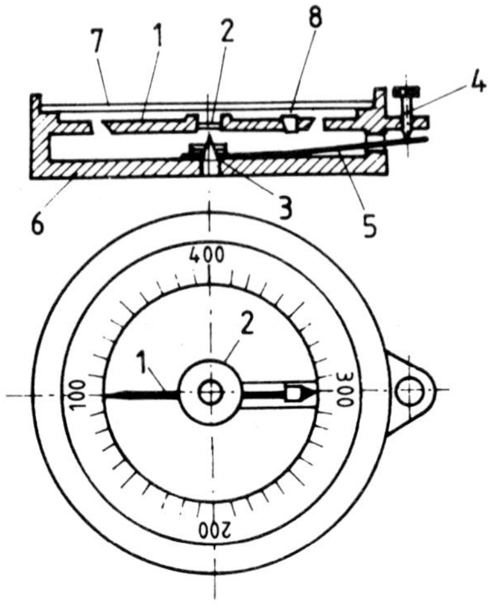

Figure 2.103. Circular compass

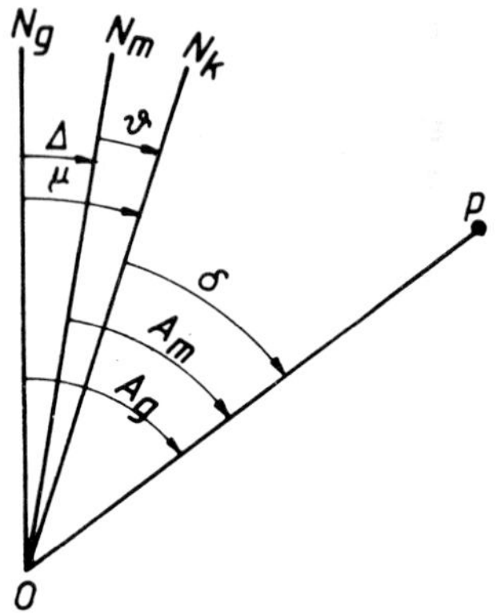

Figure 2.104. Connections between geographic north N_g, map (grid) north N_k and magnetic north N_m

from magnetic north towards the east/west. The declination Δ is the difference between the convergence of the meridian and the magnetic orientation angle:

$$\Delta = \mu - \vartheta.$$

The direction angle δ of the line connecting the station point O and a terrain point P is the difference between the magnetic azimuth A_m and the magnetic orientation angle ϑ:

$$\delta = A_m - \vartheta \tag{2.114}$$

or by using Eq. (2.113):

$$\delta = A_m + \Delta - \mu. \tag{2.114a}$$

Magnetic declination has several types of variation; the daily variation and the so-called secular variation are more or less regular, the period of the latter being several hundred years. In addition to these the geomagnetic field is frequently disturbed, and even geomagnetic storms appear from time to time. Thus the declination has local changes with shorter or longer periods. Magnetic declination (and the magnetic orientation angle) can be determined by compass, if geographic (astronomical) north (or grid north, respectively) is known. For the continuous recording of magnetic declination special recording instruments, magnetic variometers, are used which are set up in geomagnetic observatories.

The vertically directed deviating force of the magnetic inclination is balanced by a small adjustable weight *8* on the magnetic needle (see *Fig. 2.103*).

For surveying purposes, circular box compasses and trough (orienting) compasses are used.

2.5.1.2 Circular Box Compass

The usual diameter of circular box compasses is 10 to 15 cm, and the edge of the housing has a counterclockwise graduation in grades. If the instrument is rotated clockwise, the readings at the ends of the needle increase and give immediately the magnetic azimuth A_m of the direction parallel to the graduation lines $0° \ldots 180°$, with an accuracy of $0.1°$ by estimation.

The *Schmalcalder circular box compass (Fig. 2.105)* has an open sight attached to it. The graduated circle is made of a light material (cardboard, aluminium), has graduations of $1°$ and is attached to the flat magnetic needle so that the zero graduation coincides with the northern end of the needle. The numbers on the circle graduation increase clockwise. When observing, the circle graduation is read through the prism on the eyepiece sight at the narrow slit which is used as an index on the mirror coating of the prism. The bottom, convex surface of the prism is used as a reading magnifier.

The circular box compass with a telescope is fixed on a base which can be set on a tripod, and the telescope has a support on one side of the compass. The box level on it is used to set the instrument into the horizontal.

Box compass theodolites usually also have a vertical circle, and their horizontal circle is at the same time the graduated circle of the circular box compass on which the magnetic azimuth of the observed point is read, using the northern end of the magnetic needle as index. The horizontal graduated circle of the theodolite is often made attached to the magnetic needle according to the Schmalcalder principle, and the horizontal circle swings together with the needle. The horizontal circle and needle can be lifted from the pivot and pressed against the edge of the bottom part of the

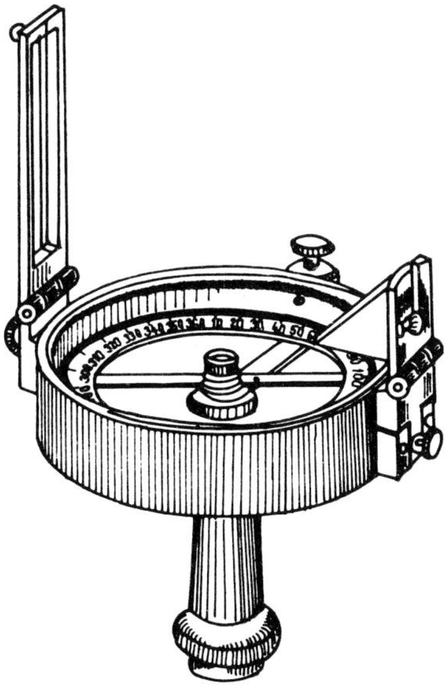

Figure 2.105. Schmalcalder compass

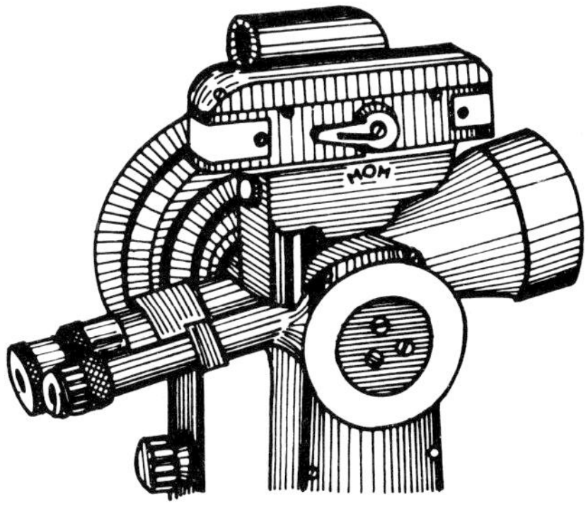

Figure 2.106. Theodolite *Te D2* with compass attachment (MOM, Budapest)

instrument. In such a case the compass theodolite can be used as a normal theodolite. Both the horizontal circle and the vertical circle are read by a microscope. The telescope is also provided with stadia hairs in addition to the cross-hair, therefore these instruments are usually classified with compass tacheometers which are treated in Point 6.1.1.2. Prismatic circular box compasses attached to one of the telescope supports of the theodolite *(Fig. 2.106)* are also often used.

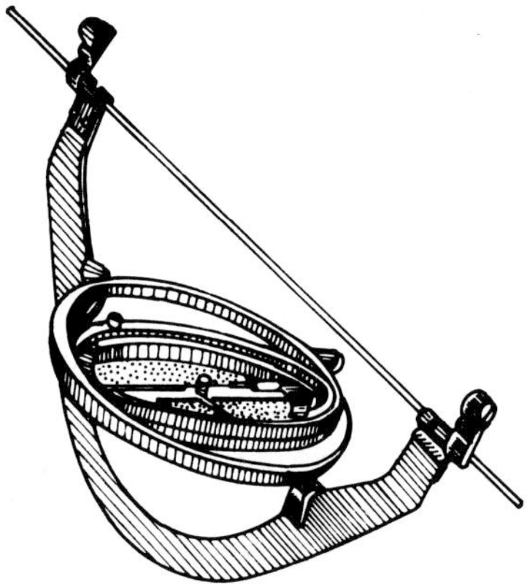

Figure 2.107. Suspension compass (Fennel, Kassel)

The compass plays an important role in mine surveying even today. The suspension compass consists of the compass arc, the compass ring and the compass housing. In the *Freiberg* type the compass ring and suspension arc are rigidly fixed to each other and only the compass housing can be tilted. In the *Fennel* type *(Fig. 2.107)* both the compass housing and the compass ring can be moved and, for transportation, the ring should be folded into the plane of the arc.

2.5.1.3 Trough (Orientation) Compass

Compass attachments used for the orientation of theodolites and tacheometers contain only a small part of the graduated circle of the compass, consisting of short segments with a few graduations at both ends of the diameter 0° . . . 180° (0^g . . . 200^g) in a long narrow rectangular trough. The compass attachment can be placed between the two telescope supports or on one of the telescope trunnions *(Fig. 2.108)*. The tubular compass attachment has no graduation at all. At both ends of the magnetic needle there is a fork with prongs bent upwards by 90° *(Fig. 2.109, a* and *b)* which should be brought into coincidence using the magnifier of the compass and the horizontal slow motion screw of the theodolite. Coincidence is facilitated by a fixed mirror *2 (Fig. 2.110)* above the pivot of the magnetic needle *1* which projects the image of the fork nearer to the observer *3* onto the image of the other fork *4*. When in coincidence *(Fig. 2.109, b)* the line of sight of the telescope lies in the magnetic meridian. If the horizontal circle of the theodolite is read in this position, the magnetic meridian has been fixed as an orientation to measure the magnetic azimuth of the

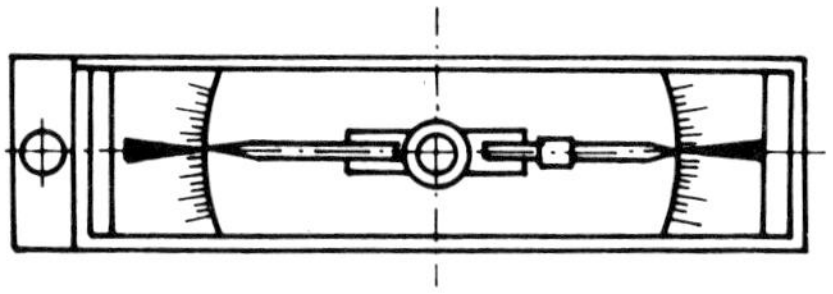

Figure 2.108. Trough compass

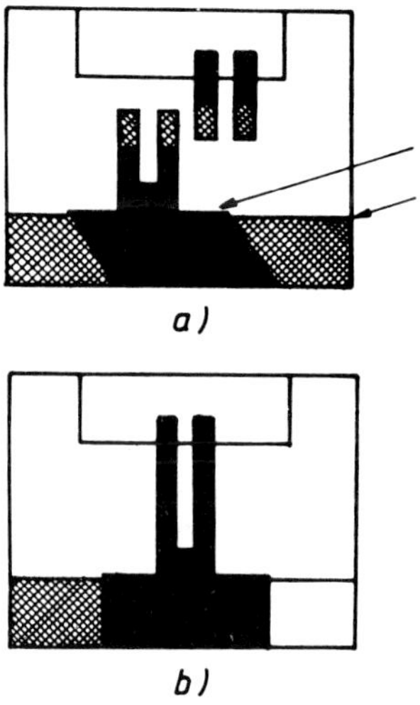

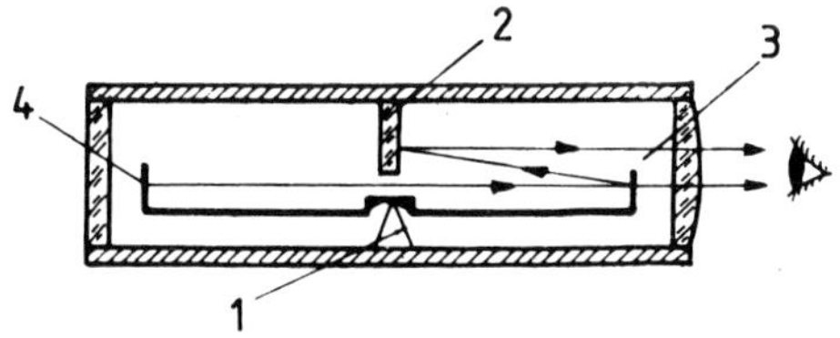

Figure 2.110. Setting of coincidence by tubular compass

Figure 2.109. Forks of a tubular coincidence compass *a)* before coincidence; *b)* in coincidence

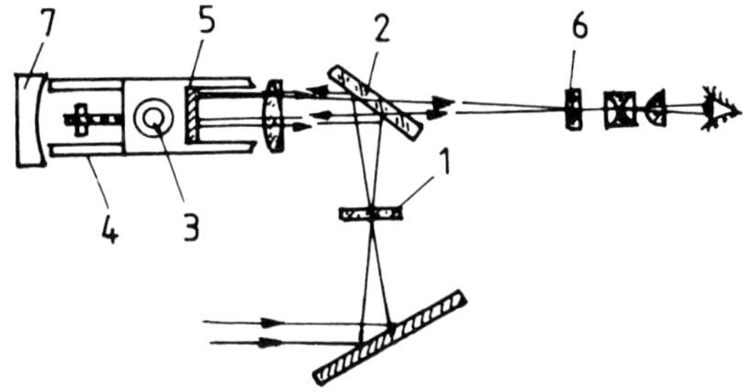

Figure 2.111. Ray path of the tubular compass with collimator setting (Rank Precision Ind., UK)

points to be determined. The accuracy of magnetic north determined by a tubular compass is about 3 to 5 minutes of arc ($\pm 6^c$ to 10^c).

In another type *(Fig. 2.111)* a collimation mirror *5* is fixed on the body of the magnet *4*, the axis of which rotates in upper and lower sapphire bearings *3*. The beam of the image of the illuminated index *1* passes through the half-silvered mirror *2* and, after reflection from the collimation mirror *5*, it is projected onto the graduated reticule *6*. The small deviations from the magnetic meridian can be read by the image of this index *1*. The oscillation of the magnet is damped by eddy currents in the copper block *7*. The compass can be used on both faces of the telescope.

2.5.1.4 Test and Correction of Compasses

1. The centre of the circular graduation should lie in the rotation axis of the magnet. This condition is fulfilled if the readings at the opposite ends of the magnetic needle differ by 180° (200^g) in different positions of the magnet. If they do not, the average of the readings at both ends is free from eccentricity error.

2. The magnet should be strongly magnetized.

3. The bearings of the magnet must not be damaged.

If conditions 2 and 3 are fulfilled, then by turning the magnet from its rest position with a small piece of iron, e.g. with a key, it must return to its rest position after the key has been removed.

4. The magnetic axis of the magnet should not deviate from its geometric axis. The effect of axis deviation would be that the ends of the magnetic needle point in a direction deviating by a constant angle from the magnetic meridian. The value of the deviation can be determined by a reversal of the magnet if both of its sides are provided with a sleeve.

Circular box compasses rigidly fixed to angle measuring instruments are tested by setting different readings on the horizontal circle of the instrument, e.g. 0°, 30°, 60°, ... and the position of the magnet is read in each setting. The compass readings should deviate by a constant angle from the horizontal circle readings. If angular differences are obtained between different settings, then the material of certain parts of the instrument may be insufficiently non-magnetic.

Care should be taken during the use of instruments with a compass that the instrument station should not lie in the vicinity of iron fences, railway lines, reinforced concrete buildings, iron constructions or DC electricity lines. The iron feet of the tripod have no significant effect on the magnetic needle as they are about 1.5 m lower and roughly symmetrical about the centre. People in the vicinity of the compass must not carry any iron objects (keys, pocket knives, etc.).

In addition to the periodic changes of declination there are sometimes sudden changes in it whose effects cannot be detected during measurement. Therefore if there is some suspicion that such changes have occurred, e.g. if a traverse measured with a compass has a large closure error, the nearest geomagnetic observatory should be asked if any geomagnetic storm had occurred on that particular day. The results of compass measurement on geomagnetically disturbed days are mostly unusable.

When compasses are used, the following general rules should be kept: 1. prior to measurement, the instrument should be checked; 2. all measurements in a series should be made by the same instrument; 3. directions in a traverse should always be measured from both end points; 4. the magnet should be lifted off its pivot after every reading; 5. the instrument should be protected from shock, impact and high temperature (shocks and high temperature deteriorate the magnetic parameters of the magnet).

2.5.2 Gyroscopic Theodolites

The accuracy of the determination of North by compass is usually insufficient, furthermore, the irregular changes of magnetic declination, the vicinity of iron and metal masses and electricity lines, make the measurement results unstable. Determination of astronomical north is accurate, but it is quite lengthy, depending on weather and locality. Some other method is therefore necessary for azimuth measurement. Pendulum experiments made by Foucault in 1852 showed that rigid rotating bodies suspended like pendulums (gyroscopes) have a rotation axis which can be used for orientation purposes.

2.5.2.1 The Gyroscope

The gyroscope is a rigid body of arbitrary shape and mass distribution which can rotate in any direction around one of its points which is fixed or can be supposed to be fixed. In practice the gyroscopes used have a mass distribution of circular symmetry, and nearly friction-free bearings at the centre of gravity. Such gyroscopes are called force-free, as the weight of the gyroscope exerts no angular momentum on the support point. The force-free, circularly symmetric gyroscopic body has three mutually perpendicular principal moments of inertia of which the two perpendicular to the spin axis (axis of symmetry) are equal and have a value B, and the third in the direction of the spin axis is different: $\Theta_{\omega} \neq B$.

In the general case the path of the force-free gyroscope can be described as follows *(Fig. 2.112)*. If a rotation around a straight line *1* through the centre of gravity S (central line) is transferred to the gyroscopic body — a torque having a spatially fixed axis — then on the one hand the gyroscope rotates around its axis of symmetry *2* with a uniform velocity, and on the other hand the axis of symmetry is rotating in the same sense around the axis of momentum *1*, maintaining a direction fixed on the surface of the so-called nutation cone *I* with vertex S. As the resultant of these two rotations, the gyroscope will be rotating around the instantaneous axis *3* through the centre of gravity, but having a direction change with a velocity which results from the sum of the two rotational velocities. The instantaneous position of the axis *3* is determined by the revolution of two cones which have the common vertex S. The generatrices of one of the cones, namely of cone *II* with a spatially fixed direction,

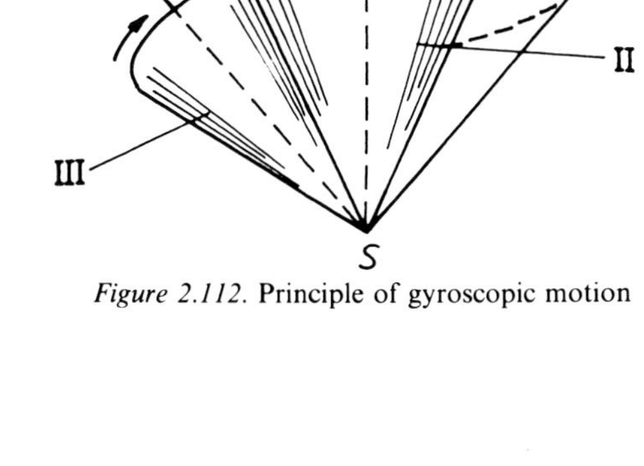

Figure 2.112. Principle of gyroscopic motion

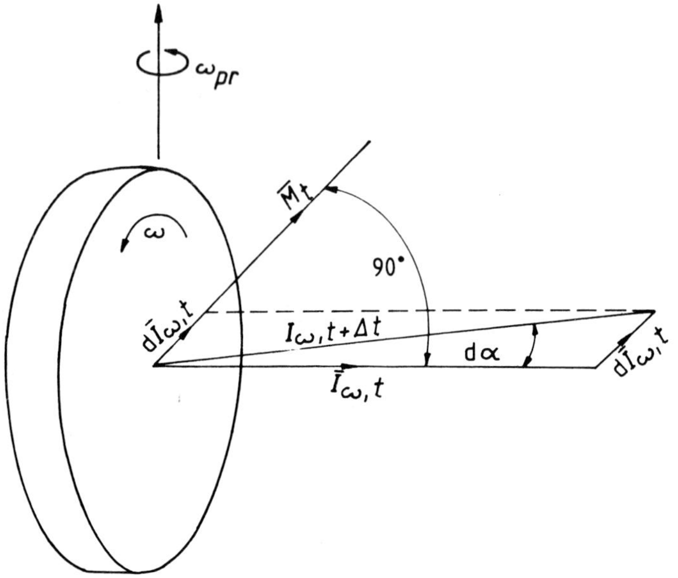

Figure 2.113. The effect of an extraneous torque on the rotation axis of a gyroscope

i.e. of the so-called hyperboloid cone, give the positions of the instantaneous axis with respect to the surroundings. The generatrices of the other cone, i.e. the revolving cone *III* with the moving axis *2*, or of the so-called polhodie-cone, are the positions of the instantaneous axis with respect to the gyroscopic body. This phenomenon is called nutation. At any point the instantaneous axis is the common straight line (common generatrix) of the two conic surfaces revolving on each other. During motion the three axes (*1*; *2* and *3*) are always in the same plane, which is rotating around the fixed axis of momentum *1* with the velocity of the nutation.

It is easily found from *Fig. 2.112* that, if the axis of momentum *1* coincides with the axis of symmetry *2*, then the instantaneous axis *3* also coincides with them, and it always remains in the same direction.

If a gyroscope with an axis of momentum nearly coincident with the axis of symmetry, and with a rotational velocity ω and a principal moment of inertia Θ (i.e. its angular momentum is $I_a = \Theta\omega$) is under the influence of a continuous external torque M of a constant absolute value $\overline{M}$, which has a changing vectorial direction always perpendicular to the instantaneous vectorial direction of the angular momentum, then the motion of the gyroscope will also change. *Figure 2.113* shows the situation at the moment t. The vector of the torque $\overline{M_t}$ is perpendicular to the angular momentum vector $\overline{I}_{\omega,t}$. During an infinitesimal time interval dt the torque M gives to the gyroscope a small angular momentum of value

$$\mathrm{d}I_{\omega,t} = M \cdot \mathrm{d}t \tag{2.115}$$

and of direction $\overline{M_t}$. The momenta $I_{\omega,t}$ and the perpendicular to it $\mathrm{d}I_{\omega,t}$ result in the momentum $I_{\omega,t+\mathrm{d}t}$. Due to the resultant momentum, the rotational axis of the gyroscope rotates by an angle $d\alpha$, lying in the plane of the vectors $\overline{I}_{\omega,t}$ and $\overline{M_t}$ which is perpendicular to the rotation plane of the gyroscope. The axis of momentum of the gyroscope tries to turn into the direction of the external torque M_t. From the figure one has:

$$dI_{\omega,t} = I_{\omega,t} d\alpha. \tag{2.116}$$

The vector of the resulting momentum $I_{\omega,t+dt}$ has, however, a constant absolute value $|I_{\omega,t+dt}| = |I_{\omega,t}|$, as the direction of the small additional vector $d\overline{I}_{\omega,t}$ has a direction perpendicular to $\overline{I}_{\omega,t}$. At the instant $(t+dt)$ the vector of the torque M_{t+dt}, of an absolute value supposed unchanged, remains perpendicular to the vector of the momentum $I_{\omega,t+dt}$ referring to this instant. Therefore, in the next infinitesimal time interval dt, the rotation axis of the gyroscope will rotate by the same small angle $d\alpha$ in the plane of the vectors $\overline{I}_{\omega,t}$ and $\overline{M_t}$, which is identical with the plane of the vectors $I_{\omega,t+dt}$, M_{t+dt}. Thus the rotation axis rotates with the constant rotational velocity

$$\frac{d\alpha}{dt} = \omega_{pr} \tag{2.117}$$

which is the precessional angular velocity in the plane of $\overline{I}_{\omega,t}$ and $\overline{M_t}$. This movement is precession. The torque M causing it is, from a comparison of Eqs (2.115) and (2.116):

$$M = I_\omega \frac{d\alpha}{dt}$$

or $I_\omega = \Theta \cdot \omega$ and by taking Eq. (2.117) into account:

$$M = \Theta\omega \cdot \omega_{pr}, \tag{2.118}$$

from which the angular velocity of precession is:

$$\omega_{pr} = \frac{M}{\Theta\omega}. \tag{2.119}$$

Thus this angular velocity is proportional to the extraneous torque M and inversely proportional to the principal moment of inertia of the gyroscope perpendicular to the spin axis (symmetry axis), also to the angular velocity.

Such precession arises e.g. if a toy top has a supporting point below its centre of gravity, at a distance s. The angular velocity of precession is determined in such a case by the expression:

$$\omega_{pr} = \frac{mgs}{\Theta\omega}, \tag{2.120}$$

where

m = the mass of the top,
g = the acceleration of gravity.
Θ = the principal moment of inertia of the top with respect to the symmetry axis (spin axis).

A similar precessional motion is made by the rotation axis of the Earth: this axis makes an angle of about 23.5° with the normal to the ecliptic, i.e. the plane of the Earth's orbit. Therefore the Sun (and also the Moon) exerts on the rotating Earth a torque (due to the flattening of the Earth), and under the influence of this torque the axis of the Earth describes a precessional cone with a semi-vertex angle of 23.5° in 25,800 years. It should be noted, however, that neither the toy top nor the Earth's axis have a pure precessional motion, as the conditions that their axis of momentum should coincide with their axis of symmetry are not exactly fulfilled, and the precessions due to the Sun and to the Moon are added together. Therefore in both cases, for the toy top and for the Earth's axis, nutational movements appear in addition to precession.

The velocity of the Earth's rotation is:

$$\omega_E = \frac{2\pi}{24 \times 60 \times 60}\,\mathrm{s}^{-1} = 7.292 \times 10^{-5}\,\mathrm{s}^{-1},$$

and is less by many orders of magnitude than the rotational velocity ω of the gyroscope of $2\pi \times 333$ to $2\pi \times 500\ \mathrm{s}^{-1}$, corresponding to 20,000 to 30,000 revolutions/min.

If a torque characterized by a vector of constant length and constant spatial direction exerts an influence on a gyroscope, then its component perpendicular to the spin axis of the gyroscope causes precession. The result is that the gyroscope rotates in the plane determined by the spin axis and the component of the torque-vector perpendicular to it as long as this component does not become zero (when the direction of the spin axis coincides with the direction of the vector of the extraneous torque). In this case the precession effect ceases but the precessional motion continues due to the inertia of the gyroscope.

There are many applications for gyroscopes in instrument production and, within it, in surveying practice.

Free gyroscope. In order to enable the symmetry axis of the gyroscope to have any direction in space, the gyroscope should be suspended so that it can move freely in all three cardinal directions *(Fig. 2.114)*. These three directions may be the directions of the axes of a right-handed system of coordinates (x, y, z). Such a gyroscope is called a gyroscope with three degrees of freedom. Its technical realization is by the cardan suspension. The rotation axis of the gyroscope rotates, possibly free of friction, within an inner gimbal ring, which can wobble, being similarly rotatable within an outer gimbal ring. The gyroscope with cardan suspension has a spin axis in the x direction which is rotatable about the axes y and z. The z axis is vertical, the x axis lies in the horizontal plane in the North-South direction, and the y axis also lies in the horizontal plane but towards East-West. This system of axes rotates in the inertial system together with the Earth. The free gyroscope can rotate with three degrees of freedom: it can rotate around the x axis, the y axis and the z axis; its spin axis may point in any direction in space. Such a so-called direction gyroscope can be compared to a theodolite: the outer gimbal ring corresponds to the vertical axis of the theodolite in the direction z, around which rotations in the horizontal plane (e.g. bearing or azimuth) can be determined in the plane xy. The second inner ring contains the y axis corresponding to the horizontal axis of the theodolite around which the gyroscope can be tilted in the plane xz, i.e. the zenith angle or inclination can be measured. The x axis of the gyroscope corresponds to the line of sight of the telescope of the theodolite.

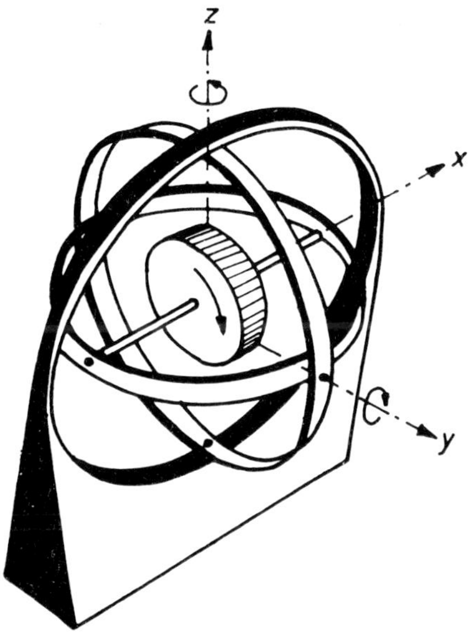

Figure 2.114. Horizontal (direction) gyroscope

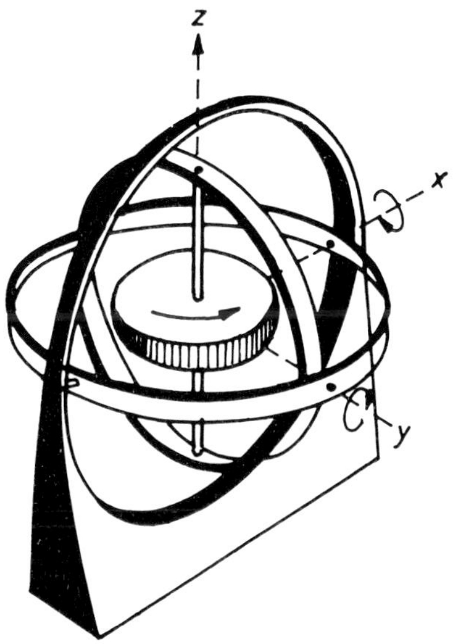

Figure 2.115. Gyroscope with a vertical axis

If the rotational plane of the gyroscope is in the plane xy *(Fig. 2.115)*, i.e. the spin axis coincides with the z axis, it is called a vertical gyroscope. By combining a direction gyroscope with a vertical gyroscope, two mutually perpendicular planes can be determined, or they can be stabilized in space. Their line of intersection is only parallel shifted in space, irrespective of the movements of the tripod on which both gyroscopes are mounted with cardan suspension. The two gyroscopes can be therefore used to set out constant directions in space, or in navigation to maintain a constant direction.

Gyroscopes for surveying. If the outer gimbal ring of the horizontal gyroscope is omitted in the situation shown in *Fig. 2.114* (Foucault's second experiment), then the gyroscope has only two degrees of freedom and the gyroscope can be tilted only around its axis y, and can make precessional movements *(Fig. 2.116)*. This is the basis of precessional gyroscopes. If such a gyroscope is set up at a station P at a geographical latitude φ, in the direction North-South, then it rotates with the Earth. It begins to wobble around the plumb-line and it begins to make a precessional movement. The precession lasts until the momentum vector I_ω of the gyroscope becames parallel to the rotation vector ω_E of the Earth. In such a case the spin axis of the gyroscope makes an angle φ with the horizontal North-South direction. Thus such a gyroscope can be used for the determination of the geographical latitude φ, or if it is known, then of the horizontal plane. Such a gyroscopic instrument with two degrees of freedom yields a so-called horizon-determining instrument which maintains its spatial direction, having a role in flight, space flight and in navigation. In surveying it is applied in aerial photography.

If the inner gimbal ring of the horizontal gyroscope is omitted (Foucault's third experiment), i.e. the gyroscope is fixed along the y axis, and the spin axis can rotate

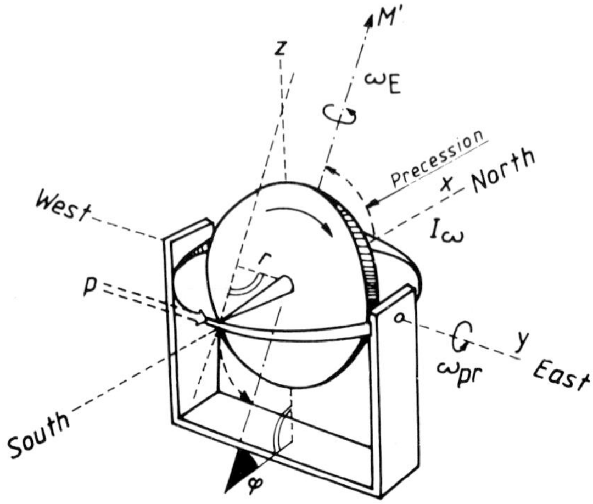

Figure 2.116. Inclination gyroscope

only around the z axis *(Fig. 2.117)*, then the north-seeking gyroscope, the gyro-compass or, if built together with a theodolite, the gyro-theodolite is obtained. At a station P at geographical latitude φ, the angular momentum vector I_ω of the north-seeking gyroscope can move from its original position only around its vertical axis due to the rotation velocity ω_E of the Earth; therefore it detects only the effect of the horizontal component of the rotation vector of the Earh, ω_E, and the rotating gyroscope makes a precessional movement in the horizontal direction. This effect lasts until the rotation axis of the gyroscope lies in the meridian plane through the observation point. Due to being fixed in the direction of the y axis, the north-seeking

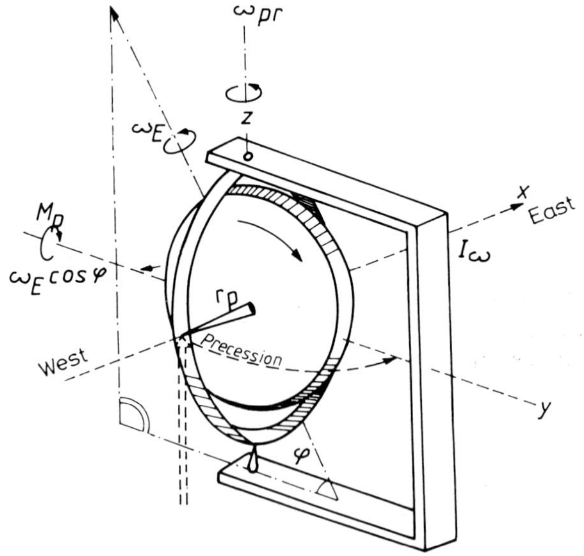

Figure 2.117. North-seeking gyroscope

gyroscope can have only a horizontal forced motion due to the gravitational force of the Earth.

If the point of suspension is above the point of intersection of the axis of the gimbal ring and the spin axis of the gyroscope, then a vertical self-aligning gyroscopic pendulum is obtained (a gyroscope forced into the vertical by the gravitational force). The elastic suspension and the inertia of the gyroscope prevent the precession from ceasing in the plane of the meridian, but the gyroscope swings in an elongated ellipse to the right and left of the meridian, as well as slightly upwards and downwards from the horizontal.

If the vector of the angular momentum I_ω makes an angle α with the meridian plane (the direction of North) then the torque due to the spin of the gyroscope and the Earth's rotation is:

$$M(\alpha)=\Theta\omega\cdot\omega_E\cos\varphi\sin\alpha=I_\omega\cdot\omega_E\cos\varphi\sin\alpha. \qquad (2.121)$$

The torque is greatest at the Equator ($\varphi=0$). At points of higher geographical latitudes the torque and the orientation accuracy of the gyroscope decrease and it swings more slowly. At the poles ($\varphi=90°$) the gyroscope does not work (here all directions are south and north respectively).

The differential equation of the precessional motion is:

$$B\frac{d^2\alpha}{dt}+\Theta\omega\cdot\omega_E\cos\varphi\sin\alpha=0 \qquad (2.122)$$

(where B is the principal transversal moment of inertia of the gyroscope), and its swing time is in the case of small angular deviations α:

$$T=2\pi\sqrt{\frac{B}{\Theta\omega\cdot\omega_E\cos\varphi}}=2\pi\sqrt{\frac{B}{I_\omega\omega_E\cos\varphi}}, \qquad (2.123)$$

which increases with increase of the geographical latitude φ.

Equations (2.121) and (2.123) show that, due to the very small rotational velocity of the Earth ω_E, the angular momentum of the gyroscope I_ω should be very large in order to obtain a sufficiently large momentum M and a sufficiently short swing time. As the increase of the rotating mass is limited by the bearing friction and the tension of the suspension, therefore in order to reach a sufficiently large I_ω, the rotational angular velocity of the gyroscope should be as high as possible.

The gyro-compass so constructed can be used to determine the direction of North or the angle between the direction of the line joining two arbitrary points and True North, i.e. the azimuth, over nearly the whole surface of the Earth with the exception of areas of very high latitude, independently of weather or time, both on the surface and at sub-surface points. The effect of smaller local anomalies of gravity can mostly be neglected, or it can be taken into account in the form of corrections.

2.5.2.2 Types of Gyro-theodolites

The requirements for gyro-theodolites are as follows:

The torque on the gyroscope should be large, the mass of the instrument as small as possible. It is an advantage if the instrument is easy to operate similarly to theodolites. Bearing friction, bearing wobble and disturbing heat generation should be small. With band suspended instruments the torsional momentum of the suspension band should be small, and the disturbing effect of the torsion must be eliminated by as exact a correction as possible. The instrument should retain its adjustment and safety of operation for a long time, even under field conditions.

There are systems operating on the principle of liquid support, band suspension and momentum adjustment, and systems with rigid and flexible clamping. These systems are also used in combination.

There are two possible ways for the determination of True North: by observation of swing and subsequent calculation, and immediate north seeking by electronic damping.

Gyroscopes with liquid support. The spin axis of the gyroscope is set into the horizontal, and it floats lightened together with the gyroscope housing in a liquid. It is held centred by a very thin suspension band or by a trunnion, the latter being loaded with only a very small weight. The advantage of this solution is that the dimensions of the suspension band can be small, therefore the disturbing effect of the band torsion remains negligible even in the case of swings with greater amplitudes. The disadvantages are the longer swing time, the liquid makes the instrument heavier and in low temperatures the instrument must be heated up.

Liquid support with band suspension is used in the instrument Girolit of Anschütz, developed at the Mining Academy in Clausthal at the beginning of the fifties, also in the instrument type MW (Meridianweiser) of the Mining Survey Institute in Bochum. Liquid support with a trunnion is used in the instrument of the All-Union Research Institute of Mining Survey in Leningrad.

Gyroscopes with band suspension. In these instruments the total weight of the gyroscope loads the suspension band. The dimensions of the band are greater, therefore corrections must be made. The torsional effect of the band depends on the amplitude (on the maximum deviation from the rest position): therefore the amplitude during measurement must be made small and, prior to measurement, the instrument must be set to North with a good approximation. A great proportion of modern instruments belong to this group.

Gyroscopes based on the principle of momentum equalization. The solution of problems of the mechanical and optical reduction of the gyroscope to centre, as well as of the technological problems of metal band production, led to the construction of other types of gyroscope, mainly in the UK and US. Let us set the north-seeking gyroscope (the gyro-theodolite) so that its z axis *1* (the output axis) is vertical and rotatable *(Fig. 2.118)*. The y axis *2* (input axis) is under the influence of the horizontal component $\omega_E \cos \varphi$ of the Earth's rotation ω_E, which induces a precessional motion of angular velocity ω_{pr} in both the output axis and, together with it, in the cylinder-shaped pendulum system. The precession induces an alternating current which is measurable with the detector *3*. The information so obtained can be used in two different ways.

In the case of the first solution *(Fig. 2.118)* the alternating current induced by the precession and detected by the detector *3* is amplified by the amplifier *4* then

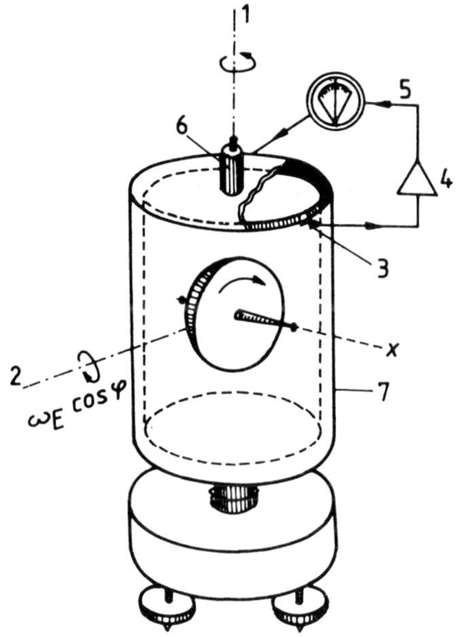

Figure 2.118. Gyroscope operating on the principle of momentum equalization

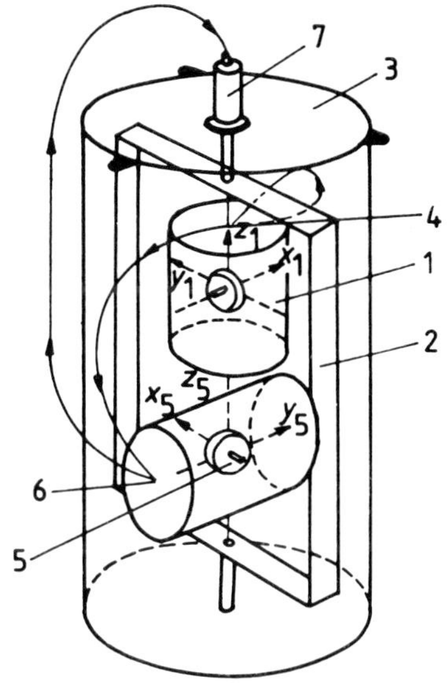

Figure 2.119. Gyroscope operating on the principle of momentum equalization with regulator (Kearfott, USA)

transformed into direct current which is brought by the electrical zero device *5* to the contra-rotating motor *6*. This motor exerts a torque on the output axis *1* which is equal in absolute value, but of opposite sign, to the precession, and which compensates the precessional motion of the gyroscope. The electrical zero device *5* gives the current intensity for the compensation and its polarity. During measurement the outer housing of the gyroscope *7* is rotated on the instrument base with a slow motion screw until the zero device reads zero. Then the gyroscope is freed from precessional motion and no current flows through the detector; the deviation of the rotation axis of the gyroscope from North is zero. The line of sight of the telescope of the theodolite above the gyroscope housing points towards North if it is parallel to the spin axis of the gyroscope. This north-seeking process is made by repeated approximations due to the inertia of the gyroscope and to other disturbing effects, and North can be found accurately only by making several corrections. Error in north-seeking due to wobble of the bearing can be eliminated by repeating the observation with opposite rotation of the gyroscopic motor, when the error is of the same absolute value but with opposite sign. The average of the two measurement results is free of the effect of this error.

The momentum of the gyroscope is given by Eq. (2.121) and a momentum of equal absolute value, but of opposite sign, is created by the contra-rotating motor. As the electrical signal is proportional to the angular deviation, this gyroscope is called a rate gyroscope. In this instrument the principle of momentum equalization is used. Such an instrument is e.g. the precise meridian seeker *PIM* (UK) and the automatic direction displaying gyro-theodolite *ABLE* (US).

The other solution *(Fig. 2.119)* takes into account the error δ in the vertical position of the standing axis z. Due to this error the projection $\omega_E \cos \varphi$ of the Earth's rotation velocity changes and an error of $\delta \tan \varphi$ is being brought into the azimuth determination. This error is eliminated by placing the north-seeking gyroscope *1* in a frame *2* suspended by a band, so that the axis y_1 is always horizontal and the input axis z_1, being always perpendicular to it, is vertical. The rotatable frame *2* is mounted in the external housing *3* and its vertical axis rotates on bearings. The precession is electronically recorded by the detector *4* and the electrical signal is projected on the output axis y_5 of another azimuth stabilizing gyroscope *5*. The axis x_5 of the second gyroscope is horizontal, its input axis z_5 is vertical, and the output axis y_5 is also horizontal. The electronic momentum brought to the gyroscope causes a precession of the axis y_5 around the axis z_5. This is also detected electronically by detector *6*. The amplified current controls a readjusting motor *7*, mounted on the housing, which rotates the frame according to the readjustment. The controlled readjustment lasts until the gyroscope points North, i.e. when the signal disappears on the detector. This solution together with the first is a combination of momentum adjustment and of band suspension.

Rigidly clamped north-seeking gyroscopic systems are not treated here in more detail as they are not used in surveying due to their lower accuracy.

2.5.2.3 Construction and Operation of Gyro-theodolites

The majority of gyro-theodolites presently used for surveying have pendulum-type suspensions. Therefore the construction of this type of gyroscopes will be described in the main.

Complete gyro-theodolites consist of four main constructional elements. These are, as seen in *Fig. 2.120*: the instrument tripod *A*, the sensor gyroscope *B* (north-seeking unit), the theodolite *C* (direction measuring unit), and the supply unit *D*, the current transformer generator.

The instrument tripod is either a tripod generally used for surveying instruments (Subsection 2.2.1), or a reinforced type of this. The reinforcement becomes necessary as the mass of the instrument can be significantly higher than that of traditional theodolites, and the pendulum system is sensitive to both vibration and torsion.

The main part of the gyroscopic sensor unit is the high revolution motor (20,000 to 30,000 revolution/min) with a large moment of inertia. The rotation axis of the motor (spin axis of the gyroscope) makes slightly damped swings about the plane, containing the vertical at the station and parallel to the rotation axis of the Earth, i.e. around the meridian plane. Gyroscopic motors are three-phase asynchronous electric motors. Recently synchronous motors are also beginning to be used. The axis rotates on ball bearings. The balls should be selected with great care to make the rotation axis free of wobble. The outer diameter of gyroscopic motors in present use is 20 to 50 mm, and their interior space is a vacuum or is filled with helium. The section of such a gyroscopic motor is presented in *Fig. 2.120, b*. In *Fig. 2.120, a* the axis of the gyroscopic motor *1* is nearly horizontal, depending on the previous setting; it is suspended above its gravity centre by a torsional tape *2*. The tape is made of special steel and has a nearly oblongue cross-section which is most advantageous for

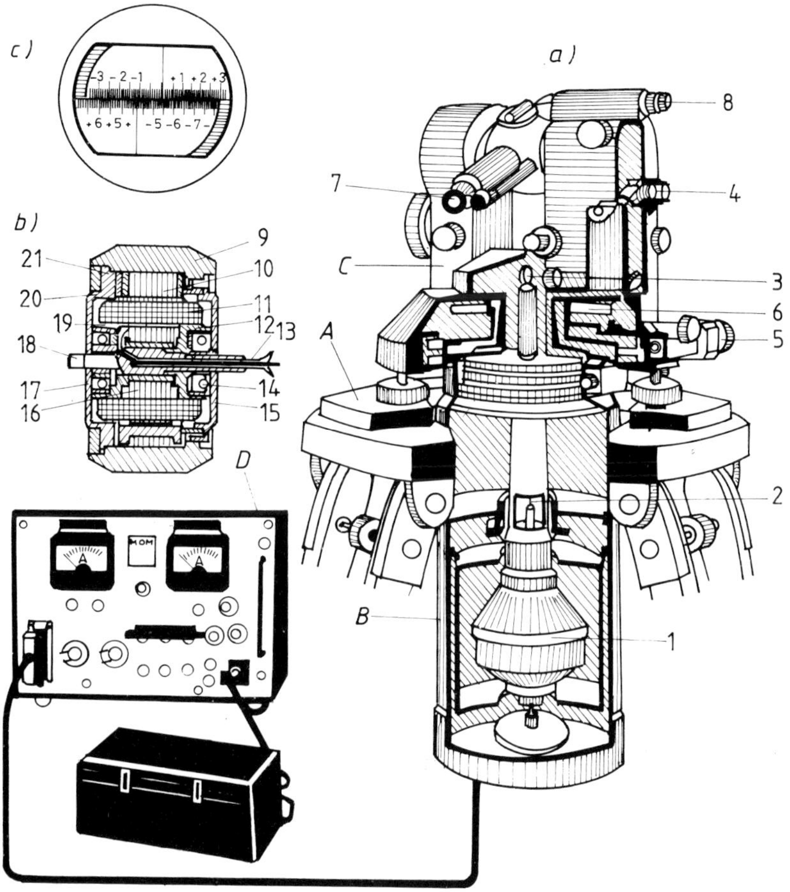

Figure 2.120. The gyro-theodolite *Gi-B1* (MOM, Budapest): *a)* cross section (with generator and power supply); *b)* section of the gyroscope motor; *c)* reading graduation; *A* tripod; *B* sensor gyroscopic unit, *C* theodolite, *D* power supply unit and generator; *1* motor housing; *2* suspension tape; *3* pendulum column with mirror; *4* autocollimator; *5* slow motion screw; *6* horizontal circle; *7* theodolite telescope; *8* auxiliary circle reading optics; *9* flywheel; *10* rotor ring; *11* three-phase coil; *12* and *19* outer rings of the rotor bearings; *13* three-phase conductor, *14* and *17* ball bearings; *15* and *20* motor housing; *18* rotation axis of the motor; *21* clamping screw; *16* stator

this purpose. The torsional momentum of the tape influences the swing of the gyroscope. A mirror *3* is mounted at the upper end of the attachment of the suspension tape to the swinging column; it reflects the image of the graduation of the autocollimator *4* into the field of view of the collimator for swing observations *(Fig. 2.120, c)*. The images of the graduations in the field of view of the autocollimator are needed for tracking the pendulum system, and to read the reversal points. Autocollimation optics with a split field of view is advantageous as, with it, the accuracy of the coincidence of the two signals with opposite motion, and of the determination of the pendulum position, can be significantly increased. The main graduation of the autocollimator is also used for the elimination of the changing torsion of the suspension tape, due to a rotation of the rotation axis of the pendulum, and for the resetting by rotation of the upper suspension point of the tape by the slow motion drive screw *5* of the upper part of the instrument. The movement of the pendulum is maximum at transit of the meridian plane, and its velocity decreases continuously when approaching the reversal points. The movement of the pendulum and of the graduation stops for some moments at the reversal points. During this standstill the position of the reversal point must be read on the horizontal circle *6* of the theodolite. It is advantageous if the reading on the circle can be made from the same body position as when the swing observations are made with the autocollimator. For this purpose an auxiliary reading device *8* is used, adjacent to the theodolite telescope *7*, in addition to the reading microscope of the horizontal circle.

The resetting of the upper suspension point of the tape by manual rotation (according to the rotation of the spin axis of the gyroscope) is very uncomfortable, moreover it may cause vibration. That is why an automatic photoelectric following device is used in the gyro-theodolites *Gi-B2* (MOM, Budapest), in the Soviet *MT-2* and in the *Lear* and *Siegler LMK 604.*

If the torsion-free position of the suspension tape does not coincide with the rest position of the gyroscopic swing, then a north position is obtained which deviates somewhat from True North, as it is burdened by the effect of the torsion. By rotation of the upper suspension point (with measurement by tracking) a constant tape torsion can be ensured during the measurement. If the tracking method is not used then the measurements are influenced by changing torsion. In both cases corrections for the torsion are being used if the torsion-free position of the suspension tape does not coincide with the zero graduation of the following autocollimator.

On the basis of the ratio of the torsional momentum M_{tors} and the gyroscopic momentum M_{pr} *(Fig. 2.121)*, the torsion correction $\Delta N = \sphericalangle 1, 2$ for a tracking

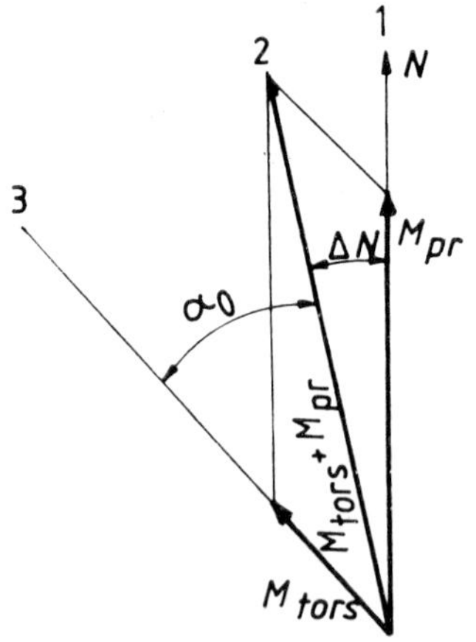

Figure 2.121. Ratio of momenta in the case of a tape suspension gyroscope

measurement, taking into account that the angles ΔN and α_0 are small, is:

$$\Delta N = \frac{M_{tors}}{M_{pr}} \alpha_0, \qquad (2.124)$$

where α_0 is the deviation of the torsion-free position *3* from the position *2* of the tracking (of the zero graduation). The latter can be determined by observing the so-called free swings (with the gyroscope not spinning).

For non-spinning measurements the corresponding correction is to be used:

$$\Delta N' = \frac{M_{tors}}{M_{pr} + M_{tors}} (\alpha_0 + \Delta N). \qquad (2.125)$$

If α_0, and thus also ΔN, are expressed in graduation units, then — to transform into seconds of arc — the value is to be multiplied by the value k of the graduation unit in seconds of arc (graduation constant). The ratio of the momenta M_{tors}/M_{pr} and $M_{tors}/(M_{pr} + M_{tors})$ is constant for a particular geographical latitude φ (if thermal and other disturbing effects are neglected); it can be computed if the momenta are known, or it can be determined experimentally. The latter method is the more advantageous as in that case temporal variations are included. For a tracking measurement the gyroscopic moment M_{pr} is the essential one, with time of swing T_1; for a measurement without tracking the essential momentum is $M_{pr} + M_{tors}$, with swing time T_2. The swing times are as follows:

$$T_1 = \frac{A}{\sqrt{M_{pr}}}; \qquad T_2 = \frac{A}{\sqrt{M_{pr} + M_{tors}}}, \qquad (2.126)$$

where A is a constant. From these the ratio of the two momenta is:

$$C_1 = \frac{M_{tors}}{M_{pr}} = \frac{T_1^2}{T_2^2} - 1; \qquad (2.127a)$$

$$C_2 = \frac{M_{tors}}{M_{tors} + M_{pr}} = \frac{T_1^2 - T_2^2}{T_1^2}. \qquad (2.127b)$$

For azimuth determinations of medium accuracy, the ratio of the momenta should be determined after each change of position of more than 100 km north or south.

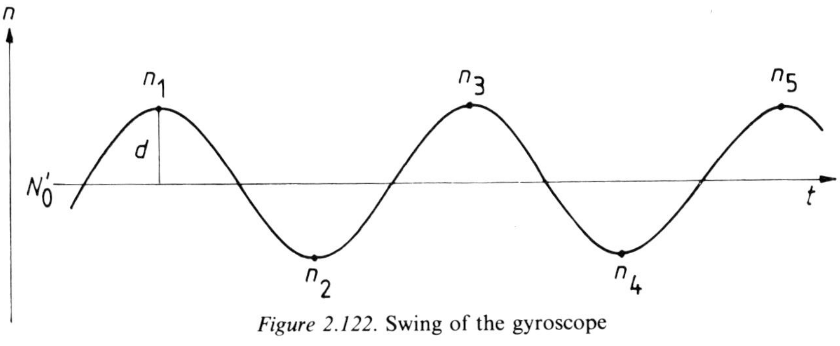

Figure 2.122. Swing of the gyroscope

For high precision measurements the best way is to use both methods at each station so that the swing times necessary for the determination of the momentum ratio are available directly from the measurements themselves.

Swing observations and their evaluation. The rotation axis of the gyroscope makes harmonic swings damped due to air resistance and bearing friction *(Fig. 2.122)*. If friction is neglected in the case of tracking measurements, then the motion of the gyroscope axis can be described by the following differential equation in the horizontal plane:

$$\frac{d^2\alpha}{dt^2}+K\alpha=0, \tag{2.128}$$

where α is the deviation from the meridian plane (for small deviations $\sin\alpha\approx\alpha$); t is measured from the instant of the start of the swing and

$$K=\left(\frac{M\omega_E\cos\varphi}{A}\right)^{1/2}=2\pi T_1^{-1}, \tag{2.129}$$

where M is the momentum causing the precession given by Eq. (2.118). The instantaneous angular deviation of the pendulum is, from Eq. (2.128):

$$\alpha=\alpha\sin(Kt+v)=a\sin\left(2\pi\frac{t}{T_1}+v\right), \tag{2.130}$$

where v is the initial phase.

At the reversal point the sine-term in Eq. (2.130) equals 1 and the amplitude a is the angular deviation of the reversal. It is advantageous to construct the instrument so that the ratio of the semi-major and semi-minor axes a and b of the flat ellipse described by the gyroscopic axis is: $a/b\approx200$ (e.g. as is the case for MOM gyro-theodolites). If the swing is undamped then the minor axis is in the meridian plane, supposing that the torsional effect of the tape is there zero. Due to the precessional swings the pendulum stops at the extreme positions, i.e. at the reversal points, for an instant and from the readings n_i on the horizontal circle corresponding to these positions *(Fig. 2.122)* the zero (rest) position N'_0 can be computed. If the positions of three consecutive reversal points are read, then

$$N'_0=\frac{1}{2}\left(\frac{n_1+n_2}{2}+\frac{n_2+n_3}{2}\right). \tag{2.131}$$

In case of more than three reversal points one has:

$$N'_0=\frac{1}{r-2}\left[\frac{1}{4}(n_1+n_r)+\frac{3}{4}(n_2+n_{r-1})+\sum_{i=3}^{r-2}n_i\right], \tag{2.132}$$

where the values n_i are the readings on the horizontal circle corresponding to the reversal points and r is their number.

The torsion-free position of the suspension tape α_0 can be also computed using Eqs (2.131) and (2.132) by reading the reversal points of free swings in the non-spinning mode. This is taken into account by Eq. (2.124) as a correction to a north determination by gyroscope. If the determination is made at a constant position of the upper suspension point of the tape, then the correction given by Eq. (2.125) is also to be applied. In this case the reversals are not read on the horizontal circle of the theodolite, but on the graduation of the autocollimator. The reading corresponding to the zero position can again be computed using either Eq. (2.131) or Eq. (2.132), giving a value n_0' differing from N_0', but this computed value must be multiplied by the graduation constant k in order to get a value which can be correctly added to the basic reading on the horizontal circle (to the reading on the horizontal circle corresponding to approximate North). The value k is the graduation unit expressed in seconds of arc.

If the line of sight of the telescope is parallel to the rotation axis of the motor (to the axis of momentum), then the azimuth A is obtained from the result l of the direction measurement to an external point and from the zero position N_0 corrected by the necessary corrections:

$$A = l - N_0. \tag{2.133}$$

As the parallel condition is usually not valid, a certain instrumental constant Δ must also be taken into account, thus:

$$A = l - N_0 + \Delta. \tag{2.134}$$

This constant Δ consists of two parts: the angle Δ_1 between the rotation axis of the motor and the normal to the tracking mirror, and the angle Δ_2 between the optical axis of the collimator and the line of sight of the telescope of the theodolite *(Fig. 2.123)*. In order to keep these values constant, the gyro-theodolite must be handled with care, it should be protected against shock and vibration, and it should be shaded from solar radiation. The instrumental constant can be determined or checked by calibration measurements carried out at geodetic stations with known astronomical azimuths.

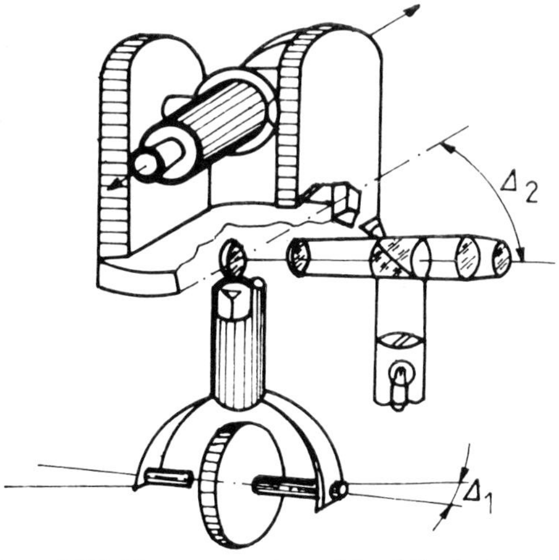

Figure 2.123. Components Δ_1 and Δ_2 of the instrument constant Δ

The power supply unit, the transformer and the regulating generator. Power for the gyroscope motor and for the necessary lights is supplied from batteries (in field work) or mains current is used with a transformer (e.g. in mines where there is a mains supply). In order to ensure the regulated three-phase current for running up the motor and for its operation, the current is fed from the current source through a transformer and a regulating generator to the gyro motor. The generator must provide voltage and frequency stability of the current, and quick running up of the motor as well as braking it after the end of measurement, and it should also regulate the automatic tracking.

The three-phase current of the necessary output is generated in semi-conductor transformers in three steps *(Fig. 2.124)*. The DC *1* is transformed by the low-output transmitter generator into AC, then by the phase selector *2* into three-phase current. The three identical amplifiers *3* amplify the output of the three-phase current up to the level necessary for the fast running of the gyro motor *4*.

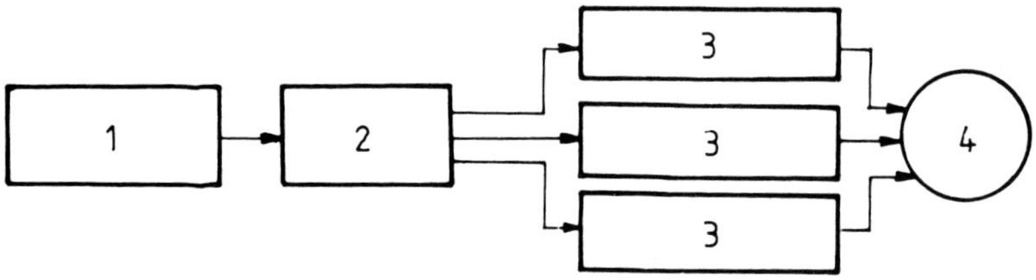

Figure 2.124. Blok diagram of the generation of three-phase current (in three stages) as used in the gyro-theodolite *Gi-B2* (MOM, Budapest): *1* transmitter genarator; *2* phase divider; *3* power amplifier; *4* gyroscope motor

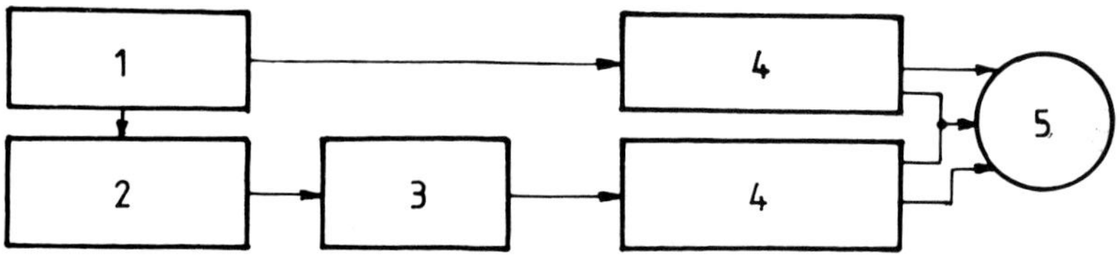

Figure 2.125. Another means of current transformation; *1* transmitting generator; *2* phase shifter ($\varphi = 90°$); *3* preamplifier; *4* power amplifier; *5* gyroscope motor

In several instruments the current is transformed into three-phase current at the output circuits of the power amplifier by short-circuiting the secondary coils of the output transformers *(Fig. 2.125)*. In simpler instruments, instead of three amplifiers only two are used. The generators of medium and low accuracy instruments operate on this basis.

The accuracy of the voltage and frequency regulation of the generator strongly influences the measurement accuracy of the instrument. Therefore the frequency of the transmitter generator is quartz-stabilized in high precision gyro-theodolites. As an example, the block scheme of the gyro-theodolite *Gi-B2* (MOM, Budapest) is presented *(Fig. 2.126)*. The vibrations of the two quartz generators operating at

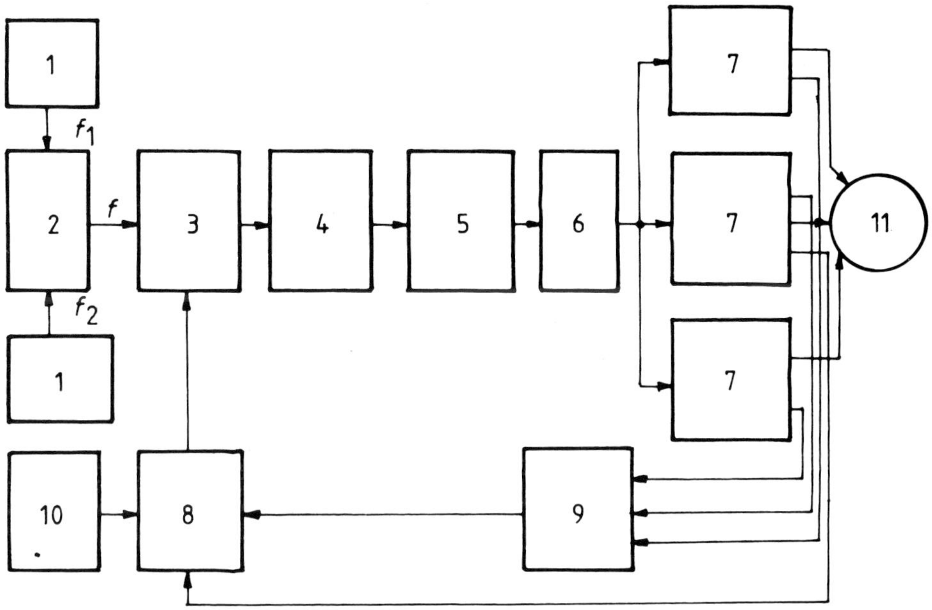

Figure 2.126. Quartz-stabilization of the frequency of the transmitter generator: *1* quartz generator; *2* mixer; *3* regulating amplifier; *4* two-stage middle amplifier; *5* power preamplifier; *6* phase divider; *7* final power amplifier; *8* voltage regulator; *9* three-phase rectifier; *10* voltage stabilizer; *11* gyroscope motor

frequencies of $f_1 = 8333.3$ Hz and $f_2 = 7916.6$ Hz *1* after mixing in the mixer *2* produce a differential frequency of $f = f_1 - f_2 = 416.7$ Hz. This voltage passes through the amplifier *3* and reaches a two-stage middle amplifier *4* and the power amplifier *5* from which it is led to the phase splitter *6* where it is transformed into a three-phase voltage. The three final amplifiers *7* amplify the power of the three-phase voltage to the necessary level. The three-phase voltage is kept at the necessary level by the voltage regulator *8*. After rectification *9* part of the output voltage reaches the voltage regulator *8* where its value is compared to the constant stabilizing voltage *10*. If the voltage deviates from the prescribed value, a signal is transmitted by the voltage regulator which affects the regulating amplifier and the gain is changed in the necessary direction. Thus a constant voltage is ensured on the stator coils of the gyro-motor *11*.

A precondition for the correct operation of the generator is to keep the operating temperature at a constant level. Modern generators are manufactured with printed integrated circuits giving smaller dimensions and an increase of accuracy.

2.5.2.4 Some Typical Gyro-theodolites

For surveying practice at present gyro-theodolites are usually of the pendulum type or of the momentum adjusting type, or both principles are combined. From the point of view of the methods used for the determination of North the following types are distinguished: 1. systems based on the observation of swings, where the direction of true North is obtained from the observed values of the swings, 2. north-seeking systems where the gyroscopic axis aligns itself into the north direction by an aperiodic damping of the motion, and 3. systems based on the principle of momentum adjustment.

The gyroscopic sensor and the theodolite can also be connected by several methods: in gyro-theodolites both units form a single rigidly mounted instrument, whereas gyroscope attachments can be fitted on theodolites which can be used individually. There are also so-called direction gyroscopes.

Pendulum-system gyro-theodolites and gyroscopic attachments for swing observations. (These instruments often have similar form, but slightly differing construction. Only a few types will be discussed here in more detail.) The principle of the construction of the gyro-theodolite *Gi-B1* (MOM, Budapest) (see *Fig. 2.120*) is similar to that of *Gi-B2* (MOM), but only the latter will be described. The gyro-theodolite *Gi-B2* is provided with an automatic tracking device *(Fig. 2.127)*. The gyroscopic motor is driven by a transistorized generator, or rather a generator with printed circuits,

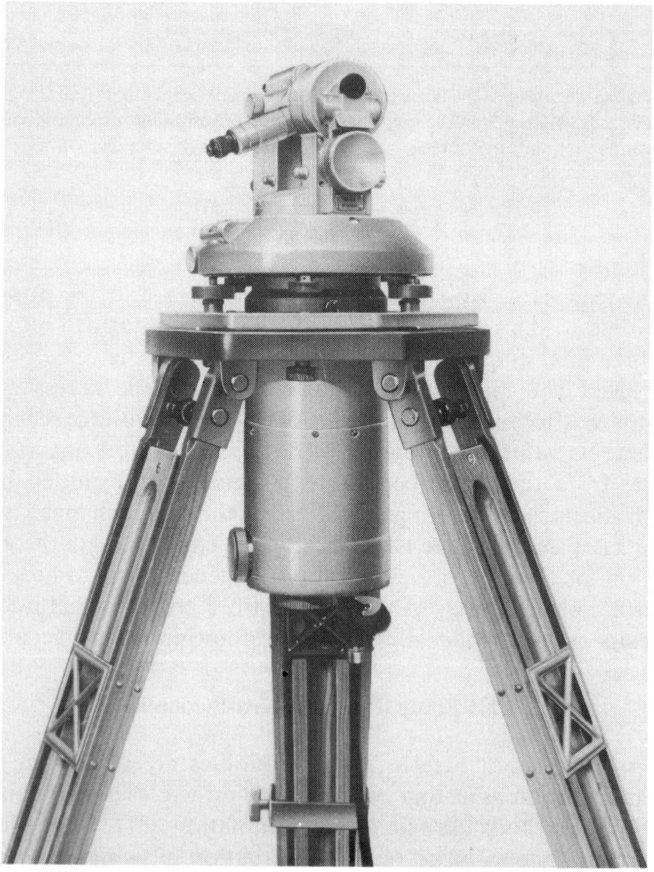

Figure 2.127. The gyro-theodolite *Gi-B2* (MOM, Budapest)

a)

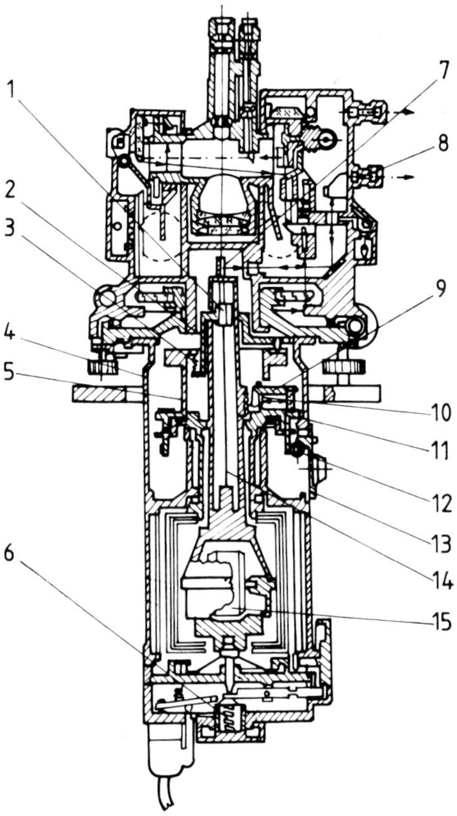

Figure 2.128. a) section and automatic tracking system; *b)* partial view of the gyro-theodolite *Ci-B2* (MOM, Budapest). *1* upper suspension point of the suspension tape; *2* tape holder and tensioner; *3* tape tensioning spring; *4* housing of the sensor unit; *5* the tape holder with braking disc; *6* security caging device; *7* upper swing mirror; *8* autocollimator to observe the swings of the pendulum; *9* casing of the autocollimator with tumb wheel; *10* autocollimator for tracking; *11* lower mirror for tracking; *12* wheel to set the zero point with a worm screw; *13* lever and arm to adjust the zero position; *14* torsional suspension tape; *15* gyroscope motor; *16* light transmitting slit; *17* photo resistors; *18* light projecting prism of the automatic tracking system

b)

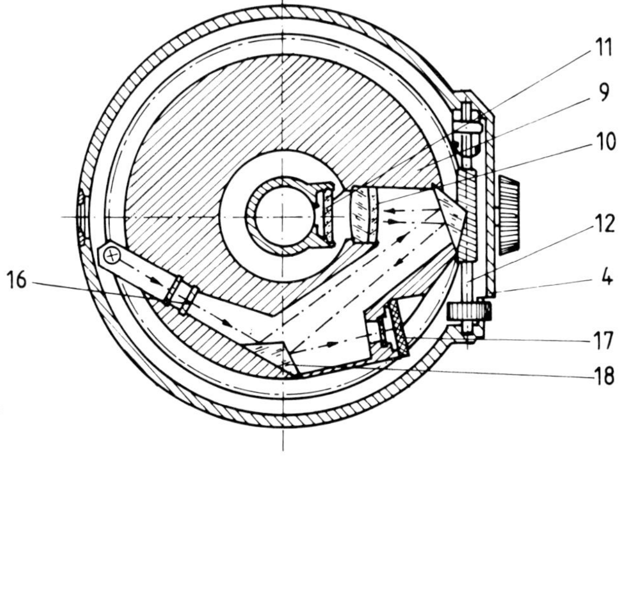

controlled by two quartz crystals in a thermostatically controlled oven, operating at a frequency of 416 Hz with a frequency stability of 3×10^{-5} Hz/1 °C. The power is supplied from 12 V batteries or by transformed mains current. The motor of the gyroscopic unit has a tape suspension and runs at 24,000 revolutions/min. An autocollimation optical system connects the reversals of the swings with the horizontal circle of the theodolite. The image of the graduations in the autocollimator is split and the two halves have opposite motions to increase the accuracy of reading.

The instrument has a reinforced lower part compared with traditional second reading theodolites. It is hollow to enable the mirror mounted on the suspension system of the gyroscope to be observed through the telescope of the autocollimator. The theodolite has a two-stage horizontal slow-motion screw system with an infinite worm thread which is adjustable on both sides.

The gyroscopic unit is enclosed by permalloy screening plates to eliminate disturbing magnetic effects. The gyroscope motor has a life of at least 500 hours of operation. The swinging part consists of the gyroscope motor *15*, which produces the orientation momentum, and of the suspension tape *14 (Fig. 2.128)*. The upper suspension point *1* of the tape can be slightly rotated by the thumbwheel *12* in order to regulate the zero point: it is advisable to determine the torsion-free position of the suspension tape prior to measurement. If it deviates by three or more divisions from the zero division of the autocollimator, then the zero point should be adjusted. The upper suspension point of the tape has a friction connection to the autocollimator which is stable during measurement. The normal to the mirror on the swinging part *11* and the normals to the tape at its upper suspension point and to the collimator *7* do not change their relative position during measurement. Friction ensures the adjustment of the zero-position of the torsion.

If the upper suspension point of the tape *1* is rotated continuously, simultaneously with the motion of the spin axis of the gyroscope, then the changing torsional effect of the band can be eliminated. It is preferable to use automatic tracking rather than manual tracking. The mirror *11* of the automatic tracking system projects the light signal controlling the tracking *16* to the photo resistors *17* by the optical system of the autocollimator *10*. The former change their resistivity depending on the intensity of the light falling on them. Depending on the position of the mirror *11*, the right or the left resistor receives light. The electronic system thus becomes asymmetric and controls a servo-motor. The servo-motor rotates the optical system, together with the upper suspension point *1* of the tape, by the worm wheel *9* to make the normal to the lower mirror *11* parallel to the axis of the optical system *10* (to the collimator axis), i.e. the photo resistor system becomes symmetrical once again.

If the tape is deformed due to the current, to ageing or to the dynamic load, the torsion zero point is moved: the normal to the upper swinging mirror *7* is no longer parallel to the optical axis of the autocollimator *8*. The parallel position can be restituted by rotating the worm wheel *12*. For this, the framework *5* is fixed to the outer housing *4* by the lever *13*. Then the cylinder *5* fits rigidly with the tape clamping *2* and the zero point of the tape torsion can be regulated by the worm wheel *12*.

To arrest the motion, the swinging part is caged between two cones and two spherical surfaces by a lever and a spiral disk system. In order to hold the tape in a constant tension, the spring *3* lifts the upper suspension point of the tape *1*. The current is supplied through two spiral springs and the suspension tape. At start the current supply is about 500 mA, therefore the instrument must be caged before starting and before braking. During this time auxiliary contacts unload the sensitive parts (spirals anf tape). Braking is only possible when caged; this is ensured by the safety caging device *6*.

The gyroscope motor rotates in a vacuum, thus the necessary driving force is small and the deformation, and the shift of the centre of gravity due to ageing, is also minimum.

In the case of automatic and manual tracking, the two parts of the image of each division are moving in opposite directions (see *Fig. 2.120, c*), and the main graduations must be brought continuously into coincidence. The reversal points must be read when the images of the graduations are completely stationary.

By switching off the automatic tracking system and not tracking by hand, the instrument can also be used for non-tracking measurement. The condition for this is that the amplitude of swing should be less than 40 min of arc. The positions of reversal are read on the graduations. Similar readings are made for the determination of the zero position of the tape torsion (by observing free swings). The optical axis of the autocollimator and the line of sight of the theodolite telescope are nearly perpendicular, thus the instrumental constant is about 90°.

For measurements free of magnetic disturbances the orientation of the instrument toward North is significantly assisted by the compass, supplemented by optical centring. If the compass is not used, the results of previous measurements must be taken into account or the approximate direction of North is determined by preliminary gyroscopic measurement. In addition to reversal point methods with and without

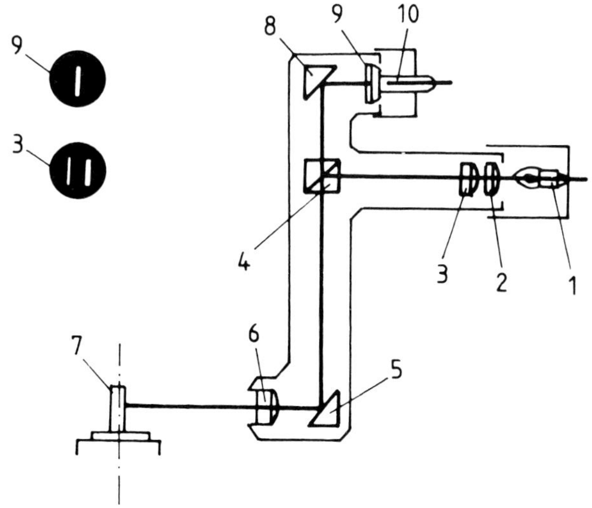

Figure 2.129. Principle of automated transit time measurement: *1* illuminating lamp; *2* condenser; *3* reticule; *4* semi-transparent prism; *5* mirror; *6* objective lens; *7* pendulum mirror; *8* mirror; *9* reticule; *10* light sensor

tracking, the *MOM Gi-B* gyro-theodolites can also be used for transit method measurements: the transit times of the pendulum through the divisions of the autocollimator are measured by stop-watch or by automatic timing, from which the reading on the horizontal circle corresponding to North can be computed. If the time necessary for the right-hand path is equal to the time for the left-hand path, then the direction of true North has been found. (Transit times can be measured by stop-watch with an accuracy of 0.1 s at best.)

The principle of *Pusztai–Beskó method* of measuring transit times *(Fig. 2.129)* is as follows. There are two transparent strips on the opaque reticule *3* of the autocollimator. Their images, as strips of light, are projected by the light source *1* of the autocollimator, via a light beam made parallel by the lens *2*, through the mirror *5* to the mirror *7* of the pendulum, and from there back to the elements *6*, *5*, *4* and then via the mirror *8* to the blackened reticule *9* which has a transparent slit. If the axis of the gyroscope crosses the meridian, the light strips projected onto the reticule *9* coincide with the slit, and light falls on the light detector *10*. The electric pulses generated by it cause the transit times to be recorded by the pulse counter, which is stabilized by the quartz-clock with an accuracy of one hundredth of a second. From these times the zero position of the pendulum can be computed. The *Gi-B21* (MOM) and the *GYMO Gi-B1A* (manufactured jointly by MOM, Budapest and Tellurometer) are instruments which can be operated in this manner.

A mean square error of ± 8 to 12 seconds of arc can be attained by the gyro-theodolite *Gi-B1* (MOM) in the determination of azimuth either by four reversal readings or by using the simple transit method with corresponding repetitions. The mean square error of the gyro-theodolite *Gi-B2* is ± 6 to 10 sec of arc. If reversal measurements and transit measurements are combined, and both methods with and without tracking are used, the correction coefficients most exactly corresponding to the measurement conditions can be determined. This means an increase in accuracy of about 50 percent. The accuracy attained with automatic transit time measurement is 2 to 3 sec of arc (by measuring 6 reversals and the corresponding transit times).

The "Kreiseltheodolit" *K1-1* (Fennel, FRG) also has a gyroscopic pendulum suspended by a metal tape. The current is supplied by spirals located horizontally around the cylindrical column of the gyroscope. The swings of the pendulum can be observed optically through an autocollimator. The amplitude of the swings can be regulated by a pneumatic brake. The instrument has a manual tracking device. Two kinds of gyroscope motors are used, with different accuracies in the determination of North (± 15 sec of arc and ± 30 sec of arc).

The gyro-theodolite *MRK 2* (Freiberger Präzisionsmechanik, German Democratic Republic) *(Fig. 2.130)* is a modified version of the earlier type *MRK 1* (Meridianrichtungskreisel) with a tape suspended pendulum gyroscope. Amplitudes are reduced by eddy-current braking. The swing time is relatively short (about 5 minutes). The generators of the gyro-theodolite *MRK 2* require a 24 V battery supply.

The gyro-theodolite *MW10* (Meridianweiser) was constructed cooperatively by the Mine Survey Institute in Bochum and Fennel *(Fig. 2.131)*. It has a relatively small gyroscope motor with a cylindrical housing floating in a liquid. It is held to the centre of the gyro housing by a suspension tape of very small dimensions — due to its lack of load — nevertheless it still has a torsional effect. The liquid support was used

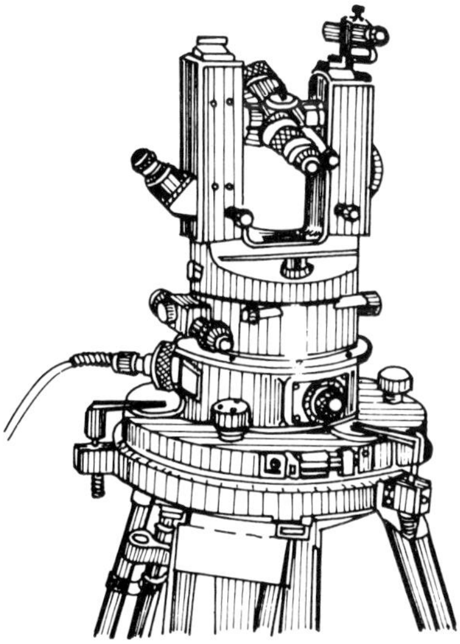

Figure 2.130. Gyro-theodolite *MRK 2* (Freiberger Pränzisionsmechanik, GDR)

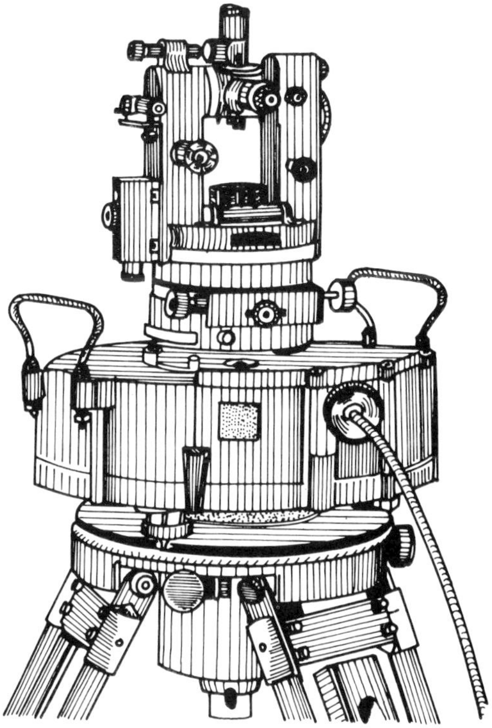

Figure 2.131. Gyro-theodolite *MW 10* (Meridianweiser, Fennel, FRG)

mainly for damp-proofing, necessary in mine surveying (spark-free connections). This instrument has the first sign of integration as the gyro-theodolite and the generator are in a single unit. The gyro-theodolite *MW 10* has not come into wide use. Among the many predecessors of this instrument the north-seeking gyro-theodolite *MW 4a* made in Bochum has a mean square error in north determination of ± 10 sec of arc. A later development is the instrument *MW 7*.

The gyro-compass *VNIMI MV 2* (Mine Surveying Institute, Leningrad) has liquid support on a different principle. Pin bearings are used to maintain centrality.

Some types of gyro-attachments to be attached to theodolites, to enable the theodolites to be used as gyro-theodolites, are now presented.

The greatest problem with these instruments is the stability of the instrumental constant. It is quite difficult to eliminate the error effects due to metals and other optical material, which have different thermal expansion coefficients, providing rather complicated mechanical and optical connections between the spin axis of the gyroscope and the line of sight of the theodolite telescope. In case of the gyro-attachment *Gi-C2* (MOM, Budapest) *(Fig. 2.132)*, the optical axis of the collimator telescope, for observation of the gyroscopic swings, and the line of sight of the theodolite telescope make an angle which is controlled by a base prism made of quartz, because this angle is one of the elements of the instrumental constant *(Fig.*

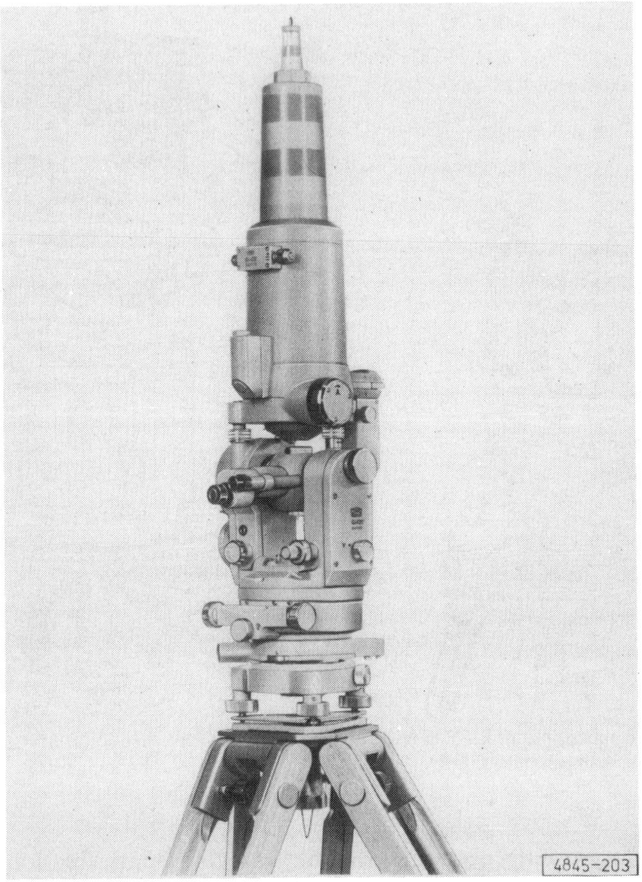

Figure 2.132. Gyro-attachment *Gi-C2* (MOM, Budapest)

2.133). This prism enables the optical axis of the collimator and the line of sight of the telescope to be made parallel within a few seconds, using the adjusting screw *5*.

The construction of the *Gi-C2* is similar to that of gyro-theodolites *(Fig. 2.133).* The swinging system in the caged position is lifted from the field of view of the glass prism *7* and of the objective lens *6*. In such a case only the image of the reticule of the autocollimator *1* can be seen in the telescope. When the optical axes of the collimator and of the theodolite telescope are set parallel, the direction pointing system is rotated by the screw *5* until both cross-hairs and, with a telescope with a split field of view, the upper and lower images of the cross-hairs coincide. If the swinging system is uncaged by the manual caging screw *3*, then the glass prism *7* is lowered between the reticule *1* and the objective lens of the collimator *6*, and the image of the graduation *2* comes into view. The movement of the pendulum can thus be tracked by the theodolite telescope.

The power supply for the motor of the gyro-attachment *Gi-C2* comes from a three-phase frequency stabilized generator of 410 Hz, supplied by 24 V batteries.

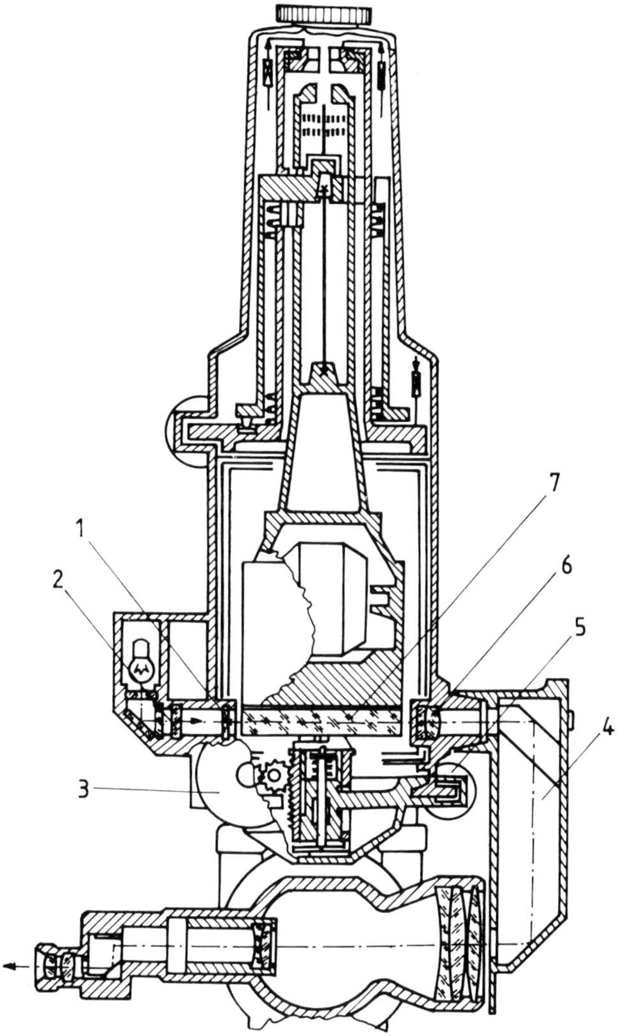

Figure 2.133. Construction of the gyro-attachment *Gi-C2:* *1* reticule of the collimator for parallel setting; *2* graduated reticule, *3* caging thumb-wheel; *4* base prism; *5* screw to adjust the instrumental constant; *6* objective lens of the sensor unit; *7* glass prism

On the theodolites *Te-B* (MOM, Budapest) the ease of use of the attachment of the gyroscope and of the tracking is enhanced by a horizontal slow-motion screw system with an infinite thread, which is advantageous for approximate north-seeking (no stops). In normal conditions with four reversals the mean square error of azimuth determination with the *Gi-C2* gyro-attachment is ± 15 to 20 sec of arc.

The gyroscope attachment *Wild GAK 1 (Fig. 2.134, a)* has a slow-motion horizontal screw with a range of $\pm 5°$. The telescope and the autocollimator are independent.

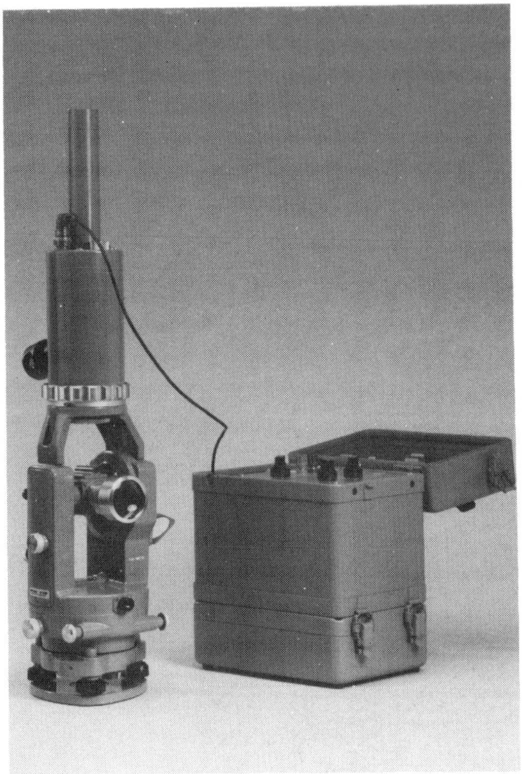

a)

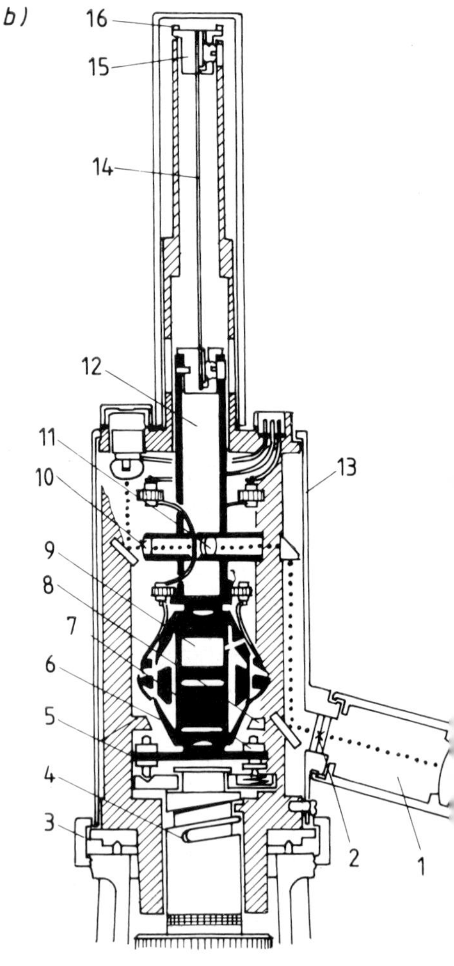

Figure 2.134. a) the gyro-attachment *GAK 1* (Wild, Heerbrugg) for the theodolite *Wild T16*; *b)* section of *GAK 1*; *1* reading telescope; *2* graduation of the autocollimator; *3* supporting bridge; *4* manual caging device; *5* swing attenuator; *6* spring; *7* arresting pin; *8* support for the caging pin; *9* gyroscope motor; *10* reticule plate of the gyroscope; *11* collimator lens; *12* framework column; *13* window; *14* suspension tape; *15* upper suspension point of the tape; *16* regulating screw

The Ni–Cd battery (12 V) and the current transformer generator (115 V; 400 Hz) are in a single unit from which the battery can be removed for charging. The generator and the sensor are connected directly by a cable (and not through the theodolite).

The caging device *4 (Fig. 2.134, b)* of the gyroscope attachment *GAK 1 (Fig. 2.134, a)* is on the base of the gyroscopic sensor unit containing the motor *8*, between the two support points of the connecting bridge. The objective lens of the autocollimator can be unscrewed and removed. The autocollimator very much simplifies the optical path necessary for tracking.

In pendulum gyro-theodolites — mainly in types of lower accuracy — a most advantageous integration process has begun. Modern elements of precise engineering, printed and integrated circuits, low power gyroscope motors and small-sized Ni–Cd bar batteries allow a significant decrease of dimension and mass as e.g. in the case of the low accuracy instrument *MW 50* (Bochum) *(Fig. 2.135)*. By using AC asynchronous motors as gyroscopes the current supply is so low that low capacity batteries can also be used and there is no danger of sparks. The theodolite together with the gyroscope, the generator and the current supply make a closed unit. The mean square error of a north determination is ± 1 min of arc. The very low mass of the instrument should be mentioned (about 7 kg). Recently a more accurate small-sized gyro-theodolite has been developed, type *MW 77* (± 5 sec of arc).

The gyro-theodolite *MVT-3* (VNIMI, Leningrad) *(Fig. 2.136)* has an integrated

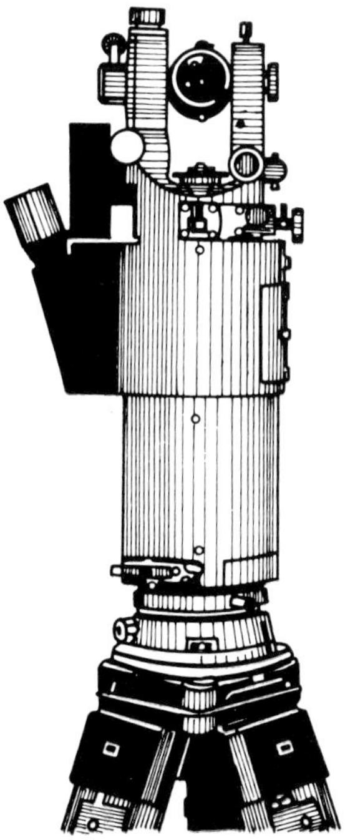

Figure 2.135. Gyro-theodolite *MW 50* (Mine Surveying Institute Bochum, FRG)

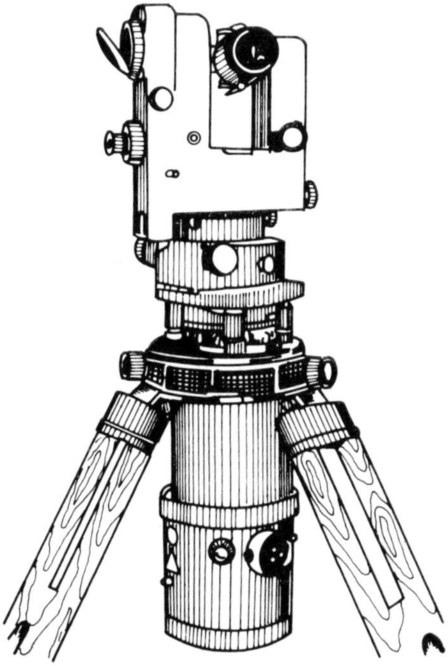

Figure 2.136. Gyro-theodolite *MVT-3* (VNIMI, Mining Research Institute, Leningrad)

generator and supply unit. The mean square error of the determination of North is ±30 sec of arc. The swing time of the pendulum is only 2 minutes (the free swing also lasts 2 minutes). The instrument is damp-proof.

Pendulum gyro-theodolites using the principle of momentum equalization. The north-seeking instrument *ALINE* (Lear–Siegler, US) is an automatic, low mass inertial system *(Fig. 2.137/a)*. Its gyroscope motor is suspended on a thin tape and the torsional effect of the tape, through which the current is fed, is negligible. The effect of temperature changes and of ageing are compensated by electrical counter-torques. In normal conditions the spin axis of the gyroscope is swinging symmetrically with respect to the meridian plane. The gyroscope is connected to a spin adjusting device. An inductive detector and torque generating system is built in. The detector transforms the angle between the pendulum body and the spin adjusting device into an electric signal which is then amplified, fed to a servo-motor and to the spin adjusting device. The motor is connected to a tacho-generator, the latter to a torque-generating unit *(Fig. 2.137/b)*. The sense of rotation is determined by the direction of the current flow, the current itself yields the angle of rotation and is continuously decreasing. Together with this, the amplitude of swing also decreases. This situation lasts until the spin axis of the gyroscope motor is aligned into the north direction when the counter-torque becomes zero (north-seeking by continuous approximation). The direction-measuring part is a theodolite Kern DKM 1. The mean square error of azimuth determination is ±25 sec of arc which will be decreased to ±10 sec of arc in the near future. The net time of a measurement is 12 minutes.

A similar principle is also used in the *PIM* precise meridian-determining instrument (UK).

The meridian gyroscope *NK 10* (Bodenseewerk, FRG) has a gyroscope with a tape suspension and with an electronic detector and torque-generating device. The torque on the gyroscope due to the Earth's rotation is detected by a torque detector and then compensated electronically. In a stationary state the current on the torque detector is proportional to the directional momentum, and thus to the deviation of the gyroscope axis from North. The electronic measurement of the gyroscope swing consists of the detection of the precessional movement, of its transformation into an electronic signal, of amplification and generation of the counter-torque. The latter has an absolute value corresponding to that of the precessional torque, but with opposite sign. By transforming the torque into an angular rotation the deviation of the pendulum from North is obtained, then the instrument is rotated through this angle by the theodolite, to point to true North. The mean square error of the simple azimuth determination is ±1 min of arc.

The instrument *Girolit II* (Anschütz, FRG) is a tape-suspended gyroscope floating in a liquid. The upper suspension point of the tape is contra-rotated by means of the directional momentum of the gyroscope, with a certain phase shift. The gyroscope sets to North rather quickly. The swing time is short (6 min) due to the relatively high mass of the gyroscope. The upper part of the theodolite, giving the reading on the horizontal circle, is rotated by a servo-motor by the same amount as the movement of the gyroscope axis so that, when the gyroscope axis lies in the north direction after damping, the telescope of the theodolite points in the same direction.

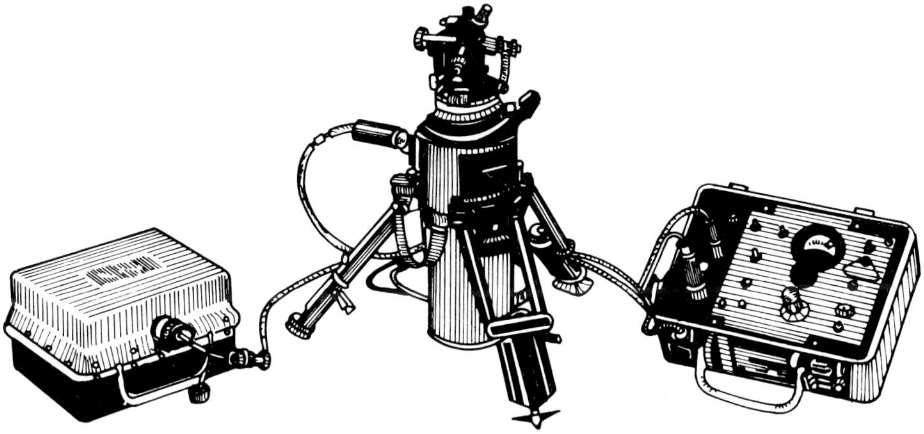

Figure 2.137/a. Gyro-theodolite *ALINE* (Lear–Siegler, USA)

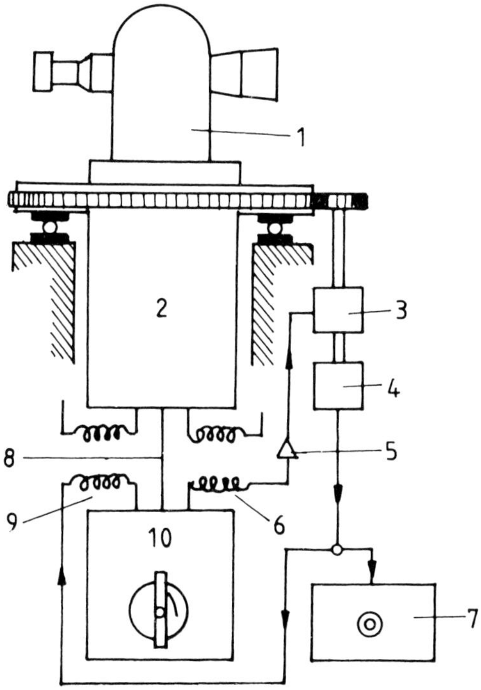

Figure 2.137/b. Construction of the gyro-theodolite *ALINE*: *1* theodolite; *2* rotation adjusting cylinder; *3* servo-motor; *4* tachogenerator; *5* amplifier; *6* detector; *7* north adjustment; *8* suspension tape; *9* torque generation; *10* housing of the gyroscope

The instrument *NSK 4* (Teldix, FRG) has electronic damping. The gyroscope is stiffened by an upper tension tape and by an automatic clutch, thus it is a perfect system with two degrees of freedom. The instrument seeks North with an accuracy of some tenths of degrees, and it is used mainly for navigational purposes.

Experiments were made to damp the axis of the gyroscope aperiodically and thus to reach true North quickly, similarly to a compass. The damping was made at first by a liquid (mercury), then by eddy currents. The time of the measurement decreased considerably, but the mean square error increased (±5 min of arc).

The most important parameters of some well-known gyro-theodolites are presented in *Table 2.5*.

Table 2.5. Characteristic data of some models of gyro-theodolites

Type	Construction*	Mass (kg)	Measuring time (min)	Oscillation time at $\varphi = 50°$ (min)	Momentum of impulse 10^6 g cm^2 s^{-1}	Mean square error of north-finding
			Gyro-theodolites			
MOM Gi–B3	1; 2	17	25	9	4.1	±3...5″
MOM Gi–B11	1	17	25	9	4.1	±3″
GYMO–Gi–B1A	1	17	40	11	4.1	±2...3″
Fennel KT1	1	14	25	8	3.7	±15...30″
LMK 604	1; 2	20	20	8	8.4	±10...15″
VNIMI MT1	1; 2	16	25	10	10.2	±10...15″
FPM–MRK 2	1	18	20	4	2	±15″
Fennel MW 10	1; 3; 4	33	25	8	1.8	±15″
VNIMI MV2	3; 4; 6	23	35	14	24	±30″
Bochum MW 50	1; 3	7	15	11	1	±50″
VNIMI MVT–4	1; 3	12	10—16	2	1.1	±30″
			Gyro-attachments for theodolites			
Wild GAK 1	1	6	20	8	2	±15...20″
Wild ARK 1	1	3	20	8	2	±1′
Fennel TK–4	1	6	20	8	2	±30″
			North-seeking gyro-theodolites			
Lear–Siegler ALINE	1; 5	14	12	—	8	±20″
Eng. Electr. Aviation PIM	1; 5	15	90	—	2	±25″
Bodenseewerk MK 10	1; 5	7	6	—	—	±1′
Anschütz Girolit II	1; 5	52	20	6	87	±3′
Teldix NSK 4	5	10	3	—	—	±0.5°

* *Remark*: In this column 1 = band suspension, 2 = automatic tracking, 3 = damp-proof, 4 = in liquid, 5 = torque-balancing, 6 = needle-bearing.

2.5.2.5 Future Developments of North-seeking Gyroscopic Instruments

The development of gyro-theodolites with traditional construction tends towards automatization together with increase in measurement accuracy, the elimination of subjective errors and making measurements less laborious. Two possible development paths are expected: the first is automatic recording at different points of the swing path, the second is the further development of north-seeking systems.

By automatic recording of the data of the swing path the maximum attainable accuracy is estimated to be ± 2 sec of arc.

From north-seeking gyroscopic systems, the development of systems on the principle of momentum adjustment is expected on the one hand to increase the accuracy of measurement (± 5 sec of arc), on the other to shorten the time necessary for measurement (5 to 6 minutes of time). Such an instrument could be used together with a distance meter for the simultaneous measurement of directions and the corresponding distances, i.e. for the determination of vectorial polar coordinates.

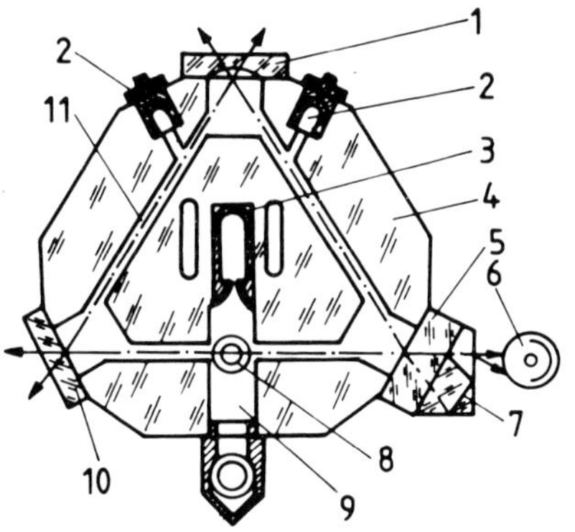

Figure 2.138. Laser gyroscope: *1* concave mirror; *2* anode; *3* cathode; *4* quartz block; *5* light outlet mirror; *6* photomultiplier; *7* uniting prism; *8* diaphragm; *9* laser tube; *10* plane mirror; *11* light path

North-seeking instruments with aperiodic damping are low accuracy instruments for supplementary measurements (gyro-compass). The mean square error of North determination is ± 3 to 5 min of arc, thus these instruments can hardly be used in surveying, except for some subsurface measurements. A special role is played by one-phase, low mass, slow revolution, so-called "cold" motors. With modern electronics and low current motors, this type of instrument can be manufactured with small size and very short observation time. The optical or laser gyroscopes being presently developed suit medium accuracy requirements. *Sagnac* has proven experimentally that lasers detecting angular rotation can be used as gyroscopes *(Fig. 2.138)*. The three or more mirrors and light conductors of the optical or laser gyroscope make a closed plane loop of light. In the closed loop, two infrared monochromatic and coherent light beams propagate in opposite directions. The energy loss due to reflection from the mirrors and to the output signal is continuously made good by the quantum generator built into the light loop. If the system is rotated around an axis perpendicular to the plane of the loop, the light path in the direction of the rotation is longer than the opposite light path. The frequency in the beam with the longer path decreases, in the other it increases, and the resulting frequency shift $\Delta\nu$ can be measured optically. The body of the gyroscope contains a quartz monolithic block with three fused glass channels *4*, with the axes in a plane, forming an equilateral triangle. At the vertices of the triangle are mounted two plane mirrors (*5* and *10*) and one spherical mirror with a long radius of curvature *1* for the regulation of the light loop. The mirror surfaces *1* and *10* have a high light reflection, using dielectric multilayer coatings; mirror *5* is a half-silvered mirror so that radiation energy may be removed from the loop. The light signal passing through this mirror falls on the prism *7*, connected to the mirror, and thence to the photoelectric device *6* which produces interference fringes. The inner cavities of the quartz block are filled with a pressurized mixture of helium and neon gases. The electrodes for the excitation of the quantum generator are in the quartz block: two anodes *2* and two cathodes *3*. One of the resonator channels contains the diaphragm *8*.

In order to determine North by optical gyroscope, the plane of the light loop of the resonator must be brought into the vertical, and it must be rotated around the vertical until the frequency difference $\Delta\nu$ becomes zero. The projection of the vector of the Earth's rotation velocity, ω_E, on the axis of the optical gyroscope (on the direction perpendicular to the plane of the light loop) is zero if the plane of the loop coincides with the meridian plane through the station.

The greater are the area F of the light loop and the optical length l of the resonator, the more sensitive is the optical gyroscope. The accuracy of measurement depends inter alia on the stability of the plane of the light loop with respect to the collimation plane of the telescope of the theodolite (i.e. on the error in the setting into the vertical). The laser gyroscope *Honeywell* (USA) has a mean square error of North determination of ± 15 sec of arc at the Equator.

The elimination of fast rotating mechanical parts and the very short preparation time (running up) has made possible the production of relatively reliable laser gyro-theodolites of high stability.

A further possibility for development is given by the gyromagnetic properties of the elementary particles of molecules (atoms, electrons, etc.) resulting in the production of atomic gyroscopes.

The output signals of laser gyroscopes operating on an optical principle can be recorded in a numerical form applicable to computer processing, thus the automatic measurement and control of the azimuthal position of both stable and moving objects will be possible. Such instruments will presumably play a significant role in the orientation of inertial measuring systems, in inertial navigation and in space flight.

The other possible direction of development of north-seeking instruments without fast rotating parts, of small size and low energy consumption, is the vibration gyroscope. In such instruments the kinetic momentum of piezoceramic cylinders of rapid swing are used for north-seeking. Present instruments offer an accuracy of some minutes of arc and therefore, from the point of view of surveying, only very low accuracy requirements can be fulfilled by them.

3 Instruments for Measuring Elevation

Instruments for measuring elevation are mostly used for the measurement of elevation differences. The instruments used for different methods can be divided into two main groups:

a) Instruments using *geometrical methods* are used to measure geometrical quantities (such as lengths and angles) related directly or indirectly to elevations or elevation differences. Instruments for levelling and trigonometric levelling (trigonometrical heighting) belong to this group.

b) Instruments using *physical methods* are used to measure physical quantities (such as atmospheric pressure or liquid pressure) which have a functional connection with height difference. Instruments for barometric and hydro-manometric height measurements belong to this group.

Levelling, trigonometric levelling (trigonometrical heighting) and barometric altimetry (barometric heighting) are the main methods used in surveying.

In the case of levelling a level surface, or a horizontal plane tangent to the level surface, is defined from which the vertical distances of the points are measured by means of levelling staffs or other measuring devices. From the measured data the elevation differences between the points can be calculated.

In the case of trigonometric levelling (trigonometrical heighting) the height difference between two points is calculated from the vertical angle (or zenith angle) of the line of sight measured from either point to the other, and from the distance between them, by means of trigonometric formula.

In case of *barometric altimetry* (barometric heighting) the elevation difference between two points is determined from the difference between the atmospheric pressures measured at the two points.

3.1 Instruments for Levelling

There are two main types of levelling methods. In the case of the seldom used *hydrostatic levelling* a level surface is produced above the points to be measured from which vertical distances to the points are measured. The level surface is produced on the basis of intercommunicating tubes: in the tube system of the instrument the free surfaces of a liquid come to the same level due to the pull of gravity — if no other effect influences the liquid.

In the case of the commonly used *optical* or *spirit levelling*, called in practice simply levelling, a plane tangent to the level surface is produced optically above the points to be measured, i.e. the line of sight of an open sight, telescope, etc. or a narrow light beam emitted from the levelling instrument, is adjusted into the horizontal by elements of the instrument such as a spirit level or pendulum. The horizontal line of sight defines the tangent plane to the level surface (or the horizontal plane of the instrument) by swinging the levelled instrument around its vertical axis. The vertical distance to the points is measured from this horizontal plane.

3.1.1 Instruments for Hydrostatic Levelling

The *intercommunicating tube level*, which is the simplest instrument, consists of a U-shaped glass tube half filled with a coloured liquid and embedded into a support. The horizontal plane is defined by the surfaces of the calm liquid in the two vertical parts of the tube.

In the *hose level* two vertical glass cylinders are connected with a longer flexible plastic or rubber hose called a hydrostatic tube. The cylinders of the simpler levels are graduated in millimeters for reading the water level. The elevation difference

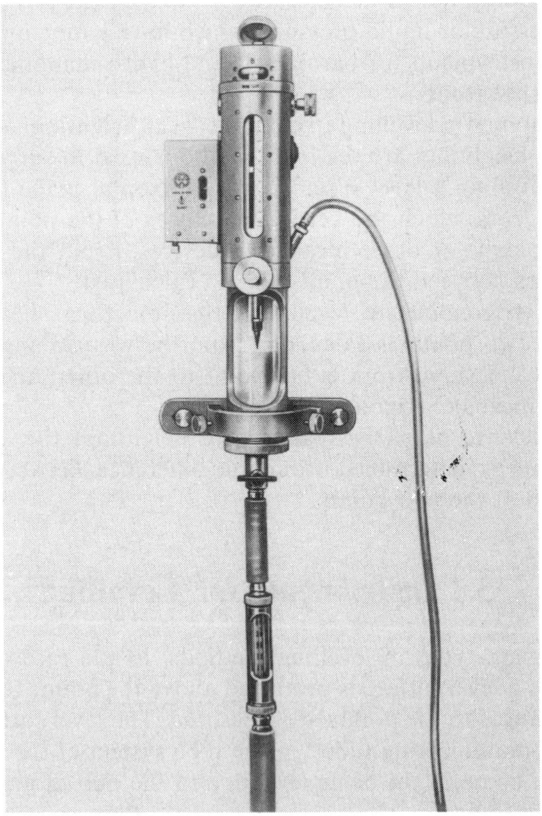

Figure 3.1. Precision hydrostatic level (VEB Freiberger Präzisionsmechanik)

between the zero points of the two graduations, that between the suspension points or foot points of the two metal frames, can be determined within the range of the graduations.

For the exact observation of the positions of the two liquid surfaces, measuring spindles are used which can be moved vertically by a micrometer screw. The micrometer reading of the spindle position, when the spindle tip is set to the surface of the liquid, gives an exact determination of the position of the liquid surface.

Elevation difference can be determined to high accuracy by, inter alia, the *Meisser* precision hydrostatic level manufactured by VEB Freiberger Präzisionsmechanik (GDR) *(Fig. 3.1)*. It consists of two measuring devices and a 25 to 50 m long pressure-tight connecting hose with 10 mm interior diameter. Before use the instrument must be filled with a liquid (water) free of bubbles. The measuring devices can be suspended by their supports onto wall plugs, or they can be placed on their own portable tripods.

In order to suspend the support on a wall, three plugs are built into the wall *(Fig. 3.2)* the upper of which has a round head *2* and is coated with a special anti-corrosive material. The measuring element has a polished agate plate *1* with a horizontal surface which rests upon the head, thus ensuring a precise contact. The other two plugs *3* in the wall are used to fix the lower stirrup *4* of the measuring element.

The most important part of the instrument is the observing tube *11*, which is a glass cylinder of 42 mm interior diameter with a spindle tip *12* movable vertically by a screw spindle with 1 mm pitch *9* and by a measuring drum *10*.

Before measuring, the observing tubes as well as the whole measuring instrument must be adjusted into the vertical by means of the screws of the stirrup *15* and by means of a circular bubble *6* which is on the upper part of the instrument. The bubble can be observed in a mirror *5* fixed above it. When measuring by rotating the measuring screw, the spindle tip is brought into contact with the surface *13* (visible in the observation tube) of the liquid *14* in the hose connecting the two measuring elements. The displacement of the spindle tip can be read on the scale with a measuring range of 100 mm *19* and on the micrometer drum *7* which has a graduation of 100 with an accuracy of 0.01 mm (and estimation to 0.001). If the tip of the spindle is set to the water surface, an electrical circuit is closed and the light of a pilot lamp *17* on the side of the measuring element is extinguished, being observable through a window *18*. The pilot lamp requires a 3 V battery.

From one reading on each of the observation tubes the elevation difference between the suspension points of the two measuring elements can be determined, if the suspension point coincides with the zero point of the graduation of the measuring screw. If there is a difference, the height determined is erroneous due to the foot point error (zero point error); the value of the error is the difference between the foot point errors of the two measuring elements. For one single elevation difference the value of this error can be determined by exchanging the two measuring elements.

Precision hydrostatic levels are mainly suited for determining vertical displacements of structures (bridges, buildings, dams, machine foundations, etc.), but they can also be used for observing the vertical component of different ground movements, e.g. for mines, hill slides, rock movement, etc.

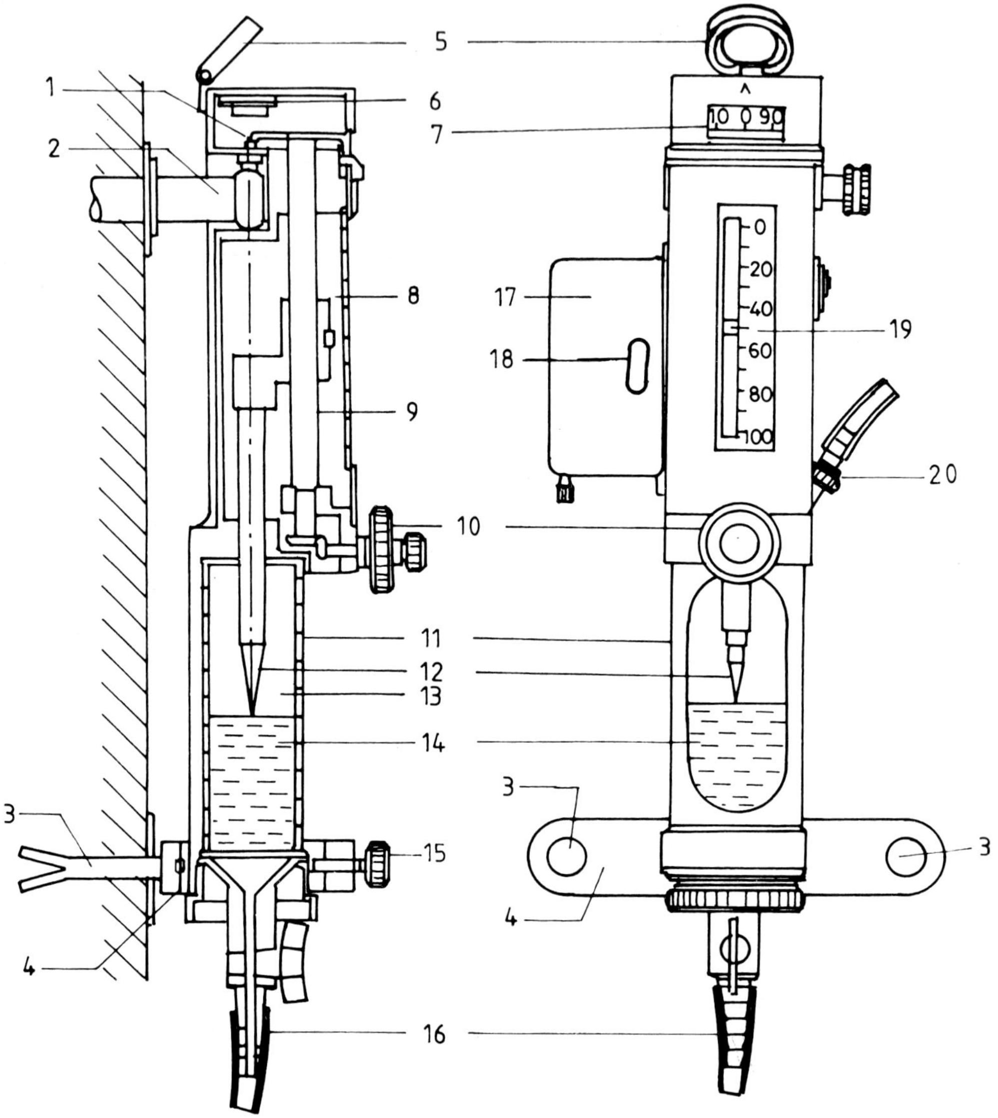

Figure 3.2. Diagram of the precision hydrostatic level showing the various necessary parts reduced to their simplest form (VEB Freiberger Präzisionsmechanik)

The measured results are influenced by physical conditions which cause the liquid surfaces in the two observation tubes not to come exactly to the same level. One frequent source of error is caused by the temperature of the liquid not being constant along the hydrostatic tube *16*. In order to correct this the instrument is provided with two thermometers which can be placed into the tubes. The temperature error can be decreased by using a heat-insulating hose or by arranging the measuring elements so that the vertical parts of the hose are as short as possible. Another source of error is the difference between the barometric pressures at the two observation tubes. In the *Meisser level* this effect is eliminated by connecting the air spaces of the two observation tubes with a second plastic tube *20*. There is only one hole at the centre of this tube, consequently the barometric pressure is equal in both observation tubes.

3.1.2 Instruments for Optical Levelling

Instruments used for optical levelling can be classified with respect to various aspects. Referring to the instrumental elements which define a horizontal line of sight, one can distinguish between simple instruments or devices, tilting levels and automatic levels.

Those instruments or devices belonging to the first group have no telescope and the horizontal direction (or horizontal plane) is defined by simple instrumental elements or devices.

The line of sight of the telescope of a *tilting level* can be set horizontal by means of a spirit level mounted on the telescope.

The line of sight of an *automatic level* (or compensator level) is set horizontal in general by means of an automatically acting element built integrally with the telescope, namely the compensator.

Levels can be grouped according to their precision on the basis of their *a posteriori* mean square error per km ($\mu_{(km)}$) as follows:

very high precision levels:	$\mu_{(km)} \leqq 0.5$ mm,
high precision levels:	$0.5 \text{ mm} < \mu_{(km)} \leqq 2.0$ mm,
levels of medium accuracy:	$2.0 \text{ mm} < \mu_{(km)} \leqq 6.0$ mm,
levels of low accuracy:	$6.0 \text{ mm} < \mu_{(km)} \leqq 20$ mm.

The *a posteriori* mean square error per km depends not only on the instrument used, but also on the field conditions and on the measuring method used.

Levels can also be classified on the basis of their intended purpose: instruments of very high and highest accuracy, high accuracy, medium and low accuracy are called in general *precise level*, *engineers' level*, and *architects' level*, respectively. It must be mentioned, however, that there is no common agreement on the question of the borders of the instrument classes.

Levels constructed for special purposes and their auxiliary devices, as well as those where the main constructional elements differ from those in conventional instruments, belong to the group of special levels.

As regards additional equipment, a tripod and levelling staffs also belong to the level.

3.1.2.1 The Tripod

In order to provide ease and convenience for measurement with a level and to secure it free from movement, various types of tripods have been constructed. Theodolite tripods are also often used for levels. One such type is the disc-head tripod used for theodolites with a base plate and foot-screws (see Subsection 2.2.1).

Centring is not necessary when setting up a level and the vertical axis needs to be adjusted only approximately, but quickly, to the vertical: tripod heads with a ball-and-socket joint are therefore often used. As a variant of these the upper surface of the head fixed to the tripod is spherical and a level with a conical or concave base plate can be moved on it. The vertical axis of the instrument is tilted by such movement and can be brought approximately to the vertical. In this position the level can be

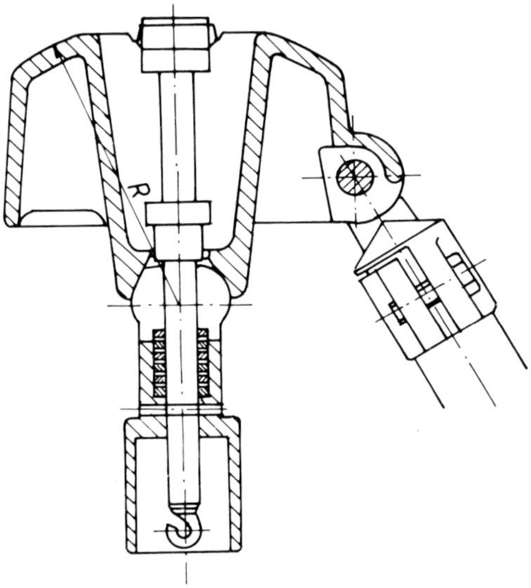

Figure 3.3. Spherical tripod head with ball-and-socket joint (Kern and Co. AG, Aarau)

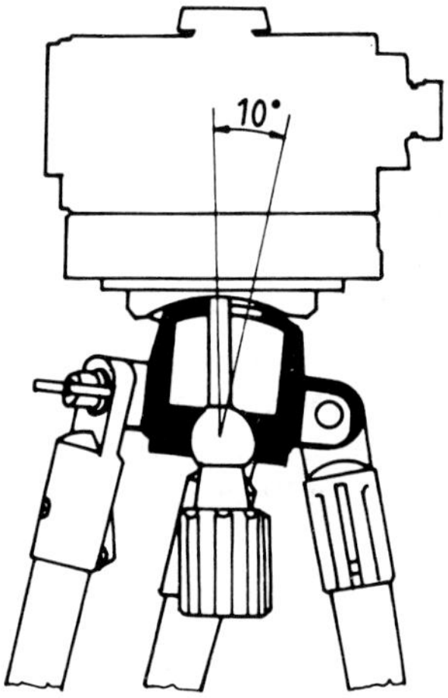

Figure 3.4. Spherical tripod head with ball-and socket joint (Ertel Werke, München)

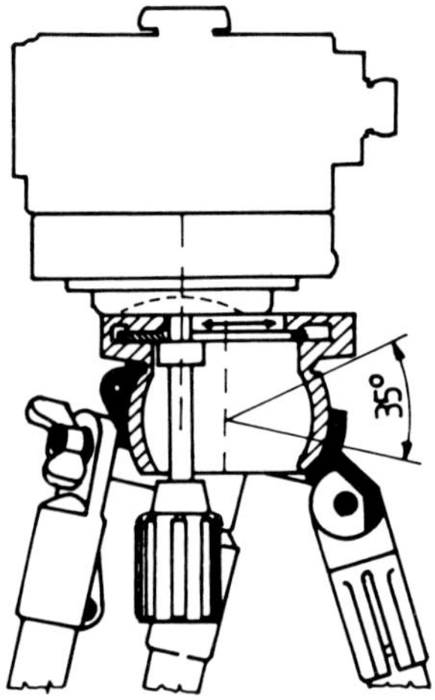

Figure 3.5. Plane tripod head with ball-and-socket joint (Ertel Werke, München)

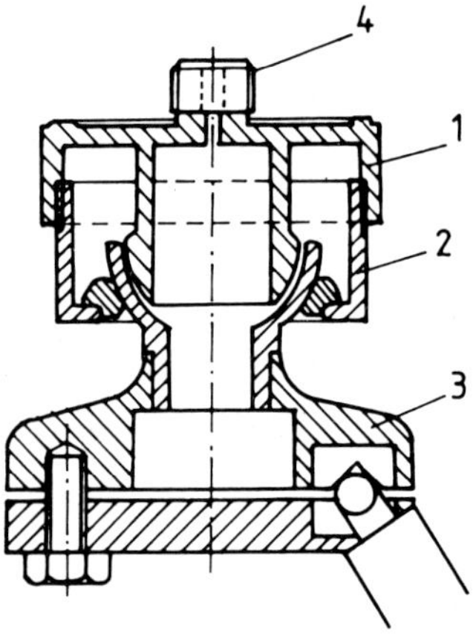

Figure 3.6. Plane tripod head with ball-and-socket joint (MOM, Budapest)

clamped by means of the central clamping screw which is formed as a ball-and-socket joint. The *Kern (Fig. 3.3)* and *Ertel* ball tripod heads *(Fig. 3.4)* are of such a type. Another variant of ball tripod heads is a plane tripod plate with spherical zones at the sides *(Fig. 3.5)* so that it is tiltable with reference to the legs of the tripod. The level can be moved on the tripod head if necessary for centring over a point.

The tripod of the compensator level *MOM Ni–E1* is of similar principle, but of different construction *(Fig. 3.6)*. The level is tightened (*4*) to the disc-shaped zone *1* of the central clamping screw *3*, which is at the same time the hilt of the screw. The hemispherical zone of the clamping screw fits the corresponding concave part of the tripod head and operates as a ball joint. The level attached by the clamping screw is tilted until its vertical axis is normal to the horizontal plane. Now the ring *2* clamping the ball joint is screwed into the inner threaded part of the hilt of the clamping screw, thus fixing the level in the vertical position.

The light, pinned tripod, similar to the tripod for a photo-camera, is used for simple levels as it is light in weight.

3.1.2.2 Simple Instruments and Devices for Optical Levelling

In the case of simple instruments and devices the tangential plane to the level surface can be defined by the unaided eye, i.e. without a telescope.

The simplest devices are *pendulum instruments* provided with a coarse sight (sighting tube) or a sighting collimator, the line of sight of which settles automatically into the horizontal plane. Such an instrument is the *NCOL 20* made by SLOM, where a signal can be moved up and down on the staff and adjusted to the horizontal through the instrument. The level *NCOL 400* used with a pinned tripod is of similar construction, but it works on the principle of coincidence *(Fig. 3.7)*. A mirror fixed on a pendulum and tilted at 45° to the horizontal, also a reversing prism fixed to the instrument housing, are placed into the cylindrical body of the level. Two split half images of the target, which is movable up and down on the staff, are visible in the window on the top of the level. The target must be moved up or down the vertically-held levelling staff until the split images are brought into coincidence. Now the target is in the horizontal plane through the instrument (i.e. at the height of the rays passing in). The standard deviation for heighting the target is ± 5 mm/30 m.

The *SL 200 Cowley Automatic Level* also works according to the principle of coincidence *(Fig. 3.8)*. There are two mirror systems in the instrument which can be put on a pinned tripod; the mirrors produce two images reversed to each other, of the right and the left hand sides respectively of the vertical plane halving the field of view. The mirrors *1* and *2* of the first system, fixed to the housing of the instrument, produce a normal image on the left hand side. Mirrors *3* and *4* of the second system are also fixed to the instrument but the mirror *5* of this system is connected to a pendulum which is damped by a magnet. Due to the force of gravity the pendulum *6* hangs vertically when the instrument is mis-levelled up to a maximum of $\pm 6°$ in the vertical plane of the line of sight. The reversed image on the right side of the field of

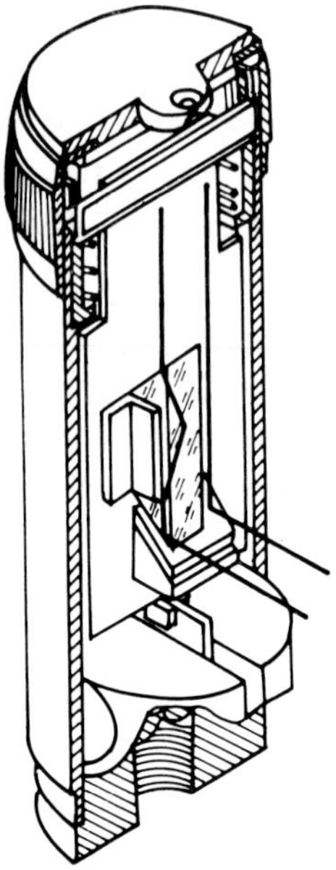

Figure 3.7. The coincidence-level *NCOL 400* (SLOM, Paris)

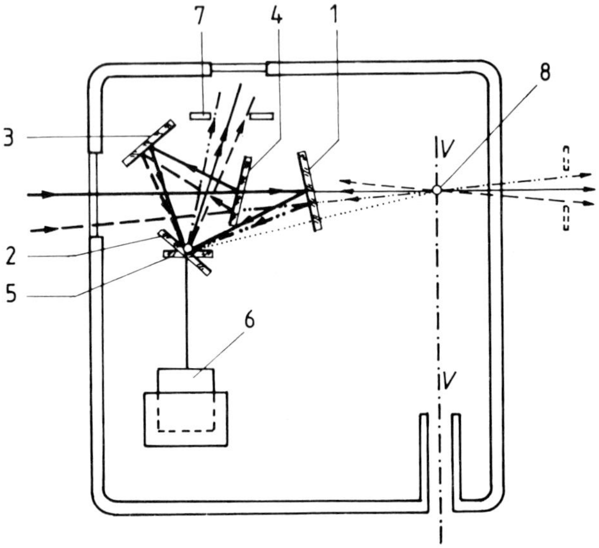

Figure 3.8. Diagram of the "Cowley Automatic Level" *SL 200* (Hilger and Watts Ltd, England)

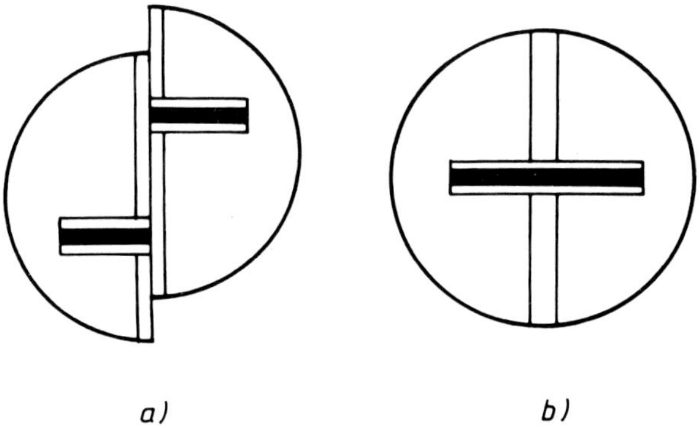

Figure 3.9. Double image in the "Cowley Automatic Level": *a)* staff below the horizon; *b)* staff in the horizon

view is projected by this second mirror system. The two split half images of the target (or staff) can be observed by looking into the instrument from the top through the diaphragm *7 (Fig. 3.9, a)*. The two half images of a horizontal staff will coincide only when the staff is in the horizontal plane through the level *(Fig. 3.9, b)*. This plane is defined as that which passes through the virtual image *8* of the axis of the tilting mirror *5*, given by reflection from mirror *1*, and the vertical axis *V—V* intersects this image. The plane is maintained horizontal by the mirror *5* which is mounted on the pendulum. A measuring accuracy of ± 6 mm/30 m can be attained by this level.

Levelling, or defining the tangential plane of the level surface, can easily be made by the so-called *planoscope (Fig. 3.10)*. The device is in the form of a box, the front and back panels of which make a certain angle and are close to vertical after setting up the device. These panels consist of translucent grids of lines parallel to the edges of the lower and upper sides. Hence, after setting up the instrument, the planes defined by the corresponding lines of the grids are horizontal. In the central horizontal section of the panels the lines of both grids are omitted or covered with a narrow strip. When looking at it from the front this is seen as a white median line.

Looking towards the device from a certain distance one sees the images of the front and back grids projected onto each other. Due to the *Moiré* effect one sees a regular series of black and white lines or strips which have a defined position with reference to the horizontal median line, depending on the point from which they are

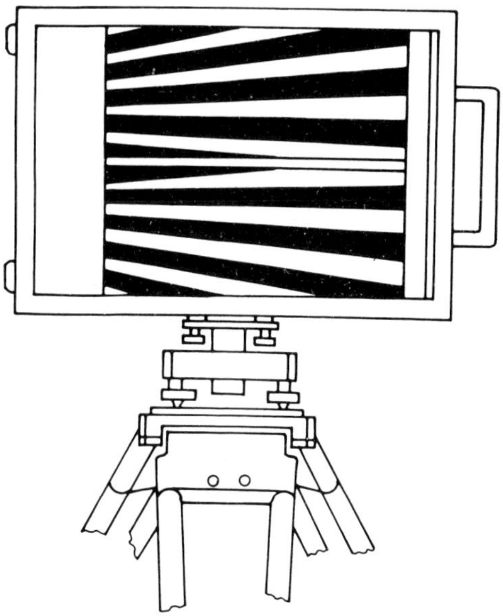

Figure 3.10. Planoscope (N. V. Optische Industrie "De Oude Delft", Delft)

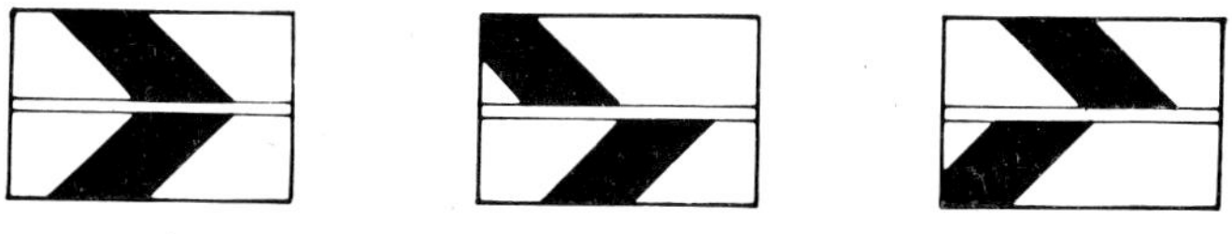

Figure 3.11. The image visible in the planoscope from a distance *a)* eye in the horizon; *b)* eye too low; *c)* eye too high

observed. If the observer's eye is in the horizontal plane of the median line of the grids, i.e. in the horizontal plane of the instrument, the Moiré fringes are symmetrical with respect to the horizontal median line *(Figs 3.10* and *3.11, a)*. If the observer's eye is above or below the horizontal plane, the Moiré fringes become asymmetric *(Fig. 3.11, b* and *c)*. The light illuminating the grids (natural daylight or artificial light at night) enters into the device through the back grid.

The density of the Moiré fringes depends on the distance of the observation: by increasing the distance the density of the visible fringes decreases; when observing from a short distance (about 5 m) one can see the fringes with a density as shown in *Fig. 3.10*. From a longer distance (about 100 m) the Moiré pattern is as shown in *Fig. 3.11*.

The Moiré pattern can be observed from any direction within a sector of 45° to the right and to left of the normal to the front, therefore it is advisable to set up the device on the edge of the area to be measured in such way that all the points to be measured lie within this angle. The sides with the grids must be vertical during the measurement, made so by means of a spirit level mounted on the device. By fitting the spirit level to the device in different positions the horizontal plane or reference plane of the instrument can be made to have a defined slope, up or down, enabling the instrument to trace a plane with a planned inclination.

After having correctly set up the instrument the measurement can be simply made by a single observer without assistance: holding the levelling staff on a point, he looks for the symmetrical position of the visible Moiré pattern by moving his eye up and down, and repeats this procedure at the other points. When finding a symmetrical pattern the observer's eye is in the horizontal plane of the instrument, and he can read the height of his eye on the levelling staff. The reliability of the measurement is two or three cm over a distance of 80 or 100 m. Depending on the weather, the visibility distance is from 100 to 180 m and by means of a telescope it can be extended to 400–700 m. A disadvantage of the instrument is that it can only be used in a nearly horizontal field, since for a measurement the eyes must always be moved into the horizontal plane of the instrument.

Simple instruments, which allow a constant angle to be set out, can be used as *simple devices for levelling* (such as right-angled mirrors, or a prism). As shown in *Fig. 3.12*, if a simple right-angled prism *1* is mounted on a plumbing rod *2*, in such a way that the hypotenuse surface of the prism is parallel to the rod and the intersection of the reflecting surfaces is perpendicular to the rod, then the prism allows a right angle to be set out in the vertical plane. If the plumbing rod *2* is held hanging free in the vertical, the shortened image of the plumbing rod visible in the prism gives the horizontal line of sight *3*. Approximate levelling with a prism can be used advantageously for e.g. calculating the horizontal equivalent of a measured slope distance.

The levelling of cross sections, especially on steep and difficult terrain, can be simply made by means of a horizontal straight edge used for step measuring *(Fig. 3.13)*. The 4 m long straight edge, which is graduated every 10 centimetres, can be set into the horizontal by means of a tubular bubble mounted on it. One end of the straight edge is put on the starting point and the other end is held by a support which is movable along a measuring rod, which is graduated in centimetres and set into the vertical with a bull's eye bubble. The elevation difference can be read on the vertical rod. The

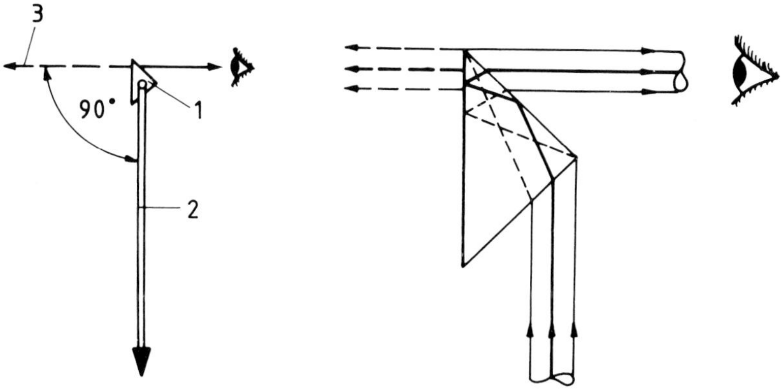

Figure 3.12. Levelling with a prism square

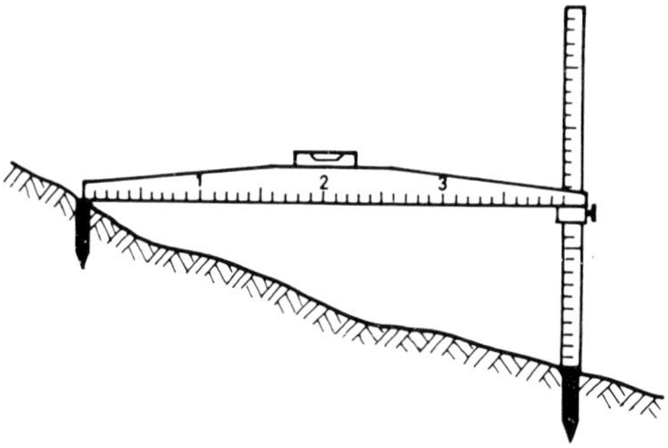

Figure 3.13. Horizontal straight edge

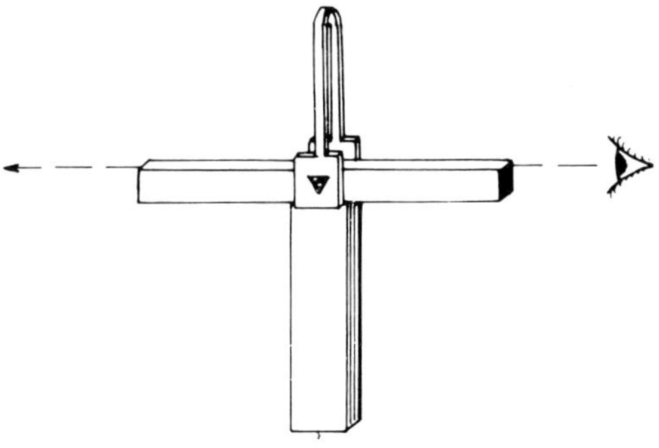

Figure 3.14. Levelling cross (compensator with a free-swinging pendulum)

axis of the tubular bubble must be parallel to the bottom of the horizontal straight edge.

The levelling cross, which is in fact a very simple pendulum instrument is suitable for approximate levelling *(Fig. 3.14)*. When the freely swinging cross comes to rest, a nearly horizontal line of sight can be set by looking along the upper surface of the horizontal rod.

3.1.2.3 Tilting Levels

The majority of the levels used today can be put into the following three groups:

1. levels with a rigid telescope and telescope bubble (or spirit level) (the telescope is attached to its support but it can usually be given a slight tilt in the vertical plane by means of a tilting screw, the bubble tube being fixed to the telescope);
2. levels with a telescope rotatable around its longitudinal axis, and usually with a reversible bubble (or spirit level) fixed to the telescope;
3. levels with a reversible telescope rotatable around its longitudinal axis and with a reversible bubble which acts as a striding level.

Today tilting levels belong to the first group. The telescope bubble is levelled by the tilting screw, and the vertical axis of the instrument can be set vertical using a circular bubble mounted on the top of the instrument.

The level consists of three main parts: levelling head (tribrach with socket), vertical axis system and top part (alidade). The main elements of the top part are: the telescope (see Point 2.2.5.2), the telescope bubble (spirit level) and the tilting screw *(Fig. 3.15)*.

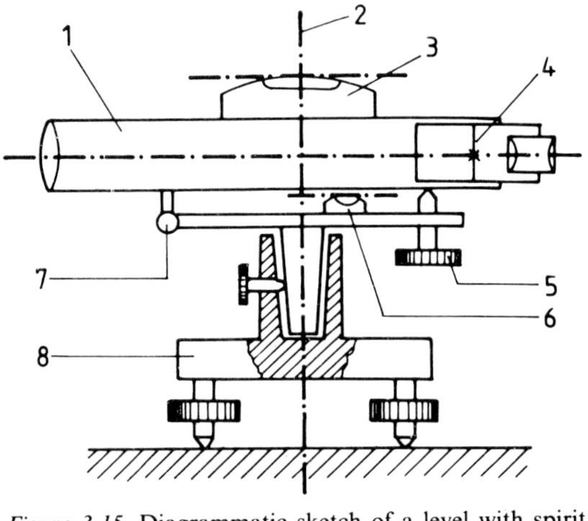

Figure 3.15. Diagrammatic sketch of a level with spirit level: *1* telescope; *2* vertical axis; *3* spirit level; *4* cross-hair; *5* tilting screw; *6* circular bubble for the vertical axis; *7* horizontal axis; *8* base plate

Levelling head. The most frequently used levelling heads have three foot screws *(Fig. 3.16)*, where the foot screws rest either on a foot plate (or base plate) or directly on the plate of the tripod head, and are tightened by the central clamping screw through a spring plate. Sometimes the base plate is also the base of the housing (bell-shaped container) *(Fig. 3.17)*.

For lower precision levels which operate with a spirit level or more usually with a compensator, this type of levelling head with foot screws is used, with a foot plate which has a spherical surface on the central part of its lower side and which can be clamped to a tripod with a spherical head. The *Miller* and *Sokkisha compensator levels* have this type of levelling head (see *Fig. 3.72*). The foot plate of the *Fen I*

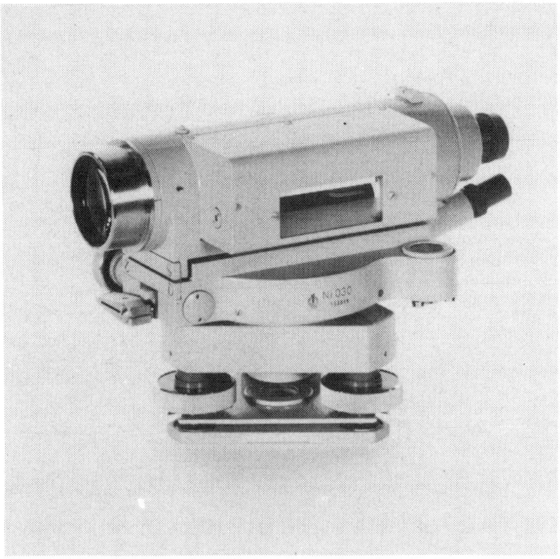

Figure 3.16. Level Jenoptik *Ni 030* (VEB Jenoptik, Jena)

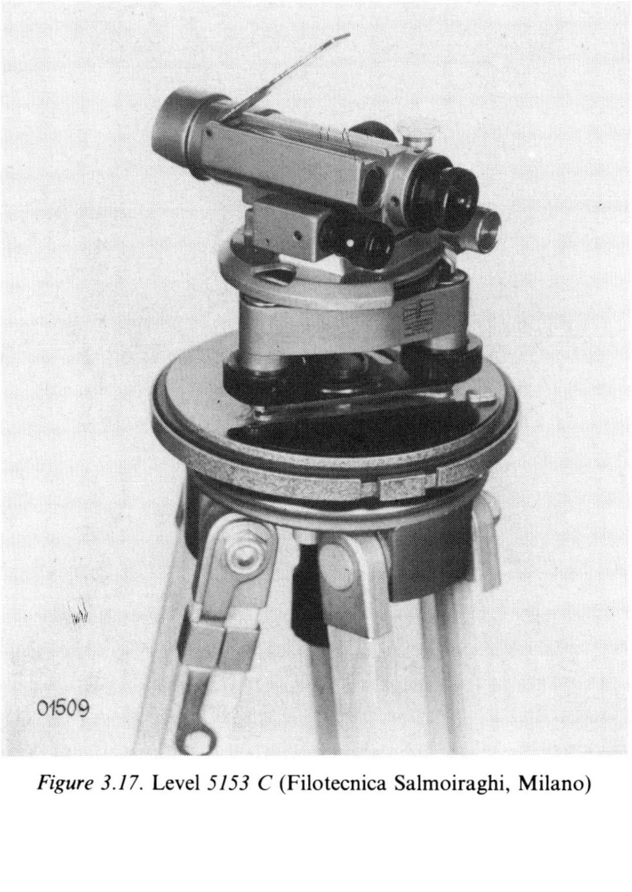

Figure 3.17. Level *5153 C* (Filotecnica Salmoiraghi, Milano)

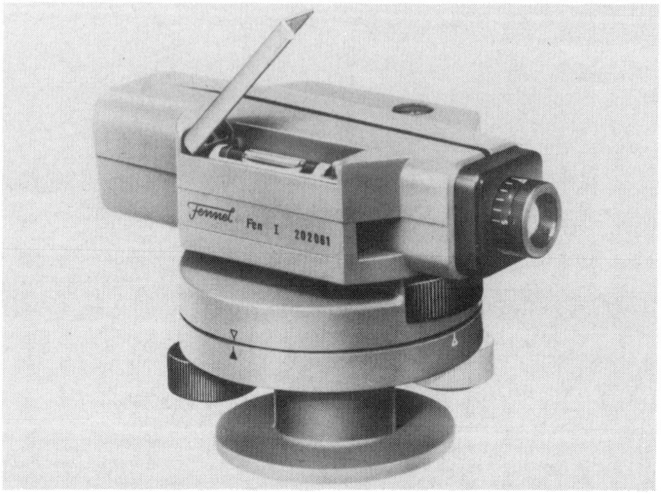

Figure 3.18. Level *Fen I* with a spirit level (Otto Fennel GmbH, Kassel)

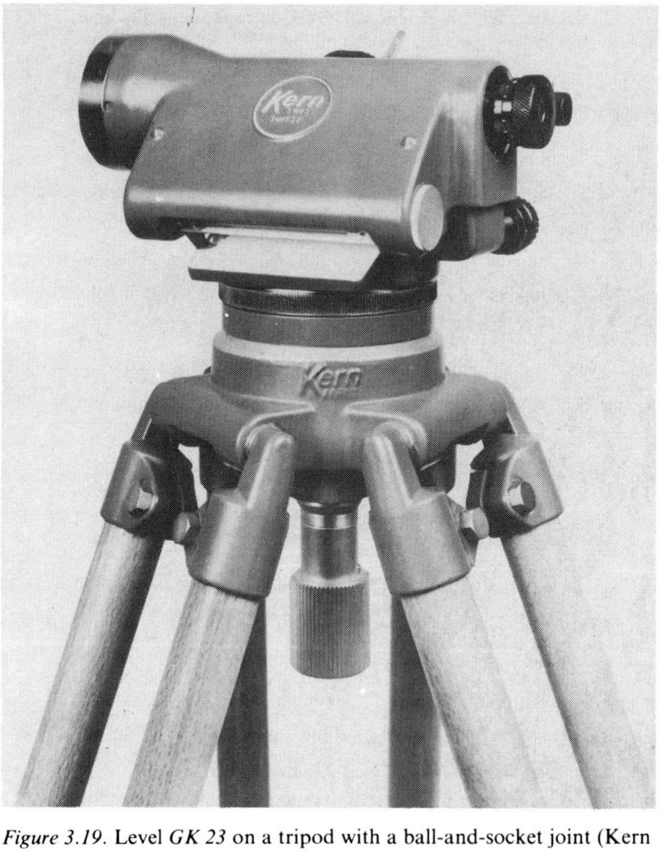

Figure 3.19. Level *GK 23* on a tripod with a ball-and-socket joint (Kern and Co. AG, Aarau)

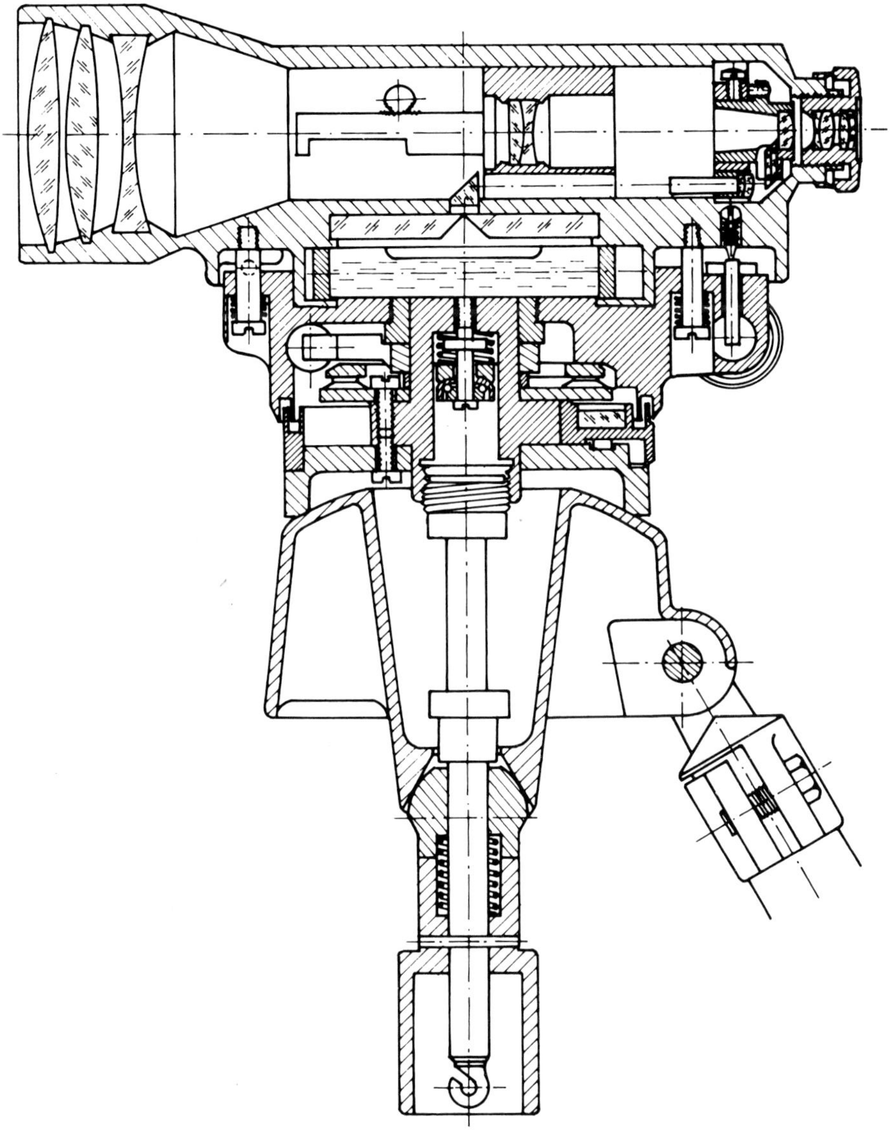

Figure 3.20. Diagrammatic sketch of the level *GK 23* (Kern and Co. AG, Aarau)

(Fennel, Kassel) differs somewhat from the usual *(Fig. 3.18)* in that the upper part can be tilted around a ball joint by means of two foot screws in two mutually perpendicular directions.

There are levelling heads without foot screws, where a concave or conic bottom surface fits to the spherical head of the tripod *(Figs 3.19* and *3.20)*. This ball and socket joint provides a faster setting up and this advantage is mainly useful for compensator levels. The levelling head of the level *Jenoptik Ni 050* has no foot screws

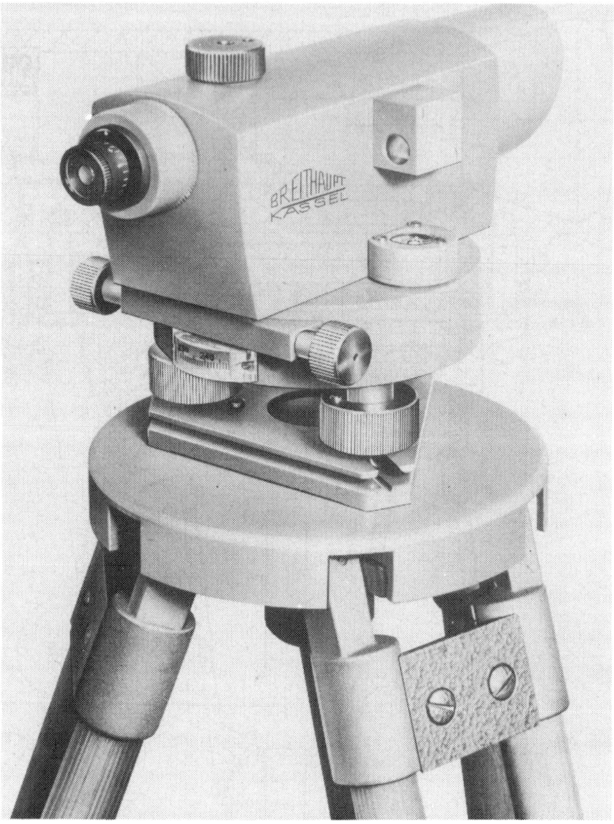

Figure 3.21. Architect's level "Nifix" (F. W. Breithaupt und Sohn, Kassel)

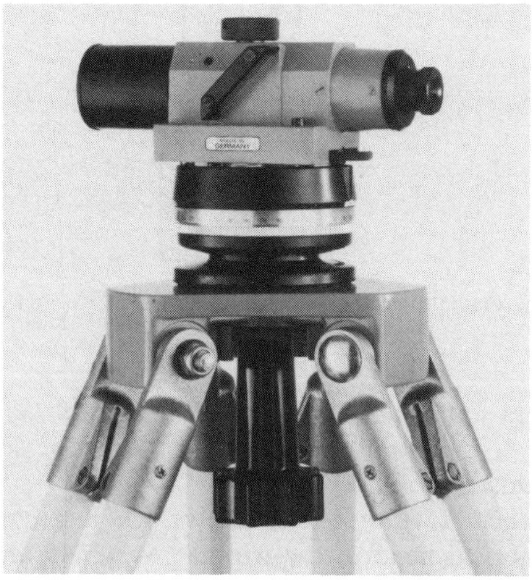

Figure 3.22. Architect's level BNL "Baunivellier" (Ertel-Werk, München)

Figure 3.23. Level *NAKO* (Wild, Heerbrugg)

and consists of two discs with a wedge-shaped cross section (see *Fig. 3.87)*. Rotating the discs, one over the other, the vertical axis can quickly be set to the vertical.

Many levels are equipped with a horizontal circle, hence they can also be used as tacheometer-levels. The horizontal circle most frequently has half-degree or one-degree graduations; these are engraved either on the side of the cylindrical base or on a glass circle *(Figs 3.21*, *3.22* and *3.23)*.

The top part of the level (alidade). This can be rotated around the vertical axis which is encased in the socket of the levelling head (see *Fig. 3.15*). Two types of construction exist: either the axis shaft is part of the top part (telescope or telescope support) with the socket in the levelling head or vice versa. The axis system can be conical or cylindrical *(Fig. 3.24)*. Axis systems with ball bearings are also used (see Subsection 2.2.3).

For holding the rotating top part and for fine setting, a clamping screw and slow-motion screw described under theodolites (in Point 2.2.5.4) are used. A clamping lever is sometimes used instead of a clamping screw as e.g. on the level *Jenoptik Ni 030* (see *Fig. 3.16*).

A friction clamp without a clamping screw is widely used for newer instruments (see Point 2.2.5.4). A rough pointing can be made when rotating the top part against slight friction, but for fine setting friction clamps also have a slow-motion screw. By means of the slow-motion screw, with an endless thread for infinite fine movement, the top part can be turned through any arbitrary angle.

The vertical axis can be set to the vertical by means of a circular bubble but, on a few levels, two perpendicular tubular bubbles are used for this purpose. On the top part of the precise level *Jenoptik Ni 004 (Fig. 3.25)* besides the circular bubble a tubular one is also mounted perpendicular to the longitudinal axis of the telescope.

Circular bubbles have in general a sensitivity between 6 and 20 min of arc, but sometimes they have a sensitivity of 2 or 60 to 80 min of arc.

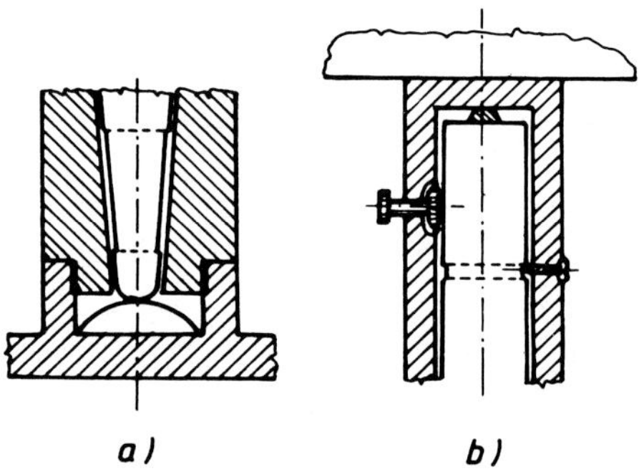

Figure 3.24. Diagram of axis systems: *a)* conical axis with spring support; *b)* cylindrical axis

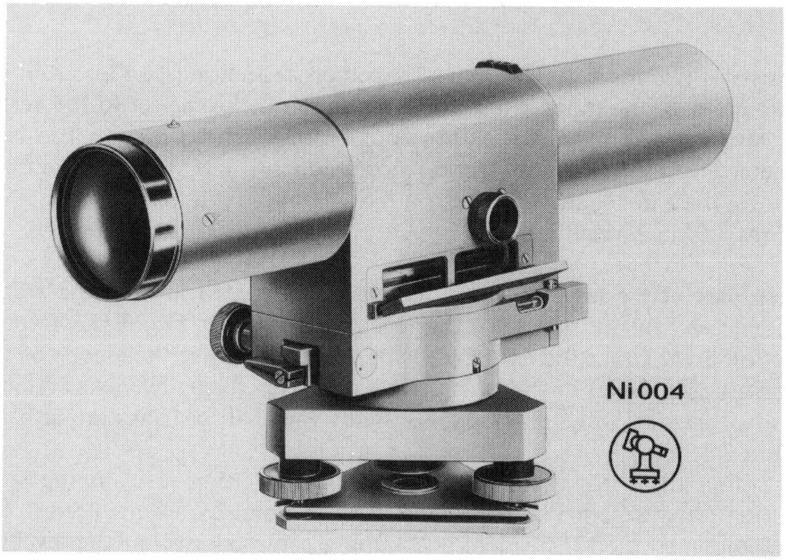

Figure 3.25. Level *Jenoptik Ni 004* (VEB Jenoptik, Jena)

The telescope of the level. In general a geodetic telescope with an anallactic and internal focussing lens is usually used for modern levels; it is fixed to the top part (see Point 2.2.5.2) and by means of a tilting screw *(Fig. 3.15)* it can be tilted around a horizontal axis normal to the vertical axis. On most levels of very high precision the horizontal axis intersects the vertical axis, but for levels of lower accuracy the horizontal axis is eccentric with respect to the vertical axis. The axis is usually cylindrical, but the pivot is often substituted by a flexible metal plate.

In exceptional cases such instruments are manufactured with a telescope which is fitted with a reversible spirit level (see Subsection 1.6.1) and which can be rotated around its longitudinal axis. The telescope of the level *Wild N2* is of this type *(Fig. 3.26)*, held in position by pressure springs, it rests upon V-shaped supports and can be rotated around its longitudinal axis through 180°. In the course of measuring, the levelling staff can be read simultaneously with centring the telescope bubble; the telescope is then rotated and the reading on the staff is repeated. If the level is well adjusted, the two readings are identical; if it is not, the arithmetic mean of the two readings gives the value assigned to the horizontal line of sight.

For levels of lower accuracy the telescope has a 15 to 25 times magnification, while for higher accuracy a magnification from 30 to 60 times is used. The manufacturers of the telescopes make efforts to minimize eyepiece motion as well as looseness of the focussing lens guide, and to make the line of sight, i.e. the optical axis of the objective system, and the geometrical axis of the telescope coincide as exactly as possible.

The focussing screw often has a coarse and a fine focussing function by double transmission; the image can be fine focussed by about a quarter or a half revolution

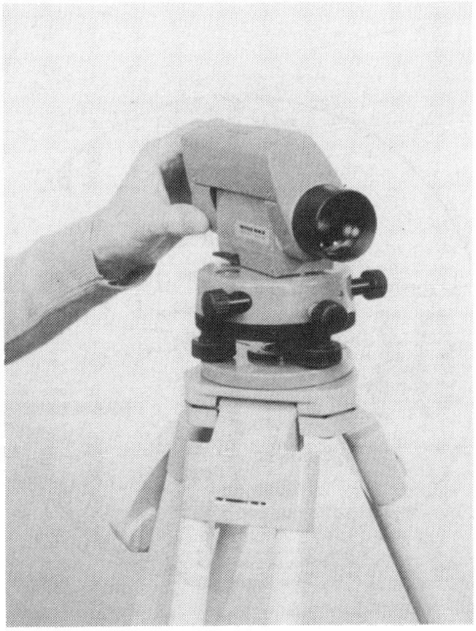

Figure 3.26. Level *N2* (Wild, Heerbrugg)

Figure 3.27. Level *N3* (Wild, Heerbrugg)

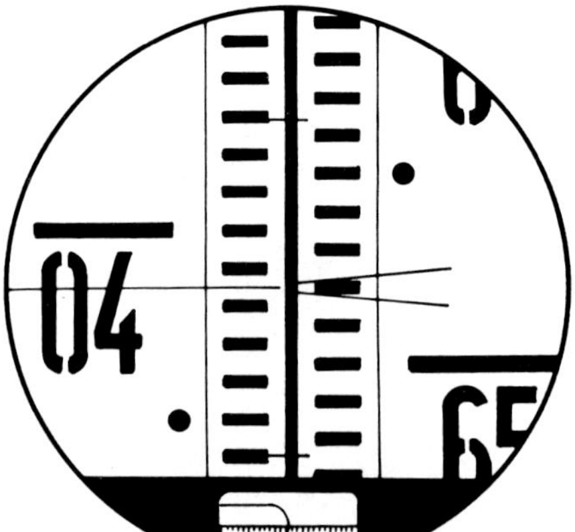

Figure 3.28. The field of view in the telescope of the level *Jenoptik Ni 004*

of the screw. When turning the screw somewhat beyond the sharpest image, the image can be fast focussed against a somewhat greater friction.

The shortest observing distance of the telescope lies between 1 and 3 m, depending on the magnifying power. It is worth mentioning that the minimum observing distance of the newest model of the level *Wild N3 (Fig. 3.27)* is 0.45 m, made possible by the panfocal lens system (see Subsection 1.5.3).

On the cross-hair plates of the levels there are usually simple cross-hairs with a vertical and a horizontal hair, and stadia lines are added. In levels with an optical micrometer one half of the horizontal hair is usually substituted by a line wedge. This wedge refines the readings on the levelling staff *(Fig. 3.28)*.

For precision levelling the telescopes are equipped with an optical micrometer for increasing the reading accuracy *(Fig. 3.29)*. This is usually a plane parallel glass plate *1* attached in front of the objective of the telescope, which can be rotated around a horizontal axis perpendicular to the line of sight by means of a rack and pinion connection *2* and a micrometer drum *3*. When rotating the drum the line of sight of the telescope is raised or lowered in the vertical plane, parallel to itself. The separation between the line of sight and the nearest graduation line of the levelling staff can be measured with the optical micrometer. The parallel shift of the line of sight can be read on a micrometer graduation connected with the plane parallel glass plate. The graduation is either on the micrometer drum *(Fig. 3.30)* or on the glass plate *4* fixed to the rack *2* which rotates the plane parallel glass plate *1*. Thus the error caused by the clearance of the micrometer screw is eliminated. Using a deflecting glass prism *5* the graduation, illuminated from below in the majority of instruments, is imaged in the field of view of the telescope, or can be observed beside the telescope eyepiece. The rack and pinion system must engage the rotating lever of the plane parallel glass plate without clearance. This is ensured by the construction shown in the figure by which the spherical end of the rotating lever fits into a fork at the end of the rack. Frequently the spherical body is at the end of the rack and the fork is on the rotating lever, as e.g. in the optical micrometers of the levels *Wild N3* and *MOM Ni A1*. Clearance due to possible wear can be taken up by adjusting the prongs of the fork (with an adjusting screw).

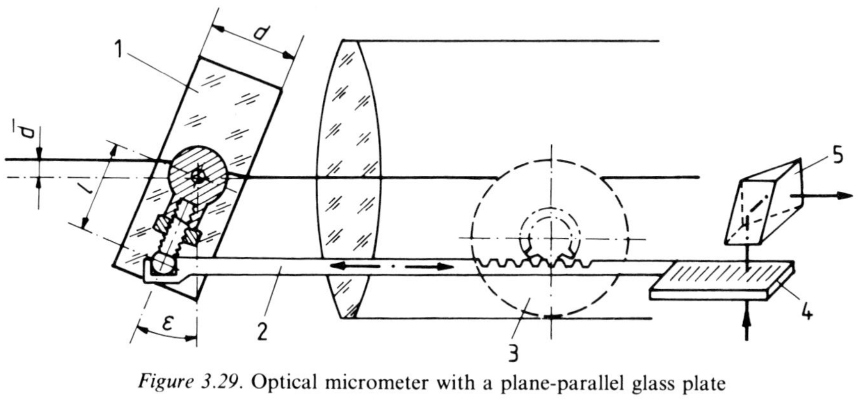

Figure 3.29. Optical micrometer with a plane-parallel glass plate

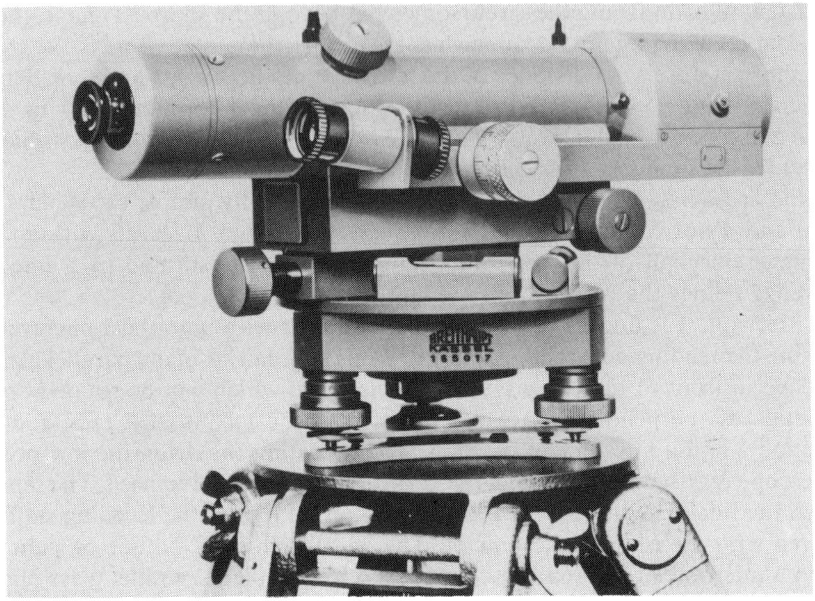

Figure 3.30. Precise level "Nabon" (F. W. Breithaupt & Sohn, Kassel)

The line of sight can usually be raised or lowered by the micrometer by an amount corresponding to the graduation unit of the staff, with a maximum shift of 5 mm, sometimes 10 mm. The micrometer drum is usually subdivided into 100 parts, hence a micrometer graduation unit corresponds to a shift of 0.05 mm for a 5 mm staff graduation. The optical micrometer system is usually either an integral part of the telescope, or it is manufactured as an auxiliary device that can be attached in front of the objective lens. The plane parallel glass plate of the micrometer can also be placed inside the telescope, as e.g. in the Soviet precise level *N-1 (Fig. 3.31)*. The lenses *1* and *2* are not movable and form a telescopic system. The rays are parallel between the negative lens *2* and the positive lens *4* which can be moved. With a rotation of $\pm 15^\circ$ the plane parallel plate *3*, placed between the two latter lenses, moves the line of sight parallel to itself by ± 2.5 mm. The image is visible on the fixed cross-hair plate *5*. The advantage of having the micrometer inside the telescope is that it can operate with a simpler transfer mechanism, while its drawback is that special care must be taken to make the focal planes of the lenses *1* and *2* coincident.

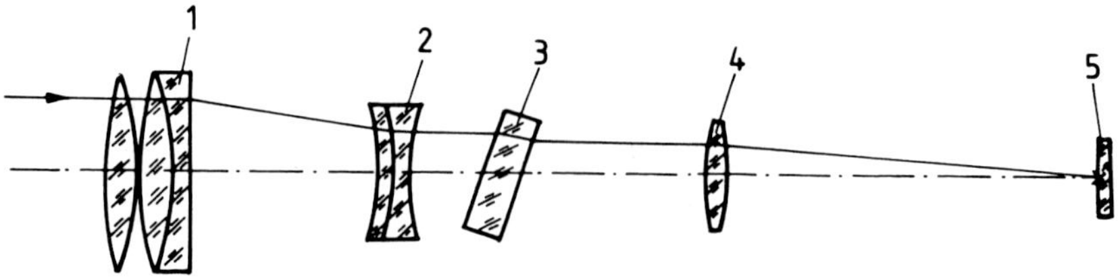

Figure 3.31. The optical train in the Soviet level *N-1*

The spirit level. The spirit level is placed longitudinally beside the telescope, and in up-to-date levels it is rigidly attached to the telescope. By means of vertical and horizontal adjustment screws the bubble axis can be made parallel to the line of sight of the telescope. The bubble sensitivity of a level of low and medium accuracy lies between 30″/2 mm and 60″/2 mm, while that of a level of high or highest precision is 10″/2 mm to 30″/2 mm.

Bubble mirrors are used in order to observe the level bubble from the eyepiece of the telescope; this shortens the time of observation and increases the accuracy of the levelling. The simplest construction is a swing mirror placed above the bubble and rotatable around a horizontal transverse axis (see *Figs 3.17* and *3.18*).

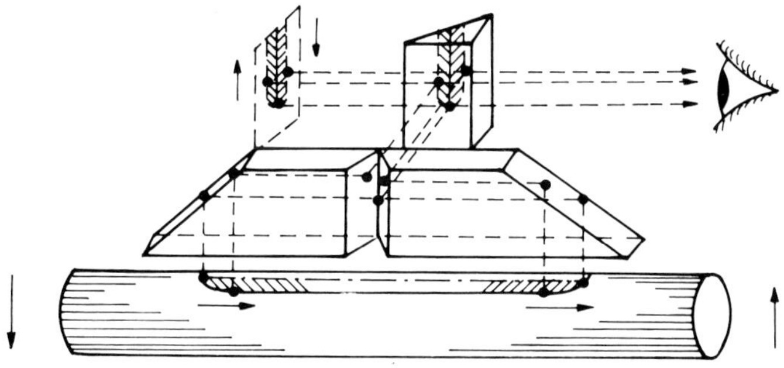

Figure 3.32. Coincidence bubble reader

For medium and high precision levels a coincidence bubble reader for a split bubble is used *(Fig. 3.32)*. This images one half of each of the bubble ends side by side, and both bubble ends can be observed either in the field of view of the telescope or through a magnifier placed beside the eyepiece. The bubble is central if both ends coincide. Using a coincidence bubble reader the mean square error of setting the bubble is about 1/50th part of the spirit level sensitivity. In some levels the two ends of the split bubble image are projected onto a graduation (see *Fig. 3.28*) on which the separation of the ends can be read. If the sensitivity of the spirit level is known, the inclination angle of the line of sight can be calculated from the separation of the bubble ends and, if the staff distance is also known, the horizontal equivalent of the staff reading can be calculated. The graduation is observed with a magnifier, hence its interval is usually less than 1 mm. In the level *Jenoptik Ni-004*, and in some other precision levels (such as the Soviet level *N-1* and *NA-1* and the *Fennel* "*Plani*") the graduation interval is 0.8 mm. Consequently a separation of the bubble ends of one unit, with a 10″/2 mm sensitivity of the spirit level as is in common use, corresponds to a 2″ tilt of the line of sight.

Tilting screw. With the exception of low or medium precision levels whose line of sight is usually levelled by means of footscrews, all up-to-date levels equipped with a spirit level are provided with a tilting screw, by means of which the telescope can be slightly tilted around the horizontal axis of the instrument (see *Fig. 3.15*). In spite of the fact that the vertical axis has been brought only approximately into the vertical by the circular level, the line of sight of the telescope is set horizontal if the bubble of the spirit level is centred with the tilting screw before each staff reading. The pitch of the screw is low (less than 0.5 mm); it usually works with a lever transmission which makes possible a quick, comfortable and precise centring of the bubble. The horizontal axis of most instruments has an eccentricity towards the object lens of the telescope which increases the active arm of the tilting screw. A disadvantage of this arrangement is that, with a slightly tilted vertical axis, the instrument horizon very slightly changes between foresight and backsight when levelling. This is not permissible for precise levels, therefore they are manufactured with a central horizontal axis.

Tilting screws of precise levels are sometimes provided with graduations. If the sensitivity of the spirit level is known, these graduations allow the measurement of small tilt angles, and facilitates the determination and setting of the normal position of the tilting screw. This normal position of the tilting screw is denoted by σ_n; in this position the axis of the spirit level is perpendicular to the vertical axis. To determine σ_n the vertical axis of the instrument should be set approximately to the vertical by means of a circular level, then the bubble of the spirit level should be centred with the tilting screw and the screw position σ_1 read. After rotating the spirit level (the telescope) through 180°, the bubble of the spirit level should be again centred with the tilting screw and the screw position σ_2 read. From these two readings the reading of the normal position is:

$$\sigma_n = \frac{\sigma_1 + \sigma_2}{2} . \qquad (3.1)$$

For one level σ_n is always a constant as long as the position of the spirit level does not change with respect to the telescope.

If there is no graduation on the tilting screw, graduations should be scratched on either the socket or the driving knob of the tilting screw, depending on its construction, and the positions σ_1 and σ_2 should be scratched on the opposite side by lengthening the corresponding graduations. The midpoint between the two scratches corresponds to the normal position σ_n, which should be marked.

Determining the normal position of the tilting screw facilitates, indeed, makes possible, the setting of the vertical axis of the instrument into the vertical by means of the spirit level.

3.1.2.4 Compensator Levels

Apart from the built-in element (called a compensator) which adjusts the line of sight into the horizontal, the construction and main characteristics of compensator levels are similar to those of levels with a spirit level. Since the compensator works automatically under the influence of gravity, these instruments are also called automatic levels.

Principle of operation and classification of compensator levels. A horizontal line of sight h (horizontal principal ray), passing through the optical centre of the objective lens of a simple geodetic telescope focussed to infinity, meets a very distant imaginary levelling staff ($d = \infty$) at the point P. The line of sight c, supposed to coincide with the optical axis, is inclined to h by the angle α *(Fig. 3.33, a)*. The optical image P'_0

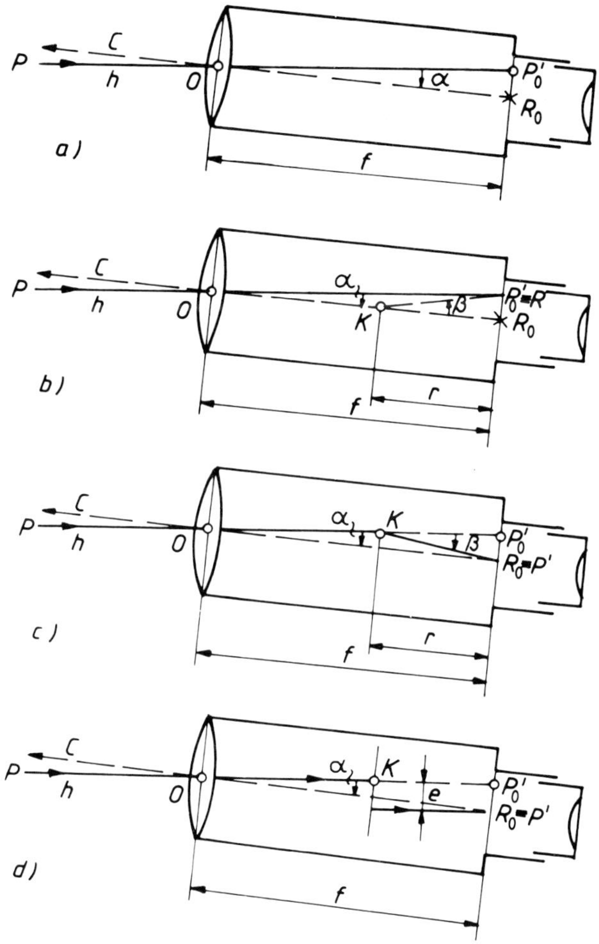

Figure 3.33. Operating principle of compensator levels; *a)* the horizontal principal ray of a telescope with a tilt α; *b)* compensator monitoring the line of sight; *c)* compensator monitoring the principal ray by rotating the beam of rays; *d)* compensator monitoring the principal ray by parallel shifting the beam of rays

of the point P is generated by the objective lens in its image-side focal plane at distance R_0P_0' from the centre of the cross-hair R_0 which is also in the same focal plane. The image P_0' and the centre R_0 of the cross-hair must coincide when levelling. This can be achieved in two ways.

The essence of the first compensating method is that either the cross-hair plate moves automatically with respect to the objective, or the objective moves with respect to the cross-hair plate, so that the points P_0' and R_0 coincide. *Compensators* based on this principle deflect the line of sight by monitoring either the cross-hair or the objective, hence these are called line-of-sight-deflecting compensators. An automatic mechanism (the compensator) moves either the point R_0 relative to the centre O, or the point O relative to the point R_0, in such a way that the point R_0 (or O) rotates about the bending point K by an angle β, where K lies on the optical axis at a distance r from the cross-hair *(Fig. 3.33, b)*. Within the compensation range of about 8 to 10 minutes of arc the condition

$$f\alpha = -r\beta \tag{3.2}$$

must be satisfied, where f is the focal length of the objective, α the tilt of the telescope. The negative sign on the right hand side of Eq. (3.2) shows that the bending direction of the angle β is opposite to that of the angle α.

The principle of the second compensating method is that the compensator placed at the beinding point K_0 deflects the horizontal principal ray h in such a way that the deflected point P' coincides with the centre R_0 of the cross-hairs. These are the principal ray-monitoring compensators. In one type the horizontal principal ray h is rotated by the angle β *(Fig. 3.33, c)*, in the other type the ray is parallel-shifted by the distance e *(Fig. 3.33, d)*. Within the operating range of compensation the following equation should be satisfied, for the first type:

$$f\alpha = r\beta, \tag{3.3}$$

for the second type:

$$f\alpha = e. \tag{3.4}$$

The principal ray-monitoring compensator can be placed either in front of the objective lens or between the objective and the cross-hair plate.

According to the physical principle of compensation the compensators are classified as compensators with a pendulum, a spirit level or with a free liquid surface; according to their technical construction they can be classified as mechanical or optomechanical (or, in abbreviation, optical) compensators.

By introducing the angular magnification of the compensator from Eqs (3.2) and (3.3):

$$n = \frac{\beta}{\alpha}, \tag{3.5}$$

the basic equation of the compensator is:

$$f = \pm nr. \tag{3.6}$$

In Eq. (3.6) the negative sign is valid for line-of-sight-deflecting compensators, while the positive sign is valid for the principal ray-monitoring compensators. If the inclination angle of the principal ray is β (denoted below by β_{op}), then the quotient $n_{op} = \beta_{op}/\alpha$ is called the *optical angular magnification* of the compensator. The swing angle of a certain mechanical element of the compensator, where the angle is considered in relation to the telescope tube (and to the optical axis), is denoted by β_r; the absolute tilt of the element with respect to the horizontal (or to the vertical) is denoted by β_a. Consequently, the *mechanical angular magnification* of the compensator can be either relative: $n_r = \beta_r/\alpha$ or absolute: $n_a = \beta_a/\alpha$.

The position of the bending point K depends on the angle magnification: $\beta_{op}/\alpha = n_{op}$. If $n = 1$, then $r = f$, i.e. the bending point K must coincide with the centre of the objective (more exactly with its principal point from the image side). In case of $n_{op} = 2$, $r = f/2$. As n_{op} is usually greater than 2, hence the point K is placed closer to the image plane and compensators with smaller dimensions and masses can be used, moreover there is more space for the focussing lens.

The basic equation Eq. (3.6) is rigorously valid only if the telescope is focussed to an infinite object distance, and if it is tilted around a horizontal axis which intersects the optical centre of the objective lens and is perpendicular to the optical axis. The formula is even valid for finite object distances if the tilting point is at a certain position which depends on the type of telescope and on the position of the compensator (see below).

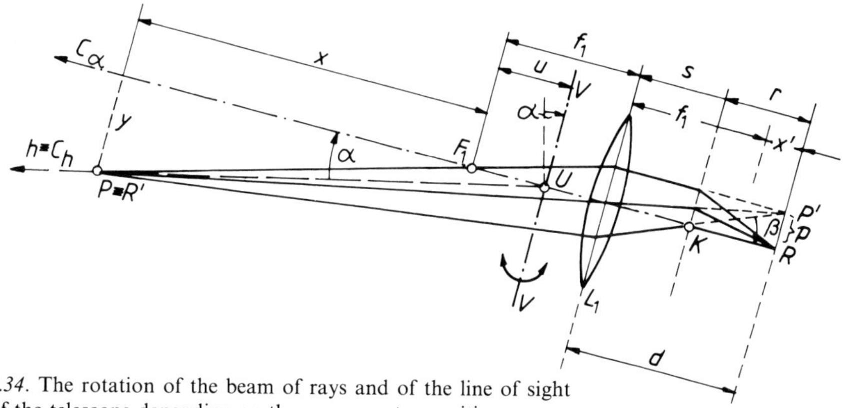

Figure 3.34. The rotation of the beam of rays and of the line of sight of the telescope depending on the compensator position

The compensator must have an arrangement and manner of operation such that the line of sight rotates around a fixed point U which is always on the optical axis, independently of the distance and the tilt of the telescope *(Fig. 3.34)*. This point U should also lie on the vertical axis. In a simple geodetic telescope the line of sight c_α, shown in the figure in coincidence with the optical axis of the telescope, is inclined to the horizontal plane by the angle α. If the compensator is inserted at any point K of the optical train, the image R' of the cross-hair lines R comes to the position P, i.e. it is moved by y from its original position. If the point U is on the vertical axis of the instrument, it is obvious from the figure that

$$\frac{y}{x+u} = \tan\alpha \approx \alpha. \tag{3.7}$$

Depending on the type of telescope and the position of the compensator, the following relationships are valid for the value u determining the position of the point U:

a) If the compensator is in front of the telescope, point U coincides with point K, around which the compensator deflects the line of sight:

$$K \equiv U. \tag{3.8}$$

b) If the compensator is inside the telescope at the constant distance s from the objective lens, then for both the simple telescope and the telescope with internal focussing:

$$u = \frac{f_1^2}{f_1 - s} = f_1 \frac{\beta_{op}}{\alpha} = n_{op} f_1, \tag{3.9}$$

where u is the distance of the point U from the first focal point F_1 of the objective lens L_1; f_1 is the focal length of the objective lens; s is the distance of the point K from the objective lens and n_{op} the optical angle magnification of the compensator. The value of u is positive if the point U is separated from the focal point F and is moved towards the telescope.

c) If the compensator is at a constant distance r from the cross-hair, then

$$u=0 \tag{3.10}$$

for the simple telescope, and

$$u=x\left\{\frac{f_2 f_0}{f_1(f_2-b)}\right\}-1 \tag{3.11}$$

for the telescope with internal focussing, where

$$x=\frac{f_1^2(f_2-b)}{b^2-(f_1-d)(f_2-b)}; \tag{3.12}$$

here f_1: the focal length of the objective lens,
f_2: the focal length of the internal focussing lens,
f_0: the resultant focal length at infinite focus,
e: the distance between the objective and the internal focussing lens,
d: the distance between the objective lens and the cross-hair,
$b=d-e$: the distance between the internal focussing lens and the cross-hair, depending on the object distance.

The value $b=d-e$ can be calculated suitably from Eqs (3.38) and (3.39). The point U determined by the value u is identical with the anallatic point of the telescope and its position changes with change of object distance, as can be seen from the equations. With a suitable choice of telescope characteristics, however, the shift of point U is small, consequently, in practice, it can usually be regarded as fixed.

d) If the compensator is at half the image distance in a simple telescope (without a focussing lens), then

$$u=f, \tag{3.13}$$

i.e. the point U coincides with the object-side principal point of the objective lens.

If these conditions are fulfilled, the basic equation (3.6) of the compensator is not only valid in the simplified case shown in *Fig. 3.33* and for infinite object distance, but also for a telescope with internal focussing lens and for any object distance. In the case of a compensator fixed at a constant distance r from the cross-hair, the focal length f in Eq. (3.6) must be substituted by the resultant focal length of the objective system; in other cases it should be substituted by the focal length of the fixed lens of the objective system.

Cross-hair monitoring compensators. In the case of cross-hair monitoring compensators the bending point K must be, from Eq. (3.6), at the distance

$$r=-\frac{f}{n_r} \tag{3.14}$$

from the cross-hair plate. According to the construction of the compensators such instruments can be subdivided into different groups.

The compensator is the cross-hair plate itself, suspended by vertical wires, in the *Filotecnica Salmoiraghi* precise level *5190 (Fig. 3.35/a)*. The cross-hair plate *4*, onto which the cross-hairs and a supplementary fine circular line have been engraved, is attached to a simple physical pendulum suspended on three wires *(Fig. 3.35/b)*. Since $\alpha = -\beta_r$ and $n_r = -1$, the rotation point K of the pendulum has been placed at the image-side principal point O of the objective lens. Parallax can be eliminated by moving the objective *3* vertically. The optical elements *5, 7, 9* create the image of the cross-hair *4* on another reticule *8* fixed in front of the eyepiece lens before ocular *12*. Thick circular lines are engraved on the latter reticule *(Fig. 3.35/c)*, by means of which the verticality of the vertical axis of the instrument can be checked independently of the circular bubble. If the images of the cross-hair *4* and of the supplementary fine circular line coincide with the marks on the reticule *8*, or if they lie within the limits given by these marks, then the vertical axis and the plumb-line are sufficiently near to coincidence, i.e. the deviation is within the operating range of the compensator, namely ± 5 minutes of arc. The pair of mirrors *2* can be moved vertically by rotating a fine drive with a range of 10 mm, consequently not only do they deflect the vertically stabilized line of sight through 90°, but also they play the role of an optical micrometer at the same time. The vertical motion of the mirror pair as well as that of the line of sight can be read on the micrometer scale *6* which is rigidly connected to the pair of mirrors. This scale has 100 graduation intervals with an accuracy of 0.1 mm (this can be increased by estimation to 0.01 mm). The micrometer graduation is also imaged on the reticule *8* by the lens *10* and the projecting prism *11*. The cover glass *1* of the telescope is in fact an optical wedge, by means of which the line of sight can be adjusted. The mean error of the compensator oscillation is ± 0.2 second of arc.

Fennel used a cross-hair plate hanging from crossed bands for the compensator element in the levels *AUING* and *AUPRE*. The operation of the compensator is outlined in *Fig. 3.36*. The two pairs of metal wires fixed to the telescope body at points A and B suspend the frame of the cross-hair plate from points C and D. The distance $\overline{CD}$ is shorter than $\overline{AB}$, consequently the tilt angle β_r is greatly magnified compared to the tilt α of the telescope, which results in an increase in the angular magnification and in a shortening of the distance r. Hence the pendulum system is small, light and protected against shock. With a telescope tilt α and at the same time imagining a motionless pendulum, *Fig. 3.36* shows the position of the pendulum marked by dashed lines; the continuous lines show the pendulum in its normal position.

A horizontal cross-hair line monitoring compensator with a spirit level is used on the *Stodolkewich–Romanov* level type *NS-2*, in which the ends of the bubble of a tubular spirit level are projected into the field of view and used as a horizontal cross-hair *(Fig. 3.37)*. The images of the bubble ends *1* are projected by the pentagonal prisms *2* and *3*, by the splitting prisms *4* and *5*, by the prisms *6* and *7* as well as by the cylindrical lens *8*, onto a quadrangular field on the left hand side of the image plane. The right hand side of the field of view is covered by the diamond-shaped prism *9* of the telescope. The bubble ends appearing in the quadrangular field, being coincident and imaged by the cylindrical lens as a straight line, serve as the horizontal

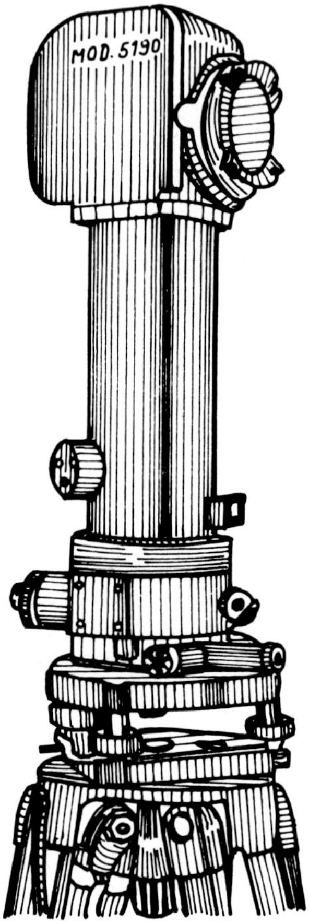

Figure 3.35/a. Automatic level 5190 (Filotecnica Salmoiraghi, Milano)

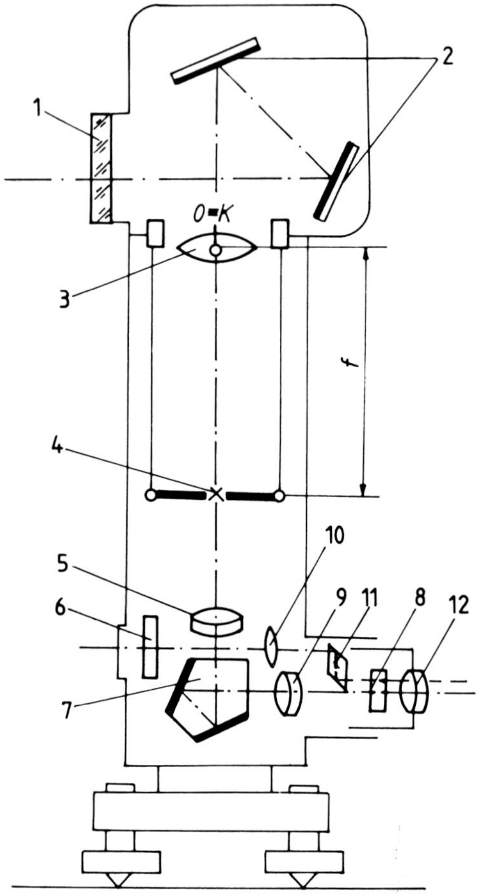

Figure 3.35/b. Diagram of the automatic level *5190* (Filotecnica Salmoiraghi, Milano)

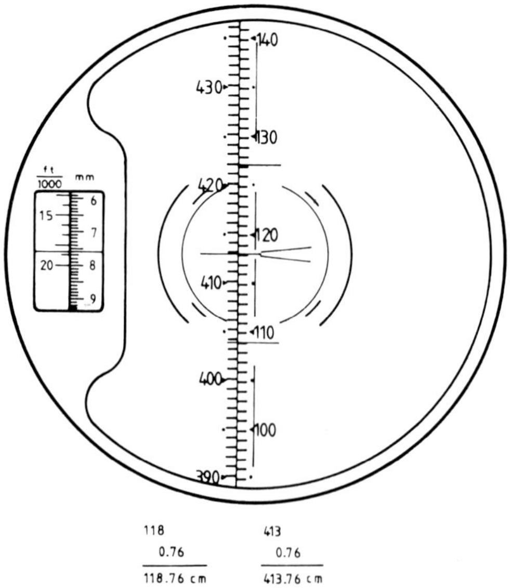

Figure 3.35/c. The field of view in the automatic level *5190* of Filotecnica Salmoiraghi

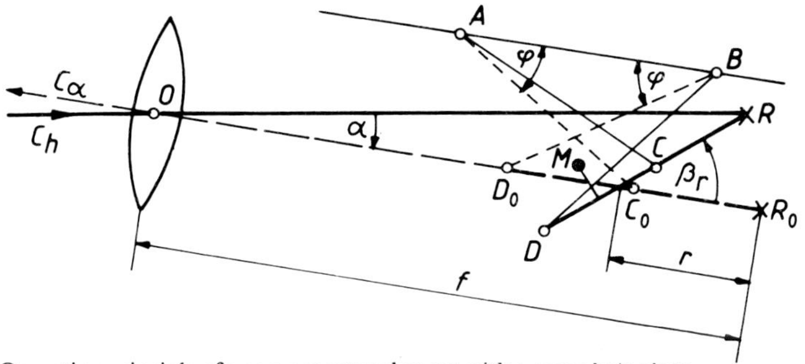

Figure 3.36. Operating principle of a compensator element with a cross-hair plate suspended by crossed bands

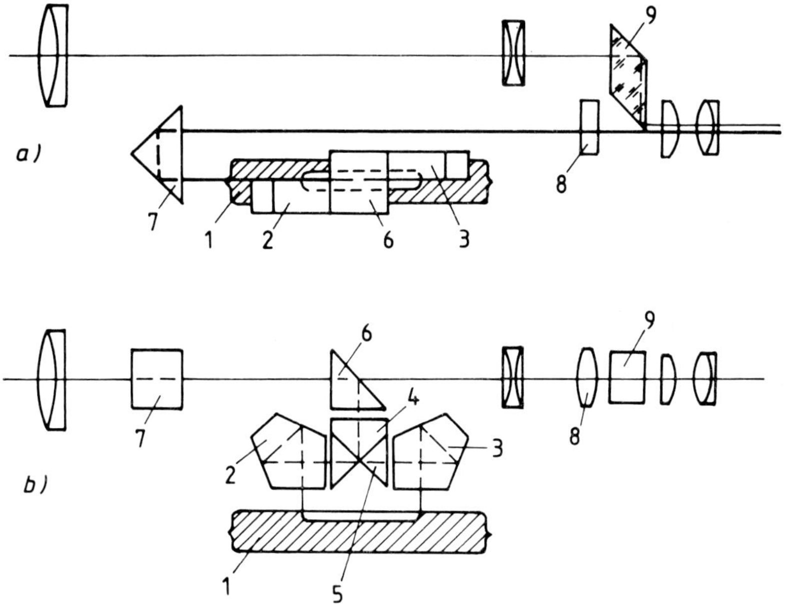

Figure 3.37. The optical system of a compensator with a spirit level monitoring the cross-hair: *a)* view from above; *b)* side view

cross-hair of the telescope; they automatically define the horizontal line of sight of the telescope within a tilt of ± 1.5 minutes of arc. The two adjusting screws of the level are used for the adjustment of the cylindrical lens, and for the coincidence setting of the bubble ends.

Objective monitoring compensators. This type of compensator has been used by the firm Filotecnica Salmoiraghi in the levels of medium accuracy *No. 5173-5175 (Fig. 3.38)*, and by the factory Franke in the levels *FN2* and *FN3 (Fig. 3.39)*. The objective lens is suspended on three wires in a periscope type telescope in such a way that the length of the equivalent pendulum is equal to the focal length of the objective

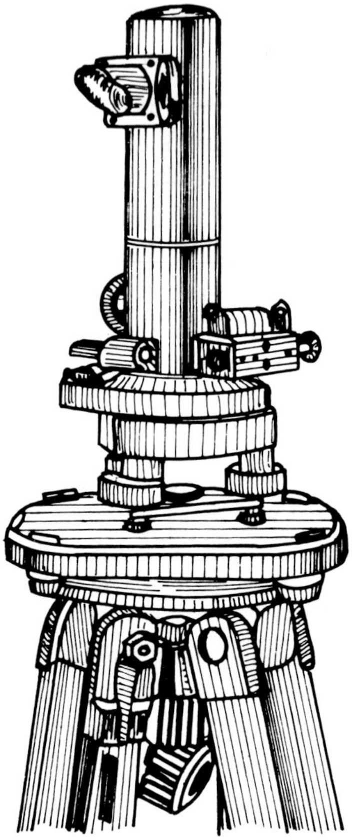

Figure 3.38. Automatic level *5174* (Filotecnica Salmoiraghi, Milano)

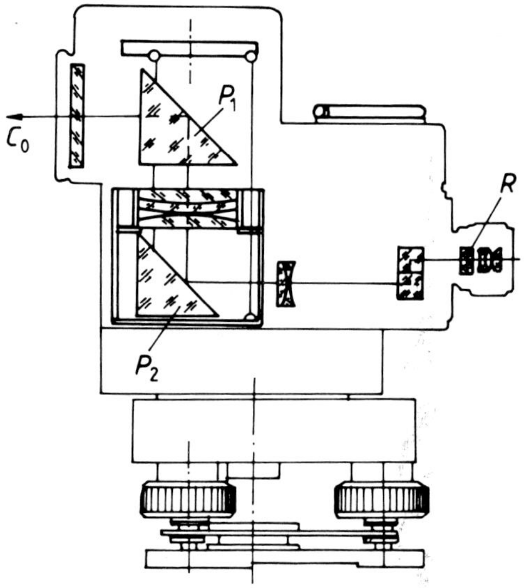

Figure 3.39. Diagrammatic sketch of the objective-monitoring compensator of the level *FN2* (Franke et Co. Optik GmbH, Giessen)

lens. The principle of the compensator is shown in *Fig. 3.40*, where a vertically positioned simple telescope is delineated instead of the bent telescope. If the tilt of the telescope is zero ($\alpha = 0°$), the suspension point of the objective lens K_0 (the rotation point of the pendulum), its optical centre O_0 and the cross-hair centre R_0 all lie in a common plumb-line. If the telescope is tilted by an angle α around the point O_0, the centre of the cross-hair, the suspension point of the objective lens and the pendulum-suspended objective lens wander to points R, K and O, respectively. Since $\overline{K_0O_0} = \overline{O_0R_0} = f$, the line of sight $RO \equiv C_{2\alpha}$ makes an angle of 2α with the vertical within the operating range of the compensator. The reflecting surface of the prism P_1, which is in front of the objective lens and is fixed to the housing of the telescope (see *Fig. 3.39*), brings the inclined line of sight to the horizontal by a single reflection.

The compensator in front of the objective. One type of the principal ray-monitoring compensators fixed in front of the objective acts with an optical wedge. For the deflection of the beam of rays (line of sight) the ray-deflecting effect $\beta = \tau(n-1)$ of an

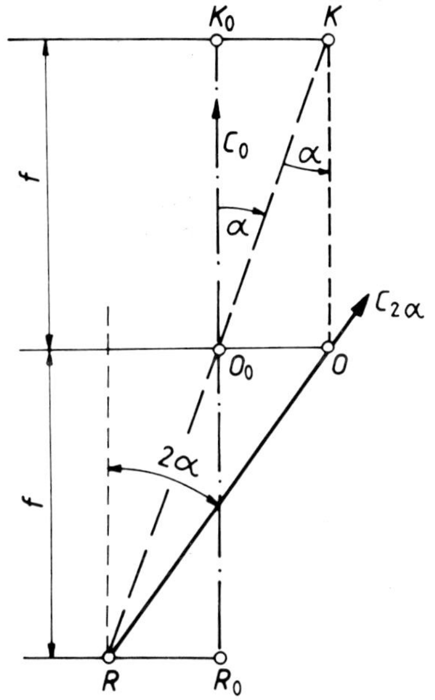

Figure 3.40. Operating principle of an objective-monitoring compensator

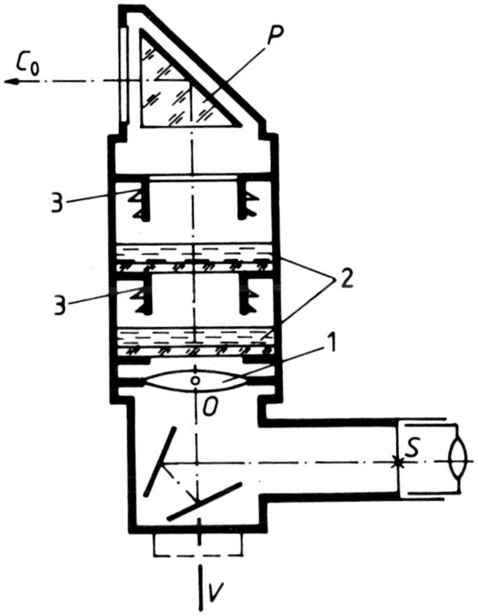

Figure 3.41. Diagram of the Gusev liquid wedge compensator (Soviet Union)

optical wedge with angle τ can be used, where n is the refractive index of the material of the optical wedge. For compensators fixed in front of the objective lens the necessary optical angle magnification of the compensator is:

$$n_{op} = \frac{\beta_{op}}{\alpha} = \frac{\tau}{\alpha}(n-1) = 1. \tag{3.15}$$

The wedge angle τ must change proportionally to the angle α within the operating range of the compensator. One solution is to use liquid, when

$$\tau = \alpha \tag{3.16}$$

hence

$$n_{op} = (n-1). \tag{3.17}$$

Since the refractive index of liquids varies between 1.3 and 1.8, at least two liquid wedges must be used in order to reach $n_{op} = 1$.

Such a compensator with a two-chamber liquid wedge is used in the *Gusev level (Fig. 3.41)*. The compensator with the two-chamber wedge is placed above the objective lens *1* of the bent telescope. The refractive index of the dibutilphtalate liquid in metal cups with a glass bottom is 1.4960 at +20 °C, its boiling points is +289 °C, its freezing point about −60 °C. In both cups *2* there is a liquid layer about 1.5–2 mm thick. The cylindrical rings *3* in the liquid chambers prevent the liquid from coming through to the upper glass plate when roughly tilting the instrument. The two liquid wedges of the compensator, operating within a compensation range of ±10–15 minutes of arc, deflect the line of sight (tilted by the angle α) again by the angle α, corresponding to the tilt of the telescope. Hence the line of sight is inclined to the vertical by an angle of 2α. It is brought back into the horizontal by a rectangular prism placed above the compensator and tilted together with the telescope housing by angle α.

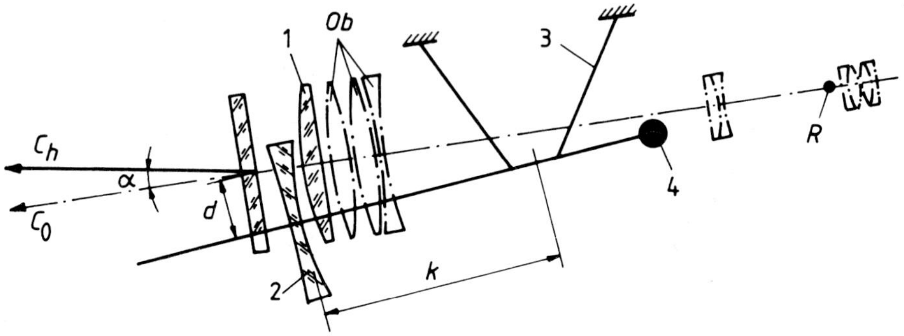

Figure 3.42. Principle of the lens compensator of the level *NSM-2* (Soviet Union)

The Kharkow Works for Mining Instruments (USSR) used a *lens compensator* in the level *NSM-2 (Fig. 3.42)*. The compensator lens consists of two parts: a plano-convex lens *1* fixed in the casing of the objective system in front of the objective lenses *Ob*, and a plano-concave lens suspended on a jointed quadrilateral *3*. The mass *4* is used for weight compensation. By suitable dimensioning (by an appropriate design of the jointed quadrilateral, by the proper selection of the focal length f of the compensator lens *2* and of the pendulum length k) it can be arranged that, with a tilt α of the telescope the ray deflection $\beta = d/f$, caused by a displacement d to the lens *2*, should be just equal to α. Consequently, the optical angular magnification for all values of α is $n_{op} = \beta_{op}/\alpha = 1$.

The basic principle of the frequently used *mirror compensator* is that a mirror (or a mirror surface of a prism), attached to a simple or combined pendulum and placed in the image-forming beam path, deflects the line of sight or the horizontal principal ray of the telescope by the angle β_{op}. With a tilt α of the telescope the mirror makes a relative and mechanical swing of angle β_r. Due to this the reflected beam is deflected by

$$\beta_{op} = 2\beta_r \tag{3.18}$$

in comparison to its original direction. With a single reflecting surface the optical angle magnification is from Eqs (3.5) and (3.18):

$$n_{op} = 2\frac{\beta_r}{\alpha} = 2n_r. \tag{3.19}$$

For a compensator placed in front of the objective lens the condition $n_{op} = 1$ must be fulfilled, therefore a relative mechanical angular magnification

$$n_r = \frac{1}{2} = \frac{\beta_r}{\alpha} \tag{3.20}$$

is required.

The requirement $n_r = 0.5$ has been achieved in the *MOM Ni E1* level *(Fig. 3.43)* by attaching the compensator mirror to a pendulum oscillating on a leaf spring above the objective lens of the vertical bent telescope. The physical parameters of the spring pendulum and the suspension point of the mirror were so selected that, with a telescope tilt of α, the mirror swings through the angle $\alpha/2$. Hence $\beta_{op} = \alpha$, and the optical angular magnification is $n_{op} = \beta_{op}/\alpha = 1$. The telescope of the instrument is a simple geodetic telescope which has 6.25 times magnification and an objective lens with 16 mm free diameter. The erect image produced by the compensator mirror and the roof prism placed inside the telescope can be focussed by rotating the cross-hair tube. The operating range of the compensator is ± 30 minutes of arc and the pendulum is air damped. A standard deviation for 1 km of levelling of ± 4–15 mm can be achieved, depending on the graduation and the observing distance to the staff.

The *MOM Ni D1* level has a similar compensator, by means of which a standard deviation of ± 3–4 mm/km can be attained.

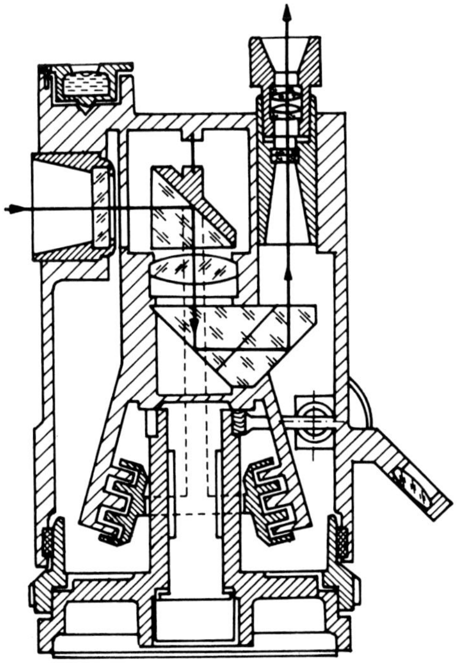

Figure 3.43. Diagrammatic sketch of the architect's level *Ni E1* (MOM, Budapest)

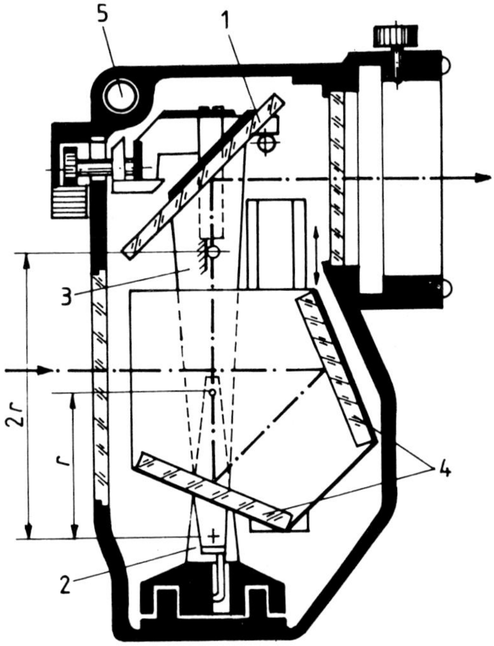

Figure 3.44. Auxiliary compensator *VK 05/1* attachable in front of the objective lens (VEB Feinmess, Dresden)

An *auxiliary compensator (Fig. 3.44)*, attachable in front of the objective lens, was manufactured by VEB Feinmess, Dresden; it can be used on both the *IN 30G* and the *Jenoptik Ni 030* levels. The tilt of the compensator mirror *1* is controlled by the simple physical pendulum *2*, of length r, by rotating the $2r$ long cantilever *3* which holds the compensator mirror, using a mechanical transfer. Hence the tilt of the mirror is $\beta_r = \alpha/2$. The line of sight is brought in vertically by the compensator mirror and then deflected to the horizontal by the mirrors *4*. The latter can be moved vertically by a slow-motion screw and can thus be used as an optical micrometer. The auxiliary compensator can be plumbed by the tubular level *5* mounted across the auxiliary instrument. The tubular level has a sensitivity of 6′/2 mm.

Compensators placed within the telescope either deflect the beam of rays transmitting the image (horizontal principal ray) or they give them a parallel shift. Compensators deflecting the beam may act with a lens or with a mirror, according to the optical element used. The material for the lens can be either glass or liquid, and the surface of a plain mirror, prism or liquid can be used equally as a reflecting surface.

The *Kern GKO-A* level is manufactured with a compensator lens *(Fig. 3.45)* in which the lens system *1* is attached to the horizontal arm of the pendulum *3* which has an axis supported by ball-bearings *(Fig. 3.46)*. The swinging of the compensation mass *2*, which is part of the pendulum, is damped by the magnet *4*. The image of the levelling staff is generated in an intermediate image plane *8* by the objective lens system *5*, the movable internal focussing lens *6* and the lens *7* rigidly fixed to the telescope, and an aperture made of red plastic is placed in the intermediate image plane. This inverted image is finally re-inverted by the compensator lens to become an erect image in the plane of the reticule *9*. The image of the aperture *8* appears as a red warning disc in the upper or lower part of the field of view if the vertical axis

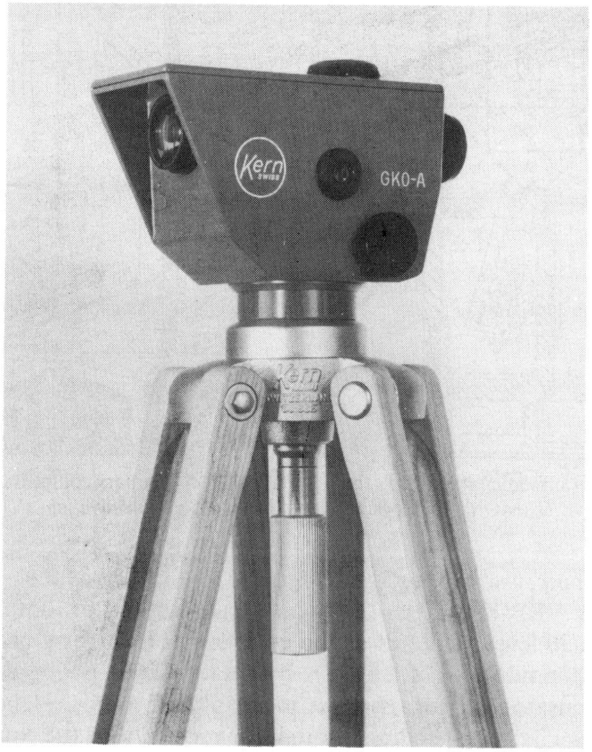

Figure 3.45. Level *GKO-A* (Kern et Co. AG, Aarau)

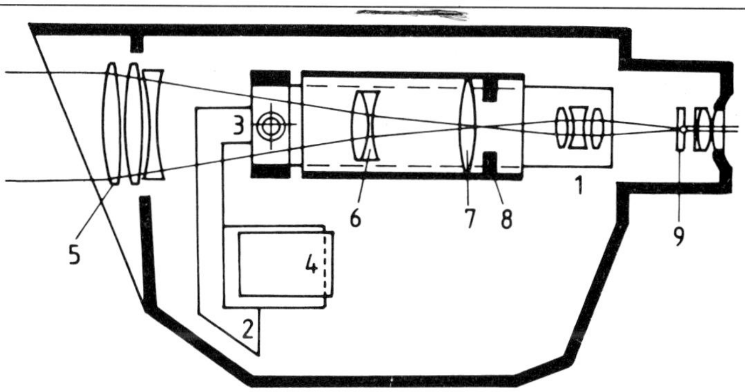

Figure 3.46. Diagrammatic sketch of the level *Kern DKO-A*

is not plumbed sufficiently well; this checks whether or not the tilt of the telescope lies within the operational range of the compensator of ± 30 min of arc. The standard deviation of the compensator stabilization is ± 3 sec of arc, and that of direct and return levelling is ± 5 mm per km.

The *Opton Ni 4* level *(Fig. 3.47)* also uses a compensator lens. The lens of the compensator is the circular level itself, operating as a concave dispersing lens which is placed in the vertical part of the optical train. One surface of the dispersing lens is the bottom plane surface inside the cylindrical glass container, while the other is the lower concave surface of the bubble. The horizontal axis is pre-levelled by the same circular level; for this purpose the circular level can be observed through the

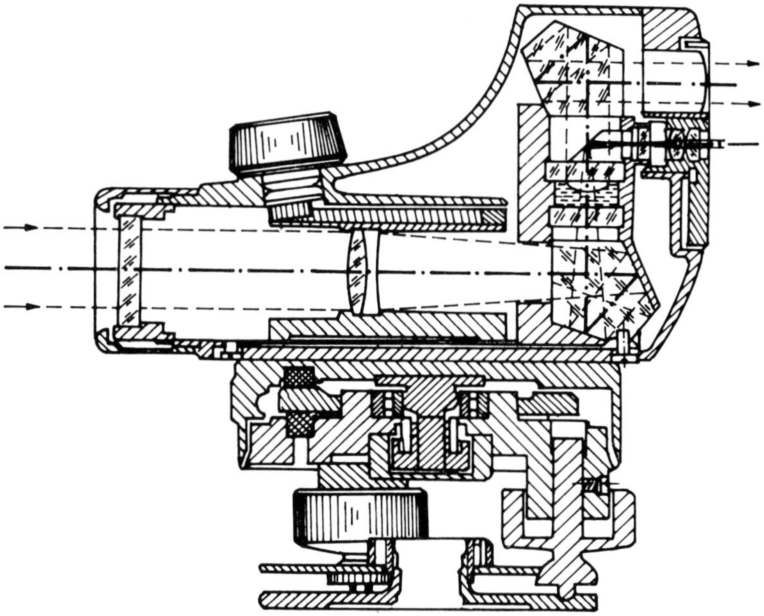

Figure 3.47. Level *Ni 4* (Opton Feintechnik GmbH, Oberkochen)

viewer above the eyepiece. The image of the levelling staff can be focussed by moving the objective longitudinally. The tilt of the line of sight can be eliminated by the optical wedge in front of the objective lens, whose focal length is 110 mm. There is a separation of 130 mm between its first focal point and the vertical axis of the instrument, which should coincide with each other according to Eq. (3.10). The measuring error arising from this separation can, however, be neglected in practical measurement (where the mm values of staff readings are estimated). The compensator is very sensitive to temperature change, but appropriate dimensioning and special materials have ensured that the effect of temperature change on the compensator is very small. The standard deviation of direct and return levelling is ± 8 mm/km.

Mirror compensators placed inside the telescope rotate the image-generating beam of rays and act in fact in the same way as mirror compensators placed in front of the objective. Their construction can be very different according to

— the position of the compensator in the telescope;
— whether a simple mirror or a reflecting prism is used as reflecting surface;
— whether the mirror is mounted on a simple or a composite pendulum;
— how many reflecting surfaces are used and in what arrangement.

Figure 3.48 shows the basic principle of the operation of the compensator. The mirror M is within the bent telescope at the point K of the optical axis, attached to a pendulum swinging about this point. The mirror is so attached to the pendulum that, after reflection, the horizontal optical axis intersects the centre of the cross-hair R. If the pendulum G with the attached mirror M were fixed to the telescope then, after tilting the telescope about the point O by the angle α, the horizontal principal ray arriving from the point P on the staff would be reflected from the mirror (in position M_0) into the direction $K_0P'_0$. Since the pendulum is not fixed, however, it swings from position G_0 to G, and the mirror swings from position M_0 to M and rotates through the angle $\beta_r = -\alpha$. From Eq. (3.18) now $\beta_{op} = -2\alpha$, and from Eq.

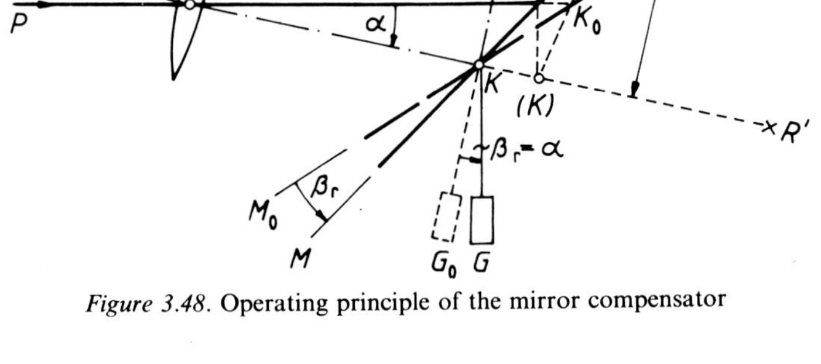

Figure 3.48. Operating principle of the mirror compensator

(3.15) the optical angular magnification is $n_{op}=-2$. The position of the compensator is determined from Eq. (3.6) as

$$r=\frac{f}{n_{op}}=\frac{f}{2}. \tag{3.21}$$

The mirror also reflects the sense of rotation of the beam, therefore the sign in the basic equation (3.6) reverses and the value of r becomes positive.

With a compensator situated between the objective lens and the inner focussing lens at distance s from the objective lens, the bending point of the line of sight is from Eq. (3.9) at a distance $u=n_{op}f_1$ from the inner focal point of the objective lens. If the compensator mirror is attached to a simple physical pendulum, then $n_{op}=2$,

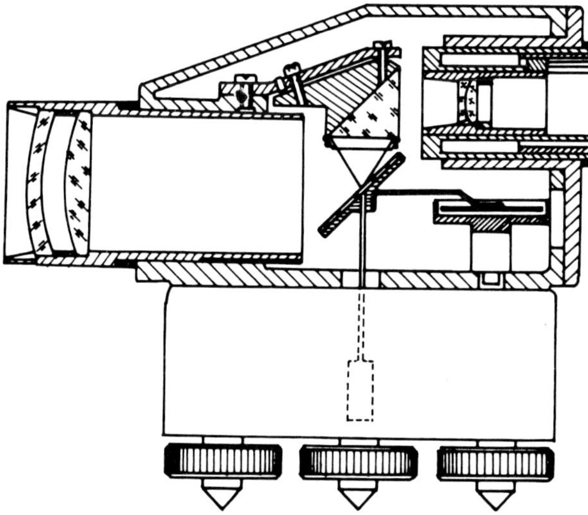

Figure 3.49. Engineer's level *Na 1* (Franke et Co. Optik GmbH, Giessen)

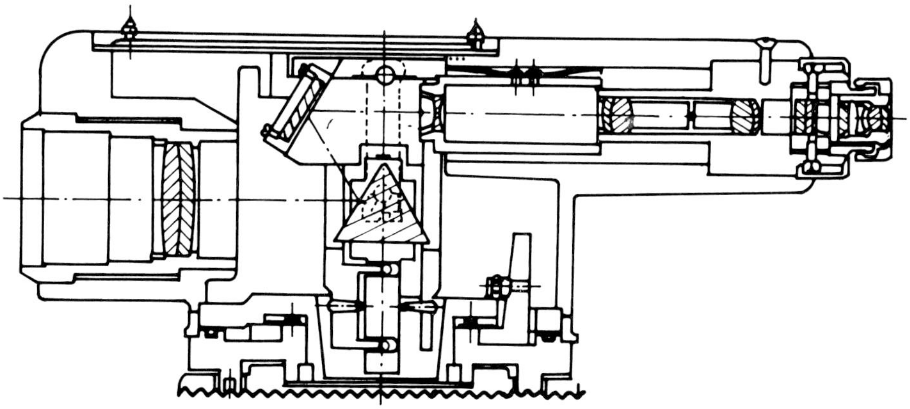

Figure 3.50. Diagrammatic sketch of the level Nikon *AP-2* (Nippon Kogaku K.K., Tokyo)

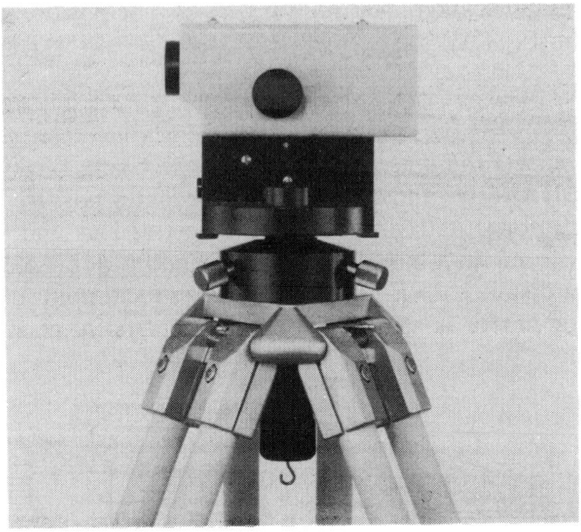

Figure 3.51. Level *Ni 42* (Opton Feintechnik GmbH, Oberkochen)

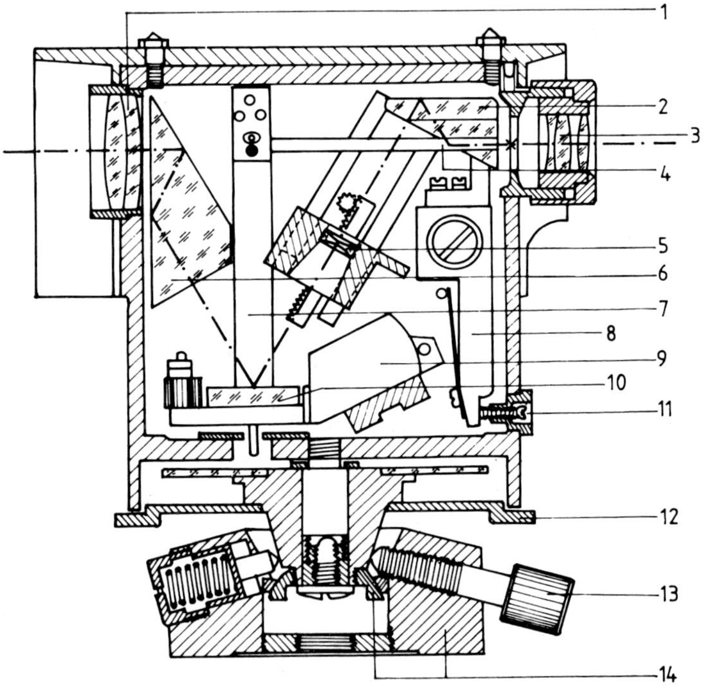

Figure 3.52. Diagrammatic sketch of the level *Opton Ni 42*: *1* objective lens; *2* roof prism; *3* eyepiece; *4* indicator of compensator position; *5* focussing lens; *6* prism with reflecting surface; *7* compensator pendulum; *8* adjusting device; *9* damping device; *10* compensator mirror; *11* screw for adjusting the line of sight; *12* exposed rim for turning the limb; *13* foot screw; *14* instrument base with ball joint

consequently $u=2$. The position of the compensator is also defined by Eq. (3.9), where $s=f_1(n_{op}-1)/n_{op}$. When taking the optical magnification $n_{op}=2$ into account, then $s=f_1/2$. Hence the reflecting surface of the compensator must be placed at this distance from the objective in order to give a correct operation independent of object distance; further, the vertical axis theoretically must be at the distance $u-f_1=f_1$ from the objective.

The compensator mirror (reflecting prism) attached to a simple physical pendulum is placed between the objective lens and the inner focussing lens for the *Franke* level *Bauart Na 1 (Fig. 3.49)* as well as for the similar level *Nikon AP-2 (Fig. 3.50)*. An additional reflecting surface is placed within the optical train, attached to the telescope housing. In this way the telescope has a horizontal construction.

The *Opton Ni 42* level *(Fig. 3.51)* has a similarly placed compensator, but its telescope has a quite different arrangement. The construction of the instrument also differs from that of conventional levels. It is worth mentioning that the ball and socket joint type of base plate, with the vertical axis of the ball joint supported by a spring, can be levelled by means of two inclined foot screws *(Fig. 3.52)*. The field of view of the anallatic telescope, with 22 times magnification and with a 30 mm objective diameter, is 3.60 m at a distance of 100 m. The operating range of the compensator attached to the pendulum and swinging on ball bearings is $\pm 1°$, the standard deviation of stabilization is ± 5 sec of arc, and the standard deviation of direct and return levelling is ± 5 mm/km. The mass of the instrument is 1.6 kg. The index *1*, visible in the field of view of the instrument and indicating the position of the compensator *(Fig. 3.53)*, shows whether the longitudinal component of the tilt of the vertical axis is within the operating range of the compensator or not.

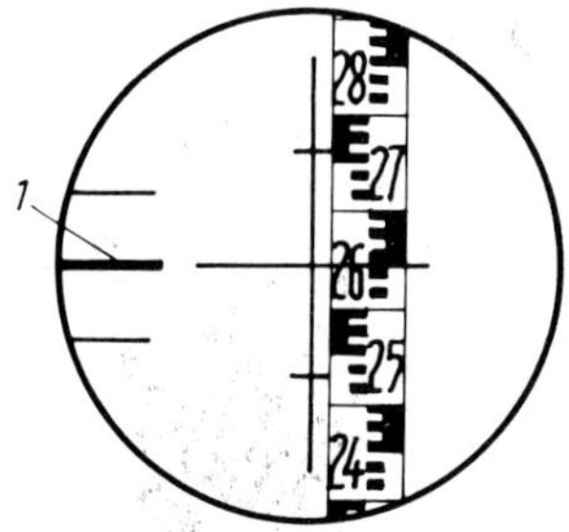

Figure 3.53. Field of view in the telescope of the level *Opton Ni 42*: *1* indicator of the compensator position

The *Kern GK1-A* level has a different construction and arrangement again *(Fig. 3.54)*, where the compensator in the telescope with an inner focussing lens is again at a constant distance from the objective lens. A roof prism is used as a reflecting surface of the compensator which is attached to a simple physical pendulum *(Fig. 3.55)*. The air-damped pendulum holding the compensator mirror has a magnetic axis with low friction. The magnetic field of the permanent magnet holding the axis *(Fig. 3.56)* is closed by the horizontal compensator axis. The ductile iron pendulum

Figure 3.54. Level *GK1-A* (Kern et Co. AG, Aarau)

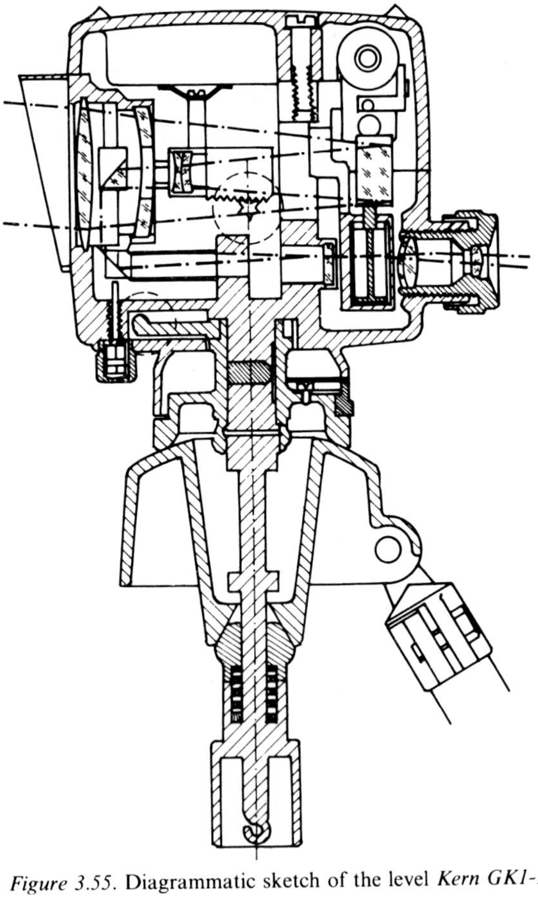

Figure 3.55. Diagrammatic sketch of the level *Kern GK1-A*

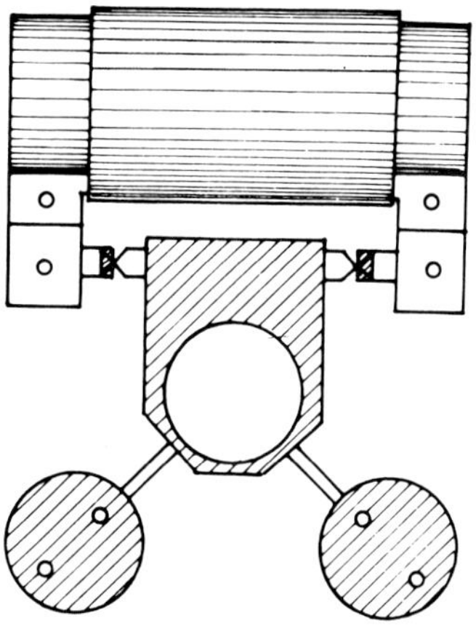

Figure 3.56. Compensator with a magnetic bearing in the level *Kern GK1-A*

Figure 3.57. Level *GK2-A* (Kern et Co. AG, Aarau)

axis is situated between the conical poles of the magnet; the poles have plane end-surfaces covered with a sapphire-cap and the conical ends of the pendulum axis are made of hardened metal. The tilt of the line of sight can be eliminated by moving the lower rectangular prism *1*. The *Kern GK2-A* is similar, but it is supplied with an optical micrometer and has a somewhat different optical construction *(Fig. 3.57)*.

In another group of compensators acting with relative mechanical angular magnification $n_r = -1$, the compensator is positioned so that its reflecting surface always lies at half the image distance in a simple telescope. In this case — from Eq. (3.13) — the tilting point *U*, i.e. the vertical axis of the instrument, must coincide with the object-side principal point of the objective lens.

Such a compensator is used in the precise level *MOM Ni A31 (Fig. 3.58)*. The cross-hair plate *1* of the instrument is cemented to the central element of the three-lens objective system in the principal plane of the objective *(Fig. 3.59)*. The image can be focussed by longitudinally moving the pair of mirrors *2*, *3*. Due to the even number of reflections of the beam of rays, even irregular motion of the image-adjusting mirror

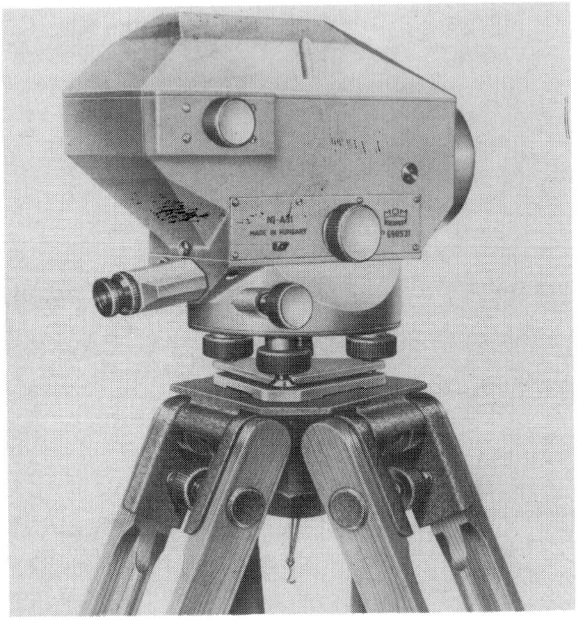

Figure 3.58. Level *Ni A31* (MOM, Budapest)

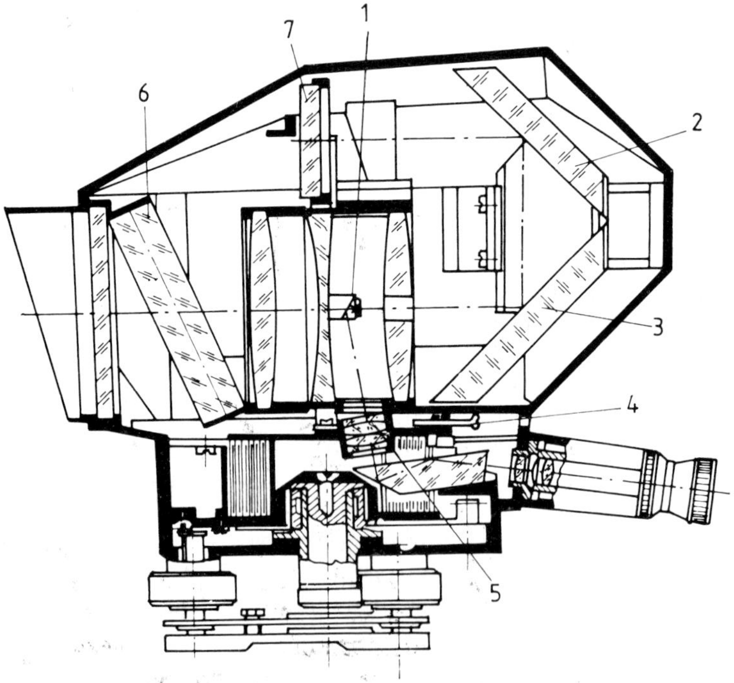

Figure 3.59. Diagrammatic sketch of the level *MOM Ni A31*

pair does not cause a wobble of the line of sight. The compensator mirror *7* attached to the pendulum always bisects the image distance due to the movement of the mirror pair *2, 3*. The lens system *5* generates the image of the cross-hair and of the observed staff, which is projected here on the second reticule placed in the second image plane in front of the eyepiece. If the projected images of the horizontal and the vertical hair lie within the pairs of thick lines engraved on the second reticule *(Fig. 3.60)*, then the vertical axis of the instrument is within the operating range of compensation. This is because one element of the lens system *5* is rigidly fixed to the compensator pendulum and therefore the image of the cross-hairs observed through the eyepiece changes its position depending on the tilt of the instrument. The instrument is then ready for measuring, or in other words, the vertical axis then has a tilt within the compensation range of ± 6 min of arc.

The vertical axis is eccentric with respect to the object-side principal point of the objective. Thus the systematic error, caused by the plane parallel glass plate *6* (see *Fig. 3.59*) in front of the objective lens, being used as an optical micrometer, is eliminated. Otherwise a systematic error would arise since, with a tilted vertical axis, the plane parallel plate *6* shifts the line of sight vertically and this shift is proportional to the longitudinal component of the tilt of the vertical axis. The measuring range of the optical micrometer *6* is ± 2.5 mm. One unit of the micrometer graduation as seen through the telescope corresponds to a shift of the line of sight of 0.05 mm.

The free aperture of the objective lens is 67 mm. The telescope gives an erect and side-correct image. It has changeable eyepieces giving a 30, 40 or 50 times magnification. The telescope is not anallatic, hence, for distance measurement with stadia hairs with a multiplying constant of 100, an additive constant of about 360 mm must be taken into consideration.

The compensator has two possible adjustments: firstly, the tilt error of the line of sight can be eliminated by moving the centre of gravity of the pendulum, which can be done by means of the screw *4* placed behind the small hole above the eyepiece; secondly, the lines on the second reticule can be adjusted by moving the reticule to its correct position after removing the eyepiece.

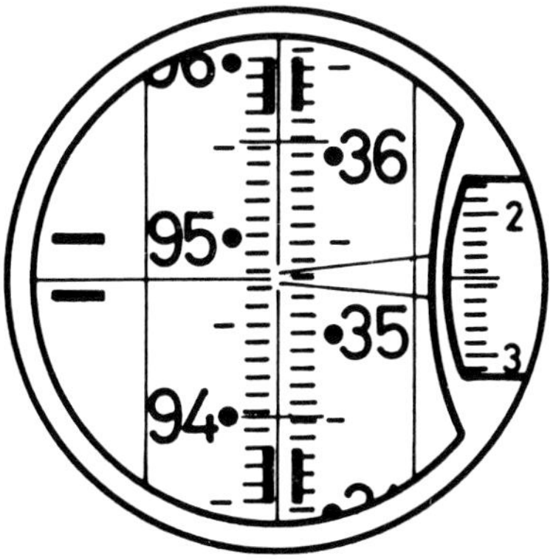

Figure 3.60. Field of view in the telescope of the level *MOM Ni A31*

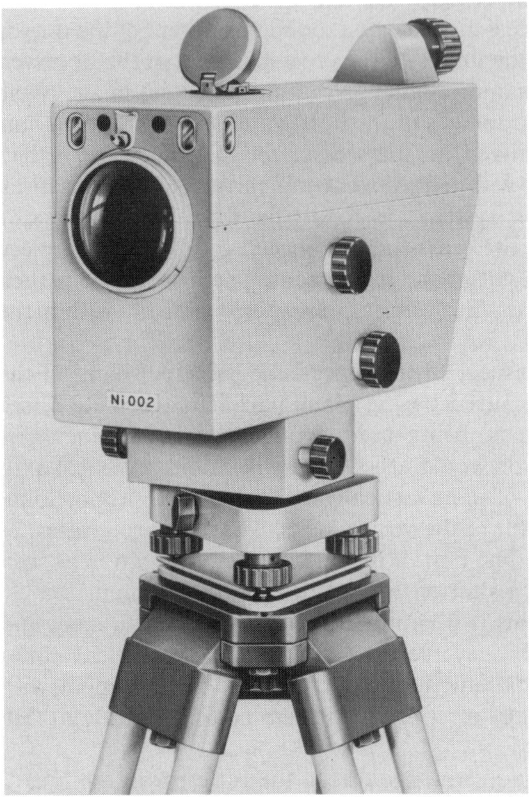

Figure 3.61. Percision level *Jenoptik Ni 002* (VEB Jenoptik, Jena)

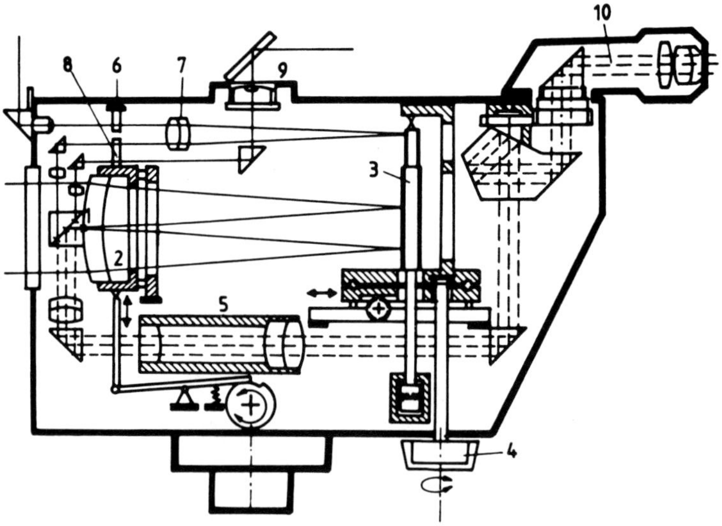

Figure 3.62. Diagrammatic sketch of the level *Jenoptik Ni 002*

The standard deviation for 1 km levelling direct and return is $\pm 0.2 \ldots \pm 0.3$ mm/km.

In the high precision level *Jenoptik Ni 002 (Fig. 3.61)* a pendulum-suspended mirror *3* is used as a compensator. It is placed behind the objective *2 (Fig. 3.62)* at half the image distance. Further, the cross-hair *1* is engraved on a prism which is cemented to the objective lens. The compensator mirror can be moved longitudinally along the telescope in order to focus the image. When measuring, the compensator mirror can be reversed around its vertical axis by 180° by knob *4* and so, by measuring in both compensator positions, an absolute horizon is given within ± 1 sec of arc within the compensation range ± 10 min of arc at all temperatures occurring in practice. The optical elements *5* project the image of the cross-hairs *1* and also the image generated in the plane of the cross-hairs onto the focal plane of the eyepiece *10*. In this same plane the circular level *9* and the graduations of the optical micrometer *8* are also imaged. The optical micrometer is not a plane parallel glass plate as usual, but the line of sight of the telescope can be moved vertically by the micrometer which raises or lowers the objective lens system and the cross-hairs together, within its measuring range of 5 or 10 mm. The shift of the line of sight can be read on the micrometer graduations *8*, attached to the housing of the objective lens system, with the aid of the index *6*; the images are unified by a second mirror fixed on the pendulum above the compensator mirror and by the optical system *7*. By this construction the error caused by the eccentricity of the object-side principal point of the objective lens in relation to the vertical axis has been eliminated. The eyepiece *10* is mounted on the top part of the instrument; it can be swivelled together with its prism system around a vertical axis through $\pm 90°$ from the optical axis of the telescope, the image always remaining erect. Consequently the surveyor can observe foresight and backsight readings on the staff without changing position. The standard deviation of stabilization of the air-damped compensator is ± 0.5 sec of arc, and that of direct and return levelling over 1 km is ± 0.2 to ± 0.3 mm.

The compensator of the *Sokkisha AL-2* level is a pendulum-suspended plane mirror placed at the bisector of the image distance *(Fig. 3.63)*, while the compensator of

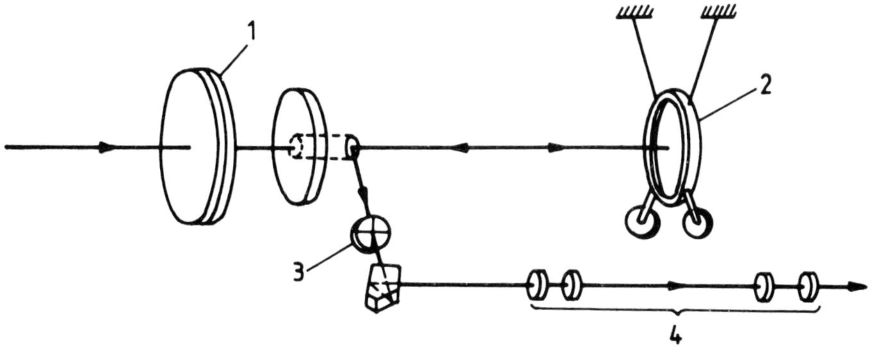

Figure 3.63. Compensator and optical train of the level *Sokkisha AL-2: 1* objective lens; *2* compensator mirror; *3* cross-hair; *4* eyepiece

the *Topcon U2* level is a horizontal *Porro* prism system of the first type *(Fig. 3.64)*. The image can be focussed for both by moving the compensator mirror.

In order to place the compensator nearer to the cross-hairs — for this the optical angular magnification $n_{op} = \beta_{op}/\alpha > 2$ is necessary — *astatic pendula* are frequently used by means of which a mechanical angular magnification $n_r > 1$ can be achieved. A pendulum is called astatic if its rest position with respect to the vertical depends on the tilt of the pendulum base. E.g., the tilt γ of the mirror or prism which is

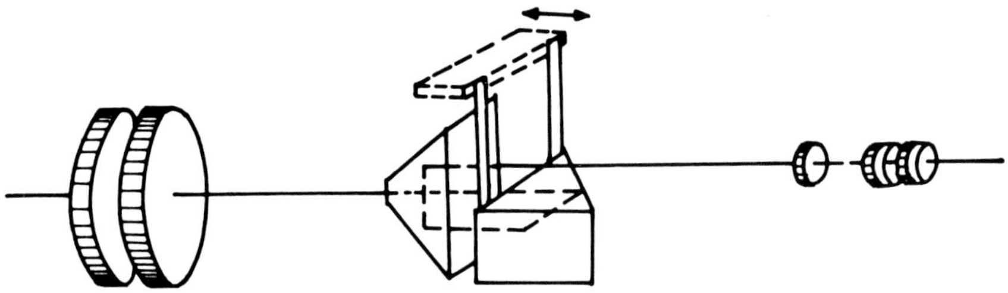

Figure 3.64. Compensator and optical train of the *Topcon* auto-level

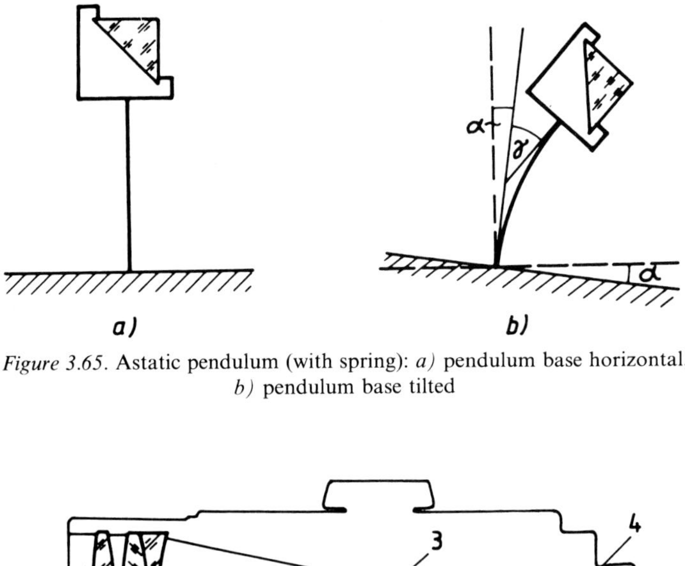

Figure 3.65. Astatic pendulum (with spring): *a)* pendulum base horizontal, *b)* pendulum base tilted

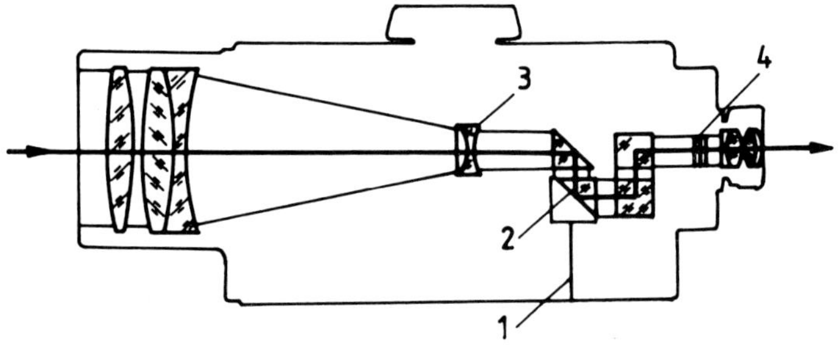

Figure 3.66. Compensator and optical train of the level *INA 65* (Ertel Werke, München)

attached to the top of an elastic spring made of invar steel *(Fig. 3.65)* is greater than the tilt α of the telescope and hence $n_r > 1$.

In the levels *Ertel INA 65* and *BNA 65 (Fig. 3.66)* the reflecting prism *2* mounted on the spring *1* is placed between the inner focussing lens *3* and the cross-hairs *4*.

In the *Wild NA2* level *(Fig. 3.67)* the body of the pendulum is connected to the housing through a pair of pre-stressed leaf springs *(Fig. 3.68)*. The swing of the

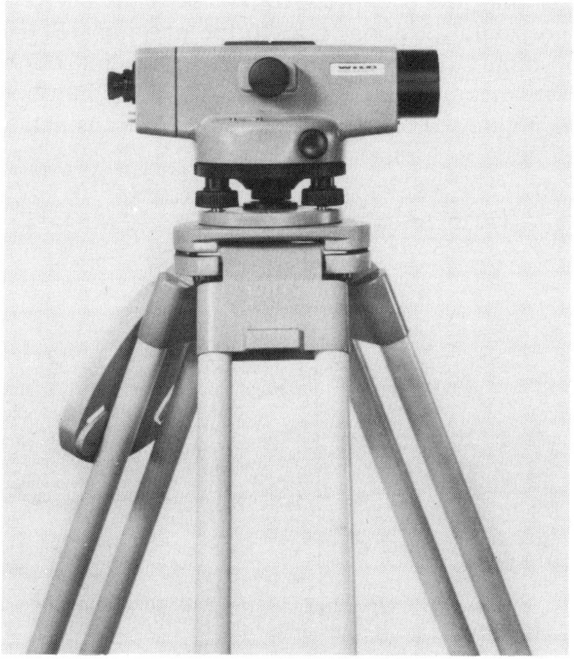

Figure 3.67. Level *NA2* (Wild, Heerbrugg)

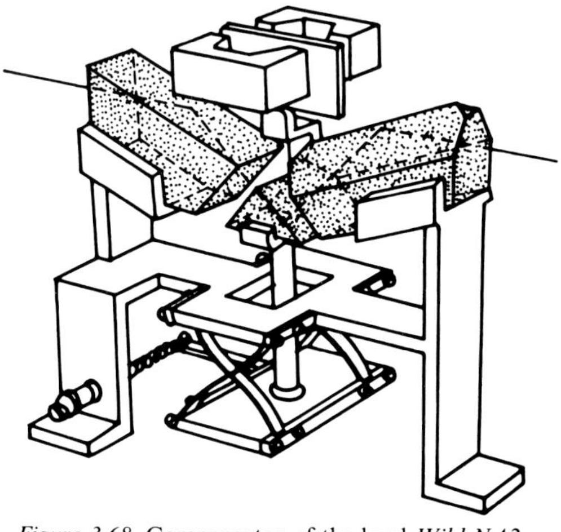

Figure 3.68. Compensator of the level *Wild NA2*

pendulum is damped by a pair of magnets. The compensation range of the instrument is ± 10 min of arc, its standard deviation of stabilization is ± 0.3 sec of arc. The standard deviation of direct and return levelling is ± 1.5 mm per km, which can be improved to ± 0.4 mm/km by using an optical micrometer.

The effect $n_r > 1$ can also be realized by a jointed quadrangular suspension *(Fig. 3.69)*. The swinging part CD is suspended on four wires of equal length. The distance $\overline{A_0B_0} = \overline{AB}$ is longer than the distance $\overline{C_0D_0} = \overline{CD}$. If the telescope is tilted by the angle α from the horizontal position A_0B_0 to the position AB, the pendulum body moves from the position C_0D_0 into the equilibrium position CD by tilting through the angle $\alpha + \beta_r$. The angle $\beta_r = n_r\alpha$ is the tilt of the pendulum body relative to the telescope tube AB. Within the compensation range n_r = constant.

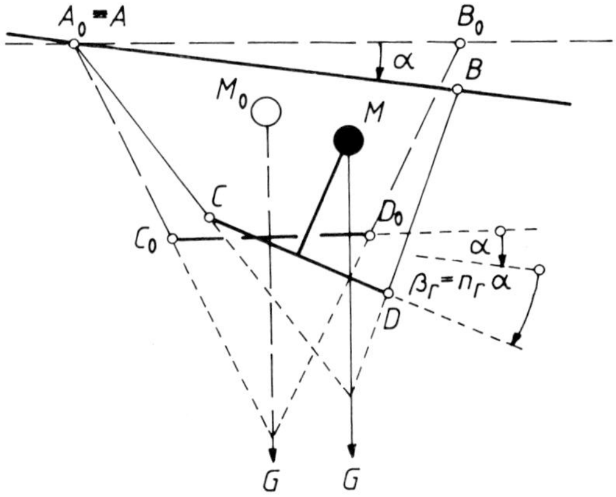

Figure 3.69. Operating principle of the jointed quadrilateral

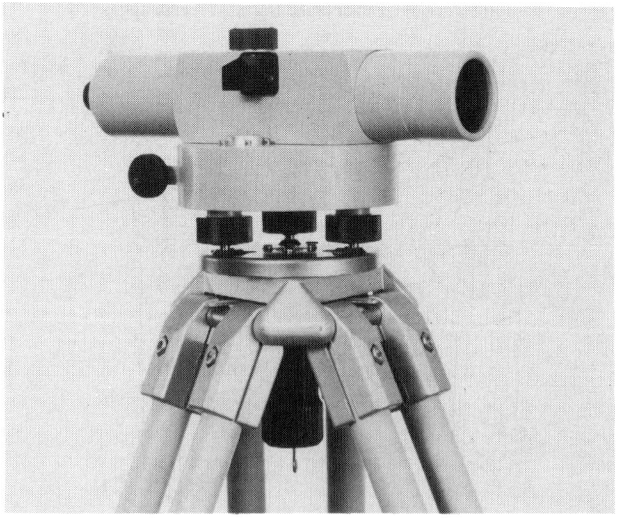

Figure 3.70. Level *Ni 2* (Opton Feintechnik GmbH, Oberkochen)

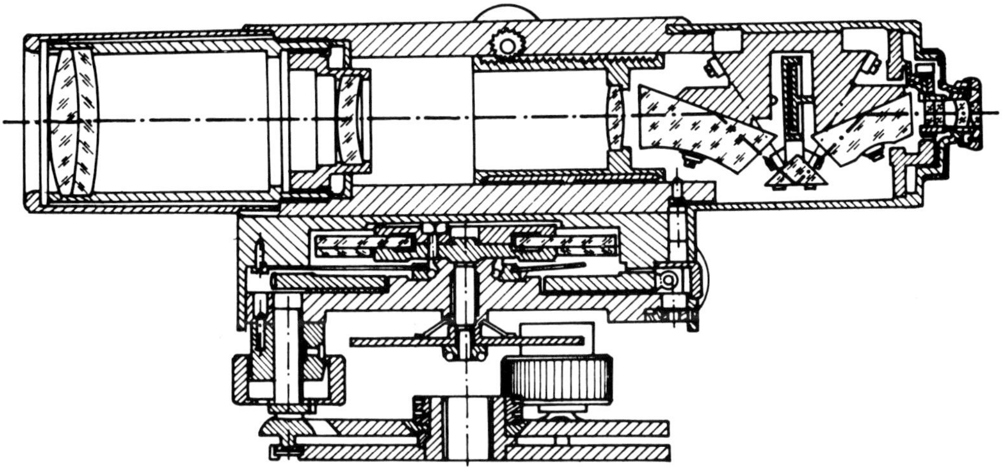

Figure 3.71. Diagrammatic sketch of the level *Opton Ni 2*

The levels *Opton Ni 2 (Fig. 3.70)* and *Nikon AE* are constructed with a compensator which includes three prisms, the two outside of which are attached to the telescope tube while the central one is attached to the jointed quadrilateral *(Fig. 3.71)*. The pendulum system has been dimensioned so that $n_r = 3$ and the optical angular magnification amounts to $n_{op} = 6$. The prism on the eyepiece side forms a roof edge which provides the side correspondence of the upright image generated by the reflecting surface of the compensator. Jointed quadrilaterals are also used in the levels *Miller Ni 15*, *Sokkisha B2* and *Pentax AL-2 (Figs 3.72* and *3.73)*. Levels of the *Pentax PAL type* also have mirror compensators, but these are newly developed *(Fig. 3.74)*.

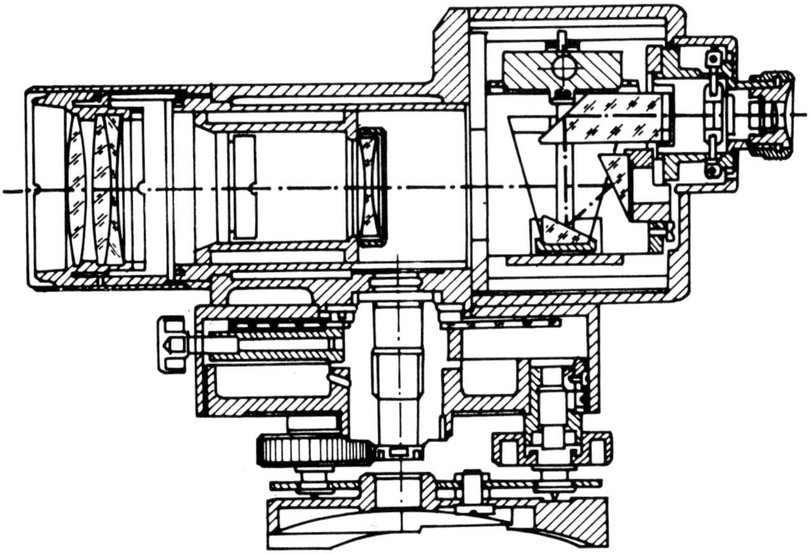

Figure 3.72. Diagrams of the levels *Miller Ni 15* and *Sokkisha B2*

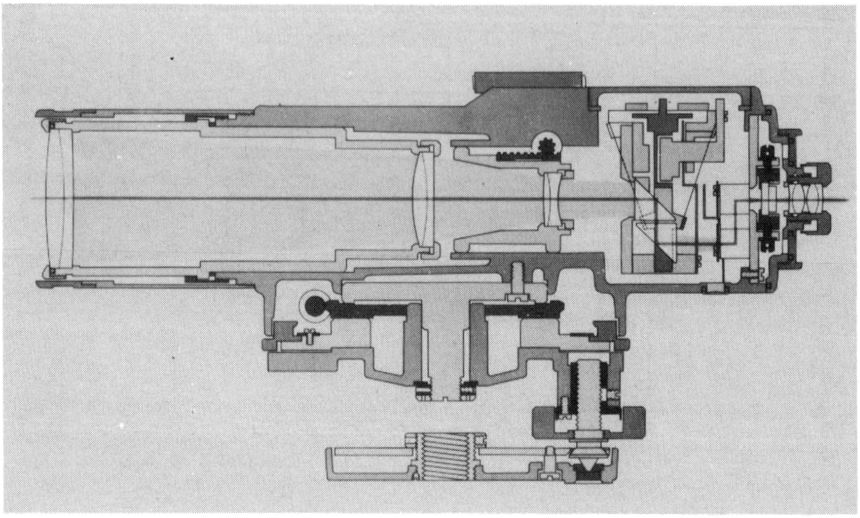

Figure 3.73. Diagrammatic sketch of the level *Pentax AL-2*

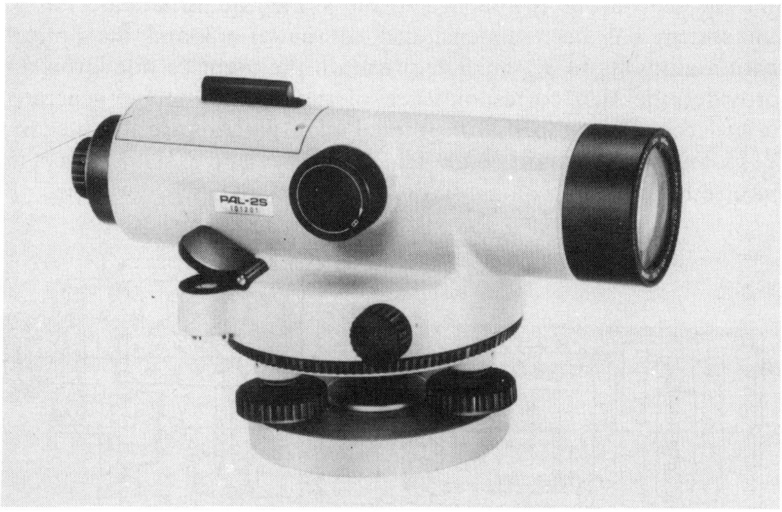

Figure 3.74. Level *PAL-25* (Pentax Asahi Precision Co. Ltd, Tokyo)

The optical angular magnification can also be increased by the use of a simple physical pendulum if the reflecting surface (or surfaces) on the pendulum reflect the light rays at least twice, while they are also reflected by at least one mirror (or by an odd number of mirrors) fixed to the telescope. From *Fig. 3.75* showing the path of the horizontal principal ray in a simplified manner, as well as from Eq. (3.6), the basic equation has the following form:

$$f = n_{op1}(r_1 - r_2) + n_{op2} r_2. \tag{3.22}$$

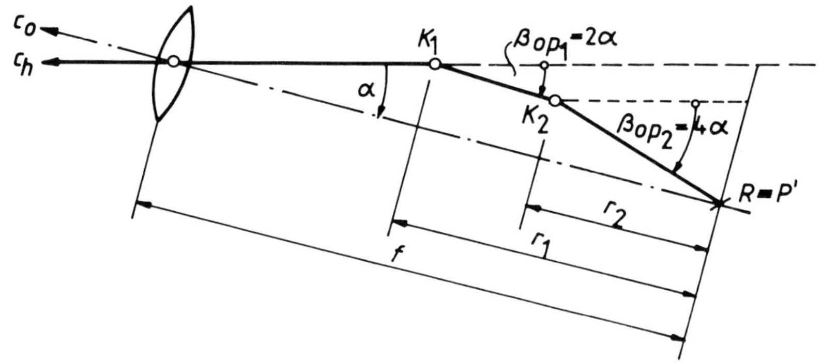

Figure 3.75. Increase in optical angular magnification of a compensator by multiple reflection: c_o the optical axis of the telescope; c_h the line of sight of the telescope coinciding with the horizontal principal ray after the rotation of the telescope; K_1, K_2 bending points of the horizontal principal ray at the compensator mirrors; α the tilt of the telescope; β_1, β_2 the rotation angles of the horizontal principal ray; R the centre of the cross-hair; P' the image of the sighted staff point; r_1, r_2 the distances of the compensator mirrors from the cross-hair; f the focal length of the objective

Since the mirror at the point K_1 and the reflecting surface at the point K_2 deflect the beam of rays from their to original direction through the angles $\beta_{op1}=2\alpha$ and $\beta_{op2}=4\alpha$ respectively, therefore $n_{op1}=2$ and $n_{op2}=4$. Substituting these values, from Eq. (3.22):

$$r_2=\frac{f-2(r_1-r_2)}{4}. \tag{3.23}$$

It is practical to make the distance (r_1-r_2) short in order to make the more effective r_2 relatively large.

The compensator in the *MOM Ni B5* level *(Fig. 3.76)* is a Porro prism system of the second type, where the two outside prisms, connected rigidly to each other, are attached to the pendulum suspended on two pairs of thin metal wires arranged in the form of a reversed V *(Fig. 3.77)*. The air-damped pendulum has a mechanical axis and an air bearing. This means that there is an air layer of appropriate size between the mechanical axis and its bearing when the pendulum swings about the cross-over point of the suspending metal wires. That is why skin friction does not exist. The mechanical axis is necessary to avoid damage due to dynamic stresses (shocks) during transport (it prevents the breaking or permanent deformation of the wires).

The operating range of the compensator is ± 8 min of arc, the standard deviation of the stabilized line of sight is ± 0.4 sec of arc; the standard deviation of direct and return levelling is ± 2 mm per km.

A similar compensator is used in the *FNAK 1* level *(Fig. 3.78)*.

Two rectangular prisms *1 (Fig. 3.80)* of the compensator in the *Jenoptik Ni 025* level *(Fig. 3.79)* are fixed to the pendulum which is air-damped and suspended on thin metal wires which cross each other. The rays are triply reflected within the third prism *2* which is placed centrally between the other two and rigidly fixed to the telescope tube, consequently the image is erect. The third reflecting surface of this

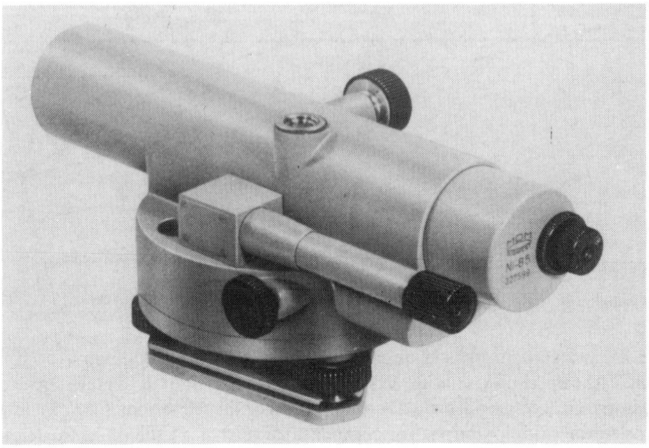

Figure 3.76. Level *Ni B5* (MOM, Budapest)

prism, forming a roof edge, ensures the side correspondence of the image. The operating range of the compensator is ± 10 min of arc, the standard deviation of stabilization of the line of sight is ± 0.5 sec of arc, and the standard deviation per km of direct and return levelling is ± 2.5 mm.

The *Autoset-level 2* of Rank Industries *(Fig. 3.81)* is of similar construction; here the prisms B_1 and B_2 are pendulum-suspended in a common housing. The third

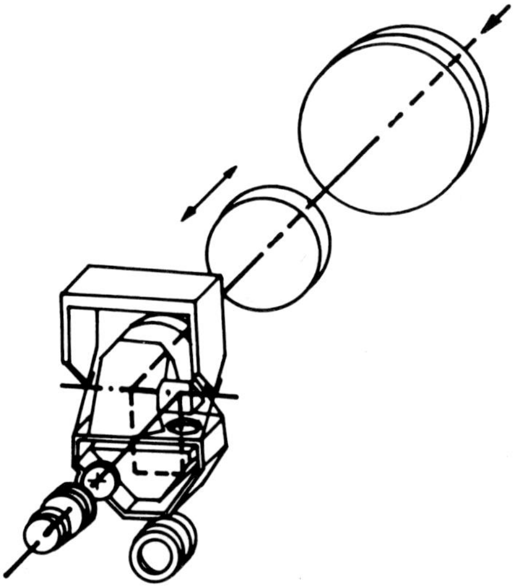

Figure 3.77. Compensator of the level *MOM Ni B5*

Figure 3.78. Level *FNAK 1* (Otto Fennel GmbH, Kassel)

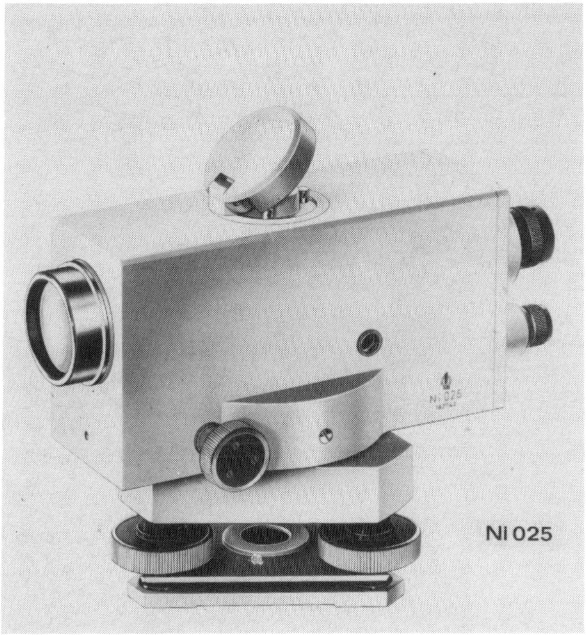

Figure 3.79. Level *Jenoptik Ni 025* (VEB Jenoptik, Jena)

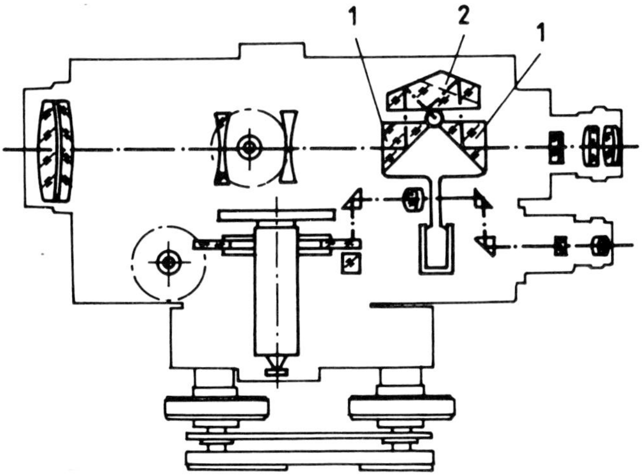

Figure 3.80. Optical train of the level *Jenoptik Ni 025*

prism *A* with a roof edge is rigidly fixed to the telescope. Due to the triple reflection and the roof prism *A* the image is erect and side-corresponding.

Compensators placed inside the telescope, and operating by parallel shifting the beam of rays, must fulfil Eq. (3.4), following *Fig. 3.33, d.* A simple method for parallel shifting the beam of rays (principal ray) is to move the rectangular prism in the

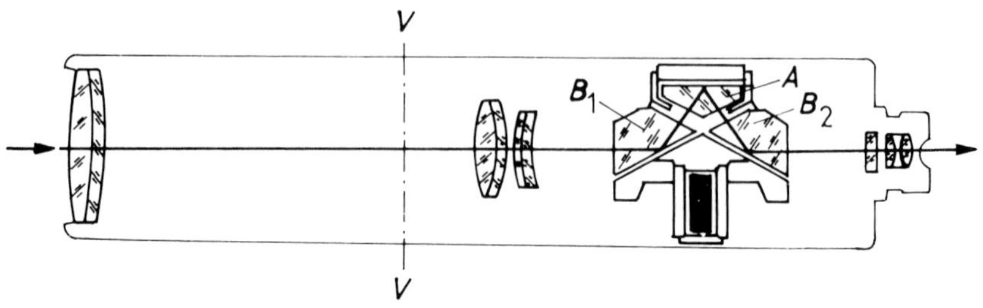

Figure 3.81. Optical train of the *Autoset-level 2* (Rank Precision Industries, England)

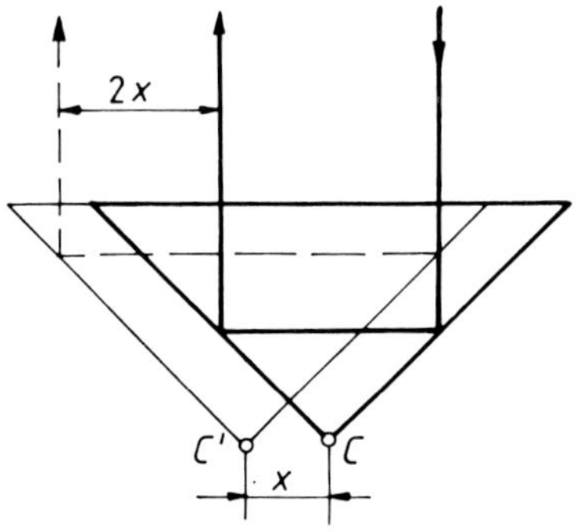

Figure 3.82. Parallel shift of rays by a prism

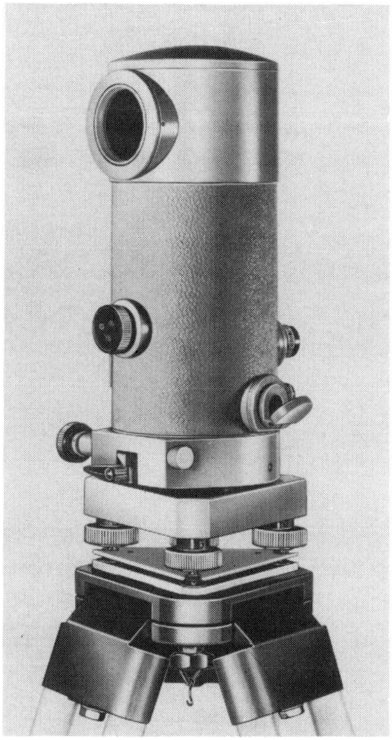

Figure 3.83. Level *Ni 007* (VEB Jenoptik, Jena)

direction perpendicular to the rays *(Fig. 3.82)*; this prism is also within the optical train and two of its sides are reflecting surfaces. If the prism is shifted by an amount x, the rays are parallel shifted by $2x$. Rotation of the prism around the intersection line C of the reflecting surfaces does not affect the direction of the rays. This solution has been used in the *Jenoptik Ni 007* level *(Fig. 3.83)*, where the distance between the refracting edge of the compensator prism fixed to the pendulum and the pendulum axis is $f/2$ *(Fig. 3.84)*. When tilting the instrument the prism and the principal ray are moved by $x = f/2$ and by $2x = f$, respectively. Due to the compensation the line

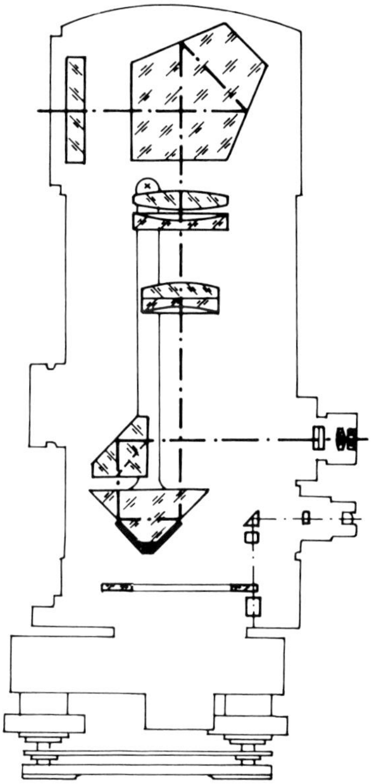

Figure 3.84. Optical train of the level *Jenoptik Ni 007*

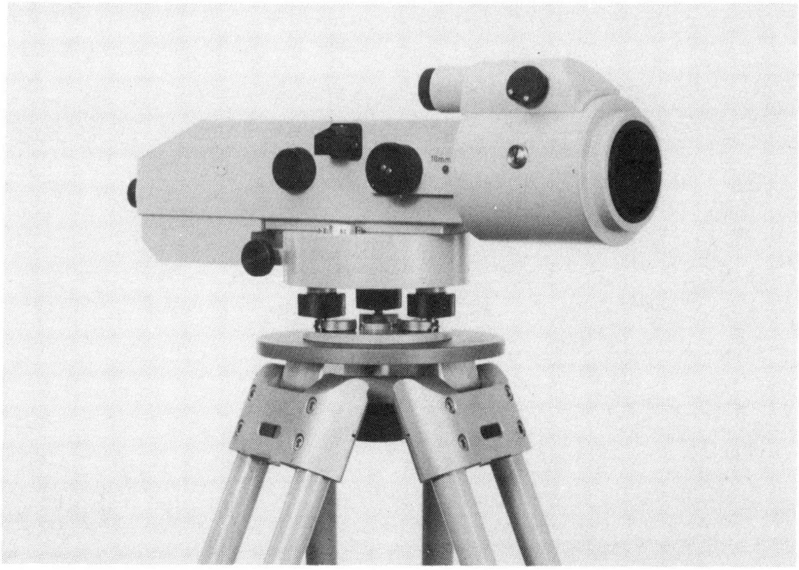

Figure 3.85. Level *Ni 1* (Opton Feintechnik GmbH, Oberkochen)

of sight of the instrument becomes vertical and it is deflected horizontally by the pentagonal prism placed above the objective lens. This prism functions at the same time as an optical micrometer, by means of which the line of sight can be moved vertically within a range of 5 mm. The anallatic point for an infinite object distance lies at the centre of the pentagonal prism at the crossing point of the principal rays passing in and out and, even for the shortest sighting distance, the point is moved by only 12 mm. The operating range of the compensator is ± 10 min of arc; the standard deviation of stabilization is ± 0.15 sec of arc; the standard deviation of direct and return levelling is ± 2 mm per km using a staff graduated in cm and without the optical micrometer, while, with it, it is ± 0.5 mm per km.

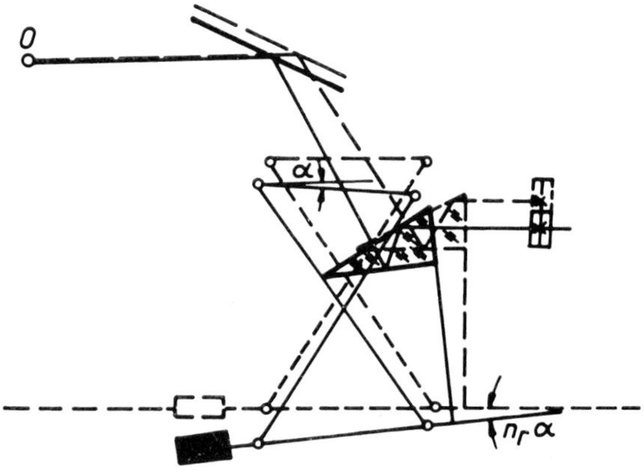

Figure 3.86. Operating principle of the compensator of level *Opton Ni 1*

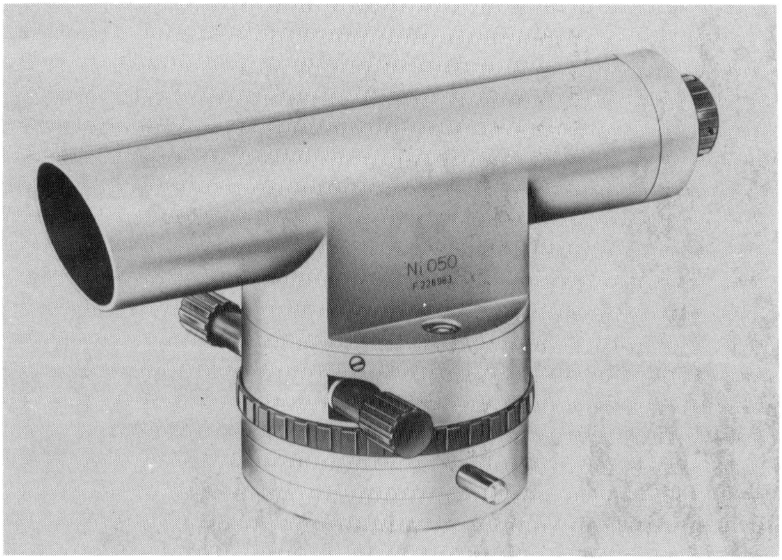

Figure 3.87. Level *Jenoptik Ni 050* (VEB Jenoptik, Jena)

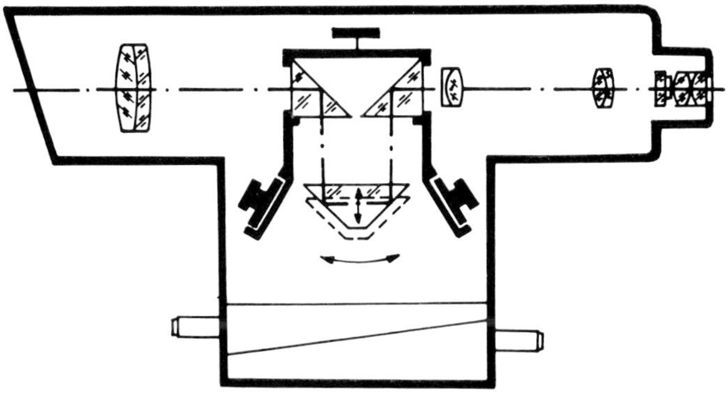

Figure 3.88. Optical train of the level *Jenoptik Ni 050*

In the first order level *Opton Ni 1 (Fig. 3.85)* the principal ray is moved by a triangular prism attached to a jointed quadrilateral; the rays are double reflected within the prism *(Fig. 3.86)*. The operating range of the compensator is ±8 min of arc, the standard deviation of the stabilized line of sight ±0.2 sec of arc, and the standard deviation of levelling ±0.2 mm per km. The telescope, fitted with an optical micrometer, has interchangeable eyepieces giving 30, 40 or 50 times magnification.

In the *Jenoptik Ni 050* level *(Figs 3.87* and *3.88)* the diagonal sides of the two pendulum-suspended, rectangular prisms are the reflecting surfaces of the compensator. The third rectangular prism placed between the two compensator prisms partly participates in the compensation, partly serves as a means of focussing the image by moving it vertically. The operating range of the compensator is ±30 min of arc, the standard deviation of the stabilized line of sight ±1 sec of arc, and that of direct and return levelling ±5 mm per km.

3.1.2.5 Special Levels, Auxiliary and Additional Devices for Levels

Some levels are developed for special purposes or are fitted with auxiliary devices.

The *MOM Ni C3* level with its auxiliary device can be used for rail settlement measurement and for levelling. The complete levelling instrument consists of a level, a vertically movable target supplied with a special graduation and a special tripod fixable to the rail.

Opton made the *Ni 2* level suitable for levelling across rivers or valleys. Two instruments can be placed on a common base plate mounted on one single tripod. A rotatable optical wedge can be attached in front of the objectives of both instruments by means of which their lines of sight can be slightly but measurably deflected from the horizontal plane. Prior to commencing measurement the two instruments are mutually collimated, in other words, their lines of sight are made parallel. With the optical wedge the line of sight of one instrument is deflected upwards by the same angle as that by which the other instrument is deflected downwards. There are special targets for the river crossing equipment, which can be placed on a tripod one above the other *(Fig. 3.89)*. When levelling, the two targets on the opposite side of the river

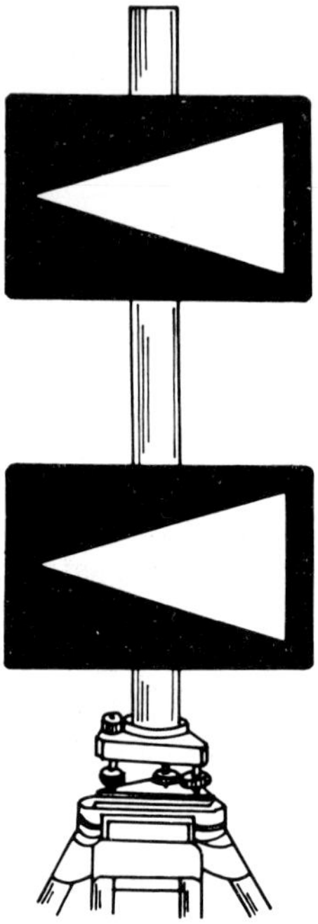

Figure 3.89. Opton targets for levelling across a river

Figure 3.90. The level *Geoplane 300* (AGA Geotronics AV. Lidingö)

should be sighted and the lines of sight of the telescopes set successively on them by rotating the optical wedges. The elevation differences of the targets with respect to the instrument can be computed from the tilt angle measured by the optical wedges and from the known height difference between the targets. The adjustment error of the line of sight is eliminated by calculating the mean value of the elevation differences obtained by the two instruments.

The *Opton Ni 2* has been fitted with other additional devices in order to make levelling possible under exceptional conditions or to carry out special tasks. By means of a frontal lens, placed in front of the objective lens, observing at distances from 1.8 m to 3.6 m is also possible. By means of a frontal pentagonal prism attachable to the telescope, the instrument can also be used as an optical plummet. By means of two further special frontal prisms, the instrument can also be used as an astrolabe.

For the *Topcon Auto level* a special additional device is used to illuminate the cross-hairs, where the illuminating light is conducted by fibre optic from the light source fixed to the tripod.

The *AGA Geoplane 300* level *(Fig. 3.90)* is essentially different from instruments with a spirit level or compensator. Its fundamental principle is that a horizontal plane is defined by a rotating laser beam, the height of which can be read on the vertically graduated staff. The He–Ne laser beam, generated in the vertical cylindrical head of the instrument (see Subsection 1.5.3), is divided into two parts by the optical system in the head and is emitted from the rotating part in two directions differing by 180°. However, the axes of the emitted beams of rays are not horizontal: one of them is tilted slightly below the horizontal, the other slightly above it, and the corresponding extreme rays from both beams touch the horizontal. Since the moving part of the head rotates at 10 revolutions per second, the observer standing in the horizon of the instrument sees 20 light pulses per second, which can be perceived as a steady and continuous light. Above and below the horizon of the instrument, however, flashing light pulses 0.1 s apart can be individually perceived.

The position of the instrument horizon can be determined with the naked eye from behind the graduated staff by reading the eye position on the staff when continuous light is perceived. For more accurate measurements or for greater distances photo-electric equipment can be used for positioning and reading the instrument horizon.

The horizon of the instrument up to a maximum distance of 250 m is automatically horizontal within the ± 12 min of arc operating range of the instrument, while the accuracy up to 60 m is within ± 2 mm.

The laser level *Modell 1054 DSL* (Spectra-Physics, Darmstadt) is the first level with a rotating laser that can operate by remote control. The rotation plane of the laser beam is automatically horizontal within 6°. The stabilization range of the tilt angles in the directions of the two axes is 9.99 percent. The laser beam rotates at 300 or 600 revolutions per minute.

In the case of the *EL-1* level (Electronic-Level-System, Spectra-Physics, Darmstadt) invisible electronic signals are radiated by a rotating device and recorded by a hand-held detector. The detector indicates the recording of the signal by a sound signal, by a visible signal (a direction arrow is displayed) or by both. If the detector is moved vertically over the point to be measured, a pulse tone can be heard immediately when the electronic signals are detected. The pulse tone changes into a whistle at the centre of the wave band of the electronic signals. This means that the detector is in the horizontal plane of the level. If the detector is used only with the visible signal, three horizontal lines appear on the display when the electronic signal is detected. When the detector is in the instrument horizon, the two outside lines disappear and only the central one remains visible. The detector can be used with either a simple or a telescopic levelling staff. The indicator can be implemented by a small lamp so that the detector may be used for overhead heights or in the dark. The rotation plane of the electronic signal is automatically horizontal within the operational range, while the rotating signal is automatically extinguished if the tilt of the instrument is not within this range.

Using electronic and laser levels for setting out, the vertical alignment of different planned structures and buildings can be performed advantageously, and their use can also be very economical for grid levelling.

3.1.2.6 Test of Tilting Levels

Levels must be tested periodically and adjusted if necessary.

Testing the circular bubble (bull's eye bubble). The bubble tube axis of the circular level, or rather the tangent plane at the centre which implies this axis, must be at right angles to the vertical axis of the level. The test must be carried out in the following steps:

— the normal position of the tilting screw is determined as described in Point 3.1.2.3;

— the tilting screw is turned to the normal position;

— the vertical axis is plumbed by means of the tubular level and the foot screws.

Then the adjustment consists merely of bringing the bubble back to its central position by means of the capstan adjusting screws.

Test of the horizontal cross-hair. The horizontal line of the cross-hair must be perpendicular to the vertical axis. For the test a clearly visible point is set onto the horizontal line of the cross-hair at either the right or the left edge of the field of view and then the telescope is turned around the vertical axis. If the horizontal line is in adjustment, the image of the point will remain on it. If it does not, the error can be eliminated by delicately turning the reticule frame.

Test of the tubular level. The axis of the tubular level must be parallel to the line of sight of the telescope, i.e. it must be parallel

— to the vertical plane which contains the line of sight (there should be no cross divergence of the bubble) and

— to the horizontal plane containing the line of sight (there should be no collimation error).

Test of the cross divergence. Let φ be the cross divergence angle of a spirit level mounted on a telescope whose line of sight is C and which is set horizontal. Let k be the rectangular projection onto the line of sight of the axis of the level $AB=l$ (*Fig. 3.91, a*). When the telescope is rotated around the line of sight by the angle β (*Fig. 3.91, b*), the ends of the axis of the level A and B move to A' and B' whose height difference is

$$h=l\sin\delta, \tag{3.24}$$

where δ is the angle between the axis of the spirit level and the horizontal.

From Eq. (3.24) and the figures

$$k=l\sin\varphi \quad \text{and} \quad h=k\sin\beta, \tag{3.25}$$

whence

$$\sin\delta=\sin\varphi\sin\beta \tag{3.26}$$

or, since the angles φ and δ are small:

$$\delta=\varphi\sin\beta. \tag{3.27}$$

From the angle δ and from the constant ε of the spirit level one obtains the graduation value of the bubble displacement:

$$p=\frac{\delta}{\varepsilon}. \tag{3.28}$$

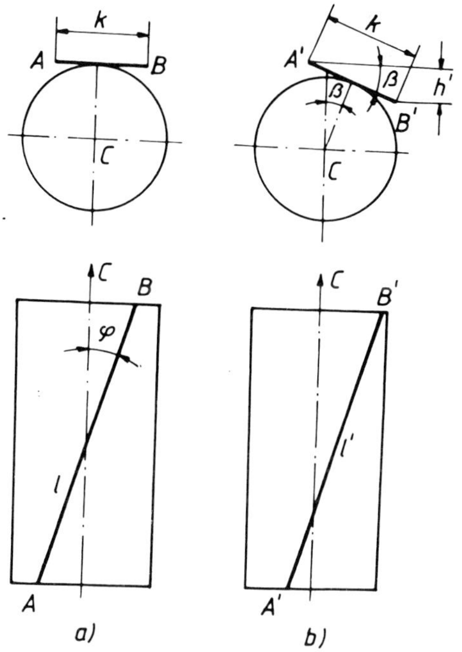

Figure 3.91. Cross divergence of the spirit level

If e.g. $\varphi = 10$ min of arc, $\beta = 5$ min of arc, then $\delta = 0.87$ sec of arc, which is considerably greater than the bubble accuracy of ± 0.25 sec of arc for spirit levels mounted on precise levels.

With an inclined vertical axis, a cross divergence of the spirit level causes a horizontal inclination of the line of sight of the telescope, and this can be a dangerous source of error as it is systematic. This error is tested in the following steps:

a) The level is set up so that one foot screw points towards a levelling staff held at a distance of about 20–50 m. The vertical axis of the level is now made vertical by means of the tubular level.

b) The bubble of the tubular level is centred by means of the levelling screw and a reading is made on the levelling staff.

c) One of the two foot screws perpendicular to the telescope should be turned two or three times and then the other must be turned in the opposite direction until the line of sight intersects the previously observed value of the staff. (The vertical axis is thereby tilted from its originally vertical position only in the plane perpendicular to the line of sight.) If the bubble of the tubular level has now moved off centre, the spirit level has a cross divergence.

d) The cross divergence can be eliminated by centring the bubble by means of the horizontal capstan adjusting screw. Since the manufacturer tests and adjusts the levels appropriately, such an adjustment is not generally necessary.

Test of the inclination of the line of sight. Tests in the field by readings on the levelling staff:

a) On a nearly horizontal field two points A and B 30 to 50 m apart are marked either by change-plates or by round head nails in pegs *(Fig. 3.92)*.

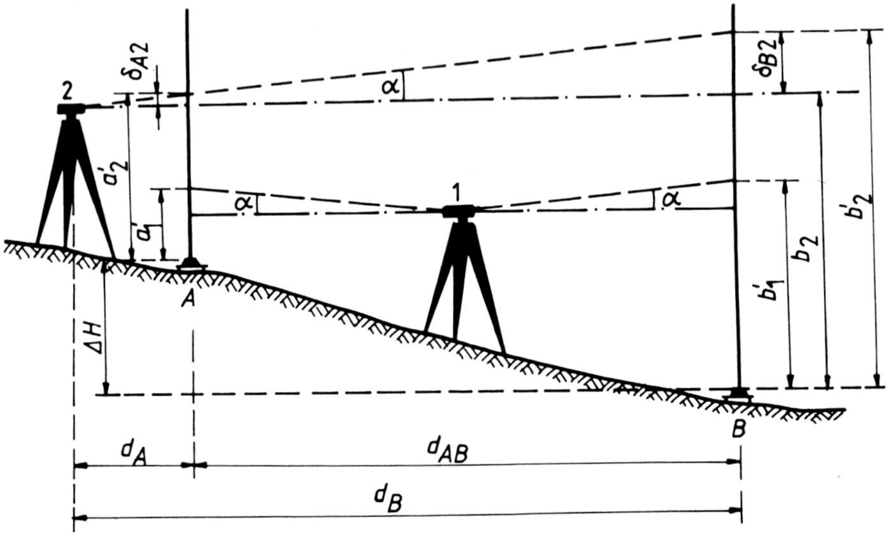

Figure 3.92. Test and adjustment of the tilt of the line of sight in the field

b) The level is set up at point *1* equi-distant from A and B, and the difference of height between these two points:

$$\Delta H_{AB} = a'_1 - b'_1 \tag{3.29}$$

is measured.

c) The level is now moved to point *2* which is at the minimum observing distance of $d_A \approx 2$–5 m from the point A, and the staff is observed at A and then at B. After centring the bubble for each, the readings a'_2 and b'_2 are obtained.

If a numerical determination of the inclination of the line of sight is required, then the measurements must be repeated several times with slight changes of instrument height.

d) The height difference is computed:

$$\Delta H'_{AB} = a'_2 - b'_2, \tag{3.30}$$

where the values are erroneous due to the inclination α of the line of sight.

e) From Eqs (3.29) and (3.30) the value of the inclination of the line of sight valid for the distance d_{AB} is calculated:

$$\alpha'' = \frac{\Delta H_{AB} - \Delta H'_{AB}}{\alpha_{AB}} \cdot \rho''. \tag{3.31}$$

If the sign of α is positive (negative) the line of sight is inclined upwards (downwards) to the horizontal.

f) For the adjustment the error-free reading b_2 on the farther point B, at distance d_B, is computed:

$$b_2 = b'_2 - \delta_{B2},$$

where

$$\delta_{B2} = \frac{\alpha'' d_B}{\rho''}. \tag{3.32}$$

g) The line of sight of the instrument is set to the reading b_2 on the staff by means of the tilting screw, and then the bubble of the spirit level is centred by means of its vertical capstan adjusting screw. If there is a tiltable glass wedge in front of the objective of the level, the bubble should first be adjusted to centre by the tilting screw and then the line of sight should be adjusted to the reading b_2 by rotating the glass wedge. If possible, adjustment by moving the cross-hair plate should be avoided for fear of making the cross-hairs eccentric.

When testing and adjusting precise levels, the effect of the curvature of the level surface and refraction must be taken into account.

According to Kukkamäki the correction to the staff reading (in metres) due to the curvature of the level surface and refraction is as follows:

$$\delta_{s,r} = -\Delta \frac{d^2}{2} = -d^2\, 0.168 \cdot 10^{-6}, \tag{3.33}$$

where Δ is the difference between the curvature of the level surface $1/\rho_s$ and that of the line of sight $1/\rho_r$, and d is the staff distance in m. The curvature of the line of sight consists of two parts: one part $1/\rho_t$ depends on the vertical temperature gradient $\mathrm{d}t/\mathrm{d}h$, and the other part $1/\rho_B$ on the vertical barometric pressure gradient $\mathrm{d}B/\mathrm{d}h$. The total curvature difference is:

$$\Delta = \frac{1}{\rho_s} - \left(\frac{1}{\rho_t} + \frac{1}{\rho_B}\right). \tag{3.34}$$

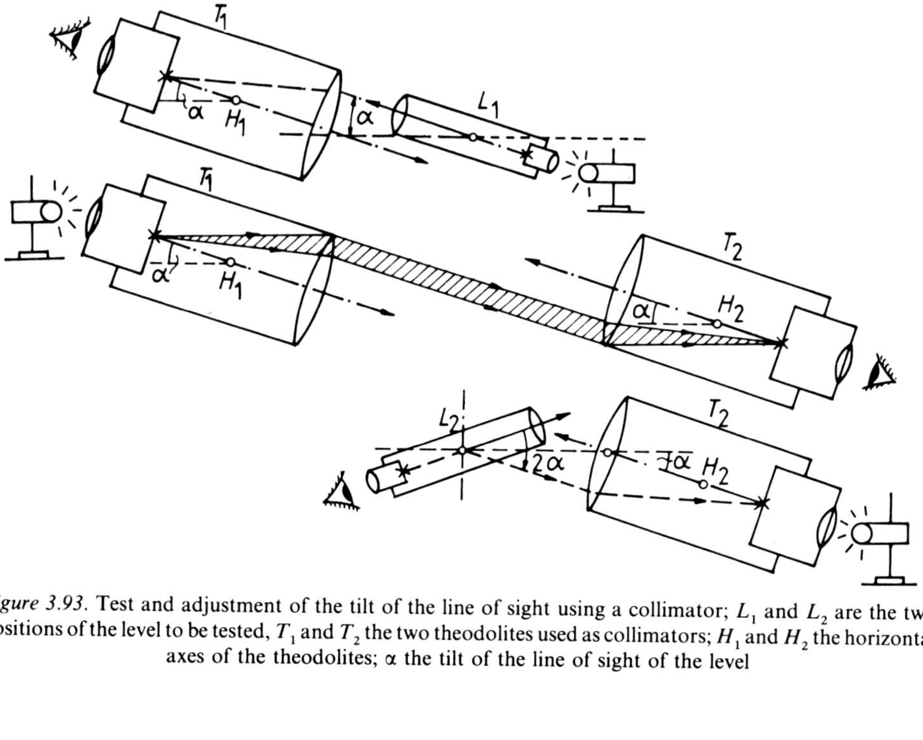

Figure 3.93. Test and adjustment of the tilt of the line of sight using a collimator; L_1 and L_2 are the two positions of the level to be tested, T_1 and T_2 the two theodolites used as collimators; H_1 and H_2 the horizontal axes of the theodolites; α the tilt of the line of sight of the level

If the sign of curvature is positive (negative), the corresponding curvature as seen from above is convex (concave). The curvatures can be computed by the formulae

$$\frac{1}{\rho_s}=0.157\cdot 10^{-6},$$
$$\frac{1}{\rho_t}=0.933\cdot 10^{-6}\frac{dt}{dh}, \qquad (3.35)$$
$$\frac{1}{\rho_B}=0.385\cdot 10^{-6}\frac{dB}{dh}.$$

According to the data given by Kukkamäki, if the average daytime value of $dt/dh=-0.23$ °C/m at 1.5 m above the ground is taken as the temperature gradient, and if $dB/dh=-1$ mm/11 m is taken as the pressure gradient, then one obtains from Eq. (3.34) the numerical value which is given in Eq. (3.33). (If the pressure gradient is taken as $dB/dh=-1$ mbar/8.25 m, then the multiplying factor $1/\rho_B$ in Eq. (3.35) must be taken as -0.289×10^{-6}.)

Some computed values of the correction $\delta_{s,r}$ are:

If	$d=$	10	20	30	40	50	60	70	80 m
then	$\delta_s=$	−0.01	−0.03	−0.07	−0.13	−0.20	−0.28	−0.38	−0.50 mm
	$\delta_r=$	−0.01	−0.04	−0.08	−0.14	−0.22	−0.32	−0.44	−1.07 mm
and	$\delta_{s,r}=$	−0.02	−0.07	−0.15	−0.27	−0.42	−0.60	−0.82	−1.57 mm

The temperature gradient depends on the local geographical conditions and on the time variation of meteorological conditions, therefore, the correction δ_r may deviate considerably from the value given.

The readings in Eq. (3.30) must be corrected by the correction $\delta_{s,r}$ in order to compute a $\Delta H'$ height difference free of the effects of the curvature of the level surface and of refraction.

Test using collimators. As opposed to the previous method, in the method using a collimator the focussing lens of the telescope must not be moved. For this method two laboratory collimators are needed (Subsection 1.5.4) but, if none are available, two theodolites or, if the numerical value of the inclination of the line of sight has to be determined, two levels can be used as collimators. The test is made in the following steps:

a) The level to be tested is set up between the two theodolites, with their telescopes focussed to infinity, in such a way that the vertical axes of the three instruments lie in the same vertical plane within a few millimetres and their telescopes are at the same height within some millimetres *(Fig. 3.93)*.

b) The cross-hair of the level is illuminated through the eyepiece by a diffuse light source, the telescope is focussed to infinity and the telescope bubble is centered. The horizontal cross-hair is then sighted by theodolite T_1 *(Fig. 3.93)*. The lines of sight of the theodolite and of the level now make the same angle α with the horizontal, but in opposite senses.

c) Removing the level, the horizontal cross-hair of the undisturbed theodolite T_1 is sighted by the theodolite T_2.

d) The level is put back, i.e. set up in its previous position and its telescope is turned toward theodolite T_1. Its telescope bubble is now adjusted to centre by the tilting screw and the position σ_1 of the tilting screw is either read or marked. Now the horizontal cross-hair of the theodolite T_2 is sighted by the level, tilting the telescope by the tilting screw, and the position σ_2 of the screw is either read or marked.

For adjustment one must compute the position

$$\sigma_n = \frac{(\sigma_1 + \sigma_2)}{2}$$

of the tilting screw which gives a horizontal line of sight: the screw is set to this position and the bubble of the spirit level is then centred by means of its capstan adjusting screw.

If the numerical value of the inclination of the line of sight is also required, then the horizontal cross-hair of the level must be sighted by theodolite T_2 in both positions σ_1 and σ_2 of the tilting screw. From the two readings the double value 2α of the inclination of the line of sight can be computed.

Test of a level with a telescope rotatable about its longitudinal axis and with a spirit level reversible with respect to the inclination of the line of sight. The axis of the reversible spirit level must be parallel to the longitudinal axis of the telescope and the line of sight of the telescope must coincide with the longitudinal axis.

For testing the parallelism of the axis of the spirit level and the longitudinal axis of the telescope, the vertical axis is set approximately vertical. The bubble of the spirit level is centred by means of the tilting screw or the foot screws. Now the telescope is reversed around its longitudinal axis. If the spirit level is not in adjustment, the bubble moves from the central position. To adjust, half of this bubble displacement must be eliminated with the tilting screw (or with the foot screws), half with the vertical bubble capstan adjusting screw.

For testing the coincidence of the line of sight and the longitudinal axis of the telescope, again the vertical axis must be set approximately vertical and a graduated staff at a distance of 100 m from the level must be sighted.

In this position of the telescope the line of sight, determined by the optical centre of the objective lens O_1 and the centre of the cross-hairs R_1 intersects P_1 on the staff *(Fig. 3.94)*. If the telescope is reversed around its longitudinal axis $A-A$, the centre of the objective lens moves to the point O_2 and the cross-hair centre to the point R_2; the line of sight now intersects P_2. In this case the cross-hair must be moved by its vertical adjusting screws to the position R_2, such that the line of sight bisects the distance P_1P_2 at the point P_0. If the centre of the objective lens lies on the longitudinal axis $A-A$, and consequently its separation e from the longitudinal axis is zero, the cross-hair is set onto the longitudinal axis by its adjustment. If $e \neq 0$, after moving the cross-hair plate the line of sight makes an angle α with the longitudinal axis. Now a near point (e.g. a graduation line on the staff) should be sighted and the telescope

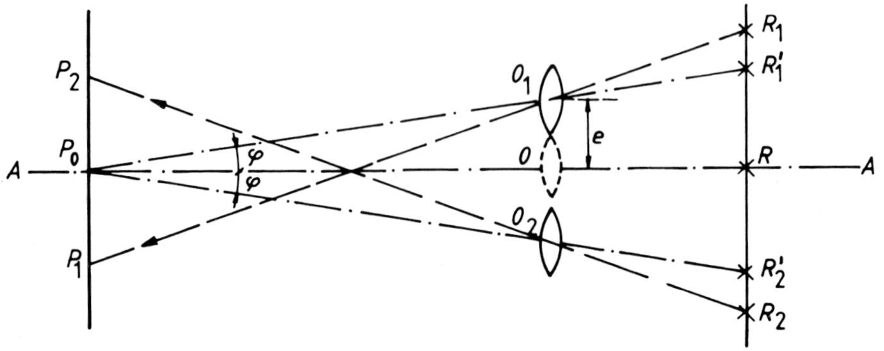

Figure 3.94. Test and adjustment of the tilt of the line of sight of a telescope reversible around its longitudinal axis

reversed around its longitudinal axis. If the centre of the cross-hairs is moved off the image of the point sighted, the objective system is eccentric.

If the line of sight of the telescope is not correctly adjusted, the arithmetical mean of the two staff readings P_1 and P_2, read in the two telescope positions, gives the correct reading for the point P_0 on the longitudinal axis of the telescope.

Test of the form of the line of sight. According to Helmert's definition the geodetic line of sight of a geodetic telescope is the locus of points of which the objective lens (lens system) produces an image on the cross-hair at different sighting distances. If, in the case of the modern telescope with an internal focussing lens, the centre of the cross-hairs lies in the straight guide of the focussing lens, then the line of sight as determined by the above definition is also a straight line connecting the point, where the line of the guide meets the first principal plane of the objective, and the first focal point of the objective. If the centre of the objective lens also lies on this straight line, the straight line of sight also coincides with this and it is now identical with the optical axis of the telescope. Due to technological and manufacturing reasons, however, the guide line of the centre of the focussing lens does not in general exactly meet the centre of the cross-hairs. Consequently, if the path of the focussing lens is

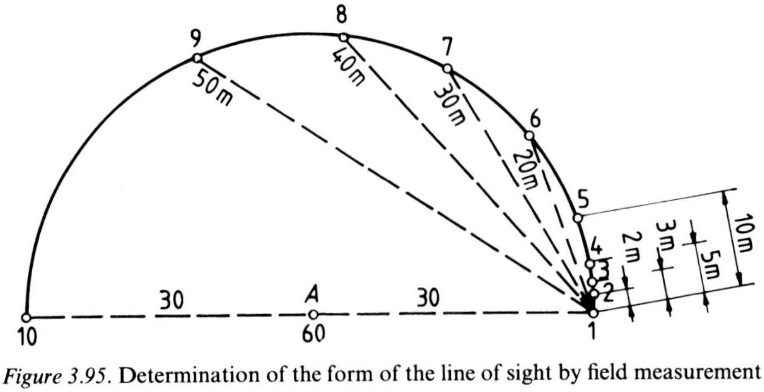

Figure 3.95. Determination of the form of the line of sight by field measurement

otherwise a straight line, the line of sight is a flat half hyperbola. If the path is not straight, the line of sight is an irregular curve. For levelling, the vertical component of the deflection of the line of sight from a straight line is important. The form of the line of sight in the vertical plane can be tested *by means of field measurements* in the following way:

a) Ten stable points should be marked in a semicircle *(Fig. 3.95)* and their elevation differences should be measured by a level set up at the centre *A* of the semicircle. The measurements should be made at a relatively constant temperature and under stable refraction conditions. Due to the equal staff distances the internal focussing lens must not be moved.

b) Readings are made with the level set up over the point *1* and the staff held at points *2* to *10*, one after another.

c) The measurements made from point *A* should be repeated for a check and in order to increase accuracy.

d) The staff readings made with the instrument at point *1* must be corrected in order to eliminate both the residual error of the line of sight and the effects of the Earth's curvature and refraction [Eq. (3.33)].

e) The deviations of the corrected elevation differences, computed from the corrected staff readings, from those measured from point *A* give information about the height errors caused by the looseness of the inner focussing lens guide, and about the form of the line of sight which can be plotted on a graph.

If the measurements are repeated from station point *1* several times, the resulting variations of the line of sight, due to irregular movement of the internal focussing lens, can also be determined. Because of the temperature sensibility of the spirit level, the instrument and the spirit level must be carefully protected during measurement against the effect of temperature changes.

The form of the line of sight can also be tested *in the laboratory* by the *Glissmann* method. Preconditions of such a test are: the level should have an optical micrometer, an attachable autocollimating eyepiece should be added to the telescope (Section 1.5.4), and a plane mirror of high quality should be available.

The basic principle of the test is that, in an arbitrary but unmoved position of the level and with the telescope focussed to infinity, the line of sight $\overline{RR'_\infty}$ is produced as

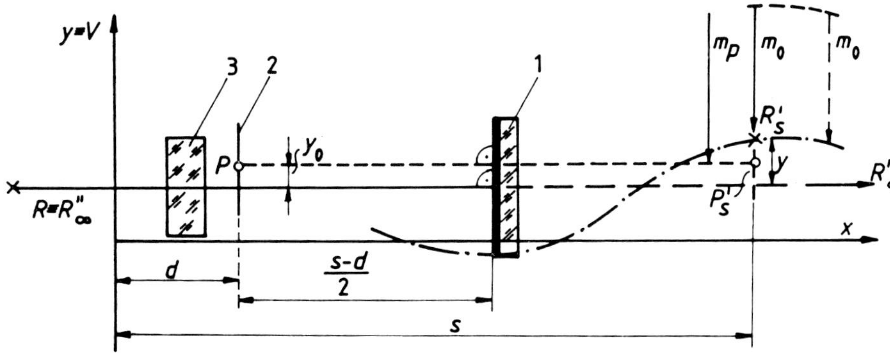

Figure 3.96. Test of the form of the line of sight in the laboratory (autocollimation position)

a straight line of reference *(Fig. 3.96)*, and the deviation y of the geodetic line of sight of the telescope, focussed to finite object distances s, is measured with the optical micrometer. In the figure the geodetic line of sight (the line determined by points at different distances s which are imaged by the objective at the centre of the cross-hairs) is denoted by the chain-dotted line, and the point at distance s from the vertical axis of the instrument, which is imaged by the objective at the centre of the cross-hairs R, is denoted by R'_s.

In order to carry out the test the high quality plane-mirror *1* is placed at such a distance from the instrument that the mirror image R'_s of the cross-hair centre R should be generated at distance s from the vertical axis of the instrument; the plane of the mirror should be perpendicular to the line of sight of the telescope when focussed to infinity. The mirror must thus be placed at distance $(s-d)/2$ from the glass cover plate *2* of the telescope, where d is the distance between the glass cover plate and the vertical axis. The mirror should be tilted and rotated around its vertical and horizontal axis respectively, using the mirror support, until its plane is perpendicular to the line of sight. This is achieved when the cross-hair centre R coincides with the image R''_∞ reflected from the mirror *1*, using the autocollimation eyepiece attached to the instrument.

Then the mirror image P'_s of the mark P, fixed on the cover plate *2* of the telescope, before starting the test, must be focussed by means of the focussing lens, without moving the mirror and telescope. It is advisable to make the mark P similar to the graduation lines used on invar staffs and it should then be fixed on the glass plate (e.g. the mark is drawn on paper and stuck to the plate) in such a position that its separation in height y_0 from the reference line of sight should be not more than a few mm, but it must be within the measuring range of the optical micrometer. When focussing the mark P and simultaneously adjusting the plane parallel glass plate *3* perpendicular to the line of sight, the image R of the cross-hair centre, generated by the objective lens in the image plane, appears on the geodetic line of sight (chain-dotted line) at the point R'_s. In order to sight the mirror image P'_s vertically, the plane parallel glass plate *3* must be rotated *(Fig. 3.97)*. If the reading of the optical micrometer when the plane parallel glass plate is perpendicular to the line of

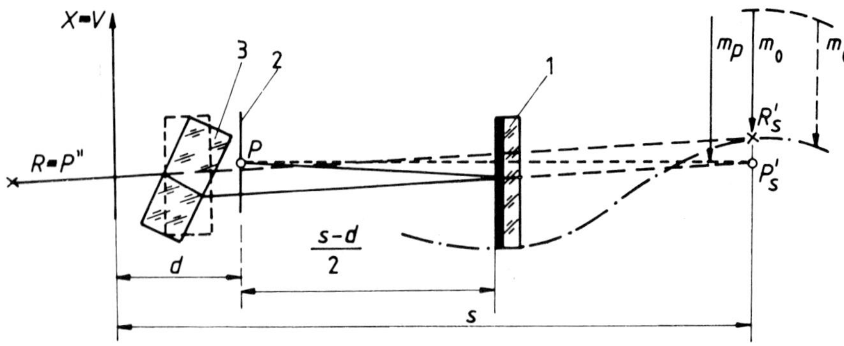

Figure 3.97. Test of the form of the line of sight in the laboratory (pointing to the mark P)

sight is m_0, and if it is m_p after rotation, the principal ray $\overline{RR'_s}$ is moved by the amount $(m_p - m_0)$. From *Fig. 3.96* it can be seen that

$$(m_p - m_0) + y_0 = y,$$

whence

$$m_p = y + (m_0 - y_0). \tag{3.36}$$

If the mirror is placed at various distances from the instrument and both the adjustment and the measurement are repeated then, for each micrometer reading m_p of the mirror images P_s generated at various distances s, the value $(m_0 - y_0)$ remains constant, because m_0 at any distance s remains unchanged. Hence, plotting the m_p values as ordinates versus s values as abscissae in a Cartesian coordinate system, one obtains the plot of the geodetic line of sight of the instrument, where the y axis of the coordinate system is the vertical axis of the instrument and the x axis is a straight line parallel to the line of sight $\overline{RR'}$ which was chosen as the reference straight line at distance $(m_0 - y_0)$ from the x axis.

With this method the determination of the line of sight is not influenced by refraction. The beam of rays passing from the telescope and reflected from the mirror traverses the same path during the autocollimating adjustment as when sighting the mark P: therefore, if refraction remains constant during the test in each mirror position, its effect is eliminated from the difference $(m_p - m_0)$.

During the measurement (or measurement series) in one mirror position, the mirror must remain motionless, therefore it must be protected against external effects (primarily those of temperature). The result is also highly affected by the flatness of the mirror, whose deviation from plane causes a systematic error. In order to decrease this error care must be taken that the beam of rays should fall on the same surface element of the mirror in each mirror position. This can be achieved by ensuring that the telescope always points at the centre of the mirror (a target should be fixed there) before the autocollimating adjustment as well as before the measurement, and the autocollimating adjustment is always made by rotating and tilting the mirror. The amount of surface error can be controlled if not one mark P but two are stuck to the telescope within the measuring range of the micrometer. The difference between the measurements made to both marks gives information not only about the errors of reading and adjustment but also about the surface quality of the mirror.

Because of possible looseness of the focussing lens guide the focussing screw must always be turned in the same sense when focussing the image of the mark P. The form of the line of sight determined in this way is only valid for this sense of turning.

If the test is made so that at the same mirror position the focussing lens is moved several times in both positive and negative directions and a micrometer reading is made after each focussing of the mark P, the variance of results gives information about the instability of the line of sight due to the irregular motion of the internal focussing lens.

The vertical instability d_y of the line of sight caused by the vertical looseness p_y of the focussing lens guide can also be computed numerically from the connection

$$\Delta\alpha = \frac{p_y}{f f_2}(d - e), \tag{3.37}$$

the formula being analogous to the horizontal instability of the line of sight caused by a loose focussing lens [Eq. (2.17)]. In Eq. (3.37) the resulting focal distance of the objective system of the telescope corresponding to a finite object distance can be computed by means of the connection

$$f=\frac{f_1 f_2}{f_1+f_2-e}. \tag{3.38}$$

Here e is the distance between the focussing lens with a focal distance f_2 and the objective lens with a focal distance f_1 (more exactly, the distance between their principal planes which face each other) in the case when the sighted point is at a finite distance s from the objective. It can be computed from the equation

$$e=\frac{s'+d}{2}-\sqrt{s'\left(f_2-\frac{d}{2}+\frac{s'}{4}\right)+d\left(\frac{d}{4}-f_2\right)}, \tag{3.39}$$

where

$$s'=\frac{sf_1}{s-f_1} \tag{3.40}$$

is the image distance from the objective lens, where the image is generated by the fixed objective lens; d is the optical length of the telescope (i.e. the distance between the rear principle plane of the objective lens and the plane of the cross-hairs).

If the straight line of the focussing lens guide does not pass through the centre of the cross-hairs, then the line of sight is a slightly curved line. The separation of the curvature points of the line of sight (from the asymptote of the hyperbola) is most noticeable for short distances (within 5 to 10 m). Conclusions about the quality of the cross-hair adjustment can be drawn from the 2 to 10 m section of the plot compiled from the results of testing the line of sight. By a vertical shift of the cross-hair it is possible to set its centre onto the straight line of the focussing lens guide with sufficient accuracy. In this case, apart from its irregular instability, the line of sight is a straight line.

If the plot shows a curvature or irregular character of the focussing lens guide, only an optimum straight line can be attained.

Test of the optical micrometer. It is a requirement that a tilt of the plane parallel glass plate gives a reading of the micrometer graduation identical with the vertical shift of the line of sight, and that the micrometer must not have any lost motion. A calibrated glass scale with millimetre graduations is used for testing, set vertically and 10 m from the level. The vertical axis of the instrument is made vertical using the spirit level which is then centred. The cross-hair is now set to each millimetre graduation line of the glass scale sequentially, forewards and then backwards, each time reading the micrometer drum. From the differences between the micrometer readings and the intervals of the glass scale the errors of the micrometer graduation can be computed. Several series of measurements will test whether or not the optical micrometer has any lost motion.

If a turn of one unit of micrometer graduation is not equal to one interval on the glass scale, which corresponds to an elementary shift of the line of sight, the optical micrometer has an error of run. This can be corrected by modifying the arm length (l) as shown in *Fig. 3.29.*

3.1.2.7 Test of Compensator Levels

Test of the circular level. The axis (or the plane of the axis) of the circular level (or alidade level) must be normal to the vertical axis. For testing a tubular level is fixed on the upper part of the instrument, or on the telescope, by means of which the vertical axis should be carefully set vertical. If now the bubble of the alidade level is not central, it should be centred by the bubble adjusting screws.

The error of tilt in other directions (see below under Test of the compensator) depends on the tilt of the vertical axis, therefore it is advisable to determine individually both the component of the residual error in the longitudinal axis of the telescope, from the adjustment of the alidade level, and the component perpendicular to it. For this purpose the level is put on a level trier (see Subsection 1.6.3) and its vertical axis is made vertical by the tubular level fixed on it or the telescope, which is turned into the direction of the tilting table of the level trier. In this position a reading is made on the graduation of the level trier (m_1). Next the bubble of the alidade level of the vertical axis is centred by means of the measuring screw of the level trier and a new reading (m_2) is made on the level trier.

The error of the alidade level in the direction of the telescope, which is to be eliminated, is

$$\vartheta = (m_2 - m_1)\alpha'' \tag{3.41}$$

where α'' is the constant of the level trier in seconds of arc.

When repeating this procedure with the telescope perpendicular to the tilting table of the level trier, one obtains the error component of the alidade level perpendicular to the longitudinal axis of the telescope.

Test of the horizontal cross-hair. Testing and adjustment must be carried out in the same way as described for the testing of tilting levels.

Test of the horizontality of the principal line of sight. Among the lines of sight of compensator levels there is a particular one for an instrument with its vertical axis vertical, which is called the principal line of sight. If the principal line of sight makes an angle with the horizontal, it is called the tilt of the principal line of sight, which corresponds to the tilt of the line of sight of tilting levels. The principal line of sight must be normal to the vertical axis.

The tilt of the principal line of sight is determined almost in the same way as that of the tilting level but with the difference that, instead of centring the bubble of the spirit level, the vertical axis of the instrument must now be set vertical very carefully, if possible by means of a tubular level mounted on the instrument for this purpose. The test can also be carried out with a collimator, similarly to the test of tilting levels, with the difference that here again the vertical axis must be vertical instead of centring the bubble of the spirit level.

The adjustment is made in general by the vertical adjusting screw of the cross-hair. Rarely a special adjusting screw is used by means of which the basic position of the pendulum compensator, or the position of one of the optical elements placed within the optical train, can be changed.

Test of the compensator. If the principal line of sight is horizontal but the compensator does not operate correctly, then the inclination α of the telescope is compensated by the different angle α'. If $\alpha' > \alpha$ or $\alpha' < \alpha$, we speak about overcompensation or undercompensation, respectively. In both cases, with a tilted vertical axis, there is an angle $\delta\alpha = \alpha - \alpha'$ between the horizon of the instrument and the horizontal *(Fig. 3.98)*.

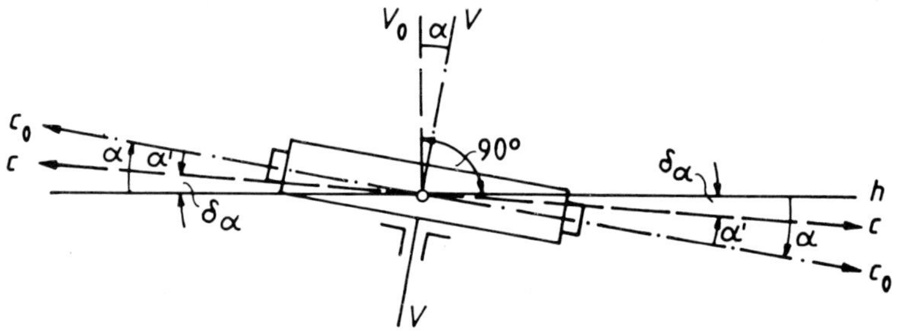

Figure 3.98. Tilt of the horizon with incorrect operation of the compensator: h = the horizontal; V-V = the vertical axis; α = tilt angle of the vertical axis; c_0 = the principal line of sight; c = the position of the line of sight rotated by the angle α' (compensated position); δ_α = the tilt of the horizon due to the compensation error

The horizon tilt $\delta\alpha$ is composed of the effects $\delta\alpha_A$ and $\delta\alpha_B$, caused by the components ζ_A in the direction of the longitudinal axis of telescope and ζ_B perpendicular to it. These two components give the obliquity ζ of the vertical axis:

$$\delta\alpha = \delta\alpha_A + \delta\alpha_B = k_A\zeta_A + k_B\zeta_B, \tag{3.42}$$

where

$$k_A = \frac{\delta\alpha_A}{\zeta_A} \tag{3.43}$$

is the relative instability error in the direction of the telescope, which is to be compensated, and

$$k_B = \frac{\delta\alpha_B}{\zeta_B} \tag{3.44}$$

is the error in the direction perpendicular to the telescope. These generally constant values specify the instrument and are usually given as a fraction with unit numerator. Equation (3.42) gives the tilt of the horizon if the values ζ_A and ζ_B are substituted by the error components of the alidade level which are to be eliminated.

The relative instability error in the telescope direction is determined as follows:

a) The level is set onto the plate of the level trier, its vertical axis is made vertical and its telescope is turned into the direction of the tilting table of level trier.

b) A staff with mm graduation is held at a distance of 10 m from the level.

c) The position l_0 of the principal line of sight is read on the staff and, simultaneously, the initial value m_0 of the measuring screw.

d) The vertical axis of the instrument is tilted through small but increasing angles in several steps, once forewards and once backwards, within the operating range of the compensator, the corresponding readings $m_1, m_2, \ldots, m_i$ of the measuring screw of the level trier are taken, and simultaneously the positions $l_1, l_2, \ldots, l_i$ of the line of sight are read on the staff. From this the component of the horizon tilt in the telescope direction is:

$$\delta''_{\alpha A} = \frac{l_i - l_0}{d} \rho'' \tag{3.45}$$

and the component of the tilt of the vertical axis in the telescope direction is:

$$\zeta''_A = (m_i - m_0)\alpha'', \tag{3.46}$$

where d = the distance between the vertical axis of the level and the staff, which must be graduated in mm,
α'' = the constant of the level trier.

The sign of the readings of the measuring screw of the level trier is positive if the numbers increase with elevation of the tilting table, negative if they decrease.

If the vertical axis of the instrument is not in the vertical plane containing the horizontal rotation axis of the level trier then, when turning the measuring screw of the level trier, the level is also moved vertically; this motion is contained in the reading, together with the effect of the error of instability. Therefore one obtains a reading free of the effect of this motion by subtracting the vertical shift from the erroneous reading l'_i:

$$l_i = l'_i - \frac{r\zeta''_A}{\rho''} \tag{3.47}$$

where r is the distance of the vertical axis of the instrument from the rotation axis of the level trier, and ζ''_A is the change of tilt of the level trier in seconds of arc, as determined by Eq. (3.46).

e) The calculated values $\delta\alpha_A$ can also be plotted against the angles ζ_A in a graph. The rise of the nearly straight line obtained for the instability errors (i.e. its mean slope) gives the relative instability error $k_A = \delta\alpha_A/\zeta_A$.

The relative instability error in the direction perpendicular to the telescope can be determined in the same way, but the telescope must be initially set perpendicular to the tilting table of the level trier.

The relative instability error k_A in the telescope direction can be changed or decreased in several types of instrument by moving the compensator longitudinally with respect to the telescope. The relative instability error k_B perpendicular to the telescope can only be eliminated by very careful manufacture. With instruments in which the compensator pendulum is not suspended on metal wires with circular cross-section but on metal bands, the cross-telescope relative instability error can be neglected in general.

3.1.2.8 Accuracy of Levels

The reliability of levelling depends primarily on the accuracy of the level. This, together with the reliability of levelling, is jointly characterized by the *a posteriori* standard deviation for 1 km. In order to determine this, a levelling line of about ($L=$) 5 km is chosen on a slope which should not exceed 3 percent. The line is made up of n sections of approximately equal length t_i ($i=1, \ldots, n$), where n is about 10. After measuring the elevation difference for each section both direct (1) and return (2), the differences δ_i of the elevation differences of the direct and return levelling of the sections are given by:

$$\delta_i = \Delta H_{i(1)} - \Delta H_{i(2)}. \tag{3.48}$$

From them one obtains the *a posteriori* standard deviation for 1 km of a twice (direct and return) measured levelling line:

$$\mu_{(\mathrm{km})} = \pm \sqrt{\frac{1}{4n}\left[\frac{\delta_i \delta_i}{t_i}\right]_{i=1}^{n}}. \tag{3.49}$$

The formula contains the joint effect of random and systematic errors. The standard deviation of levelling for 1 km almost free of systematic error is:

$$\mu_{(\mathrm{km})} = \pm \sqrt{\frac{1}{4n}\left[\frac{\delta_i \delta_i}{t_i}\right] - \frac{[\delta_i]^2}{[t_i]}}. \tag{3.50}$$

The efficiency of the levelling equipment can, however, be also characterized in a more simple way. If the same surveyor uses the same instrument and the staff distances are as usual for levelling, then the standard deviation of staff readings can be taken as proportional to the staff distance d:

$$\mu = \alpha d \tag{3.51}$$

from which

$$\alpha = \frac{\mu}{d}. \tag{3.52}$$

This angle is called the *mean wobble of the line of sight*. Its value is usually given in seconds of arc:

$$\alpha'' = \frac{\mu}{d} \rho'' \tag{3.53}$$

and this depends mainly on the telescope magnification, on the sensitivity of the spirit level (with compensator levels, on the sensitivity of the compensator), on the quality of the tilting screw and optical micrometer, on the form of the graduation of the levelling staff and on the measuring skill of the surveyor.

The numerical value α of the mean wobble of the line of sight can be determined in practice from the standard deviation $\mu_{\Delta H}$ of the elevation difference ΔH, instead of from Eq. (3.52). This is given by:

$$\mu^2_{\Delta H} = 2\mu^2 \tag{3.54}$$

and from this one obtains from Eq. (3.53):

$$\alpha'' = \frac{\mu_{\Delta H}}{d\sqrt{2}} \rho''. \tag{3.55}$$

The value of $\mu_{\Delta H}$ can be determined by n measurements of the elevation difference between two points (A and B):

$$\Delta H = \frac{[\Delta H'_i]_{i=1}^{n}}{n} \tag{3.56}$$

where

$$\Delta H'_i = (l_A - l_B)_i$$

and

$$\mu_{\Delta H} = \sqrt{\frac{[v_i v_i]_{i=1}^{n}}{n-1}}, \tag{3.57}$$

in which $v_i = \Delta H - \Delta H'_i$.

The distance d between the instrument and the staff must approximate the average distance used during the levelling. During the test the height of the instrument horizon must be changed between measurements in order to eliminate the systematic error of the staff reading.

When using a compensator level the bubble of the alidade spirit level is moved by means of the foot screws after each measurement and then centred again. The centring of the bubble must always be made for the same telescope setting in order to free the determined standard deviation from the effect of possible horizontal tilt.

The mean wobble of the line of sight is valid only within certain distance limits because, in fact, the standard deviation of the staff reading has a non-linear connection with the distance.

3.1.2.9 Levelling Staffs

Simple levelling staffs. In general a levelling staff is made of dry, straight and clear-stemmed pine or fir wood. The length of the staff is usually 3 or 4 m, but staffs with different lengths are also made for special purposes. Staffs are usually 6 to 10 cm wide. The staff is firstly impregnated in order to avoid changes in its size due to humidity, then it is covered several times with layers of oil paint. These layers serve as a base for graduation which is super-imposed, usually in black.

The *rigid staff* is made of one piece of wood and is normally 3 m long. For the 3 or 4 m long *folding staff* the graduated side is protected in the folded position. The transportation of such staffs is more convenient.

For low order levelling a 2 m long "pocket staff" or an extension staff (telescope staff) is used. The pocket staff is in fact a folding scale consisting of 20 cm long sections. The *reversible staff* is graduated on both its front and back surfaces with an uneven displacement and is produced as a rigid 3 m long staff. The readings on both graduations serve as a safeguard against blunders as well as biased estimation errors, and increase the reliability of the measurements.

The bottom end of the staff has a fitting of hardened steel with a plane lower surface which is perpendicular to the longitudinal axis of the staff.

The origin of the staff graduation (for staffs with dual graduation, the origin of one graduation) usually coincides with the lower plane surface. Any deviation is called the *zero error of the staff*. If one works with one single levelling staff the effect of the zero error is eliminated from the elevation difference determined from one instrument set-up. When using two levelling staffs, the elevation difference measured at one set-up is in error by the difference of the zero errors of the two staffs. This error has opposite sign if the two staffs are interchanged. Since the staffs are interchanged at sequential instrument set-ups, over a levelling with an even number of instrument set-ups elevation differences are free of zero error.

The staffs can have simple area graduations *(Fig. 3.99)*, checquerboard graduations *(Fig. 3.100)*, or line graduations *(Fig. 3.101)*.

Simple area graduations can have centimetre units (sometimes with half-centimetre divisions) which are 10 mm wide and about 20 mm long, alternately black and white stripes. The graduation is numbered at every 10 centimetres and normally the numerals are placed within the corresponding 10 centimetre section. An exception is the Kern reversible staff with line graduations *(Fig. 3.101, c)*, where the numerals are in the preceding 10 centimetre section.

The numerals of staffs made for longer sighting distances, mainly for topographical purposes, may be as tall as almost the whole 10 cm section in order to provide easy

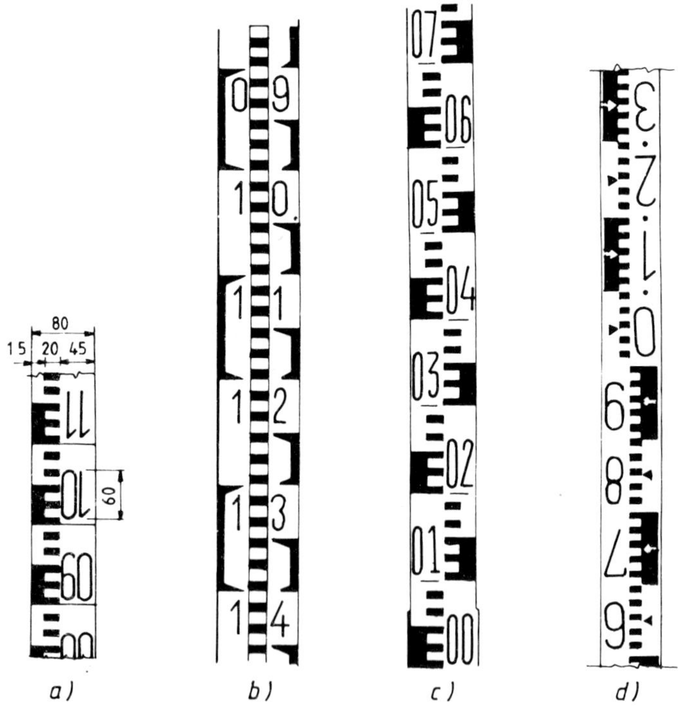

Figure 3.99. Levelling staffs with simple area graduation: *a)* staff made according to the standard DIN 3003; *b)* Kern folding staff, 3 (4) m long; *c)* Opton folding staff, 4 m long; *d)* 3 m long Wild folding staff for topographic surveying

recognition, also for distances longer than 150 m *(Fig. 3.99, d)*. Levelling and tacheometric staffs have smaller figures (40 to 60 mm) and they are not placed at the centre of the 10 cm section, but nearer to the corresponding line *(Fig. 3.99a, b* and *c)*. For the sake of easier reading additional bands of 10 centimetres, alternately black and white, are placed beside the centimetre graduations.

The graduation is often placed between the numerals in the longitudinal axis of the staff, but it can be on one side of the staff with the numerals on the other. In the latter case graduation and numerals sometimes interchange position every metre or 10 centimetres *(Figs 3.99, c* and *d)*.

The majority of staffs used to be made with numerals upside down, because the telescope of the level reversed the image. Staffs made for up-to-date levels, whose telescopes generate erect images, have erect numerals.

For *checquerboard graduations* each black centimetre unit has an adjacent white one above and below, and vice versa, thus seeming to form another identical graduation with a displacement of 1 centimetre. The centimetre units of such graduations are about 10 to 15 mm long *(Fig. 3.100, a)*. Besides the cm graduation there is often an additional banding, alternately black and white and of 10 centimetres or 5 centimetres, either only on one side of the graduation *(Fig. 3.100, a, d, e)* or on both sides *(Fig. 3.100, b, f)*. Compared with the simple area graduation, the

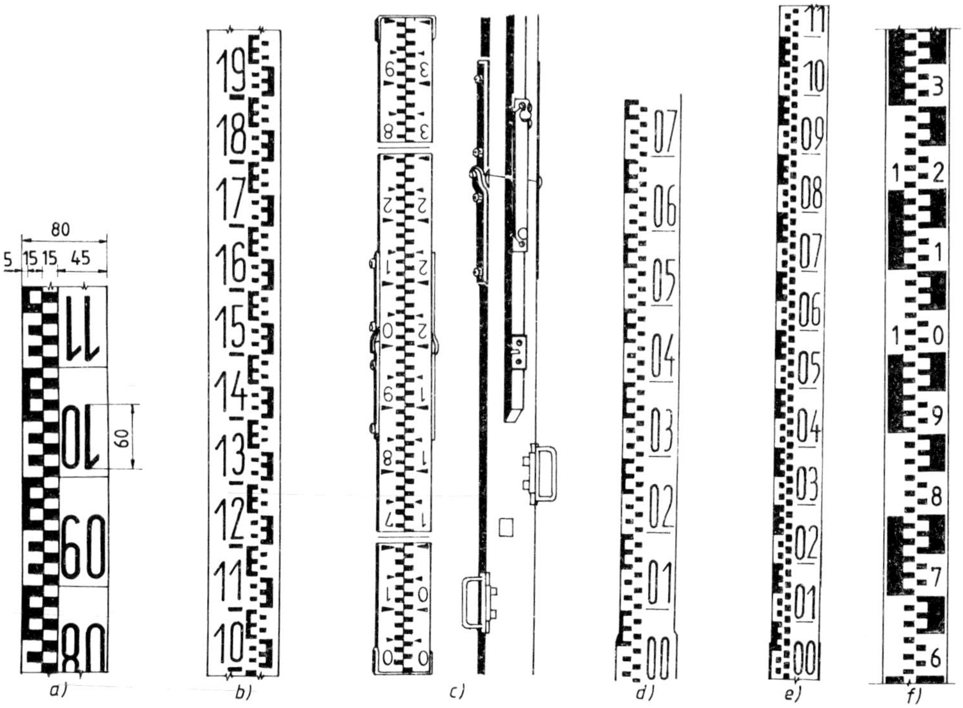

Figure 3.100. Levelling staffs with checquerboard graduation; *a)* staff made according to the standard DIN 303; *b)* MOM folding staff, 3 (4) m long; *c)* Kern folding staff, 3 (4) m long; *d)* Opton folding staff, 3 (4) m long; also with reflecting coating; *e)* Opton folding staff, 0.5 m and 3 m long; *f)* Wild folding staff, 3 (4) m long

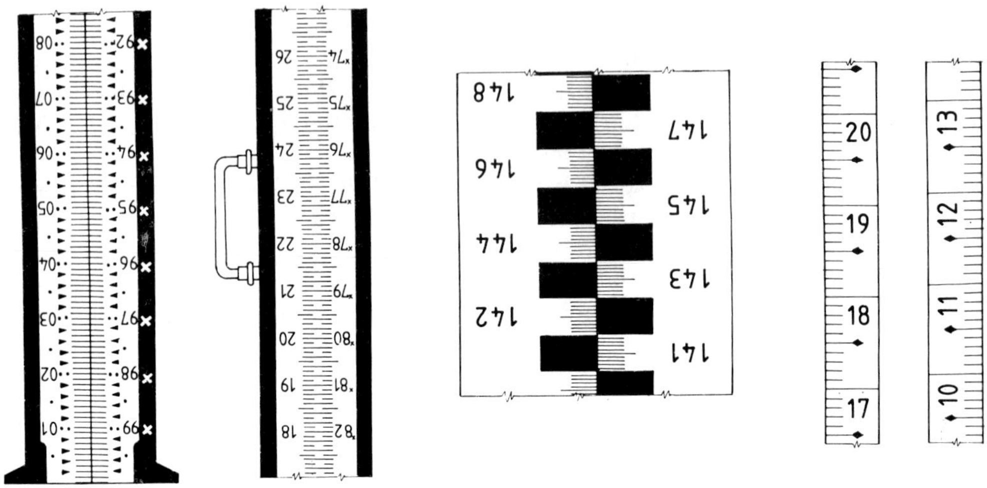

a) b) c) d)

Figure 3.101. Levelling staffs with line graduations: *a)-b)* staffs with dual graduation and numbering (used in early Prussian land surveying); *c)* staff with cm wide banded graduation, supplemented with mm lines (used in early land surveying in Switzerland); *d)* Kern staff with 1 cm divisions on both sides, 3, 4, 5 m long; (the 10 cm numerals are in the previous field);

checquerboard graduation has the advantage that the estimation of fractions of centimetres is not so tiring for the eye, because estimation can be made on a white strip at any position of the horizontal cross-hair. Sometimes, mainly on staffs used for tacheometric measurement, the graduations are made in two colours (black and red) alternating every metre.

Staffs with *line graduations (Fig. 3.101)* are good if the staff is always set up nearly at the same distance from the instrument. In that case, by choosing the width of the staff graduations to be the same as the apparent width of the cross-hair, precise readings can be made on the staff. Levelling staffs with graduation lines are generally used for precise levelling. Readings on a staff with two sets of relatively displaced graduations should be made on both sets *(Fig. 3.101, a, b)* in order to check the readings, since the sum of the two readings is 100. Using a staff with a chequerboard graduation combined with line graduations in between *(Fig. 3.101, c)*, even tenths of mm can be estimated if the staff distance is short and the telescope has a large magnification. Graduations are added to the simpler staffs by stencil while, for higher accuracy requirements, they are made by means of a graduation machine and the bands are painted later.

For wooden staffs any graduation unit is allowed to deviate from an exact centimetre by a maximum of ± 0.2 mm, while the total length of the staff graduation can deviate from the nominal value by ± 0.8 mm. The thermal change of length of wooden staffs is about 4 μm/°C for 1 m. The change of the humidity content of the staff can cause an irregular change of length by as much as several tenths of mm per m, therefore for precise levelling a wooden staff must be calibrated immediately before and after measurement.

Invar staffs. Levelling staffs fitted with an invar tape are now used for precise levelling *(Fig. 3.102)*. The staff is usually rigid, 3 m long, made of wood or light metal and has a hollow into which an invar tape about 25 mm wide *1* is placed under tension. The invar tape at its lower end is fastened to the bottom of the staff, while at the upper end it is pulled by a spring-actuated apparatus with a tensile force of 20 daN. The tape rests on supports placed every 50 cm within the body of the staff. It is graduated by an invar stencil used under tension *(Fig. 3.103)*. Usually there are two graduations displaced relatively to each other *(Fig. 3.104)*.

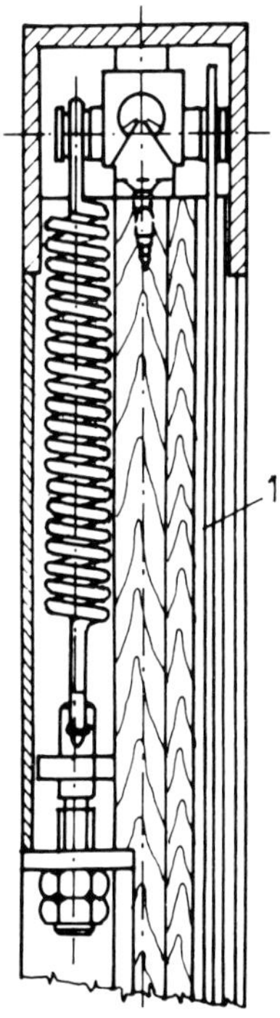

Figure 3.102. Device for tensioning an invar tape in an invar staff

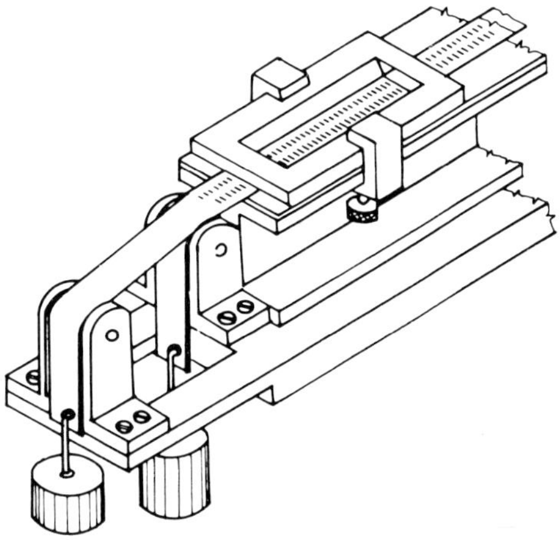

Figure 3.103. Stencil for engraving an invar tape

The two graduations relatively displaced by a non-integer *(Fig. 3.104, e)*, and the different numbering on both sides of the graduation, aid the avoidance of blunders as well as the decrease of random and systematic reading error, provided that both graduations are always used.

On the Jenoptik and MOM staffs 0.5 cm graduation units and 1 mm wide lines are used *(Fig. 3.104, a)*, on the Wild staff a line width of 1.8 mm is used with a 1 cm graduation unit *(Fig. 3.104, e)*. With increasing observing distances the narrow lines can be observed only with decreasing reliability, therefore on the Opton staff *(Fig. 3.104, b)* 3 mm wide graduation lines follow each other alternately on the right and left sides of the invar tape, forming individually a cm graduation but together a graduation with 0.5 cm separation.

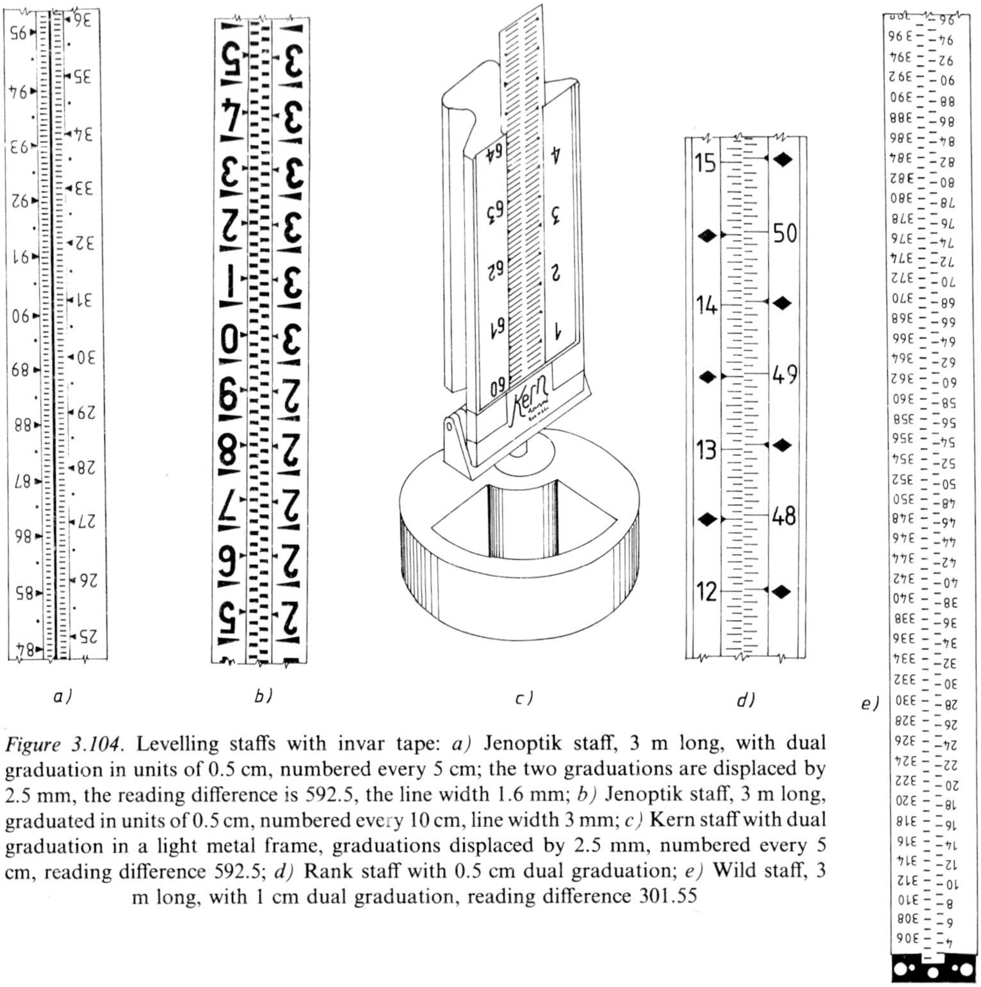

Figure 3.104. Levelling staffs with invar tape: *a)* Jenoptik staff, 3 m long, with dual graduation in units of 0.5 cm, numbered every 5 cm; the two graduations are displaced by 2.5 mm, the reading difference is 592.5, the line width 1.6 mm; *b)* Jenoptik staff, 3 m long, graduated in units of 0.5 cm, numbered every 10 cm, line width 3 mm; *c)* Kern staff with dual graduation in a light metal frame, graduations displaced by 2.5 mm, numbered every 5 cm, reading difference 592.5; *d)* Rank staff with 0.5 cm dual graduation; *e)* Wild staff, 3 m long, with 1 cm dual graduation, reading difference 301.55

The invar tape of the Kern invar staff *(Fig. 3.104, c)* is covered with a light-reflecting coating, the staff thus being very useful for night measurement.

The invar staffs should be calibrated only at the beginning and end of a measurement campaign, because the thermal change of the invar tape is only about 1 μm/°C per m. Normally the standard deviation of the random error of the graduation is less than ±0.01 mm, and that of the systematic error is less than 0.01 mm per metre.

Special levelling staffs. On *target staffs*, used for levelling over rivers, a target with a point marker can be moved up and down and its position can be read on the staff graduation by means of its index. The disadvantage of the target staff is that measuring with it is complicated and the readings must be made at the staff.

On levelling staffs used for precision levelling in Hungary, black bands of width 2 cm alternate with white bands of width 3 cm. When observing over a longer distance,

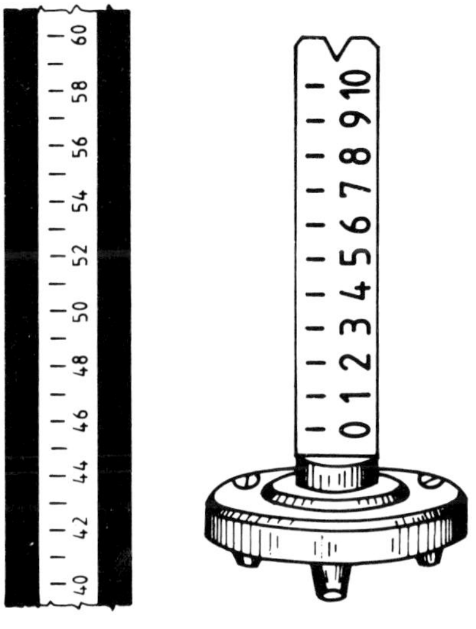

Figure 3.105. Special levelling staffs: *a)* Wild invar staff for surveying for civil engineering, 92 cm or 182 cm long, with shoes with different profiles, numbered sideways every 2 cm; *b)* Wild 10 cm long invar staff with a base for forced centring

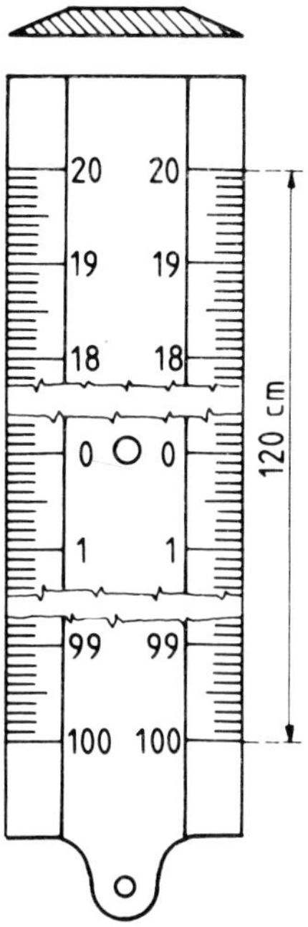

Figure 3.106. Hanging staff

this alternating graduation appears in the field of view similar to the image of the invar staff. Using a *Wild N3* level for the measurement, the 2 cm black strip must be flanked by the cross-hair wedge: the principal reading is obtained from reading the staff and the secondary reading is made by means of the micrometer scale. Since the line of sight of the instrument can be raised or lowered by the micrometer only within a range of 1 cm, five additional change points (pegs with a nail on top) are placed near to each other in such a way that the height difference between two adjacent pegs is always 1 cm. Now there will be certainly one on which the staff will have one 2 cm wide black band which can be flanked by the wedge, and thus the reading can be obtained.

The Wild invar staff with cm graduation is mostly used for industrial surveying. This staff is either 92 cm or 182 cm long and it is provided with three fittings on the bottom, each having a different profile. The Wild 10 cm long staff mounted on a forced centring tripod *(Fig. 3.105, a, b)* is also convenient in industrial surveying.

Out-of-date levelling marks, consisting of a hole in a wall plate, can only be levelled using so-called hanging staffs *(Fig. 3.106)*. The hook of the 120 cm long metal staff, which has mm graduations and cm numbering, fits into the hole in the wall plate.

Staff accessories. Staffs must be held vertical, therefore a circular bubble (bull's eye bubble) with a sensitivity of 5 to 20 min of arc/mm is fitted as a *staff level* to the side or rear of the staff at about 1 m above the zero, in order to ensure verticality. For low order levelling the staff level consists of a circular bubble mounted on top of a metal angle plate which is held by hand against the edge of the staff. For precise levelling the staff is supported by two *staff braces.* Staff braces manufactured for some types of staffs can be fixed on the side or upper end of the staff, and their lengths or their attachment point on the staff can be adjusted in order to adjust the staff into the vertical. Straight bars are used for staffs which have no suitable manufactured braces; they are attached to the staff and held by hand during measurement.

For marking turning points as well as for unambiguous and stable support of the levelling staff in precise levelling, *ground pegs* made of hard metal (or pegs made of any other material of equivalent quality) are used. For low order levelling *change-plates* (base plates) are used. A possible form of ground peg is shown in *Fig. 3.107.* Important components are a cylindrical tip ending in a regular hemisphere and the 2.5 to 3.0 cm wide rim by means of which the peg can be driven into the ground.

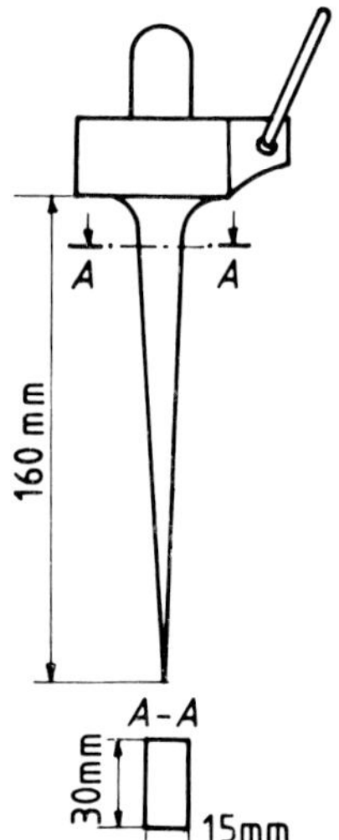

Figure 3.107. Iron peg for marking turning points (type used in geodetic levelling in Hungary)

Figure 3.108. Change-plates for marking turning points

The commonly used change-plates are made of cast iron *(Fig. 3.108)* with a hemispherical upper surface or with a cylindrical tip ending in a curved surface. If the levelling staff is set up on the cylindrical tip of the change-plate, a metal staff shoe is usually fixed to the bottom of the staff; on the one hand it ensures that the centre of the staff bottom lies on the change-plate, on the other it restrains the staff from slipping on the change-plate.

Test of the levelling staff. The plane of the axis of the staff level fixed to the levelling staff (or held against it by hand) must be perpendicular to the longitudinal axis of the staff. For testing the staff level the staff must be set vertical using a plumb bob suspended on a 3 m long string. In the field the levelling staff can be set up vertically using the vertical cross-hair of the level, by sighting the staff from two perpendicular directions. If the bubble of the staff level is not central, it must be adjusted by its adjusting screws.

The surface of the staff shoe must be plane and perpendicular to the longitudinal axis of the staff. For testing it is advisable to use a nail with spherical head in a wooden peg. The staff is set up several times on the nail in such a way that each time a different point, from points of equal distribution on the bottom surface including the centre, is set onto the nail, and a reading is made at all set-ups. If there is a difference between the reading using the centre and those using the other points, then the surface of the staff shoe is either not plane or not perpendicular to the longitudinal axis of the staff. The differences should not exceed the reading accuracy.

The zero point of the graduation (in the case of dual graduations, one of them) must coincide with the bottom plane of the staff shoe in order to avoid zero error. If only one staff is used for levelling, the zero error has no influence on the elevation difference. When two staffs are used, each elevation difference also includes the difference of the two zero errors. In practice it is sufficient to determine the difference between the zero errors of the two staffs. For this purpose both staffs are set up on the same point; the difference between the level readings on them gives the difference of their zero errors.

In certain cases (e.g. when using wall plate with a hole) the zero errors of the staffs must be individually known. In such a case the zero errors can be determined by using a hanging staff whose zero error is already known (determined in the laboratory with a glass scale). The level should be set up at a point where there is a rise roughly half a metre above the ground 10 to 15 m away from the level. Now a peg should be driven into the ground on this rise and a nail with a spherical head is driven into the peg. The hanging staff is set up vertical but upside down on the nail and a reading is made with the level after its telescope bubble has been centred. Now the hanging staff is replaced on the nail by the first and then by the second ordinary staff, and readings are made with the level on both graduations on both staffs. Lastly, for a cheek, the hanging staff is reinstated and the reading is repeated. Knowing the zero error of the hanging staff, the zero errors of both graduations of both staffs can be computed. For computation the arithmetical mean of the first and last reading on the hanging staff is taken.

The length unit of the staff graduation, i.e. the *staff metre*, is usually not identical with a true metre and it may change with temperature. Therefore, the elevation difference of a levelled section computed from the readings must be corrected with correct sign by the calibration correction

$$k = c_t(\Delta H), \tag{3.58}$$

where the calibration coefficient c_t is the average difference of the staff metre from a true metre, i.e. the calibration correction of the staff metre at the temperature t. The

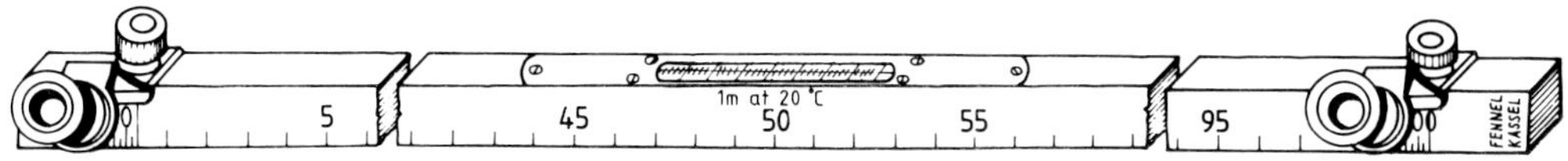

Figure 3.109. Fennel standard metre for the calibration of levelling staffs (Otto Fennel GmbH, Kassel)

value of c_t is determined by a comparator device. The most simple device is a standard metre supplied with a magnifying viewer or a microscope. The Fennel standard metre *(Fig. 3.109)* is of such a type, having scales of 0.2 mm or 0.1 mm at both ends. The scales can be observed either with a magnifying viewer or with a microscope. The temperature of the standard metre is taken by an inner thermometer.

In order to determine the calibration factor c_t the standard metre is placed upon the levelling staff which is laid horizontal on the comparator bench; the longitudinal axes of the standard and the staff must be parallel. Now the graduation lines i and $i+1$ of the levelling staff are read on the graduations at both ends of the standard metre. From these readings the difference between the length l_k of the staff metre assigned to the temperature t_k and the length L_k of the standard metre is computed:

$$\Delta = l_k - L_k. \tag{3.59}$$

The length of the standard metre at the temperature of calibration is

$$L_k = 1\ \mathrm{m} + c_{n0} + \alpha_n(t_k - t_0) \tag{3.60}$$

and the length of the staff metre is

$$l_k = 1\ \mathrm{m} + c_{l0} + \alpha_l(t_k - t_0), \tag{3.61}$$

where c_{n0} = the deviation of the standard metre from 1 metre at the standard temperature t_0,
c_{l0} = the calibration correction of the staff metre at the standard temperature,
α_n and α_l = the temperature coefficients of the standard metre and of the pair of staffs.

Substituting Eqs (3.60) and (3.61) into Eq. (3.59) one obtains the following equation:

$$\Delta = c_{l0} - c_{n0} + (\alpha_l - \alpha_n)(t_k - t_0), \tag{3.62}$$

where the calibration correction of the staff metre at standard temperature is:

$$c_{l0} = \Delta + c_{n0} + (\alpha_n + \alpha_l)(t_k - t_0). \tag{3.63}$$

Since the length of the staff metre usually varies along the staff, the calibration must be repeated at different points on the staff or the staff pair (usually at every half metre) and the average difference

$$\bar{\Delta} = \frac{[\Delta_i]_{i=1}^{n}}{n} \tag{3.64}$$

between the staff metre and the standard metre is computed at the temperature of calibration. Substituting this into Eq. (3.63) one obtains the average calibration correction of the staff metre at standard temperature:

$$c_{l0} = \bar{\Delta} + c_{n0} + (\alpha_n - \alpha_l)(t_k - t_0). \tag{3.65}$$

Knowing the calibration correction at the standard temperature and the thermal expansion coefficient of the staff pair, the calibration coefficient (the average calibration correction of the staff metre) at any temperature t is:

$$c_t = c_{l0} + \alpha_l(t - t_0). \tag{3.66}$$

From Eqs (3.65) and (3.66) a series of values is computed and compiled in practice in tabular form for each pair of levelling staffs, from which the calibration coefficient can be taken for the temperature of the measurement and the calibration correction of the elevation difference of the levelled length can be computed from Eq. (3.58).

A more reliable determination of the calibration correction can be made by using a levelling staff comparator with a double image, operating on the principle of coincidence *(Fig. 3.110)*. The different types operating on the basis of the same principle are more or less similar. E.g., the majority of the optical elements of the comparator shown in *Fig. 3.111* are placed in a steel tube. The images of the two scale lines showing the distance on the invar staff *9* to be calibrated are projected into the tube by the pentagon prisms *1* and *1a*. Lenses *2* and *3*, and *2a* and *3a*,

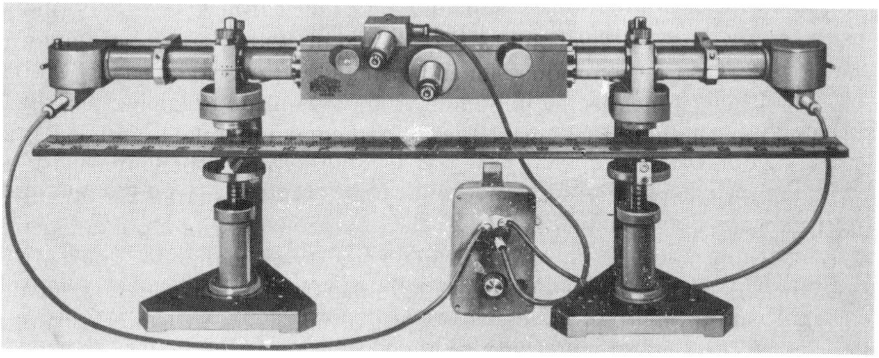

Figure 3.110. Breithaupt levelling staff comparator (F. W. Breithaupt, Kassel)

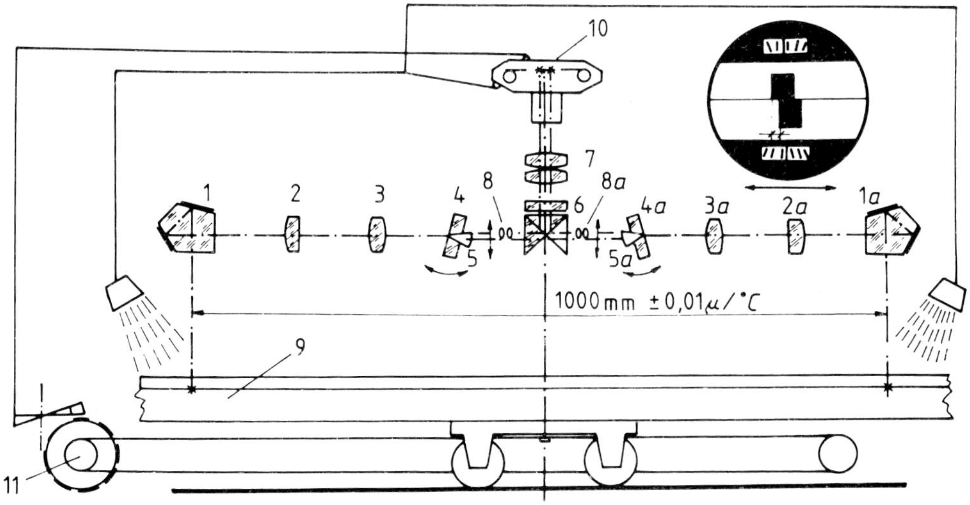

Figure 3.111. Diagrammatic sketch of the Pusztai staff comparator based on coincidence reading (BGTV Laboratory, Budapest)

respectively, form a once magnified image of the scale lines on the surface of the image-forming prism *6*; this surface is marked by a cross in the figure. Part of the beam of rays coming from the scale lines passes through the micrometer scales *5* and *5a* after passing through the plane parallel glass plates *4* and *4a*. The micrometer scale lines are also reflected onto the surface of prism *6* by the optical elements *8* and *8a*. The image generated on the surface of prism *6* (shown in the upper right corner of the figure) can be observed by the interchangeable eyepiece *7* (8 times, 12 times or 16 times magnification).

To calibrate the levelling staff firstly the standard metre is put on the rolling mechanism under the comparator tube and the plane parallel plate *4* is rotated until the micrometer scale is in its zero position. Now the images of the two end lines of the standard metre are brought into coincidence by means of the glass plate *4a*, then the micrometer graduation *5a* is read. The known deviation from 1 m of the standard metre used is added to the value read, with correct sign, and the value obtained is set on the micrometer graduation *5a*. This way the optical base line of the comparator is adjusted to exactly 1 metre. Then the levelling staff (*9*) is placed on the comparator instead of the standard metre and, by gradually moving it, the images of two division lines of the staff metre, which are nominally 1 m apart, can always be brought into coincidence by turning the plane parallel plate *4*.

The readings on the micrometer scale give the difference from 1 m of the distance defined by the scale lines. The value 0.01 mm can be read directly on the micrometer and 0.001 mm can be estimated.

The comparator can also be used automatically. In this case the eyepiece must be substituted by a photo-camera *10* after adjustment of the optical base line. The levelling staff is moved by the electromotor *11* in 1 cm steps. When motion-less the camera is exposed electromagnetically by a delayed relay. The images recorded on the film can be evaluated with a microscope.

The graduation error of a levelling staff can be suitably tested by means of a 0.1 mm glass scale. The permissible differences between the distances between two adjacent graduation lines are about ± 0.2 mm for wooden staffs and ± 0.02 mm for invar staffs.

3.2 Instruments for Trigonometrical Heighting

3.2.1 Simple Instruments

Simple instruments such as clinometers with a base, suspended clinometers without a telescope or simple levels with a pointing device on simple legs are usually used for less accurate height measurement on building sites, for vertical or horizontal positioning of building elements, for approximate measurement of profiles, for reduction to the horizontal of inclined lengths measured by measuring bars or measuring tapes. In such instruments the position of an index mark brought into the horizontal by means of a spirit level or a vertical index (plummet) can be read on a graduated arc. Hence the slope angle between the instrument base and the suspended device — both rigidly fixed to the graduated arc — can be determined. This type of instrument includes the clinometer with a spirit level for determining the tilt of bars

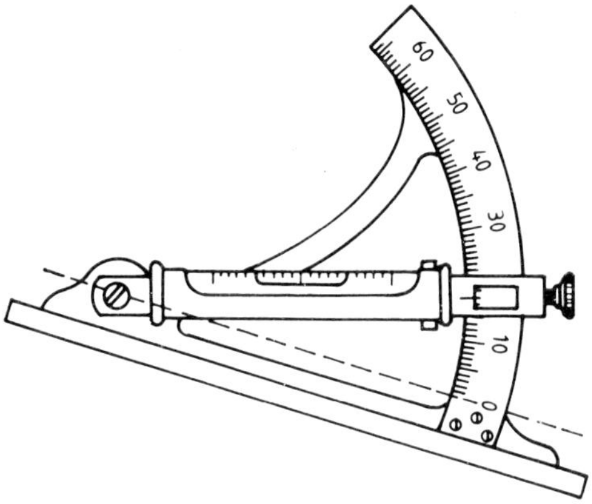

Figure 3.112. Clinometer with a spirit level for measuring the tilt of a staff

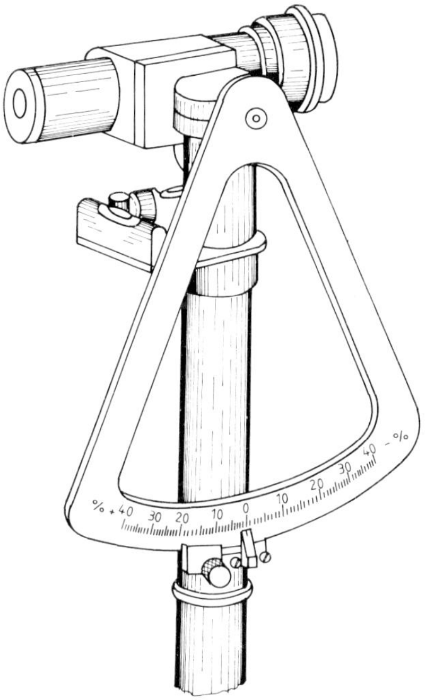

Figure 3.114. Clinometer with a telescope (Otto Fennel GmbH, Kassel)

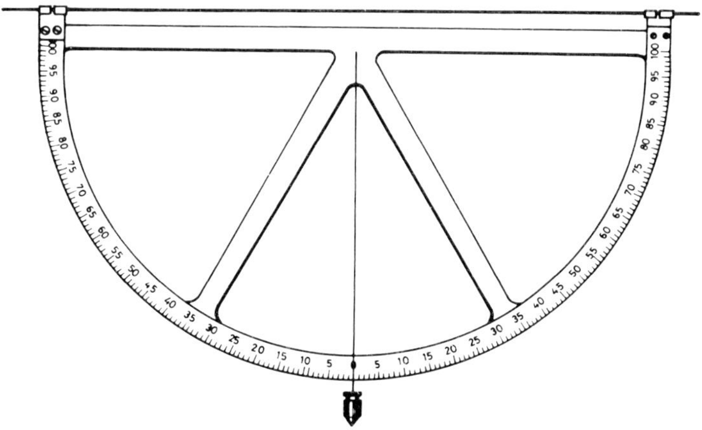

Figure 3.113. Clinometer suspended on a wire (Otto Fennel GmbH, Kassel)

or staffs *(Fig. 3.112)*, the suspendible clinometer used in underground surveying *(Fig. 3.113)* and the clinometer with a telescope used on simple legs *(Fig. 3.114)*.

The mean square error of measurement by a clinometer with a pendulum or plummet is ± 5 min of arc, by a clinometer with a spirit level ± 1 min of arc.

3.2.2 The Theodolite as an Instrument for Trigonometrical Heighting

3.2.2.1 The Construction of Vertical Circles

In order to determine vertical angle or zenith distance a graduated vertical circle is attached to the horizontal axis of the theodolite; it is perpendicular to the horizontal axis and concentric with it. The vertical circle rotates together with the telescope and the vertical angle or zenith distance of the line of sight of the telescope can be read by means of an index mark placed beside the circle; the horizontal (vertical) position of the index mark is stabilized by means of an index level or a compensator.

The numbering on the vertical circle can be:

— in *quarters of the circle*: four numberings from 0° to 90° (see *Fig. 3.115, a*);

— *in half circles*: two numberings from 0° to 180°; the numberings can be in opposite sense *(Fig. 3.115, b)* or in the same sense *(Fig. 3.115, c)*;

— in a *whole circle*: the numbering goes from 0° to 360°.

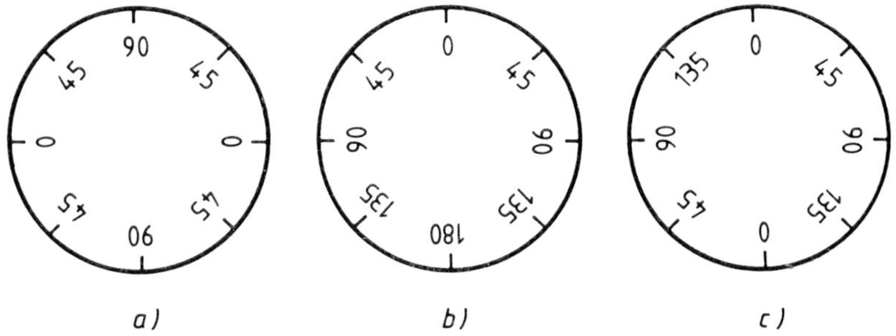

Figure 3.115. Vertical circles with numbering in each quarter and half of the circle

The vertical circle can be fixed on the horizontal axis:

— in the sense of the *elevation angle*, i.e. the reading on the circle l_{I} gives the elevation angle α,

— in the sense of the *zenith distance*: the reading l_{I} on the circle gives the zenith distance (zenith angle) z.

Most theodolites have a vertical circle with whole circle numbering fixed in the sense of the zenith distance. In this case the zenith distance can be obtained from the readings l_{I} and l_{II} of the two measurements in two telescope positions as:

$$z = \frac{l_{\mathrm{I}} + 360° - l_{\mathrm{II}}}{2}. \tag{3.67}$$

If the circle is fixed in the sense of the elevation angle, Eq. (3.67) gives the elevation angle from the two readings. An exception is the *Wild T3* theodolite, where the readings on the vertical circle with two half-circle numberings in the same sense give half the actual elevation angle *(Fig. 3.116)*, hence the zenith distance is:

$$z = l_{\mathrm{I}} + 90^\circ - l_{\mathrm{II}} \tag{3.68}$$

or the elevation angle is:

$$\alpha = l_{\mathrm{II}} - l_{\mathrm{I}}, \tag{3.69}$$

where l_{I} and l_{II} are the readings in the first and second telescope positions respectively (in the "left" and "right" positions of the vertical circle).

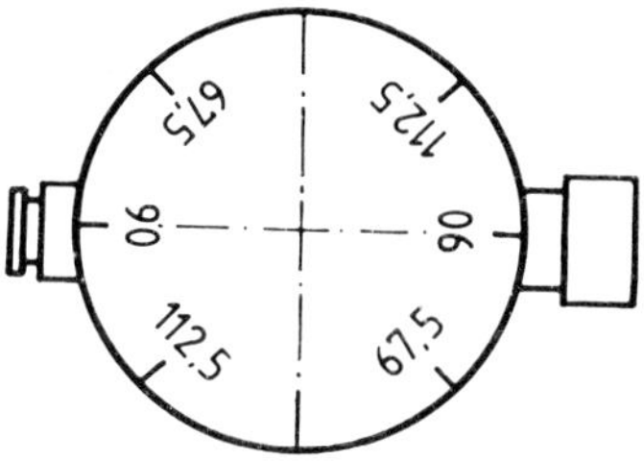

Figure 3.116. Numbering of the vertical circle of the theodolite *Wild T3*

In theodolites of low or medium precision usually only one reading index mark is provided, but in precise theodolites two diagonally opposite index marks are provided and the two readings are unified in one reading device. The device is now usually a reading microscope fixed alongside the telescope. Both the horizontal and the vertical circles can be read in the same microscope.

3.2.2.2 The Index Mark of the Vertical Circle

The straight line perpendicular to the horizontal axis which contains the index mark(s) for vertical reading must always form the same angle with the horizontal when reading. This can be attained in two ways. In *theodolites with index levels* the level is fixed to the arm containing the index mark in such a way *(Fig. 3.117)* that the axis of the spirit level must be parallel to the plane of the vertical circle and it should be rotatable, together with the arm of the index mark, through a small angle about the horizontal axis H. On up-to-date theodolites the index level and the index mark are usually set simultaneously before measurement by means of a coincidence observation device. If the instrument is in adjustment, the spirit level axis must be parallel to the straight line connecting the index marks when horizontal; for index marks in a vertical position it must be perpendicular to the connecting line.

Today a built-in *index compensator* (also called *automatic index*) is widely used. This automatically eliminates the effect of the tilt of the vertical axis in the telescope direction. In the microscope, which is rigidly fixed to the telescope *(Fig. 3.118)*, the image of the index mark i_0, generated by the objective lens O_M of the microscope, will come to the position l_i' instead of l_i if the vertical axis is tilted by the angle ϑ, thus having an error of $r\vartheta$, where r is the radius of the horizontal circle *(Fig. 3.118)*. The compensator must bring the image l_i' of the index automatically to l_i.

In one type of automatic index the compensator built into the microscope rotates the image-generating rays of the microscope in the plane of the vertical circle around the point K by the angle $\beta = f(\vartheta)$ *(Fig. 3.118)*. The error-free reading l_i is obtained if equation

$$\frac{r\vartheta}{b} = \frac{s\beta}{a} \tag{3.70}$$

holds. From this the optical angular magnification is

$$\frac{\beta}{\vartheta} = \frac{a}{b}\frac{r}{s}. \tag{3.71}$$

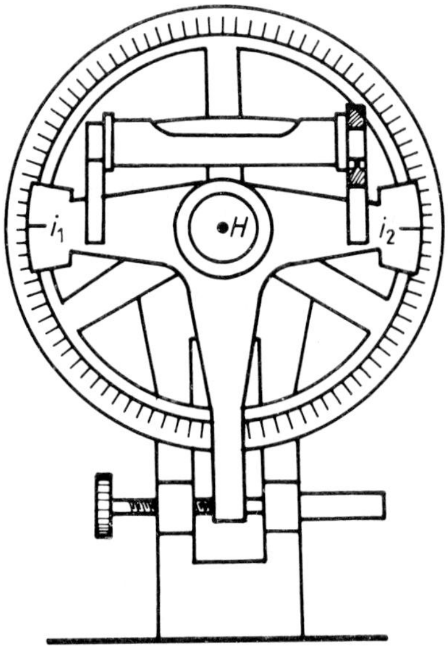

Figure 3.117. Vertical circle index mark with index spirit level

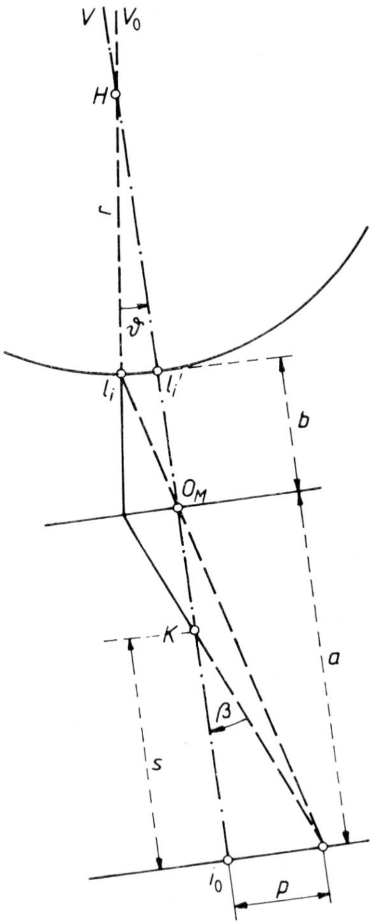

Figure 3.118. Operating principle of an automatic vertical circle index mark (with compensator)

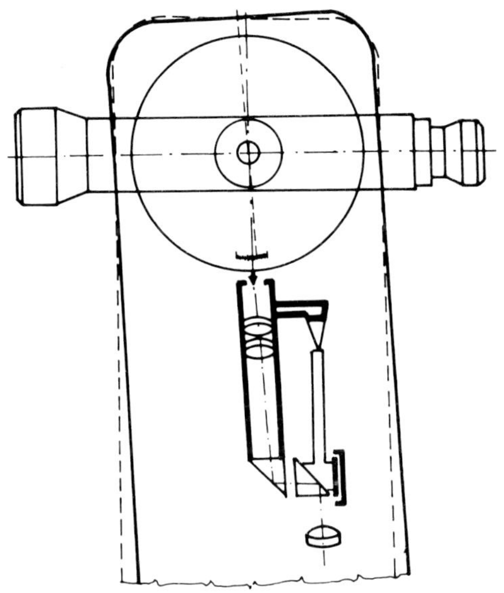

Figure 3.119. Automatic vertical circle index mark on the theodolites *Tu* and *Tt* (Franke et Co. Optik GmbH, Giessen)

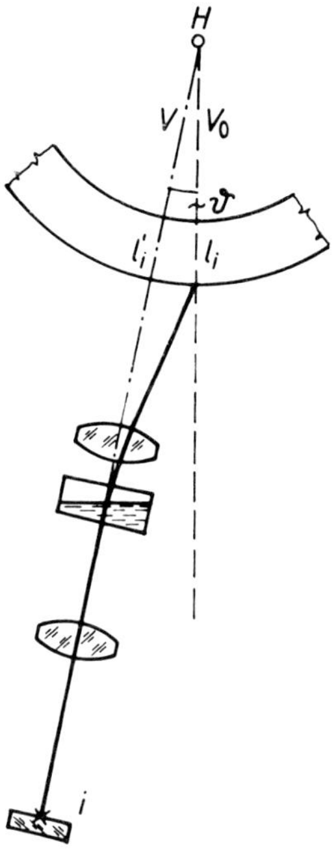

Figure 3.120. Operation of the liquid wedge compensator (*Wild T1-A* theodolite)

Such a compensator can be found in the theodolites *Franke Tu* and *Tt,* where a prism with a reflecting surface and suspended on a simple pendulum is placed in the ray path of the microscope *(Fig. 3.119).*

In the *Wild T1-A* theodolite the compensator is a liquid wedge inserted between the two parts of the microscope-objective; it moves the image l_i' of the index i automatically to l_i if the vertical axis V and the real vertical V_0 form the angle ϑ *(Fig. 3.120).* The *Kern DKM 2A* theodolite also has a liquid compensator *(Fig. 3.121),* but the compensation is made by the reflecting horizontal surface of the liquid.

In other types of compensator the reading index is automatically moved along the telescope axis proportionally to the tilt ϑ of the instrument. This type is used in the *Opton Th 3* theodolite, where the end of the bubble of the index level is the index mark of the reading on the micrometer. If the instrument is tilted by an angle ϑ, the bubble of radius R is moved by

$$p = R\vartheta. \tag{3.72}$$

From *Fig. 3.118* the error-free reading is obtained if the equation

$$\frac{r\vartheta}{b} = \frac{p}{a} \tag{3.73}$$

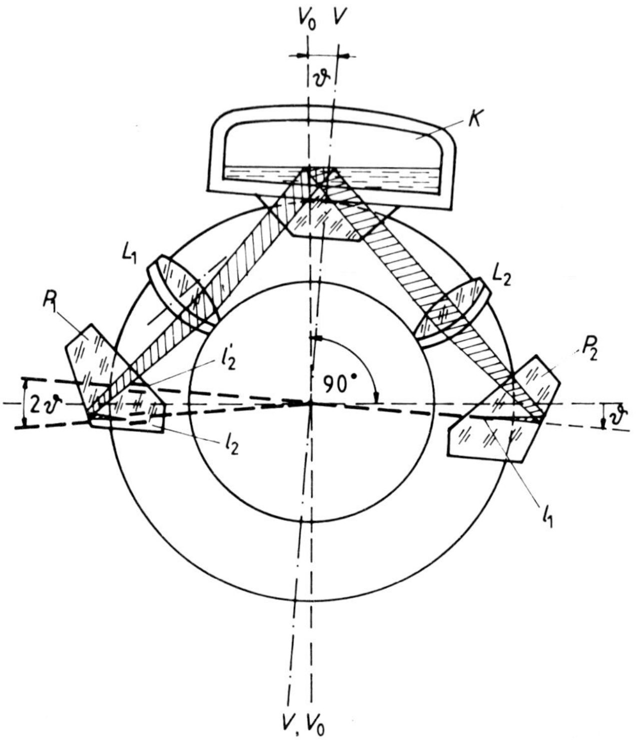

Figure 3.121. The liquid compensator of the theodolite *Kern DKM 2A:* V_0-V_0 = the vertical axis of the theodolite in the vertical position; V-V = the tilted vertical axis of the theodolite; ϑ = the tilt angle of the vertical axis; K = the liquid compensator; P_1 and P_2 the prisms deflecting the light; L_1 and L_2 projecting lenses; l_1 the point on the numbered circle graduation corresponding to the index mark; l_2' the point on the unnumbered double graduation projected to l_1 if the vertical axis is vertical; l_2 the circle graduation projected to l_1 if the vertical axis is tilted by the angle

holds. From Eqs (3.72) and (3.73) one obtains:

$$R = \frac{a}{b} r. \tag{3.74}$$

The 10 min of arc divisions of the vertical circle can be made to coincide with the vertical index mark by means of the optical micrometer connected to the level, and now readings can be made with an accuracy of 0.1 min of arc on the graduation of the optical micrometer over the range of 10 min of arc.

The effect of a tilt ϑ of the vertical axis can also be eliminated by an automatic movement of the objective lens of the microscope. The movement necessary is:

$$d = \frac{a}{a+b} r\vartheta. \tag{3.75}$$

In the instruments *Jenoptik Theo 020A* and *Dahlta 010A* the objective lens of the microscope is suspended on a spring pendulum *(Fig. 3.122)*. The two objective lenses *3* and *4* of the reading microscope are fixed to the pendulum *10* which is suspended by the leaf spring *11* and damped by the air-damper *12*. The objective generates the image of the divisions of the vertical circle *1* through the prisms *2* and *5* onto the index plate *6* (a glass plate with graduations).

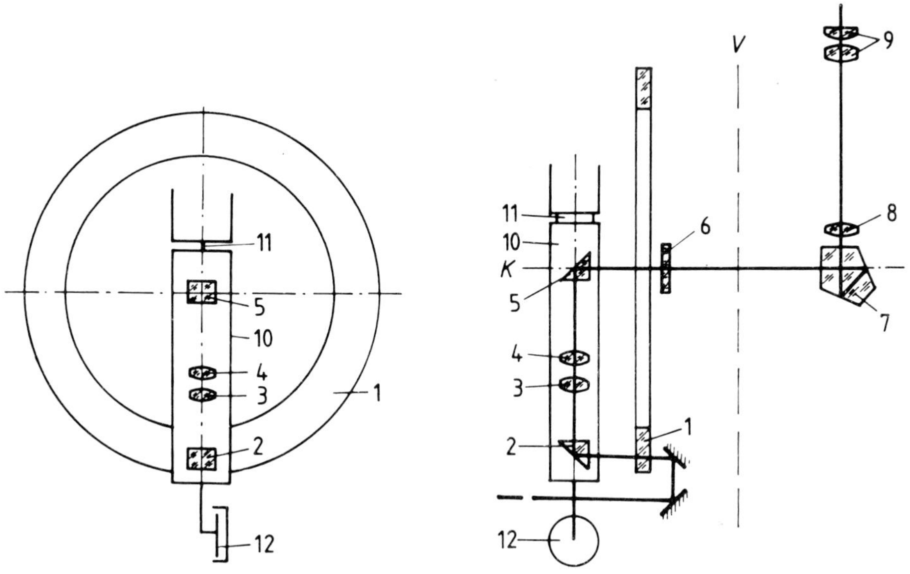

Figure 3.122. The automatic vertical circle index mark of the instruments *Jenoptik Theo 020A* and *Dahlta 010A* (objective-suspended compensator)

By proper dimensioning of the pendulum the optical centre of the objective lens of the microscope can be made to come into the vertical plane through the horizontal axis of the instrument when there is a tilt ϑ of the vertical axis. Consequently the reading point, which lies in the vertical plane containing the horizontal axis, is always imaged on the reading index plate *6*, and this image is observed through the prism *7* by means of a second microscope objective lens *8* and the eyepiece *9*.

The index marks of the vertical circles of the theodolites *Filotecnica Salmoiraghi 4149 A*, as well as of the *Soviet OMT 3* and *T 5*, are similarly controlled by the objective lens of the microscope. In the latter, one of the lenses is suspended on four 42 mm long fibres, the other is fixed to the telescope support *(Fig. 3.123)*.

In the *Kern* theodolites *K1-A* and *K1-RA* the compensation is made again by an objective lens controlled by a simple physical pendulum *(Fig. 3.124)*. The length of the pendulum is identical with the radius *r* of the vertical circle.

The effect of a tilt of the vertical axis in the telescope direction on the reading of the vertical circle can be compensated by a parallel shift of the beam of rays passing through the microscope. The automatic vertical index of the theodolite *Wild T2* operates on the basis of this principle; here a plane parallel glass plate mounted on

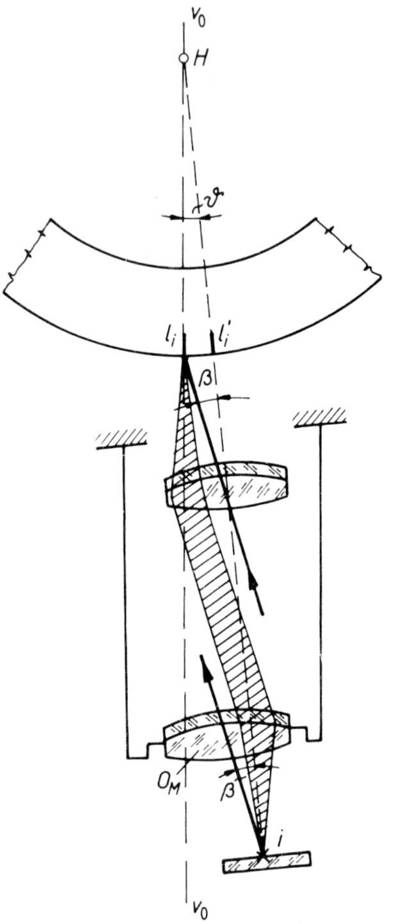

Figure 3.123. The automatic vertical circle index mark (objective-suspended compensator) of the theodolites *OMT 3* and *OMT 5* (Soviet Union): *H* the horizontal axis, i.e. the centre of the vertical circle; V_0-V_0 the vertical; ϑ the tilt component of the vertical angle; *i* the index mark; O_M that part of the microscope objective which is suspended on a pendulum; l_i' the erroneously positioned image of the index mark on the vertical circle with fixed O_M; l_i the error-free image of the index mark produced after the movement of the compensator objective O_M; β the deflection angle of the ray

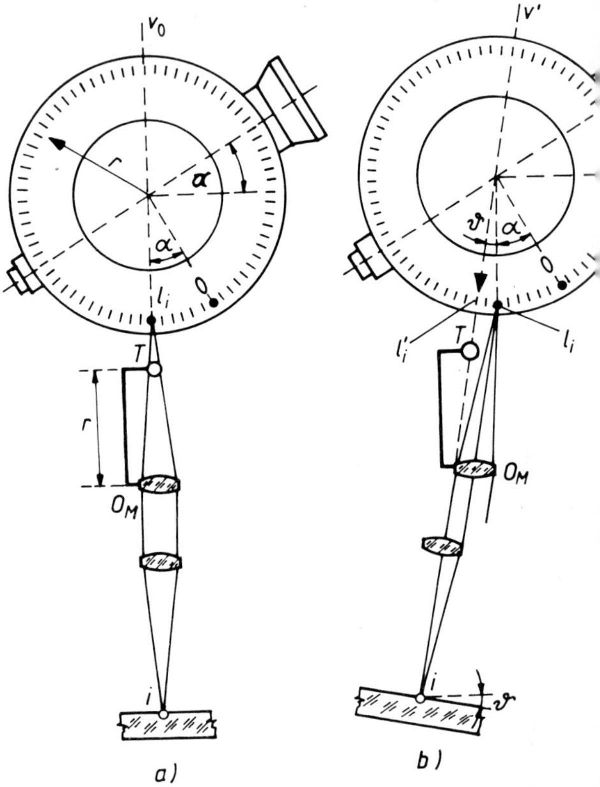

Figure 3.124. The automatic vertical circle index mark (objecti suspended compensator) of the theodolites *K1-A* and *K1-RA* (K et Co., Aarau, Switzerland). *a)* with the vertical axis V_0 vertical: l_i image of the index mark *i* on the vertical circle produced by microscope objective O_M; *T* the rotation axis of the compensa pendulum; α the elevation angle read; *b)* with oblique vertical a V' with a tilt of angle ϑ: l_i' the erroneously positioned image of index mark *i* on the vertical circle for fixed O_M; l_i the error-free ima of the index mark *i* produced after the movement of the compensa objective O_M

an astatic pendulum suspended on crossed tapes ("X" compensator) gives a parallel shift to the beam of rays generating the image of the circle graduation *(Fig. 3.125)*. In the index compensator of the theodolite *Wild T16* the principle is the same but the construction is rather different.

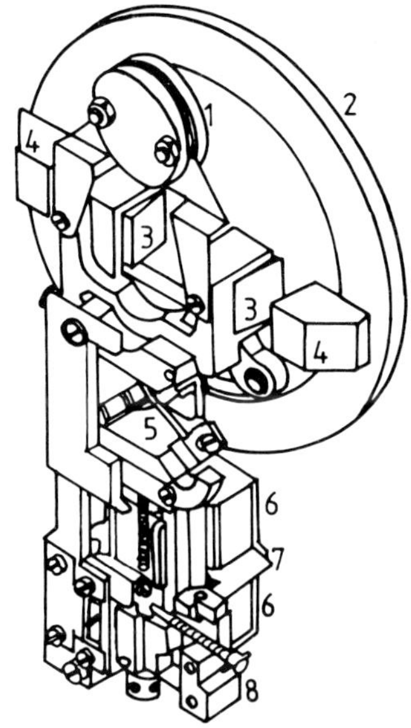

Figure 3.125. The automatic vertical circle index mark of the theodolite *Wild T2*: *1* cross-damping; *2* vertical circle; *3* plane-parallel glass plate on the compensator pendulum; *4* prism for reading the circle; *5* suspension bands; *6* magnet; *7* damping plate; *8* adjusting screw

The beam of rays is also parallel shifted by a plane parallel glass plate with an "X"-shaped suspension of the pendulum device in the *TDJ2* theodolite, and with a "V"-shaped suspension in the *TDJ6* theodolite (Beijing Optical Instruments, Beijing). Similar index compensators are used in the theodolites *Opton Th 2* and *Th 4*.

3.2.2.3 Test of the Vertical Circle and of the Reading Device

The reading on the vertical circle of the theodolite gives the correct elevation angle or zenith distance only if:

1. the vertical plane of pointing is perpendicular to the horizontal axis,
2. the horizontal axis is perpendicular to the vertical axis,
3. the vertical and horizontal axes intersect each other,
4. the line of sight of the telescope intersects the horizontal axis,
5. the plane of the vertical circle is perpendicular to the horizontal axis,
6. the centre of the vertical circle lies on the horizontal axis,
7. the reading l_I in the first telescope position gives directly the angle of elevation (or the zenith distance), i.e. there is no index error,
8. the errors in the graduation of the vertical circle can be neglected,
9. the reading device has no systematic error.

If the listed conditions 1, 2, 3, 4 and 5 are only slightly not satisfied, they have such a small effect that they can be neglected in practice. If condition 6 is not satisfied, its effect cannot be neglected even in the case of a very small eccentricity. Instruments with two diagonally opposite index marks are exceptions because the effect of eccentricity becomes zero in the arithmetical mean of the readings on both index marks. In theodolites of small or medium precision, however, usually only one index mark is used. Here the effect of the eccentricity of the vertical circle is not eliminated by the mean value of the measurements in two telescope positions. The effect of the error depends on the angle of elevation and it is a maximum if the pointing direction is horizontal.

If the centre O of the vertical circle of radius r has an eccentricity e with respect to the horizontal axis H, and if the latter forms an angle ω with the line of sight of the telescope (as in *Fig. 3.126*), then the following relationships are valid for pointings to the same point in the first and second telescope positions (using the notation of the figure):

$$l_{\mathrm{I}}+\beta+\gamma+\varepsilon_{\mathrm{I}}+L+\alpha=180° \tag{3.76}$$

$$l_{\mathrm{II}}+\beta+\gamma+\varepsilon_{\mathrm{II}}+L-\alpha=360°, \tag{3.77}$$

where l_{I} and l_{II} are the readings on a vertical circle with whole circle numbering in the first and second telescope positions; L is the deviation from the horizontal of the straight line connecting the end of the horizontal axis H with the prescribed position i' of the index; β is the alignment error of the circle; γ is the error of the angular adjustment of the index; ε_{I} and $\varepsilon_{\mathrm{II}}$ are the angular errors due to the eccentricity of

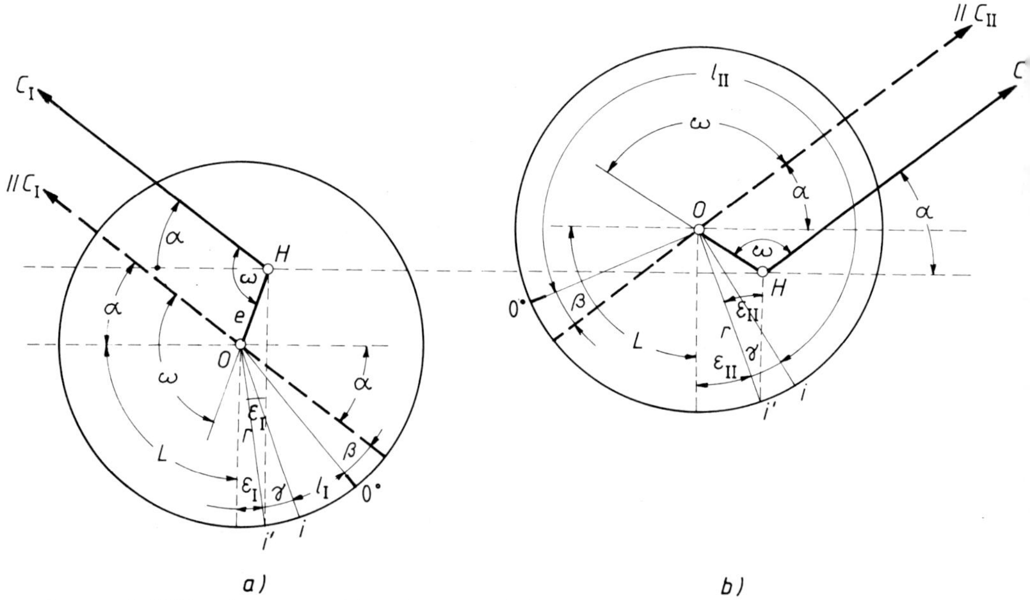

Figure 3.126. Effect of the eccentricity of the vertical circle: *a)* first telescope position; *b)* second telescope position

the vertical circle (the corrections to the readings) in the first and second telescope positions.

After reducing the equations and taking into account that $180° - L - l_I = \alpha_I$ and $l_{II} + L - 360° = \alpha_{II}$ are the vertical angles with reading errors in the first and second telescope positions respectively, and after introducing the index error $\delta = \beta + \gamma$, then Eqs (3.76) and (3.77) can be written as:

$$\alpha = \alpha_I - \varepsilon_I - \delta, \tag{3.78}$$

$$\alpha = \alpha_{II} + \varepsilon_{II} + \delta. \tag{3.79}$$

The angular error ε_I and ε_{II} can be determined from the triangle *OHi'* as seen in the figure. As $\sin \varepsilon_I \approx \varepsilon_I$ and $\sin \varepsilon_{II} \approx \varepsilon_{II}$, one obtains:

$$\varepsilon_I = \frac{e}{r} \sin(L - \omega + \alpha) = \frac{e}{r} \sin(L - \omega) \cos \alpha +$$

$$+ \frac{e}{r} \cos(L - \omega) \sin \alpha, \tag{3.80}$$

$$\varepsilon_{II} = \frac{e}{r} \sin(L - \omega + 180 - \alpha) = -\frac{e}{r} \sin(L - \omega) \cos \alpha +$$

$$+ \frac{e}{r} \cos(L - \omega) \sin \alpha. \tag{3.81}$$

Introducing the notation:

$$K_1 = -\frac{e}{r} \sin(L - \omega), \tag{3.82}$$

$$K_2 = -\frac{e}{r} \cos(L - \omega) \tag{3.83}$$

also called *eccentricity constants*, Eqs (3.80) and (3.81) may be used to write Eqs (3.78) and (3.79) in the following form:

$$\alpha = \alpha_I + K_1 \cos \alpha + K_2 \sin \alpha - \delta \tag{3.84}$$

$$\alpha = \alpha_{II} + K_1 \cos \alpha - K_2 \sin \alpha + \delta. \tag{3.85}$$

The arithmetical mean of the two equations obtained for the first and second telescope positions is:

$$\alpha = \frac{\alpha_I + \alpha_{II}}{2} + K_1 \cos \alpha, \tag{3.86}$$

from which it is seen that the mean value of the vertical angles α_I and α_{II} obtained (with reading error) in the two telescope positions does not give the error-free vertical angle, but it is influenced by the *first eccentricity constant* K_1. The effect of the eccentricity error is a maximum at $\alpha = 0°$, i.e. in a horizontal pointing when its value is K_1. From Eq. (3.86) the vertical angle determined in the two telescope positions is free of the eccentricity error of the vertical circle if $K_1 = 0$ or, by taking Eq. (3.82) into account, if $L-\omega=0$. This happens if the direction of the eccentricity, when pointing horizontally, intersects the reading index mark.

If the measurement of the vertical angle is made only in one telescope position, it can be corrected using Eqs (3.84) and (3.85), where the constants K_1, K_2 and δ must be known. Measurement in both telescope positions is always possible in practice, therefore Eq. (3.86) shows that it is sufficient to determine only the eccentricity constant K_1.

This constant K_1 can be determined by field measurement as follows.

Change-plates are placed on the two points A and B which are at a distance of about 30–40 m from each other and at about the same height *(Fig. 3.127)*. The small height difference Δ between the graduations a and b of a levelling staff set up on the points A and B is measured by a level. Now the theodolite to be tested is set up at the minimum observing distance from A, in such a way that its horizontal axis is nearly at the height of the nearly horizontal straight line $\overline{ab}$.

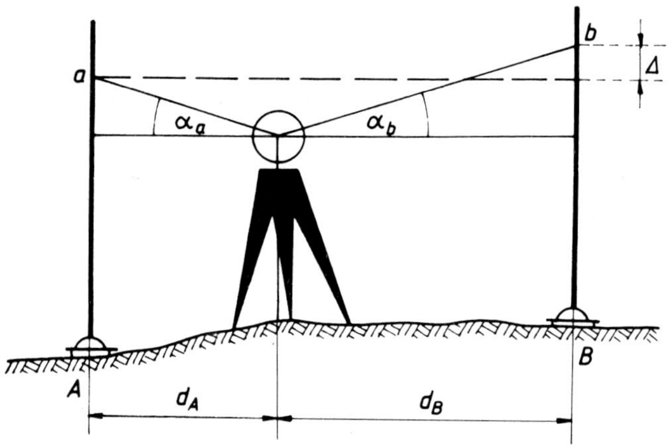

Figure 3.127. Determination of the eccentricity constant K_1 of the vertical circle

Measurements are made of the vertical angles α_a and α_b to the points a and b in two telescope positions, as well as of the horizontal distance d_A and d_B. From these data the height difference between a and b is, using Eq. (3.86):

$$\Delta=d_B \tan(\alpha_B+K_1 \cos\alpha_B)-d_A \tan(\alpha_A+K_1 \cos\alpha_A). \tag{3.87}$$

Since α_a and α_b are small angles, one has:

$$\Delta=d_B(\alpha_B+K_1)-d_A(\alpha_A+K_1), \tag{3.88}$$

from which the *first eccentricity constant* of the tested instrument in radians is:

$$K_1=\frac{\Delta-(d_B\alpha_B-d_A\alpha_A)}{d_B-d_A}, \tag{3.89}$$

where Δ is the height difference measured with a level.

If geometric condition *7* is not satisfied, i.e. if the reading l_I of the circle in the first telescope position is not equal to the elevation angle (or zenith distance) (see *Fig. 3.126*), then the error of the vertical circle when elevation angles are read is:

$$\delta = \alpha - l_I \tag{3.90}$$

or when zenith distances are read:

$$\delta = z - l_I. \tag{3.91}$$

This consists of two parts. One is the alignment error β of the vertical circle, i.e. the deviation of the direction of the vertical circle reading 0° or 180° from the horizontal observing plane of the telescope. The other part is the adjustment error γ of the index level (or compensator), which means that the plane determined by the index mark and the horizontal axis is not vertical (or horizontal) when the index level has been correctly set or the compensator operates correctly.

With measurement in both telescope positions the index error vanishes from the mean value of the two elevation angles (zenith distances) as can be seen from Eq. (3.86). Further, from Eqs (3.84) and (3.85) the index error is:

$$\delta = \frac{\alpha_I - \alpha_{II}}{2} + K_2 \sin \alpha. \tag{3.92}$$

In other words, if the vertical angle of a point with an elevation $\alpha \approx 0°$ is measured with a theodolite with one reading index mark in two telescope positions, then half the difference between the vertical angles α_I and α_{II} gives the index error free from the influence of the eccentricity error. With the usual whole circle numbering the index error can be computed from the readings l_I and l_{II} measured to a point with an elevation angle $\alpha \approx 0°$ in two telescope positions:

$$\delta = \frac{360° - (l_I + l_{II})}{2}. \tag{3.93}$$

The tests for conditions 8 and 9 are similar to the tests of the horizontal circle and the horizontal reading device (Point 2.2.8.4). The vertical circle, however, cannot be brought into different positions during the repetitions of the measurement, therefore a prism of known refracting angle is fixed in front of one part of the telescope objective. Now the targets (collimators) placed at different elevation angles are observed firstly through the prism and then directly. In this way a constant angle can be measured at an arbitrary part of the vertical circle.

The fulfilment of conditions 8 and 9 is seldom tested, because the errors when they are not satisfied are generally less than those due to the external conditions of the measurement of elevation angles.

3.3 Instruments for Barometric Heighting

3.3.1 Basic Principles of Barometric Heighting

Atmospheric pressure decreases with elevation as well as with other parameters (meteorological conditions), consequently the atmospheric pressure at points at different heights gives information about the height differences between these points.

Due to the continuous change of atmospheric conditions, however, atmospheric pressure has a continuous but non-uniform change at the same observation point, and at points far from each other atmospheric conditions may be different at the same time. Therefore, it can be assumed only for relatively near points (at a distance of maximum 5–10 km) that atmospheric conditions are nearly identical at the same time. Consequently height above sea level cannot be determined directly by measuring atmospheric pressure.

If the height difference between two points P_1 and P_2, relatively not far from each other, is to be determined, and if for this purpose the atmospheric pressure (B_1 and B_2) as well as the temperature (t_1 and t_2) and the water vapour pressure (e_1 and e_2) are measured at the same instant, then the height difference between the points P_1 and P_2 can be computed in metres by means of the modified *Laplace* formula:

$$\Delta H = K'(1+\beta\cos 2\alpha)(1+\alpha t)\left(1+0.377\frac{e}{B}\right)\left(1+\frac{2}{R}H\right)\log\frac{B_1}{B_2}, \tag{3.94}$$

where

$$K' = \frac{0.76}{\mu}\,\frac{13.59593}{0.00129277}\,\frac{1}{1.00021} = 18{,}400$$

is the barometric constant (0.76 is the height in metres of the mercury column counter-balancing standard atmospheric pressure; 13.59593 is the density of mercury, and 0.00129277 the density of air with respect to water; 1.00021 the reduction coefficient with respect to carbon dioxide);

μ = 0.4342945, the modulus of common logarithme,
β = 0.00264, the constant in the formula for the force of standard gravity,
φ = the mean geographical latitude of the points P_1 and P_2,
α = 0.003665, the expansion coefficient of air for 1 °C,
t = $(t_1+t_2)/2$, the average air temperature,
e = $(e_1+e_2)/2$, the mean water vapour pressure of air,
B = $(B_1+B_2)/2$, the mean air pressure,
R = the radius of curvature of the terrestrial ellipsoid at the mid-point and in the direction of the line joining the two points,
H = the mean of the approximate heights above sea level of the points P_1 and P_2.

Requirements in practice allow the use of the average value of each of the quantities in the formula except air temperature and air pressure. In this case the simplified version of the modified *Laplace* formula can be used:

$$\Delta H = K(1+\alpha t)\log\frac{B_1}{B_2}, \tag{3.95}$$

where K = the barometric constant for the area in question (if e.g. $B = 720$ Torr, $e = 7.2$ Torr, $H = 500$ m, $\varphi = 47°$ are the average values, then $K = 18{,}469$); the other quantities are identical with those in Eq. (3.94).

It must be mentioned that the barometric constant K is basically different if the dimension of B is different.

On the basis of their operating principle there are three types of instruments for measuring atmospheric pressure: mercurial barometer, aneroid barometer and boiling-point thermometer or hypsometer. Aneroid barometers supplied with a linear height scale corresponding to air pressure are called altimeters.

In survey practice aneroid barometers are used in general, usually *precise barometers* reading to an accuracy of 0.1 to 0.2 Torr or 0.1 to 0.2 mbar. Relative heights of points lying in a small area and within 200 m of height difference can be determined by them with a reliability of ± 1 m to ± 2 m. Barometers with a reading accuracy of 0.005 to 0.03 Torr (0.01 to 0.02 mbar) are called *micro-barometers*. Relative heights determined by them have a reliability of ± 0.3 m to ± 0.8 m.

3.3.2 Mercury Barometers

3.3.2.1 The Construction of Mercury Barometers

The mercury barometer operates on the principle of a pair of scales. One scale is affected by the air pressure, the other by the pressure of the mercury column. The air pressure can be determined from the height of the mercury column counter-balancing it. There are three types of mercury barometers, depending on their construction.

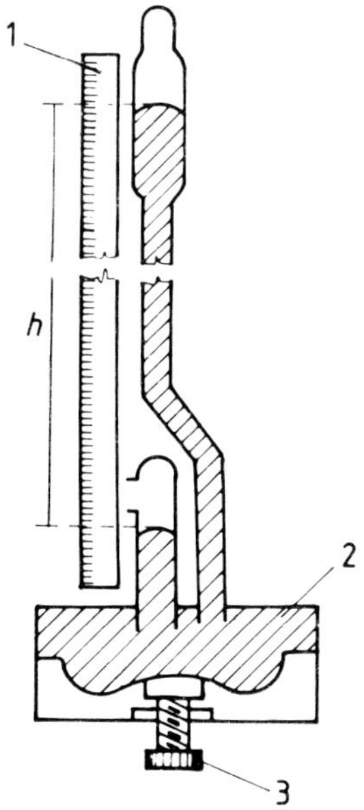

Figure 3.128. The siphon-cistern barometer

1. The *cistern barometer* consists of a cistern filled with mercury into which a glass tube with a closed upper end and open lower end is immersed and the glass tube itself is also filled with mercury. The pressure of the atmosphere acts on the surface of the mercury in the cistern and this is in balance with the mercury column in the glass tube. The height of the column above the surface of the mercury in the cistern can be read on the graduation beside the tube. When reading the starting or zero point of the graduation must be at the height of the mercury level in the cistern. This can be realised by raising or lowering either the graduation or the bottom of the cistern.

2. The *siphon barometer* is a vertical glass tube with its upper end closed and its lower end open and U-shaped. The tube is filled with mercury in such a way that a vacuum arises below the upper closed end. The height of the mercury above the free surface can be read on a mm scale.

3. The combined *siphon-cistern barometer (Fig. 3.128)* is similar to the previous type, but the lower U-shaped part of the tube is substituted by a cistern *2* with a movable bottom. Thus both surfaces of the mercury can be raised or lowered by means of a screw *3*. For repeated measurement the height of the mercury column can be read at different parts of the graduation *1*.

3.3.2.2 Reading Correction of Mercury Barometers

By international agreement the pressure of a 760 mm high mercury column at geographical latitude 45°, at sea level and at temperature 0 °C, has been taken as the standard atmospheric pressure of the physical atmosphere (1 atm = 760 Torr = 1.01325 bar). The height of the mercury column balancing the pressure of the atmosphere depends not only on the atmospheric pressure, but also on the gravitational acceleration which varies with height above sea level, on the temperature of the mercury and on capillary action in the tube. Therefore measured values must be reduced to a geographical latitude 45°, to sea level and to temperature 0 °C, and it must also be corrected for the capillary depression.

The reduction v_φ due to the geographical latitude φ is:

$$v_\varphi = -\beta B \cos 2\varphi, \tag{3.96}$$

the reduction v_H due to the height H above sea level is:

$$v_H = \frac{2BH}{R}, \tag{3.97}$$

and the reduction v_t due to the temperature t is:

$$v_t = -(\gamma - \alpha)Bt. \tag{3.98}$$

In these formulae $\beta = 0.00264$, B = the measured atmospheric pressure, R = the Earth radius, $\gamma = 18 \times 10^{-5}$ the thermal expansion coefficient of mercury, α = the thermal expansion coefficient of the material of the graduated bar.

The decrease v_k of the mercury level caused by capillary action *(capillary depression)*, which depends on the inner diameter d of the tube containing mercury and on the meniscus height m of the mercury surface *(Fig. 3.129)*, can either be taken from tables or determined from a comparison with a so-called standard barometer

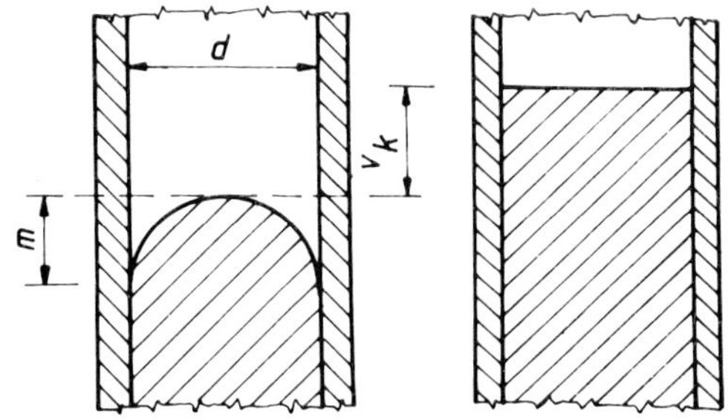

Figure 3.129. Capillary depression

which has a large tube diameter. If the diameter of the tube is greater than 20 mm, capillary depression can be neglected. Capillary depression can also be neglected with siphon barometers and siphon-cistern barometers if the diameter of the tube is everywhere the same.

Using the above corrections the reduced atmospheric pressure is:

$$B_0 = B + v_\varphi + v_H + v_t + v_k. \tag{3.99}$$

3.3.3 Aneroid Barometers

The operating principle of aneroid barometers is that the atmospheric pressure acting on the sensor element of the barometer is balanced by the tensile force of a spring or by the pressure of gas in a closed space. The variation of the atmospheric pressure can be determined by measuring the elastic deformation of the sensor element of the barometer. This deformation must be measured with an accuracy of 0.5 to 1 μm. The problem is solved in different types of instrument by mechanical magnification of the displacement, by measuring with a micrometer screw or by optical magnification.

3.3.3.1 Aneroids with an Elastic Membrane

Most of the currently produced aneroid barometers with an elastic membrane have basically the same construction as the aneroid manufactured by *Vidi* in 1847. In the *Vidi* aneroid barometer, improved by *Naudet (Fig. 3.130)*, the sensor element is a box *1* with a corrugated elastic top plate (membrane) which is evacuated to low

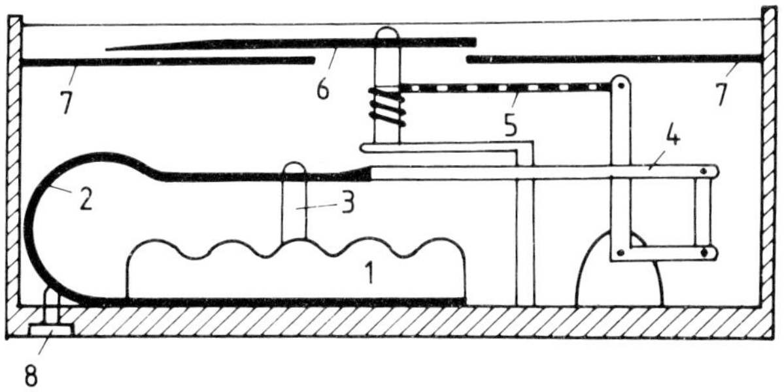

Figure 3.130. Naudet–Vidi barometer

pressure. The elasticity of the top plate is increased by the bent leaf spring *2* which continuously strains the plate through the pin *3* which is rigidly attached to it. The displacement of the top plate is mechanically magnified by the pivot mechanism *4* and transmitted through a chain *5* to the points *6*, by which the atmospheric pressure can be read to an accuracy of 1 Torr (by estimation to 0.1 Torr) on the graduation *7* attached to the casing. The pointer can be adjusted by the adjusting screw *8*. The aneroid is usually supplied with a thermometer in order to determine the reductions of the readings. Recently developed aneroids have automatic temperature compensation, thus no thermometer is used with them.

In the *Paulin* barometer *(Fig. 3.131)* the deformation of the two membrane-topped boxes *10* is transmitted without play or friction by the rigid arm *3*, which is mounted on a leaf spring *1* and a spiral spring *2*, to the torsion balance *4* which magnifies the movement and transforms it into a rotation of the pointer *5*. The deviation of the pointer *5*, caused by the change of atmospheric pressure, can be cancelled by the adjusting screw *6*. The eccentric disc *7* connected to this screw compensates the

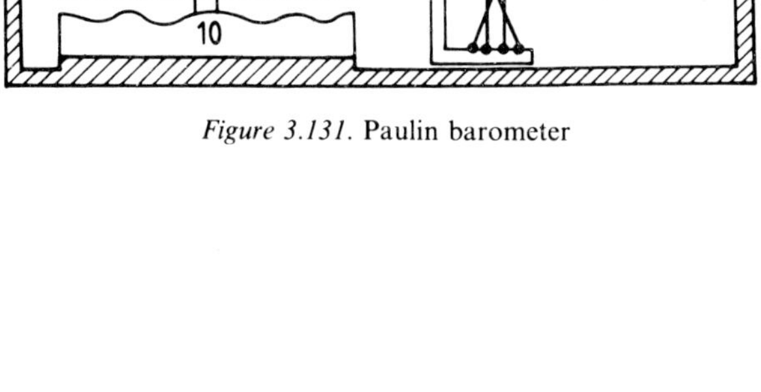

Figure 3.131. Paulin barometer

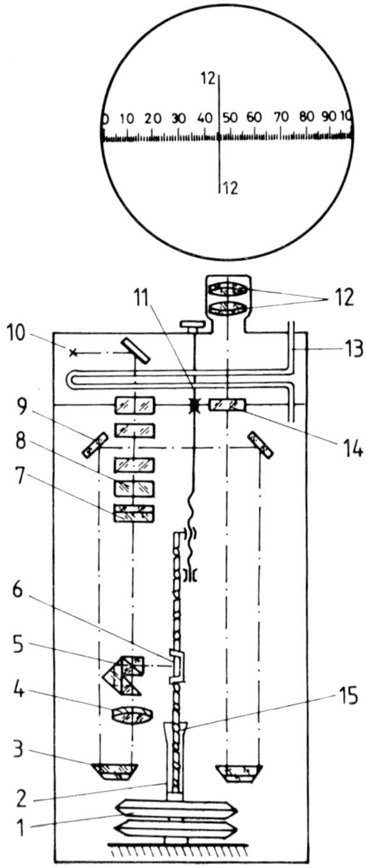

Figure 3.132. The micro-barometer *OMB-3P* (Soviet Union)

movement of the arm *3* by straining the spring *9*, by means of the lifting mechanism *8*. Another pointer *11* also rotates with the adjusting screw; by means of this either the difference of atmospheric pressure or the direct height difference can be measured on the graduated scale under the pointer *11*.

In the Soviet optical micro-barometer *OMB-3P (Fig. 3.132)* the sensor element is an aneroid component with two boxes *1* to which is fixed one end of the spiral spring *2*. A mirror *6* is mounted on the centre of the spring. The other end of the spring is connected to the instrument housing by the micrometer screw *11*; the measuring range of the barometer can be changed by this screw. The circular graduation of the glass plate *7*, consisting of 20 divisions, is illuminated by the light source *10* through an optical system. The light passing through the graduation goes through the prism system *5* onto the mirror *6*; it is reflected from this and deflected downwards by the prism system *5*, and meets the objective *4* which generates an image of the graduation *7* on the micrometer graduation *14* (of 100 divisions) after a multiple ray deflection. The micrometer graduation and one division line of the graduation *7*, after displacement by the mirror *6*, can be observed through the eyepiece *12*. Geometrical and optical parameters are chosen so that 0.03 to 0.04 mbar corresponds to one micrometer division. The accuracy of reading by estimation is 0.01 mbar. The barometer is placed in a Dewar vacuum flask (insulated thermoflask), whose interior is connected with free air only through the spirally bent tube *13*, hence it is well protected against sudden temperature changes and gusts of wind. Part of the spiral spring is within the cylindrical cistern *15* — in order to decrease its mechanical oscillations — and the cistern is filled with a special liquid. The inner temperature can be read on an internal thermometer of accuracy 0.1 °C by estimation. The reliability of the measured atmospheric pressure difference of up to 20 mbar is ± 0.02 mbar. The measuring range is between 60 and 80 mbar or 180 to 240 mbar with extension. The weight of the instrument is 3.1 kg.

3.3.3.2 Aneroids with an Elastic Tube

The sensor element of this aneroid is the *Bourdon*-tube of elliptical cross section and spiral shape. The distance between the ends of the tube changes if the air pressure in the tube changes due to the forces acting on the inner and outer surface of the spiral. The elastic deformation between the ends of the tube is measured by means of mechanical or other magnification.

The sensor element of the micro-barometer *Askania Gb5* is a cylindrical spiral *Bourdon*-tube with 9 turns *(Fig. 3.133, a)*; its upper end is attached to the torsion head *2*, which is rotatable by the micrometer screw *1*, while its lower end is attached to the metal tube *4* which is attached to the torsion axis *3*. The torsion axis itself is a steel wire strained by a spring and it is rigidly attached both to the torsion head and to the suspended tube. The suspended tube ends at its lower end in the disc *5*, the oscillations of which are damped by a magnet *6*. When the atmospheric pressure changes, the lower end of the Bourdon-tube rotates and the metal tube, which is rigidly connected to it, is rotated around the torsion axis. The rotation of the suspended tube, which is proportional to the change of atmospheric pressure, is determined optically by means of the mirror *7* by the method of autocollimation.

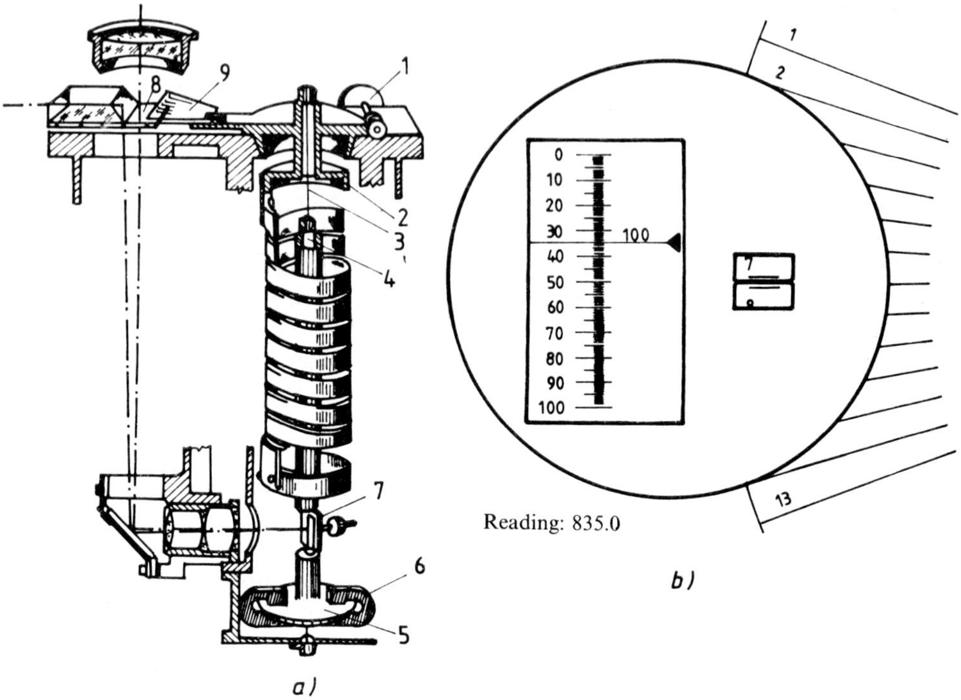

Figure 3.133. The micro-barometer Askania *Gb5* (Siemens AG, West-Berlin)

In the focal plane *8* of the reading device *(Fig. 3.133, b)* the moving autocollimation index, marked 0, 100, 200 or 300 appears between divisions 0 and 100 of the micrometer graduation, according to the swing of the mirror. The distance between the magnified images of adjacent divisions of the index is identical to the distance between the division lines 0 and 100 of the micrometer graduation. The greatest measurable swing of the mirror is 400 micrometer division. One division corresponds to 0.1 Torr and 0.01 Torr can be read by estimation, corresponding to a height difference of 10 cm. In order to increase the measuring range of the instrument, the torsion head can be rotated by means of the micrometer screw (see *Fig. 3.133, a)*. On the glass plate *9* connected to the torsion head there are numbered divisions from 1 to 30; one of them must be made to coincide with the index in the field of view *(Fig. 3.133, b)*, in which one of the autocollimation indices must also appear. Rotation of the torsion head by one division exactly corresponds to 100 micrometer units.

The reading consists of three parts:

— the number of the division line adjusted on the torsion head multiplied by 100:	700
— the number of the autocollimation index:	100
— and the reading of the autocollimation index on the micrometer graduation:	35.0
Total:	835.0

The total measuring range of the instrument is 300 Torr (400 mbar); this is divided into 30 subranges by turning the torsion head, adjacent sub-ranges always overlapping by 30 Torr. The range of atmospheric pressure corresponding to the instrumental range depends on the adjustment of the manufacturer. If the aneroid is adjusted for a range of 600 to 900 Torr (800 to 1200 mbar), then a basic atmospheric pressure 600 Torr (800 mbar) must be added to the value obtained when the reading is divided by 10. Hence in respect of *Fig. 3.133, b* the approximate atmospheric pressure is 683.5 Torr (911 mbar).

The aneroid is calibrated by the manufacturer for each sub-range in a pressure chamber and the calibration data are given in tables. The tabulated data also contain the errors of the index and refer to an instrument temperature of 20 °C. Due to the high precision of the barometer the temperature correction must be taken into account, in spite of the fact that the instrument is temperature compensated and therefore the temperature coefficient is small: 0.005 to 0.01 Torr/°C.

The instrument is very sensitive to a change of tilt of the torsion axis, therefore it must be adjusted before measurement by means of a built-in spirit level.

3.3.3.3 Aneroids with Compressible Gas

The operating principle of a compressible gas aneroid is that it measures the change of volume of a defined quantity of gas which varies with atmospheric pressure. The instrument developed by *Stribling Broding* and *Wilhelm* (USA) is of this type: the gas chamber is connected with the outside free air by a graduated capillary tube and a drop of oil moves in this tube in accordance with change of atmospheric pressure. The measuring reliability is ±0.03 Torr (0.0 mbar). The instrument is very sensitive to temperature, therefore much supplementary equipment is necessary; consequently it is not suitable for use in the field.

3.3.3.4 Reading Corrections of Aneroid Barometers

The readings B' obtained with aneroids must be given different corrections in order to obtain the true atmospheric pressure B. In surveying practice the following linear relation is usually used for the computation of atmospheric pressure:

$$B = B' + \delta_t(t - t_0) + \delta_B(B_0 - B') + \delta_i, \tag{3.100}$$

where

B' = the reading on the aneroid,
δ_t = the temperature coefficient of the aneroid,
t = the temperature of the aneroid,
t_0 = the temperature of calibration (usually 0 °C or 20 °C),
δ_B = the graduation coefficient of the instrument,
B_0 = the atmospheric pressure corresponding to a reference graduation line (usually 760 Torr or 1000 mbar),
δ_i = the index error $= B - B'$, if $t = t_0$ and $B' = B_0$.

The aneroid is usually calibrated in a pressure chamber or by comparing it with a mercury barometer, and the coefficients of Eq. (3.100) are determined from this.

The graduation correction is, however, often non-linear; in such cases the corrections of the different divisions are tabulated in the calibration certificate. The index error can also change as a result of mechanical effects occurring during transport from station to station. Therefore, it is advisable to determine the index error often by comparison with a mercury barometer or a hypsometer.

3.3.4 Hypsometers

With hypsometers the temperature of boiling water is measured. Since the boiling point depends on atmospheric pressure, the latter can be determined from the temperature measured. One tenth of a degree centigrade change in the boiling point corresponds to a 2.7 Torr (3.6 mbar) change of atmospheric pressure. The temperature of the boiling point can be determined by special thermometers with an accuracy of 0.01 °C. The boiling point thermometer (hypsometer) consists of a special metal vessel and a thermometer graduated in units of 0.1 °C *(Fig. 3.134)*. Distilled water must be put into the metal vessel and then the vessel together with the thermometer is placed in a metal box standing on three legs. The water is brought to boiling by means of an alcohol burner below the vessel. Only that part of the thermometer which is necessary for reading the temperature protrudes from the vessel, the greater part being protected by a metal cover. The atmospheric pressure corresponding to the boiling point is taken from a table, or the thermometer is graduated directly in units of atmospheric pressure.

The mean error of the determination of atmospheric pressure is about ± 0.2 Torr ($\approx \pm 0.3$ mbar).

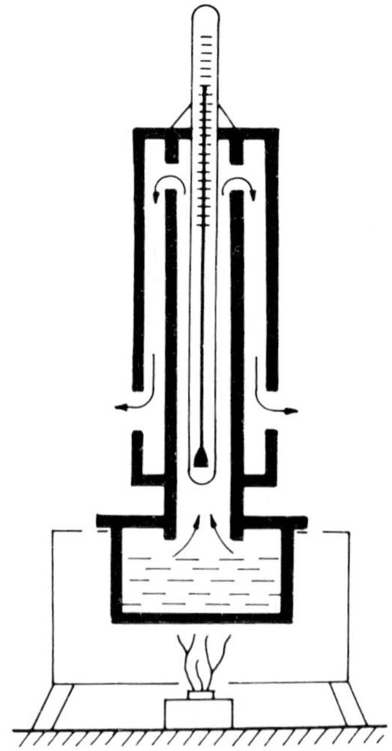

Figure 3.134. Hypsometer or boiling-point thermometer

4 Precision Instruments for Horizontal and Vertical Setting Out

4.1 Precision Instruments for Setting Out

The process of setting out can be reduced to establishing angles and distances in a manner similar to the basic operations of surveying. Consequently instruments for angle and distance measurement are equally suitable to carry out the practical tasks of setting out, and their applicability is determined by the accuracy that can be attained by them. The following instruments are classified as precision instruments:

— those which are manufactured so that they fulfil the special aims and often demanding requirements of setting out,

— those which fulfil the requirements either by means of the basic instrument (theodolite, level, etc.) or by means of the instrument and an attachable auxiliary device.

Precision instruments for setting out are always optical instruments. Using the line of sight of the telescope a certain part of a vertical or horizontal (sometimes sloping) plane is established with the required accuracy. Practical setting out can be made considerably easier with those instruments which visibly mark this line of sight (or a line parallel to it) at any point on it. These are the laser instruments, which are made either as separate instruments or as a laser device attachable to surveying instruments.

4.1.1 Alignment Instruments

Alignment is a method to determine the position of nearly vertical or linear construction, or rather a group of points well approximating a generally vertical plane. The main principle of the method is that a vertical (nearly parallel) plane, namely a reference plane, is established near to the group of points to be fixed and then the normal distances of the points from the reference plane are measured.

Any straight line in the vertical plane can be established by the line of sight of a theodolite telescope if the theodolite has been set up and the telescope is rotated around its horizontal axis into the slope direction in question. This principle is followed by the theodolite-like instruments for alignment measurement. Since these instruments are used to establish the vertical plane containing a straight line defined by two points, they need neither a horizontal nor a vertical circle graduation for the alignment. On certain instruments, however, a horizontal circle is also supplied for a possible orientation of the established straight line. In order to obtain a high accuracy in the measurement of normal distances, (at least 0.5 mm/100 m) the instruments have a telescope with large magnification ($G > 40$) and a very sensitive striding level that can be reversed on the horizontal axis. Otherwise their construction is identical with that of a theodolite.

The old-fashioned alignment instrument "Taspe" (Fennel, Kassel) has a 440 m long telescope with an 80 mm diameter objective lens and a magnification of 60 times. For setting the vertical axis vertical there is a circular bubble with a sensitivity of 8 sec of arc, while for levelling there is a reversible striding level with a sensitivity of 15 sec of arc. There is neither a horizontal nor a vertical circle. The weight of the instrument is 7.9 kg including the striding level.

The telescope of the alignment instrument manufactured by VEB Freiberger Präzisionsmechanik (Dresden) has a 65 times magnification *(Fig. 4.1)*. There is a circular bubble with a sensitivity of 8 sec of arc for setting the vertical axis vertical and a reversible striding level with a sensitivity of 20 sec of arc for levelling the horizontal axis. The basic instrument is supplied with several auxiliary devices; they are mainly used for more precise setting out of constructions and for deformation surveys.

Figure 4.1. Alignment instrument, VEB Freiberger Präzisionsmechanik (Dresden, GDR)

Breithaupt (Kassel) has marketed its alignment instrument under the name *NITAL*; it is recommended primarily for control surveying of hydro dams. The total equipment consists of the basic instrument, a fixed reference mark and movable (adjustable) targets *(Fig. 4.2)*. The telescope of the basic instrument has a 42× magnification and it can be rotated around its horizontal axis within a range of ±30 degrees. The telescope gives a reversed image and up to 200 m readings can be made with an accuracy of order mm, up to 450 m with an accuracy of order 10 mm. The shortest observing distance is 2 m. A spirit level with a sensitivity of 1 min of arc is used for setting up the instrument which is fitted with a forced centering device, and a striding level with a sensitivity of 20 sec of arc can be mounted on its horizontal axis. The targets are equipped with pin-type forced centring devices and their spirit levels have a sensitivity of 45 sec of arc. The movable target can be moved on a forced path by a spindle on the side towards the water through 50 mm and on the other side through 150 mm, and it can be fixed in position. Readings with 1 mm precision can be made

Figure 4.2. Instrument *NITAL*, Breithaupt (Kassel, FRG), for the control survey of hydro dams

directly on the graduation which must be set perpendicular to the line of sight, while the normal distance between the aligned target and the instrument station can be read directly on the graduated drum of the spindle with an accuracy of 0.1 mm (and 0.01 mm by estimation).

A considerable advantage of the alignment method is its simplicity and the rapid solution of the measurements, therefore time intervals between successive measurements (e.g. measurements of movements) can easily be decreased.

The test of alignment instruments must be made for the same inner geometrical conditions as for theodolites. First of all the error of the perpendicularity of the line of sight must be eliminated, in order that the effect of the collimation error should be negligible in the results of measurements observed only in one telescope position (as is often the case). The perpendicularity error of the horizontal axis is eliminated in practice by carefully levelling it by means of a striding level after setting up and approximate pointing. If e.g. the collimation error is $\varepsilon_k = 10$ sec of arc and the two points defining the plane of reference are at the same height, and if the distance to the observed point is 100 m, then the linear error:

$$\Delta_k = \varepsilon_k \left(\frac{1}{\cos \alpha} - 1 \right) \frac{100}{\rho} \tag{4.1}$$

occurs for a target observed with a telescope vertical angle of α. With the same conditions a horizontal axis tilt of $\varepsilon_H = 10$ sec of arc causes the linear error:

$$\Delta_H = \varepsilon_H \tan \alpha \frac{100}{\rho}. \tag{4.2}$$

Table 4.1 shows numerical results of Eqs (4.1) and (4.2) for some values of α. It may be seen from the table that the effect of the tilt of the horizontal axis on the results

Table 4.1. The values of Δ_K and Δ_H when $\varepsilon_K = \varepsilon_H = 10$ sec of arc

α°	5	10	15	20	30	40
Δ_K (mm)	0.0	0.1	0.2	0.3	0.8	1.5
Δ_H (mm)	0.4	0.9	1.3	1.8	2.8	4.1

of the measurements is several times higher than the effect of a collimation error of the same magnitude. Therefore the accuracy of the alignment method depends primarily on the exact adjustment of the horizontal axis.

Under favourable circumstances measurements of reliability ± 0.1 to ± 0.5 mm can be made with well-adjusted alignment instruments.

4.1.2 Laser Alignment Instruments and Devices

Instruments emitting a laser beam parallel to the line of sight of the telescope have recently come into general use. Since the laser beam makes the line of sight (or a line parallel to it) visible along its full length, the process of setting out can be carried out with simple manual positioning of the target, while with a traditional instrument the alignment process is directed from the instrument which can be difficult for a long distance. The use of the laser line of sight is especially favourable if e.g. staff readings must be made at various distant points along a reference line, or if the path of a moving target has to coincide with a reference line.

Special laser levels are used for vertical setting out, laser theodolites for aligning horizontal or slope lines and laser plummets for precision plumbing.

The basis of surveying instruments equipped with a laser is some type of conventional optical instrument. The laser can be attached to the basic instrument in different ways:

— the laser emitter (laser tube) can be mounted on the telescope of the instrument, in such a way that the axis of the emitted beam is parallel to the line of sight of the telescope,

— the laser tube is positioned independently of the telescope (e.g. fastened to one leg of the tripod) and the laser beam is introduced into the surveying telescope by means of fibre optics,

— the laser tube is rigidly connected to the telescope and the beam is conducted into the telescope through a prism system, thus providing a concentric exit.

Due to its bundling sharpness and beam density a gas laser is generally used in surveying instruments. For instance, a laser beam of 2 mm diameter, with 1 min of arc divergence ($\approx 2^c$), gives a light spot of about 30 mm diameter on a screen at a distance of 100 m. The divergence, however, can be further decreased by the telescopic optics and the centre of the light spot can be determined in general with a maximum error of 1 mm.

Laser theodolites are the laser instruments most frequently used for setting out. Laser theodolites are advantageously used for alignment measurements in the construction industry, for establishing straight lines and directions, for the control of the marking of inaccessible points on building constructions and on the ground surface (as e.g. on bluffs).

Wild (Heerbrugg) markets its *laser attachments* under the trade names *GLA1*, *GLA2* and *GLA3*; they can be attached equally to the theodolites *T1A*, *T2* and *T16*. *Figure 4.3* shows the adaptor *GLA3* attached to the theodolite *Wild T2*; it has a *Spectra Physics Model 120* laser beam source. The adaptor is composed of two parts: the laser tube and the support. The relative position of these two parts can be altered by means of adjusting screws so that the axis of the beam can be adjusted parallel to the line of sight of the telescope. The adaptor can be removed from the telescope (or fixed to it) by the simultaneous rotation of two clamping levers; this will not change the adjusted position of the laser beam. The adaptor also includes a spring support which inactivates the effect of the harmful torque of the adaptor on the horizontal axis in the case of steep pointings. An eyepiece with a bent ray path must be used for pointing to a theodolite provided with an adaptor, because otherwise the adaptor impairs the use of the telescope.

Wild (Heerbrugg) also produces laser eyepiece devices under the trade name *GLO* for the theodolites *T1*, *T16*, *T2* as well as for the theodolite *T3*. In addition, the laser eyepiece *GLO2* of this series also serves as an auxiliary device for the levels *NA2*, *NAK2*, *N3* and for the optical precision plummets *Wild NL* and *ZL*. The laser eyepiece can be fixed on the telescope with a bayonet joint after the removal of the standard eyepiece. A red beam is emitted by the laser tube which is attached to the leg of the tripod *(Fig. 4.4)*. This beam is conducted into the laser eyepiece by a flexible light

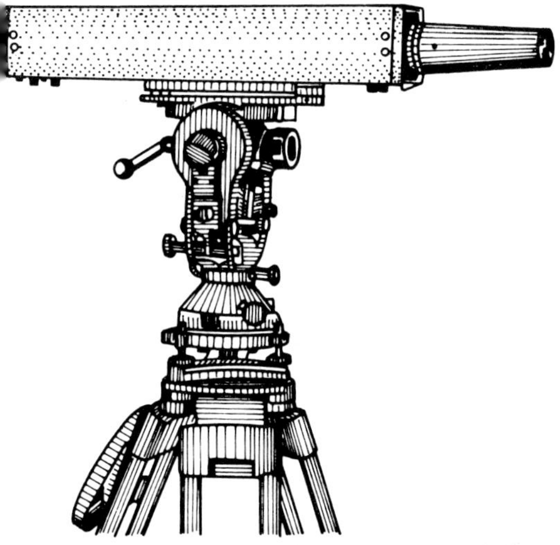

Figure 4.3. Laser adaptor on the *Wild T2* theodolite

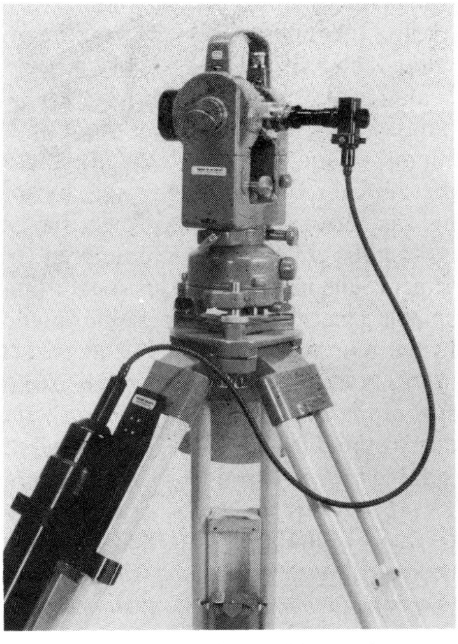

Figure 4.4. Laser eyepiece and laser tube *GLO2*, Wild (Heerbrugg, Switzerland), on the theodolite *Wild T2*

conductor in such a way that the image of the light spot of the beam, focussed by the focussing lens of the telescope, falls on the centre of the cross-hair. The intensity of the light spot can be controlled by a light filter placed in the eyepiece tube. When using a laser eyepiece the cross-hair plate of the instrument must be replaced by one through which the laser beam may pass undisturbed *(Fig. 4.5)*. As additional devices for the *adaptor GLO2 Wild* produces a converter, a telescope spirit level and an objective prism of 90°. The objective prism placed on the telescope can be rotated around the axis of the telescope and it deflects the laser beam in a plane perpendicular to the line of sight of the telescope. In this way the laser theodolite can be used as an optical plummet or for direction transfer. Particulars of the adaptor *GLO2* are listed in *Table 4.2*.

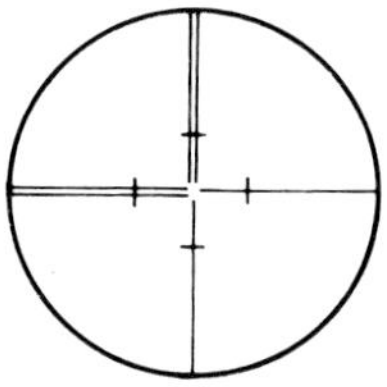

Figure 4.5. Cross-hairs in the laser eyepiece, Wild (Heerbrugg, Switzerland)

The laser theodolite *DKM 2-AL* (Kern, Aarau) is a combination of the one second (or arc) theodolite *DKM 2A* and the *Spectra Physics Gas-laser* attached to the tripod of the theodolite *(Fig. 4.6)*. The connection between the theodolite and the laser tube is also made by flexible fibre optics *(Fig. 4.7)*. The beam, emitted by the laser *1* as parallel rays, is focussed by the lens *2* onto the diameter of the fibre optics *3*. The beam, conducted through the fibre optics and the cross-hair plate *4*, is directed along the optical axis of the telescope by the dividing prism *5*. The high density of laser light is necessary only to project the cross-hair and give a well-defined image of the cross-hairs on the laser-illuminated field of the observed surface (e.g. on the target screen). The image of the cross-hair (and its environment), projected to the laser light spot, is generated in the cross-hair plane of the telescope. The light intensity is harmful to the human eye but it is decreased to an innocuous value by a light filter placed in front of the eyepiece. The cross-hair *6* of the telescope can be found on the eyepiece side of the dividing prism. When the laser is in operation, this cross-hair is invisible due to the diffused laser light propagating toward the eyepiece.

Kern (Aarau) produces its laser eyepiece *LO* for the theodolites *Kern KO-S*, *KL-S*, *K1-M* and *DKM 2-A* as well as for the level *GK 2-A*. The small laser eyepiece can be easily exchanged with the standard eyepiece of the basic instrument and does not need a counterweight on the telescope. Nevertheless, measurement even in two telescope positions can be easily carried out while the laser is attached to the eyepiece. If the telescope axis coincides well with the axis of the laser beam, there is no need for any extra adjustment of the laser beam with the focussing screw. Then, by focussing the image of the target in the telescope the laser beam is also focussed and a sharp

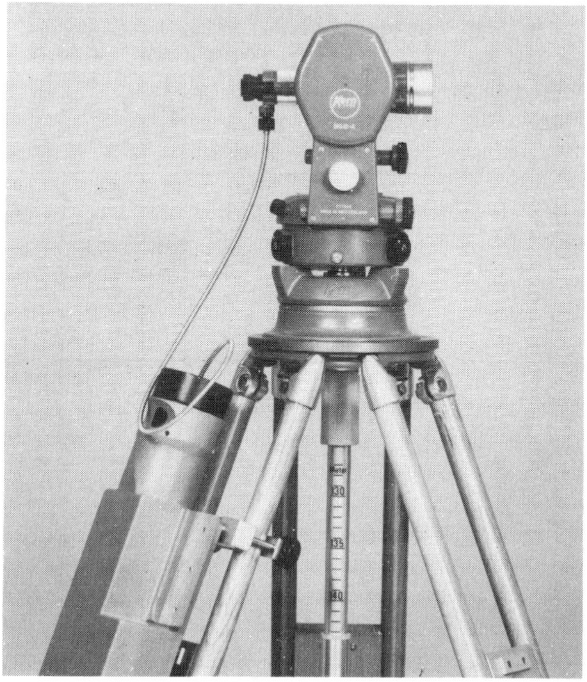

Figure 4.6. Laser theodolite *DKM 2-AL* with a laser tube, Kern (Aarau, Switzerland)

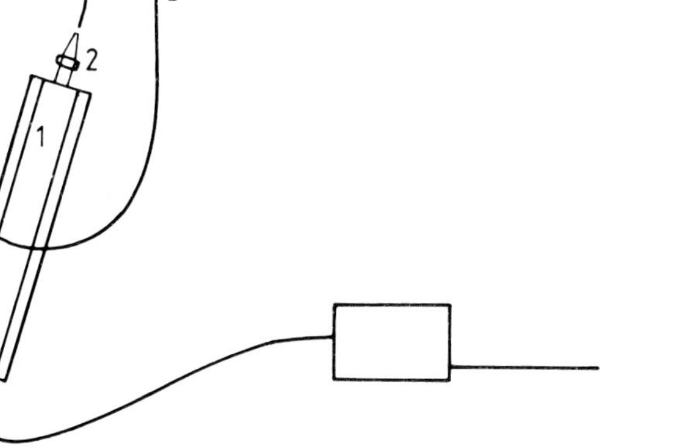

Figure 4.7. Optical train in the laser eyepiece for the *DKM 2-AL* (Kern, Aarau)

Table 4.2. Instrumensts for setting-out with a laser and laser eyepieces

Instrument name	DKM 2-AL	SLT 20	ZLA	LF 1	LFG 1	GL 0	L0
Manufacturing firm	Kern Co.	Sokkisha Ltd	Ertel-Werk	Jenoptik		Wild AG	Kern Co.
Instrument type	Laser theodolite	Laser theodolite	Optical plummet and level with laser	Level for setting out with laser	Instrument for setting out with laser	Laser eyepiece	Laser eyepiece
Basis instrument	DKM 2-A	SLT 20	ZLA	LF 1	LFG 1	T1, T16, T2, T3, NA2, NAK2, N3, NL, ZL	K0-S, K1-S, K1-M, DKM 2-A, GK2-A
Laser	Spectra-Physics gas-laser 120 T	He–Ne gas-laser	He–Ne gas-laser	Gas-laser	Gas-laser	He–Ne gas-laser	He–Ne gas-laser
Power supply	12 V	12 V	12 V	$12\ V^{-1+2}$		12 V	12 V
Output power	5 mW	2 mW	2 mW	0.8 mW		1 mW	1 mW
Laser spot diameter	3 mm at 50 m 7 mm at 100 m 14 mm at 200 m 20 mm at 300 m 27 mm at 400 m	3 mm at 50 m 7 mm at 100 m 17 mm at 200 m	12.6 mm (parallel up to 100 m)	About 15 mm at 200 m with a collimator of $\Gamma = 15$; about 30 at 1000 m with a collimator of $\Gamma = 30$		5 mm at 100 m 10 mm at 200 m 15 mm at 300 m	2 mm at 50 m 4 mm at 100 m 7 mm at 200 m 10 mm at 300 m
Range of measurement day-time night-time	400 m	200 m		About 0.5 km 1.0 km		200 m 400 m	300 m 600 m
Wavelength	632.8 nm	632.8 nm		633 nm		630 nm	

Current consumption				1.7 A	2.1 A	
Temperature interval	− 20 to + 50 °C	− 20 to + 50 °C		−25 to +45 °C	− 20 to + 50 °C	
Important accessories	Theodolite accessories, interchangable fibre optics (100 and 200 cm long)	Zenith eyepiece	Rotatable pentagonal prism detector	Cylindrical lenses for the laser light 1, 5, 10, 20 m per 100 m Targets illuminable Optical plummet for setting-up targets	Rectangular objective prism with tubular level	

light spot can be seen in the image of the observed target. The characteristics of the laser eyepiece *Kern LO* are listed in *Table 4.2.*

The *Sokkisha* (Tokyo, Japan) laser theodolite *SLT 20 (Fig. 4.8)* is produced mainly for measurement and setting out in underground work as well as in the construction industry, civil engineering and hydraulic engineering. The He–Ne gas-laser including the current supply is rigidly fixed to the telescope. The stabilization error of its automatic vertical index is ± 1 sec of arc, therefore the instrument can also be used to set out individual points in the horizontal plane. Its standard eyepiece, giving an upright image, can be exchange with a zenith eyepiece (further details are given in *Table 4.2*).

As in the instrument *SLT 20,* the Japanese laser alignment instruments *Nikon NT-2* and *Topcon TL-2* have a laser device mounted on the theodolite telescope whose beam is guided by prisms to coincide with the optical axis of the telescope.

The *Fennel* (Kassel) laser tube *FVL* is mounted perpendicularly to the line of sight of the telescope. The laser beam emitted from the centre of the tube has a propagation path which is 10 cm above the line of sight of the telescope. Its effective measuring range is 250 m and the diameter of the laser spot is 26 mm at this distance.

On the Fennel (Kassel) laser alignment instrument "Vermessungslaser", again the telescope is substituted by a laser tube. The tube can be rotated both horizontally and vertically. Both rotations can be read on metal circles with an accuracy of 0.1^{g}. A finder telescope mounted on the laser tube is used for sighting.

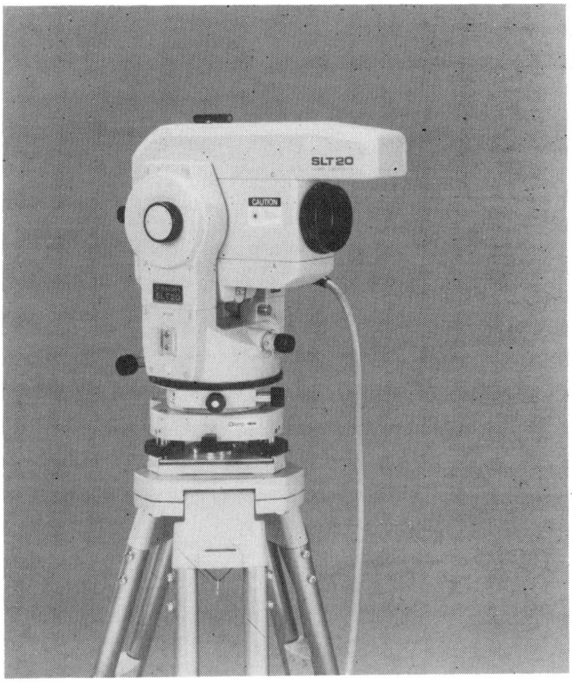

Figure 4.8. Laser theodolite *SLT 20,* Sokkisha (Tokyo, Japan)

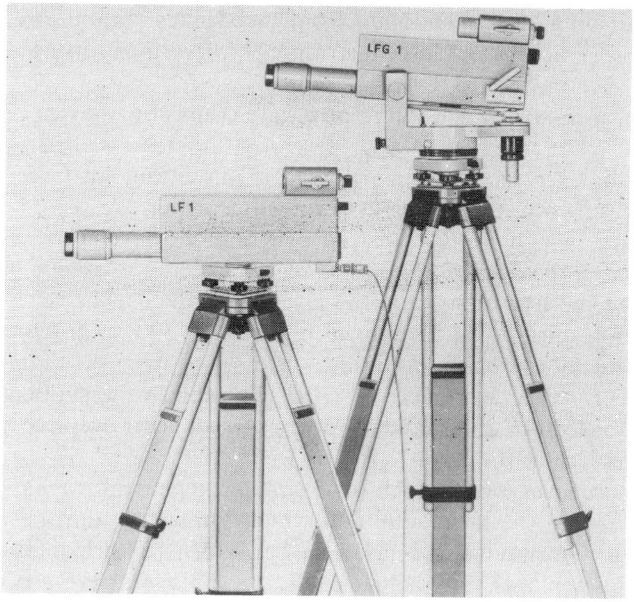

Figure 4.9. Laser alignment instruments *LF 1* and *LFG 1*, (VEB Jenoptik, Jena, GDR)

The *Jenoptik* (Jena) instruments *LF1* and *LFG1* (*Fig.* 4.9) are two laser alignment instruments with substantially the same construction. The gas-laser, the electronics, the finder telescope and the collimator for bundling the laser beam form a common unit. Both alignment instruments have a 34 mm pin that fits into the socket of every instrument base manufactured by Jenoptik. The instrument *LF 1* is constructed only for setting out levels, *LFG 1* also has a graduated horizontal circle and a measuring screw for setting the beam in a slope expressed as a percentage. Upon request the firm delivers the instrument with a collimator of $\Gamma = 15$ or $\Gamma = 30$ magnification power.

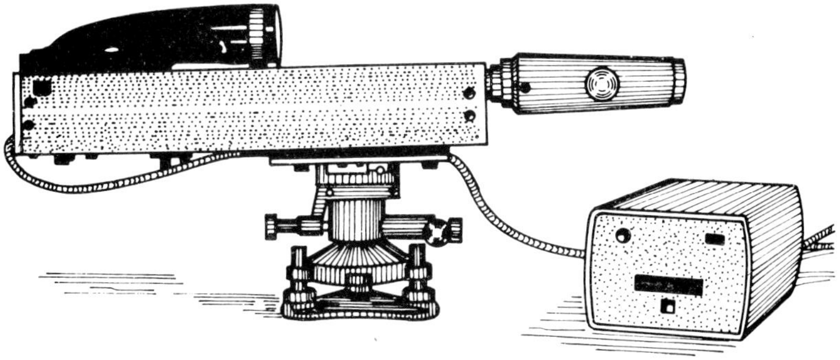

Figure 4.10. Laser telescope *Spectra Physics Model 133 P* on a Wild instrument base

As in distinguishing laser theodolites from theodolites with laser attachments, so one can specify several types of laser instruments for vertical setting out (setting out levels):

— laser adaptor mounted with an appropriate connecting device on the telescope of the level;

— independent laser telescope which has a pin fitting into the socket of the instrument base, utilizing the principle of forced centring;

— laser eyepiece which can be replaced by a standard eyepiece.

Figure 4.10 shows the *Spectra Physics Model 133 P* laser telescope placed in a *Wild* instrument base. The line of sight of the laser telescope can be set horizontal by a tubular bubble (or spirit level). Its maximum range is 300 m, at which distance the light spot diameter of the beam is 30 mm.

The Wild (Heerbrugg) laser eyepiece *GLO* can replace the standard eyepieces of the levels *Wild NA2, NAK2, N3*, and the Kern (Aarau) laser eyepiece *LO* can replace the standard eyepiece of the level *Kern GK 2-A*.

In practice those laser levels which have come into general use have a laser beam which is adjusted into the horizontal and which rotates around the vertical axis of the instrument. By continuous revolution the beam forms a horizontal reference plane (or a so-called laser plane). The height of the laser plane above ground points can be read on a graduated staff (e.g. on a levelling staff) within the effective range of the laser. Readings can be made by some-one standing near the staff. In certain cases special targets are used which can be moved up and down the staff. The target gives an acoustic (or optical) signal when the centre of the target (or that target point which is used as an index mark) moves into the laser plane.

The basic constructional element of the *Ertel* (München) laser instrument *ZLA* is a special laser instrument with a compensator *(Fig. 4.11)*. The laser beam emitted by the laser tube first passes through a magnetically damped pendulum and prism system, which directs the axis of the beam into the vertical with a stabilization accuracy of ± 4 sec of arc. The compensator also ensures that the exit point of the laser beam axis from the objective lens does not depend on the preliminary adjustment of the laser tube. The 65 cm long cylindrical device, of diameter 13 cm, is connected to the tripod by a ball joint. In order to carry out levelling tasks, a 90° deflecting prism can be placed in the path of the emitted beam. An adaptor head, with an electrically driven rotating pentaprism, can also be attached to the special laser instrument *ZLA*. The angular velocity of the prism, rotating around a vertical axis, can be chosen between 0 and 800 revolutions per minute. The laser beam passing through the rotating prism forms a continuous laser plane within the effective range of the laser. An additional device *ZLA* is an optical and acoustic detector, which can be moved upwards or downwards on the vertically held staff and which indicates the correct position of the index mark by a light or an acoustic signal. The levelling (or vertical alignment) accuracy obtained with a special laser instrument acting with a laser bundle of 12.6 mm diameter is ± 3 mm/100 m.

A similar laser setting-out instrument is produced by AGA in Sweden under the name *Geoplan* (see *Fig. 3.90*).

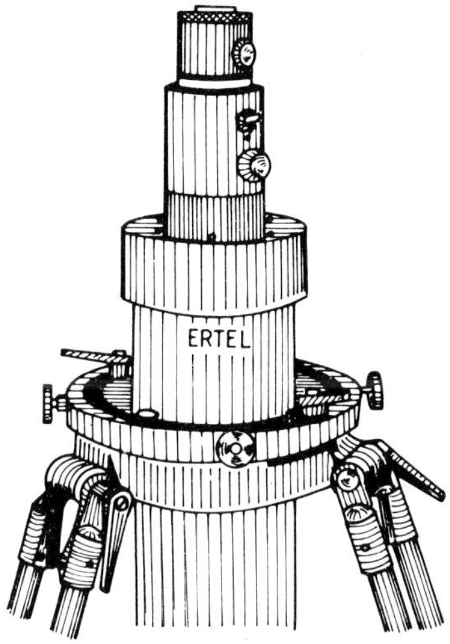

Figure 4.11. Special laser instrument *ZLA*, Ertel (München, FRG), for vertical setting-out, with attached rotating prism

4.2 Optical Precision Plummets

In the course of practical surveying (building construction, civil engineering, mine surveying) it is often necessary to set out points with high accuracy at great heights or depths below the ground surface.

The mechanical tools of vertical projection are plumb bobs with a suitable mass, the instruments are optical plummets. From optical plummets used for vertical setting out those used for great height differences, which in general require a coordinate transfer within some mm, are called optical precision plummets. The optical setting out of long vertical distances is also possible by means of surveying instruments (theodolites, levels) where the vertical section of the line of sight is generated by means of a deviating prism placed in front of the telescope of the instrument (e.g. auxiliary eyepiece, objective prism).

The main construction element of optical precision plummets used primarily (or solely) for setting out vertical straight lines is a geodetic telescope usually set vertical. The telescope can be rotated around a central vertical axis embedded in the instrument base. The majority of optical precision plummets can only be used for setting out one direction: either the zenith or the nadir. A few such plummets, however, are capable of setting out both directions. There are two types of construction for such instruments:

— the observation direction is determined by a deviating prism, which is centred on the vertical axis,

— two concentric telescopes are mounted on a common instrument base, one for zenith, the other for nadir observation.

Optical plummets using a spirit level and a compensator can be categorized separately. Firstly the line of sight is adjusted into the vertical by a spirit level. Secondly the approximately vertical line of sight is automatically stabilized to be exactly vertical.

The majority of optical plummets with a compensator bring the line of sight into the vertical only in one plane, namely in the so-called plane of compensation. Therefore the vertical through a point is set out with such instruments as the intersecting line of two vertical planes at right angles to each other. Recently such compensated optical precision plummets have been constructed with a dual compensator which simultaneously sets out both vertical planes of sight. In other words, when the vertical axis is set approximately vertical by a spirit level of low sensitivity (usually a circular bubble), then the line of sight of the telescope is automatically vertical in all telescope positions.

As with level setting-out instruments which incorporate a laser, optical plummets can also be fitted with a laser. After levelling the instrument, the light spot of the laser beam can be read or measured on a horizontal signal target which is in the horizontal plane through the light spot. When using a light-sensitive target the result can be obtained as a digital output by means of electronics, and this can be the input of a computer or automatic plotting machine.

The *Wild* (Heerbrugg) optical plummet *ZNL (Fig. 4.12)*, controlled by a spirit level, is suitable for setting out both the zenith and the nadir. Its telescope is horizontal and the one projecting direction can be changed to the other by switching a rectangular deviating prism. This prism is concentric with the vertical axis, hence it bends the line of sight as if it were an extension of the axis. The telescope has 10 × magnification,

Figure 4.12. The optical plummet *ZNL*, Wild (Heerbrugg, Switzerland), with a spirit level

the minimum observing distance is 0.6 m. The tubular level adjusting the line of sight has a sensitivity of 20 sec of arc. According to the technical data given by the factory, the relative error of a projection made with the optical plummet *ZNL* is 1 : 30,000.

Breithaupt (Kassel) produces a nadir plummet under the name *TELIM*, in which the nadir can be set out by electric illumination *(Fig. 4.13)*. The objective is in the lower part of the vertical telescope, consequently the eyepiece must be viewed from above. An auxiliary eyepiece is provided as an accessory of the instrument; it turns the line of sight through 90° thus enabling observation from the side. A circle graduated every 30° rotates together with the telescope, enabling the projection planes to be oriented. The two tubular levels, of sensitivity 20 sec of arc, are attached to the graduated circle with their axes perpendicular to each other. The magnification power of the telescope is 40 ×, the minimum projection distance 2 m. Targets of dimensions 300 × 300 mm or 360 × 360 mm are delivered by the factory for this instrument. The former is divided into 1 cm bands, the latter into 2 cm bands, and they can be used up to a distance of 200 m and 400 m respectively.

Figure 4.13. Nadir plummet *TELIM*, Breithaupt (Kassel, FRG)

Figure 4.14 shows the *Fennel* (Kassel) nadir plummet *FL1* on the lateral tribrach arrangement used in mining survey. The telescope construction is similar to that of *TELIM*. The tubular level used for the adjustment of the line of sight has a sensitivity of 20 sec of arc.

An accuracy of setting out of 1 mm/100 m can be attained with the Breithaupt *TELIM*, as well as with the *Fennel FL1*.

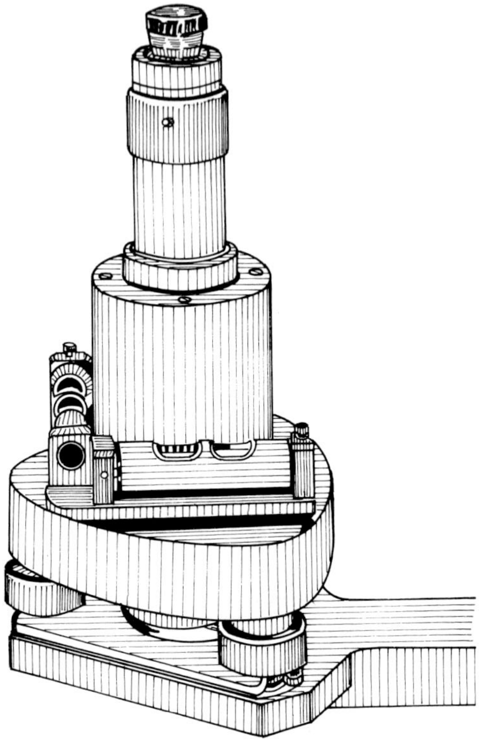

Figure 4.14. Nadir plummet *FL1,* Fennel (Kassel, FRG)

Figure 4.15. Optical precision plummet *OL,* Kern (Aarau, Switzerland), which gives the directions of both the zenith and the nadir

The optical system in the *Kern OL* (Aarau) optical precision plummet *(Fig. 4.15)* consists of two telescopes which are independent of each other. Both have their ray paths bent vertical and made concentric with respect to each other, and they have a 22.5 × magnification. The diameter of the telescope field of view at 100 m is 3 m. The minimum observing distance is 0.8 m. Kern supplies the *OL* plummets either with a tubular level of sensitivity 20 sec of arc, or with a split bubble of sensitivity 30 sec of arc. For setting out the former gives an accuracy of ±2 mm/100 m, the latter ±1 mm/100 m.

The construction of the compensators in optical plummets is the same as that of the compensators in levels. Thus the compensator of the *Jenoptik PZL* (VEB Jenoptik, Jena) optical precision plummet *(Fig. 4.16)* is the same as that of the *Jenoptik Ni 007* precise level (see *Fig. 3.84*). If the vertical axis of *PZL* has been set approximately vertical, the pendulum compensator automatically sets out a correct vertical plane perpendicular to the compensation plane *(Fig. 4.17)*. Since compensation is made only in one vertical plane, the vertical above the point of a set-up can be set out as

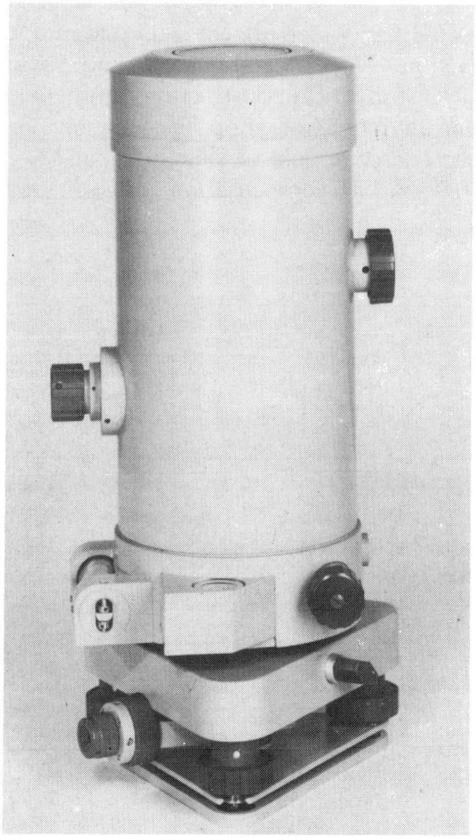

Figure 4.16. The precision zenith plummet *PZL*, Jenoptik (VEB Jenoptik, Jena, GDR), with a compensator

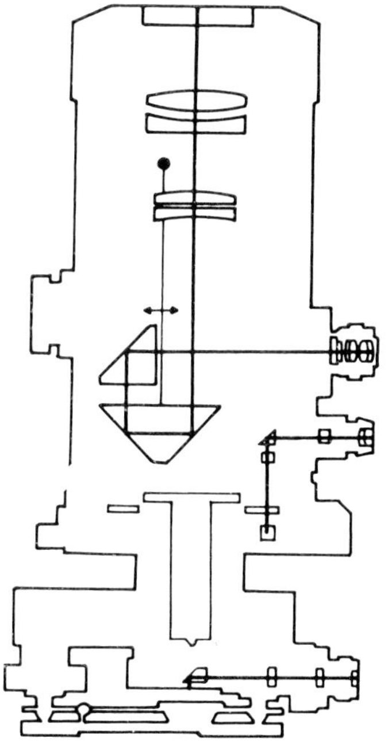

Figure 4.17. Diagrammatic sketch of the optical train of the optical precision plummet *PZL*

the intersection line of two planes perpendicular to each other. The magnification of the telescope of *PZL* is 31.5 ×, its field of view 1.3° and the air-damped compensator comes to rest within 1 sec. The maximum stabilization error of the compensation plane is ±15 sec of arc. The instrument has a circular bubble of sensitivity 8 min of arc for setting up the instrument, and a tubular level of sensitivity 30 sec of arc whose axis is perpendicular to the compensation plane. The instrument is manufactured with a pin-bush forced centring device, standardized for all Jenoptik instruments. An optical plummet of 2.8 × magnification and a prism rotate together with the optical precision plummet, the former being for setting up over a point. The reliability of measurement is ±1 mm/100 m.

The optical precision plummet *Watts Autoplumb* (Rank Precision Ltd, Great-Britain) can also be used simultaneously as a zenith and a nadir plummet. The compensator of the zenith plummet is the same as that of the *Autoset-level* (see *Fig. 3.81*). The line of sight is made vertical by the compensator only in one vertical plane. The accuracy of setting out in this compensation plane is ±0.5 mm/100 m. The line of sight of the nadir plummet can be adjusted into the vertical by a tubular level of sensitivity 45 sec of arc.

Wild (Heerbrugg) makes two instruments, both of which are compensated in both vertical planes, denoted *ZL* for zenith projection *(Fig. 4.18* and *Table 3)* and *NL* for nadir projection. The two instruments, of almost identical construction, can be built together on request into a single instrument in any Wild workshop. The magnification of the telescope is 24 ×, the diameter of the telescope field of view is 3.2 m/100 m and the minimum observing distance is 0.9 m. The compensation range in both

Table 4.3. Optical

Type	Manufacturing firm	Projection direction	Direction setting device	Telescope		
				magnification	diameter of objective lens (mm)	minimum sighting distance (m)
ZNL	Wild AG	zenith-nadir	spirit-level	10	12	0.6
OL	Kern Co.	zenith-nadir	spirit-level	22.5	30	0.8
OL	Kern Co.	zenith-nadir	spirit-level	22.5	30	0.8
TELIM	F. W. Breithaupt	nadir	spirit-level	40	50	2.0
Ferlo	Otto Fennel GmbH	nadir	spirit-level	30	40	1.6
ZBL	Wild AG	zenith	spirit-level	5	8	0.3
FL1	Otto Fennel GmbH	nadir	spirit-level			
Autoplumb	Rank Precision Ind.	nadir	spirit-level	17	25	0.6
		zenith	compensator	30	40	1.8
ZL	Wild AG	zenith	compensator	24	36	0.9
NL	Wild AG	nadir	compensator	24	30	0.9
PZL	Jenoptik	zenith	compensator	31.5	40	2.2
ZLA	Ertel Werke	zenith	compensator			
OZP	USSR	zenith	compensator	31	34	3.0

compensation planes is ± 10 min of arc, and the stabilization accuracy is ± 0.3 sec of arc. The weight of the ZL instrument is 3.1 kg, that of the NL 3.0 kg. The two precision plummets assure an accuracy of ± 0.5 mm/100 m for setting out with repeated readings in diametrically opposite positions. The *standard eyepieces* of both *Wild* ZL and NL, which have bayonet connections, can be substituted by an auxiliary eyepiece attachable in front of the eyelens which provides a 90° deflection for the

Figure 4.18. Optical precision plummet ZL, Wild (Heerbrugg, Switzerland), with a compensator

precision plummets

Spirit-levels		Mass		Projection accuracy of a single setting		Reference to figure
tubular	circular	instrument (kg)	box (kg)			
20″		2.6	2.2	1 : 30,000	prism with changeable projecting direction	4.12
20″		3.7	1.8	±2 mm/100 m	two telescopes	4.15
30″		3.7	1.8	±1 mm/100 m	coincidence spirit-level	
20″		7.1	6.5	±1—2.5 mm/100 m		4.13
20″		3.2	2.8	1 : 100,000		
60″		1.5	0.5	1 : 5000		
20″				±1 mm/100 m		4.14
45″					two telescopes	
	20′			1 : 200,000		
	4′	3.1	2.2	±0.5 mm/100 m	compensating in space	4.18
	4′	3.0	2.2	±0.5 mm/100 m	compensating in space	
	8′	4.8	2.8	±1 mm/100 m		4.16
				±3 mm/100 m	laser-determined line of sight	4.11
	10′	5.0		1 : 10,000		

collimation axis, or by a *laser eyepiece* of the type *GLO 2*. The optical precision plummet with a laser eyepiece can be used very advantageously for surveys and setting-out for buildings as well as for supervision of construction.

Tilting or compensator levels can also be used as optical precision plummets if a prism, deflecting the line of sight by 90° into the vertical, is attached in front of the telescope. Such an objective prism is usually made by all firms manufacturing surveying instruments at least for one of their types of level. The *objective prism* made for the *Ni 2* compensator level (Opton, Oberkochen) is shown as an example in *Fig. 4.19*. The vertical section of the line of sight of levels with an attached objective prism is always eccentric. The eccentricity effect on the result of a coordinate transfer can be eliminated either by a reduction of eccentricity (by computation) or by a measuring method (measurement repeated in diametrically opposite telescope positions).

In optical precision plummets the spirit level — or the compensator as the case may be — must be carefully tested and adjusted. The test must be further extended for both the concentric and the parallel positions (with respect to the vertical axis) of the line of sight.

The in-built eccentricity of optical precision plummets is generally very small and can be neglected (e.g. in the case of the instrument *Kern OL* it is a maximum of 0.02 mm). However the coordinate transfer error occurring if the line of sight and the vertical axis are not parallel can depend on the position of the focussing lens. This error can be eliminated by the measuring method (by taking the mean value of diametrically opposite readings). Therefore, with instruments compensating only in one vertical plane, it is advisable to make repeated readings (or repeated optical

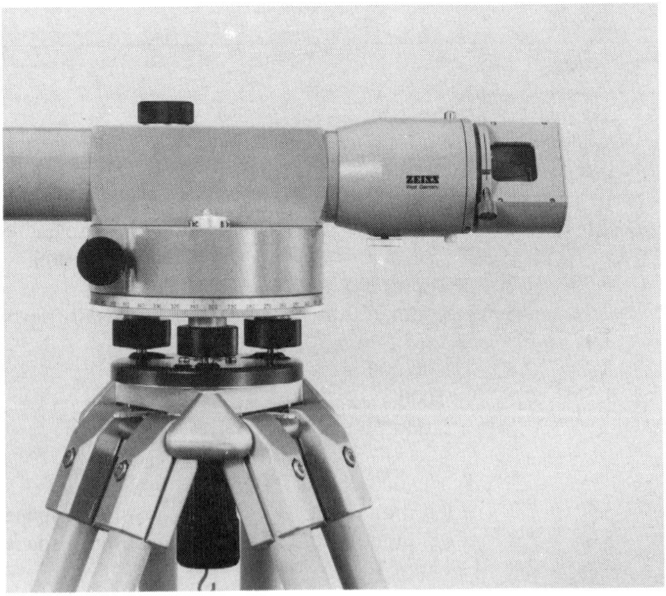

Figure 4.19. Adaptor prism for optical plumbing on the level *Opton Ni 2* (Oberkochen, FRG)

setting-out) in telescope positions reversed by 180° (200^g) for setting-out in both the two mutually perpendicular vertical planes. Setting out or coordinate transfer made this way can only be affected by a systematic error caused by a possible tilt of the vertical axis, and this can also be influenced by the random error of a wobble of this axis. Thus a tilt of the vertical axis can be the deciding factor for the frustration felt if a low accuracy is obtained which otherwise is not characteristic for optical precision plummets. With optical plummets using a compensator, a systematic tilt of the line of sight can occur due to the instability of the compensator. According to practical experience, however, this source of error does not significantly affect the results of the measurement.

5 Instruments and Equipment for Distance Measurement

Both direct and indirect methods are used for distance measurement in surveying. Instruments and equipment are correspondingly divided into two main groups: instruments for direct measurement (or length measurement) and instruments and devices for indirect measurement (or distance measurement).

The basic principle of measurement by length measuring devices is that a basic length from a zero mark to an end mark is directly compared with the distance to be measured. The basic length marked on the measuring device is usually an integral multiple of the measuring unit of length measurement.

Instruments for distance measurement can be again divided into two groups: distance meters operating on geometrical (optical) pricniples and those operating on the principle of wave-physics (electronic or physical distance meters).

In the case of the geometrical distance meter the determination of the distance is generally reduced to the solution of a triangle. Electronic distance meters use either the phenomenon of light interference or electromagnetic waves; they are described in the second part of the book.

In a wider sense one can speak of distance measurement in the case of an indirect distance determination, when the distance is computed from the measured data of an established geometric framework in the field. In this case instruments for measuring angles are also used as tools for distance determination.

The internationally accepted unit of length is the metre. The reference standard of the metre used to be an iridium-platinum bar at Sèvres near Paris. The metre is defined as the distance between the centre marks of the groups of marks at the two ends of the bar, with the bar in a horizontal position and at a temperature of 0 °C. Copies of the original reference standard metre are in the possession of the member-countries of the metre convention. The length of a metre bar, however, changes with time to a degree that cannot be predicted — presumably due to the slow change in the position of the atoms as a result of internal contraction. Therefore, the metre was defined at the 11th International Conference on Weights and Measures as a multiple of the constant wavelength of an exactly defined radiation. According to the definition:

$$1\ \text{m} = 1{,}650{,}763.73\ \lambda,$$

where λ is the wavelength in vacuo of the radiation emitted by krypton of mass number 86 for the transition between energy levels $2p_{10}$ and $5d_5$.

A new definition was accepted by the General Committee of Weight and Measure in 1983, according to which one metre is the distance which light travels in vacuo during 1/299,792,458 s.

5.1 Instruments for Length Measurement

5.1.1 Simple Equipment and Methods for Length Measurement

Long distances can be adequately measured from the observed time of travel by foot, knowing the speed. A simple method of measuring shorter distances is the determination of the distance from the number of paces. According to *Jordan* the length of a natural pace is equal to half the height of a person to eyelevel, and this is a good approximation. The result is even more accurate if the length of the pace is determined from the number taken to traverse a longer known distance.

The *hodometer* and *field caliper* are worth mentioning among simple length-measuring equipment. Both are used mainly in agriculture (horticulture) for the determination of approximate lengths. The hodometer *(Fig. 5.1)* is made of wood or metal and supplied with a revolution counter and a handle. Its basic length is the length of one revolution (usually 1 or 2 m). The field caliper is usually made from hard wood (or light metal) and has a defined leg length, but there are field calipers with a variable leg length.

Figure 5.1. Measuring wheel *M10* of Fennel (Kassel) and its carrying bag *M 10K*

The practical accuracy of both the hodometer and the field caliper is strongly affected by the ground conditions (quality, flatness, hardness, etc.), so that they can only be used for approximate measurement.

5.1.2 The Surveyor's Equipment for Length Measurement

5.1.2.1 The Measuring Bar

The 3 to 5 m long measuring bars used for length measurement in surveying are made from dry soft-wood. In order to prevent moisture absorption, the wood is initially impregnated (with oil, creosote, paraffin or zinc sulphate). The shape of their

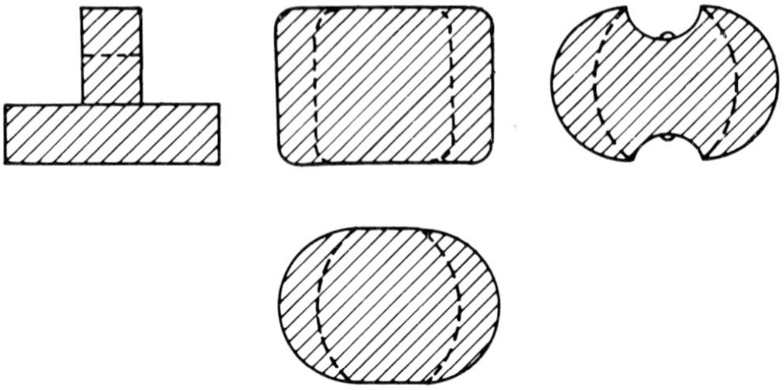

Figure 5.2. Cross sections of the central part of measuring bars

cross section has been changed in the course of time *(Fig. 5.2)*; 5 m long bars now usually have an oval cross-section. The major axis of the ellipse is about 5 cm along the central part of the bar, and about 3 cm at its ends. The bars are covered by two coats of oil paint. Measuring bars are usually used in pairs with alternating colour combinations at metre intervals: black and white on one, red and white on the other. Steel plates are fixed to the ends of the bars. The basic length is defined by the end-surfaces of the steel plates (bars with end-plates), by an index mark on the plates (bars with end marks) or by the edges of the wedge-shaped plates (bars with end-edges) *(Fig. 5.3)*. When measuring by means of bars with end-edges, a horizontal edge should be placed against a vertical one and a vertical against a horizontal one.

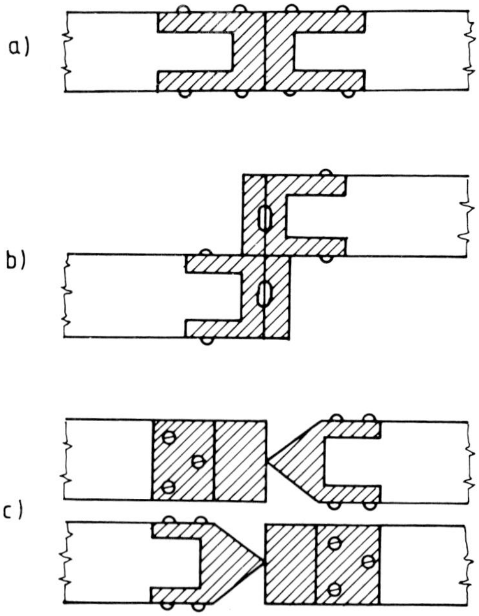

Figure 5.3. End-marks of measuring bars: *a)* end plate; *b)* end mark; *c)* end edge

The thermal expansion coefficient of wooden measuring bars is small (3×10^{-6} to 9×10^{-6}), consequently the change of length due to thermal expansion is usually neglected in practice. The effect of humidity can be considerably, however; one can calculate a change of length of between +0.6 mm/m and −0.3 mm/m, depending on the moisture absorption.

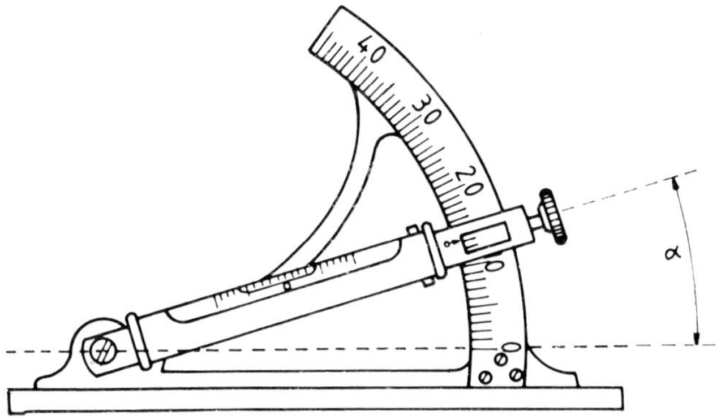

Figure 5.4. Bar clinometer with a spirit level

The horizontal projection of the slope length measured with measuring bars is deduced from the inclinations of the different bar positions, which are measured with a clinometer. With the bar-clinometer with a spirit level, the lever containing the index-mark is brought into the horizontal by means of a spirit level *(Fig. 5.4)*. When the index mark reads zero the adjusted axis of the spirit level (i.e. the tangent through the zero graduation) is parallel to the plane of the base of the clinometer. On the bar-clinometer with a pendulum, the lever of the index-mark is substituted by the cord of the pendulum which, when at rest, defines the vertical when it crosses the centre of the graduated circle. If the base plane is horizontal, the index-cord points towards the zero of the circle graduation.

The horizontal projection length l' of a bar of length l for a slope angle α is:

$$l' = l \cos \alpha, \tag{5.1}$$

and the projection reduction is:

$$\Delta l = l(1 - \cos \alpha). \tag{5.2}$$

The accuracy of the reduction computed from the inclination depends on both the accuracy of the length measurement and the angle of inclination. If a relative accuracy of $1/v = \Delta l/l$ is sought, then the inclination α must be measured with a mean square error:

$$m_\alpha = \frac{\rho}{\sin \alpha} \frac{1}{v}, \tag{5.3}$$

where ρ is the analytical angle unit (1 radian).

If, for example, a relative accuracy $1/v = 1/5000$ is required from the length measurement and the average slope of the field is 10° ($\approx 11^g$), then the angle α must be measured with an accuracy:

$$m_\alpha = \frac{3438'}{\sin 10^\circ}\,\frac{1}{5000} = \pm 4'.$$

Comparators are used to determine the basic length of the measuring bar. The comparator is a timber balk or iron bar somewhat longer than the length of the measuring bar; it is placed horizontally on a suitable support. One of the horizontal and vertical steel edges at both ends of the comparator can be moved along the length of the comparator after slackening the fixing screw. The distance between the edges is chosen so that the length between them is greater than that of the bar but small enough to be measurable by means of geodetic wedges *(Fig. 5.5)*.

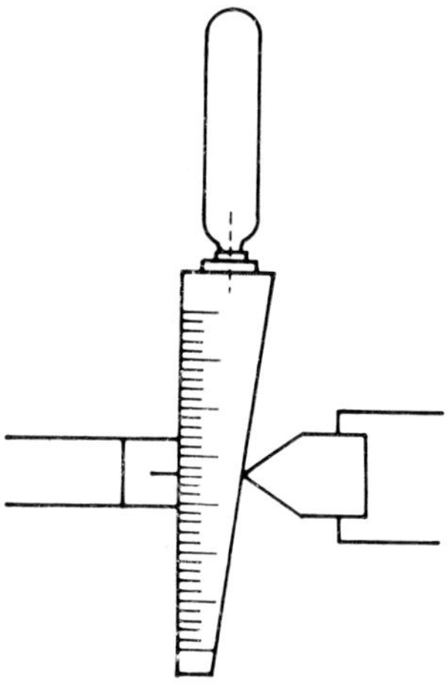

Figure 5.5. The geodetic wedge and its application to the measurement of small distances

The geodetic wedge is an aid for the precise measurement of small distances; it is made of steel. The edges of its surface, one of which is graduated, converge at a small angle (usually $5^\circ\,42'\,38'' = 6^g\,34^c\,51^{cc}$).The graduation unit is usually 1 mm, so that the convergence of 1/10 gives units of 0.1 mm along the distance to be measured. Thus distances can be measured with a mean square error of ± 0.1 mm. The width of the wedge at the different graduation lines is given in its calibration certificate. For the calibration of the edge-distance of a comparator a so-called standard metre-pair is used: this is a pair of steel metre bars with edge ends (usually of cross section 10×10 mm). These standard bars are previously calibrated at the Institute of Weights and Measures and the result of the calibration is expressed in equations of the following form:

$$M_1 = 1\,[\mathrm{m}] + c_1\,[\mathrm{mm}] + \alpha(t - t_0)\,[\mathrm{mm}]$$

$$M_2 = 1\,[\mathrm{m}] + c_2\,[\mathrm{mm}] + \alpha(t - t_0)\,[\mathrm{mm}]. \qquad (5.4)$$

In the equations c_1 and c_2 are the deviations of the two metre bars from the legal metre with an accuracy of 0.01 mm at the calibration temperature t_0, t is the temperature at the comparison and α is the thermal expansion coefficient of the material of the bar.

In the course of the calibration of a bar, firstly the edge distance on the comparator is measured by means of the standard metre-pair. At the same time the temperature of the bars (in simpler cases the air temperature) is also measured. The lengths M_1 and M_2 of the standard bars during calibration are computed from Eq. (5.4), and then the distance b between the edges of the comparator is obtained from:

$$b = n_1 M_1 + n_2 M_2 + \delta, \tag{5.5}$$

where n_1 and n_2 are the number of total bar lengths, and δ is the remaining distance measured by the geodetic wedge. When the comparator length is known, the bar to be measured is placed in the comparator, its horizontal edge is placed against the vertical edge of the comparator and the surplus distance is measured by means of the geodetic wedge. The length of the bar is:

$$l_0 = b - \Delta .$$

It is worth mentioning that the length of the comparator and the length of the bar are both determined as the mean of two measurements. In this case the first measurement of the comparator length should be made by bar No. 1 and the second by bar No. 2. When calibrating a measuring bar the vertical edge of the bar should be placed against the horizontal edge of the comparator when repeating the calibration.

Direct measurement of horizontal distances is also possible on steep terrain by measuring in steps *(Fig. 5.6)*. When used as illustrated in *Fig. 5.6, b*, the measuring equipment consists of one (or two) 4 to 5 m long measuring bars, a projection bar

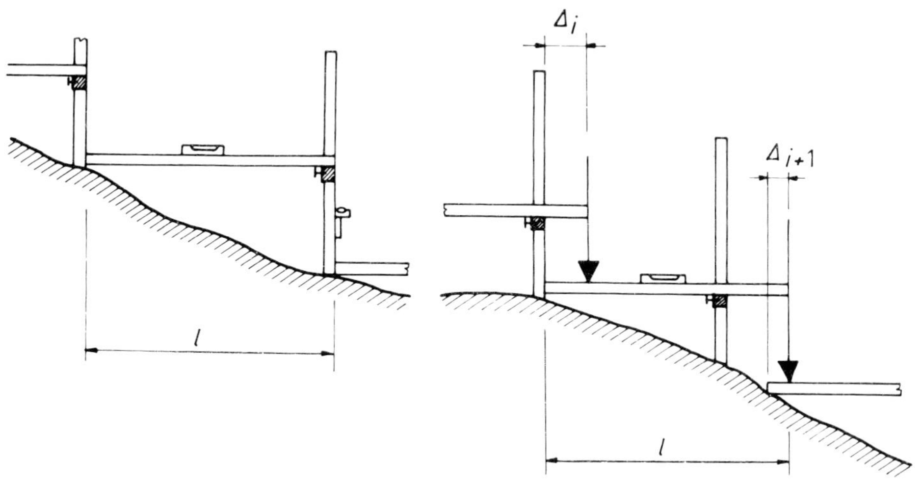

Figure 5.6. Stepwise measurement by measuring bars: *a)* by a projection staff; *b)* by projection with a plummet

with a support and a clampable slide, a bar level for setting the bar into the horizontal and a plummet. Measuring bars are generally fitted with a fixed level with an accuracy of some secs of arc. The bars are graduated every 10 cm, sometimes every 1 cm, along the whole length. The projection bar usually has a cm graduation along its whole length, hence the height difference between the end points can also be determined when measuring in steps.

Due to their short base-length the accuracy of a length measurement using measuring bars change considerably depending on the quality of the terrain or of the measurement path:

— The relative accuracy of a measurement with the bar laid on the ground on a surface of medium quality is about 1:2500 to 1:3000.

— In case of measurement on prepared smoothed ground or on the shoulder of a road, and by setting out the straight line with a cord and calibrating the bar from time to time, a relative accuracy between 1:10,000 and 1:20,000 can be achieved.

— If the bar is calibrated before and after the measurement, if a reciprocal observation is made and if the bar is used on solid pavement or iron rail, then a relative accuracy between 1:50,000 and 1:100,000 can be attained, depending on the number of repetitions.

5.1.2.2 The Measuring Tape (Steel Band)

A measuring tape is usually used for the measurement of short distances in the field and of lengths of structures. In surveying practice it is now usual to use 20 m and 50 m long field tapes, while for the measurement of structures or other lengths 10 to 100 m long hand measuring tapes with a handle (10, 15, 20, 25, 30, 50, 100 m) are commonly used. The field tape is from 12 to 20 mm wide, from 0.3 to 0.4 mm thick, and is made of steel or invar (an alloy of 64 p.c. steel and 36 p.c. nickel). The thermal expansion coefficient of invar ($1.6 \times 10^{-6}/1\,°C$) is less than that of steel ($1.2 \times 10^{-5}/1\,°C$) by one order of magnitude, consequently its expansion due to temperature is negligible in the majority of practical measurements.

Field measuring tapes are graduated on both sides with 10 cm divisions and the direction of increasing numbers is opposite. Whole metres are generally marked by embossed numbers on small coloured metal plates *1*, half metres by small copper plates of 6 to 10 mm diameter *2*, 10 cm-s by holes of 1.5 to 2 mm diameter *(Fig. 5.7)*.

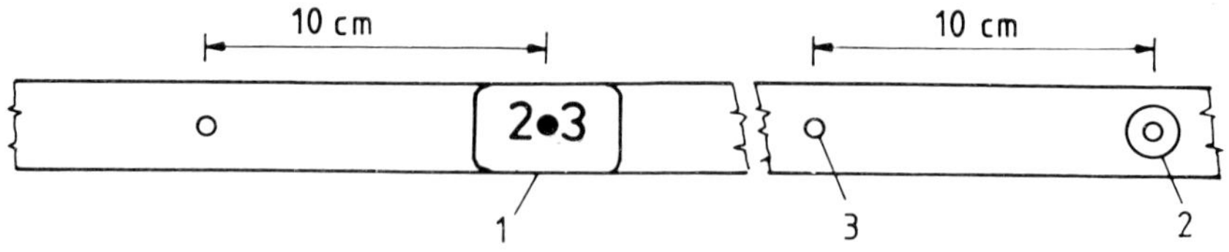

Figure 5.7. Graduation of a field measuring tape

According to the marking at their ends, tapes can be distinguished between those for use with ranging poles or those for use with marking pins *(Fig. 5.8)*. Tapes with end lines or end graduations are used for precise length measurement *(Fig. 5.9)*. Measuring tapes with end marks for use with ranging poles end in metal rings attached to the tape by a hinge. The ranging poles fit into the 30 mm diameter rings. The ends of the tape are fixed during measurement by the vertical axis of the ranging poles stuck into the ground. Measuring tapes for use with marking pins have hinged handles attached at both ends of the tape, and the basic length of the tape is marked by end lines at the cuts for the marking pin.

The marking pin is a pointed metal nail of length 30 cm and 5 to 6 mm diameter. A set of 11 pins and two metal rings belong to each measuring tape *(Fig. 5.10)*. The marking pins are used both for marking the end of the tape in the field and for

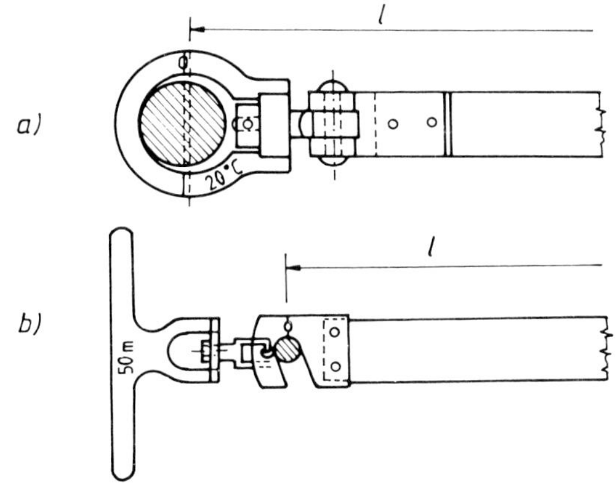

Figure 5.8. Ends of field measuring tapes: *a)* for survey poles; *b)* for marking pins

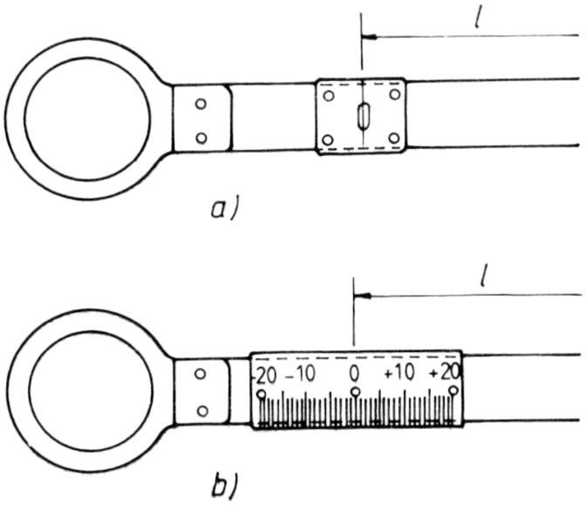

Figure 5.9. The ends of a precise measuring tape: *a)* end line; *b)* end graduation

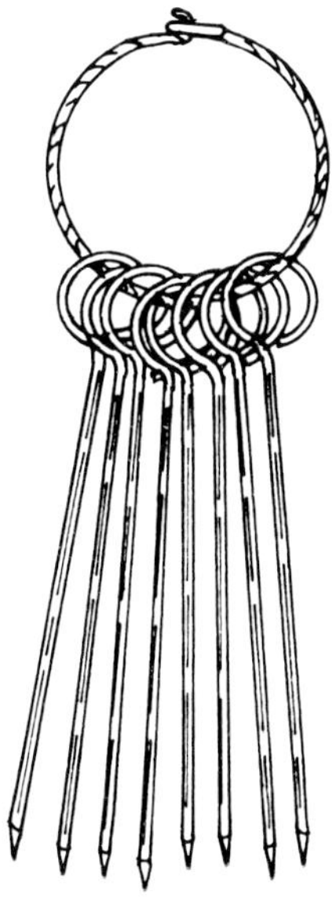

Figure 5.10. Metal marking pins with keeper ring

counting the number of whole tape lengths within the measured distance. Field measuring tapes are wound on metal frames of about 20 to 24 cm diameter for transport and storage *(Fig. 5.11)*.

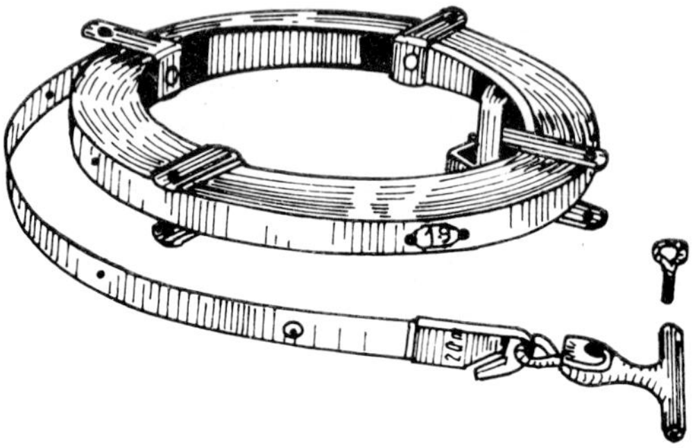

Figure 5.11. Measuring tape on a metal frame

Field measuring tapes used for precise measurement also have hinged handles outside the end points of the tape. The basic length is marked here by an end line *(Fig. 5.9, a)*, in some cases by the zero line of a short graduation *(Fig. 5.9, b)*. In the former case the end points of the tape are marked in the field by index plates.

The *Löschner* index foot plate *(Fig. 5.12)* is a metal device for marking the end point; it weighs from 3 to 5 kg and is fitted with a handle and three pointed feet. Its folding tongue embodying the index mark can be moved in the direction of the tape by a slow-motion screw. The index mark of this device, placed near the end of the tape, can be made to coincide with the end line by the slow-motion screw.

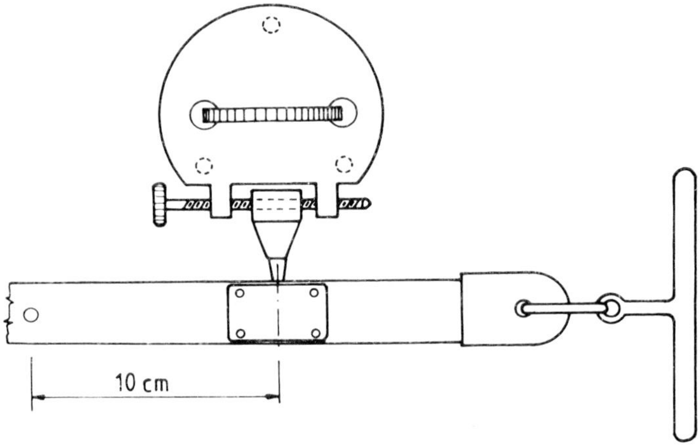

Figure 5.12. The use of an index plate for precise length measurement

A spring-balance or dynamometer is used to produce the tension required for the tape *(Fig. 5.13)*. The graduated spring-balance has a handle and hook, and a tension of between 5 and 20 daN can be set with an accuracy of ± 0.5 daN.

Hand tapes with handle *(Fig. 5.14)* are steel tapes from 12 to 14 mm wide and from 0.25 to 0.35 mm thick. Recently they are usually provided with a white (or

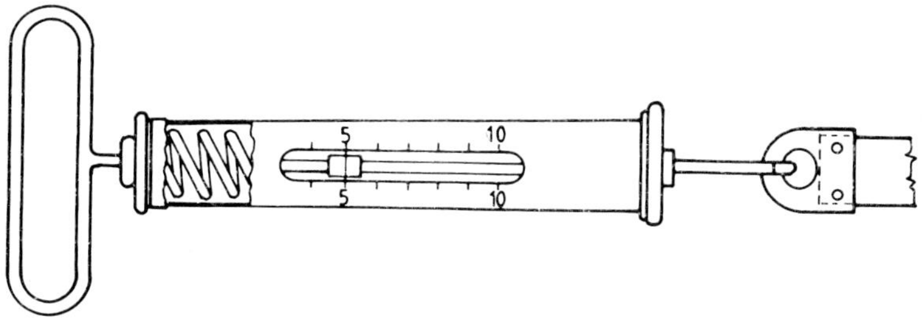

Figure 5.13. Dynamometer or spring-balance

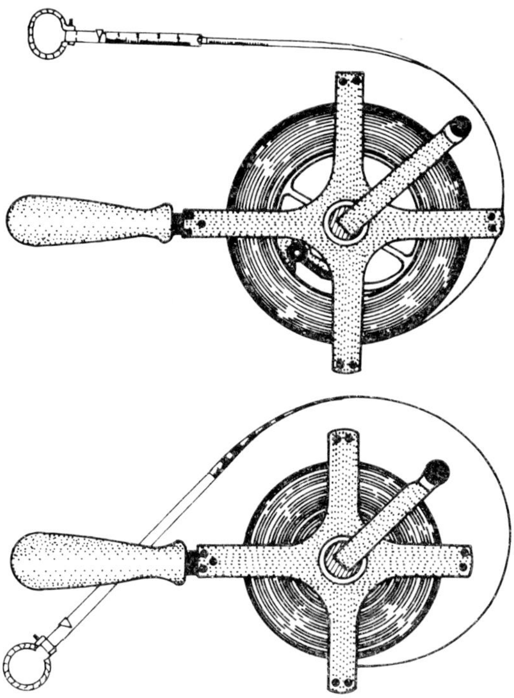

Figure 5.14. Hand tapes with a handle

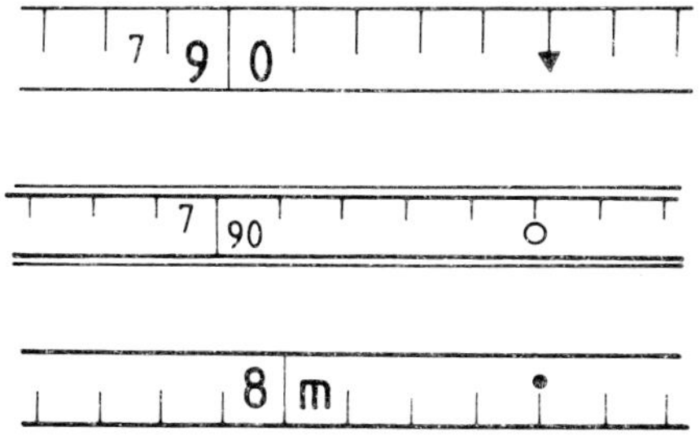

Figure 5.15. Graduation and numbering on hand tapes

coloured) plastic cover. The wooden or plastic handle is mounted on a metal frame. The tape can be quickly wound off and on the frame by the crank-handle on the centrally rotating drum. The tapes are usually graduated in cm along their whole length and numbered every 10 cm (sometimes every cm) *(Fig. 5.15)*. Certain hand tapes have mm graduations along their whole length. The zero line of the graduation is either on the metal tape itself or refers to a certain point on its end *(Fig. 5.16)*.

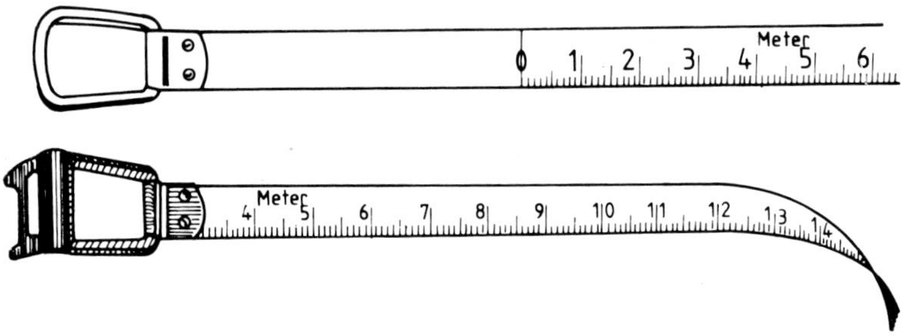

Figure 5.16. The end of a hand tape

Recently hand tapes, 50 and 100 m long, 4 to 6 mm wide and 0.3 mm thick, have also been produced in steel or invar *(Fig. 5.17)*. These are usually used freely suspended and under great tension.

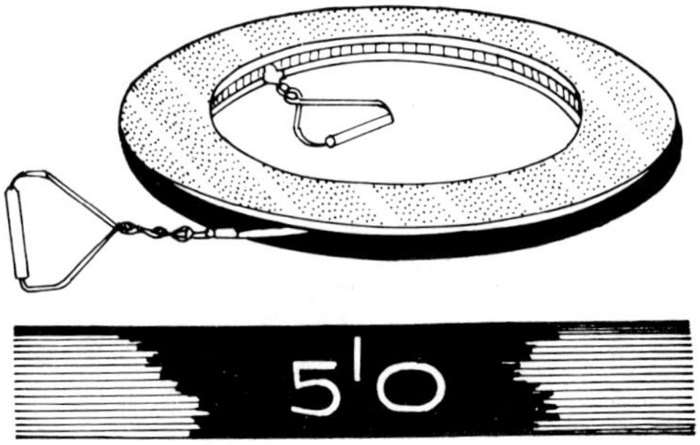

Figure 5.17. 6 mm wide invar measuring tape

For precise length measurements with a steel tape the change of length due to temperature change must be taken into account. The temperature of the tape during measurement is measured by a tape thermometer.

A tape thermometer consists of a 30 cm long steel tape in a protective case with an open top, and of a mercury thermometer connected to it. If it is exposed to the same temperature as the tape, it will show a temperature approximating that of the tape. If a tape thermometer is not available, measurement of air temperature must suffice. The reduction is then computed from the average of the temperatures measured at the beginning and end of the measurement.

The horizontal projection of slope lengths measured by a tape is computed from the height differences between the points at which the slope changes. The reduction of the i-th section, of length d_i and elevation difference Δh_i, is:

$$\Delta_i = \frac{\Delta h_i^2}{2d_i}. \tag{5.6}$$

The reduction of a total distance consisting of n sections is:

$$\Delta l = \sum_{i=1}^{n} \Delta_i = \sum_{i=1}^{n} \frac{\Delta h_i^2}{2d_i}. \tag{5.7}$$

For a required relative accuracy of $1/v$, the admissible mean square error $m_{\Delta h_i}$ of the elevation difference Δh_i is:

$$m_{\Delta h_i} = \frac{d_i^2}{\Delta h_i v}. \tag{5.8}$$

The basic length marked on the measuring tape is calibrated with a tape comparator — similarly to that of measuring bars. This comparator consists of two metal rules fixed on the even surface (e.g. a solid floor-covering) of a gallery or corridor. The rules are generally 10 cm long and the distance b between their zero lines is nearly equal to the nominal length of the measuring tape. The zero line is at the centre of each graduation, consequently both graduations have positive and negative division lines. The length b of the comparator is determined by precise measurement (e.g. with standard metres or calibrated measuring bars). The tape is laid along the base of the comparator and the sign-correct readings Δ_1 and Δ_2 of the two end marks of the tape give the calibrated length of the measuring tape as:

$$l_k = b + \Delta_1 + \Delta_2. \tag{5.9}$$

The tape is maintained at constant tension during calibration and the temperature of the tape (in simpler cases the air temperature) is measured. The length of the tape can be computed from the formula:

$$l = l_k + \Delta l_t + \Delta l_p = l_k + [\alpha(t_m - t_k) + \beta(p_m - p_k)]l_k, \tag{5.10}$$

where Δl_t is the temperature correction and Δl_p is the elongation correction. These two corrections can be determined from the thermal expansion coefficient α and the elongation coefficient β, also from the average temperatures t_m observed during the measurement, and from the tension p_m applied, with respect to the calibration values t_k and p_k.

The thermal expansion coefficient α is $11.5 \times 10^{-6}/1$ °C for steel tapes. The alongation coefficient β is obtained from the elasticity coefficient E and the cross section F of the tape. The area F can be substituted in practice by the quotient of the tape mass m per running metre and the density γ, i.e.:

$$\beta = \frac{1}{F \cdot E} = \frac{\gamma}{m \cdot E}. \tag{5.11}$$

Since the tension differs in practice from the 10 daN tension at calibration by at most a few daN-s, it is sufficient for tapes of usual cross section to compute the reduction for field tapes with $\beta = 7 \times 10^{-6}$/daN, and for hand tapes with a handle with $\beta = 16 \times 10^{-6}$/daN.

For a freely suspended tape the measured length must be further corrected by the sag correction in order to get the highest accuracy. The form of the measuring tape is substituted by a flat parabolic arc and the correction Δl can be computed from the sag h of the arc with sufficient precision as:

$$\Delta l = -\frac{8h^2}{3l}. \tag{5.12}$$

For calculating the reduction (5.12) the measurement of the sag is generally difficult, therefore, the reduction is taken in practice from graphs or tables constructed or listed for certain tensions.

The reliability of measurement with a tape strongly depends on the quality of the device, on the method of marking the end points, on the preparation of the ground along the line and on the external circumstances of the measurement.

— For field tapes with marking pins, on medium quality terrain and neglecting the temperature effect, the accuracy of one single measurement can be characterized by relative errors between 1:1000 and 1:2000.

— In field conditions, after smoothing the ground along the line, using a tape with end lines and index plates (or index pins) and taking the temperature effect into account, a relative accuracy of between 1:10,000 and 1:20,000 can be attained from repeated measurements.

— For measurement on road pavements or rail tracks the use of a precise measuring tape with end marking and a tape thermometer, and with at least reciprocal measurements will give a relative accuracy between 1:50,000 and 1:100,000 — depending on the number of repetitions.

5.1.2.3 The Measuring Wire

The measuring wire is a precise instrument for length measurement in which the distance between the end marks, which determine the basic length, is maintained nearly constant when the invar wire connecting them is under tension.

A wire used to be used to measure longer distances with extreme accuracy requirements (e.g. the base line of a triangulation), but the high accuracy of electronic distance meters has now considerably narrowed its area of economical application. Recently a new field of application has been found for them in special measuring tasks where the variation of distances of a maximum of 50 m must be determined with very high accuracy (within 0.1 mm).

Conventional measuring wires for the measurement of long distances are 24 m long and have a diameter between 1.6 and 1.7 mm. At each end of the wire there is a metal prism of triangular cross section with an 8 cm long graduation *(Fig. 5.18)*.

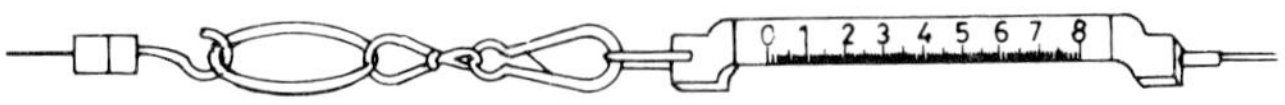

Figure 5.18. Graduated metal prism on a measuring wire

The basic length of the wire is defined by the distance measured along the stretched wire between the zero lines of the two prisms. For transportation and storage it is rolled on a metal drum of 50 cm diameter fixed in a box.

The measuring wire is used for measurement in a freely suspended state, the ends of the wire being suspended on straining trestles, such as the *Witram* trestle *(Fig. 5.19)*. The tension wires, which have a hook, are placed over the pulley wheels of the trestles and strained by cylindrical weights of 10 kg (± 1 kg). The Kern (Aarau) straining device is mounted on a pillar together with the support for the index mark. The wire is here also strained by cylindrical weights *(Fig. 5.20)*.

Before using a measuring wire, the distance to be measured is divided into 24 m long partial distances; they are the so-called index-distances. These index distances

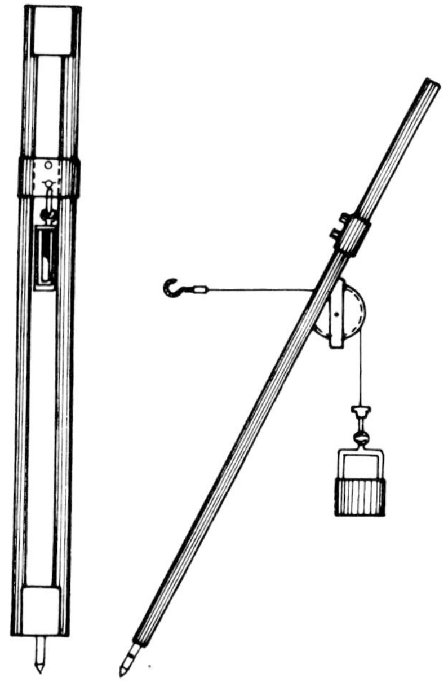

Figure 5.19. Witram straining trestle

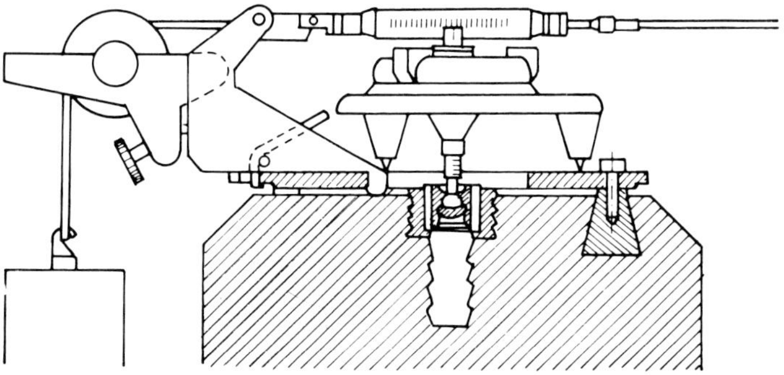

Figure 5.20. The Kern (Aarau) straining device

are marked by index pins carefully set out on the straight line between the end-points of the line. The index pins with index marks are mounted on bases with three foot screws on an index tripod, or by means of the threaded shaft *(Fig. 5.21)* on the index pegs hammered into the ground.

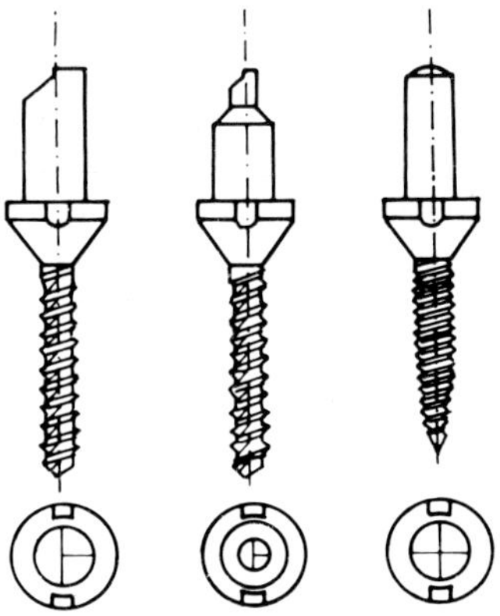

Figure 5.21. Index pin with threaded shaft

When measuring the wire is strained between adjacent index tripods (or index pegs) and the index lines are read on the scales of the metal prisms, with estimation to 0.1 mm. The readings are repeated several times, moving the wire a little along the line after each. The temperature of the wire (or the air), also wind-direction and velocity, are measured at each wire position.

The basic length of the wire is determined by calibration before and after measurement, either in the laboratory or in the field on a comparative base line. The laboratory calibration of the wire is the same as the calibration of the measuring tape. The exact length $b \approx 24$ m of the comparator is very carefully measured by a pair of standard metre bars or by the light interference method. The distance between the index marks defining the length of the comparator is determined by readings with microscopes on the mm graduations of the wire.

If it is measured in only one wire position the effects of the systematic errors do not vanish, but they are eliminated by measuring in more wire positions. Very precise results can be achieved if the calibration is made on a comparative base line that is 30–50 times the length of the wire. In such a case the distance to be measured is directly compared with the comparative base line. The main point of the method is that the base line d_0 known with a high accuracy is measured several times by the wire and its length d'_0 is computed with the nominal length of the wire. The preliminary value d' of the length to be measured is determined in the same way, and the factor $q = d'/d'_0$ is computed. The corrected length from this is:

$$d = q \cdot d_0. \tag{5.13}$$

Hence for this method the true length of the measuring wire need not be known. The constant errors and the mean of the systematic errors of the measurement with the wire does not affect the result in practice.

By repeated measurements of the comparative base line with different calibrated measuring wires at different times, more and more reliable values of the comparative base line can be obtained.

The comparative base line can also be measured by the method of light interference, by means of the *Väisälä* device. These measurements have shown that the unit of length of the base line can be made equal to the international metre with a precision of less than 1 : 10,000,000.

A special piece of equipment for civil engineering has been developed by the Institute for Road Construction of the Federal Technical College Zürich (Institut für Strassenbau der Erdgenössischen Technischen Hochschule: ISETH). It is manufactured by the Swiss firm Kern (Aarau) under the name *Distometer ISETH*. The Distometer can be used very advantageously for the precise determination of length variations in the study of motions and deformations. The equipment is also suitable to measure 1 m to 50 m distances of arbitrary slope. The range of the possible length variations is 100 mm.

In the *Distometer ISETH* two main elements, i.e. the tension meter and the dial gauge for reading the length variation, are unified in a hand device about 1 m long *(Fig. 5.22)*. (A complete measuring outfit consists of three main elements, i.e. invar wire, straining device and reading device.) The *Distometer* is the reading device which can be connected to a measuring wire of standard cross section as an extension. The invar wire has different lengths depending on the distance to be measured. Further accessories of the total outfit are the pins marking the ends of the line and the connecting elements which assure a connection between the Distometer and the end points of the invar wire.

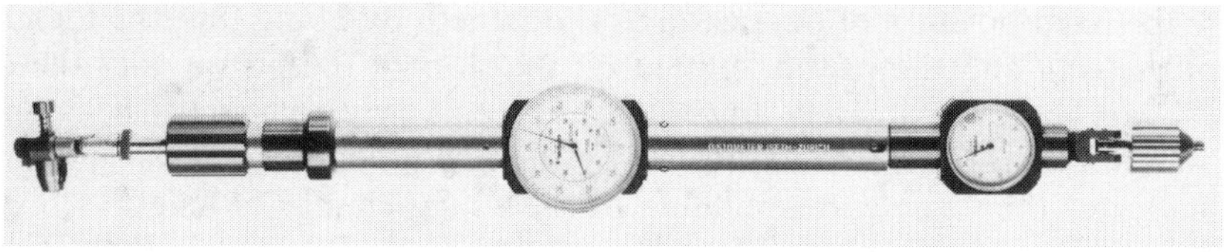

Figure 5.22. The *Distometer ISETH* wire measuring device

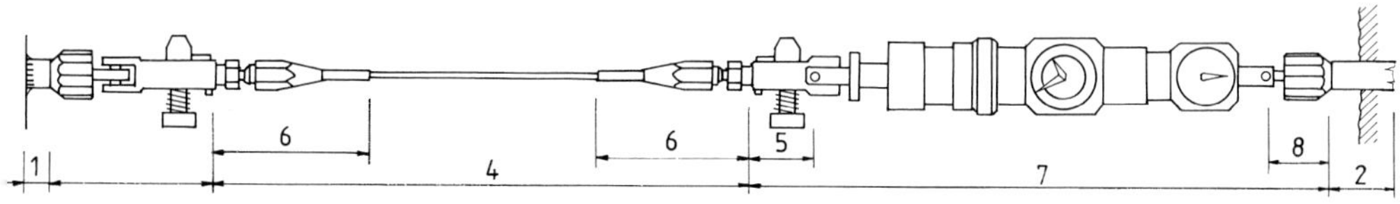

Figure 5.23. Distometer and measuring wire assembled for measurement

The instrument is schematically shown assembled for measurement in *Fig. 5.23*. The link assemblies *3* and *8*, screwed to the pins *1* and *2* which are let into concrete (or fixed by welding), are attached to the invar wire *4* through the junction piece *5* and the Distometer *7*. The invar wire, cut to the desired length, is connected at one end to the Distometer by the wire clamp *6*, and at the other end by the link assembly clamp. The link assemblies which can be screwed to the pins, enable the measuring device to be used between end points of different heights.

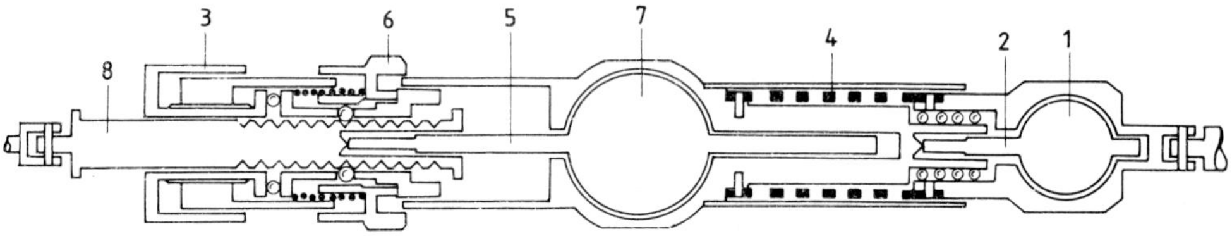

Figure 5.24. Construction of the *Distometer ISETH*

An important element of the Distometer is the precisely manufactured steel spring *4 (Fig. 5.24)*. The tensile force of the spring, to ensure the required tensioning of the wire, can be adjusted by the fine adjustment screw *3*, and the tension can be read — transmitted by the push bar *2* — on the dial of the tension gauge *1*. If the wire tension changes, the push bar *5* also moves the pointer of the dial gauge *7*. Since a certain reading on the dial gauge indicates a given tensile force at an unchanged measured distance, the difference between the dial gauge readings at different epochs indicates the measure of the change of distance. The dial of the gauge has two scales, both divided into 100 units: the unit of the outer scale is 0.01 mm, that of the inner one 1 mm. Hence a maximum of 100 mm variation of length can be measured on the dial gauge with a direct reading accuracy of 0.01 mm. A more comfortable use is possible by using the ring for coarse adjustment *6*; if it is pulled in the direction of the wire the fixed position of the piston *8* is loosened. By doing this, the locking ball between the ring and the piston is lifted off the slots of the piston, and the change of length can be approximately adjusted in advance by displacing the piston. Now the tensile force on the wire is adjusted by the fine adjustment screw and the dial gauge indicates the value corresponding to the actual change of length.

A calibrating device is also delivered with the *Kern Distometer ISETH* — it is packed in the common transporting box. This device consists of two horizontal end-plates at a constant separation which is ensured by three connecting invar bars. Both tension gauge and dial distance gauge can be calibrated by them. The tension gauge can be tested by an 8 kg weight hung on the Distometer suspended from the upper end-plate *(Fig. 5.25)*. If the pointer of the tension gauge does not read zero, it must be turned by hand to the zero mark. For testing the distance gauge, the Distometer, assembled with the link assemblies, must be placed between the measuring pins fixed to the two end-plates. Having adjusted the corresponding tension with the fine adjustment screw, the pointer of the dial gauge must show the reading corresponding to the known distance between the two end-plates. Any difference can either be eliminated when measuring by turning the pointer, or it can be taken into account by computation.

The calibration of the Distometer on a test stand is applicable only for the repeated detection of distance changes. For the measurement of absolute lengths the total length of the assembled equipment must be determined by careful measurement on a proper base length.

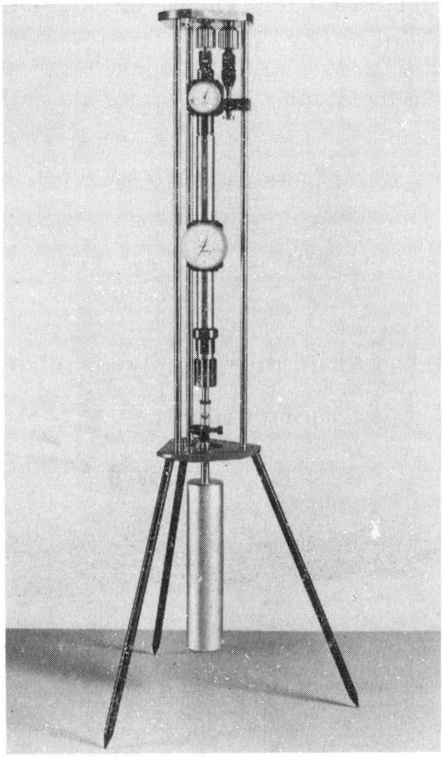

Figure 5.25. The *Distometer ISETH* on a calibrating device

A considerable advantage of the determination of length variations by Distometer is the high accuracy that can be achieved by the measurement. According to the data given by the factory a standard deviation of $m = \pm 0.02$ mm can be achieved over distances not longer than 20 m. If the distance d is longer than 20 m the reliability of the result is:

$$m = \pm 1 \times 10^{-6}\, d.$$

5.2 Geometrical (Optical) Distance Meters

Surveying instruments for geometrical distance measurement for a more precise determination of distance are always optical instruments. The distance measuring system is usually built into the geodetic telescope or optically projected into it. Therefore instruments for geometrical distance measurement are also called optical distance meters.

The determination of a distance by optical distance meters can always be reduced to the solution of one or two triangles. As regards the known and measured quantities in the triangle of distance measurement there are two basic cases:

— the side b, called the base line, is known and the stadia (parallactic) angle φ opposite to it is measured;

— the stadia angle φ is known and the base line b is measured.

If b is perpendicular to the distance d and forms one single triangle with it *(Fig. 5.26, a)*, then:

$$d = b \operatorname{cotan} \varphi \tag{5.14}$$

and if b forms with d two congruent triangles *(Fig. 5.26, b)*, then:

$$d = \frac{b}{2} \operatorname{cotan} \frac{\varphi}{2}. \tag{5.15}$$

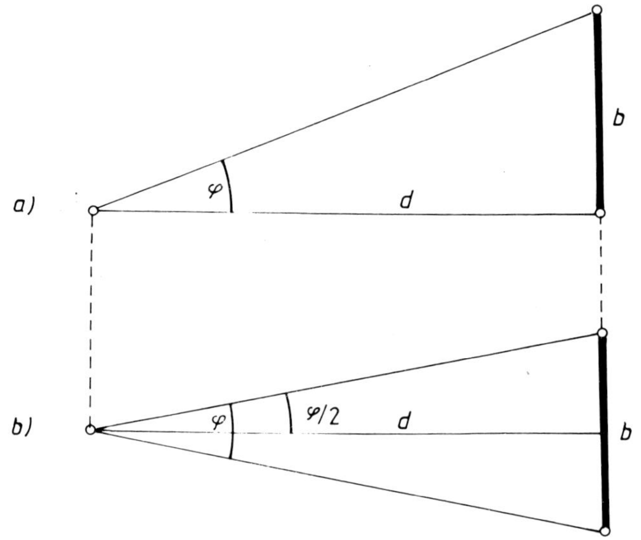

Figure 5.26. The parallactic triangles of geometrical distance measurement

If the error of the base b is denoted by Δb, the error of the parallactic angle φ by $\Delta\varphi$, and the error of the distance d by Δd, then it is easy to find the following connection by differentiating Eq. (5.14)

$$\frac{\Delta d}{d} = \frac{\Delta b}{b} + \frac{\Delta \varphi}{\varphi}. \tag{5.16}$$

For the determination of a distance of e.g. 100 m to an accuracy of 0.10 m (1 : 1000 relative error) it is advisable to control the error in both b and φ in such a way that the effect of each does not exceed a relative error of 1 : 2000 in the distance d (in our example 0.05 m). Hence, if $b = 2$ m, then $\Delta b = 1$ mm. Consequently the measurement of the angle

$$\varphi'' = \frac{b}{d} \rho'' = \frac{2}{100} \rho'' = 4125.3'' \qquad (12{,}732.4^{cc})$$

requires that the condition

$$\Delta\varphi \leqq \frac{4125.3}{2000} = 2.1'' \qquad (6.5^{cc})$$

should be satisfied in order that the relative error of d be less than 1 : 1000. If, however, the relative error of b is not higher than 1 : 5000, the former condition is:

$$\Delta\varphi = \frac{4125.3}{1250} = 3.3'' \qquad (10.2^{cc}).$$

In order to decrease the relative error of the distance d, a high accuracy must be sought in the measurement of both b and φ (or, if their values are known, in their invariability).

In both groups of optical distance meters two further cases are distinguished, according to the position of the base line b with respect to the instrument.

— The distance meter has an external base-line if a base independent of the instrument (e.g. a stadia rod) is required at the opposite end of the distance to be measured.

— A distance meter is of the inner base-line type if there is a base-line of limited length within the meter itself. In this case it is only necessary to be able to point to the opposite end of the distance.

The stadia rod for distance meters with an external base-line must be set up either horizontally or vertically, depending on the measuring system. A horizontal stadia rod usually gives a higher accuracy because the effect of vertical refraction may be assumed to be the same on both legs of the horizontal stadia angle, and thus not to change it significantly. In spite of this the unequal effects of refraction, or differential refraction, acting on the legs of the stadia angle in the vertical plane is a considerable source of error in distance measurement. Experiments have shown that differential refraction can reach a relative error of as much as 1 : 300 of the measured distance in extreme cases, depending on external circumstances. One advantage of the vertical base, however, is that it can always be set up centrally over the end-point of the line, and its verticality can easily be assured with a spirit level attached to the stadia rod; on the other hand a horizontal base, which must be placed on a tripod, requires to be centred over the point like a theodolite, and distances to certain field objects (such as corners of buildings) can be often only measured with great difficulty, if at all.

The measurement error of distance meters with a constant stadia angle (or varying base-line) increases proportionally to the distance, while the measurement error of distance meters with a varying stadia angle (or constant base-line) increases quadratically with the distance measured. This quadratic error increase has a very unfavourable effect on the maximum distance that can be measured by distance meters with a constant base-line.

Horizontal projections of slope distances measured by optical distance meters are computed using the vertical angles measured together with the distance. Sometimes the reduction is made by the distance measuring system itself. Self-reducing type distance meters automatically change the stadia angle as a function of the inclination of the telescope in such a way that the measurement directly yields the horizontally projected distance. However, self-reducing systems also have other solutions; their use is wide-spread in tacheometers (Subsection 6.1.2).

5.2.1 Distance Meters with an External Base-line and a Constant Stadia Angle

5.2.1.1 Simple Distance Meters with Fixed Stadia Hairs

The main constructional element of simple distance meters with fixed stadia hairs is a geodetic telescope. In the plane of the telescope graticule there are, as well as the cross-hairs, two stadia hairs, both parallel to one of the cross-hairs (usually the horizontal one), symmetrical about the centre and apart *(Fig. 5.27)*. These stadia hairs with the constant separation p define a constant stadia angle. Such stadia hairs are now provided in the telescopes of all surveying instruments (theodolites, levels, etc.), even of those which are not intended primarily for distance measurement. It is also necessary to have a graduated staff which is set up vertically at the end of the line. The simple stadia rods used in practice have cm divisions, numbered every 10 cm. Their length is usually 3 or 4 m. The levelling staffs discussed in Chapter 3 can e.g. also be used as stadia rods.

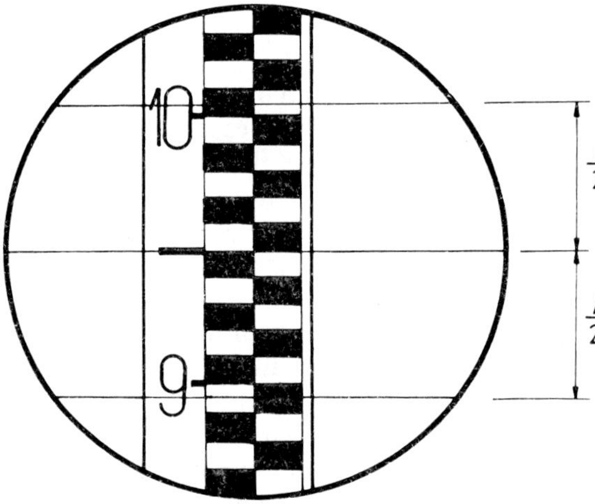

Figure 5.27. Field of view of a simple distance meter with stadia hairs and the image of a stadia rod

Distance observation includes taking the readings l_1 and l_2 of the stadia hairs on the rod. The difference Δl between them depends on the distance between the instrument and the rod. With the telescope horizontal and the rod vertical *(Fig. 5.28)*, the distance d can be derived from the proportion:

$$\frac{d-(a+f)}{f}=\frac{\Delta l}{p} \tag{5.17}$$

to be

$$d=a+f+\frac{f}{p}\Delta l, \tag{5.18}$$

where a is the distance of the objective from the horizontal axis H and f the focal length of the objective.

Since a, f and p are constants for a telescope, consequently

$$a+f=c,$$

and

$$\frac{f}{p}=k$$

are also constants. Substituting them into Eq. (5.18) gives:

$$d=c+k\Delta l, \tag{5.19}$$

where c is the additive constant and k the stadia constant of the distance meter.

Equation (5.19) is also valid for slope distances if the stadia rod is set up in the vertical plane of the telescope and perpendicular to the line of sight. The additive constant can be even 50 cm in simple distance meters. The stadia constant is $k=100$ for a hair-separation of $p=0.01\times f$, and now the stadia angle φ can be obtained from the expression

$$\frac{2f}{p}=\operatorname{cotan}\frac{\varphi}{2} \tag{5.20}$$

as

$$\varphi=34'\,22.6'' \qquad (\approx 63^{c}\,66.0^{cc}).$$

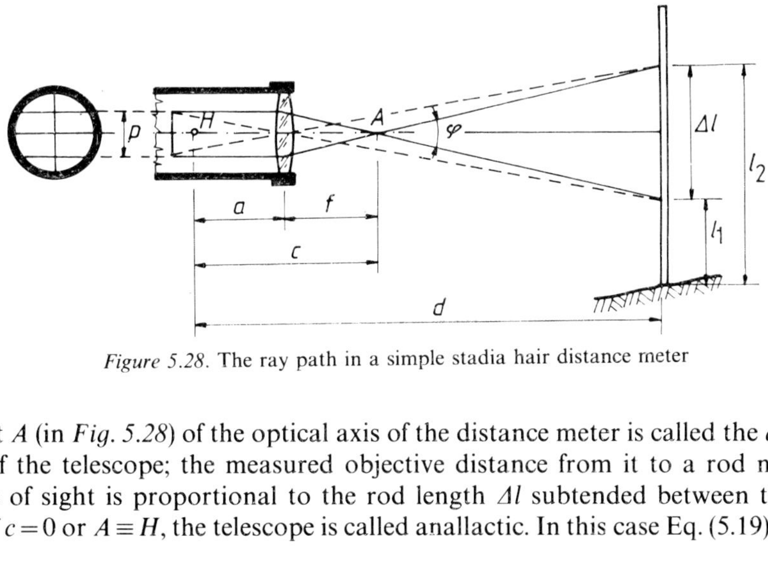

Figure 5.28. The ray path in a simple stadia hair distance meter

Point A (in *Fig. 5.28*) of the optical axis of the distance meter is called the *anallactic* point of the telescope; the measured objective distance from it to a rod normal to the line of sight is proportional to the rod length Δl subtended between the stadia hairs. If $c=0$ or $A\equiv H$, the telescope is called anallactic. In this case Eq. (5.19) becomes simply

$$d=k\Delta l. \tag{5.21}$$

For example the *Porro* telescope is anallactic *(Fig. 5.29)*, its objective lens consisting of two condensing lenses with a considerable air-space between them. By proper choices of the physical length s, of the focal lengths f_1 and f_2 of the lenses L_1 and L_2 and of the lens separation e, point A can be made to e.g. bisect the physical length s. In order to satisfy this condition, however, the telescope would be rather long and therefore the *Porro* anallactic telescope never came into general use.

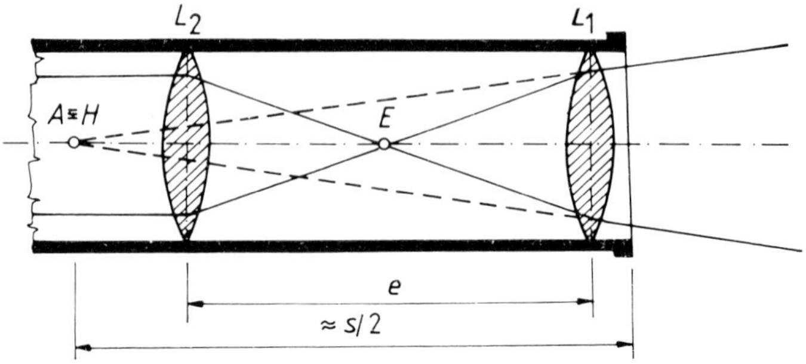

Figure 5.29. The ray path in the *Porro* anallactic telescope

Telescopes with internal focussing, now widely used in surveying instruments, are nearly anallactic (see *Fig. 1.55*). Here the objective lens system is composed of the condensing lens L_1 and the dispersing lens L_2; the equivalent (resulting) focal distance of the objective varies with the distance of the object observed. By changing the distance between the two lenses, however, the equivalent focal length and the magnification of the telescope also changes. Due to focussing, the anallactic point is moved and the line of sight of the stadia hairs takes a hyperbolic form. Consequently the constants c and k of a telescope with internal focussing change slightly, and the anallatic point is displaced depending on the distance. Therefore, the flat hyperbolic line can be substituted by a properly chosen straight line, thus obtaining a practically constant stadia constant k. Either a chord of the hyperbola or its asymptote can be taken for this line. In the former case when the substituting chord is chosen, it must be remembered that the combined error effect of the changes in c and k should not exceed values acceptable in practice. In the case of a substitution by the asymptote, the error due to the shift of the anallactic point is significant over short distance, but it decreases over increasing distances and becomes zero over infinitely long distances.

The intersection of the substituted straight line and the optical axis is called the *quasi-anallactic* point.

Roelofs' equation for the hyperbolic line of sight is:

$$f_2 d^2 + f_1(s - f_1 - 2f_2)d\frac{y}{p} + f_2 f_1^2 \frac{y^2}{p^2} -$$

$$-2f_1 f_2 d + f_1^2(2f_2 - s)\frac{y}{z} + f_2 f_1^2 = 0. \tag{5.22}$$

Roelofs' equation is in content identical with *Fialovszky*'s equation [Eq. (2.20)], noticing that the equivalents for the symbols d, s and p in Eq. (5.22) are S_1 with negative sign, d and q_2 in Eq. (2.20).

Deriving y from either Eq. (2.20) or Eq. (5.22) gives the ordinates y of the hyperbola at different distances d. The difference between the ordinate y_d computed thus and the staff length Δl_d, which is proportional to the distance d, is:

$$\delta y_d = \delta \Delta l_d = y_d - \Delta l_d,$$

and the error in length, i.e. Δd, is:

$$\Delta d = k(y_d - k\Delta l_d). \tag{5.23}$$

It is worth mentioning here that Eq. (2.20) is also suitable for determining the resulting focal length $f_0 = \Gamma \times f_{0k}$ which depends not only on the required magnification Γ of the telescope and on the chosen focal length f_{0k} of the eyepiece, but also on the optimum focal lengths f_1 and f_2 of the two parts of the objective lens system and on the stadia separation p. The requirement is that the defined position of the quasi-anallactic point should lie e.g. at the centre of the instrument, and at the same time that the displacement of the anallactic point should remain within the required measurement range below a defined error limit.

Table 5.1. Distance measurement errors due to the shift of the anallactic point (in thousandths of the distance)

Distance d (m)	5	10	20	50	100
Δd (mm)	24	9	4	2	1
p ‰	4.8	0.9	0.2	0.05	0.01

From Eq. (5.22) *Table 5.1* gives the distance measurement errors for a telescope with the following characteristics: $s = 154$ mm, $f_1 = 136.156$ mm, $f_2 = -101.51$ mm, $f_0 = 206.45$ mm, $p = f_0/100$ and the distance of the quasi-anallactic point from the objective lens $= s/2$.

Since the telescope tilt α of distance meters with fixed stadia hairs can usually be measured, the horizontal distance can be computed from the slope distance d_s by the formula:

$$d = d_s \cos\alpha \tag{5.24}$$

and by substituting Eq. (5.21) one obtains

$$d = k\Delta l \cos\alpha. \tag{5.25}$$

With the usually horizontal stadia hairs, a vertical stadia rod must be used for distance measurement. It would be very difficult to obtain a position of the stadia rod normal to the line of sight (for a vertical rod), therefore the rod is set up vertically at the end of the line. When the line is sloping, however, the rod is not perpendicular to the line of sight, therefore the basic equation of distance measurement (5.25) changes, as may be deduced from *Fig. 5.30*. The stadia angle φ is bisected in the figure by the line of sight which has a tilt of α, and for an anallatic telescope:

$$\Delta l = d\left\{\tan\left(\alpha + \frac{\varphi}{2}\right) - \tan\left(\alpha - \frac{\varphi}{2}\right)\right\}.$$

After transformation

$$\Delta l = d\frac{2\tan\frac{\varphi}{2}}{\cos^2\alpha\left(1 - \tan^2\alpha\tan^2\frac{\varphi}{2}\right)},$$

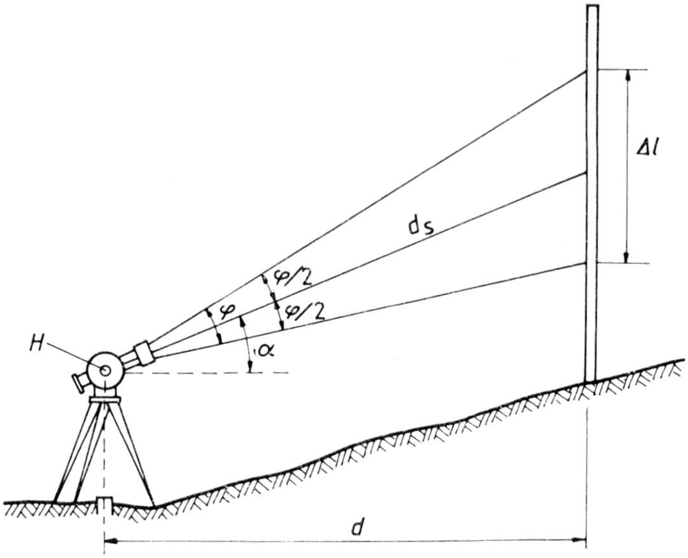

Figure 5.30. Distance measurement with a vertical stadia rod

and the horizontal distance d is:

$$d=\frac{\Delta l}{2}\operatorname{cotan}\frac{\varphi}{2}\cos^2\alpha\left(1-\tan^2\alpha\tan^2\frac{\varphi}{2}\right),$$

or by taking into account that $\operatorname{cotan}\frac{\varphi}{2}=\frac{f}{\frac{1}{2}p}=2k$:

$$d=k\Delta l\cos^2\alpha\left(1-\tan^2\alpha\tan^2\frac{\varphi}{2}\right). \tag{5.26}$$

Since the slope angle never exceeds 45° (50^g) in practice and with the usual multiplying factor $k=100$,

$$\tan^2\alpha\tan^2\frac{\varphi}{2}=1\times0.000025\approx0,$$

whence the horizontal distance can be computed by the simplified formula:

$$d=k\Delta l\cos^2\alpha. \tag{5.27}$$

If the telescope is not anallactic, the horizontal projection of the additive constant c must be added to Eq. (5.27) and the formula now becomes:

$$d=c\cos\alpha+k\Delta l\cos^2\alpha. \tag{5.28}$$

The constants c and k of distance measurement can be determined separately or together.

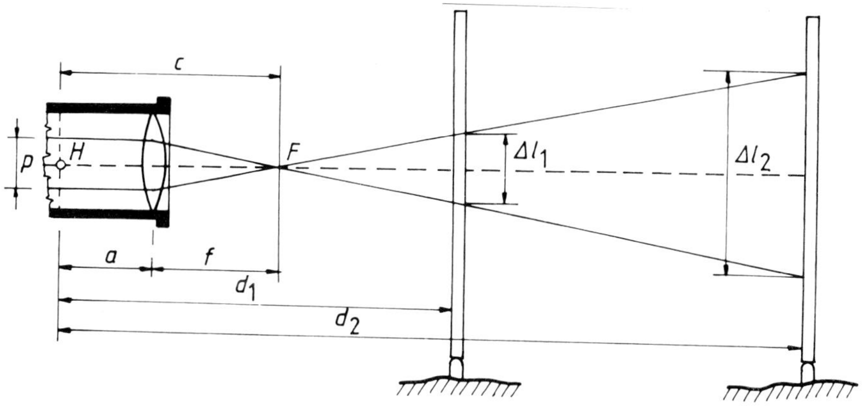

Figure 5.31. Determination of the distance measurement constants c and k

A separate determination of the constant c is possible in a laboratory. Since from *Fig. 5.31*

$$c = a + f, \tag{5.29}$$

the distance a can be measured on the instrument itself, f can be determined on an optical bench.

For the determination of the stadia constant k, uniformly divided distances d_i are set out very carefully and distance measurements are carried out for each with the distance meter to be tested. From the measurements, with telescope horizontal, the stadia constant

$$k_i = \frac{d_i - c}{\Delta l_i} \tag{5.30}$$

is computed for each measurement; the mean value gives the most probable value of k.

The two constants can be determined together in a similar manner. Repeated measurements are made of the distances d_1 and d_2 set out (e.g. 30 m and 50 m) and two equations are obtained for the means Δl_1 and Δl_2 obtained from them:

$$\begin{aligned} d_1 &= c + k\Delta l_1 \\ d_2 &= c + k\Delta l_2, \end{aligned} \tag{5.31}$$

from which c and k can be computed.

The mean error of one single measurement by a distance meter with fixed stadia hairs is, for $k = 100$ and under favourable circumstances, about ± 0.15 m/100 m. The accuracy decreases with increasing distance, therefore the simple distance meter with stadia hairs (and in general distance meters with an external base-line) are not used for measuring distances longer than 150 m. When carrying out a survey for small scale maps (e.g. topographic maps), however, the longest distance used can be as much as 300 to 350 m. At such distances stadia rods with 10 cm graduations are used, with estimation of cm-s.

5.2.1.2 Simple Distance Meters with a Prism

The operational principle of distance meters with a prism is based on the light-deflecting effect of an achromatic glass prism in front of the telescope objective.

On the most simple distance meter with a prism the light-deflecting prism can be swung off the line of sight when measuring. The reading on the stadia rod is made by the respective hair of the telescope, i.e. the horizontal hair is used for a vertical rod and the vertical hair for a horizontal rod. The reading l_1 on the stadia rod is made with a direct line of sight, and the reading l_2 with a line of sight deflected by the prism. The distance is computed by a formula similar to that used for the simple distance meter with stadia hairs.

If the rod is perpendicular to the line of sight and the deflecting angle of the prism is φ, then with a horizontal line of sight:

$$d = c + \Delta l \operatorname{cotan} \varphi \tag{5.32}$$

(see *Fig. 5.32*), or by substituting $\operatorname{cotan} \varphi = k$:

$$d = c + k\Delta l.$$

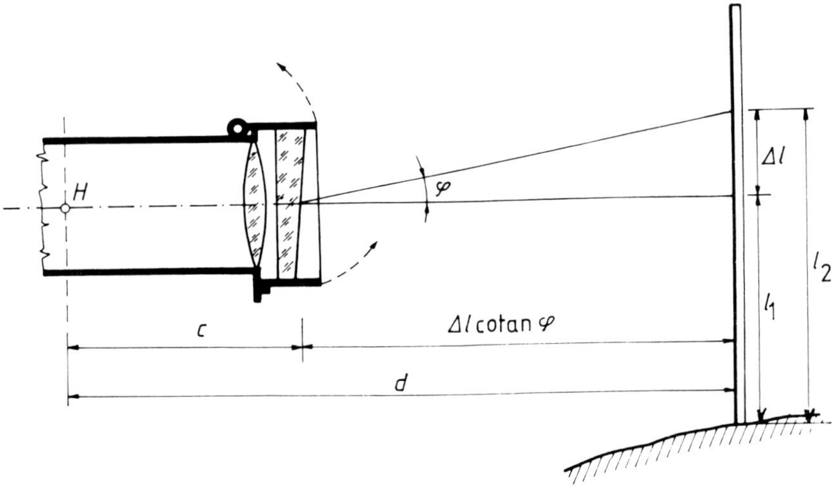

Figure 5.32. Basic principle of a simple distance meter with a prism

The deflecting angle of the prism for a stadia constant $k = 100$ is $\varphi = \operatorname{arc\,cotan} 100 = 34' \, 22.57''$ $(= 63^{c}6596^{cc})$.

For a sloping line of sight and a vertical stadia rod, the horizontal distance is obtained from Eq. (5.28). If the meter uses a horizontal stadia rod which is always set normal to the line of sight, then:

$$d = c \cos \alpha + k\Delta l \cos \alpha.$$

A tripod or folding legs are used for setting up the horizontal rod. The rotation axis of the rod fits into the bush of the instrument base, hence the set up is the same as for a theodolite. The rod is adjusted into the horizontal by means of a box level with a constant of 8 to 10 min of arc, and it is adjusted to be perpendicular to the line of sight by means of a simple pointing device (open sight) which is mounted perpendicularly to the stadia rod. By means of this sight an accuracy of about ± 2 min of arc ($\approx \pm 4^c$) can be achieved, and this is sufficient in practice to ensure the required perpendicularity.

Let $\Delta l = 2.000$ m, then a perpendicularity error of 5 min of arc corresponds to a relative error of the rod length of 1:480,000 which, for a stadia constant $k = 100$, causes a relative error of 1:4800 of the distance. This agrees with the relative accuracy of 1:5000 which is required from distance meters with prisms, even in extreme cases.

The stadia rod can usually be moved along its longitudinal axis and then fixed. Thus, if one leg of the stadia angle points to the zero of the rod graduation (so that $l_1 = 0$), then the division d/k is almost in the vertical through the station mark.

5.2.1.3 Distance Meters with a Prism and Split Image

Distance meters with a prism and split image use an optical wedge, which is sometimes attachable in front of the telescope objective lens. However, the wedge, with a deflecting edge normal to the horizontal sighting plane, covers only half the objective. Hence two images of the rod graduation, one corresponding to the direct line of sight and the other to the deflected one, can be observed together.

In order to achieve the relative accuracy of 1:5000 which is usually required from precise optical distance measurement, the error of the deflecting angle of the wedge-prism must not exceed a value of ± 0.43 sec of arc ($= \pm 1.33^{cc}$). In order to decrease the colour error, an achromatic prism composed of flint- and crown-glass is usually used.

The horizontal stadia rods required for distance meters with a prism and split image have two graduations. One is the base graduation (with 1 or 2 cm division intervals) and the other is a vernier graduation (of 10 or 20 divisions). When observing the rod a mixed image of the two graduations, shifted towards each other, can be seen but the relative displacement Δl of the two graduations cannot yet be read. One possible way to separate the two images is to use colour separation on the stadia rod. This is mainly used on rods required for instruments with attachable prisms. Another solution is the image-separating biprism mounted on the graticule plate of the telescope (Point 6.1.2.3), which produces one image in the field of view above the division line, the other below it.

For colour separation the corresponding combination of black and white is most frequently used; e.g. the two graduations are marked on the rod by black lines on a white background, but the fields below and beyond them are black *(Fig. 5.33)*. Now the super-imposed images are slightly grey but the black lines are easily distinguishable. In another solution white division lines are used on a black background (e.g. the stadia rod in *Fig. 5.34*). Short distances, however, cannot be measured by image separation using colour separation because — with black division lines — the field below the vernier must also be painted black. Short distances (in general $d < 20$ m) are measured with a measuring tape or with an auxiliary staff.

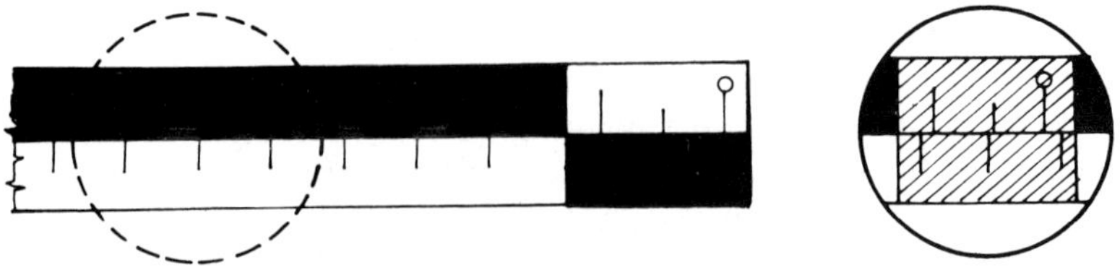

Figure 5.33. Colour separation on the stadia rods for distance meters with a split image

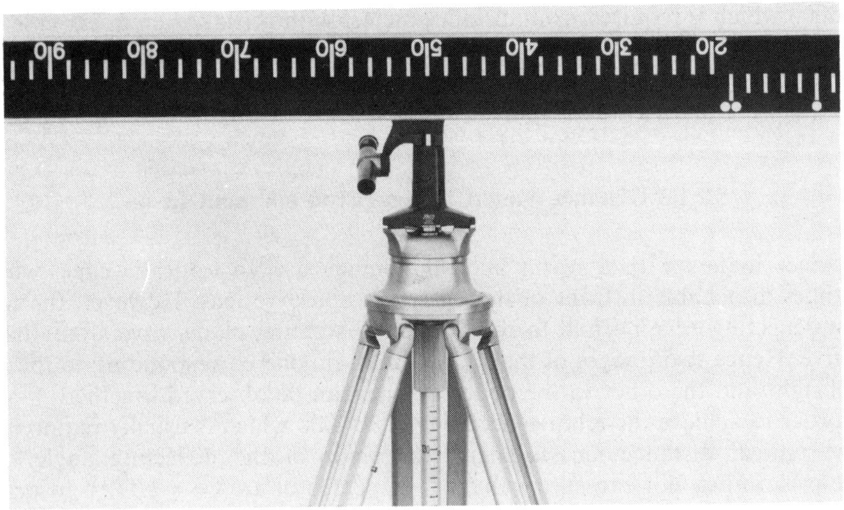

Figure 5.34. Stadia rod for the *DM-M* objective prism (Kern, Aarau)

A vernier of 10 divisions, often used on stadia rods, makes possible readings to one tenth of the base-graduation interval. For e.g. $k = 100$ and a base graduation of 1 cm, the vernier directly yields distance with an accuracy of 0.1 m. The accuracy of reading can be further increased by means of a plane-parallel plate placed in the housing of the prism together with a connected micrometer. With this system one division line of the vernier is to be made coincident with the corresponding line of the base graduation, using the micrometer which is then read. If there are 100 micrometer divisions in one revolution of the micrometer screw, the optical micrometer gives a direct reading on the rod with an accuracy of 0.1 mm and 0.01 mm can be obtained by estimation. The plane-parallel plate, however, causes a shift of the anallatic point, due to which the additive constant c also changes. Since in any case c is not zero in distance meters with a prism, the reduced value of the components is regarded as their c. In the computation of distance, however, the value of c is usually not present because the resultant constant c is taken into account, either by a shift c/k of the rod index or by a correspondingly rotated zero-position of the optical micrometer.

Distance measuring systems of distance meters with a prism and split image are frequently manufactured as self-reducing systems (Point 5.2.2.4).

5.2.1.4 Distance Meters with an Objective Prism

The basic instrument of distance meters with an objective prism is the theodolite. The distance measuring prism, encased in a short metal tube, can be pushed onto the objective end of the theodolite tube and fixed. Horizontal stadia rods are necessary for the majority of these distance meters, but there are some objective prisms which require a vertical stadia rod. Image separation by objective prisms with a split image varies from type to type; length, base-graduation as well as the optical micrometer are all adapted to the requirements of different objective prisms.

The *DM-M* objective prism (Kern, Aarau) manufactured as an attachment to *Kern* theodolites (see *Fig. 5.34*) is used for distance measurement of medium accuracy. A 1.7 m long horizontal stadia rod is associated with the prism, yielding a slope distance. The back-ground colour of the rod is black, the graduation lines every 2 cm are white. Two verniers with 10 divisions have the same colour as the back-ground and they are displaced with respect to each other by 30 cm. Distances between 20 and 100 m can be measured with the inner vernier, between 50 and 150 m by the outer one. For the measurement of distances shorter than 20 m a 25 cm long auxiliary rod is available; it can be attached to the stadia rod. A tripod or folding legs are used for setting up the rod.

The Kern objective prism *DR* gives the horizontal distance directly *(Fig. 5.35)*. The self-reducing method of its distance-measuring system is the same as that of tacheometers with a reducing prism which operate on the basis of the *Bosshardt* principle (see Point 6.1.2.3). It differs from the prism-systems included in tacheometers in that here the prism, consisting of two members, is not automatically rotated by the rotation of the telescope around its horizontal axis. After pointing, the axis of the tube level of the objective prism on the *DR* must be adjusted to the horizontal by its adjusting screw; by this operation the prisms are turned by an amount that depends on the telescope tilt. There is no optical micrometer; the stadia rod adopted

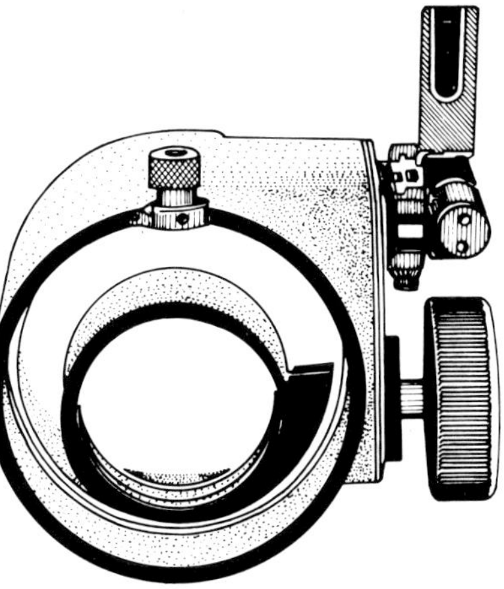

Figure 5.35. The *DR* self-reducing objective prism (Kern, Aarau)

is 1.5 m long, and it has a 1 cm graduation and two verniers with 20 divisions, the latter being displaced with respect to each other by 30 cm, as on the rod of the *DM-M* (*Fig. 5.36*). The nominal mean error is ± 5 cm/100 m.

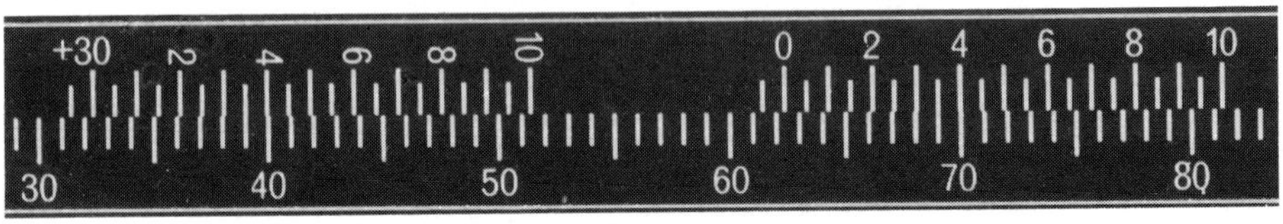

Figure 5.36. Stadia rod for the *DR* objective prism

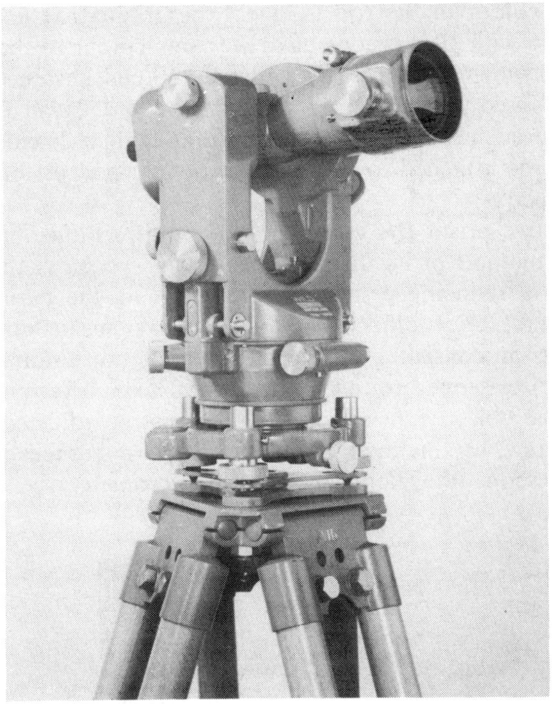

Figure 5.37. The *DM-1* objective prism (Wild, Heerbrugg) on a theodolite

The *Wild* (Heerbrugg) objective prism *DM-1* with an optical micrometer has a system similar to that of the *Kern DM-M* prism (*Fig. 5.37*). It can be pushed onto the telescopes of the *Wild* theodolites *T1-A*, *T16* and *T2*. The stadia rod adopted is 1.6 m long and has a base-graduation of 1 cm, numbered every 10 cm (*Fig. 5.38*). It has two verniers with 10 divisions, giving measurement ranges between 10 and 100 m and between 60 and 150 m respectively. The stadia rod can be folded for transportation and storage. Folding legs or a tripod are used for setting up; forced centring can also be used with the latter. Slope distance can be determined by its optical micrometer

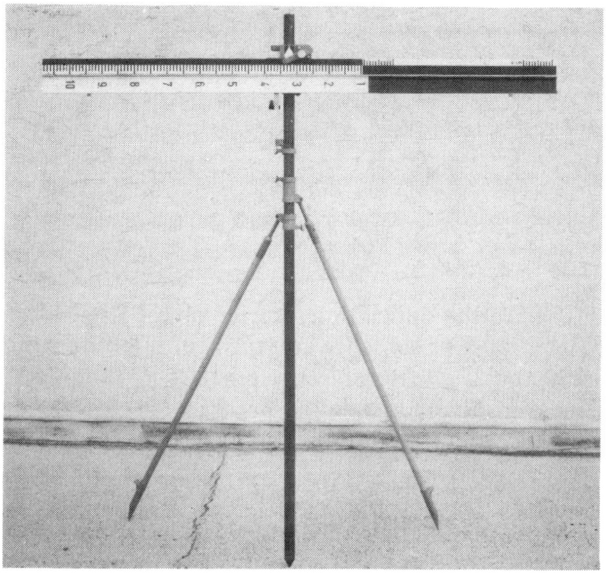

Figure 5.38. The stadia rod of the *DM-1* objective prism with folding legs

Figure 5.39. The *K-320* distance measuring objective prism (MOM, Budapest)

with a relative accuracy of 1 : 5000 under favourable circumstances. (Wild has ceased production of this objective prism *DM-1*).

The *K-320* objective prism with optical micrometer is manufactured by MOM (Budapest) as an accessory for the *B* types of MOM theodolites. The achromatic prism covers an 18 mm wide central strip of the objective *(Fig. 5.39)*. The split image can be eliminated without moving the objective prism by swinging a cover plate of the same width in front of the prism. The micrometer has 20 divisions. The effective length of the associated stadia rod *K-086* is 160 cm *(Fig. 5.40)*. There is a 2 cm graduation with 10 cm numbering on the face of the rod which has a triangular cross-section. The zero lines of the two 10 division verniers are 50 cm from each other, but there is no base-graduation between them. In order to separate the double image the rod is black both below the verniers and above the sections of the main graduations. The 110 cm long base-graduation makes possible the measurement of slope distances of up to 90 m with the inner vernier and up to 150 m with the outer vernier. The rod can be set up using the socket which guides the forced centring of

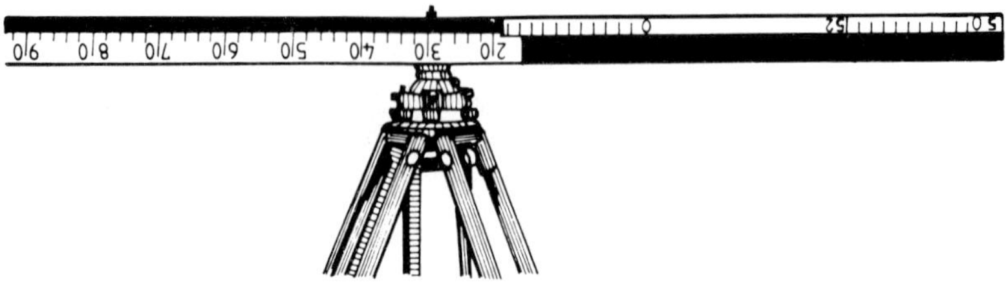

Figure 5.40. The *K-086* stadia rod (MOM, Budapest) on a tripod

MOM instruments. Since the optical micrometer of *K-320* gives a halved reading, the optical micrometer must be set twice and the sum of both readings added to the vernier reading on the rod. Using the objective prism *K-320* and the stadia rod *K-086*, distance can be measured with a relative accuracy of 1:5000 under favourable circumstances and taking the temperature effect into account. The reduction formula

$$\Delta t = 1.15 \times 10^{-5}\, d(t^\circ - 20\,^\circ C) \tag{5.33}$$

is used for the computation of the temperature effect, where t° is the temperature during measurement.

The Rank Precision Industries (Leicester) objective prism with an optical micrometer has a 1.6 m long mahogany stadia rod, with a vernier of 10 divisions. Its measuring range lies between 20 m and 140 m. The nominal relative accuracy is 1:5000.

VEB Jenoptik (Jena) manufactures objective prisms named *Dimesskeil 002* and *Dimesskeil 004* which can be used on several Jena instruments *(Fig. 5.41)*. The two prisms differ only in the diameter of this mount. The former can be fitted to the instruments *Theo 020*, *Theo 030*, *Dahlta 020* and the earlier instruments *Th IV* and *Th II*, while the latter fits the telescope of the theodolite *Theo 120*. The optical micrometer for *Dimesskeil 002* gives a measurement with a mean square error of ±2

Figure 5.41. Dimesskeil objective prism (VEB Jenoptik, Jena) and its counterweight

cm/100 m, and for *Dimesskeil 004* one with a mean square error of ±4 cm/100 m. Both prisms are central and cover astrip across the objective of area about half the free objective aperture. The additive constant of the instrument with prism attached must be regarded as zero, since the rod reading Δl is increased by c/k due to a displacement of both the rod index and the optical micrometer.

There are three stadia rods of different sizes for the *Dimesskeil* type distance measuring prisms *(Figs 5.42, 5.43, 5.44)*. Their basic material is wood and the graduations are on a metal band. Graduation errors at a temperature of +20 °C are less than 0.05 mm per 1 m rod length. The temperature effect can be taken into account by Eq. (5.33) for the distance measured, or the length of the rod at the temperature of the measurement can be taken from a table provided with it. The separation of the double image is made by a combination of black and white. Each of the three rods has an auxiliary rod of appropriate size for the measurement of short distances. The lengths of the rods and their measurement range in m are:

Rod length	Range of measurement		Auxiliary rod
	Dimesskeil 002	Dimesskeil 004	
2.09	22...172	22...100	2...22
1.52	22...128	22...100	2...22
0.77	6... 66	6... 55	2...10

On the 2.09 m stadia rod a base-length of 2 m ±0.1 mm is also marked for distance measurement with a constant base-line. Since the rod constant (the horizontal distance between the front graduation plane) of the rod and the vertical through the station point is not taken into account in the 2 m long base, the measured distance d must be increased by 4.5 cm. The 0.77 m so-called hand-rod can be very useful for measuring distances shorter than 66 m (e.g. for polar distances).

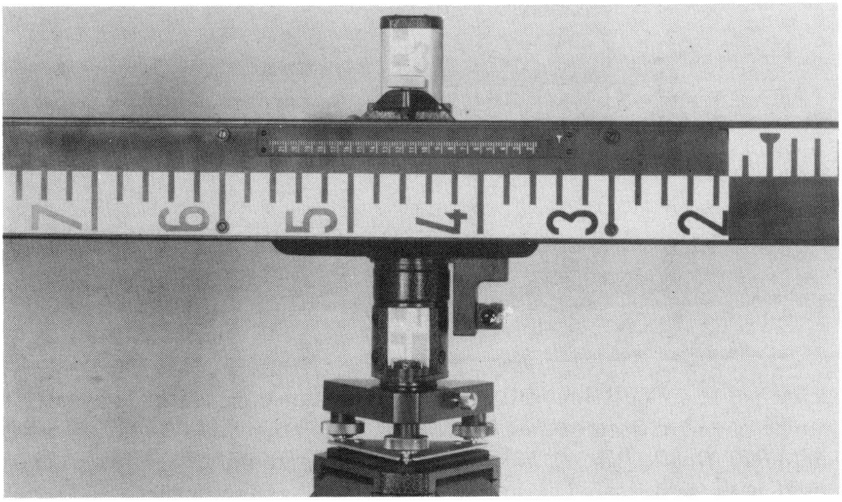

Figure 5.42. The 2.09 m long stadia rod for the *Dimesskeil* distance measuring prisms

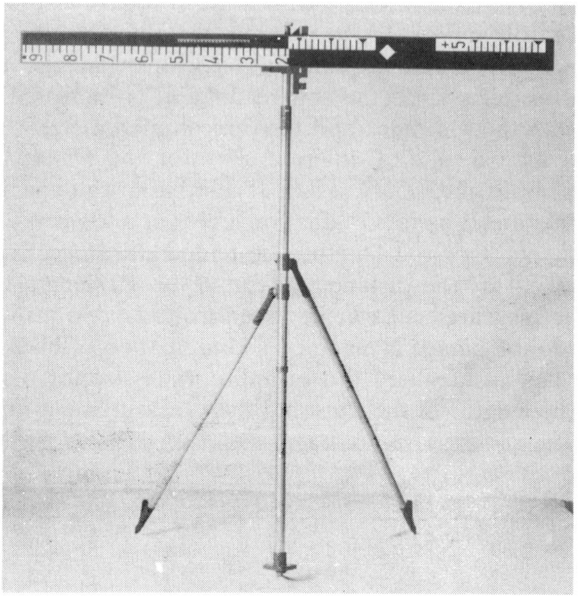

Figure 5.43. The 1.52 m stadia rod for the *Dimesskeil* distance measuring prisms on folding legs

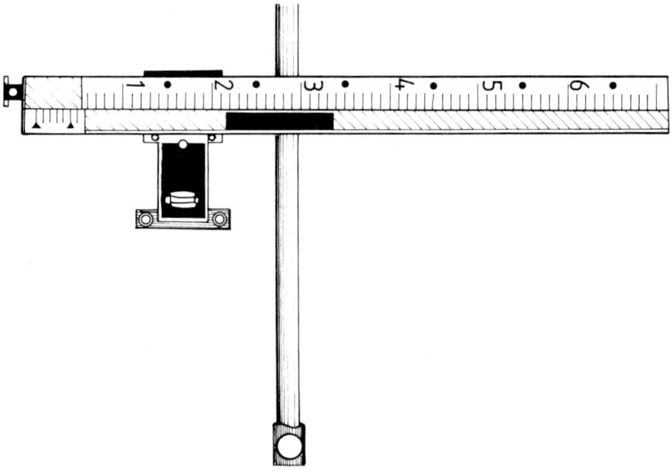

Figure 5.44. The 0.77 m stadia rod for the *Dimesskeil* objective prisms

The *Lotakeil 004* (VEB Jenoptik, Jena) is a distance measuring objective prism which can be placed in front of the objective on the instruments *Theo 020, Theo 020 A,* and *Dahlta 020, Dahlta 020 A (Fig. 5.45).* A 2 m horizontal or a 3 m vertical stadia rod is used with the prism. The measurement range of the former is 400 m, of the latter 600 m. The wide measurement range is realized by a special, logarithmic

Figure 5.45. The *Lotakeil* distance measuring prism (VEB Jenoptik, Jena) with counterweight

graduation of the rod. With this graduation distances between two adjacent divisions vary to make the logarithmic differences between the reading in constant. Consequently the distance between the first division (l_0) and those following in turn are:

$$l_1 = l_0 10^a, \quad \text{i.e.} \quad \lg l_1 = \lg l_0 + a \cdot \lg 10,$$

$$l_2 = l_0\, 10^{2a}, \quad \text{i.e.} \quad \lg l_2 = \lg l_0 + 2a \cdot \lg 10,$$

$$\ldots\ldots\ldots \qquad \ldots\ldots\ldots\ldots\ldots\ldots$$

$$l_n = l_0\, 10^{na}, \quad \text{i.e.} \quad \lg l_n = \lg l_0 + na \cdot \lg 10,$$

where a is a suitably chosen constant.

The i-th division separation is:

$$b_i = l_i - l_{i-1} = l_0 10^{ia} - l_0 10^{(i-1)a} = l_0 10^{ia}(1 - 10^{-a})$$

or

$$b_i = l_i \frac{10^a - 1}{10^a}. \tag{5.34}$$

From Eq. (5.34) each division separation is proportional to the reading l_i on the rod and hence also to the distance to be measured. On the stadia rods for the *Lotakeil 004*, $a = 0.02$. The value of the stadia constant k depends on the angle ε_i at which the graduation separation b_i is visible:

$$\varepsilon_i = \frac{b_i}{d_i} = \frac{b_i}{k \cdot l_i} = \frac{10^a - 1}{k \cdot 10^a}; \tag{5.35}$$

in other words, ε_i is independent of distance.

A distance measurement can be carried out with a stadia rod with a logarithmic graduation in the same way as with a stadia rod with a linear graduation if the division lines are numbered with the mantissae of the Briggsian logarithm. An advantage of the logarithmic rod graduation is that, when reading the rod at any distance, the division separation subtends the same constant angle at the observer. Division separations which increase with distance also enable longer division lines

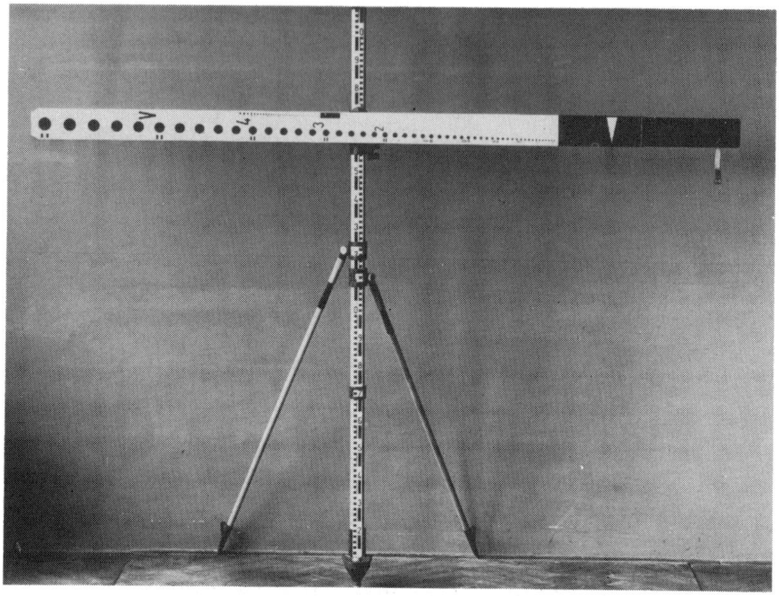

Figure 5.46. Stadia rod with logarithmic graduation for a *Lotakeil* objective prism

and increased numbering to be used for longer distances *(Fig. 5.46)*. The numbering on the Lotakeil rod is the logarithm of k-times l_i. With a rod normal to the line of sight the reading gives the logarithm of the slope distance. The zero mark on the rod is a rhombic symbol, the characteristics 1 and 2 are one and two short lines respectively. Mantissae can be read on the rod directly with an accuracy of two decimals. The division marks are black circles of increasing diameter, the index mark is a white wedge on a black back-ground. This mark appears white between two division marks. The extra distance to a division mark is measured with an optical micrometer. The optical element of the micrometer is a rotating prism behind the distance-measuring prism, rotated by a micrometer screw. The measuring range of the rotating prism is chosen in such a way that, over four whole revolutions of the micrometer screw, the stadia angle changes by t, the distance-independent angle subtended by the stadia

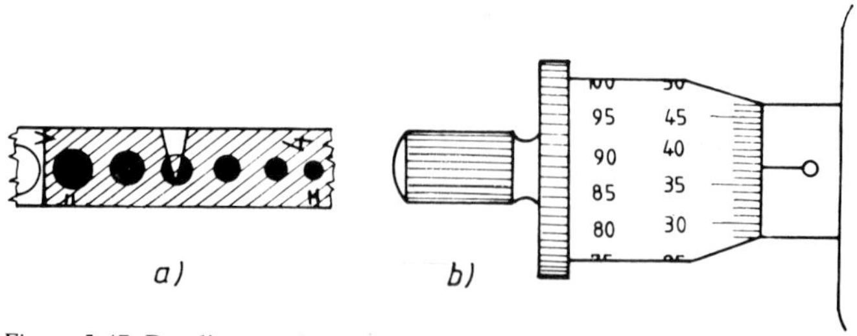

Figure 5.47. Reading on the stadia rod for the *Lotakeil* prism and on its micrometer drum *a)* on the rod: 2.46; *b)* on the optical micrometer: 374, lg $d_s = \log k \cdot l = 2.46374$

distance which increases with the measured distance. The micrometer drum has 50 divisions, hence the stadia distance on the rod can be read on it with an accuracy of 1/200. This reading gives the third and fourth decimals of the mantissae, and the fifth decimal can still be estimated *(Fig. 5.47)*. The logarithm of the horizontal distance can be obtained as a function of the zenith distance angle z, measured at the same time as the distance, in the form:

$$\lg d = \lg (k \cdot l) + \lg \sin z. \tag{5.36}$$

Equation (5.36) is modified for a vertical rod as follows:

$$\lg d = \lg (k \cdot l) + \lg \left(\sin^2 z + \frac{\sin z \cdot \cos z}{k} \right), \tag{5.37}$$

or

$$\lg d = \lg (k \cdot l) + A. \tag{5.38}$$

The value A as a function of z can be taken from a table provided with the instrument.

A theodolite supplied with a Lotakeil prism can also be used as a tacheometer. In this case the logarithm of the height difference between the axis H and the observed point on the stadia rod is:

$$\lg \Delta h = \lg (k \cdot l) + B, \tag{5.39}$$

where the value B can also be taken from the above-mentioned table or computed by the formula:

$$B = \lg \left(\sin z \cdot \cos z + \frac{\cos^2 z}{k} \right). \tag{5.40}$$

With a Lotakeil prism a reliability of ± 4 cm/100 m can be attained with a horizontal rod and ± 7 cm/100 m with a vertical rod.

Objective prisms manufactured in the Soviet Union follow the principle of differential distance meters, and the role of the optical micrometer is taken over by a prism system. This systen is composed of two prisms with different deflection angles and they cover half the telescope field of view. Two parallel rays meeting the prism with a separation of p_1 are emitted still parallel after passing through the prism.

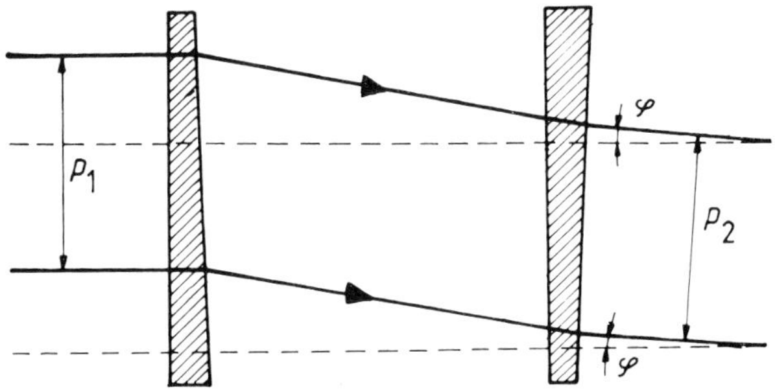

Figure 5.48. Ray path in a differential distance meter

However, their separation will be p_2 and both will be deflected by the angle φ *(Fig. 5.48)*. In the differential distance meters manufactured in the Soviet Union the prisms are chosen in such a way that the ratio of their magnifications is $\Gamma = 90/89$, and the stadia angle corresponds to a stadia constant $k = 100$ ($\varphi = 34' 23'' \approx 63.66^{cc}$). If the prism is attached in front of the telescope, one leg of the stadia angle is determined by the line of sight and the other by the direction of the light ray deflected by the prism system. The distance can be computed by Eq. (5.18). The relative position of the split image is read on the vernier. The zero line of the vernier differs from the zero mark of the rod graduation by c/k, therefore the addition constant is not to be taken into account in the measurement. The vernier has 10 divisions and its total width is 89 mm. In consequence of the 90/89 magnification ratio of the prism system, the end-line of the vernier is shifted by 1 mm with respect to the rod graduation if the line of sight of the telescope is rotated by as much as corresponds to the total length of the vernier. If, however, the vernier line is made to coincide with the nearest line of the main rod graduation by a small swing of the telescope, the reading on the vernier with the horizontal hair is proportional to the shift of the vernier. Hence the total reading on the stadia rod is composed of three parts: 1. the reading of the base-graduation of the rod, 2. the reading of the coincident vernier line, 3. the reading of the horizontal hair on the vernier graduation *(Fig. 5.49)*.

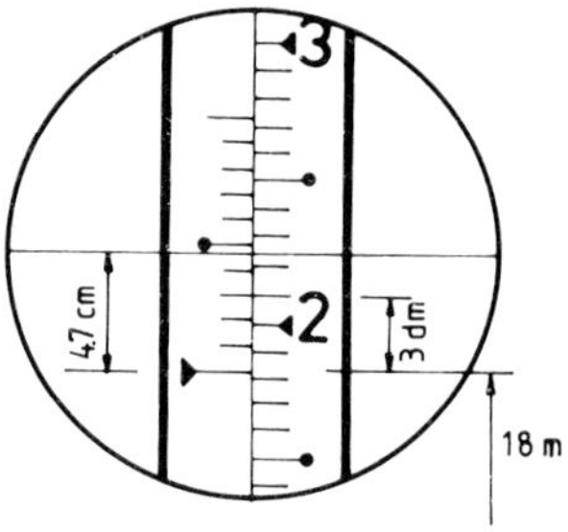

Figure 5.49. Reading on the rod of a differential distance meter ($d = 18.347$ m)

Differential distance meters can be used both with vertical and horizontal stadia rods. The result of the distance measurement is a slope distance in both cases.

In some differential distance meters the distance-measuring system consists of two half-lenses, instead of prisms, giving a constant parallactic angle (see Point 5.2.3.3), as e.g. on the attachment system of the differential distance meter *DN-10* manufactured in the Soviet Union. The Soviet attachment system *DNR-06* gives reduced horizontal distances. Its characteristic is that it consists of a fixed and of a swinging prism, the latter operating as a pendulum. The rays passing through the prism system are deflected by an amount which varies proportionally to the cosine of the telescope tilt. The optical data of the prisms ensure that the reading multiplied by $k = 100$ gives the horizontal distance directly.

A relative accuracy between 1:2000 and 1:2500 can be attained with differential distance meters, depending on the type of stadia rod and on external circumstances.

5.2.1.5 Test and Adjustment of Distance Meters with a Prism and a Fixed Stadia Angle

Distance meters with a prism must be tested not only for the errors and adjustment of the basic instrument (theodolite), but also for the constants c and k of the distance-measuring system. There may well be a considerable change in the value of the additive constant c given by the manufacturer.

The test of the stadia constant k is made by repeated measurements of different long base distances. Having substituted the most reliable value of a distance obtained from the measurement series into the formula for distance measurement, the stadia constant k can be expressed by the formula. Since displacement in the prism system is a significant source of systematic error, distance meters with a prism are supplied with an adjusting screw. The prisms can be adjusted by the screw in the following way: The instrument and the stadia rod are set up at the respective end-points of an integer distance (e.g. $d = 100$ m). The rod is sighted and then the optical micrometer is rotated to its base (zero) position. The next step is to use the prism adjusting screw to set the zero line of the vernier very carefully on the rod graduation which corresponds to the distance set out. To check the adjustment this distance is measured several times and the mean value obtained must give the known distance within the accuracy that can be achieved by the distance meter.

5.2.2 Distance Meters with an External Base and a Varying Stadia Angle

5.2.2.1 Subtense Bars for Distance Measurement with Theodolites

A very simple method of indirect distance determination by angle measurement is to use a theodolite at one end of the line and a base b of known length at the other end, set perpendicular to the line. The stadia angle φ subtending the ends of the bar is measured and, knowing the base-length b, the horizontal distance can be obtained from Eq. (5.14) or Eq. (5.15). The reliability of the distance d depends not only on the accuracy of the measurement, but also on the ratio of distance to base length. The reliability of the distance d increases with decrease of the ratio d/b.

Due to the high accuracy (and stability) required of the length of the base, manufactured base rods (called subtense bars) are used for measurement in practice. For the sake of easy handling and transport, 2 m long subtense bars are usually used. Sometimes the subtense bar can be folded once or it can be dismantled into two parts. The connection of the two parts requires, however, a precise device such that the distance between the end-points of the assembled bar should not have an error greater than 0.2 mm (relative error 1 : 10,000). The relatively short base severely limits the maximum distance that can be measured by a substense bar with the accuracy required (see Point 5.2.2.3).

With a given base-length the accuracy of distance measurement is influenced by the system of marking the ends of the base and by the stability of the distance so defined. The end points are marked in very different ways. *Figure 5.50* shows the markings used on a few products.

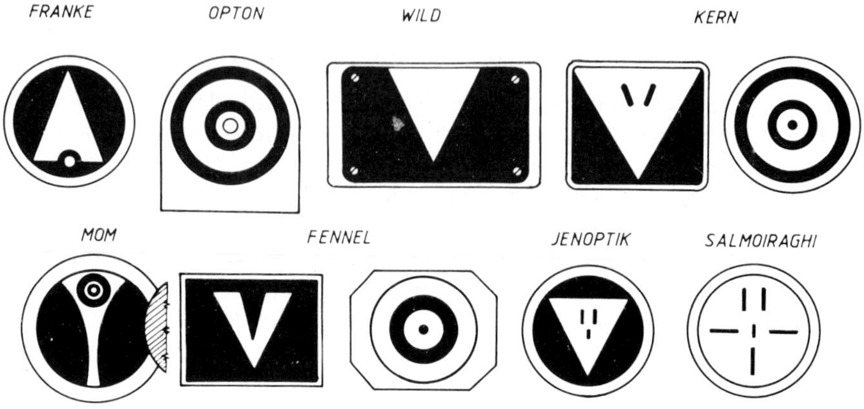

Figure 5.50. Markings of targets on different types of subtense bars

The subtense bar is usually made of light metal. For measurement it can usually be set up on a tripod by means a forced centring device *(Fig. 5.51)*. In order to ensure an unchanged distance between the marks they are often attached to an invar inset inside the casing. Furthermore a frequently used construction is the bimetal compensator; here the expanding effect of temperature is balanced by the automatic shift of one of the marks. On some substense bars the constant distance between the marks is ensured partly by an internal invar inset and partly by a compensator. The bar is adjusted into the horizontal with a box level, and set perpendicular to the line by an open sight through which the theodolite at the opposite end is sighted. It is sufficient to adjust the horizontal and the perpendicular positions with an accuracy of a maximum of 20 min of arc ($\simeq 40^c$).

The two parts of the *GBL* subtense bar (Wild, Heerbrugg) can be connected by a lever clamp. There is a free hanging invar wire in each of the two steel tubes. The

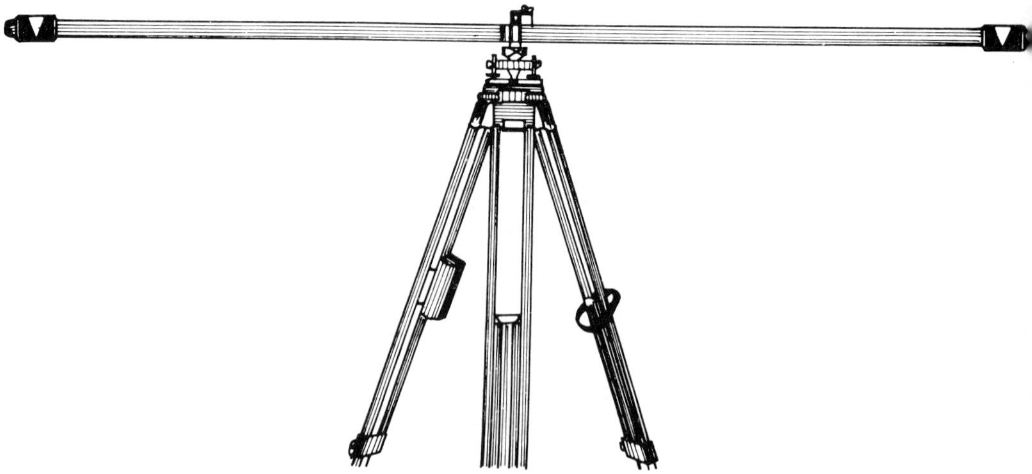

Figure 5.51. The *GBL* subtense bar (Wild, Heerbrugg) on a tripod

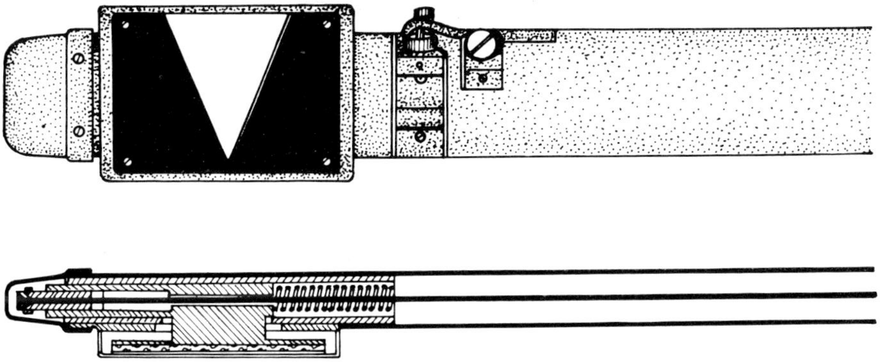

Figure 5.52. Temperature compensation of the *Wild GBL* subtense bar

inner ends of the wires are fixed, their outer ends are strained by the constant force of straining springs *(Fig. 5.52)*. While a temperature change of e.g. 10 °C alters the length of the metal bar by 0.24 mm, the sum of the lengths of the two invar wires is changed only by 0.024 mm. The spring effect is increased at the same time by 0.2 daN, which results in an additional extension of the wire of 0.004 mm, hence the invar wires become longer by 0.028 mm. An 0.018 mm temperature expansion of the brass supports of the targets in the opposite direction decreases the alteration of the separation of the marks to 0.010 mm. Hence, even for a temperature change of 20 °C, the base-length b undergoes a relative change of only 1 : 100,000.

On the *IB* subtense bars (Kern, Aarau) the two marks *8*, affected by the two end springs *7*, are only partly connected by invar *(Fig. 5.53)*. Distance compensation is

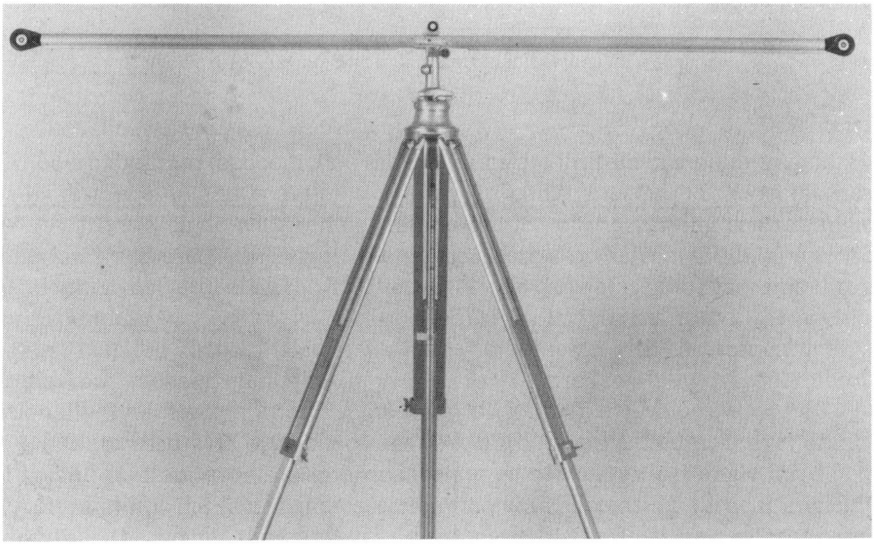

Figure 5.53. The *IB* subtense bar of Kern (Aarau) on a tripod

accomplished inside the light metal tube *2* by the steel insets *3*, *5*, *6* as well as by the aluminium inset *4* which has a greater temperature expansion; the insets are between the invar wires *1* *(Fig. 5.54)*. According to the data given by the factory, a temperature change of 30 °C causes a change of only 0.168 mm in the distance between the marks. Even this is eliminated by the inset aluminium tube-compensator *4*, whose length ensures that the 2 m mark separation should remain constant with an accuracy of ±0.03 mm.

On the *Bala-2* subtense bar (VEB Jenoptik, Jena) *(Fig. 5.55)* again a bimetal compensator is used to ensure that the length between the inner ends of the invar insets holding the base-marks should remain constant.

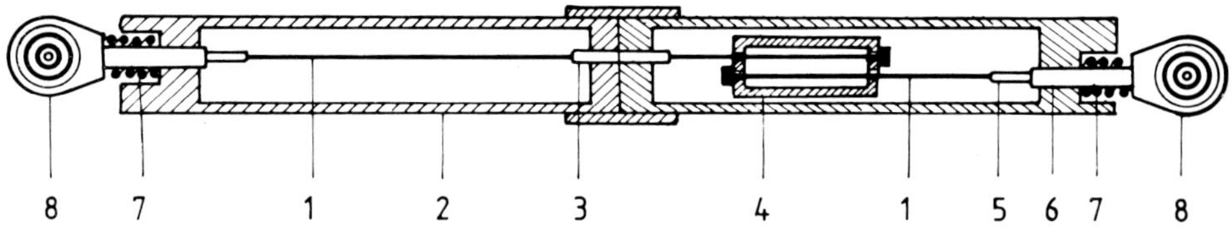

Figure 5.54. Temperature compensation of the *IB* subtense bar

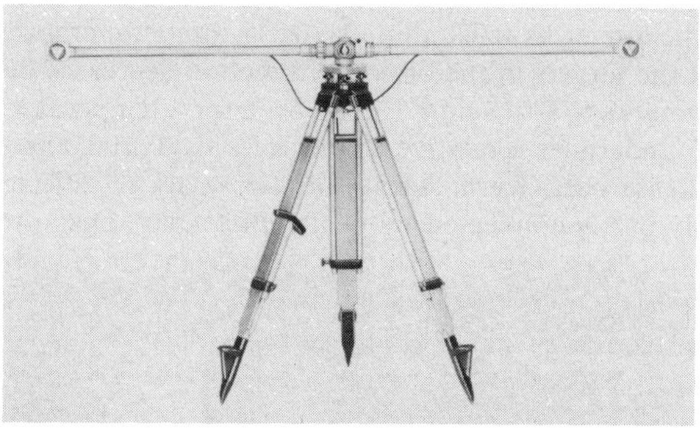

Figure 5.55. The *Bala-2* 2 m subtense bar (VEB Jenoptik, Jena)

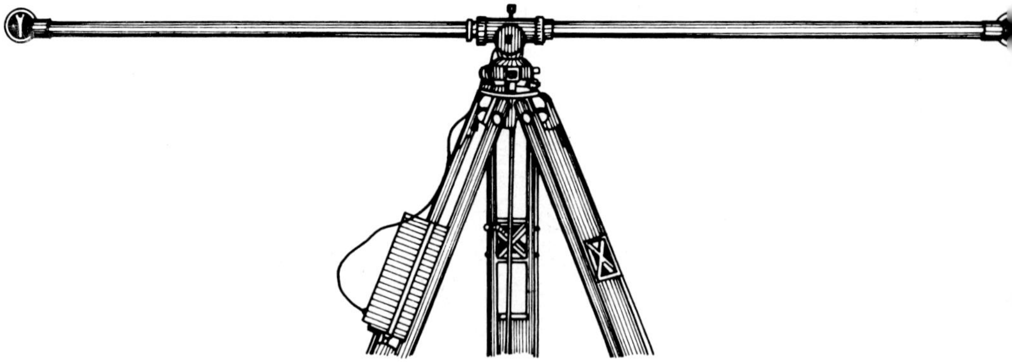

Figure 5.56. The *K-070* subtense bar (MOM, Budapest) on a tripod

The *K-070* subtense bar (MOM, Budapest) consists of two bars with an invar inset and a bayonet locking connection at the centre *(Fig. 5.56)*. A *K-810* optical plummet is used for centring over the point. The base-marks can be illuminated, if necessary, through a connection cable from a battery box type *K-150*. There is a sighting target at the centre of the subtense bar, therefore its replacement by a centred target is unnecessary when traversing at the same time as measuring distance.

In the *Askania* subtense bar (Franke et Co., Giessen) the temperature effect is compensated by the opposite extensions of the invar wire and the target support.

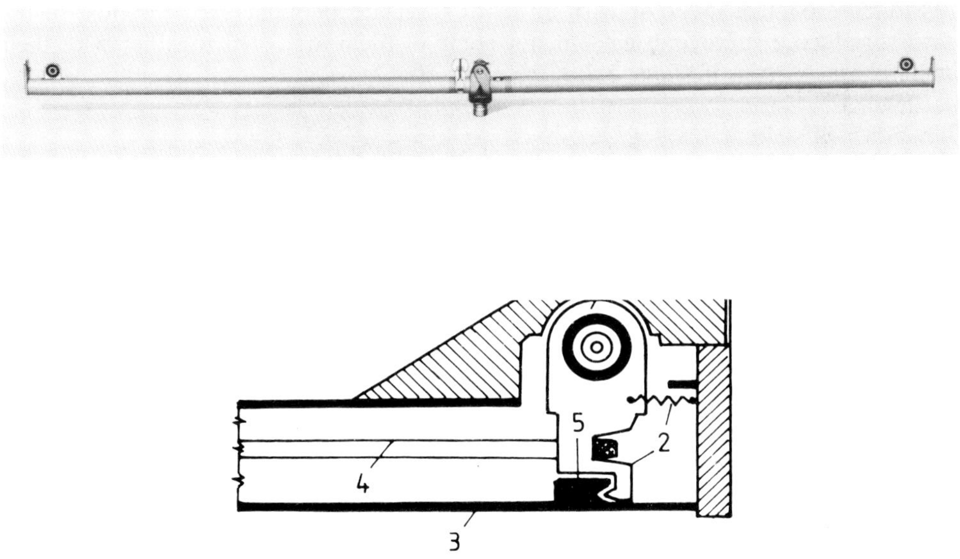

Figure 5.57. The *Opton* (Oberkochen) 2 m subtense bar and its compensation

Table 5.2. Subtense bars for distance measurement with theodolites

Manufacturing firm	Mark	Theodolite	Base length (mm)	Correction of temperature effect
Wild AG	GBL	T2	2000	invar + compensator
Kern & Co.	IB	DKM-2	2000	invar + compensator
Jenoptik	Bala-2	Theo 010/A Theo 010/B	2000	invar + compensator
Opton GmbH	—	Th-3	2000	compensator
Otto Fennel GmbH	—	FTS	2000	compensator
MOM	K-070	TeB1, TeB3, TeB41	2000	invar
Filotecnica Salmoiraghi	—	4200	2000	invar
Rank Precision Ind.	—		2000	invar
Officine Galileo	—	TG 1	2000	invar
Hilger & Watts	—	Microptic 2	2000	invar
Franke	—	Tu	2000	invar

The *Opton* subtense bar (Opton Feintechnik GmbH, Oberkochen) is contained in an aluminium tube and uses the different temperature expansions of steel and aluminium for length compensation *(Fig. 5.57)*. In the aluminium tube *3* the left-hand target *1* operates with a spring tensioner *2* as a rocker; its lower edge, ending in a wedge shape, is clamped by the steel wire *4* beyond it to the nose *5*. The tension spring *2* works opposite to it. If constructed with proper ratios, the stability of the base-length can be kept within the required limits.

Fennel (Kassel) manufactures its 2 m long subtense bar from quartz in order to compensate the temperature effect.

Subtense bars of invar or with an invar inset are manufactured by most of the firms producing surveying instruments; some of them are listed together with their main characteristics in *Table 5.2*.

5.2.2.2 Test of Subtense Bars

Systematic errors of distances measured by subtense bar have their sources in errors of setting up and in defects in manufacture as well as in errors in the measurement of the parallactic angle. The error of setting up can be eliminated in practice by being careful provided that the box level, the projecting (centring) device and the open sight are properly adjusted.

The error in the base-length of the bar, and in the relative eccentricity of the vertical target-plane and of the vertical rotation axis, are two characteristic systematic error sources in the construction of a subtense bar. Distance measurement is firstly considerably influenced by error in the length of the base, e.g. a 1 mm error (1:2000 relative error) in the base-length causes a systematic error of 5 cm over a distance of 100 m. The eccentricity e of the target plane causes a constant error of measurement, independent of distance. Since the manufacturers guarantee $e<0.5$ mm, the eccentricity influence is negligible in practice.

The error of the base-length can be determined by direct or indirect methods. A direct method is calibration with a comparator, carried out similarly to the calibration of levelling staffs with an invar inset (see Point 3.1.2.9). However, special comparators have been constructed for the calibration of subtense bars, such as e.g. the subtense bar comparators of Wild (Heerbrugg) or Breithaupt (Kassel).

One possible method for indirect determination of the base-length b is due to *Kobold (Fig. 5.58)*. A distance $d \approx 10$ m is established on a straight line set out at a distance s from a comparison base l and parallel to it. Next the angles δ_1, δ_2 and φ are measured very carefully with several repetitions.

From the measured data one obtains:

$$d = l + \frac{s}{\rho''}(\delta_1'' + \delta_2''), \tag{5.41}$$

and

$$b = 2d \tan\frac{\varphi}{2}. \tag{5.42}$$

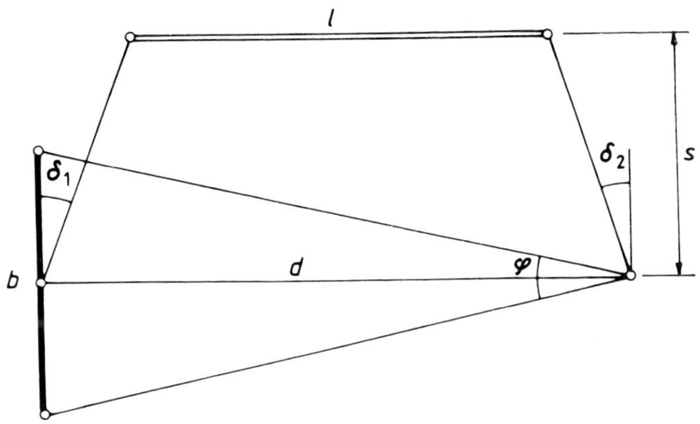

Figure 5.58. The Kobold method to determine the base length of a subtense bar

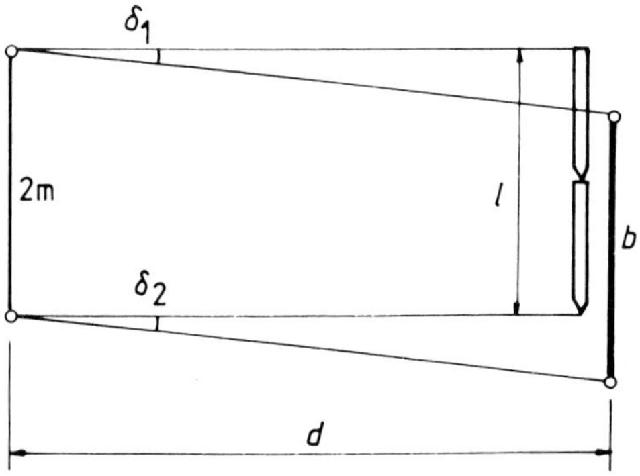

Figure 5.59. The Ochsenhirt method to determine the base length of a subtense bar

According to the *Ochsenhirt method (Fig. 5.59)* a comparison base of $l \approx 2$ m is established by a pair of standard metres (or by a calibrated bar with an invar inset) and the subtense bar is placed in the same vertical plane (above or below the comparison base). From a distance $d = 6.37$ m the horizontal angles δ_1'' and δ_2'', between the corresponding ends of the comparison base and of the subtense bar, are measured. Then

$$b = l + \frac{d}{\rho''}(\delta_2'' - \delta_1''). \tag{5.43}$$

For both methods the angles must be measured with a mean square error less than ± 1 sec of arc ($\pm 3^{cc}$) in order to obtain the base-length with an accuracy of ± 0.1 mm.

5.2.2.3 The Accuracy of Distance Measurement with a Subtense Bar

The factors influencing the accuracy of distance measurements with a theodolite and subtense bar can be divided into three groups: error of the position of the theodolite and the bar, manufacturing defects of the bar and error in the measurement of the parallactic angle.

Among the errors of position, the error of centring over the point affects the distance measurement — as in the case of every distance meter — by the corresponding projection of the eccentricity. If the centring is carefully made with an optical projector, the error in the distance measurement is very small ($e < 1$ mm). The maximum deviation of about 20 min of arc (37^c) in the horizontal and perpendicular position of the subtense bar is an error limit easy to attain by using adjusted box levels and sights.

Among the manufacturing defects the eccentricity of the plane of the targets with respect to rotation axis of the bar has no significant effect on the distance measured. The error of the length of the base however, can only be neglected if its value is less than 0.1 mm. The dilatational change of precise subtense bars with an invar inset or with length compensation theoretically does not exceed the error limit, but the stability of the length of the bars must be checked from time to time by calibration on a comparator.

The accuracy of distance measurement with a calibrated subtense bar is mainly influenced by the ratio of the measured distance to the base-length d/b (which determines the angle φ) and, within that, by the accuracy with which the angle φ is measured. If the law of error propagation is applied to the basic equation of distance measurement (5.14) (where b should also be regarded as a variable), then after reduction the reliability of measurement is characterized by the formula:

$$m_d = \sqrt{\frac{d^2}{b^2} m_b^2 + \left(\frac{d^4}{b^2} + 2d^2 + b^2\right) \frac{m_\varphi^2}{\rho^2}} . \tag{5.44}$$

Investigating Eq. (5.44) with regard to the mean square error of angle measurement $m_\varphi = \pm 1$ sec of arc ($\pm 3^{cc}$), to be achieved by an appropriate number of repetitions, one finds that Eq. (5.44) can be reduced without practical error to the form:

$$m_d = \sqrt{\frac{d^2}{b^2} m_b^2 + \frac{d^4}{b^2 \rho^2} m_\varphi^2} . \tag{5.45}$$

By separating Eq. (5.45) into the error effects of the two variables, one obtains:

$$m_{d_b} = \pm \frac{d}{b} m_b \tag{5.46}$$

and

$$m_{d_\varphi} = \pm \frac{d^2}{b\rho} m_\varphi . \tag{5.47}$$

In *Table 5.3* the mean square errors for some values of d are listed, computed from Eqs (5.45), (5.46) and (5.47) with the assumption that the mean square errors $m_b = \pm 0.1$ mm and $m_\varphi = \pm 1$ sec of arc.

It can be seen from the Table that a 0.1 mm change in base-length has no practical effect on the mean square error of distance measurement. On the other hand an error of 1 sec of arc in angle measurement — with a constant base-length — causes the mean square error of distance measurement to increase with increasing distance, proportionally to the square of the distance.

According to Eq. (5.45) the relative error is 1:5000 at a distance of about $d=80$ m, if $m_\varphi = \pm 1$ sec of arc. By increasing the accuracy of angle measurement a relative error of 1:5000 should be obtained at a distance of 100 m. Hence the accuracy of measurement with a subtense bar can be characterized by a relative mean square error of ± 2 cm/100 m.

If the distance to be measured is greater than 100 m, but does not exceed 200 m, it is advisable to measure the distance separately in two parts *(Fig. 5.60, a)*. In this case, if $d_1 \approx d_2$, $m_{d=200\,\mathrm{m}} = \sqrt{2} m_{d=100\,\mathrm{m}} = \pm 35$ mm (in spite of the error ± 97 mm given in *Table 5.3*).

Table 5.3. Accuracy of distance measurement by subtense bar and theodolite ($m_\varphi = \pm 1''$ and $m_b = \pm 0.1$ mm)

d (m)	50	100	150	200	250	300
m_{d_b} (mm)	±3	±5	±8	±10	±13	±15
$m_{d\varphi}$ (mm)	±6	±24	±55	±97	±152	±218
m_d (mm)	±7	±25	±55	±97	±152	±219

Longer distances ($d > 200$ m) can be deduced from the parallactic angle determined by the end-points of an auxiliary base b_0 measured by a distance meter *(Fig. 5.60, b)*.

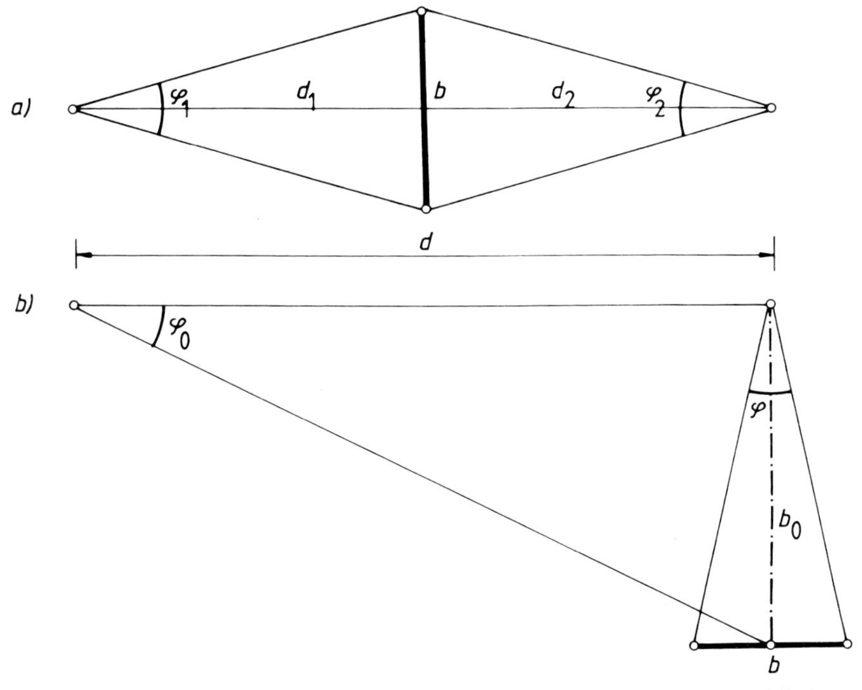

Figure 5.60. The measurement of long distances with a 2 m subtense bar: *a)* 100 m<d<200 m; *b)* d>200 m

By this method a distance of $d = 1$ km can be measured, depending on external circumstances, with a mean square error between ± 5 cm and ± 10 cm (or on average with a relative error of 1 : 15,000).

5.2.2.4 Distance Meters with a Split Image and a Varying Stadia Angle

On some distance measuring instruments the parallactic or stadia angle can be changed by an image-shifting optical system. A beam of rays comes from each of the two subtense bar targets which are at a constant separation b, and enters the telescope of the instrument. The two images are brought into coincidence by changing the stadia angle φ by an amount depending on the distance d.

The distance meter consists of two main constructional elements: the metal frame containing the optical system and the reading microscope connected to it. The meter can be attached to a telescope objective in a similar way as prism attachments.

The operating principle of a distance meter is as follows: if the condenser lens, of focal length f, is cut along its diameter and the two half lenses are given a relative displacement of Δp along the surface of the cut and perpendicular to the optical axis, then parallel beams of rays entering the two half lenses form on exit an angle of:

$$\varphi = \frac{\Delta p}{f} \rho \tag{5.48}$$

(Fig. 5.61). In the meter the angle φ, varying with the displacement Δp, is used as a varying parallactic angle. For a constant external base b the distance is

$$d = \frac{b}{\varphi} \rho . \tag{5.49}$$

Substituting Eq. (5.48) gives:

$$d = \frac{bf}{\Delta p} = \frac{k}{\Delta p} . \tag{5.50}$$

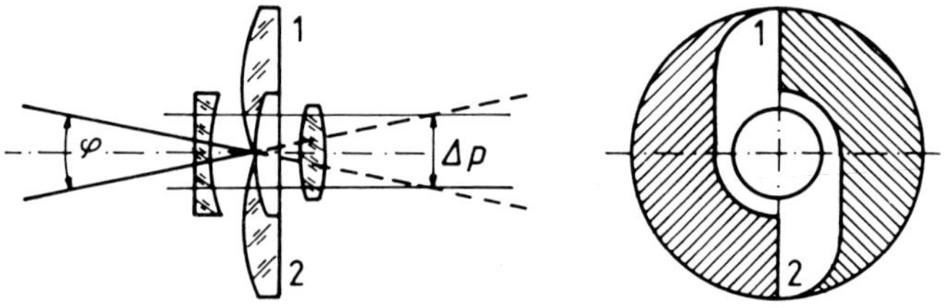

Figure 5.61. Schematic diagram of the optics of a distance meter with a split image and variable stadia angle

In front of the two half lenses a dispersing lens is attached with an identical absolute focal length. Due to it the two beams of rays leaving the optical system at an angle φ remain separately parallel. The double image formed in the telescope is separated by a biprism placed in the image plane.

Measurement with these distance meters is as follows: The end targets of the base b are made to coincide using the coincidence screw i.e. by altering the separation of the half-lenses. The movement is read in the reading microscope with an accuracy of a tenth of a division, and the slope distance is obtained from Eq. (5.50). The distance obtained must be increased by the addition constant c.

Some of the attachment systems manufactured in the Soviet Union use the above-mentioned principle of distance meters with a double image and a varying parallactic angle. The attachment systems *DN-08*, *DNT-2*, *DNT-2N* are examples; distances can be measured with them with relative accuracies from 1 : 2500 to 1 : 5000. If the distance measuring system is built into the telescope of the distance meter, it becomes a tacheometer (e.g. the tacheometer *OTD* made in the Soviet Union which ensures a relative distance measuring accuracy of 1 : 5000).

5.2.3 Distance Meters with an Internal Base

5.2.3.1 Basic Principle

A base line of constant or varying length is built into distance meters with internal base. Thus, when measuring a line of length d, the base b is at the observer and the apex of the parallactic angle φ is at the other end of the line *(Fig. 5.62)*. The measurement is of either:

1. the stadia (parallactic) angle φ if the internal base b is constant, or
2. the base length b if the stadia angle φ is constant.

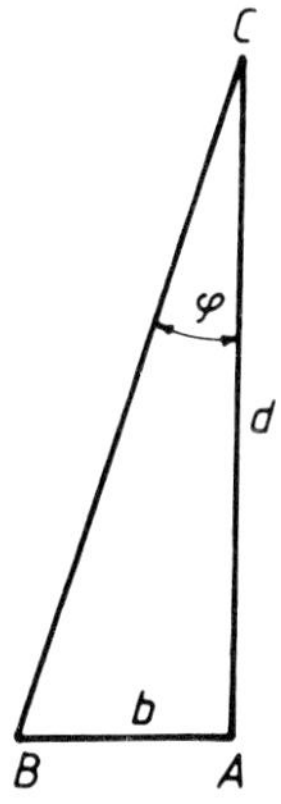

Figure 5.62. Operating principle of a distance meter with an internal base

The advantage of distance measuring instruments with an internal base is that no base (subtense bar) is needed at the far end of the line, and therefore no-one has to walk there. The instruments are useful for detail survey if the accuracy required can be satisfied by the relatively short internal base. In such distance meters made for surveying purposes the base is horizontal. The optics of these instruments form two displaced half-images or two total images of the beams of rays coming to each end of the internal base from the other end of the distance to be measured. The images are observed either in the common eyepiece (monocularly) or in two eyepieces (stereoscopically).

5.2.3.2 Distance Meters with a Monocular Internal Base

In instruments with one eyepiece (monocular) the images can be:

a) two half-images, one above and the other below the dividing line, which must be moved with respect to each other by varying the length of the base in such a way that the upper half-image should be a continuation of the lower one, i.e. the two half-images should together form a complete image *(Fig. 5.63, a)*;

b) two half-images inverted with respect to each other which should be moved horizontally to bring them exactly into one vertical line across the dividing line *(Fig. 5.63, b)*;

c) two whole images which must be brought into coincidence *(Fig. 5.63, c)*.

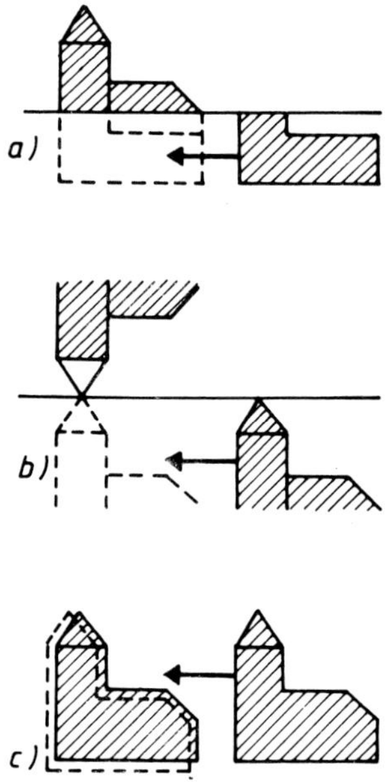

Figure 5.63. Half images or doubled images: *a)* two upright half images; *b)* two reversed half images; *c)* two whole images

From *Fig. 5.62* the distance d is:

$$d = b \operatorname{cotan} \varphi. \tag{5.51}$$

The standard deviation of the distance measurement is:

$$m_d = \sqrt{m_b^2 \operatorname{cotan}^2 \varphi + \left(\frac{m_\varphi}{\rho}\right)^2 \frac{b^2}{\sin^4 \varphi}}, \tag{5.52}$$

where b = the length of the base,

m_b = the standard deviation of the length of the base-line read on the instrument,

φ = the stadia angle,
m_φ = the standard deviation of the stadia angle,
ρ = one radian in sec of arc.

[Eq. (5.52) is essentially the same as Eq. (5.45).]

If the mean square error of coincidence of the two images due to error in the parallactic angle is m_γ sec of arc with no magnification (which can be from 15 to 30 sec of arc depending on the object observed and on the visibility), then it is m_γ/Γ for a magnification of Γ. The value of this expressed in radians i.e. $m_\gamma/\rho\Gamma$, must be substituted into Eq. (5.52) instead of m_φ/p, giving:

$$m_d = \sqrt{m_b^2 \operatorname{cotan}^2 \varphi + \left(\frac{m_\gamma}{\rho\Gamma}\right)^2 \frac{b^2}{\sin^4 \varphi}}, \tag{5.53}$$

and after some reduction and with permissible approximation:

$$m_d = \frac{10^3 D}{b} \sqrt{m_b^2 + 10^6 D^2 \left(\frac{m_\gamma}{\rho\Gamma}\right)^2} \quad \text{(mm)}, \tag{5.54}$$

where D = the measured distance in m,
b = the length of the base in mm,
m_b = the mean square error of the length of the base-line in mm,
m_γ = the mean square error of coincidence due to error in the parallactic angle in sec of arc,
ρ = one radian in sec of arc,
Γ = the angular magnification of the telescope of the distance meter.

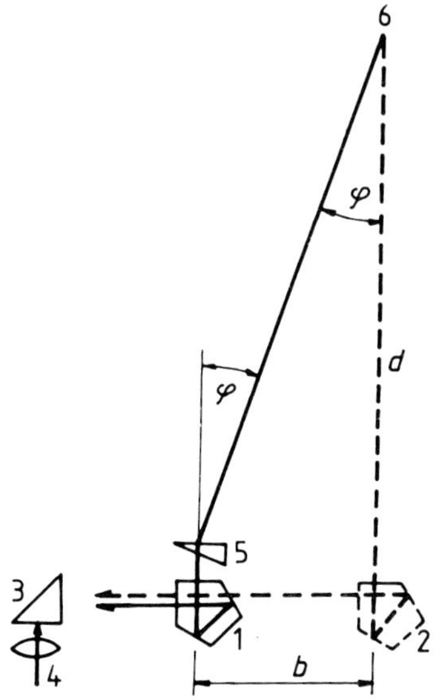

Figure 5.64. Theoretical schematic diagram of a distance meter with a variable internal base

From Eq. (5.54) it may be seen that the mean error of distance measurement increases with increase in the ratio of the distance to the base length, and additionally the mean square error of coincidence increases with increasing distance and with decreasing telescope magnification.

Distance meters with a variable internal base. In these instruments the pentagonal prisms *1* and *2*, placed at the two ends of the variable base b at slightly different respective heights *(Fig. 5.64)*, always deflect the light rays falling on them by 90° and then the rays are directed into the eyepiece *4* by the rectangular prism *3*. In the eyepiece two partial images of the observed object are visible. The glass wedge *5* placed before one of the pentagonal prisms deflects the rays coming from the object *6* sideways by a constant angle φ. Consequently, imagining the ray path to be reversed, the ray falling from the eyepiece onto the pentagonal prism *1* leaves the glass wedge *5* deflected by the angle φ. By changing the base length b, i.e. by moving one or both pentaprisms sideways, the two images of the object can be made to coincide. The base length thus established can be read on the length graduation of the measuring bar on which the prism (prisms) slide. The deflecting angle of the glass wedge is chosen in such a way that $k = \cot \varphi$ is an integer. The slope distance is:

$$d_s = k \cdot b, \tag{5.55}$$

where k is the stadia constant of the distance measurement.

The instrument is usually supplied with several wedges with different constants k.

The upper measuring limit is for a wedge of given deflecting angle φ, i.e. of given constant k:

$$d_{s,\max} = k \cdot b_0 \tag{5.56}$$

from Eq. (5.55), where b_0 is the longest base length that can be set.

For the measurement of any slope distance d_s it is advisable to use that wedge which has the smallest constant possible, i.e. that with which the base length obtained from Eq. (5.55):

$$b = \frac{d_s}{k} \tag{5.57}$$

is just shorter than the maximum base length b_0, because the error of the measurement increases with increase of the constant k.

The effective length of the measuring bar on the *Teletop* distance measuring instrument, manufactured by VEB Jenoptik, Jena *(Fig. 5.65, a)*, is 30 cm. The left hand fixed pentaprism P_1 is placed higher than the right hand one P_2 which can be moved *(Fig. 5.65, b)*. The glass wedge is in front of the fixed prism. Two half-images can be observed in the 130 mm long telescope, which has an objective lens diameter of 20 mm and a magnification of 6, and they are separated by a horizontal line. The measurement consists of moving the right-hand prism on the measuring (sliding) bar until the two half-images form a whole image in the telescope field of view. The separation of the two half-images is ensured by a *double wedge* in the telescope eyepiece (see Point 6.1.2.3) and by its *diaphragm aperture.*

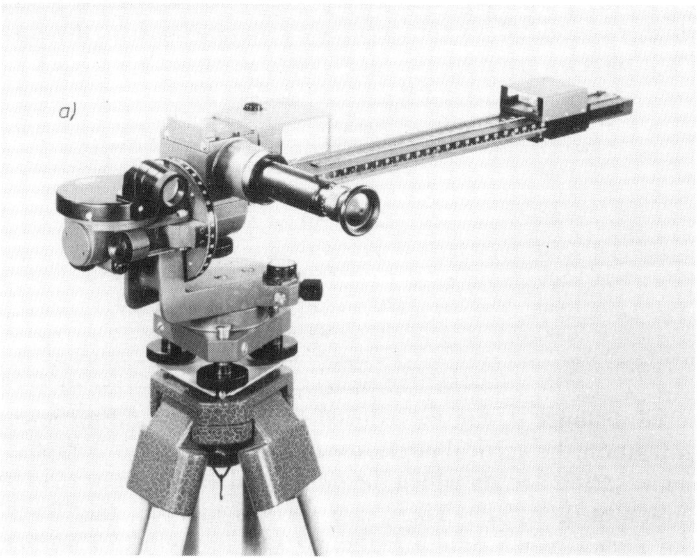

b)

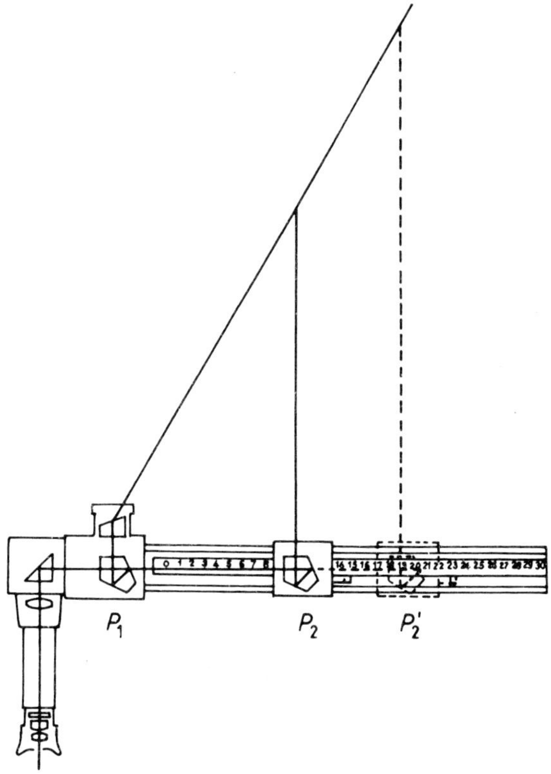

Figure 5.65. a) the *Teletop* distance meter (VEB Jenoptik, Jena). $H = 220$ mm; $W = 530$ mm; *b)* operating schematic diagram of the *Teletop*

There are several measuring wedges for each instrument; they have different constants which correspond to the accuracy required and to the distances to be measured. Their data as given by the factory are listed below:

The k constant of the measuring wedge	Range of measurement (m)	Measuring error as a percentage of the distance
100	2... 30	±0.2
250	5... 75	±0.3
500	10...150	±0.5
1000	20...300	±1.0
2000	40...600	±2...3

The measured distance d_s can be reduced to the horizontal distance d using the slope angle α read on the vertical circle, $d = d_s \cos \alpha$. A circular compass is placed to the left of the measuring bar, on which the horizontal angle can be read by estimation with an accuracy of 0.1° (0.1^g). There is a vertical mark in the lower part of the telescope field of view; when pointing, the mark is brought to a vertical line (edge, axis, etc.) of the observed object and this line is used to bring the two half images

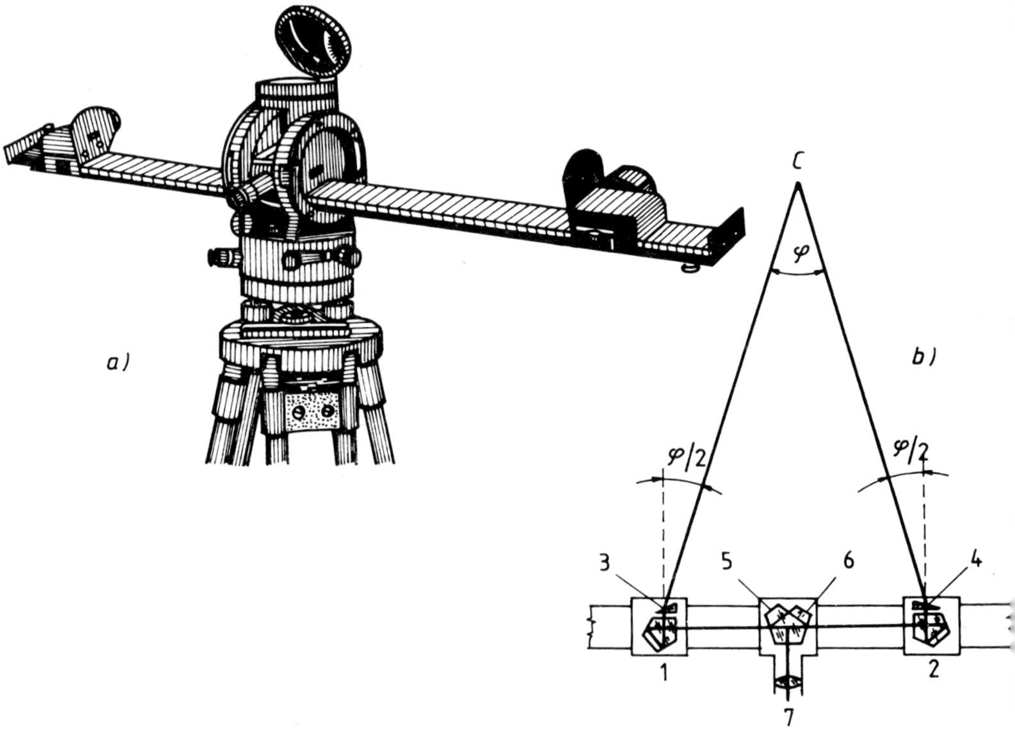

Figure 5.66. a) the *Todis* distance meter (Breithaupt, Kassel); *b)* operating schematic diagram of the *Todis* distance meter

into coincidence. If coincidence is brought about away from the vertical mark in the field of view, the base line is not perpendicular to the line of sight and a distance measuring error occurs due to the mis-orientation of the base. The diameter of the vertical circle is 85 mm, its graduation unit 1°, the compass diameter is 59 mm, its graduation unit 1°. The instrument weights 3 kg, the box 3.5 kg and the tripod 3.2 kg.

The *Todis* distance meter with an internal base has been built into the lower part of a simple vernier theodolite, with a telescope of 6 times magnification, by *Breithaupt* (Kassel) *(Fig. 5.66, a)*. The measuring bar, which is fixed on rings and consists of two parts with an effective length of 90 cm/80 cm, supports on each part one movable pentaprism *1* and *2* and one deflecting wedge *3* and *4*, the two wedges having the same deflecting angle *(Fig. 5.66, b)*. The double pentaprisms *5* and *6* rigidly mounted at the centre of the bar deflect the rays coming from the two sides towards the eyepiece *7*. Four glass wedges with different constants are supplied with the instrument. In this instrument the two images are again separated by a biprism built into the eyepiece. The instrument has a horizontal metal circle of 110 mm diameter on which direction can be read by verniers with an accuracy of 2 min of arc. On the vertical circle of 125 mm diameter, also metal, readings can be made to an accuracy of 1°/10. The compass of 80 mm diameter is fixed on the two base rings, the graduation unit of the circle is 1° or 1^g. The weight of the instrument is 5.4 kg, that of the box 8.4 kg and that of the tripod 3.5 kg.

The *Soviet* distance meter *DV-20* is constructed similarly to the Todis, with a telescope magnification of 14 ×. Base lengths from 170 mm to 600 mm can be set on the base bar. The four accompanying measuring wedges have the following parameters:

Constant *k* of the measuring wedge	Range of measurement (m)	Measuring error as a percentage of the distance
50	1... 40	±0.02
100	1... 80	±0.05
200	4...160	±0.1
500	10...400	±0.3

The instrument weighs 4.2 kg, the box 4.9 kg.

Distance meters with a variable internal base are used for topographical survey, for preparatory engineering works, and for geological and forestry survey, i.e. to determine distances, elevations and directions during such surveys. Care must be

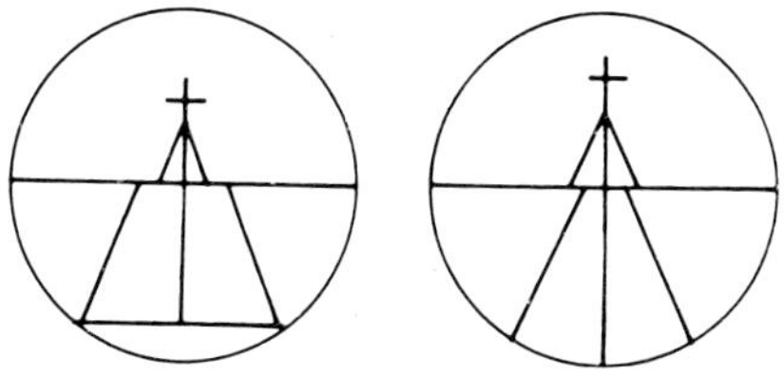

Figure 5.67. The effect of the height error

taken during measurement that there should be no part of the image of the object missing, nor should any part of the image appear doubled *(Fig. 5.67)*. The two oblique edges of the object (one going upwards to the right, the other downward to the left) should appear after coincidence as a continuous line without a break. If such an error between the two half images is visible it must be corrected by adjustment, because an erroneous distance is given.

Distance meters with a constant internal base. In monocular instruments with a constant internal base the pentaprisms *1* and *2*, placed at the ends of the base *a (Fig. 5.68)*, deflect the rays falling on them into the direction of the base and through the objectives *7* and *8*. From here the ray passes through the movable measuring glass wedge *9* or the rigidly fixed wedge *10*, then the rectangular prism *3* or the pentaprism *4*, and finally passes through the image plane *5* and enters the common eyepiece *6*. The two half images visible in the eyepiece are made to coincide by moving the distance measuring wedge *9*. The moving mechanism is coupled to a reading device (with graduations) and the distance can be read directly on it. The instrument can be adjusted by means of the wedge *10* placed on the other side. The two parts of the image visible in the eyepiece are both in the same sense vertically *(coincidence distance meter)*.

In inverting distance meters one of the two central prisms *3* and *4* is a roof edge prism *(Fig. 5.69)* which inverts the image passing through it *(inverting coincidence distance meter)*.

Equation (5.54) expressing the mean square error of distance measurement by distance meters with an internal base becomes more simple for distance meters with

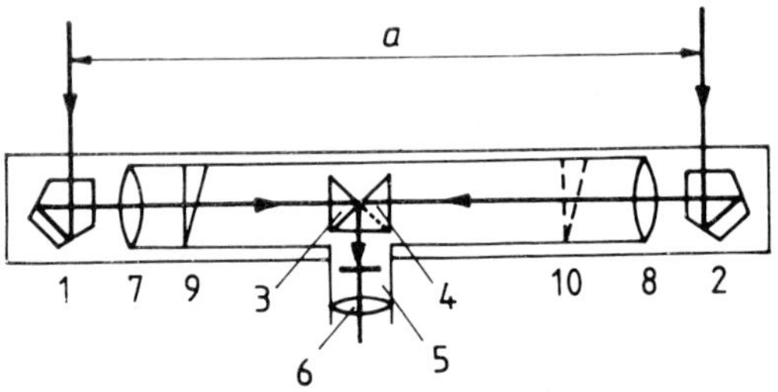

Figure 5.68. Theooretical schematic diagram of a monocular distance meter with a constant internal base

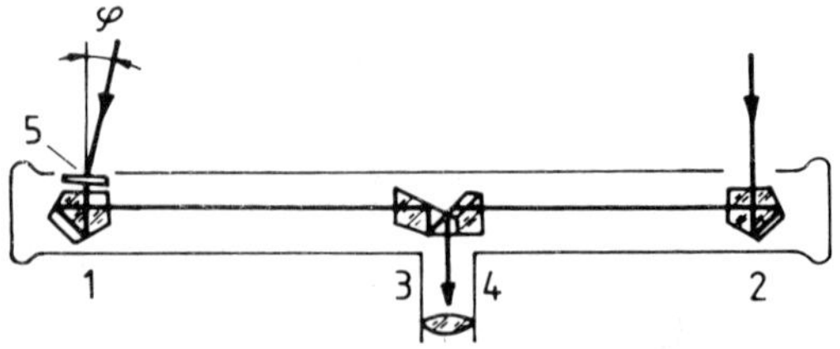

Figure 5.69. Schematic diagram of a monocular inverting distance meter with a constant internal base

a constant base, because the constant base has no random error. Hence, neglecting m_b, Eq. (5.54) can be written:

$$m_d = \frac{10^6 D^2}{b} \frac{m_\gamma}{\rho \Gamma}, \tag{5.58}$$

where the symbols D, b, m_γ, ρ and Γ are the same as for Eq. (5.54).

Among others the *Zeiss* and *Görz* distance meters with the Barr and Stroud system used to be manufactured according to the design shown in *Fig. 5.68*. Distance meters of such types are also manufactured for special purposes with a 0.5 to 10 m base and a $\Gamma = 10 \ldots 50$ magnification.

Figure 5.70. The *TM 10* distance meter (Wild, Heerbrugg)

The *TM 10* distance meter (Wild, Heerbrugg) is also made according to the design of inverting coincidence distance meters with a constant internal base as shown in *Fig. 5.69*. The roof of pentaprism *4 (Fig. 5.69)* in the distance meter *TM 10 (Fig. 5.70)* also reverses the image, besides deflecting the right-hand beam of rays into the eyepiece. The rotating pair of wedges *5*, placed in front of the pentaprism *1* at the left end of the base, must be rotated during measurement until the required stadia angle φ is obtained, i.e. until the two half images of the observed object exactly coincide in the field of view of the eyepiece *5*. The graduation of the rotating mechanism of the pair of wedges gives the distance directly. The *TM 10* can be either used in the hand or placed on a tripod. The base is 50 cm long, the magnification of the telescope 6, the diameter of the objective 21 mm. The right-hand eyepiece belongs to the telescope, while the left-hand one is used to read the distance. The horizontal and vertical circles can be read by verniers with an accuracy of 1°/10. The instrument weighs 3.4 kg, the metal box 1.5 kg, the tripod 5.7 kg. For the mean error of distance measurement the following data are given by the factory for a measuring range between 30 m and 1000 m:

Distance:	30	50	100	200	500	1000 m
Mean error:	±0.02	±0.05	±0.2	±0.8	±5.0	±20.0 m

The *Hilger and Watts Rangefinder* theodolite, with a 1 m long base and an effective range of up to 10,000 m, is similar to the *TM 10*. Hilger and Watts give for it the following mean square errors:

Distance:	250	500	1000	2000	4000 m
Mean square error:	±0.3	±1.0	±3.8	±15	±61 m

The distance meter can be attached to the bottom part of a vernier theodolite. It has a telescope magnification of 14, the readings on its horizontal and vertical circles can be made with an accuracy of 30 sec of arc. The weight of the instrument is 12.2 kg, including the tripod.

5.2.3.3 Stereoscopic Distance Meters

Binocular stereoscopic distance meters are based on the principle of stereoscopic vision with two eyes. Since the average distance of 68 mm between the human eyes is too short a stereoscopic base, prisms are built into these instruments by means of which this base is artificially increased several times *(Fig. 5.71)*. Depth perception is further increased by the magnification of the telescope. Stereoscopic distance meters for special purposes are also manufactured, with a base length between 0.5 and 4.0 m and a telescope magnification Γ between 8 and 40. When looking into the two telescopes the sharpness of the depth perception of the eyes is increased many times due to the increased stereoscopic base and the magnification Γ. There are two types of stereoscopic distance meters.

1. In stereoscopic distance meters with a *mark system* there are a series of *measuring marks* on the cross-hair planes of both telescopes. Marks determined by corresponding pairs give different parallactic angles φ relating to different distances d_s, and the

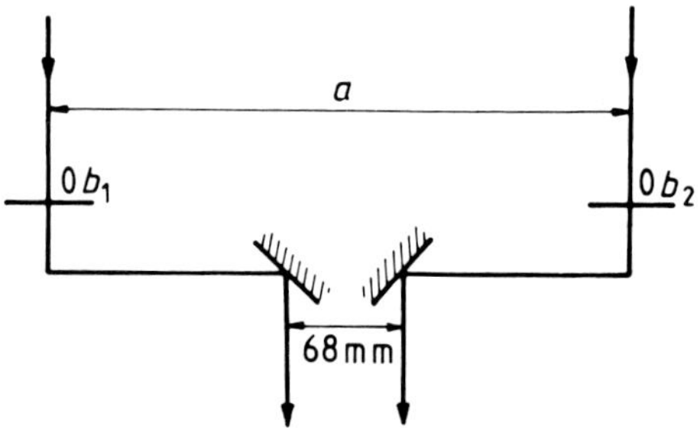

Figure 5.71. Principle of the stereoscopic distance meter

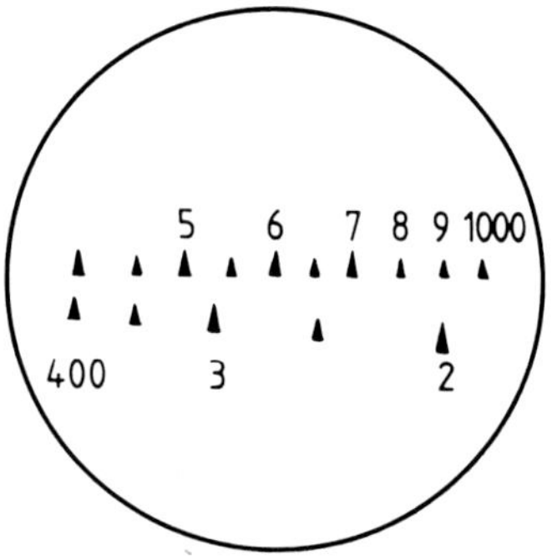

Figure 5.72. System of marks in the stereoscopic distance meter

marks are numbered with the respective distances *(Fig. 5.72)*. If one looks simultaneously into the two telescopes, the two mark systems appear in stereoscopic depth as one system. A measurement is carried out as follows. The point observed is brought between the two marks about it in depth and its distance is estimated by interpolation. The marks are often put on the cross-hair planes in a zigzag line in order to obtain more marks in the field of view.

2. In stereoscopic distance meters with a *micrometer screw* there is a fixed mark in the image plane of one of the telescope, while in the image plane of the other there is a mark that can be moved sideways by means of a micrometer screw (parallax screw). By rotating the parallax screw the *space distance index*, determined by the two marks when coincident, moves in depth. The measurement consists of moving the index onto the image of the observed object. The distance can be read directly on the graduation of the parallax screw.

The mean error of distance measured by a stereoscopic distance meter strongly depends on the *mean square error* of the *depth perception* ability of the observer. If this mean square error is m_0 for a certain distance d_0 with an unaided eye, or the mean square error of relative depth perception for the same distance is $m_{0,\,\mathrm{rel}}=m_0/d_0$, and if the base b of the meter is n times the eye base (68 mm) and the magnification of the instrument telescope is Γ, then the observer can measure with the distance meter a distance $n\Gamma$ times longer, i.e. the distance $n\Gamma d_0$, with a relative mean square error of $m_{0,\,\mathrm{rel}}$. The relative mean square error of an arbitrary distance d_s for this observer is:

$$m_{d,\,\mathrm{rel}}=d_s\frac{m_{0,\,\mathrm{rel}}}{d_0 n\Gamma}=d_s\frac{m_0}{d_0^2 n\Gamma}\,. \tag{5.59}$$

For instance, let $b=1$ m, $\Gamma=10$, $d_0=10$ m and $m_0=0.2$ m, i.e. $n=b/0.068=15$, then for the measurement of the distances $d=50$; 100; 500; and 1000 metres a relative distance measurement error of 0.0006; 0.0013; 0.006 and 0.013, or a mean square error of 0.03; 0.13; 3.0; and 13 m can be expected from the observer. By increasing the distance d_s the relative distance measurement error increases linearly and the real mean square error quadratically.

6 Instruments for Tacheometry

Tacheometry is a method that unifies horizontal and vertical surveying. In tacheometry data for the spatial determination of a point are measured with respect to a known point (the instrument station) and a known direction from it. The data to be determined are spatial coordinates: two polar coordinates in a plane (horizontal angle and distance) and a vertical distance. The latter is the elevation difference between the instrument station and the point to be fixed.

The instrument for tacheometry is the tacheometer. By means of it the horizontal distance is determined by optical or electronic (electro-optical) distance measurement, and the horizontal angle is determined numerically or graphically. A numerical angle is obtained by readings on the horizontal circle of the tacheometer, a graphical angle is obtained by pointing with a telescopic alidade on a plane table. Elevation differences are usually determined by tacheometers independently of the distance but, in some cases, it can be only obtained if the distance is known.

One group of tacheometers measures the slope distance corresponding to the tilt of the telescope; this must be reduced to the horizontal as a function of the zenith distance (or vertical angle) measured at the same time. These are called simple tacheometers. Self-reducing tacheometers yield horizontal distances directly and elevation differences can be determined by most of them without knowledge of the telescope tilt.

Tacheometric tasks can be carried out with all instruments with which the spatial polar co-ordinates of a point to be fixed can be determined. Optical tacheometers are after all geometrical distance meters combined with a theodolite, or a telescopic alidade with a plane table. Consequently their accuracy is defined primarily by the accuracy of the geometrical distance measurement. The distance meter of electronic tacheometers works mainly on the basis of wave physics (it is usually a short-distance light-wave distance meter). As well as distance, the horizontal reference angle and the vertical angle (or zenith distance) can also be measured by them. They have a higher accuracy due to the electronic distance measurement. The longest distance measurable by them is determined by the effective range of the distance meter.

6.1 Optical Tacheometers

All instruments which yield nothing but numerical data (by tacheometric methods) are classified as optical tacheometers. Graphical tacheometry also uses optical (sometimes electro-optical) distance measurement, but instruments belonging to this group are dealt with in Section 6.2, due to the particular methods and instrumentation of graphical survey.

Optical tacheometers can be tacheometer-theodolites or compass tacheometers. One of the spatial polar coordinates, the polar angle, is determined indirectly by tacheometer-theodolites (by orienting on known points) as by theodolites, but it is given directly by compass tacheometers with direct orientation (with the compass). Tacheometers of both types have a telescope with stadia hairs, a horizontal and a vertical circle. Since theodolites now have a vertical circle with an index level or an automatic vertical circle (with a compensator), there is no essential constructional difference between a theodolite and a tacheometer. They are classified according to their main function. Consequently one speaks both of theodolites — mostly precise — used mainly for angle measurement, and of *tacheometer-theodolites* having a double function and *proper tacheometers* manufactured mainly for tacheometric purposes.

The telescope of a proper tacheometer is now almost exclusively self-reducing. The possibility of measurement in two telescope positions, and two diametrically opposite reading devices on the circle, are unnecessary. The reading accuracy can also be lower; an accuracy of 1 min of arc (1^c) direct reading is usually satisfactory for tacheometric measurements.

An external base (tacheometric staff) is necessary for most optical tacheometers, but there are some with an internal base in which a base of limited length is integral with the instrument. Tacheometric staffs used for distance measuring systems are identical with the stadia rods discussed in Section 5.2. Sometimes special graduations and staff extensions are used in order to reduce computation.

6.1.1 Simple Tacheometers

6.1.1.1 Tacheometer-theodolites

Tacheometer-theodolites are theodolites of low or medium accuracy manufactured for angle measurement and tacheometric measurement. Their telescopes are usually fitted with either stadia hairs of constant separation or a simple distance meter with a prism. The reading device on earlier instruments is a 20 to 60 sec of arc vernier and, on newer instruments, a 1 min of arc (1^c) direct reading microscope. The instruments have on the vertical circle an index level, or on the new types an automatic index (compensator), for the reduction of slope distances to the horizontal. Sometimes a telescope level is used and there may be a compass placed on the instrument. The stadia rods, which must be set up vertically, have a simple or double graduation with cm division units and a length of 3 or 4 m (see Point 5.2.1.1).

The spatial data of a point are measured by a tacheometer theodolite as follows: the horizontal angle by readings on the limb or horizontal circle, distance and height difference by readings on the tacheometric staff and on the vertical circle, respectively.

The simple anallactic distance meter, with a constant stadia distance and using a vertical staff, gives the reduced distance according to Eq. (5.27). The height difference between the horizontal axis of the telescope and the point observed on the staff is obtained from the slope distance $d_s = k\Delta l \cos\alpha$ if it is multiplied by the sin of the angle α:

$$\Delta h_s = k\Delta l \sin\alpha \cos\alpha. \tag{6.1}$$

The height difference of the observed point above the station point can be obtained from the height Δh_i of the instrument axis H above the station point, and the staff reading l of the horizontal cross-hair; with an anallactic telescope. Eq. (6.1) shows that:

$$\Delta h = \Delta h_i + \Delta h_s - l = \Delta h_i - l + k\Delta l \sin\alpha\cos\alpha. \tag{6.2}$$

With a non-anallactic telescope it is:

$$\Delta h = \Delta h_i - l + c\sin\alpha + k\Delta l \sin\alpha\cos\alpha. \tag{6.3}$$

In practice it is usually sufficient to measure Δh_i and l with an accuracy of 1 cm. The height Δh_i is usually measured with a short (2 m) measuring tape with cm divisions. The observed point on the staff of height l is read together with the stadia hairs.

Special computation accessories are used for the computation of the terms in Eqs (5.27) and (6.2) which contain trigonometric functions. The most well-known of them are the different tables (e.g. those of Jordan, Jadanza, Hanisch, etc.) and stadia slide rules. Due to the wide-spread use of small electronic computers and calculators their importance has now decreased considerably.

Tacheometer-theodolites must be checked according to their dual function as both theodolites and tacheometers. Theodolite tests have been discussed in Chapter 2. The following must be tested in connection with the tacheometer function:

— the constants c and k of the distance meter telescope (Point 5.2.1.1);

— the index error of the vertical circle (Point 3.2.2.3);

— the axis of the telescope level (if there is a spirit level on the telescope). The latter test and adjustment must be carried out in the same way as the test and adjustment of telescope levels (Point 3.1.2.6).

The mean error of a measurement by tacheometer-theodolites is between ± 1.5 and ± 2.0 m with a stadia factor $k=100$, nearly horizontal sights and favourable conditions. Accuracy decreases with increasing slope, consequently the accuracy of a single measurement can be characterized by a relative error between 1:100 and 1:800.

6.1.1.2 Compass Tacheometers

The compass tacheometer is an instrument like a theodolite fitted with a distance measuring telescope, a vertical circle and a compass *(Fig. 6.1)*. Its telescope is usually a simple distance meter with a constant stadia distance. It is usually used for small scale mapping (e.g. topographic survey). For measurement it is set up on a tripod as used generally in surveying; its vertical axis is adjusted to the vertical with a box level. The vertical circle reading device is either a vernier or a simple reading microscope (a microscope with lines or a scale microscope). The telescope is usually eccentric, the use of a telescope level is rather rare *(Fig. 6.2)*. The tacheometric staff used is the same as those for tacheometer-theodolites. The compass always has a whole circle graduation of 360° or 400^g. The compass is used for a direct orientation of the measured points, otherwise measurement and computation as well as the test and adjustment of the instrument is the same as discussed in Point 6.1.1.1. A special test is necessary for the compass (Point 2.5.1.2).

Figure 6.1. TB compass tacheometer (Kern, Aarau)

Figure 6.2. BUMON compass tacheometer (Breithaupt, Kassel) with a telescope level

6.1.2 Self-reducing Tacheometers

Tacheometers, which yield directly the horizontal projection of the distance measured by the readings on the tacheometric staff, are called self-reducing tacheometers. Measurement of the inclination and the now superfluous computation of the reduction is thereby omitted.

Self-reducing tacheometers usually have an anallactic telescope. With some non-anallactic telescope (e.g. those with a prism) the addition constant $c \neq 0$ is taken into account by means of a shift c/k of the staff-index (or any other index mark).

The base length Δl_d obtained on the stadia rod by a self-reducing anallactic tacheometer must be multiplied by the integer of the stadia constant k_d to obtain the horizontal distance d. The multiplication constant k_d is 100 or 200, seldom 50.

Self-reducing tacheometers — except the so-called semi-reducing ones — yield height difference in a similar way, but by multiplying the measured base length Δl_h with the integer constant k_h. The usual values of the height factor k_h are $k_h = \pm 10$, ± 20, ± 50 or ± 100.

In practice the most frequently used tacheometers are those which use the tangent principle, varying stadia separation and the reducing system of distance meters with a prism.

Tacheometers are called semi-reducing if the reduced distance is obtained directly but the height difference is obtained only as a function of the horizontal distance, as is the case for tacheometers operating on the basis of the tangent principle and some of the self-reducing tacheometers with a prism.

Self-reducing tacheometers using the tangent principle and a varying hair separation always operate in practice with vertical staffs, while self-reducing tacheometers with a prism operate with a horizontal internal or external base.

6.1.2.1 Self-reducing Tacheometers Operating on the Basis of the Tangent Principle

Self-reducing tacheometers operating on the basis of the tangent principle use the ratio between the horizontal distance d and the base length $\Delta l = l_2 - l_1$ as determined on the vertical staff as a function of their construction. If an inclination α_1 gives a reading l_1 and α_2 gives l_2, then from *Fig. 6.3*:

$$l_2 - l_1 = \Delta l = d(\tan \alpha_2 - \tan \alpha_1)$$
$$b_2 - b_1 = \Delta b = s(\tan \alpha_2 - \tan \alpha_1), \qquad (6.4)$$

from which the horizontal distance d is:

$$d = \frac{s}{\Delta b} \Delta l. \qquad (6.5)$$

In Eq. (6.5) s is a suitable fixed dimension (e.g. 100 mm), Δb can either be a constant or must be determined from case to case. Tacheometers using the tangent principle differ from each other as to whether Δb is a constant or not and, if not, how to determine it. The value of $s/\Delta b = k$ can, however, almost always be made an integer and hence the following basic equation of distance

$$d = k\Delta l \tag{6.6}$$

is obtained for such self-reducing tacheometers.

If, e.g. $s = 100$ mm and $\Delta b = 1$ mm, then $k = 100$, i.e. the computation of d from Δl is nothing but the corresponding transposition of the decimal point.

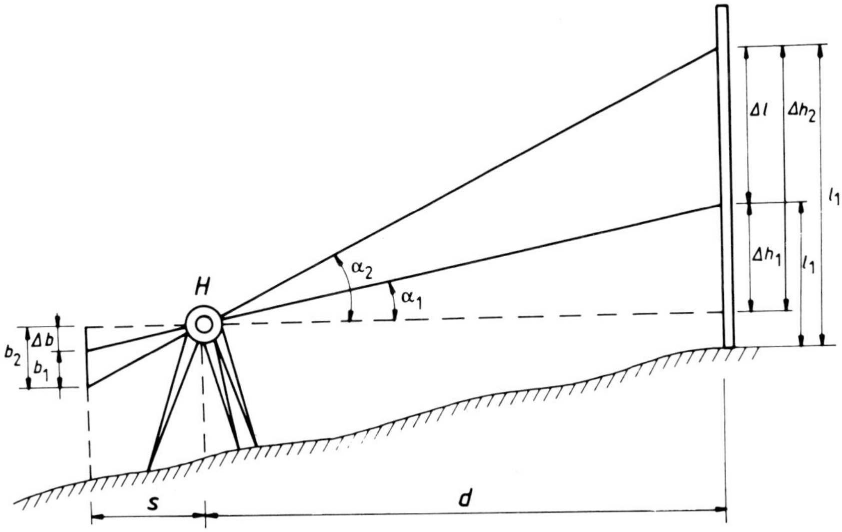

Figure 6.3. Application of the tangent principle in tacheometers

Knowing the distance d, the height difference between the instrument axis H and the point l_1 read on the staff is from *Fig. 6.3*:

$$\Delta h_1 = d \tan \alpha_1 = d\frac{b_1}{s}$$

and similarly (6.7)

$$\Delta h_2 = d \tan \alpha_2 = d\frac{b_2}{s},$$

whence the height of the ground point at the staff above the axis H is:

$$\Delta h = d\frac{b_1}{s} - l_1 = d\frac{b_2}{s} - l_2. \tag{6.8}$$

Tacheometers operating on the tangent principle are called contact tacheometers; they operate with one or more bases Δb fixed on the instrument.

Contact tacheometers. On contact tacheometers the several individual constant tilt angles of the telescope are determined by stops which impede further rotation of the telescope around the axis H. Using pairs of stops during measurement, pointings can be repeated with a constant tilt difference Δp. There is a different staff reading Δl for each tilt different Δp, each yielding the horizontal distance after being multiplied by the integer k. If, e.g. the staff readings l_1 and l_2 are obtained with telescope tilts p_1 and p_2, the horizontal distance is:

$$d=\frac{l_2-l_1}{p_2-p_1}=\frac{\Delta l}{\Delta p}. \tag{6.9}$$

Since $k=1/\Delta p$, therefore $d=k\Delta l$. Let $\Delta p=0.01$, then $k=100$ and $d=100\Delta l$.

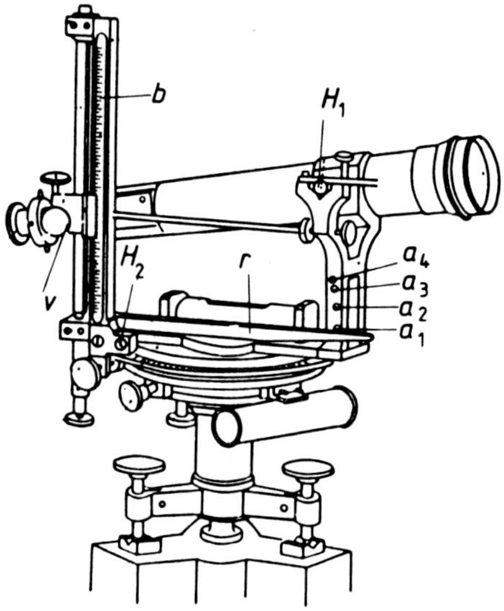

Figure 6.4. Sanguet tacheometer (Secretan, Paris) with stops

A characteristic type of contact tacheometers is the *Sanguet* (Secretan, Paris) tacheometer, mostly used in France *(Fig. 6.4)*. That part on the eyepiece side of the telescope which rotates around the axis H_1 can be moved along the vertical graduation b and fixed. The lever r rotating around the axis H_2 can be set against each of the stops a_1, a_2, a_3, a_4, thus determining four constant telescope tilts p_i depending on the position of the slide v. Out of the possible six tilt differences the following three are used in practice:

$$p_2-p_1=\frac{1}{100}; \qquad p_3-p_1=\frac{1.8}{100}; \qquad p_4-p_1=\frac{2.2}{100}.$$

Three staff reading differences Δl are determined by measurement at the three tilt differences, whence:

$$d=\frac{100(\Delta l_1+\Delta l_2+\Delta l_3)}{5}+c=$$

$$=20(\Delta l_1+\Delta l_2+\Delta l_3)+c.$$

As a final result the distance can be computed from the staff readings by means of the simple formula:

$$d=c+20(l_2+l_3+l_4-3l_1). \tag{6.10}$$

For detailed tacheometric survey it is sufficient to use one tilt difference and then:

$$d = 0.10 + 100(l_2 - l_1). \tag{6.11}$$

The elevation difference can be computed from the readings on the graduation b. Since the line of sight of the telescope is horizontal when the instrument is set to stop a_1 and the index of the vernier (which has 10 divisions) is set to the zero line b_0, any reading b_i of the graduation b gives the tangent of the telescope tilt, i.e.:

$$p_1 = \tan \alpha_1 = \frac{b_1}{100} = \frac{\Delta h_1}{d},$$

whence:

$$\Delta h_1 = \frac{b_1}{100} d. \tag{6.12}$$

Using three tilt differences, Eq. (6.10) shows that the distance can be determined with a relative error from ± 3 to ± 5 cm/100 m, and Eq. (6.11) shows that one tilt difference can give a relative error from ± 15 to ± 20 cm/100 m as obtainable from simple distance meters with stadia hairs.

Tangent tacheometers. Different constructions have been developed for setting the varying lengths b_i in tangent tacheometers. One constructional variant is a micrometer screw, on the drum of which the chosen b_i can be set. Tacheometers with a so-called tangent screw have such a construction but they are now already obsolete. As a further variant the index of the microscope, rotating together with the telescope, indicates the reading b_i corresponding to the tilt. The wear of the screw head of the former, and the image focussing corresponding to the varying graduation spacing of the latter, are considerable sources of error.

Instead of the equally spaced graduations b, *Szepessy* used on his tangent tacheometer a so-called tangent graduation i.e. the graduations were projected onto the surface of a metal disc of radius r rigidly fixed to the vertical circle *(Fig. 6.5)*. These graduations are visible in one half of the telescope field of view, while in the other half the image of the graduations of the vertical stadia rod can be adjusted *(Fig. 6.6)*. Every other line of the basic graduation $r/200$ on the tangent circle is

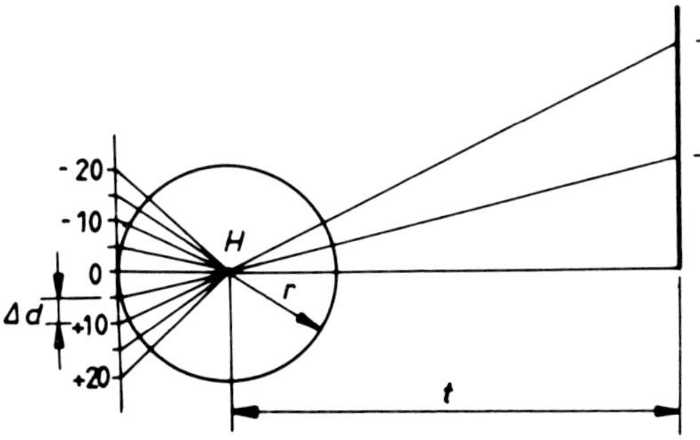

Figure 6.5. The construction of a graduation in tangents

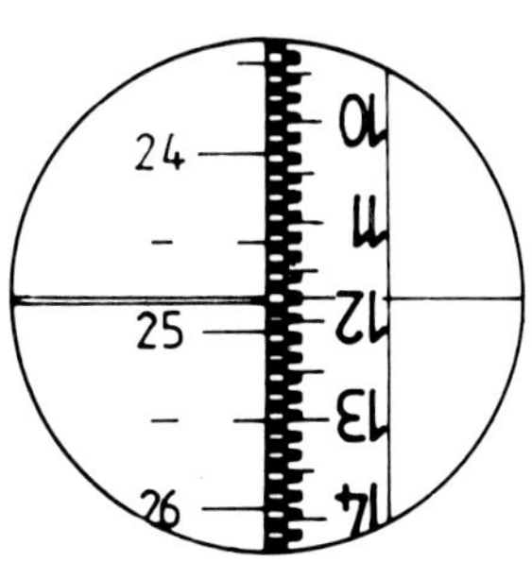

Figure 6.6 The field of view of the *Szepessy* tangent tacheometer when observing a rod

numbered, with adjacent numbers increasing by one. The doubled horizontal section of the cross-hair is the index for the tangent graduation; the chosen line of this graduation must be straddled by the double hair. If the readings on the rod of the principal hair, when set to the division lines b_1 and b_2, are l_1 and l_2 and their difference is Δl, the distance can be obtained as:

$$d=\frac{100}{b_2-b_1}\,\Delta l=\frac{100}{\Delta b}\,\Delta l\,. \tag{6.13}$$

The heights of the reading points l_1 and l_2 above the instrument axis H can be obtained from Eq. (6.12).

On the tangent tacheometers *17 S* and *17 K-S* (MOM, Budapest) *(Fig. 6.7)* a tangent graduation is engraved on the vertical circle as well as an angular one. The two graduations are projected into one microscope and their images are visible beside each other *(Fig. 6.8)*. The lines b_1 and b_2 of the tangent graduation can be adjusted by the double index hair in the image plane of the reading microscope.

In the *Tari* tacheometer (models *4180* and *4181*, *Fig. 6.9*) of Filotecnica Salmoiraghi (Milano) the tangent graduation is projected into the telescope field of view. The tangent divisions, numbered in two directions from 0° to 45° (from 0^g to 50^g) appear

Figure 6.7. The *17 K-S* tangent tacheometer (MOM, Budapest)

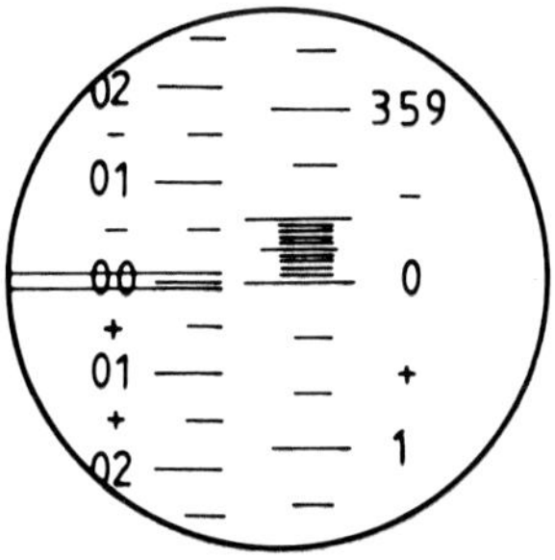

Figure 6.8. Angle and tangent graduations of the *17 K-S* in the field of view of the vertical circle microscope

Figure 6.9. The *Tari* tangent tacheometer (Salmoiraghi, Milano)

in the telescope as stadia lines with a varying separation. The stadia factor to be used with two adjacent division lines is 200; when multiplying the staff reading difference Δl by it one obtains the horizontal distance directly. In practice, however, it is advisable to set one of the projected graduation lines to an integer reading l_1 on the rod, then to read the rod l_2 at the line straddled by the double hair and to multiply their difference Δl by $k_d = 100$. By a proper choice of the spacing between the division lines used as stadia lines the distance factor can be further decreased. For the determination of height difference a smaller scale tangent graduation is projected onto the right hand side of the telescope field of view *(Fig. 6.10)*. The division spacings, numbered corresponding to a tangent difference of 0.01, are further subdivided by 10. Hence on this graduation, using the horizontal hair as an index, the tangent of the telescope tilt can be read directly to three decimals and the fourth can be estimated. Numbered graduation lines give the values of $\tan \alpha \times 100$, i.e. the engraved numbers are the first two decimals of the tangents between 0 and ± 1. From the tangent reading:

$$\Delta h_\alpha = d \tan \alpha, \quad \text{i.e.} \quad \Delta h = \Delta h_i + d \tan \alpha - l,$$

where l is the reading on the rod with the horizontal hair with a slope of α.

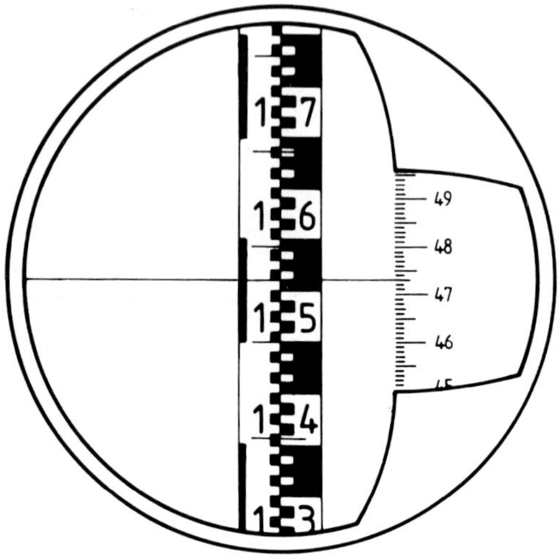

Figure 6.10. The field of view of the *Tari* tangent tacheometer when observing a rod

6.1.2.2 Tacheometers with Varying Stadia Distance

Tacheometers, in which reduction of distance to the horizontal is performed by a change of the distance $p = f_0/k$ between the stadia lines depending on the vertical angle α, are called self-reducing tacheometers with a varying stadia distance. In such tacheometers the stadia distance p for a given horizontal distance changes for a tilted pointing to

$$p_d = p \cos^2 \alpha \tag{6.14}$$

for distance measurement in accordance with Eq. (5.27), and to:

$$p_h = p \sin \alpha \cos \alpha \tag{6.15}$$

for height difference measurement in accordance with Eq. (6.1), Eqs (6.14) and (6.15) imply that the stadia lines always remain symmetrical with respect to the horizontal cross-hair of the telescope. The staff section Δl_d measured by the stadia distance of Eq. (6.14) must be multiplied by the integer factor k_d in order to obtain directly the horizontal distance d. Similarly the stadia interval Δl_h measured by the stadia distance of Eq. (6.15) must be multiplied by the integer factor k_h in order to obtain directly the height difference Δh.

In most tacheometers with a varying stadia distance the change of stadia distance is asymmetric, i.e. the reducing line distance changes with respect to the one common stadia line of fixed position. With an asymmetric change, however, the vertical tilt assigned to the half of the varying stadia distance also changes by a small amount with respect to the line of sight of the telescope. Now it can be written from *Fig. 6.11*:

$$d = d_s \cos \alpha = \Delta l_d \frac{\cos(\alpha + \varphi_d)}{\sin \varphi_d} \cos \alpha .$$

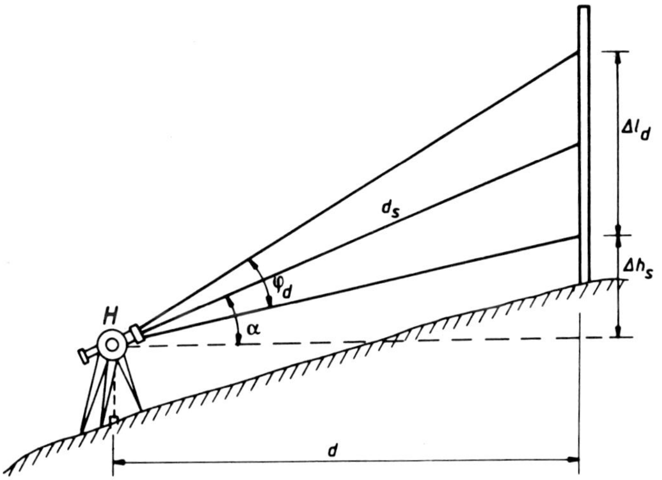

Figure 6.11. Basic principle of asymmetrical stadia diagrams

After transformation this becomes:

$$d=\cos^2 \alpha(\text{cotan } \varphi_d - \tan \alpha)\Delta l_d.$$

By making the coefficient of Δl_d equal to the integer factor k_d, as is necessary to determine the distance d, i.e.:

$$\cos^2 \alpha(\text{cotan } \varphi_d - \tan \alpha)=k_d$$

one obtains:

$$\text{cotan } \varphi_d=\frac{k_d+\sin \alpha \cos \alpha}{\cos^2 \alpha}. \tag{6.16}$$

Since the resulting focal length of the telescope is $f_0=p \text{ cotan } \varphi=p_d \text{ cotan } \varphi_d$, one can write:

$$\text{cotan } \varphi_d=\frac{p}{p_d} \text{ cotan } \varphi=\frac{p}{p_d} k_d,$$

and after comparing this with Eq. (6.16) and transforming to an explicit form for p_d, it becomes:

$$p_d=p\frac{\cos^2 \alpha}{1+\dfrac{\sin \alpha \cos \alpha}{k_d}} \approx p \cos^2 \alpha\left(1-\frac{\sin 2\alpha}{2k_d}\right). \tag{6.17}$$

When $k_d=100$ the factor in parentheses in Eq. (6.17) modifies the correct reduced distance, as given by Eq. (6.14), by some thousandths. *Table 6.1* gives the differences computed for a few slope angles.

The height difference can also be derived from *Fig. 6.11* but, from common sense, φ_h must be taken instead of φ_d and Δl_h instead of Δl_d:

$$\Delta h = d_s \sin\alpha = \Delta l_h \frac{\cos(\alpha+\varphi_h)\sin\alpha}{\sin\varphi_h} =$$

$$= \Delta l_h \left(\frac{\sin 2\alpha}{2}\operatorname{cotan}\varphi_h - \sin^2\alpha\right) = k_h \Delta l_h. \qquad (6.18)$$

The sign of the elevation stadia factor k_h depends on the sign of α, it is positive if the tilt is positive (elevation angle) and negative if the tilt is negative (depression angle).

From Eq. (6.18) by substituting

$$\operatorname{cotan}\varphi_h = \frac{p}{p_h}\operatorname{cotan}\varphi$$

one obtains

$$\frac{p_h}{pk_h}\frac{2}{\sin 2\alpha} = \frac{1}{k_h + \sin^2\alpha}$$

and

$$p_h = p\frac{k_h \sin 2\alpha}{2(k_h+\sin^2\alpha)} \approx \frac{p}{2}\sin 2\alpha\left(1 - \frac{\sin^2\alpha}{k_h}\right),$$

or

$$p_h \approx p\sin\alpha\cos\alpha\left(1 - \frac{\sin^2\alpha}{k_h}\right). \qquad (6.19)$$

Eq. (6.19) modifies Eq. (6.15) to give the correct height difference when the stadia distance varies asymmetrically; *Table 6.1* tabulates these modifications. The data in the Table refer to the most frequently used height factors. Therefore, for self-reducing tacheometers using asymmetric stadia distance variations, stadia distances are expressed by Eqs (6.17) and (6.19) if the stadia distance in question is determined by a displacement of the upper stadia line. For a fixed upper and displaced lower stadia line, the two formulae are modified so that the corrections have opposite signs, which is easy to see.

Table 6.1. Modification of height difference with asymmetric change of stadia

α	10°	20°	30°	45°
k_h	Modification in ‰			
10	3.0	11.7	25.0	50.0
20	1.5	5.8	12.5	25.0
50	0.6	2.3	5.0	10.0

In practice the stadia distance is varied in two ways:

— bent lines (diagrams) are projected onto the reticule of the telescope, assigned points on which define the stadia line distances p_d and p_h corresponding to the telescope tilt. Tacheometers using this construction are called diagram-tacheometers;

— one of the stadia lines in the telescope distance meter is a movable line, displaced in accordance with the telescope tilt by such an amount that its distance from the fixed stadia line is p_d or p_h. Tacheometers operating on this principle are called tacheometers with movable stadia lines.

Telescopes of tacheometers with a varying stadia distance usually give erect images and vertical stadia rods must usually be used for measurement. Corresponding to this the values p_d and p_h for any tilt angle are the linear distance measured along the vertical hair.

Diagram tacheometers. Their basic principle, developed by *Roncagli* and *Urbani* in 1983, is as follows. The ends of the ordinates p_d and p_h, plotted against the abscissae of a Cartesian co-ordinate system, determine the curve expressing the change of p_d (distance diagram) and a bifurcated curve expressing the change of p_h (height diagram) *(Fig. 6.12)*. The diagrams can also be constructed in such a way that the ordinates $p_d/2$ and $p_h/2$ are plotted upwards and downwards from the abscissae (symmetrical diagrams). By engraving the diagrams on the reticule (or projecting them there), it must be ensured that the ordinates of the diagrams assigned to the tilt angle α should always fall on the vertical hair. In this case the intersections of the diagrams with the vertical hair determine the intervals Δl_d and Δl_h on the image of the observed stadia rod.

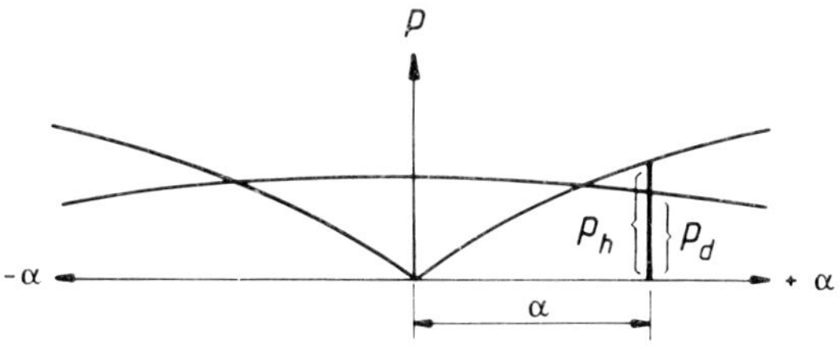

Figure 6.12. Asymmetrical diagrams in a rectangular system

On modern diagram tacheometers the diagrams are utilized in a polar system. Here the stadia distances p_d and p_h are plotted symmetrically *(Fig. 6.13)* or asymmetrically *(Fig. 6.14)* from a circle with radius r along the leg of the varying central angle α. The diagrams are used either with the vertical circle or as central diagram circles independent of the vertical circle. The vertical circle can rotate the diagrams through the corresponding vertical angle α simultaneously with the rotation of the telescope, and the moving leg of the angle can be made to coincide with the vertical hair.

On diagram tacheometers the integer factor k_d used with p_d is generally 100, exceptionally 200. Two or three, sometimes all four of the values $k_h = \pm 10$, ± 20, ± 50, ± 100 are used with p_h. This means that both (positive and negative) branches of the height diagram are used in a defined range of the vertical angle for different factors. By making the factor increase stepwise with increasing vertical angle, it is intended that the height diagram should intersect the vertical hair nearly perpendicularly at each angle α in order to have a more accurate reading on the stadia rod.

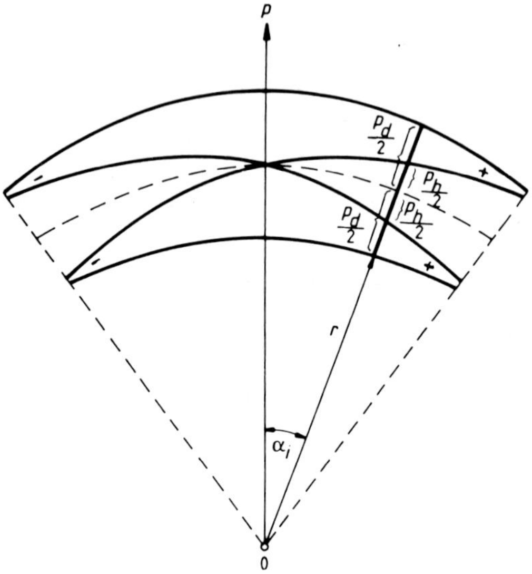

Figure 6.13. Symmetrical polar diagrams

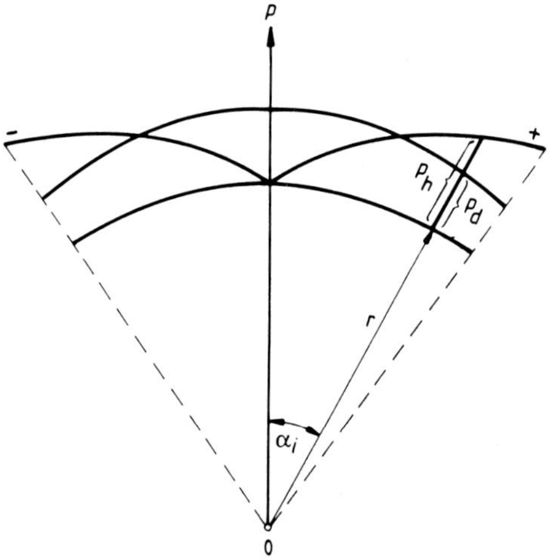

Figure 6.14. Asymmetrical polar diagrams

Kern (Aarau) used the Lehmann diagrams with a symmetrical stadia line system (see *Fig. 6.13*) on its earlier diagram tacheometer *DKR*. Such types of diagrams can be found now in the distance measuring telescopic alidade *RK* (for a plane table) manufactured by Kern (see Section 6.2).

Symmetrical distance diagrams are also used by MOM (Budapest) on its telescopic alidade *MA-5*.

In the overwhelming majority of diagram tacheometers the Hammer asymmetric diagrams are used; they require fewer readings on the rod.

In the first *Hammer–Fennel* diagram tacheometers the diagrams were photographically applied to glass plates. The plate with its mount is positioned in such a way that the centre of its base circle should lie on the geometrical centre line of the horizontal axis. The plate, however, does not take part in the rotation of the telescope around its axis H. On the side of the telescope next to the diagram D there is a glass window and the image of the diagram is projected by the mirror prism P_1 to the collecting lens L *(Fig. 6.15)*. The image of the corresponding diagram section is generated by the lens L on the surface coinciding with the image plane of the prism P_2 which covers half the field of view. The instrument is adjusted so that the horizontal stadia line touches the base circle of the diagrams *(Fig. 6.16)*, and the zero points of the diagrams coincide with the intersection of the horizontal line and the vertical edge of the prism. The parallax of the telescope has been eliminated by moving the tube of the objective lens, because the constant distance between the diagram and image plane can be ensured only in this way.

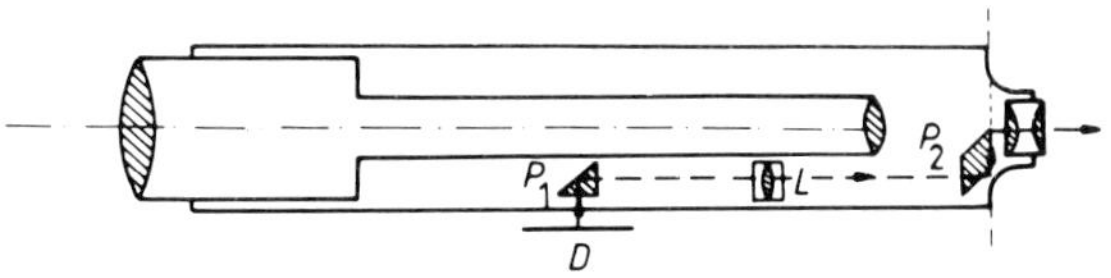

Figure 6.15. The Hammer system for projecting a diagram

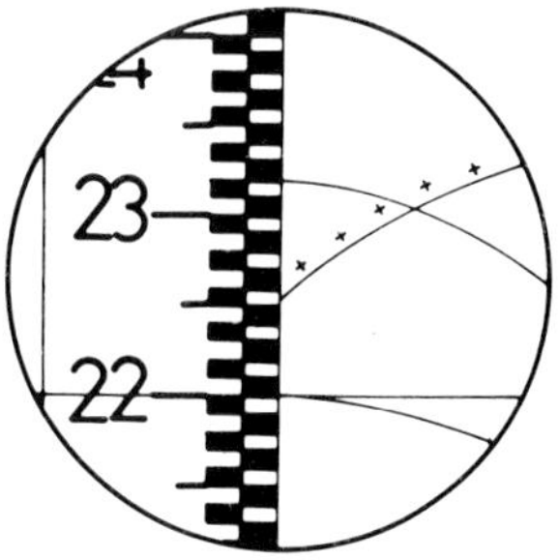

Figure 6.16. The field of view of the Hammer–Fennel (Kassel) diagram tacheometer when observing a rod

A projecting system similar to the above is used in the Soviet tacheometer *TA-2*, but here the diagrams D are on the vertical circle and they appear in the field of view together with the angular graduations L_v of the circle *(Fig. 6.17)*. The vertical circle is divided to 10 min of arc and a reading of 1 min of arc can be made by estimation.

VEB Jenoptik (Jena) marketed its first diagram tacheometer with the name *Dahlta 020*, its second as *Dahlta 010*. These were followed by *Dahlta 010 A (Fig. 6.18)*, then by *Dahlta 010 B (Fig. 6.19)*. In the *Dahlta* tacheometers the diagrams on the glass circle are fixed centrally with respect to the horizontal axis. They are visible over the whole field of view without even partly masking it *(Fig. 6.20)*. The main point of the projecting system is that the objective lens, the focussing lens and the prism system together generate an image of the rod on the diagram plane of the glass circle *(Fig. 6.21)*. When looking into the eyepiece one sees in the image plane erect images of the rod and the diagrams together with the direction hair on the reticule (the images of the rod and the diagrams are reversed by the reversing lens). The diagram system of the *Dahlta 020* includes the base circle, a distance diagram line ($k_d = 100$) and on both branches of the height diagram a three section diagram line ($k_h = \pm 10$, ± 20,

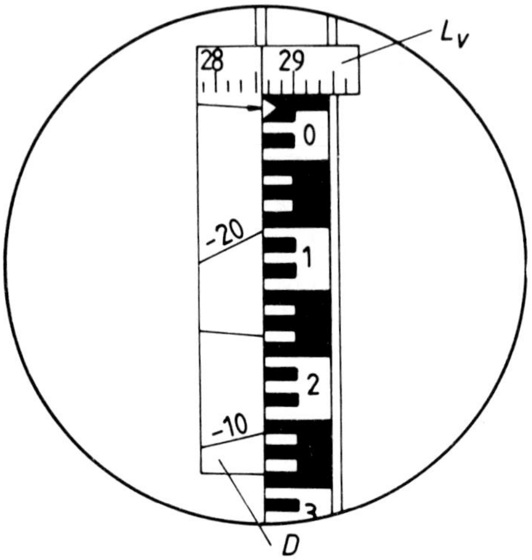

Figure 6.17. The field of view of the *TA-2* (Soviet Union) diagram tacheometer when observing a rod

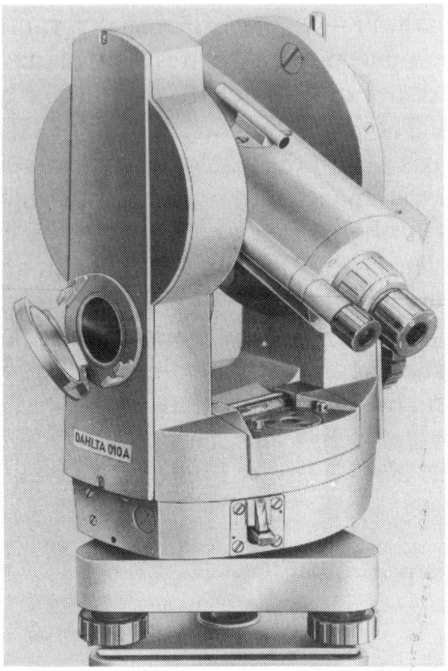

Figure 6.18. The *Dahlta 010 A* (VEB Jenoptik, Jena) diagram tacheometer

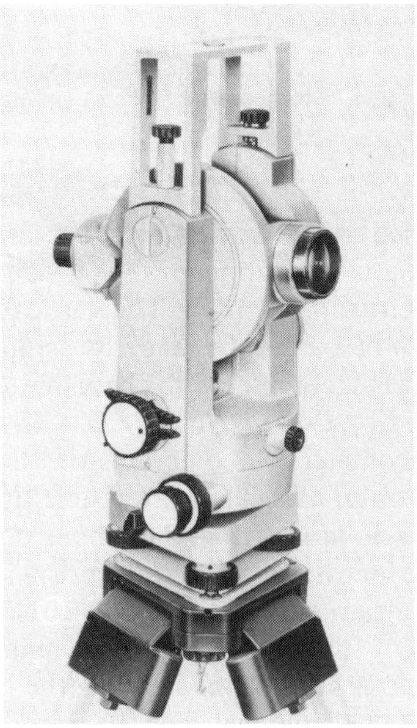

Figure 6.19. The *Dahlta 010 B* (VEB Jenoptik, Jena) diagram tacheometer

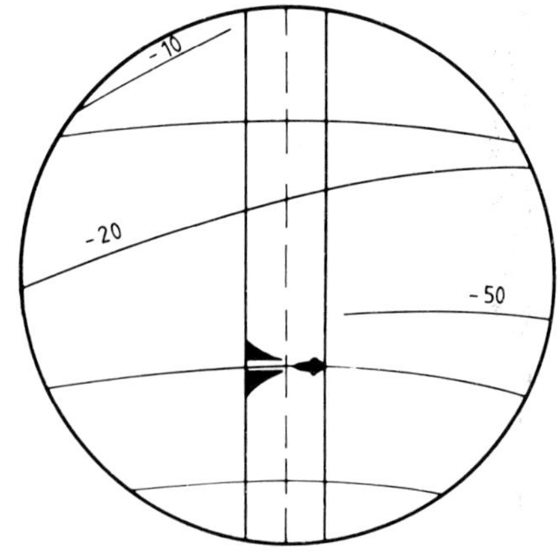

Figure 6.20. The diagrams and a rod image in the *Dahlta 010 B*

± 100). The *Dahlta 010* instrument uses two distance diagrams ($k_d = 100$, $k_d = 200$) and a height diagram similar to the *Dahlta 020*. On the *Dahlta 010 A* there is even a height diagram section of $k_h = \pm 50$. The index of the vertical circle is automatic; in older instruments an index level was used. The diagram system of the *Dahlta 010 B* is the same as in the *Dahlta 010 A*. The distance diagram with a constant of 200 is visible below the base circle, that with the factor 100 above the base circle *(Fig. 6.20)*. Its forced centring with a bush and pin enables it, among other things, to be used together with the *KARTI 250* which also has forced centring (see Section 6.2).

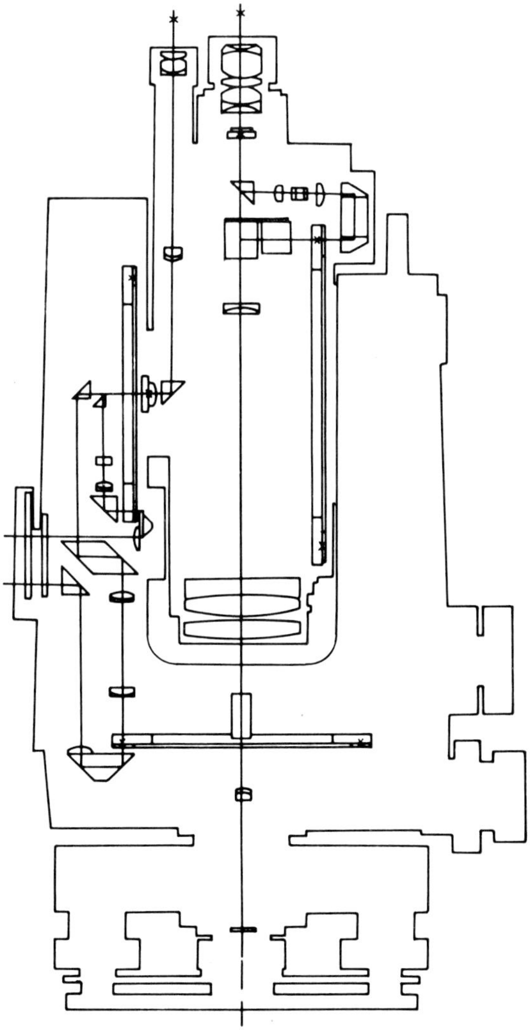

Figure 6.21. The optical system of the *Dahlta 010 A*

A 4 m long stadia rod with line graduations is used with the *Dahlta* tacheometers *(Fig. 6.22)*. The zero line of the rod is 1.40 m above its foot. Black division lines are used above the zero line and red ones below it. When measuring the base circle must be brought to the elevated zero line and the vertical hair must cut the graduation lines. In this case, instead of rod readings, the distance referring to the stadia factor constant 100, and the height difference referring to the stadia constant ± 10, can be read directly. If the stadia constant is not 100 (or not ± 10), then the corresponding factor (2.5 or 10) must also be recorded in the field book after the distance (or height difference) has been read. (Let, e.g. $k_h = -50$ and the height difference $= -2.49$ when referred to -10, then -2.49×5 must be recorded in the corresponding line of the field born.)

On the tacheometer *RDS* of Wild (Heerbrugg) the diagrams are also on a glass circle separate from the vertical circle *(Fig. 6.23)*. In order to ensure more favourable

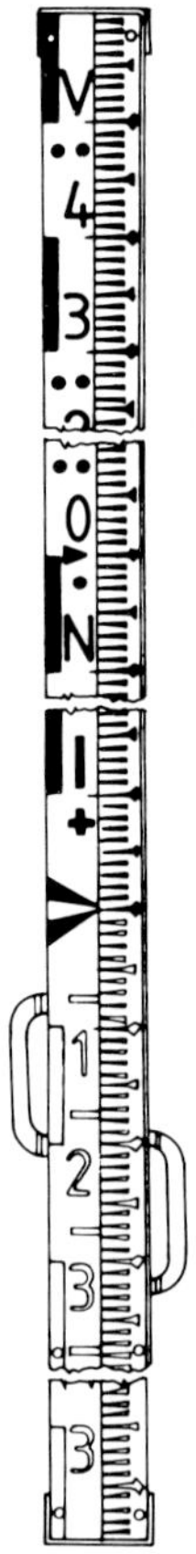

Figure 6.22. Graduation of *Dahlta* rods

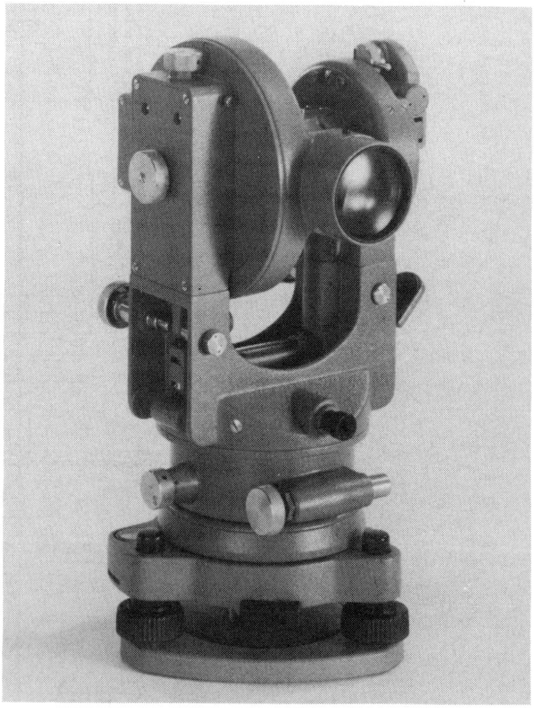

Figure 6.23. The *RDS* diagram tacheometer of Wild (Heerbrugg)

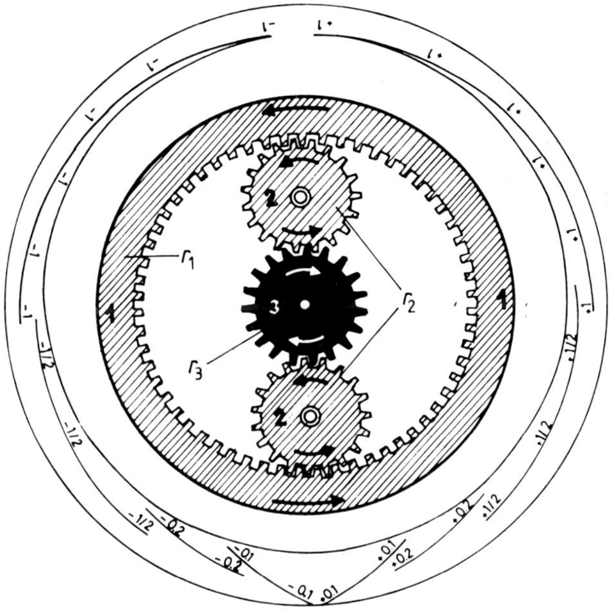

Figure 6.24. Rotating device for the diagram circle in the *Wild RDS*

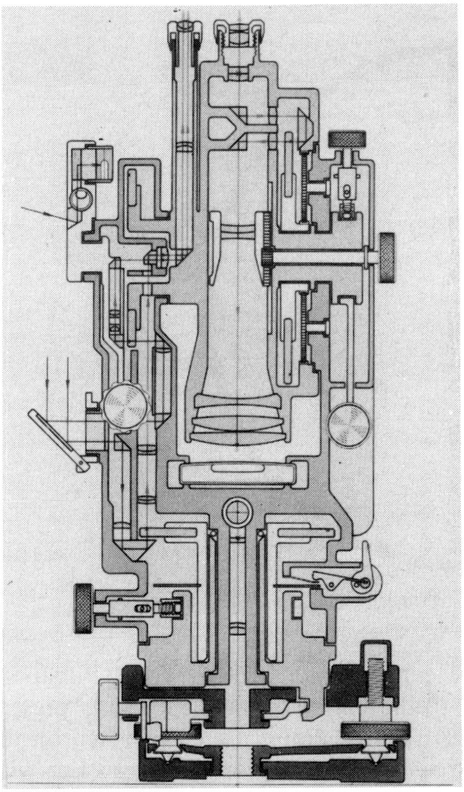

Figure 6.25. Cross section and optical ray paths of the *Wild RDS*

intersection angles between the diagram lines and the vertical hair — as well as a height diagram with more sections — the diagram circle swings through the angle 3α when the telescope swings through α. When the concentric gear wheel with radius r_1 rotates through the angle α together with the telescope, it rotates the satellite wheel with a radius of $r_2=r_1/3$ through the angle 3α *(Fig. 6.24)*. A rotation of the same angle but in the opposite sense is made by the concentric gear with radius $r_3=r_2$ and by the central diagram circle which rotates together with it. Thus the diagram circle has a four times rotation with respect to the telescope. Therefore the stadia line distances p_d and p_h computed for the angle α_i are engraved on a radius which forms an angle of $4\alpha_i$ with the zero line of the diagram circle. The projection of the diagrams into the telescope is made in the same way as with the *Dahlta* instruments *(Fig. 6.25)*. The telescope gives an erect image; the base circle can be seen in the lower part of the field of view, the height diagram between the base circle and the distance diagram *(Fig. 6.26)*. The stadia constant for distance is $k_d=100$, those for height are $k_h=\pm 10$, ± 20, ± 50 and ± 100. The height diagram, however, is engraved with the factors $k_h/100=\pm 0.1$, ± 0.2, $\pm 1/2$, and ± 1 referred to the stadia constant k_h, not with the constant itself. The zero line of the *GVLV* tacheometric staff used with this instrument is 1 m above the foot of the staff. There is a 0.70 m long telescopic staff extension in order to set the zero line of the staff to the height of the tacheometer. For topographic survey *Wild* recommends the tacheometric staff *GTLE (Fig. 6.27)*.

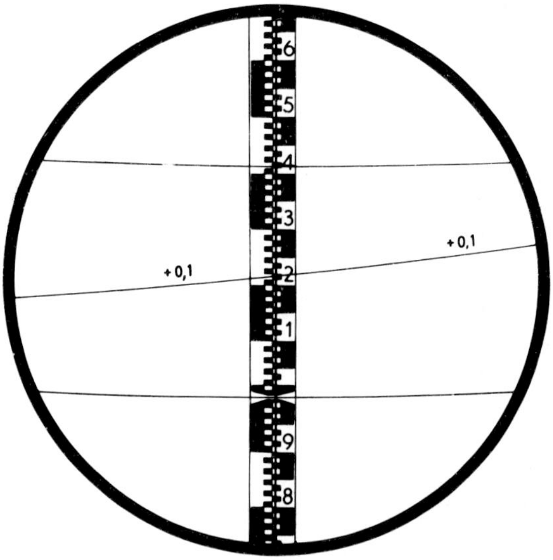

Figure 6.26. The field of view of the *Wild RDS* when observing a rod

The older tacheometer *Ta D1* and the newer *Ta D4* and *Ta D41* (MOM, Budapest) are called circle diagram tacheometers due to the shapes of the diagram. At present MOM manufactures the *Ta D41 (Fig. 6.28)*. The diagrams are constructed on the basis of the recognition that, if the glass circle with the diagrams rotates twice with respect to the vertical circle which rotates together with the telescope, then the diagram

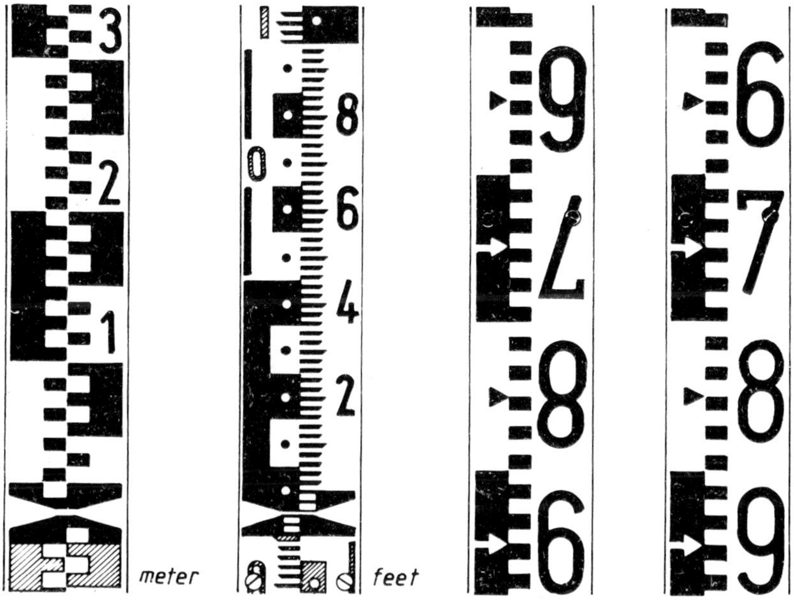

Figure 6.27. Tacheometric staffs *GVLV* and *GTLE* (Wild, Heerbrugg)

curves can be substituted by arcs of circles with different centres. This approach satisfies the necessary conditions in practice.

The diagrams on circle diagram tacheometers have a gear drive. The base circle of the diagrams lies between the two distance diagrams which have a stadia constant

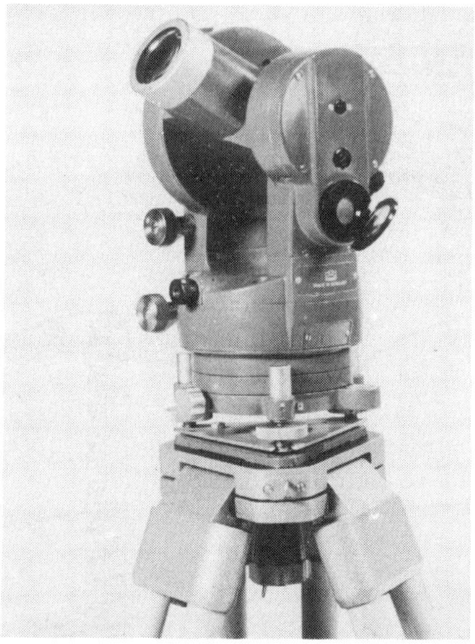

'igure 6.28. The *Ta D41* diagram tacheometer (MOM, Budapest)

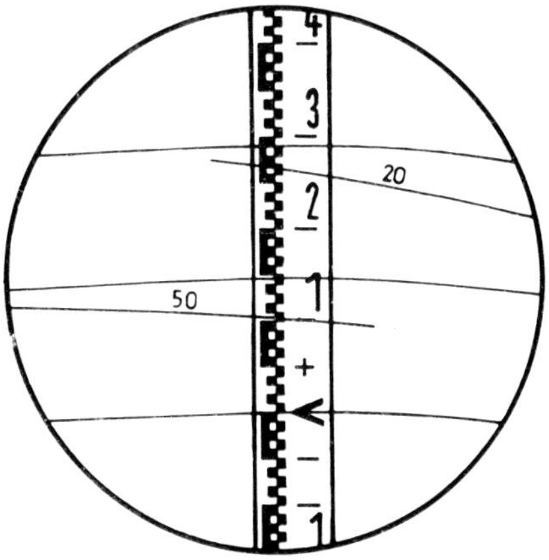

Figure 6.29. Diagram and staff image in the field of view of the *Ta D41*

of 200. Hence the varying stadia line distance between the two distance diagrams has a constant $k_d = 100$ *(Fig. 6.29)*. The stadia distances of the height diagram line are referred to the lower distance stadia line, therefore this line must be used as a base circle when measuring. The stadia constants of the height diagrams are $k_h = \pm 20$ and $k_h = \pm 50$. A 3 m long tacheometric staff is used with the instrument. The zero line of the staff lies 1 m above its foot. The staff reading $l = \Delta h_i$ can be ensured by a proper adjustment of the 0.70 m long staff extension. The tacheometers *Ta D4* and *Ta D41* have an automatic vertical circle index, the *Ta D1* has an index level with a constant of 30 sec of arc. In the *Ta D* type instruments the coincidence of the vertical hair and the diagrams with a radius angle 2α is ensured when the vertical axis is exactly vertical, provided that the instrument is in adjustment.

In the self-reducing tacheometers *FTRA* (Fennel GmbH, Kassel) the diagrams are projected into the telescope field of view in the same way as in the *Dahlta* instruments *(Fig. 6.30)*. The telescope gives an upright image. The base circle of the diagram is the lower curve in the field of view *(Fig. 6.31)*, the stadia constant of the distance

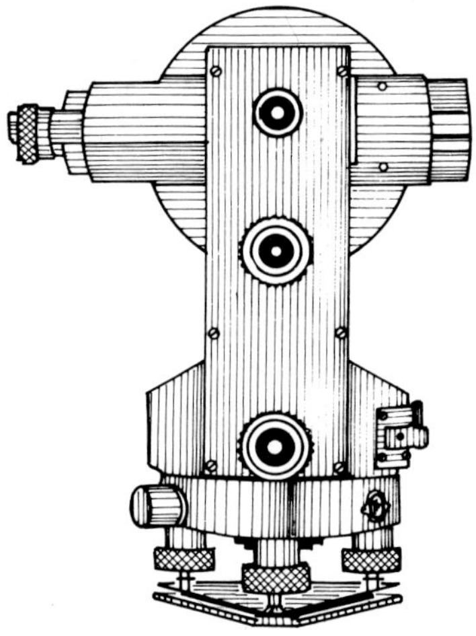

Figure 6.30. The *FTRA* diagram tacheometer (Fennel GmbH, Kassel)

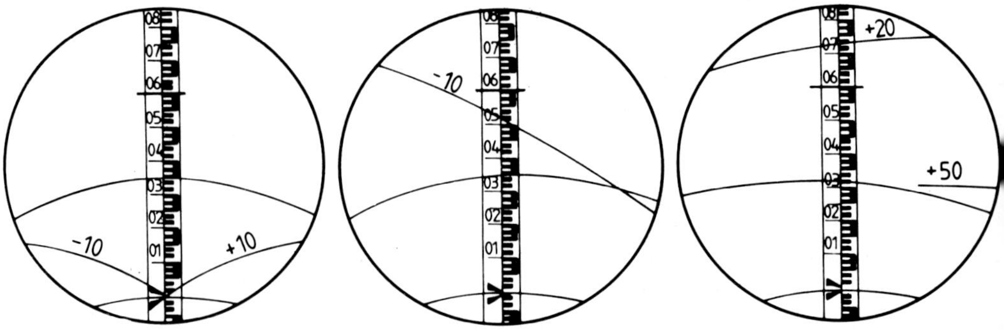

Figure 6.31. The diagrams of the *FTRA* in the field of view

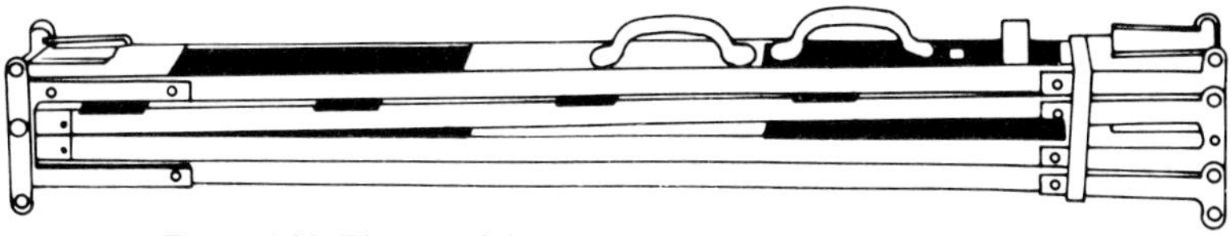

Figure 6.32. The 4 m folding tacheometric staff for the *FTRA*

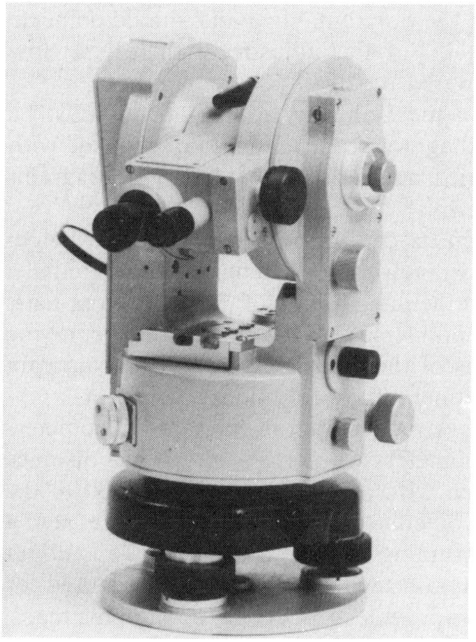

Figure 6.33. The *RTa 4* diagram tacheometer (Opton Feintechnik, Oberkochen)

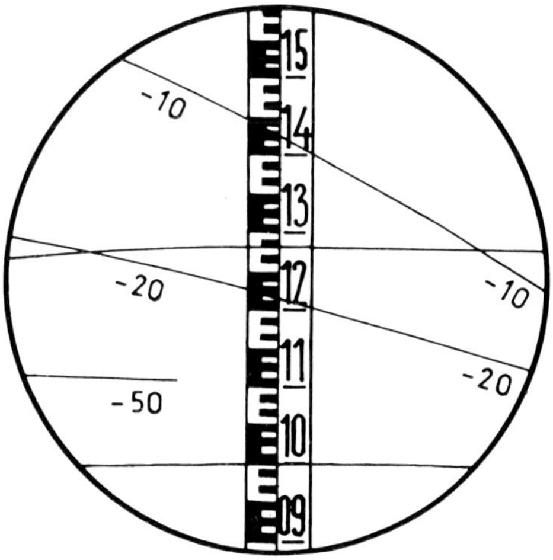

Figure 6.34. The image of the tacheometric staff *RTa 4K* in the field of view of the *RTa 4*

curve with respect to it is $k_d = 100$, and the stadia constants of the three sections of the height diagrams are $k_h = \pm 10$, ± 20, and ± 50. The zero line on the appropriate tacheometric staff is at a height of 1 m, which can be increased by pulling out the 0.60 m staff extension. The staff is hinged and ca be folded into four for transportation and storage *(Fig. 6.32)*.

On the diagram tacheometer *RTa 4* (Opton Feintechnik GmbH, Oberkochen) *(Fig. 6.33)* there is an additional curve with a stadia factor $k_d = 50$ for the measurement of short distances. A transverse graduation is used on the tacheometric staff *RTa 4K* appropriate to the instrument; the horizontal distance can be read directly from the staff *(Fig. 6.34)*.

Diagram tacheometers — as all tacheometers in general — must be tested and adjusted similarly to theodolites. The test of their distance meter systems may involve the diagrams, and their rotating and projecting systems. Since the diagrams are rigidly attached to the diagram circles by the manufacturer, they can only be tested for their required accuracy without any possibility of adjustment.

The test of the optical projecting system must be carried out by using the *Hammer* projection for the size of the projected image (i.e. for the hair distance p_d) as well as for the position of the whole of the field of view with respect to the stadia line system of the telescope. The adjustment is better left to a professional workshop. One condition for the diagram image position is that the horizontal line should touch the base circle and, when the telescope is horizontal, the common point of the two branches of the vertical curve should lie on the horizontal line and should coincide with the edge of the image-separating prism. For this adjustment there is a slow motion screw on the instrument, usually covered by a cap.

The diagram optical projecting system of the Dahlta type instruments must be adjusted if the common point of the height diagram branches does not coincide with the intersection of the base circle and the vertical hair. The field of view of the diagrams can also be adjusted by a slow motion screw.

The diagram circle of the *Wild RDS* is in its correct position if the two height diagram branches straddle the vertical hair symmetrically when pointing horizontally, and the vertical hair lies in the centre line of the interruption of the short cross-hairs visible at bottom and top of the view. The adjustment can be made after removing the cover of the diagram housing by means of the adjusting screw of the diagram circle. It would be better to make the adjustment in a professional workshop.

If the diagrams have clear intersecting images, the accuracy of diagram tacheometers is the same as the accuracy of simple tacheometers with cross-hairs. If the distance does not exceed 150 m, it can be calculated with a relative error of 1:500 of the distance. The measuring error of height difference, using diagrams with stadia constants $k_h = \pm 10$, ± 20 or greater and assuming average circumstances, is about ± 5, ± 10, ± 16 ... ± 20 cm. Therefore, diagram tacheometers are usually used for surveys for engineering location plans and maps with contours or topographic maps.

The accuracy data of the different instruments, as given by the manufacturers, and further characteristics are listed in *Table 6.2*.

Tacheometers with movable stadia lines. The reduced stadia line distances p_d and p_h changing as a function of the vertical angle, can be set by any distance meter with an eyepiece micrometer if the values p_d and p_h assigned to the elevation angle α are known, e.g. from a table. This principle was formerly used in the *Tichy* tacheometer with an eyepiece micrometer. On this the stadia line distances p_d and p_h could be read directly on the vertical circle (p_d on the graduation of the vertical circle marked *D*, and p_h on the graduation marked *H*). After having observed the staff at an angle α, the movable line was set to the line distance p_d obtained on the graduation *D* and Δl_d was then obtained directly; finally the movable line was set to p_h, and the reading interval Δl_h was obtained.

In up-to-date tacheometers with movable stadia lines the movable line is mechanically controlled and is set automatically to the correct stadia line distance, reduced with respect to the fixed line by the rotation of the telescope around its horizontal axis.

On the self-reducing tacheometer *K1-RA* of *Kern* (Aarau) *(Fig. 6.35)* the automatic shift of the movable line is controlled by the suitably shaped eccentric *4 (Fig. 6.36, a)* which meshes with the horizontal rotation axis of the telescope. A rotation of the telescope by an angle α results in a rotation of the eccentric by an angle of 2α. The

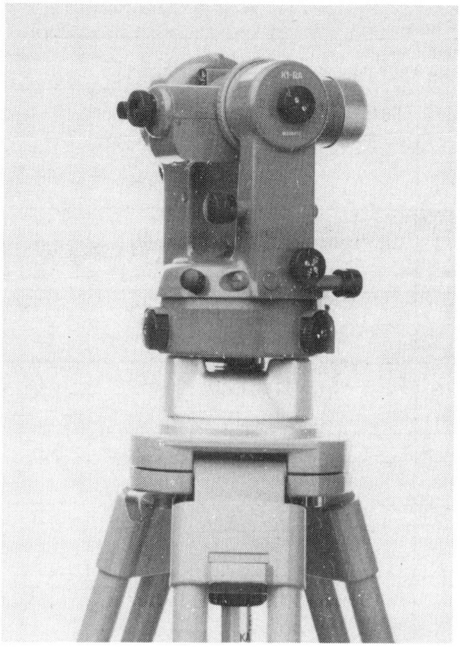

Figure 6.35. Kl-RA tacheometer (Kern, Aarau) with movable stadia lines

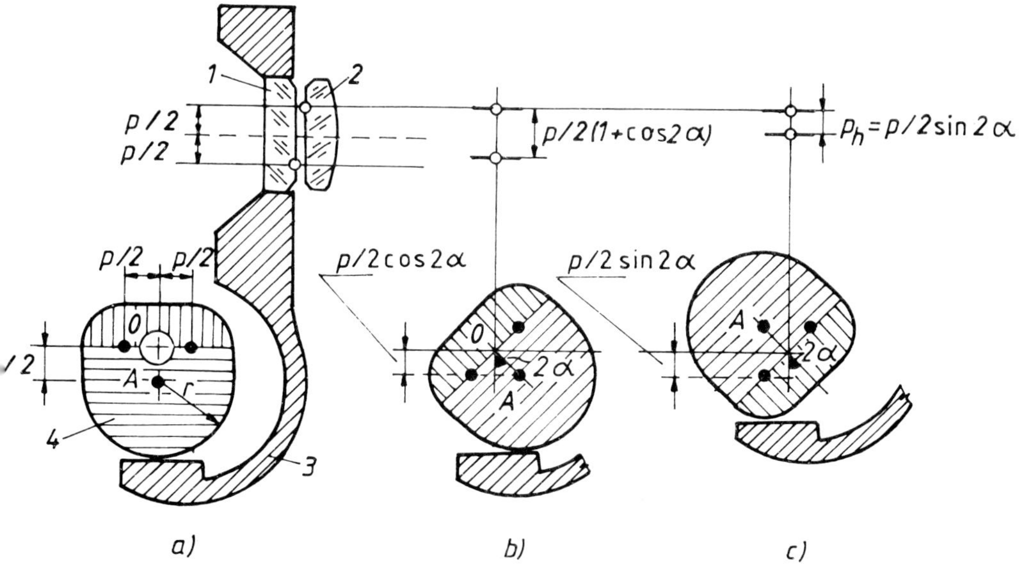

Figure 6.36. Controlling system for the movable stadia lines of the *Kl-RA*

Table 6.2. Self-redu

Type, mark	Manufacturing firm	Self-reducing system	Telescope: magnification	Telescope: diameter of the objective lens (mm)	Telescope: minimum distance (m)	Horizontal circle: diameter (mm)	Horizontal circle: reading microscope	Horizontal circle: direc reading
RDS	Wild AG	diagram	24	40	2.4	79	graduated	1^c—
FTRA	O. Fennel GmbH		25	40	1.7	90	graduated	1^c—
RTa 4	Opton Feintechnik		25	40	2.0	98	graduated	1^c—
Dahlta 010	Jenoptik		25	40	3.0	98	graduated	1^c—
Dahlta 010A	Jenoptik		25	40	3.0	86	graduated	1^c—
Dahlta 010B	Jenoptik		25	40	3.0	86	graduated	1^c—
Ta D1	MOM		24	40	3.0	77	graduated	2^c—
Ta D11	MOM		24	40	2.5	77	graduated	2^c—
Ta D41	MOM		24	40	2.5	77	graduated	2^c—
Ta-2	USSR		25	33	3.0	72	graduated	1'
K1-RA	Kern Co.	moving stadia	28	45	1.8	95	optical micrometer	1^c—
DK-RV	Kern Co.		27	45	1.8	75	optical micrometer	10^{cc}
DK-RT	Kern Co.	with split image and prism	27	45	2.5	75	optical micrometer	10^c
Redta 002	Jenoptik		25	42	2.5	114	graduated	1^c—
RDH	Wild AG		27	45	5.0	79	graduated	1^c—
TP	USSR		27	38	2.0	95	graduated	1
BRT 006	Jenoptik	basis reducing	15	25	1.8	90	estimating	5^c—
TW	USSR		16	25	2.0	95	estimating	10'
17-KS	MOM	tangential	24.5	30	2.5	116	optical micrometer	20^c 10'
Tari 4180	Filotecnica		28	40	3.0	90	graduated	1^c—
Tari 4181	Salmoiraghi		28	40	3.0	90	optical micrometer	5^{cc}—

eometers

Vertical circle			Stadia factor		Spirit level			Weight				Reference to figures
m- er m)	reading on special graduation	direct reading of angle	k_d	$\pm k_h$	alidade	height index	telescope	instrument (kg)	box (kg)	tripod (kg)	accuracy $\pm m_d$ cm/100 m	
0		1^c—1′	100	10, 20 50, 100	30″	30″	60″	7.1	2.3	6.5	10	6.23
0		1^c—1′	100	10, 20 50	40″		40″	5.4	3.4	5.5	10	6.30
5		1^c—1′	50, 100 200	10, 20 50, 100	30″			5.3	4.3		5—20	6.33
6		1^c—1′	100, 200	10, 20 50, 100	30″	30″	30″	4.7	3.1	5.8	10—20	
6		1^c—1′	100, 200	10, 20 50, 100	20″		30″	4.7	4.4	5.8	10—20	6.18
6		1^c—1′	100, 200	10, 20 50, 100	30″			4.9	4.4	6.5	10—20	6.19
8		2^c—1′	100, 200	20, 50	30″	30″		5.3	2.8	5.1	10—20	
8		2^c—1′	100, 200	20, 50	30″	30″		5.3	2.8	5.1	10—20	
8		2^c—1′	100, 200	20, 50	30″			5.3	2.8	5.1	10—20	6.28
2		1′	100	10, 20 100	45″	45″		4.0	3.0	5.0	10—20	
5	tan 1^c		100		35″			4.5	2.2	5.3	10—20	6.35
0	0.001		100		30″	30″	30″	4.9	2.7	5.3	3—5	6.38
0	0.001				30″			5.4	2.3	5.8	1—2	6.47
4	tan 1^c (1′)				30″	30″	30″	6.5	6.2	5.6	2	6.43
0		1^c—1′			30″	30″		7.1	6.7	5.2	1—2	6.46a
0		1′			30″			5.7	3.0	5.3		
0		5^c—5′	200		60″	30″		6.7	5.8	5.6	6	6.50
5		10′	20		30″			7.5	4.6	5.3		
0	tan	5^c—2′			50″	20″	20″	5.8	2.1	4.8	max. 5	6.7
0	tan				30″	30″		5.5	3.1	7.2		6.9
0	0.1%	5^{cc}—2″			30″	30″		5.5	3.1	7.2		

lever *3*, in contact with the eccentric from below, shifts the reticule *1* containing the movable line with respect to the reticule *2* containing the fixed line. If the telescope is horizontal, the two lines are separated by $p = f_0 \tan\alpha = f_0/k_d = f_0/100$. In this position the distance between the point *A* of the eccentric (the centre of the circle with radius *r*) and the centre of the axis 0 is $p/2$. For a tilt of the telescope of angle α *(Fig. 6.36, b)* *A* and the movable stadia line are raised together by:

$$s = \frac{p}{2}(1 - \cos 2\alpha).$$

Thus the separation of the two lines will be the reduced stadia line distance:

$$p_d = p - \frac{p}{2}(1 - \cos 2\alpha) = \frac{p}{2}(1 + \cos 2\alpha). \quad (6.20)$$

The stadia line distance p_h for determining the elevation difference can be obtained by a 180° reversal of the eccentric *(Fig. 6.36, c)*. Then, if the telescope is horizontal, the point *A* becomes higher in the rotation plane of the eccentric by $2p/2 = p$, and the movable line coincides with the fixed line. For a rotation of the telescope of α, the point C_1 of the arc with a radius $\rho = p/2$ rotates through an angle 2α, and the lever shifts the movable line by $p_h = p/2(\sin 2\alpha)$ with respect to the fixed line. To this stadia line distance the height factor $k_h = 100$ is assigned.

The advantage of the *Kern K1-RA* tacheometer with a movable line with respect to diagram tacheometers is that the lines always intersect the staff perpendicularly *(Fig. 6.37)*. A further advantage is that p_d and p_h do not appear in the field of view at the same time: to change from p_d to p_h it is necessary to reverse a lever. Hence the probability of a muddle of the stadia line distances (or readings) is very small. This system is, however, inconvenient with small tilts because the sign of the height difference cannot be determined visually, but must be defined from the vertical angle. On the vertical circle of the tacheometer *K1-RA* there is also a tangent graduation which can be read with an accuracy of 0.0001 by estimation.

The mean square errors of distances and elevation differences obtained by the *K1-RA* lie between ± 10 cm/100 m and ± 20 cm/100 m depending on the distance and the circumstances.

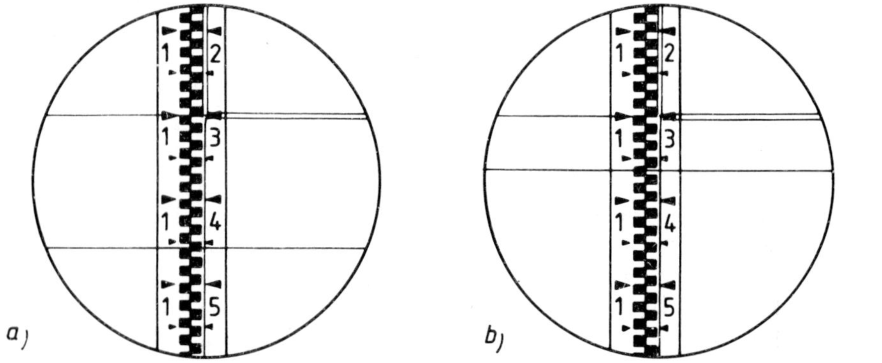

Figure 6.37. Field of view of the *Kl-RA* when observing a staff *a)* for distance measurement: *b)* for elevation measurement

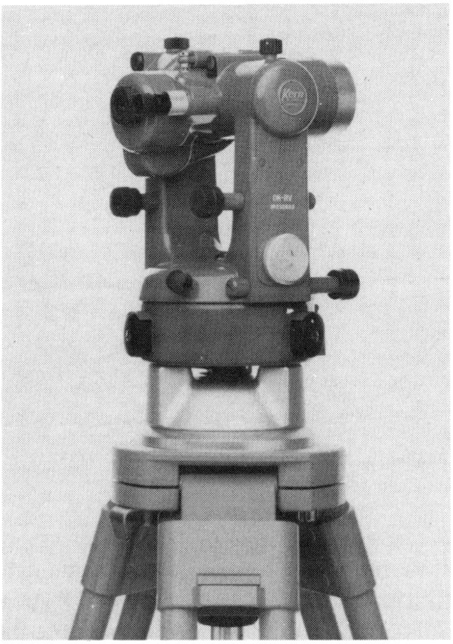

Figure 6.38. The *DK-RV* tacheometer (Kern, Aarau) with movable hairs

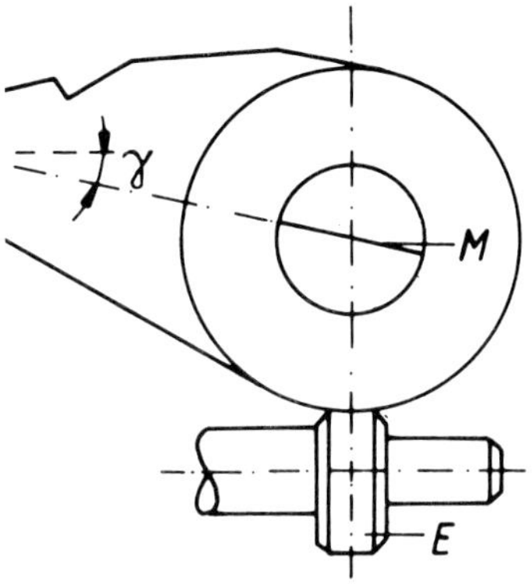

Figure 6.39. Control of the reticule in the *DK-RV*

In the tacheometer *DK-RV* (Kern, Aarau) *(Fig. 6.38)*, for the distance measurement the stadia line distance p_d corresponding to the telescope tilt is also set by the automatically controlled movable line. The elevation difference, however, can only be computed from the horizontal distance. There is a tangential graduation on the vertical circle of the instrument for this purpose. In the tacheometer *DK-RV* the movable line *M* is controlled by the eccentric *E (Fig. 6.39)*, the angular velocity of which is twice the angular velocity of the gear rotating through angle α around the instrument axis *H*. The eccentric moves the lever rotating around the axis *T*, together with the reticule fixed to it, by $p_d = \frac{p}{2}(1+\cos 2\alpha)$ as in the principle of the instrument *K1-RA*. The stadia constant k_d is 100 for this stadia line distance. The movable line is, however, not parallel to the fixed line, but forms with it the angle γ. This angle also changes somewhat with the tilt angle, due to the rotation around the axis *T*. This change corresponds to the dimension change of the staff image assigned to the different telescope tilts. The oblique line gives an increased accuracy of staff reading if the especially graduated stadia rods appropriate to the instrument are used *(Fig. 6.40)*. The unit of the main graduation on the rod is $a = 1.25$ cm and the division are marked by small white circles as well as by the usual black and white divisions. At each tenth main division there is a horizontal subscale with 10 divisions, the total length of which is $b = 6.25$ cm. The movable hair with a tilt of $\tan\gamma = a\cos\alpha/b = 0.2\cos\alpha$ — the stadia line distance of which is to be understood to be in the direction of the

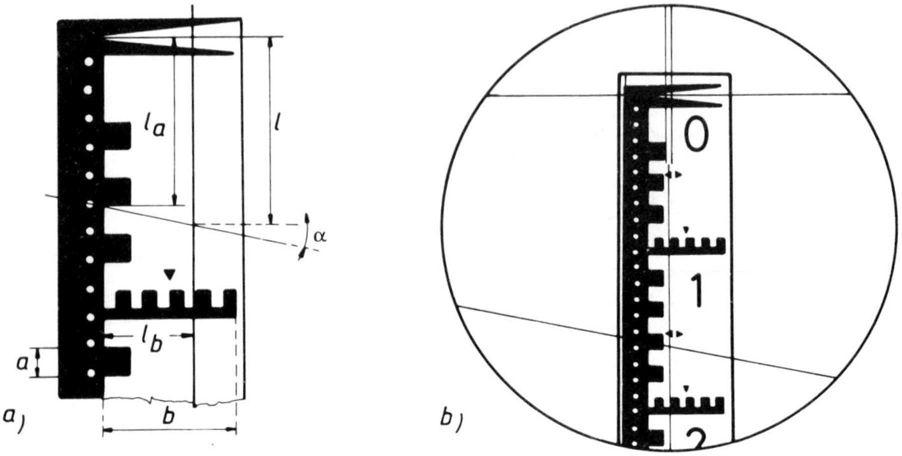

Figure 6.40. In the *DK-RV: a)* reading of the stadia line distance; *b)* the field of view when observing a staff

fixed line — must be set by the horizontal slow motion screw of the instrument to bisect a white circle of the main graduation. Now a reading l_b made by the vertical hair on the graduation b yields the values $a/10$ directly and the values $a/100$ by estimation. Due to the main graduation of $a = 1.25$ cm, $k'_d = 80$ would belong to the hair distance p_d but, when regarding the division of 1.25 cm as a division of 1 cm and with numbering in accordance with this, the actual factor for distance measurement is:

$$k_d = k'_d\, 1.25 = 80 \times 1.25 = 100.$$

For determining the elevation difference there is a tangential graduation on the vertical circle of the instrument which can be read with an accuracy of 0.0001.

According to the data given by the manufacturer horizontal distance can be obtained by the *DK-RV* with a mean square error of ± 4 cm/100 m and elevation difference with a mean square error of ± 1 cm. Further data for the instrument can be found in *Table 6.2.*

6.1.2.3 Split-image Self-reducing Tacheometers with an External Base

Basic principle. In split-image self-reducing tacheometers with an external base the *Boskovic rotating wedge pair* is used; the method of using it was developed by *Bosshardt*. If a glass wedge of small deflecting angle $\varphi/2$ (Subsection 1.3.3) is rotated around the axis $t-t$ perpendicular to the symmetry plane bisecting the deflecting angle, then the principal rays falling along the axis will describe after emergence a conical surface of half aperture $\varphi/2$ *(Fig. 6.41, a)*; the radius of the base of the cone at a distance d_s from the wedge is:

$$r = d_s \tan\frac{\varphi}{2} \approx \frac{\varphi''}{2\rho''}\, d_s. \tag{6.21}$$

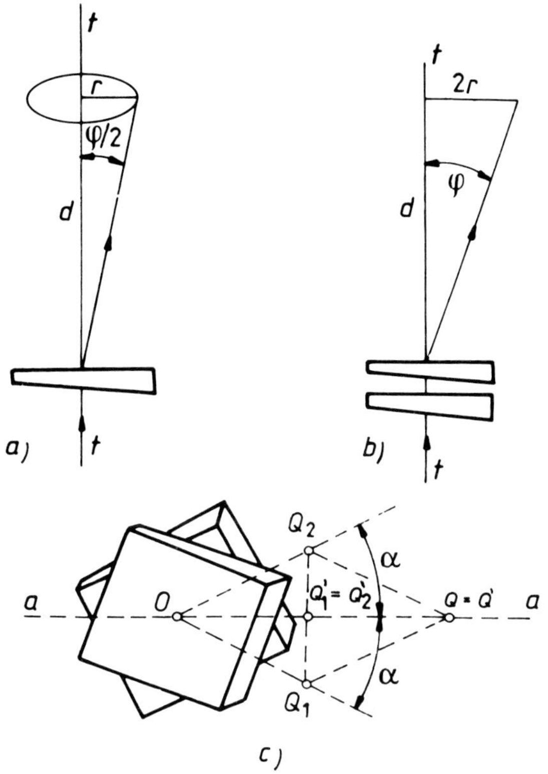

Figure 6.41. Operating principle of the reducing wedge pair

Let a second wedge with the same deflection $\varphi/2$ now be placed in front of the first and close to it. Let the two wedges be in their basic position, i.e. in the starting position where the principal planes of both wedges are in the same principal plane (the plane of the paper in *Fig. 6.41, b*) and their deflecting edges lie on the same side. The principal ray entering along the axis, after passing through the two wedges, will be deflected in the common principal intersection plane by the angle φ in the direction opposite to the deflecting edges; it will meet the plane perpendicular to the axis and distant d_s at the distance $2r = d_s\varphi/\rho$. Let the two wedges be now oppositely rotated around the axis denoted in *Fig. 6.41, c* by O, i.e. respectively by the angles $+\alpha$ and $-\alpha$ from the base plane $a-a$. In this case, due to the wedge rotated by the angle $+\alpha$, the emerging ray would meet the normal plane distant d_s (which is parallel to the plane of the paper) in the point Q_1. The projection of the distance OQ_1 onto the base plane $a-a$ is:

$$OQ'_1 = \frac{\varphi}{2\rho} d_s \cos\alpha, \qquad (6.22)$$

and the projection onto the normal to the base plane:

$$Q_1Q'_1 = \frac{\varphi}{2\rho} d_s \sin\alpha. \qquad (6.23)$$

Due to the effect of the wedge rotated by the angle $-\alpha$, the emerging ray would meet the plane distant d_s in the point Q_2. The projection of the distance OQ_2 onto the base plane is:

$$OQ'_2 = \frac{\varphi}{2\rho} d_s \cos(-\alpha), \tag{6.24}$$

and the projection onto the normal:

$$Q_2Q'_2 = \frac{\varphi}{2\rho} d_s \sin(-\alpha). \tag{6.25}$$

The projection on the base plane of the distance OQ, between the axis and the intersection Q of the normal plane by the ray formed by the two wedges, is the sum of Eqs (6.22) and (6.24), i.e.:

$$OQ = OQ' = \frac{\varphi}{\rho} d_s \cos\alpha \tag{6.26}$$

and the perpendicular projection of OQ is the sum of Eqs (6.23) and (6.25), or:

$$QQ' = 0. \tag{6.27}$$

Let us suppose that a rotatable pair of wedges — both deflecting the ray by the angle $\varphi/2$ — is attached in front of the objective lens of a geodetic telescope and they cover the upper or lower half of the objective lens. This pair of wedges should be in their basic position when the telescope is horizontal, i.e. the principal planes of the wedges should lie in the horizontal as a base plane, and the deflecting edges of both wedges should lie on the same side. It can be arranged by means of a train of gears that, by elevating the telescope by the vertical angle $+\alpha$, the first wedge is turned simultaneously around the optical axis of the telescope by $+\alpha$ and the second one by $-\alpha$. If a stadia rod is set up at the point P which is at a slope distance d_s and a slope angle $+\alpha$ from the centre of the wedge pair, the rays emerging directly from the objective lens and passing through the wedge pair will subtend on the rod

$$\Delta l_d = \frac{\varphi}{\rho} d_s \cos\alpha, \tag{6.28}$$

which is equal to Eq. (6.26). From this the slope distance is

$$d_s = \frac{\Delta l_d}{\cos\alpha} \frac{\rho}{\varphi}. \tag{6.29}$$

The horizontal distance d between the point P and the centre of the wedge pair is:

$$d = d_s \cos\alpha$$

into which Eq. (6.29) can be substituted, giving:

$$d = \Delta l_d \frac{\rho}{\varphi} = k \Delta l_d. \tag{6.30}$$

Here

$$k = \frac{\rho}{\varphi} = 100$$

is the stadia constant of distance measurement, if $\varphi = 0.01$ radian $= 2062.6$ sec of arc; in other words, if wedges with a deflecting angle of $\varphi/2 = 0.005$ radian $= 1031.3$ sec of arc are used. Equation (6.30) means that the horizontal distance d between the point P and the centre of the rotating pair of wedges is defined as k-times the rod intercept Δl_d.

If the pair of wedges is attached in front of one half of the objective lens of the telescope in a base position in which the principal planes of the wedges are vertical and the deflecting edge of the first wedge is above and that of the second below, then the wedges are turned by $-90°$ and $+90°$ with respect to their basic position. Let the telescope be now pointed to the point P which has a vertical angle of $+\alpha$. In this case the first wedge is turned by the angle $(-90° + \alpha)$ and the second wedge by the angle $(90° - \alpha)$ with respect to the basic position. Putting $(90° - \alpha)$ instead of α in Eq. (6.22) and $(-90° + \alpha)$ instead of α in Eq. (6.24), the sum of Eqs (6.22) and (6.24) will now be [instead of (6.26)]:

$$OQ' = \frac{\varphi}{\rho} d_s \sin \alpha \tag{6.31}$$

and the sum of Eqs (6.23) and (6.25) remains zero in this case also. Consequently the intercept on the horizontal stadia rod is identical with Eq. (6.31):

$$\Delta l_h = \frac{\varphi}{\rho} d_s \sin \alpha, \tag{6.32}$$

from which:

$$d_s = \frac{\Delta l_h}{\sin \alpha} \frac{\rho}{\varphi}. \tag{6.33}$$

The elevation Δh of the point P, which is at a slope distance d_s and a slope angle α from the centre of the wedge pair, is:

$$\Delta h = d_s \sin \alpha,$$

or by using Eq. (6.32):

$$\Delta h = \frac{\rho}{\varphi} \Delta l_h = k \Delta l_h. \tag{6.34}$$

Equation (6.34) means that the elevation Δh of the point P with respect to the centre of the wedge pair is determined as k-times the stadia rod intercept Δl_h.

Separation of images. The rays entering the telescope directly or through the wedge pair generate two images in the image plane of the telescope, which are displaced with respect to each other. In other words a mixed image is generated. This is disturbing, therefore the two images are separated in split-image self-reducing tacheometers in such a way that one image is visible in the lower part of the telescope field of view and the other in the upper part, and the two halves are separated by a sharp line. Three types of image separation are used:

1. There is a longitudinal dividing strip *3* inserted between the objective lens *1* and the image plane *2* or the telescope *(Fig. 6.42, a)*. In order that the back edge of this

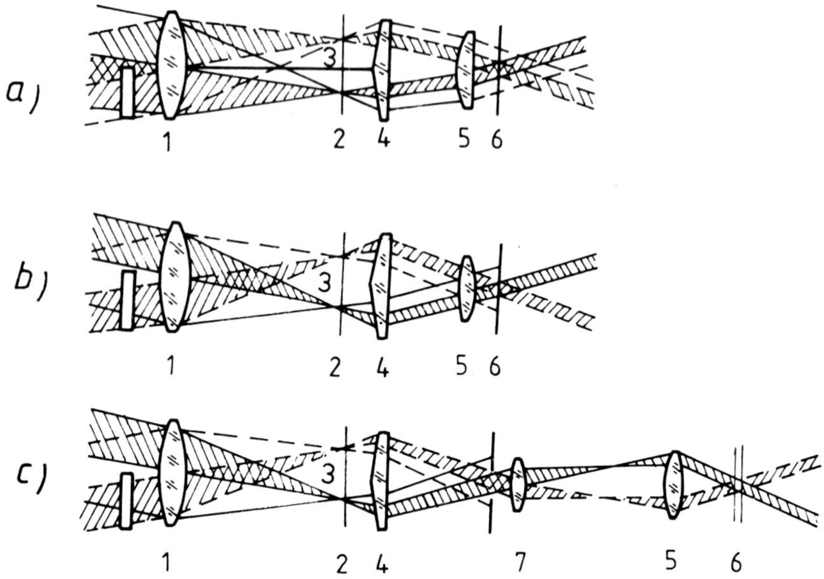

Figure 6.42. Image separation in split-image tacheometers

strip should be in sharp focus in spite of the great magnification of the eyepiece lens, also in order to eliminate the injurious effects of eye-defects due to irregular curvature of the eye lens of the observer, the two beams of rays generating the two half images are superimposed in the exit pupil *6* in front of the eyepiece *5*, by interposing the double wedge *4*.

2. The dividing strip described in point 1 can be substituted in the exit pupil *6* by a horizontal slit about 2 mm wide, or by a diaphragm with a hole of the same diameter *(Fig. 6.42, b)* which blank out the disturbing parts of the rays in the same way as the dividing strip. The disadvantage of this solution is that some observers (especially those wearing spectacles) cannot put their eye into the exit pupil, because this is occupied by the slit or hole diaphragm. Consequently the two beams of rays leaving the eyepiece *5* do not fall on exactly the same part of the eye which is in front of the exit pupil.

3. An improved variant of solution 2, in which an image-shifting lens *7* is used in front of the slit or aperture diaphragm and the eyepiece is placed before this lens *(Fig. 6.42, c)*. The observer can put his eye into the now empty exit pupil and, furthermore, this solution yields an erect image. In this case, however, the telescope is longer.

The accuracy of split-image tacheometers with an external base. In order to determine the standard deviation of the horizontal distance d measured by this type of tacheometer, Eq. (6.30) should be expanded after having substituted Eq. (6.28) into it in the form:

$$d = k\Delta l_d = kd_s \frac{\varphi \cos \alpha}{\rho}.$$

In order to determine the standard deviation of distance m_d, two independent standard deviations must be taken into account separately. Firstly there is m''_γ/Γ, the standard deviation of the reading of the stadia rod intercept Δl_d subtended by the varying stadia angle ($\varphi \cos\alpha$), due to the mean square error of image coincidence as explained in the discussion of Eq. (5.52). Secondly there is m_α, the independent standard deviation of observation of the vertical angle α. Hence:

$$m_d = k\sqrt{d_s^2\left(\frac{m''_\gamma}{\rho\Gamma}\right)^2 + d_s^2\left(\frac{\varphi}{\rho}\right)^2 \sin^2\alpha\left(\frac{m_\alpha}{\rho}\right)^2}$$

or by putting $d = d_s \cos\alpha$ and substituting $\varphi/\rho = 1/k$, one obtains:

$$m_d = \frac{kd}{\rho}\sqrt{\left(\frac{m''_\gamma}{\Gamma}\right)^2 \sec^2\alpha + m_\alpha^2\frac{\tan^2\alpha}{k^2}}. \qquad (6.35)$$

Equation (6.35) shows that the standard deviation of horizontal distance d increases with increase in both distance d and vertical angle α, and with decrease in telescope magnification Γ. It may be mentioned that, with a split-image tacheometer, the images to be coincided are precise geodetic marks of sharp contrast on the horizontal stadia rod, and the distance is always shorter than 200 m: consequently m_γ can be supposed to be ± 10 sec of arc under conditions of good visibility. Hence, by taking the stadia constant $k = 100$, the telescope magnification $\Gamma = 25$ and $m_\alpha = 30$ sec of arc, the theoretical values of the standard deviation of distance, expressed as a percentage of the distance, can be obtained from Eq. (6.35) to be as follows:

if $\alpha =$	0°	5°	10°	20°	30°
then $100\, m_d/d =$	0.020	0.020	0.020	0.021	0.024

With similar considerations one obtains from Eqs (6.34) and (6.32) the standard deviation of measured elevation Δh to be:

$$m_h = k\sqrt{d_s^2\left(\frac{m_\gamma}{\rho\Gamma}\right)^2 + d_s^2\left(\frac{\varphi}{\rho}\right)^2\left(\frac{m_\alpha}{\rho}\right)^2 \cos^2\alpha}$$

or, again putting $d = d_s \cos\alpha$ and substituting $\varphi/\rho = 1/k$:

$$m_h = \frac{k\cdot h}{\rho}\sqrt{\left(\frac{m_\gamma}{\Gamma}\right)^2 \sec^2\alpha + \left(\frac{m_\alpha}{k}\right)^2}. \qquad (6.36)$$

Eq. (6.36) shows that the standard deviation of elevation Δh increases with increase in both distance d and angle α, and with decrease in magnification Γ. For the standard deviation of image coincidence made with a single magnification, assuming $m''_\gamma = 10$ sec of arc, $k = 100$, $\Gamma = 25$ and $m_\alpha = 30$ sec of arc, one obtains the following standard deviations of height in terms of vertical angle:

if $\alpha =$	0°	5°	10°	20°	30°
then $100\, m_h/d =$	0.024	0.024	0.025	0.025	0.027

A comparison of the theoretical values computed for the standard deviations of distance and elevation shows that the accuracy of elevation measurement is less than the accuracy of distance measurement. In practice one must assume somewhat higher standard deviations due to defects of manufacture and other reasons.

Available models of split-image tacheometers with an external base. The tacheometer *Redta 002* (VEB Jenoptik, Jena) is constructed according to the system developed by *Boskovic* and *Bosshardt (Fig. 6.43)*. The instrument is basically a theodolite, with a reiterative axis system and fitted with stadia hairs, that has been further developed into a split-image tacheometer with an external base according to the principle described in Point 6.1.2.3. The rotatable wedge pair is turned by rotating the horizontal axis in the vertical plane by means of the train of gears *1*, *2*, *3*, *4*, *5*, and *6 (Fig. 6.44)*. Gear *1* is fixed to the mount of the index-bubble of the vertical circle, gears *5* and *6* to the mounts of the two wedges. Gear *7* is used only to reduce the clearance. For image separation the system shown in *Fig. 6.42*, *c* is used. The beams of rays coming through the upper glass cover plate *1 (Fig. 6.45, a)* are led in the telescope to the upper part of the objective lens *3*, after having been shifted 22 mm downwards by the rhomboid prism *2*. The second beams of rays, entering through the lower glass cover plate *4*, reaches the lower part of the objective lens after passing through the rotatable wedge pair *5*. Both beams of rays reach the double wedge *7* (biprism), which is combined with a condensing lens, after passing through the focussing concave lens *6*. The intermediate image generated here is turned upright in the second image plane *9* by the reversing lens *8*, and this is observed in the eyepiece *10*.

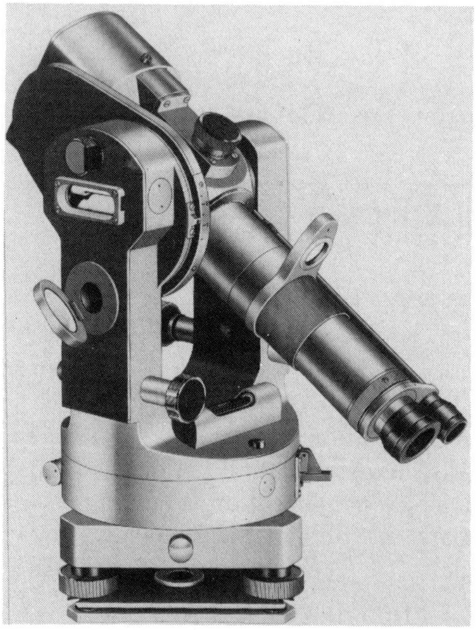

Figure 6.43. Redta 002 self-reducing tacheometer (VEB Jenoptik, Jena)

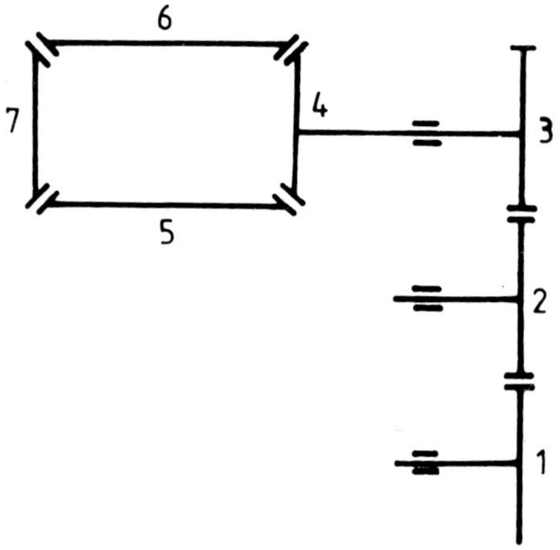

Figure 6.44. Control of the rotating wedge pair in the *Redta 002*

a)

b)

c)

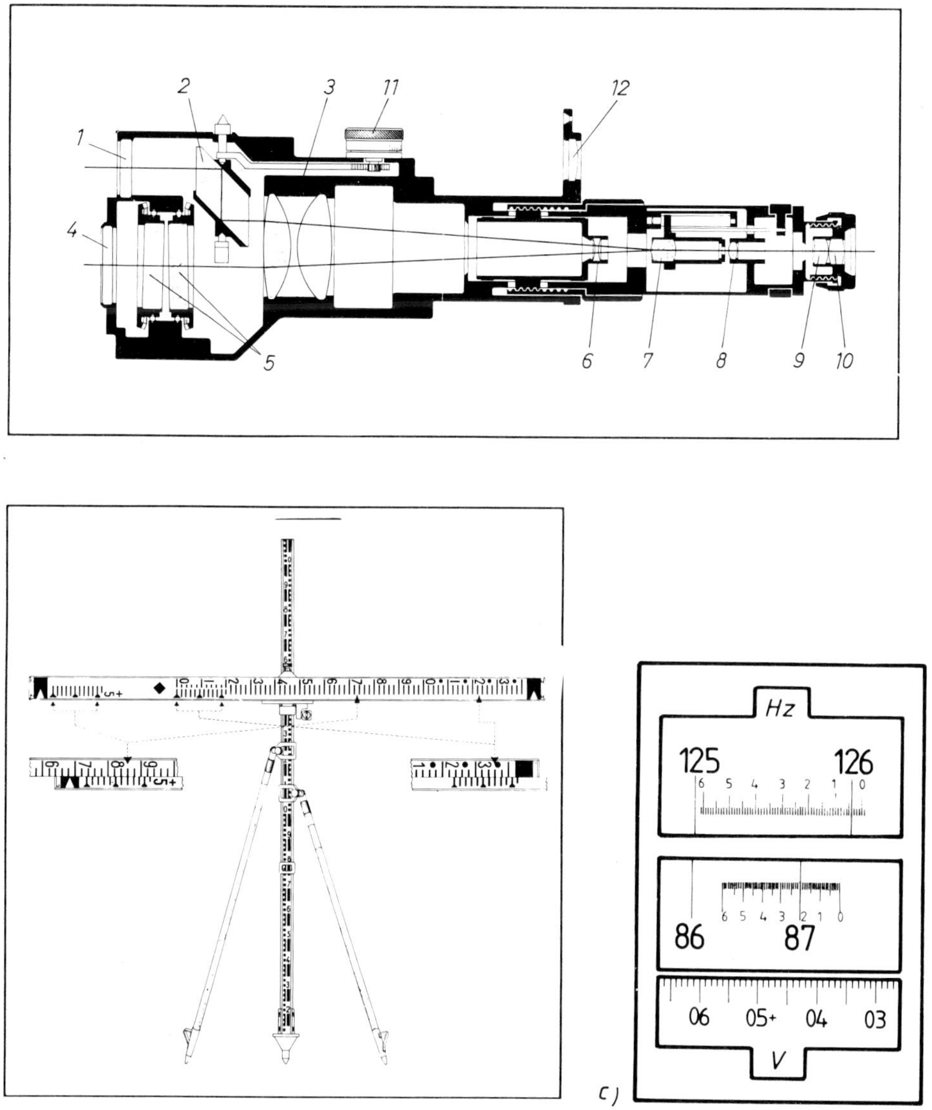

Figure 6.45. a) longitudinal section of the telescope of the *Redta 002; b)* Redta staff equipment; staff reading $l = 123.04$ m; *c)* field of view of the micrometer of the Redta; reading on the horizontal circle: 126° 04.0′; reading on the vertical circle: 87 ° 21.0′; reading on the tangent graduation: 0.0446

The interval Δl_d read on the horizontal stadia rod held at the point P to be measured *(Fig. 6.45, b)* is multiplied by $k = 100$ to give the horizontal distance d directly, the additive constant being eliminated. This constant consists of the distance between the vertex of the stadia angle (at the centre of the rotatable wedge pair) and the vertical axis of the instrument, plus the horizontal distance between the vertical graduation surface of the stadia rod and the vertical through its station mark. The constant is eliminated by moving by the required amount both the zero position of the optical micrometer used for the precise reading on the rod and the vernier of the rod. A small reduction is necessary however for steep angles; its value can be read on the cover of the vertical circle, and it has to be subtracted from the horizontal distance measured (zenith distance angles 77^g; 68^g and 58^g require a reduction correction of 0.5, 1.0 and 2.0 cm, respectively).

The rhomboid prism *2* placed in front of the upper part of the objective lens *(Fig. 6.45, a)* is not only used to shift the half ray vertically but is also at the same time an optical micrometer; therefore it can be rotated around the vertical axis normal to the line of sight. By means of this prism one division line of the vernier on the stadia rod can be made to coincide with one of the division lines on the main graduation. The plane parallel effect of the prism can shift the ray by 0.2 cm, i.e. by the graduation unit of the vernier. The scale of the micrometer drum *11* has 20 divisions, hence one division corresponds to 0.01 cm and, by multiplying it by the stadia constant $k=100$, this means a distance difference of 1 cm. The total reading is thus the reading on the stadia rod, the reading on the vernier and the reading on the drum (by means of the single microscope *12*).

If the instrument is used only as a theodolite for angular measurement, the split image can be eliminated by turning the knurled ring in front of the eyepiece, in order to make observing more comfortable. In this case one simple total image is visible in the telescope field of view, because the wedge pair together with a condensing lens and the diaphragm are rotated out of the optical path, and instead of them is interposed a lens without a wedge pair and a diaphragm covering the lower shifted image.

The instrument *Redta 002* yields only the reduced distance directly, not the height distance. On the vertical circle, however, there is also a tangent graduation with a graduation unit of 0.001 *(Fig. 6.45, c)*, and the height difference can be obtained from the reading on it and from the measured distance d as

$$\Delta h = d \tan \alpha. \tag{6.37}$$

The tangent graduation can also be used for distance measurement with vertical stadia rods, and therefore the Redta can also be used as a tangent tacheometer. From the reading $\tan \alpha_1$ and the reading $\tan \alpha_2$ made after reversal of the telescope around the horizontal axis (these are set to integer values for the sake of simplifying the calculations), and from the corresponding horizontal hair readings l_1 and l_2 on the rod, the horizontal distance can be obtained as:

$$t = \frac{l_2 - l_1}{\tan \alpha_2 - \tan \alpha_1}. \tag{6.38}$$

The distance can be measured — with lower accuracy — by the stadia hairs of the reticule on both vertical and horizontal stadia rods, when the additive constant is $C = -0.3$ cm. The Redta stadia rod has a tubular tripod of height 2.1 m *(Fig. 6.45, b)* which can be lengthened by an additional 1.40 m by its telescopic legs. For forced centring the rod can be fitted to a short (0.3 m long) pinned tubular support, which can be attached to the base part of the instrument and fixed by a clamping screw.

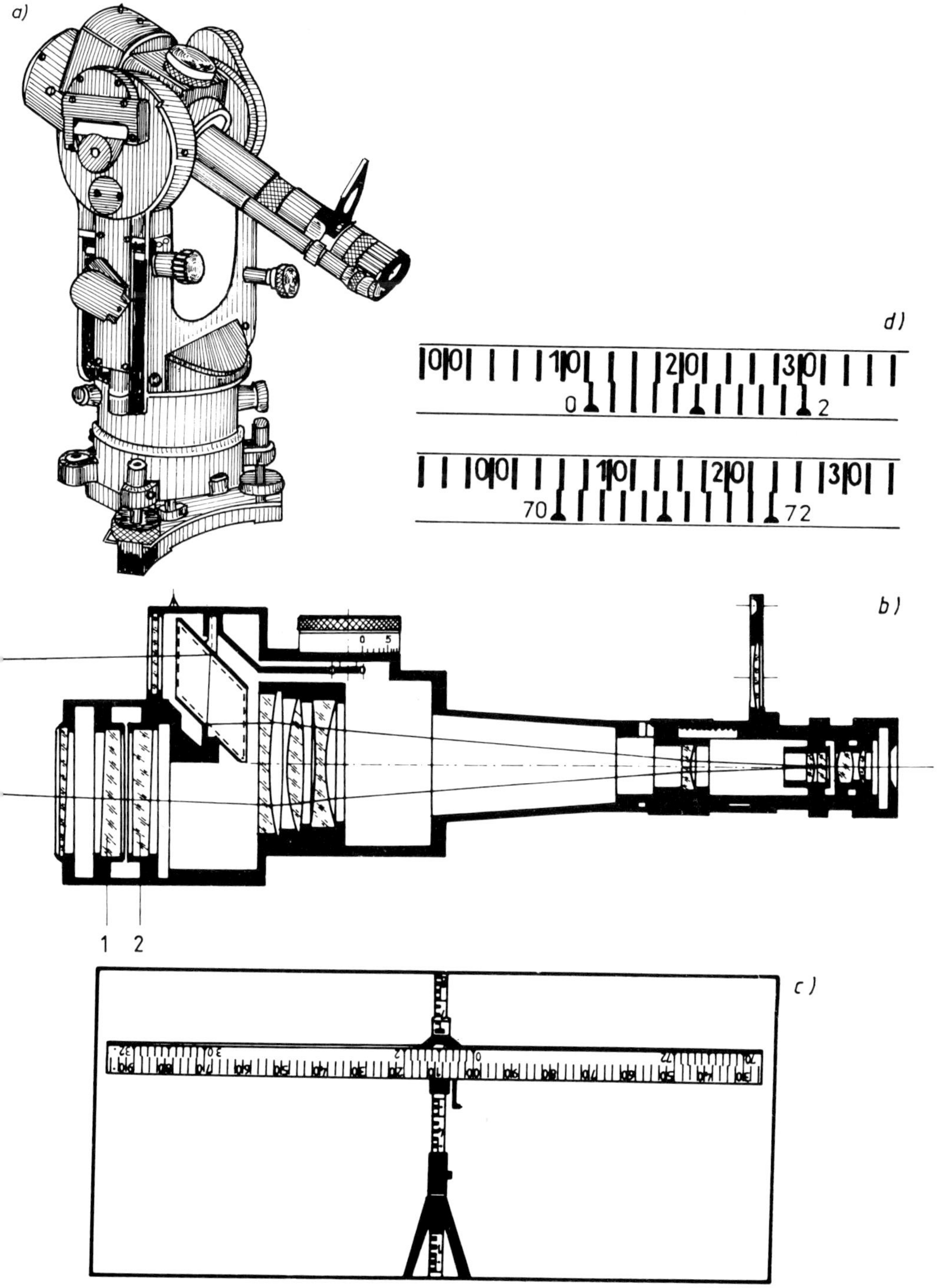

Figure 6.46. a) the *RDH* tacheometer (Wild, Heerbrugg); *b)* the optics of the *RDH* telescope; *c)* the stadia rod for the *RDH; d)* reading on the *RDH* stadia rod

There is a centimetre graduation on the tube which is used:

— to read the height above the stadia rod station, or to set the instrument height *h*;
— to enable the instrument to be used as a simple or tangential tacheometer;
— to use the tube as a levelling staff for small height differences.

The graduation is on the metal band of the 2.09 m long wooden rod (see *Fig. 6.45, b*); the cm graduations are black lines with an inner vernier with 10 divisions for the measurement of distances between 2.50 m and 122 m, and with an outer vernier — displaced by 50 cm — for the measurement of distances between 122 m (72 m + 50 m) and 172 m (122 m + 50 m). For distance measurements the telescope must be rotated around the horizontal axis until the lines of the rod graduation and the vernier become equally long above and below the line separating the telescope field of view. The zero line of the vernier marked with a triangle must be used as a reading index. There is a triangular symbol at both ends of the Redta stadia rod, the distance between them is 2 m ± 0.1 mm at a temperature of 20 °C, and they enable the rod to be used as a subtense bar of constant base length. For distance measurements with a split image the length expansion of the rod due to temperature need not be taken into account because it is compensated by the variation of the refractive index of the rotating wedges. The telescope has a magnification of $\Gamma = 25$, and a mass of 6.5 kg.

The split-image tacheometer *RDH* (Wild, Reduktions Distanz- und Höhenmesser) *(Fig. 6.46, a)* differs from the Redta in so far as both the reduced distance and the height difference can be measured with it directly. The second type of image separation has also been used here. In order to change from split image to single image (total image) the eyepiece must be turned.

The theoretical realization of height measurement as described at the beginning of this section is achieved as follows: turning the switch-over knob *10* on the right-hand side of the telescope support rotates the gear *1* shown in *Fig. 6.44* through 100^g, thus also rotating the wedges through -100^g and $+100^g$ respectively. Consequently the principal planes of the wedges are turned into the horizontal (for a telescope set horizontal) and the refracting edges of the two wedges come to the opposite side. If the rotating wedges are set into the basic position for elevation measurement, *k*-times the rod section Δl_h read on the stadia rod gives by Eq. (6.34) directly the elevation Δh of the point *P* above the centre of the wedge pair.

For a steep angle a reduction must be applied, the value of which can be read in cm-s on an auxiliary scale. The distance reduction must be subtracted from the distance, the height reduction must be added to the absolute value of the height difference.

The 2 m integers of both the distance and the height are read on the main scale of the 2 cm graduation units of the rod *(Fig. 6.46, c)*, the even 10 cm-s on one of the three verniers with 10 divisions, relatively separated by 70 cm *(Fig. 6.46, c* and *6.46, d)* and the cm-s are read on the 20 division scale of the micrometer drum *(Fig. 6.46, a)* which rotates the wedge pair *1* and *2 (Fig. 6.46, b)*. The effective range of the instrument is 164 m when using the appropriate stadia rod, and the telescope magnification is $\Gamma = 27$.

The *Kern* split-image tacheometer *DK-RT (Fig. 6.47)* differs from the *Redta* mainly in its optical system. Neither of the two beams of rays generating the half images is vertically shifted *(Fig. 6.48)*. The mechanical dividing strip *2*, running from the

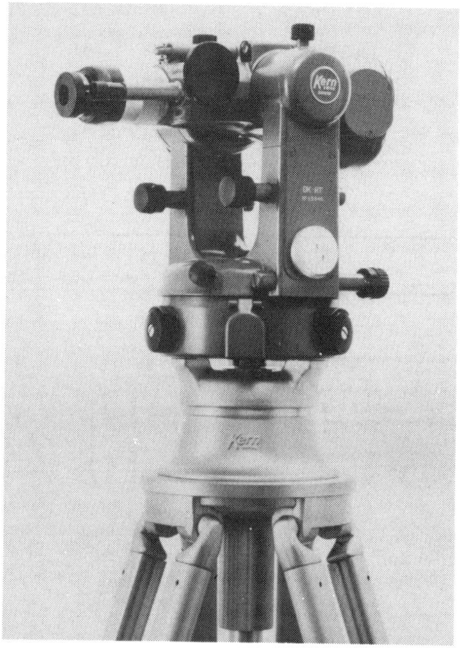

Figure 6.47. The *DK-RT* split-image tacheometer (Kern, Aarau)

objective lens *1* along the telescope, separates the two half images as in the system shown in *Fig. 6.42, a*. The roof edge prism *3* used in the neighbourhood of the image plane *(Fig. 6.48)* covers a small horizontal strip of the image; hence the disturbing effect of the dividing wall is cancelled out, and the two exit half-pupils are brought together. The rotating wedge pair *4* is placed in front of the lower part of the objective lens *1*. The optical micrometer *5*, placed between the wedge pair and the objective lens, consists here of two slightly wedge-shaped plates. By means of the two images generated by these plates the beams of rays, passing and not passing through the rotatable wedge pair, can be shifted by the same amount but in opposite directions, within the limits of ± 5 mm. The micrometer drum has a scale of 100 divisions, hence one unit corresponds to 0.1 mm on the rod or to a distance difference of 1 cm. This solution makes the use of a vernier on the rod *(Fig. 6.49, a)* superfluous and it is replaced by two index-scales of 50 cm apart, graduated in units of 2 cm and each having four division lines. If the zero line and the third (longer) line of the index scale are made collinear with one of the lines of the main graduation, then the latter is an even metre. If instead of this the second and fourth index-scale lines are made collinear *(Fig. 6.49, b)*, then the metre is odd. The reading on the control index consisting of two division lines gives a distance greater by 5.27 m *(Fig. 6.49, c)*. The rods are made from light metal. The graduation is engraved on a lacquered steel plate. The measuring range lies between 2.5 m and 166 m when using the 1.7 m normal rod, and it can be extended to 205 m when using the 2.1 m special rod.

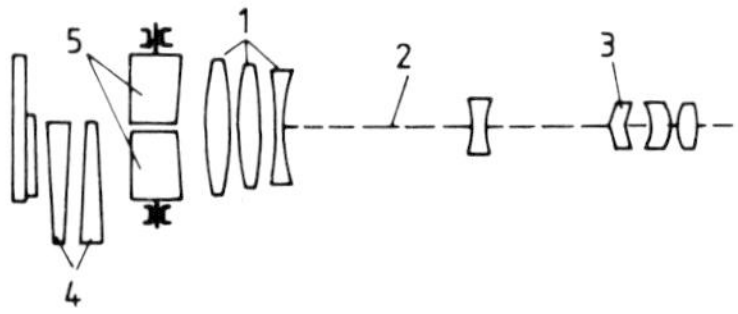

Figure 6.48. The optical system of the telescope of the *DK-RT*

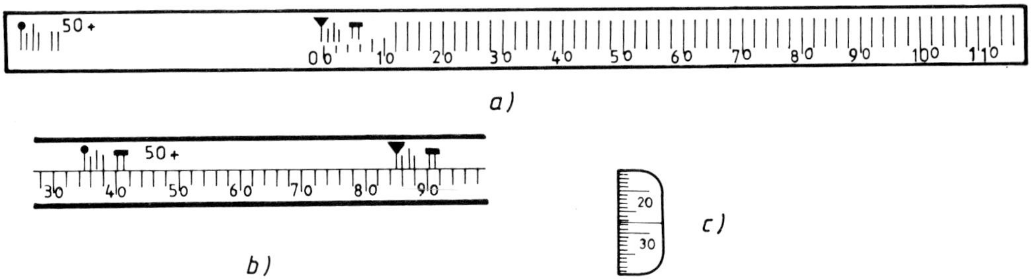

Figure 6.49, a) the stadia rod for the *DK-RT; b)* the staff index after coincidence by the micrometer, staff reading: 84.5 m; *c)* graduation of the micrometer drum, reading: 27.5 cm; the horizontal distance is 84.5+ 0.275=84.775 m

The additive constant of 73 mm (the distance between the vertex of the stadia angle and the vertical axis of the instrument) is compensated by the correspondingly shifted position of the index on the rod. Change of the stadia constant and of the additive constant cause opposite errors; below a vertical angle of 30° the remaining error is less than 1 cm, therefore no reduction need be applied.

Height is obtained by multiplying the horizontal distance by the tangent of the vertical angle. The value of the tangent can be read on the tangent graduation on the vertical circle. The telescope magnification is 27 ×.

The tacheometer *TP*, manufactured in the Soviet Union, measures horizontal distances and — after switching over the rotatable wedge pair — elevation differences. The instrument has an automatic index on the vertical circle. The measuring range is between 20 m and 180 m. The telescope magnification is 27 ×.

Test and adjustment of the split-image tacheometer with an external base. Testing and adjustment of these instruments is the same as for those which measure angles, as described in Subsection 2.2.8. The error of the reducing device, i.e. an erroneous position of the rotating wedge pair in the base position, is significant only on steep angles. For the determination of this error it is necessary to measure a slope distance several times and then interchanging the instrument and the stadia rod. The mean value of the forward and backward measurements is free of the error of the reducing device, for which the device must be adjusted. Adjusting screws are used for this purpose, their position being different from one instrument type to another. Such an adjustment, however, is only seldom necessary.

The additive constant must be zero when pointing horizontally. This can be checked by comparing measurements of a distance of 5 to 10 m made by both the instrument and a measuring tape: the error of the constant has no effect at such a short distance. The error is eliminated by a small turn of the micrometer scale, after loosening the fixing screws which must be tightened again after the adjustment is completed.

The stadia constant must be 100. It is tested by comparing the measurements of a distance between 60 and 100 m by the instrument and by a calibrated measuring tape. The instrument can be adjusted to the required value on the *Redta 002* and *Wild RDH* as follows: after adjusting the index bubble of the vertical circle, the appropriate division lines on the rod are brought into coincidence by rotating the wedge-shaped glass cover in front of one half of the objective lens. In the *Kern DK-RT* the stadia constant can be adjusted after removing the switch-over ring by means of the adjusting cross-head screws.

It is advisable to establish a comparison base line (of e.g. 100 m) for checking the addition and stadia constants from time to time, and to provide internal marks on this line at 20, 40, 60 and 80 m. It is further advisable to compile a table of corrections from the check measurements, from which the corrections to the addition and stadia constants can be taken for different distances.

The use of split-image tacheometers with an external base. They are the most precise of all optical tacheometers. The effect of near-surface refraction is eliminated by the use of horizontal stadia rods.

A reducing device for elevation measurement is only used in the *Wild RDH* and the Soviet *TP*. Distance meters with a double image and an external base are useful for traversing and precise tacheometric survey, due to their relative distance measuring error of about 1:5000. They have an average effective range of 150 m. These instruments, however, cannot rival the modern but much more expensive electronic tacheometers, either in accuracy or in effective range (see Part II).

6.1.2.4 Split-image Tacheometers with an Internal Base

In urban areas it is often impossible to set up the horizontal stadia rod of a split-image tacheometer with an external base. The internal base overcomes this difficulty. Such an instrument is the instrument *BRT 006* (VEB Jenoptik, Jena) *(Fig. 6.50)*, which is an improvement of the *Teletop (Fig. 5.65)*. On the base bar *1*, which is 30 cm long and has a graduation unit of 0.5 mm *(Fig. 6.51, a)*, the movable pentagonal prism *3* with a reading index *4* is placed in front of the lower part of the objective lens *2*, and the fixed pentagonal prism *5* is placed in front of the upper part of the objective lens. Incoming rays are deflected by both prisms to the left, towards the eyepiece, by 100^g. The reducing device *6*, establishing the variable parallactic angle φ_α, is set in the path of the rays arriving from the prism *5*. The reducing device consists of a movable half lens *(Fig. 6.51, b)*, the optical axis of which is parallel to the optical axis *t—t* of the objective lens *2*. In the basic position of the half lens *6* its centre is moved a distance e from the optical axis of the objective lens *2*, and therefore the rays arriving at *2* along its axis have been deflected by the half lens by a small angle $\rho'' e/f$ from its focal point F. The focal length f and the basic shift e are chosen in such a way that the deflecting angle is equal to the small stadia angle $\varphi'' = \rho'' b/d_s$ necessary for the measurement of the slope distance d_s with a base

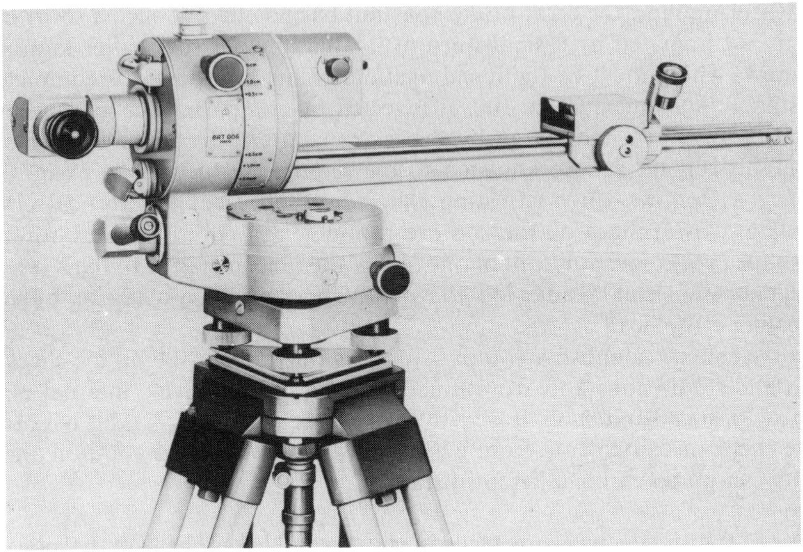

Figure 6.50. The *BRT 006* self-reducing tacheometer (VEB Jenoptik, Jena)

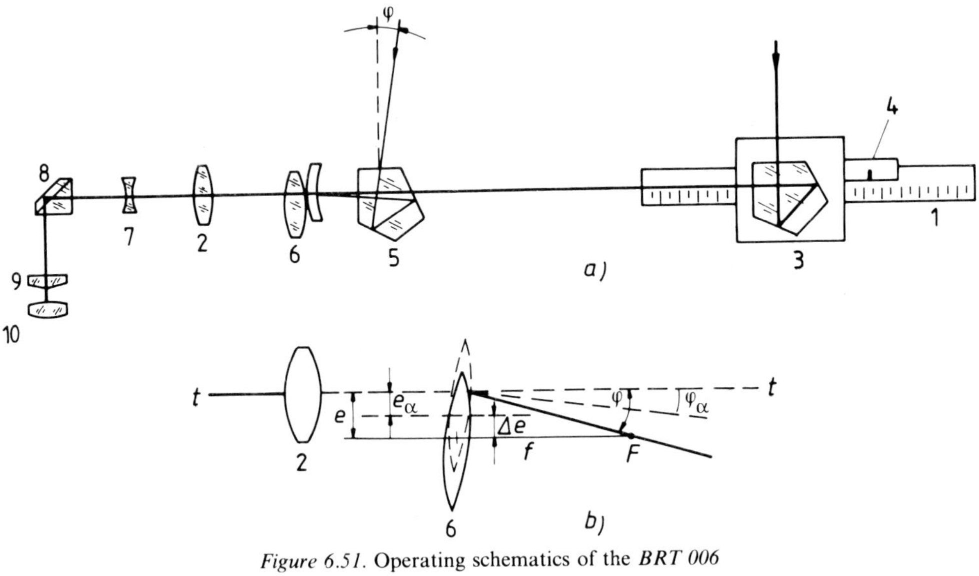

Figure 6.51. Operating schematics of the *BRT 006*

length b. If the prism *3* is moved along the base bar until the two half images, generated by the rays arriving from the observed object after passing through the two prisms, are brought into coincidence in the eyepiece, then one obtains the slope distance d_s from the base length b, read on the index *4*, multiplied by the stadia constant $\cot\varphi = f/e = d_s/b = k$; in other words $d_s = k \cdot b$. If now the half lens *6* is moved off the optical axis t—t of the objective lens *2*, not by the basic shift e but by the distance

$e_\alpha = e \cos\alpha$ corresponding to the slope angle α to the point observed, then the deflection of the rays caused by the half lens, also the actual parallactic angle, is:

$$\varphi_\alpha = \rho \frac{e \cos\alpha}{f} = \rho \frac{b \cos\alpha}{d_s} = \frac{b_\alpha}{d_s}.$$

According to this the two beams of rays, travelling from the two pentagonal prisms through the focussing lens *7* and then passing through the roof edge prism *8*, are brought to coincidence and the reduced base length $b_\alpha = b \cos\alpha$ is read on the base bar. This base length multiplied by $k = d_s/b$ yields the horizontal distance

$$k \cdot b_\alpha = k \cdot b \cdot \cos\alpha = d_s \cos\alpha = d. \qquad (6.39)$$

The movement of the half lens from the basic position with displacement e, by the changing displacement $\Delta e = (1 - \cos\alpha)e$ towards the optical axis t—t, is controlled by tilting the instrument by the slope angle α. The reducing system can be switched off; in that case the half lens *6* remains in its basic position displaced by e and independent of the slope angle α, and the slope distance d_s is measured.

The separation of the two half images is accomplished by system 2 described in Point 6.1.2.3, by means of the double wedge *9* set in the image plane and the slit diaphragm *10* in the exit pupil (keyhole diaphragm).

The additive constant of -75 mm of the distance meter is eliminated in the measurement of the slope distance d_s by displacing the position of the reading index *4*. The corrections to the additive constant, which become necessary when switching in the reducing device, can be read on the eyepiece side of the instrument cover. This correction can usually be neglected because it is less than 1 cm if $\alpha < 33^g$. In the centre of the telescope field of view there are two vertical lines which define the area (1/8 of the field of view between them) in which measurements obtained by coincidence are free of the effect of base line tilt. The stadia constant is $k = 200$. The effective range of the instrument is 60 m if image identification is made on any vertical line of the object observed or on a survey pole held at the end of the line. The effective range can be increased, however, by using the target belonging to the instrument on which there are three vertical lines 15 cm apart, or by using a survey pole with a target on which there are two vertical lines 30 cm apart and two additional vertical lines 60 cm apart. To use the target either the two adjacent lines 15 cm apart or the two outside lines are brought into coincidence, in which case 15 m or 30 m, respectively, must be added to the distance read on the base bar. The effective range is 90 m when using the target. To use the pole and target, a distance of 60 or 120 m, respectively, must be added to the distance read on the base bar if the lines 30 or 60 cm apart are brought into coincidence. Hence the maximum distance that can be measured with the instrument is 180 m when using the pole and target.

The relative mean square error of distance measurement is 1 : 1700. In the manufacturer's nomenclature BRT 006, the number 006 refers to the mean square error of measurement which is given as about ± 6 cm.

6.2 Instruments and Devices for Graphical Survey

A survey is called graphical if the measurements yield the spatial data of the points partly or wholly in graphical form, rather than their numerical values.

Two main groups of graphical survey methods can be distinguished:

— plane table survey using only surveying methods and instruments,

— photogrammetry using photographs transformed into measuring images, and processing (interpreting) them as a central projection of the objects being measured.

Photogrammetry, in respect of its special methods of data acquisition and processing, is discipline already independent, hence a description of its methods, instruments and devices belong to photogrammetry as a separate subject.

The earliest method of making a unified horizontal and levelling survey is a plane table survey. By means of its instruments and ancillary equipment a map can be drawn on the spot at a scale defined in advance. Its primary field of application is small-scale mapping and it can also be used occasionally where the lower accuracy of the graphical method is acceptable.

In topographical tacheometry even the densification of the control net is made by plane table survey. Plane table survey, however, is predominantly used for topographical surveys based on control points; in these surveys the measuring method is tacheometry and, using it, the plane table surveying method gives spatial polar coordinates for the points. The horizontal projections of the directions measured at a station are obtained graphically, but distances and elevation differences are determined by numerical distance and elevation measurements, respectively. The determination of distances is graphical if the mapping (map drawing) is done on the spot. Therefore this measuring method is often called graphical tacheometry (graphotacheometry).

6.2.1 Plane Table Equipment

6.2.1.1 The Plane Table

The two basic units of the plane table are: the plane table (with accessories) and the pointing device. The latter is now always an optical instrument (telescopic alidade), in which the telescope itself is a self-reducing distance meter.

The plane table consists of three main constructional elements *(Fig. 6.52)*: the board *1*, the tripod *3* and the instrument base *2* connecting the board to the tripod.

The board is a rectangular (sometimes circular) drawing board, usually made of wood (or light metal), and its dimensions — varying only slightly from manufacturer to manufacturer — are adapted to the size of the survey (map) sheet. The drawing surface is a very carefully manufactured plane surface. In order to avoid deformation wooden boards are laminated, with each layer having its grain at right angles to the grain of the previous one. Nuts for the connecting screws of the levelling base are mounted underneath the board.

The instrument base or central part of the plane table again consists of two parts *(Fig. 6.53)*: the instrument base in a strict sense *1* and the drawing table support *2*. The instrument base can be attached to the tripod head by the clamping screw on

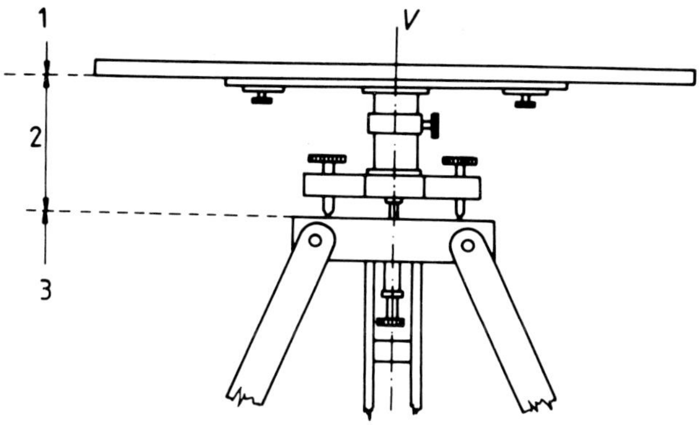

Figure 6.52. Main constructional parts of a plane table

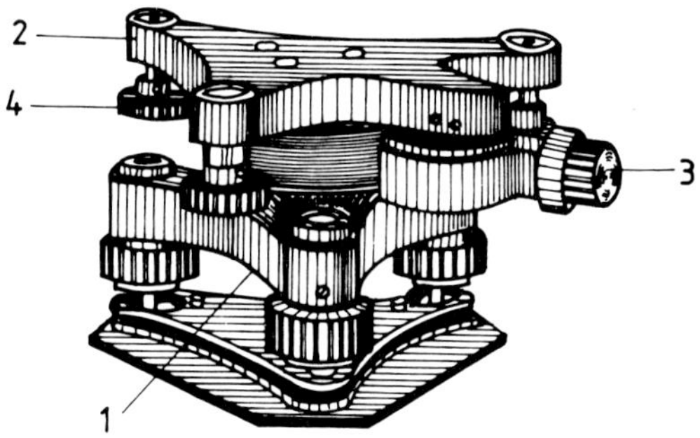

Figure 6.53. The levelling base of a plane table

the latter. From the central (usually cylindrical) part of the instrument base three arms project which hold three foot screws. The vertical axis of the rotating part of the instrument base, i.e. of the drawing board support, fits into the centred bush of the instrument base. The rotating drawing board support is connected to the immobile base by the usual clamp and slow motion screw system *3*. A rigid metal plate is mounted on the vertical axis of the drawing board support to strengthen the board. The board is attached to the plate by connecting screws *4*. The clamping heads of these screws are so made that, before tightening, the drawing table can be moved slightly in any arbitrary direction.

The tripod used for setting up the plane table is the same as those used for surveying instruments in general.

Accessories for a plane table are a spirit level for adjusting the board to the horizontal, a plumbing fork for centring over a point and a compass.

The spirit level is a box level with a constant of 8 to 10 min of arc. The plumbing fork *(Fig. 6.54)* is a *V*-shaped device of wood or metal, one fork of which is placed on the surface of the drawing board. This fork has a chamfered edge at the end, on which an index line is engraved. If the intersection of the index line with the edge is

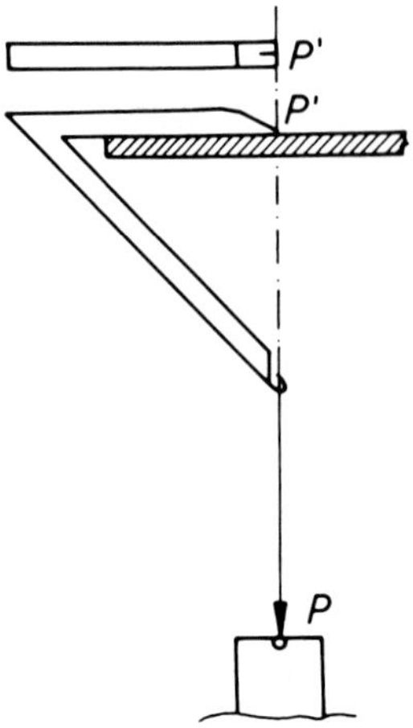

Figure 6.54. The plumbing fork

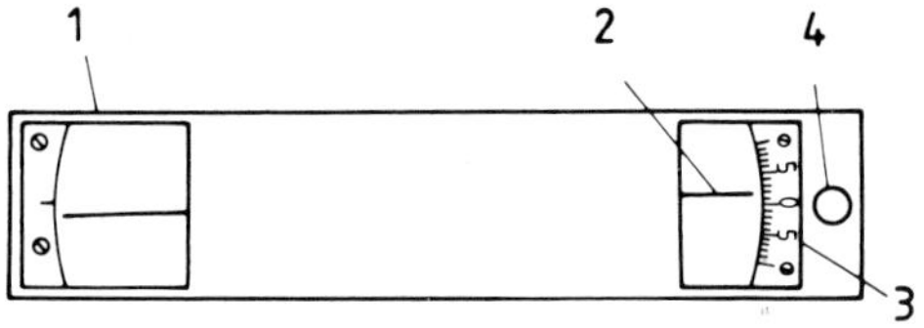

Figure 6.55. The compass for a plane table

set to the plotted position of the station, then the plumb line, attached to the end of the other fork, is plumbed. The plane table is set up if the drawing surface of the board is adjusted into the horizontal, the end of the index line of the plumbing fork is set to the plotted position of the station and the plumb line is vertically over the station mark.

The compass *(Fig. 6.55)* is usually a 10 to 15 cm long magnetized plate *2* set edgewise, rotating on an agate bush and placed in a box with a glass cover *1*. There is a graduation of $\pm 10°$ *3*, usually at the north end of the box. The division unit is usually 1°, which enables reading to an accuracy of 0.1° by estimation. The magnetic inclination is compensated by a small movable mass on the south arm of the compass. When not in use the compass is lifted off its axis by the arresting screw *4*.

6.2.1.2 The Telescopic Alidade

The telescopic alidade consists of a metal rule with a drawing edge mounted together with a geodetic telescope rotatable around its horizontal axis. The telescope is usually also a distance meter *(Fig. 6.56)*. The two parts are attached in such a way that the edge *2* of the rule *1* is parallel (nearly parallel) to the line of sight of the telescope. The horizontal axis *4* is connected to the bar *6* by means of the usual system for vertical clamping *5* and slow motion *7*. This connection, however, allows a slight rotation of the horizontal axis around an axis perpendicular to itself (parallel to the drawing edge). There is an adjusting screw for this purpose. Some telescopic alidades are supplied with a tubular level *3* for the adjustment of the instrument.

In order to simplify pointing, telescopic alidades are supplied with a parallel rule and a slow motion screw. The drawing edge *9* of the parallel rule *8* moves parallel to itself when moved away from the pricnipal rule. The two rules are connected by hinges in such a way that the four hinges form the four vertices of a parallelogram in all positions of the parallel rule. There is a slow motion screw *10* to displace the instrument on the drawing board by a small amount. This screw acts on a small

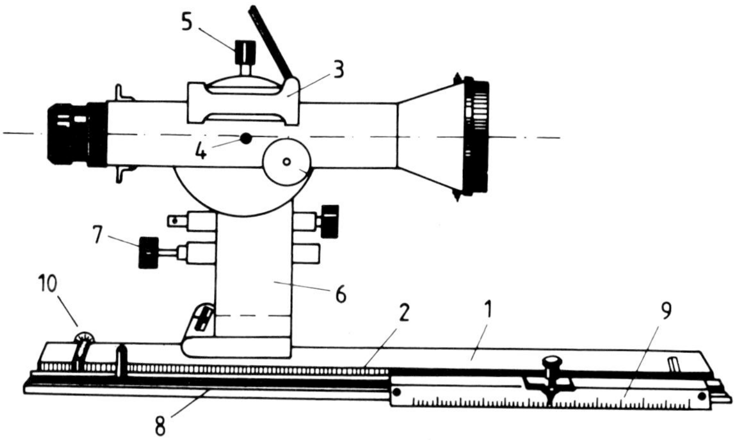

Figure 6.56. Construction of the telescopic alidade

rough plate, which is in contact with the drawing-paper, and moves the rule on the paper with respect to the plate.

Telescopic alidades which have a telescope capable of optical distance measurement (or eventually of electro-optical distance measurement) are called tacheometric telescopic alidades. Their parallel rule is supplied with interchangeable graduations at different scales for plotting distance. Most tacheometric telescopic alidades are supplied with a pricker which can be moved together with the scale in order to plot the measured distance on the drawing paper.

It follows from the construction and use of the telescopic alidade that a direction drawn along the edge of the rule is always eccentric with respect to the line of sight of the telescope. *Figure 6.57* shows the geometric relationship between the actual direction from P to Q and the corresponding line drawn. It can be seen from the figure that the plotted point Q' lies at the perpendicular distance

$$e = \Delta \frac{d'}{d}$$

from the direction r drawn. Since d'/d is identical with the scale $1/M$ of the draft (map), it can be written:

$$e = \Delta \frac{1}{M},$$

i.e.

$$\Delta = e \cdot M. \tag{6.40}$$

Let us take $e = 0.1$ mm, which can be regarded as the maximum accuracy achievable in drawing, then the maximum acceptable eccentricity Δ is 0.10 m for a scale of 1:1000, but already 2.5 m for a scale of 1:25,000. Consequently the maximum acceptable eccentricity considerably exceeds the eccentricity resulting from the

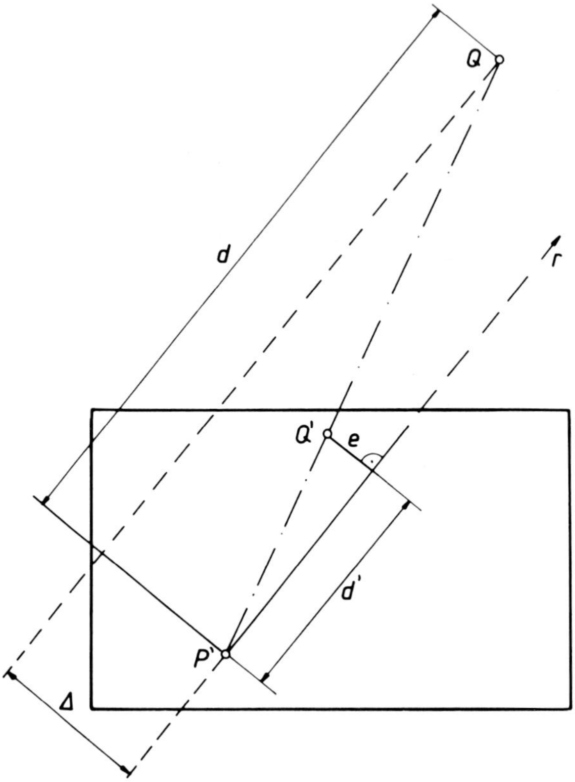

Figure 6.57. Effect of the eccentricity of the telescopic alidade

construction of the telescopic alidade and from the use of the parallel rule; this is generally less than 2 cm in practice.

6.2.2 Some Plane Table Equipment

The telescopic alidade *RK1* (Wild, Heerbrugg) *(Fig. 6.58)* has a telescope with a self-reducing tacheometric system. The diagrams, projected and moved in the same manner as in the diagram tacheometer *RDS*, are visible in both telescope positions together with the image of the stadia rod (see *Fig. 6.24* and *6.25*). In the first telescope position there is a distance diagram with a stadia constant of 100 with respect to the base circle, and a height diagram with three sections on both branches (marked by $\pm 1/5$, $\pm 1/2$ and ± 1 with the stadia constants ± 20, ± 50 and ± 100) *(Fig. 6.59)*. In the second telescope position — used for the measurement of longer distances — there is a diagram for the stadia constant 200 with respect to the basic circle (in other words it is not self-reducing) *(Fig. 6.59)*. The reduction of long distances to the horizontal is facilitated by the projection into the field of view of the telescope of an image of the vertical circle, which is graduated to 10 min of arc (10^c). The telescopic alidade has four interchangeable graduated rules at different scales, 330 mm long. For the sake of easier observation the eyepiece of the telescope is bent. The appropriate

Figure 6.58. The telescopic alidade of the set of plane table equipment *RK1* (Wild, Heerbrugg)

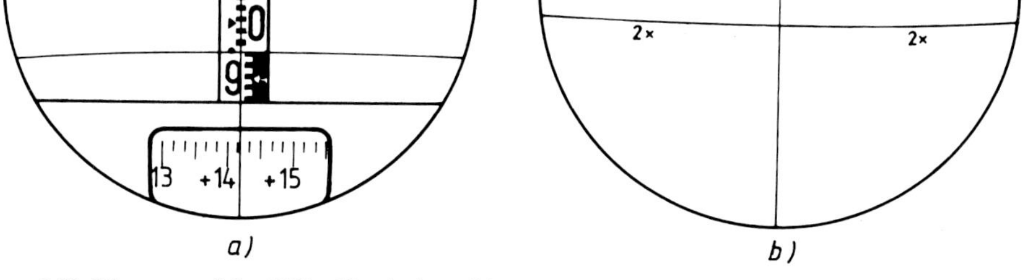

Figure 6.59. Diagrams of the *RK1* self-reducing alidade when observing a staff *a)* in the primary telescope position; *b)* in the secondary telescope position

tacheometric staffs are 3 and 4 m long with a striped graduation in cms *(GVL3, GVLV3, GVL4, GVLV4)*, or 12 feet long with graduations of 0.01 and 0.1 foot *(GVL12* and *GVLV12)*. On the rear of these staffs there are staff extensions with cm (or 0.1 foot) graduations which can be extended 0.6 m (or 2 feet) long. There are also topographic staffs for this plane table; they are 3 and 4 m long and have cm graduations *(GTLE3* and *GTL4)*.

The set of plane table equipment *RK* (Kern, Aarau) includes a telescopic alidade also tilted with reducing diagrams. The telescope constant forms an angle of 30° with the horizontal. The different elevations of the line of sight can be set by rotating the prisms attached in front of the telescope objective. There is no separate parallel rule, its role being taken over by a graduated rule connected to the base plate of the

Figure 6.60. The telescopic alidade of the set of plane table equipment *RK* (Kern, Aarau)

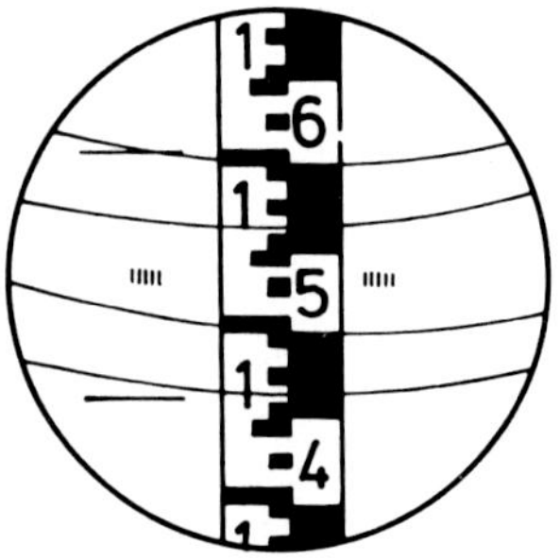

Figure 6.61. Field of view of the *RK* distance meter

alidade by a parallel mechanism *(Fig. 6.60)*. Its diagram system is symmetrical (see *Fig. 6.13*), with a distance diagram of stadia constant 100 and a height diagram having three sections on each branch, whose stadia constants are ± 20, ± 50 and ± 100, respectively. These constants are not marked on the height diagrams by numbers but by small lines between the diagrams, the number of which (two, five and one in turn) corresponds to the stadia constant *(Fig. 6.61)*. According to the data given by the manufacturer, it is possible to measure with the *RK* a slope distance of 100 m — depending on the telescope tilt — with a reliability of ± 10 to ± 20 cm for horizontal distances and ± 5 to ± 20 cm for elevation differences.

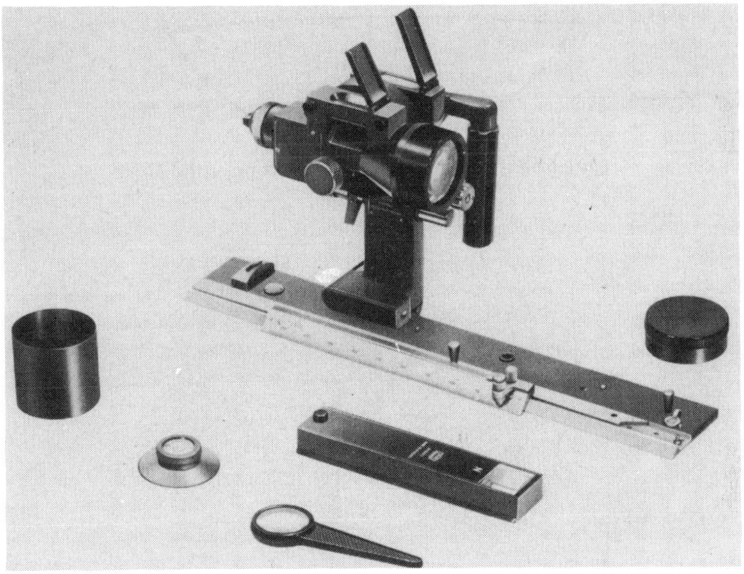

Figure 6.62. The set of plane table equipment *Ma-5* (MOM, Budapest)

For its earlier sets of plane table equipment *Ma-1*, *Ma-2* and *Ma-3*, MOM (Budapest) used the tangent principle for distance and elevation measurement with the telescopic alidade. In the sets *Ma-4* and *Ma-5 (Fig. 6.62)* recently brought to market, telescopic alidades with diagrams of varying distance are used. The base circle of the distance diagrams lies equi-distant between the two diagram lines and their stadia constant is $k_d = 200$. Hence the section Δl subtended by the two stadia hairs has a stadia constant of 100. The base curve of the height diagrams, which have two branches and three sections (with stadia constants ± 10, ± 20 and ± 50) is the lower distance stadia hair *(Fig. 6.63)* and therefore this curve must be taken as the base curve for measurement. The vertical circle has a graduation unit of 1 degree (1^g) which is divided by a 2° subscale graduated in units of $1°/6 = 10'$ (or $1^g/10 = 10^c$). The index level of the vertical circle has a constant of 50 sec of arc, the reversible level of the telescope a constant of 30 sec of arc. The appropriate tacheometric staff has cm graduations, also a striped graduation in units of 5 cm for topographical measurement. The wedge-shaped zero line of the staff graduations is 1 m above the

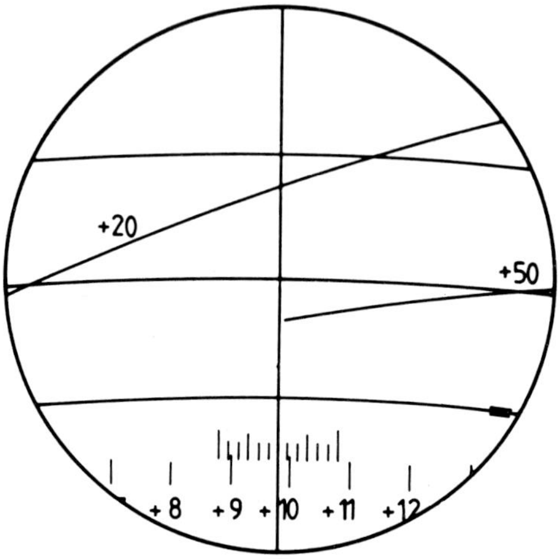

Figure 6.63. Field of view of the telescope of the *Ma-5*

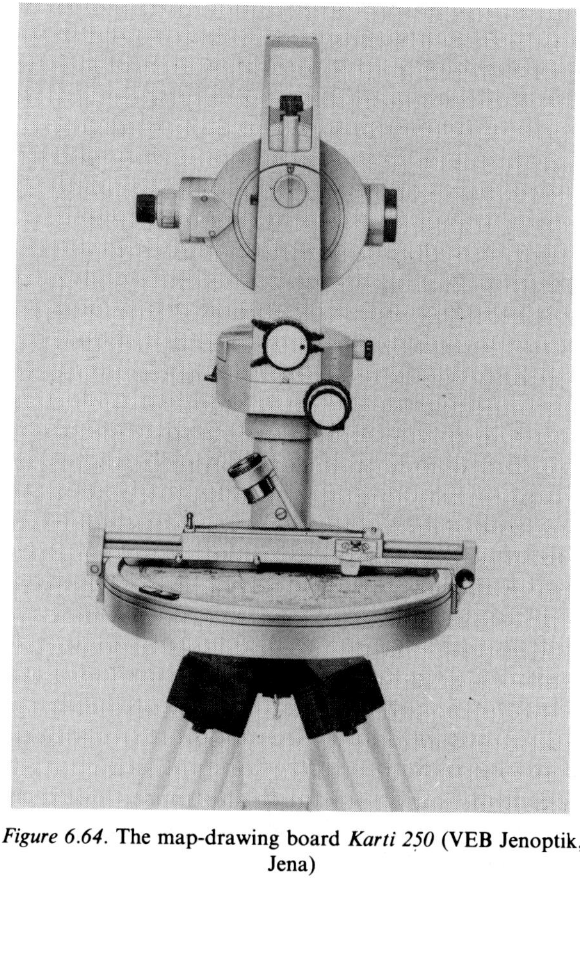

Figure 6.64. The map-drawing board *Karti 250* (VEB Jenoptik, Jena)

foot of the staff, but this height can be increased by means of the staff extension at the rear of the staff. According to the data given by the manufacturer, the distance meter of the plane table *Ma-5* provides distance measurement with a relative standard deviation of ± 20 cm/100 m, and elevation difference measurement (at a slope angle of 30°) with a relative standard deviation of ± 15 cm/100 m.

The small map-drawing board "Karti" (Kartiertisch) (VEB Jenoptik, Jena) *(Fig. 6.64)* is a rotating disc. The supporting socket of the disc together with the disc in the instrument base, can be rotated around its vertical axis. The alidade of the instrument (theodolite, tacheometer) fitted onto the instrument base of the drawing board — corresponding to the construction of forced centring — is clamped to the supporting socket which rotates with it. Due to this common rotation, the edge of the rule, which has interchangeable graduated scales and is fixed along a diameter of the socket, remains always parallel to the line of sight of the telescope (or forms a small angle with it). The actual drawing board (drawing disc) rotates by means of gearing in the opposite direction but with equal angular velocity as the supporting socket. Hence, once a drawing board is oriented to any direction, it remains continuously so oriented during rotation of the alidade.

The new map-drawing board "Karti 250" from Jena differs from the earlier Karti only in the construction of the instrument base. The rotating disc of the Karti instruments has a diameter of 250 mm, and its rule with a pricker is 125 mm long. The interchangeable graduated scales are 1 : 1000, 1 : 2000, 1 : 2500 and 1 : 5000. Since the plotted position of the instrument always corresponds to the centre of the rotating

Figure 6.65. Use of the *RED Mini* electro-optical distance meter (Sokkisha, Tokyo) as part of a set of plane table equipment

drawing disc, a separate sheet of drawing paper is needed for each point. In order to facilitate compilation of surveys from separate points, mapping on the spot is made by the Karti on a transparent sheet (plastic). Points on the final map sheet can be copied from the survey sheets by simply pricking through, after having oriented them.

Sokkisha (Tokyo) used with its electro-optical distance meter *RED Mini* not a telescopic alidade but an alidade with a rule to which the small meter can be attached and fixed *(Fig. 6.65)*. With this instrument the conventional alidade and rule used in plane tabling becomes useful as an electronic alidade with a rule for graphical surveys. Since the distance meter yields only slope distance, there is a graduation of $\pm 40^\circ$ on the vertical arc of radius 50 mm on the instrument base, which rotates together with the instrument, and direct readings to an accuracy of 5 min of arc can be made with a vernier. The effective length of the plotting rule with a point pricker is 225 mm, the interchangeable scales are 1 : 250, 1 : 500, and 1 : 1000. For the use of the distance meter *RED Mini* for plane tabling, its standard eyepiece should be interchanged with an eyepiece attachment that bends the line of sight through 90°.

Data for some further plane table equipments are listed in *Table 6.2*.

6.2.3 Test and Adjustment of Plane Table Equipment

The test of plane table equipment includes tests of the three main constructional elements, namely the drawing board, the levelling base and the telescopic alidade. The connections between these main elements must also be tested.

The requirement for the drawing board is to have a finished plane surface without a crack or split. The latter can be checked by the naked eye, but the flatness of the surface must be checked with a plane table level after it has been set exactly horizontal.

The levelling base or the table base provides the rigid connection between the board and the tripod. It is essential that the drawing board, when set up for measurement, should not suffer any irrevocable displacement. Such a displacement — supposing a stable tripod set up and a motion-free connection between the construction elements by the tightened screws — mainly results from the clearances of the connecting screw system of the table base. For testing this a distant point target is sighted by the telescopic alidade on the board, and now a slight twist is applied to the board, both vertically and horizontally. If there is only a slight displacement and the line of sight returns to the target after ceasing the twist, the connecting system of the table base is in good order. If not, the reason for the irreversible displacement must be sought. The defect is most probably caused either by play in the horizontal clamp and slow motion screw or by horizontal play in the foot screws.

The test of a telescopic alidade involves the telescope, its distance meter system, the straightness of the drawing edge of the parallel rule and the parallelity of its shift, also the relative positions of the telescope and rule.

The telescope is tested and adjusted for both collimation error and perpendicularity error of the horizontal axis in the same way as described for the test of theodolites (Subsection 2.2.8).

Test and adjustment of the distance meter can be made depending on the system, as described for the test and adjustment of the corresponding tacheometer (Subsection 6.1.1).

The straightness of the drawing edge of the rule can be checked as follows: a fine line is drawn along the parallel rule when closed against the main rule, then the alidade is reversed through 180° and its drawing edge is put along the line drawn. If the edge is straight, it fits the line along its whole length. The edge can be ground until it is straight within the limits of drawing accuracy.

An empirical check of the movement of the parallel rule into a position parallel to itself can be carried out as follows: a fine line is drawn along the edge of the parallel rule in its closed position and then an other one at its maximum extension. Now the alidade is reversed through 180° and the parallel rule is placed in its closed position against the second line. Then the rule is opened to its maximum extension, and its edge should fit the first line. If the rule does not fit, but forms an angle with it, this angle is twice the parallelity error. A small error is probably caused by an incorrect adjustment of the hinge system and can therefore usually be eliminated by loosening the screws and adjusting the hinges. If correction is impossible, the rule should always be used in the same position (conveniently in its closed position).

As regards the relative positions of the drawing edge of the plotting rule and the telescope, it is a conceptual condition that the vertical planes of sight of the drawing edge and of the telescope should be parallel to each other. The effect of any existing non-parallelity causes exactly the same error in both the measured direction and the orientation of the plane table. Therefore directions plotted from the point form angles which are free of this error, and consequently an adjustment of this source of error is unnecessary.

The complete test of a set of plane table equipment includes tests of the index bubble of the vertical circle, of the telescope level and of the compass.

Test and adjustment of the index bubble of the vertical circle should be made as described in Point 3.2.2.3, of the telescope level as described in Point 3.1.2.6; the compass is checked as described in Subsection 2.4.1.

Part II

Electronic Surveying Instruments

7 Elements of Electronic Instruments

The importance of electronic instruments in measurement techniques is ever increasing. The general principle of these instruments is that a certain sensor generates an electric signal which has a defined functional — i.e. proportional — connection with the quantity to be measured, and which is then processed in the instrument: it is amplified, stored, transmitted, modified, etc., and finally its value is displayed, recorded or processed on a computer for further purposes. Hence modern electronic instruments are complicated and often consist of many units. Each of these units is in itself an individual instrument for performing a special task. The circuits for these partial functions can easily be distinguished within the instruments. With earlier techniques the different elements of the circuits were even physically separated from each other; nowadays they are mostly integrated within a common casing and extremely miniaturized.

The information-carrying signal is the change of an electric quantity (voltage, current, frequency, or eventually charge measurable on a capacitor). Analogue technique means that information is carried in the form of variation with time, and digital technique means that a set of elementary signals carries information such that each of the signals falls into one of two distinguishable value ranges.

7.1 Elements of Electronic Circuits

7.1.1 Elements of Passive Circuits

Table 7.1 contains the names and units of the idealized elements of passive circuits together with the assigned impedance (i.e. the ratio between the voltage U arising on them and the current I flowing through them) and admittance (the reciprocal of impedance). In the case of direct current the impedance of an inductor is zero and

Table 7.1. Elements of passive circuits

Name	Letter symbol	Unit	Z (impedance)	Y (admittance)
Resistance	R	Ω (Ohm) = V/A	R	
Capacitance (condenser)	C	F (Farad) = As/V	$1/j\omega C$	$j\omega C$
Inductance	L	H (Henry) = Vs/A	$j\omega L$	$1/j\omega L$

that of a capacitor infinite (*Fig. 7.1* shows the symbols for these elements). The angular frequency of alternating current is $\omega=2\pi f$. The value $j=\sqrt{-1}$ indicates that the phase of the capacitor current leads the phase of the voltage on the poles by 90°, and the phase of the current flowing through the inductor lags the phase of the voltage by 90°. The current flowing through a resistor is in phase with the voltage.

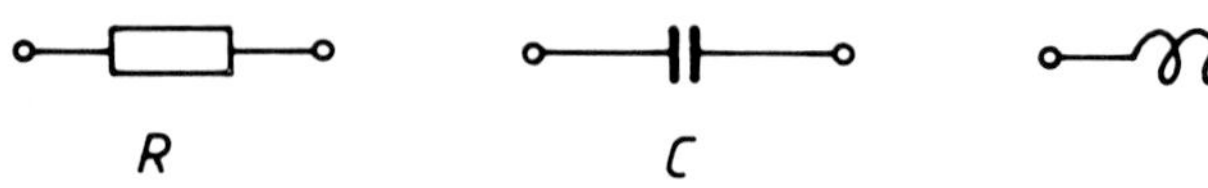

Figure 7.1. Symbols of passive circuit elements

In actual circuit elements there are always secondary (parasitic) characteristics. Resistors and inductors have parasitic capacitance; the losses of capacitors (condensers) are represented by a parallel resistance; losses of inductors by a series resistance.

Circuit elements are not always linear in reality, e.g. it is not valid for all elements that, in the case of DC, U changes proportionally to I. This means that the ratio $\frac{\Delta U}{\Delta I}$ relating to small changes depends on I, and it is not identical with the previously defined value $R=\frac{U}{I}$. Similar non-linearities also occur in the elements C and L. The value $\frac{\Delta U}{\Delta I}$ relating to a given current can be called differential resistance.

Between the poles of parallel-connected elements an equal potential difference occurs: the resultant conductance is the sum of the individual conductances. Equal current flows through elements connected in series: the impedances are summed *(Fig. 7.2)*. When planning simple circuits connections between the elements are regarded as perfect short-circuits ($U=0$ for any I flowing), and insulators are regarded as perfect cut-offs ($I=0$ at any U); sometimes, however, it must be appreciated that this is only an idealization of reality (see Subsection 7.3.1).

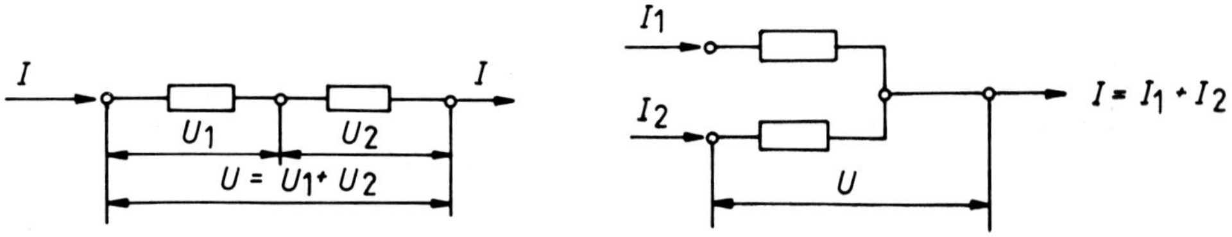

Figure 7.2. Series and parallel coupled elements

An important circuit element is the switch. An ideal switch is an element with two positions, one being a perfect short-circuit and the other a perfect cut-off, and the transition from one position to the other can be realized in an infinitely short interval of time. Real switches can, however, be characterized in such a way that they have one position of very high resistance and the other very low, and they have a certain small capacitance and perhaps an inductance to be taken into account. As a consequence of the latter, transition from one position to the other takes a finite time during which take place the processes of charge-discharge and inductive transient processes.

Transformers. A transformer consists of a primary coil with n_p windings and of one or more secondary coils with n_{s_1}, n_{s_2}, ... windings. Due to the effect of the DC voltage U_p coupled to the primary coil of n_p windings the voltages

$$U_{s_1}=\frac{n_{s_1}}{n_p}U_p=a_1U_p,$$

$$U_{s_2}=\frac{n_{s_2}}{n_p}U_p=a_2U_p \tag{7.1}$$

are induced in the secondary coils. At the same time the primary currents

$$I_{p_1}=\frac{1}{a_1}I_{s_1}, \qquad I_{p_2}=\frac{1}{a_2}I_{s_2} \tag{7.2}$$

correspond to the currents I_{s_1}, I_{s_2} flowing in the secondary coils. The total primary current is the sum of I_{p_1} and I_{p_2}.

7.1.2 Resonant Circuits

A resonant circuit consists of a capacitor suitable for storing magnetic energy and of an inductive element (coil). In the case of an ideal resonant circuit which has no loss, undamped sinusoidal oscillations are generated when energy is transferred, but in practice damped oscillations are generated due to the losses which always exist. Diagrams of series and parallel resonant circuits are shown in *Fig. 7.3*. Resistances r and R, respectively, represent the losses in both circuits. Using the notation in the figure the equations of the series-resonant and parallel-resonant circuits are respectively as follows:

$$Z_s=r+j\left(\omega L-\frac{1}{\omega C}\right),$$

$$Y_p=\frac{1}{R}+j\left(\omega L-\frac{1}{\omega L}\right). \tag{7.3}$$

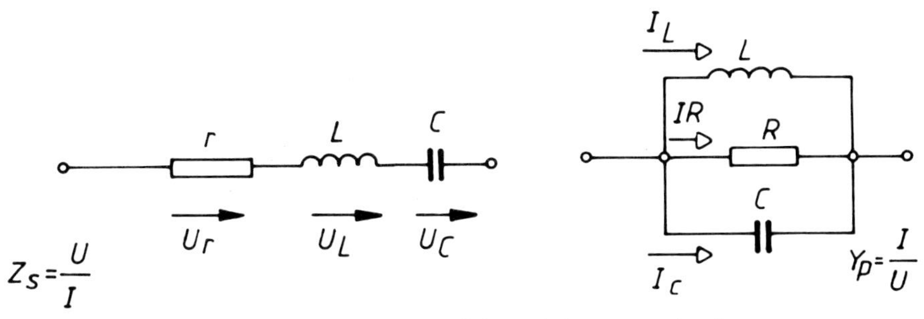

Figure 7.3. Series and parallel coupled resonant circuits

In the case of the resonant frequencies f_s and f_p, respectively, the imaginary term of Eq. (7.3) is zero,

$$f_s = f_p = \frac{1}{2\pi\sqrt{LC}} \tag{7.4}$$

and in this case Z_s and Y_p, respectively, are a minimum.

The amplitude and phase characteristics of series and parallel coupled resonant circuits are shown in *Fig. 7.4.* With series resonant circuits at the resonant frequency, the voltage U_L at the poles of the inductor is equal to the potential U_c produced at the poles of the capacitor, but they have opposite signs and their sum is hence zero. Therefore the current I_0 is determined by the small loss due to the series resistance.

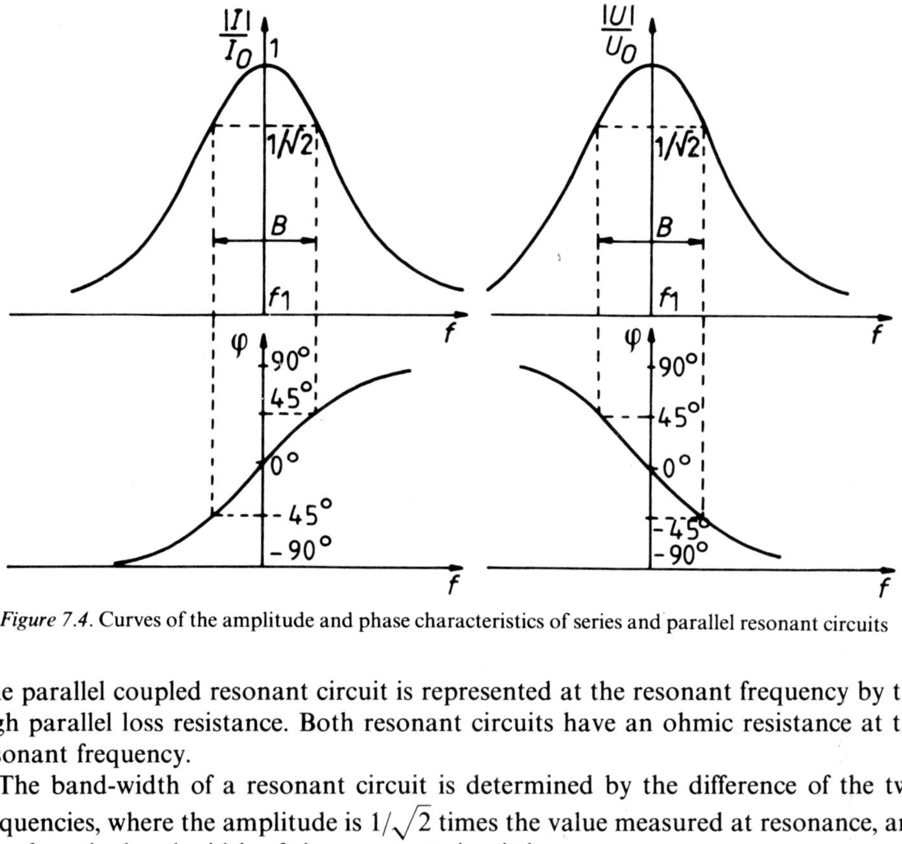

Figure 7.4. Curves of the amplitude and phase characteristics of series and parallel resonant circuits

The parallel coupled resonant circuit is represented at the resonant frequency by the high parallel loss resistance. Both resonant circuits have an ohmic resistance at the resonant frequency.

The band-width of a resonant circuit is determined by the difference of the two frequencies, where the amplitude is $1/\sqrt{2}$ times the value measured at resonance, and therefore the band width of the resonant circuit is:

$$B = \frac{f_0}{Q} = \frac{1}{2\pi RC} = \frac{\omega_0^2 rC}{2\pi}, \tag{7.5}$$

where the Q-factor of the series oscillating circuit is:

$$Q = \frac{1}{\omega_0 rC},$$

and the Q-factor of the parallel oscillating circuit is:

$$Q = \omega_0 RC.$$

The Q-factor indicates to what degree the resonant circuit approximates the "ideal" without loss. The value of Q can be several hundreds in practice.

7.1.3 Filter Circuits

Filter circuits help to select the measuring signal from the accompanying noise in such a way that they transmit a certain frequency band within the spectrum of signals of different frequencies and suppress the other bands. High-pass and low-pass filters divide the total frequency domain into two parts, band-pass and band-reject filters divide it into three parts. The theoretical basis of filtering is the frequency dependence of the circuit elements. If the corresponding elements are coupled in series the potential distribution on them will be frequency dependent and the desired frequency can be further amplified. In most simple filters there is only one frequency-dependent element *(Fig. 7.5)*. A better filtering effect is obtained, however, with two or more frequency-dependent elements. In band-pass filters series and parallel resonant circuits are used *(Fig. 7.6)*.

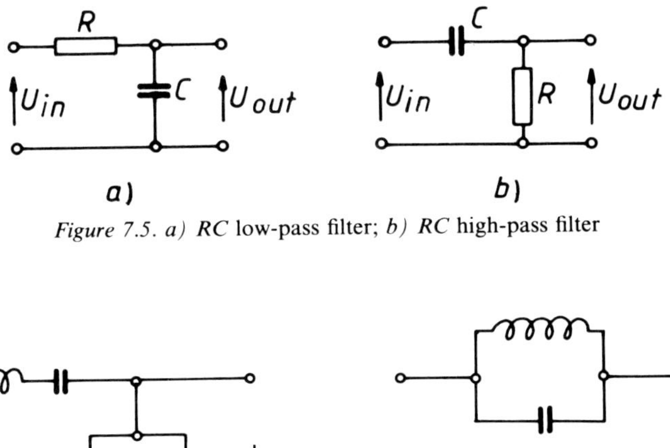

Figure 7.5. a) RC low-pass filter; *b) RC* high-pass filter

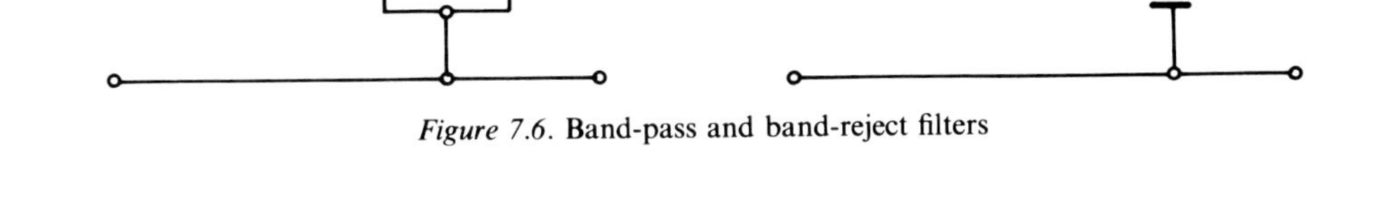

Figure 7.6. Band-pass and band-reject filters

Up-to-date filter circuits have complicated circuitry consisting of many elements and they usually include active amplifier elements. Inductors are almost obsolete; together with large resistors they are superseded by complicated solutions copying the effects of these elements.

Phase shift. High-pass or low-pass filters are also used as phase shifters, their phase-shift effect being shown in the vector diagram of *Fig. 7.7*. It may be seen that a phase quadrature cannot be attained. By the use of more elements, however, a phase-shift of even 180° can be obtained *(Fig. 7.8)*. Naturally, besides the phase, the amplitude distribution of the output potential also depends on frequency. Phase shift circuits with constant amplitude have a more complicated construction.

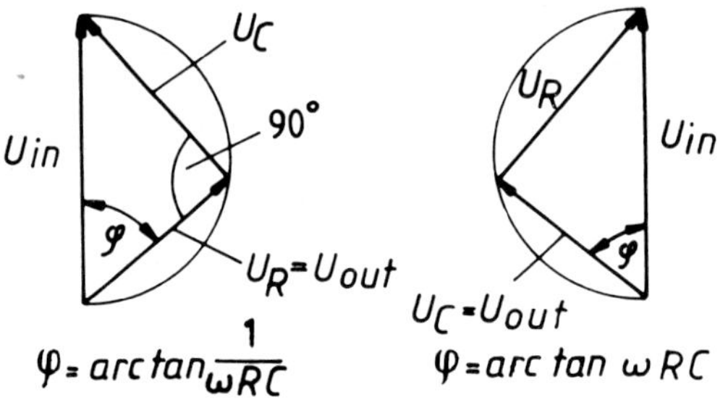

Figure 7.7. Phase shift of high-pass and low-pass filter circuits

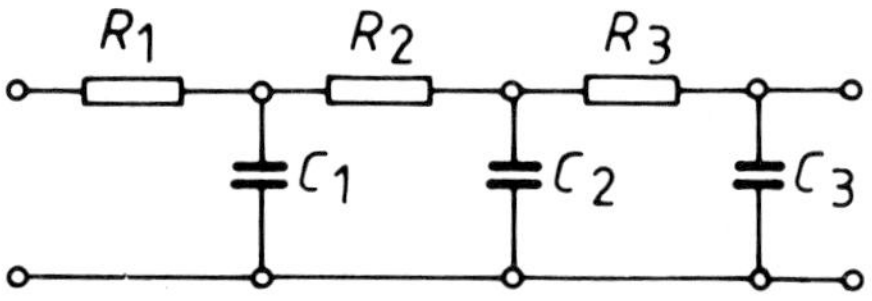

Figure 7.8. RC phase-shifting net

7.1.4 Vacuum Elements

Vacuum elements, having come into general use in the first half of the present century, have been replaced by solid elements, mainly by semi-conductors in almost all applications. Conventional grid-controlled electron tubes are used, however, even today for high-power radiation of high frequency oscillations. Special vacuum elements are used for producing very high frequency micro-wave oscillations; in these tubes the ray of electrons directly interacts with the high-frequency electromagnetic field. Cathode-ray tubes are also vacuum elements.

Cathode-ray tubes are useful for the investigation of electrical signals which vary with time. The construction of a tube is shown in *Fig. 7.9*.

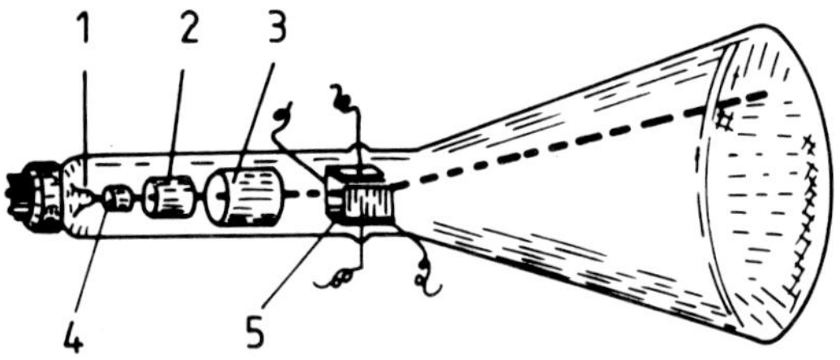

Figure 7.9. Cathode-ray tube: *1* cathode; *2* focussing electrode; *3* accelerator anode; *4* controlling electrode; *5* beam-deflecting plates

The beam of electrons emitted by the hot cathode is focussed electrostatically or magnetically. After passing through the cylindrical accelerating anode at a positive potential, the focussed beam passes between pairs of plates arranged horizontally and vertically. The voltages applied to the plate pairs produce an electrical field which deflects the passing electrons proportionally to the voltage. The deflected electron beam hits a fluorescent screen and produces a light spot on it. It is used mainly in instruments for investigating electrical signals and circuits, but it was also used for phase measurement in the first microwave distance meters (Tellurometer *MRA-1*). The television picture tube operates on a similar principle, but here the electron beam is deflected by magnetic fields.

7.1.5 Semi-conducting Elements

The discrete energy levels of individual atoms inside solid state matter are widened to a band due to the strong mutual effect of atoms in a small volume. Electrons cannot have an energy level lying between the two bands, this band being forbidden. The different possible cases of band splitting in the outside electron shell are shown in *Fig. 7.10*.

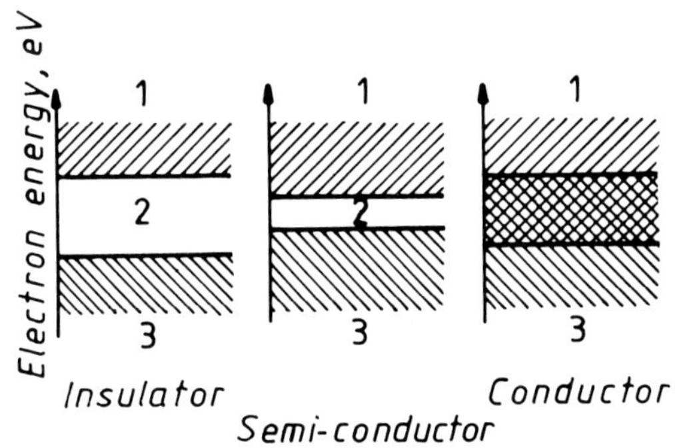

Figure 7.10. Types of band-splitting; *1* conducting band; *2* forbidden band; *3* valency band

There is no forbidden band in conductors, but in insulators the forbidden band is very wide. In semi-conductors the bandwidth is some tenths of a J, i.e. of the order of 1 eV: therefore at room temperature some electrons from the outside electron shell of the atoms can reach the conductive band, where they move as do the electrons in the half-saturated band of metals. There remains a hole in the original position of the electron, which means a non-neutralized positive charge. The hole attracts the electrons of the adjacent valency band. One of them enters the hole, so that the hole with a positive charge is displaced opposite to the electron.

In an intrinsic semi-conductor the number of holes and moving negative charge carriers is equal and this number is very small. In the most important material of semi-conductor technology, i.e. silicon, this number at room temperature is only about $10^{10}/cm^3$, whereas the order of magnitude of the number of atoms is 10^{23}.

In practice semi-conductors are always used in a doped form, in other words impurity materials are doped at a definite rate. This rate is very small. At room temperature one type of doping atoms, the donor, being wedged among the atoms of the semi-conductor, dismisses an electron, and thereby increases the number of free electrons, while the atom itself remains where it is as a positive ion. The other type of doping atoms, the acceptor, attracts to itself an electron from the adjacent atom and thereby increases the number of holes, but the atom itself remains where it is as a negative ion. In this way *N*-type and *P*-type semi-conductors are produced. If e.g. silicon is doped with phosphorus in a ratio of one in a million, the number of free electrons in 1 cm^3 (i.e. the electron concentration) will be increased 10^7 times, in other words 10 million times. The resistance is decreased by a similar ratio.

Some specific characteristics:

1. Holes and electrons continuously come into being and vanish by recombination, i.e. their concentration is in a state of equilibrium and is constant in time (with a statistical fluctuation). The lifetime of the individual charge carriers is finite (of the order of 10^{-5} to 10^{-9} s).

2. Besides the majority charge carriers produced by adding, there are also minority charge carriers always present, but their number is small with respect to the intrinsic case. (There is a greater possibility of their rapid recombination.) The product of the majority concentration and the minority concentration is equal to the square of the intrinsic concentration. Hence, in an *N*-type silicon doped as described above, the hole concentration is only $10^3/cm^3$ at room temperature.

3. Similarly to conductors, a non-neutralized charge can also be present in semi-conductors at their boundary surface. Internally they are usually neutral, the algebraic sum of the charges of the electrons and of the doping ions being zero.

4. The diffusion phenomenon in semi-conductors can cause a considerable charge flow independently of the electric field.

7.1.6 Integrated Circuits

Most of the important active and passive elements of a circuit can be produced from semi-conductors. The semi-conductive component used can perhaps consist of only one element, but it often contains more than a thousand elements (integrated circuit). The single crystal chip (mostly silicon) of the element is an extremely small plate, some tenth mm thick and with a surface of 0.5 mm^2 to 10–20 mm^2. The casing, complete with terminals, is also very small. The electronic characteristic of the single crystal is determined by the geometry of the zones into which it is divided, both on the surface and internally by the *N* and *P* type doping.

In the single crystal chip a properly doped zone of defined cross section can yield the prescribed resistance. The sides and bottom of the zone are surrounded by a closed *PN* junction (see later) and the chip is supplied at both ends with metallic connections.

For a capacitor either a metal-silicon dioxide laminated structure or the closed *PN* junction is used. Inductors are not used in semi-conductor circuits.

The different circuit elements are connected by metal strips vaporized onto the surface of the chip.

This is not the place to describe the production technology but it suffices to mention that it is extremely productive in spite its utmost complexity. This is the explanation of the price revolution of electronic products.

7.1.7 PN-Junction

The most important phenomena occur at the contact surfaces of the adjacent *P* and *N* doped zones produced inside the single crystal. At the moment they are produced, a large number of electrons diffuse from one side to the other, and a large number of holes diffuse in the opposite direction. This process is, however, short-lived. The remaining acceptor ions on the *P*-side represent negative charges fixed in position, and the donor ions on the *N*-side represent positive charges. An electric field arises between the two layers which has a repelling effect on further charge carriers. An equilibrium comes into being, i.e. there is no current flow at the junction (common contact surface), if there is no external voltage *(Fig. 7.11)*.

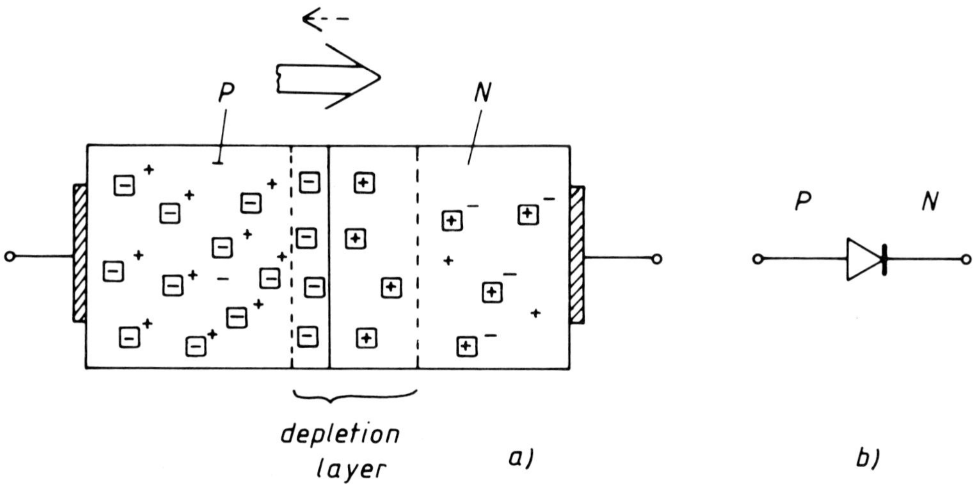

Figure 7.11. *PN*-junction

The most important characteristic of the *PN*-junction is its behaviour as a switch: its state can be changed between cut-off and short-circuit depending on the polarity of the voltage applied to it. Let this characteristic be studied in more detail. (The depletion layer arising at the junction is not taken into account.) If a positive potential is given to the *P*-side with respect to the *N*-side, this is superimposed on the former state of equilibrium and results in driving the movable electrons of the *N*-side to the *P*-side, where they move on. From the *P*-side holes carrying a positive charge flow to the *N*-side of a more negative potential. The electric current due to electrons and holes flows in the same direction, yet they move in opposite directions.

If the polarity is reversed, the electric field retains the majority charge carriers from the junction. On the other side, minority charge carriers can generate only a weak current due to the phenomenon of recombination causing the disappearance of those minority charge carriers which start farther away. Hence this weak current is only supplied by the path sections near the junction.

The state of strong current, i.e. the state approximating a short-circuit is called on-state, that approximating the cut-off state is called off-state, and the corresponding polarities on-direction and cut-off-direction, respectively *(Fig. 7.12)*.

In the on-direction the current increases according to an approximately exponential function of the voltage (a voltage increase of about 60 mV always increases the current ten times). This means in practice that, below an external potential (of on-direction) of about 0.6 V the current is negligible, but above 0.8 V it is so strong that its value is already not limited by the junction itself but by e.g. the ohmic resistors present in the circuit. This approximately verifies that the *PN*-junction corresponds to a switch which is at the state of short-circuit above a certain voltage U_0, and at the state of cut-off below it *(Fig. 7.12)*.

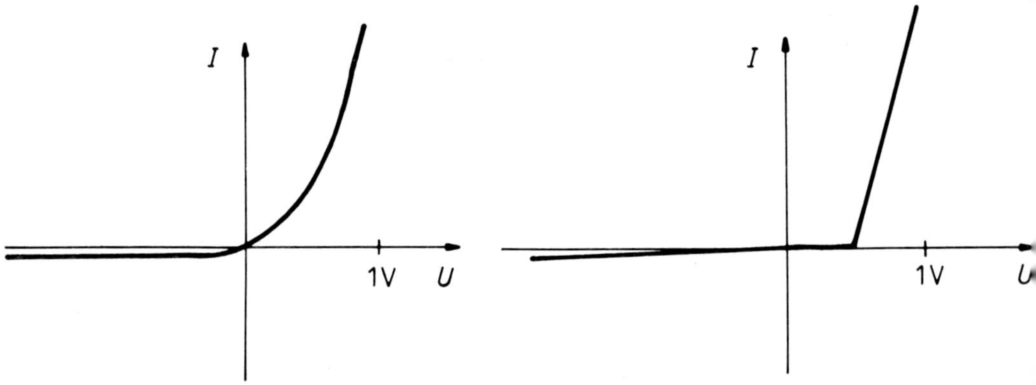

Figure 7.12. Current-voltage dependence of the idealized diode and its approximation by a linear characteristic with one suitably chosen point

The equation of an idealized junction is

$$I = I_0 \left(\exp \frac{U}{U_T} - 1 \right) \tag{7.6}$$

according to theory, where U is the external voltage applied to the P-side with respect to the N-side, $U_T = \frac{kT}{q}$ (at room temperature it is about 26 mV), and I_0 is a very small value, namely 10^{-10} to 10^{-15} A depending on the physical characteristics of the junction. Therefore the last term in the expression can usually be neglected and the voltage at the junction with a given current is proportional to the logarithm of the current.

The internal resistance of the junction (i.e. the differential resistance according to Subsection 7.1.1) at a given I is:

$$R_I = \frac{\Delta U}{\Delta I} = \frac{U_T}{I + I_0}. \tag{7.7}$$

Since in the cut-off direction the current is $I = -I_0$, the internal resistance is infinite in theory: in practice it is very high.

A very important deviation from the formula is that at a definite cut-off voltage the *PN*-junction suddenly becomes a conductor. This voltage is called the Zener potential.

The main applications of the *PN*-junction are:

a) It can be used to electrically separate the circuit elements on a single silicon chip. All the elements are enclosed in a *PN*-junction; this gives a practical means of insulating them from adjacent elements.

b) It is used as a switch. The switching voltage can be set by potential sources connected in series *(Fig. 7.13)*.

c) It can be used for rectification, since the junction allows the transit of *AC* half-waves. This pulsing signal consists, from Fourier's theorem, of a direct potential and of superimposed spherical harmonic terms of higher order; the latter can be filtered out by a corresponding low-pass filter *(Fig. 7.14)*.

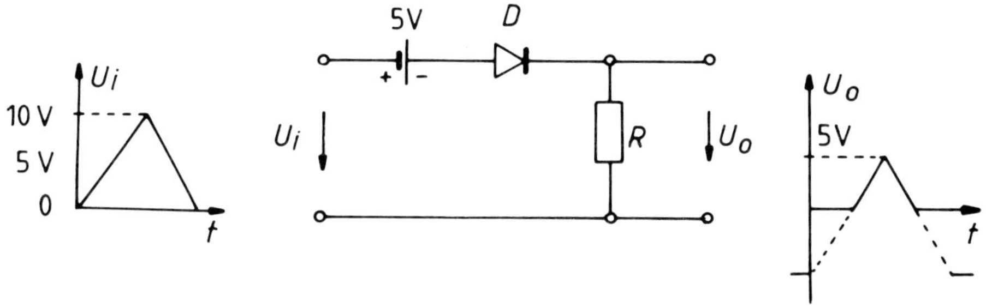

Figure 7.13. The diode as a switch

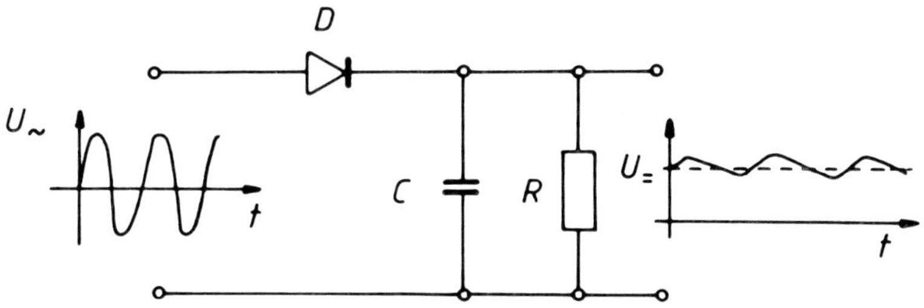

Figure 7.14. The diode as a rectifier

d) It is used as a spatially-formed Zener element for producing a stable reference voltage.

e) There are non-neutralized depletion layers at both sides of the *PN*-junction. The cut-off junction acts as a capacitor with respect to *AC* and it can be used as such an element. It is essential that the value of this capacitor be influenced by the cut-off voltage.

The realized *PN*-junction is called a diode if it provides an independent function.

Metal-semi-conductor junction. There are two types of contacts between a metal and a semi-conductor depending on the atomic characteristics of the two materials:

a) the so-called ohmic contact; charge carriers can move freely in both directions, similarly to the contact of metallic conductors; there is practically no resistance drop at the contact, *b)* the Schottky junction; it has a behaviour similar to the *PN*-junction. The value of I_0 in Eq. (7.6) is, however, much greater the current increases very rapidly and therefore the on-voltage in the switch model is smaller (0.3 to 0.4 V).

7.1.8 Transistors

The transistor is an amplifying element. It is a part of two simple circuits at once: it amplifies the signal of the input circuit in the output circuit. In analogue (linear) amplifiers it is important that the output signal be distortion-free, i.e. the output time function should correspond to the input time function as exactly as possible. One point of the amplifying element must be linked to the input circuit and the other to the output circuit, and its third point is connected to the common reference point of the two circuits (earth potential).

Two types of transistors are used today. The essential point of the first is a variable resistance. The resistance is a current path formed in the semi-conductor and belonging to the output circuit. The geometrical dimensions of the current path and the concentration of the charge carriers flowing in it can, however, be controlled by the input signal. This element is the field-effect transistor (FET) and its most frequently used variant is the metal-oxide semi-conductor or *MOS*. The direction of the controlling electric field is perpendicular to the path of the controlled current.

The second type is the bipolar transistor. Its characteristic is the PNP or NPN structure, in which the distance between the two junctions is very short. In this type the input signal at one junction affects the output current passing through the other junction. The directions of the control field and of the controlled current are the same.

MOS transistor. The metal-oxide semi-conductor structure is a small plane capacitor, one electrode of which is the isolated surface of the semi-conducting silicon chip (the "channel"), its dielectric medium is very thin (some nm-s) silicon dioxide, and its second electrode is an aluminium layer applied in vacuo, i.e. the controlling electrode (gate).

If voltage is applied between the two capacitor plates and the capacitor is charged, equal charges with opposite polarity appear at the two electrodes. Due to the very high density of free electrons in the metal a surface charge is produced. In the semi-conductor the field either attracts the electrons to the surface or drives the free charge carriers from it; in certain cases the field even attracts the minority charge carriers below the surface. If the field is further increased, the originally minority charge carriers, e.g. the electrons of a *P*-type crystal, produce down to a certain depth an *N*-type channel strongly saturated with electrons. This channel, or the resistance representing it, is a part of the output circuit. At the two ends of the channel there are the two connecting poles of the output current, i.e. small strongly *N*-doped islands on the originally *P*-type chip, which are called source and drain. They use ohmic metallic contacts to connect to external elements *(Fig. 7.15)*. If there is no control voltage, or if it is less than a defined threshold value U_{th}, no current can flow between the two poles because the source and the *P*-type semi-conductor block are coupled to the same earth potential, and they thus produce a *PN*-junction without any on-voltage. If the control voltage exceeds the threshold value, the *N*-type channel

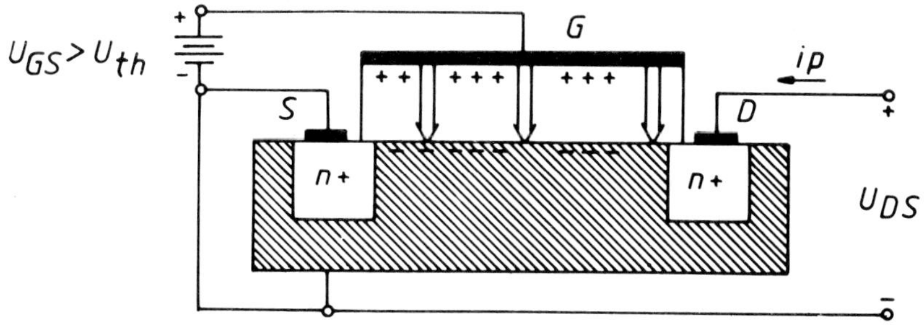

Figure 7.15. The theoretical structure of the MOS transistor

comes into being and the *PN*-junction disappears. Current can flow between source and drain if the drain has positive potential with respect to the source.

The same device can be produced on an *N*-based semi-conductor chip, but the channel is *P*-type, the charge carriers are holes and the drain is on negative potential with respect to the source.

Within proper conditions the current depends on the control voltage and is approximately:

$$I = A(U_g - U_{th})^2, \tag{7.8}$$

where A is the constant determined by the dimensions, and U_g is the voltage between the control electrode and the earth. Thus, in this rough approximation, the current is not influenced by the potential of the drain. This means that the differential resistance of the output side (output resistance) is nearly infinite, i.e. in practice very high. The proportionality factor A is determined by the geometric and doping conditions. Consequently the current variation, i.e. the output signal, depends quadratically on the input voltage. A linear amplification can be obtained in spite of this if the input signal, i.e. the variation ΔU_g, is small enough. (The short section of the parabola can be approximated by its tangent.)

If the output current flows through the load resistance R_L, the potential of the drain changes, due to the effect of the input signal voltage ΔU_g, by:

$$\Delta U_d = R_L \Delta I = 2A(U_g - U_{th}) R_L \Delta U_g. \tag{7.9}$$

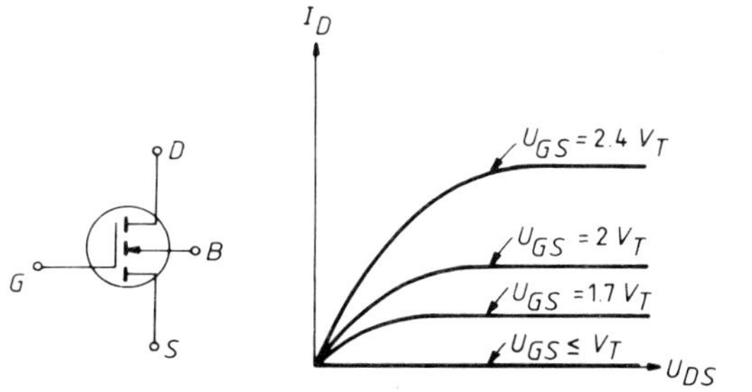

Figure 7.16. Current and voltage conditions of the MOS transistor

The use of the *MOS* transistor as a switch has even greater importance. By altering the control voltage the current can be altered within a range between a considerably high I_{max}, determined by the circuit elements, and nearly zero (the output voltage can also vary within wide limits) *(Fig. 7.16)*.

A *MOS* transistor requires very little space and is thus expedient for use as one of several thousand elements united in one complicated integrated circuit. The control electrode can be made from another conducting material instead of metal, and the dielectric medium is also not exclusively silicon dioxide.

In another group of *FET* transistors the resistance of the channel is affected by a voltage connected to the *PN*-junction. This component is seldom used.

Bipolar transistor. A frequently produced form of the bipolar transistor is shown in *Fig. 7.17* in section and in plan. The transistor in the figure is of the *NPN*-type in which the upper layer has the strongest doping. From an electronic point of view only the upper layer of a cross section of about 100 μm^2 is an *N*-layer, *H* is the emitter and the layer below this plays an active role. The layer between the two junctions is the base and it is usually $d<1$ μm thick. The lowest, thicker layer is the collector. The operation of the two junctions corresponds to that of the *PN*-junction described earlier, but due to their nearness they mutually affect each other. If an on-direction voltage is applied to the emitter-base junction and a cut-off direction voltage to the base-collector junction, a great number of electrons cross from the emitter into the base and thus the concentration of the minority charge carriers increases many times in a *P*-type base. The current at the cut-off junction depends on the number of the minority charge carriers present on both sides of the junction. Since the base is very thin, only a small number of electrons are recombined within it, while the vast majority of them diffuse to the collector junction and drift through the field of the cut-off junction into the more positive collector.

The less positive base electrode has thus only a very small current. By far the greatest part of this current is the hole current, which replaces the holes. Since only a very few of the transient electrons are recombined, the variation of the electron current is proportional to the hole current:

$$\Delta I_c = \beta \Delta I_b, \tag{7.10}$$

where I_c is the collector current, I_b the base current.

The order of magnitude of β is 100, therefore the amplification of the collector current is about 100× in case of control by the base current. In this case the base-emitter voltage is stabilized to a proper value. The collector current of the transistor can also be controlled by voltage signals, because it differs only slightly from the emitter current which is the current of a *PN*-junction operating in the on-direction, i.e. it depends exponentially on the voltage applied to the junction. However, to a first rough approximation, the cut-off voltage connected to the collector junction does not influence the flow of current. Therefore the differential resistance measured at the output side is very high. A large external resistance R_L can be connected to the collector circuit. The output signal voltage is mainly limited by the supply voltage.

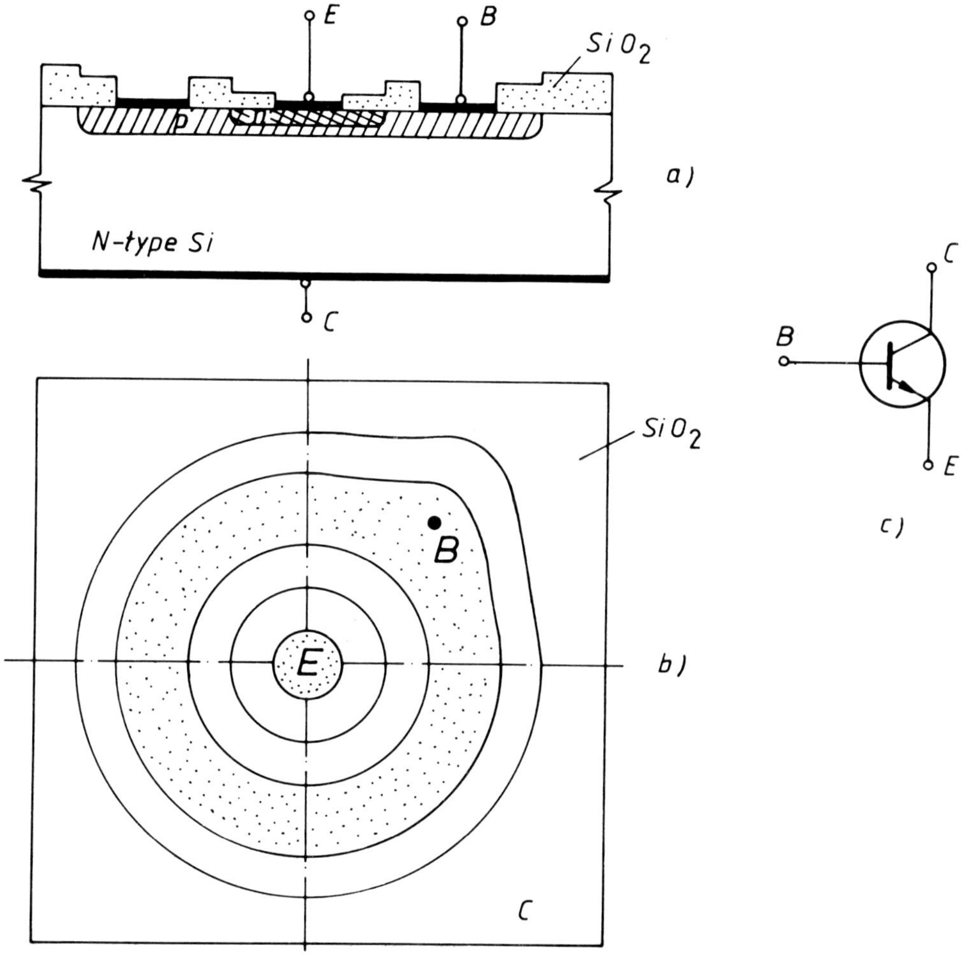

Figure 7.17. Bipolar transistor: *a)* section; *b)* plan view; *c)* symbols

The voltage variation arising due to the effect of ΔI_b, i.e. the output signal within this limit, is approximately:

$$\Delta U_{out} = R_L \Delta I_c = R_L \beta \Delta I_b. \tag{7.11}$$

From Eq. (7.7) the variation of the emitter current (and of the slightly differing collector current) is:

$$\Delta I_e \approx \Delta I_c \approx \frac{I_e}{U_T} \Delta U_{in} \tag{7.12}$$

due to the effect of the voltage signal ΔU_{in} applied between emitter and base. The output signal voltage generated by the input voltage signal is therefore:

$$\Delta U_{out} \approx R_L \frac{I_e}{U_T} \Delta U_{in}. \tag{7.13}$$

If both junctions are cut-off, only very little current flows in the transistor, while if on-voltage is given to both junctions but a little higher to the emitter side, a very large current can flow through the transistor in spite of the small emitter-collector voltage *(Fig. 7.18)*.

As a consequence of the above the bipolar transistor can also well be used as an electronically controlled switch: the section emitter-collector can be brought by base voltage from nearly short-circuit state to nearly cut-off state *(Fig. 7.18)*.

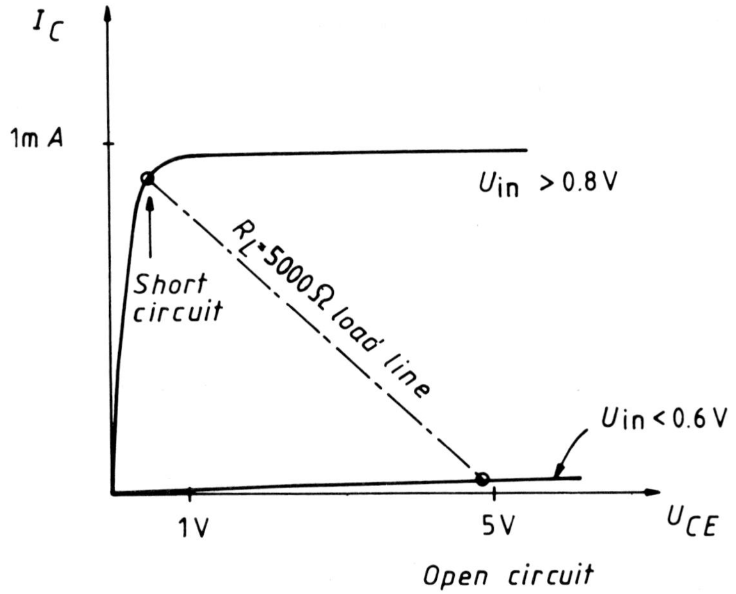

Figure 7.18. Current conditions of the bipolar transistor when operating as a switch

7.1.9 Power Sources

For field instruments dry batteries or storage batteries are the most appropriate power sources, but in exceptional cases — especially for high power requirements — motor-generators are also used.

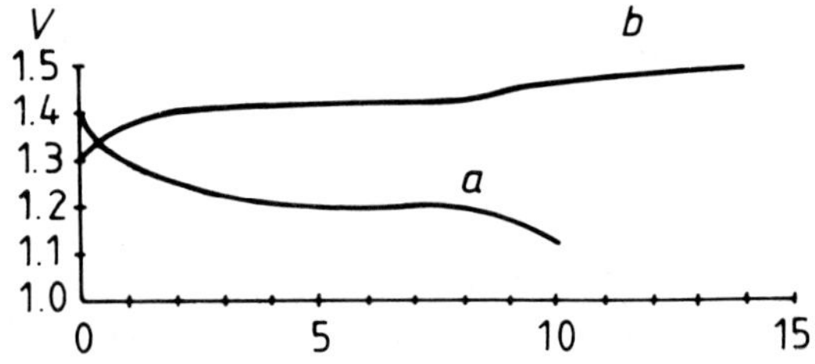

Figure 7.19. Charge and discharge curves of a Ni–Cd accumulator: a = discharge during 10 hours; b = discharge during 14 hours

Accumulators can be lead or alkaline batteries. Lead batteries are mostly used for larger equipment, since their capacitance/weight ratio is large and they are easy to handle. Ni–Fe, Ag–Zn and Ni–Cd type alkaline batteries are also widely used. In modern instruments with small dimensions the latter are usually used due to their large capacitance/mass ratio.

Accumulators can be charged several hundred times if the specifications for the type used are fulfilled. The curves of charge and discharge of a Ni–Cd type accumulator are presented in *Fig. 7.19*, from which the degree of efficiency can be read.

Controlled power sources are often used for many instruments. In such power sources the original uncontrolled supply voltage is periodically cut off by a switch and the alternating voltage obtained in this manner can be transformed to the required voltage. The rectifier and filter circuits connected to the secondary coil of the transformer produce the different *DC*-s. The error signal need for the control is the difference between a suitably chosen reference voltage and the output voltage. The reference voltage is often the break-down voltage of a Zener diode kept at constant temperature.

7.2 Electronic Circuitry

7.2.1 Noise

All changes of values of electrical quantities which are detected by instruments as signals, but do not contain useful information, are called noise.

Noise can be produced by an avoidable disturbing effect such as radiation, induction produced by other circuits, intermittent connection due to poor contacts, etc. The most important type of noise of thermodynamical origin is, however, always present at a finite temperature and is called *Nyquist noise*. The reason for it is that the charge carriers have an irregular motion due to heat as well as motion controlled by the electromagnetic field. Thus e.g. a voltage variation of

$$u_n = \sqrt{4kTR\Delta f}$$

can always be measured in any frequency band Δf with resistance R and at temperature T, where k is the Boltzmann constant ($k = 1.38 \times 10^{-23}$ J/K).

In the active components of electronics (in non-linear units which can amplify) a further type of noise occurs because the energy of the moving charge carriers has a statistically varying distribution. The relations, equations, diagrams (characteristics) describing the operation of the components therefore yield only statistical average values.

7.2.2 Amplifier Circuitry

Amplifier circuits are used to increase small electrical signals. They can be voltage, current or effective power amplifiers, depending on whether they amplify the voltage, current or power level of the input signal to the required value.

The characteristics of an amplifier are the gain, the frequency band transmitted, the unavoidable distortion and noise voltage, and the input and output impedance.

Important characteristics of power amplifiers are further the power output and the efficiency factor; the latter is the ratio between the output power and the input power taken from the supply voltage.

The amplifier circuit amplifies the electrical signal supplied to the input by means of the power gained from the direct voltage supply applied to its voltage poles. All amplifier components must be set to a suitable DC working point, around which the output voltage varies.

Amplifiers produced in practice usually consist of many circuit elements; these elements are used to set the working point quantities and combine the signals at the input and to separate them at the output. Generally a single transistor alone does not yield the gain needed and more transistors are often included in one stage; even then the operation may have to be repeated in several stages.

Part of the output signal of an amplifier or amplifier stage can be fed back to the input. It this partial signal is in phase with the original input signal, i.e. they are algebraically added, the feed-back is positive. If the partial signal is in anti-phase, i.e. it decreases the original signal, the feed-back is negative. The feed-back can be made by phase-shifting elements (see Subsection 7.1.1), hence both cases can be set independently of the phase difference between input and output signals.

In amplifiers negative feed-back is usually used. This actually decreases the gain level but it gives higher stability. For a high negative feed-back, the gain is simply:

$$\left|\frac{\text{Output signal}}{\text{Input signal}}\right| = \frac{1}{B},$$

where B is the ratio between the feed-back signal and the output signal.

Negative feed-back decreases distortion, the frequency dependence of gain and the influence of a possible shift of supply voltage or working point on the gain.

An operational amplifier is a frequently used component in up-to-date techniques *(Fig. 7.20)*. It is a transistor amplifier consisting of more stages with several internal negative feed-back loops, auxiliary current circuits and voltage stabilizing solutions. It is always produced in an integrated form. The characteristic of operational amplifiers is that they always have two input poles as well as the poles for supply voltage and output signal. The output is in phase with the signal connected to the positive input

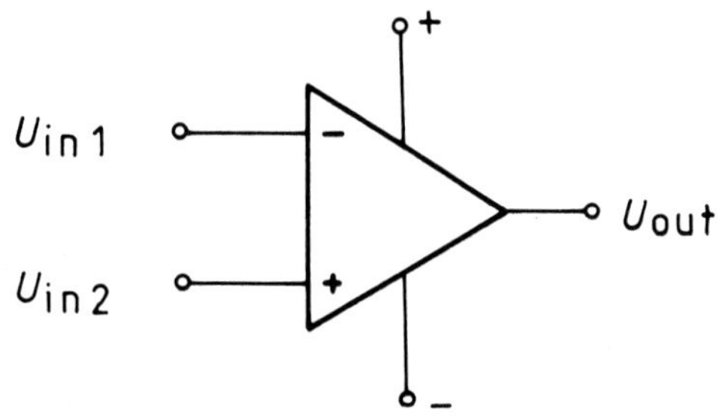

Figure 7.20. Symbol for an operational amplifier

and in anti-phase with the signal connected to the negative input, the gain in both cases being exactly the same. The output signal is thus theoretically zero if both inputs have the same signal. (In practice there always unavoidably exists a very small signal due to slight asymmetries in manufacture; this can be corrected.) The operational amplifier amplifies all signals from DC to a high limiting frequency (e.g. 10 MHz) evenly and equally, and practically without distortion, e.g. 100,000 ×. Its input differential resistance is nearly zero, its output differential resistance is very high. Gain characteristics can easily be altered by setting in negative feed-back elements. By means of a proper negative feed-back, the cut-off frequency can be increased several times or a certain frequency range can be enhanced (active filters). External elements can also be used to make the input of the operational amplifier a voltage or a current signal.

7.2.3 Oscillators

Let us now investigate positive feed-back in a voltage amplifier which has a gain of $\frac{U_{out}}{U_{in}} A_u$ without feed-back. Now let the feed-back B be gradually increased. As long as $B < \frac{1}{A_u}$, the gain increases (together with distortion and frequency dependence).

If $B = \frac{1}{A_u}$, the gain becomes infinite: there is a signal on the output even if there is no signal on the input. The output signal is naturally of finite magnitude; that is to say, the large output signal influences the working point and, together with it, the gain also. Finally the output signal becomes so large that the condition $\frac{1}{A_u} = B$ just holds and the amplifier is in sustained oscillation. The ever-present noise has a component in each frequency. With a lack of other signals, positive feed-back amplifies this component and the gain level increases until an equilibrium is reached.

It must be noted that the gain A_u is in practice not quite independent of frequency, due to the presence of the different circuit elements; this also holds for the value B. Sustained oscillation sets in at the frequency at which the condition $\frac{1}{A_u} B$ is most easily fulfilled. (This frequency, however, is not stable; further sustained oscillation can set in at several such frequencies.)

Stable oscillation can be produced by means of positive feed-back. In this case a resonant circuit must be present in the feed-back loop; its resonance defines the oscillation frequency. In a stable state the power of the output signal is equal to the loss of the resonant circuit. The whole system is called an oscillator.

The energy conditions permit another formal point of view. Since the loss of the resonant circuit is symbolized by an ohmic resistance, one can say that an amplifier with positive feed-back must cancel out the low resistance.

In a series resonant circuit the resistance r can be eliminated so that a negative resistance $-r$ is connected in series with it, the algebraic sum of the two thus being

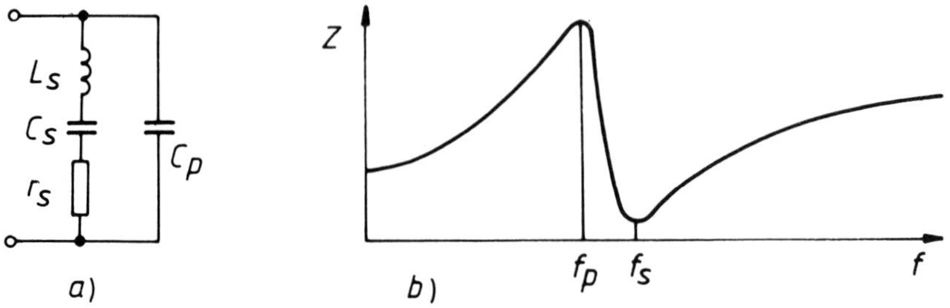

Figure 7.21. a) equivalent circuit of a quartz crystal and *b)* its impedance vs. frequency

zero. In a parallel resonant circuit the resistance $-R$ must be connected parallel to the resistance R in order to obtain $\frac{-R^2}{R-R}=\infty$.

Hence an oscillator can also be regarded as consisting of a real resonant circuit (with loss) and of negative resistance. (In practice the negative resistance is a complex system with a DC power source.) This formal consideration can refer to all oscillators, even to those in which feed-back and gain mechanisms cannot be well separated.

Quartz oscillator. If high frequency stability is required, crystal controlled oscillators are used. The oscillating quartz used as an oscillating element, is a chip cut out from a natural or artificial quartz crystal with proper orientation; it is supplied with evaporated electrodes and placed together with its glass or metal casing into an evacuated container. On the surface of the quartz chip a charge is generated by the effect of pressure and mechanical deformation, due to the effect of charge variation: consequently the chip oscillates mechanically due to the effect of the AV (Alternating Voltage) applied to the electrodes. The equivalent circuit of the oscillating crystal *Kr* *(Fig. 7.21)* consists of the inductor L_s corresponding to the mass of the crystal structure, of the capacitor C_s substituting for the elasticity of the crystal, of the resistor r_s representing mechanical friction, and of the capacity C_p between the crystal and its casing. As L_s is very high and C_s very low, therefore the Q-factor of the equivalent resonant circuit is very high, $Q>10^6$. On the impedance curve of the oscillating crystal, plotted as a function of frequency *(Fig. 7.21)*, parallel resonance occurs at the frequency f_p and series resonance at the frequency f_s.

The resonant frequency slightly depends on temperature, therefore the crystal is usually placed in a thermostatically controlled oven.

It is often the case that the frequency required is several times higher than the highest frequency f of the oscillator that can be suitably reached. The problem can be solved by frequency multiplication.

A sinusoidal signal of given frequency is distorted by the effect of an element with a non-linear characteristic; the consequence is that integer multiples of the given frequency occur in the frequency range. The amplitudes of the frequency components nf ($n=1, 2, 3, \ldots$) depend on the non-linearity of the transfer characteristic. Each of the frequency components nf can be selected by a band-pass filter and used as follows.

Microwave oscillators. Towards higher frequencies an approximate realization of idealized circuit elements, such as e.g. resonant circuits, becomes more and more difficult. At microwaves with wavelengths of some cm-s or shorter (i.e. of a frequency over 1 GHz) cavity resonators are used; in them the location of inductive and capacitive phenomena cannot be separated. A cavity resonator can resonate at several different frequencies, out of which one (usually the lowest) can be separated by means of further elements. The undamped oscillation can be generated either by positive feed-back of a special amplifier (microwave transistor), or a "negative resistance" is produced at the corresponding frequency by other electronic means.

Such an element is e.g. the Gunn diode which is essentially a very thin homogeneous crystal plate made of gallium arsenide. This semi-conductor shows a character of negative resistance in a certain DC voltage range. This behaviour can be explained in terms of quantum mechanics.

Formerly electron tubes of special construction were used for generating microwave oscillations; electrons accelerated by DV (Direct Voltage) interacted with the high-frequency field of the cavity resonator (klystron, reflex-klystron).

7.2.4 Modulation and Modulator Circuits

In a connection realized by electromagnetic waves the transfer of information (transmission of measuring signals or of the speech frequency band) is possible by modulation.

Types and characteristics of modulation. All phenomena of pure sinusoidal processes can be characterized unambiguously by their amplitude, phase and frequency. If the amplitude varies according to some time function, amplitude modulation occurs; if the frequency and the phase vary, frequency or phase modulation arise respectively.

The equation of the sinusoidal carrier frequency is:

$$U(t) = A \sin(\omega t + \Phi).$$

For amplitude modulation with a sinusoidal signal of angular frequency Ω *(Fig. 7.22)*, the amplitude varies according to the relationship:

$$A(t) = A_0(1 + m \cos \Omega t).$$

Hence the amplitude modulated function is:

$$U(t) = A_0(1 + m \cos \Omega t) \sin(\omega t + \Phi),$$

where ω = the angular frequency of the carrier,
Ω = the angular frequency of the modulation,
m = the depth of modulation.

By trigonometric transformation this can be brought into the following form:

$$U(t) = A(t) \sin \omega t = A_0 \sin \omega t + \frac{A_0 m}{2}[\sin(\omega + \Omega)t + \sin(\omega - \Omega)t]. \quad (7.14)$$

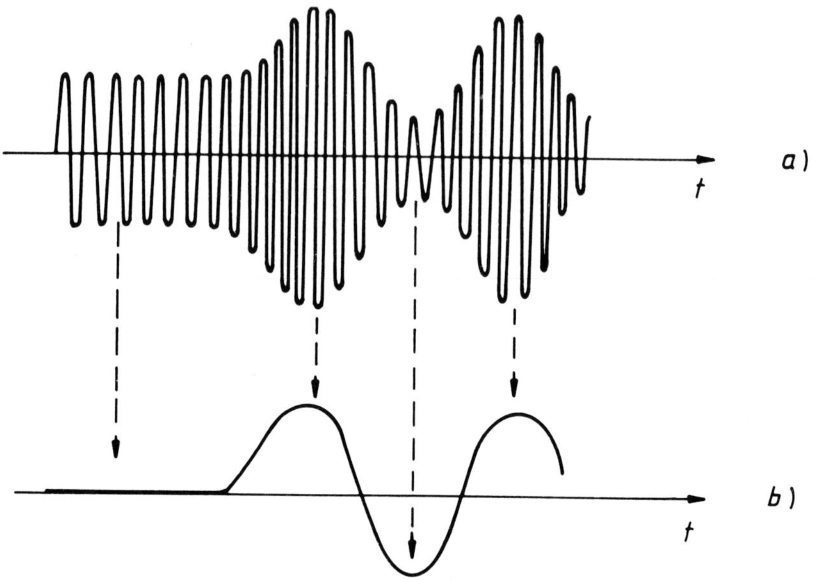

Figure 7.22. Amplitude modulation: *a)* amplitude-modulated carrier signal; *b)* modulator signal

From Eq. (7.14) the spectrum of the amplitude-modulated signal can be obtained. At both sides of the angular carrier frequency ω side-bands of amplitudes $A_0m/2$ and angular frequencies $(\omega+\Omega)$ and $(\omega-\Omega)$ appear, respectively.

For phase modulation with a sine signal *(Fig. 7.23)* the value Φ varies in the equation of the carrier frequency:

$$U(t)=A\sin[\omega t+\Phi(t)].$$

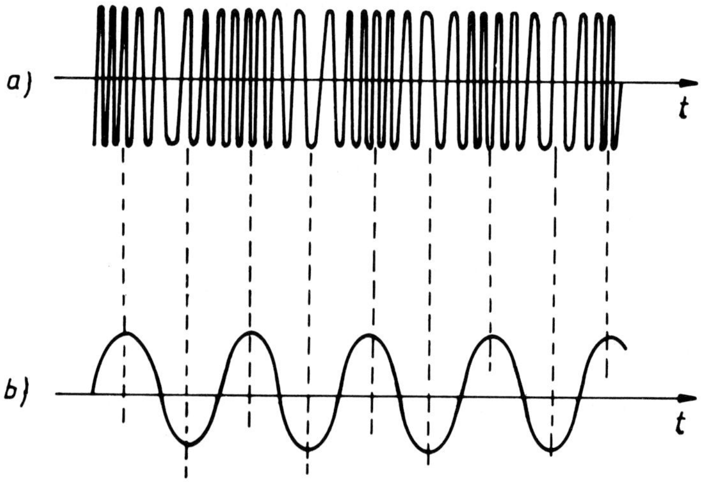

Figure 7.23. Phase modulation: *a)* phase-modulated carrier-signal; *b)* modulator signal

Let

$$\Phi(t)=m_f \sin \Omega t \qquad \text{and} \qquad \psi(t)=\omega t+m_f \sin \Omega t,$$

by means of which the phase modulated oscillation

$$U(t)=A \sin (\omega t+m_f \sin \Omega t)=A \sin \psi(t)$$

is generated.

The frequency of the phase-modulated oscillation is:

$$\frac{\mathrm{d}\psi(t)}{\mathrm{d}t}=\omega+\Omega m_f \cos \Omega t,$$

from which it may be seen that phase modulation is at the same time also frequency modulation. The modulation index m_f shows maximum oscillation of phase around the unmodulated value, i.e. the variation is $+m_f$ and $-m_f$, respectively (expressed in radians). For frequency modulation the instantanious angular frequency $\omega(t)$ oscillates around the angular carrier frequency ω within the limits $-\Omega m_f$ and $+\Omega m_f$. The quantity $\Omega m_f=\Delta\Omega$ is the angular frequency sweep, or $\Delta F=fm_f$ the frequency sweep, by which the modulation index becomes $m_f=\Delta F/f$. (Here Ω is the modulating angular frequency, f the modulating frequency.) The components of the spectrum of phase- and frequency-modulated signals, respectively, are multiples of the modulating frequency f; their amplitude distribution is determined by Bessel functions and the number of important components depends only on m_f.

7.2.5 Demodulation and Demodulating Circuits

The modulated signal carrying information is separated from the modulated carrier frequency by demodulator circuits.

For the demodulation of amplitude modulation it is advisable to use a linear demodulator characteristic with one break point, in order to achieve minimum distortion. This is best approximated by the exponential diode characteristic that becomes straight at high voltages. The principle of the demodulator is shown in *Fig. 7.24*. The modulated signal appearing at the poles of the resonance circuit is distorted by the characteristics of the diode, and it is obtained from the difference between the upper and the lower envelopes. The carrier frequency is removed by the RC network.

For the demodulation of a frequency- or phase-modulated signal, at first an amplitude modulated signal must be generated which can be demodulated by a diode demodulator. The problem is solved by the discriminator circuit preceeded by a limiter stage in order to eliminate harmful amplitude modulation from the frequency-modulated signal.

The DV component of the discriminator is also suitable for the control of an automatic frequency-control circuit. The amplified control signal controls the oscillation-generating frequency in such a way that e.g. for increasing frequency deviation the positive DV appearing in the discriminator diminishes the oscillation-generating frequency. This control system can sustain — with a small error signal

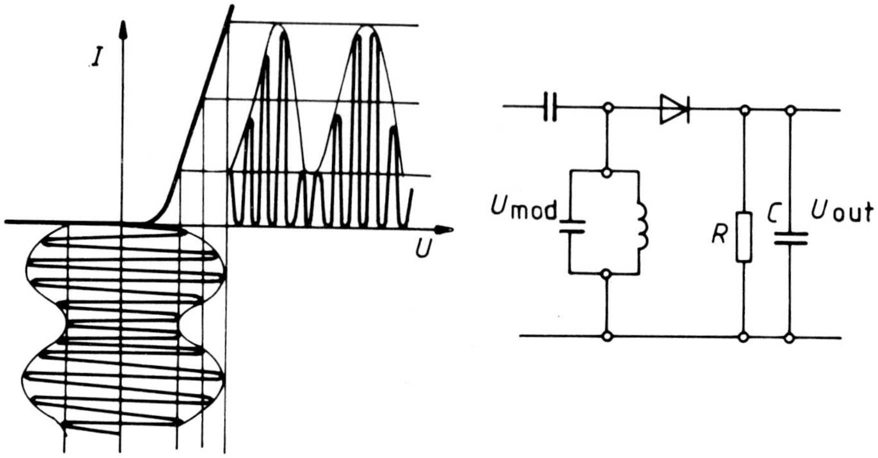

Figure 7.24. Diode demodulator

remaining — e.g. the carrier frequency difference between two microwave distance meter stations at the mean frequency of the discriminator (AFC, automatic frequency control).

7.2.6 Mixers

The aim of mixing is to change a given modulated or unmodulated frequency to another frequency in a phase correct way. Frequency change ensures the possibility of realizing further amplification in an advantageous manner. When mixing, the sum and the difference of two frequencies are produced, as well as their harmonics, from this filter circuits (resonance circuit or band filter) select the required so-called intermediate frequency.

An element of a non-linear (at least second order) characteristic is necessary for mixing.

The most frequently used mixer is the semi-conductor diode. An important requirement for the diode is to have low noise.

7.2.7 Phase-meter Circuits

Phase-meter circuits measure the phase difference of two signals of identical frequency. In general, the circuits used work optimally within a narrow frequency band; in the case of several very different measuring frequencies, the beat-frequency method can be used *(Fig. 7.25)*. A signal of identical frequency always appears at the output of the mixer, hence phase can be measured at optimum conditions. A precondition of the method is the production of a conversion signal f' which differs from the frequency of the signal to be measured exactly by the frequency difference f_m.

Phase-meter circuits can also be made by using a cathode-ray tube. The horizontal and vertical deflector plate pairs of the cathode-ray tube are controlled by the two potentials generated from the reference signal and are perpendicular to each other.

The two potential components are produced by means of a phase shifter consisting of RC elements. The signals shifted by 90°, i.e. the sine and cosine signals, draw a circle on the screen of the tube corresponding to the parameter equations

$$X = R \cos \omega t,$$

$$Y = R \sin \omega t$$

of the circle where R is the radius. The light spot moves around the circle, i.e. through the whole phase angle range of 0° to 360°. The pulses controlled by zero transits of the measuring signal, which differ from the reference voltage by the phase angle to be measured, extinguish the electron beam of the cathode-ray tube, the circle thus being interrupted for a short section. The arc between the zero point of the graduation and the interruption defines the phase angle to be read.

The most well-known phase-meter circuit is the phase-sensitive rectifier; its circuit diagram and the vector diagrams illustrating the operation are presented in *Fig. 7.26*.

Notations $i<0$ and $i>0$ in the figure mean reversal of the current direction. The phase-sensitive rectifier is used for inducing a constant shift of 90°. The phase of the signal to be measured is changed by a calibrated phase shifter until the phase difference is 90°, when the indicator (micro-ammeter) shows zero.

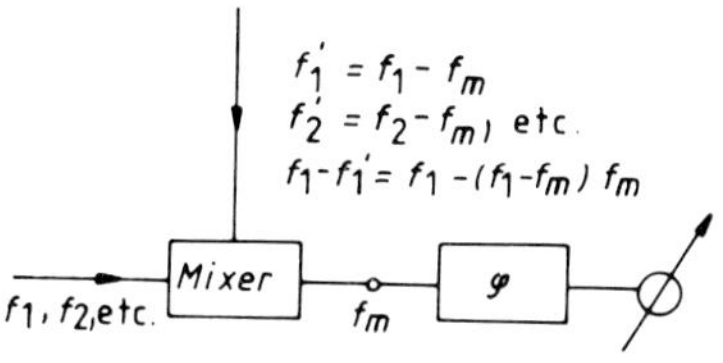

Figure 7.25. Beat frequency method of phase measurement

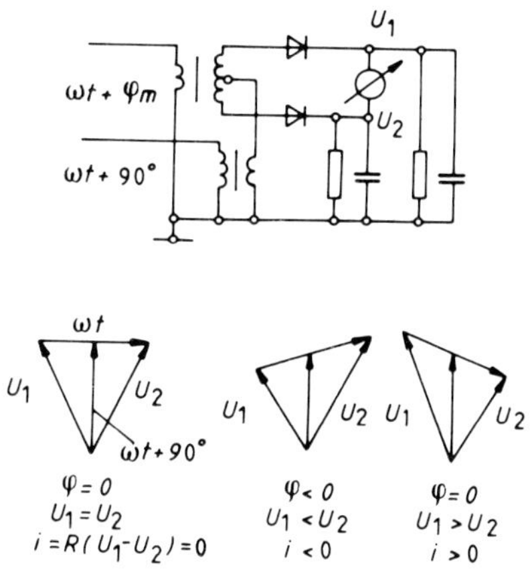

Figure 7.26. Phase-sensitive rectifier

A calibrated phase-shifter can also be made by variable RC elements, but in practice a phase shifter with a resolver is usually used due to the high accuracy required. The resolver consists of a stator and a rotor with a laminated core. The stator consists of two coils perpendicular to each other, the rotor of one coil. The two coils of the stator must be fed by a sine and a cosine voltage respectively, which can be suitably generated from the reference signal by an RC phase-shifter. The alternating flux produced by the coils induces a voltage in the rotor coil. The phase of the induced voltage is determined by the position of the rotor with respect to the stator. The phase angle corresponding to the position of the rotor can be read on a result indicating device coupled mechanically to the rotor. If a proper transmission is used (e.g. on distance meters) the value read can also be the distance.

7.3 Microwave Techniques

7.3.1 Wave Guides

In instruments high frequency electromagnetic waves must be transmitted from one circuit to another with minimum damping and minimum harmful irradiation. This can be attained by coaxial cables or by wave guides. Coaxial cable can transmit all frequencies from zero upwards, but damping increases considerably with increasing frequency.

Such guides emphatically cannot be regarded as connectors with zero impedance, but their capacity and inductance are proportional to their lengths. The wave resistance of the cable is an especially important parameter.

In order to avoid waves being reflected from the ends of the cable, terminating resistances identical with the wave resistance must be connected to both ends. The wave resistance can be calculated from the capacity C and self-induction L for unit length:

$$Z=\sqrt{\frac{L}{C}}, \tag{7.15}$$

where L has dimension H/m and C has F/m.

The propagation velocity of electromagnetic waves along a cable is:

$$v=\frac{1}{\sqrt{LC}}. \tag{7.16}$$

For the centimetre wave-band wave guides of rectangular or circular cross section are used. For low loss transmission the guides must also be matched, because energy reflexion arises at each unmatched point, creating a wave propagating backwards. Forward and backward propagating waves interfere with each other and standing waves occur.

In wave guides the electromagnetic field propagates forward in an axial direction in such a way that the electric and magnetic field vectors thereby take forms of complicated symmetry, depending on the geometry and wavelength of the guide. They are called modes.

7.3.2 Microwave Integrated Circuits

Microwave integrated circuits are not yet made of solid semi-conducting single crystal chips, but are ceramic plates of surface area one to two cm^2 which carry the laminated circuit elements and the microwave transistors. The circuits have the functions of amplifying, mixing, demodulating, phase-measuring, matching, etc.; the

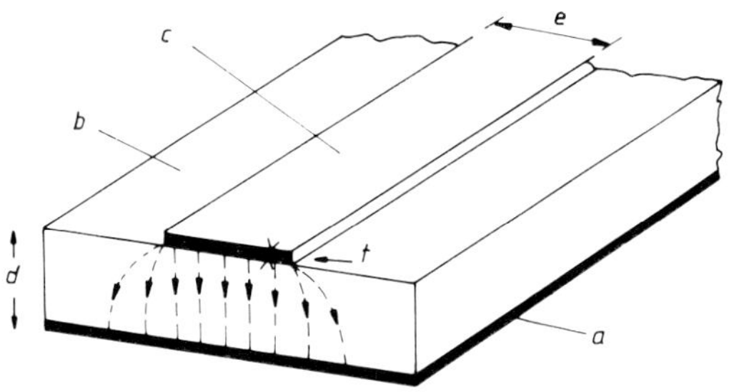

Figure 7.27. Microstrip wave-conductor; a = metal foil; b = insulator ($d \approx 1$-2 mm); c = metal band ($t \approx 0.1$ mm, $e \approx 2$ mm)

connection between the different elements is realized by small microstrips *(Fig. 7.27)*. The frequency-dependent circuit elements are sometimes microstrip stubs of defined length, representing a certain imaginary impedance.

7.3.3 Antennas

Electromagnetic waves can propagate in free space or by conduction. Conversion between the two propagation forms is made by antennas.

The main characteristics of antennas are as follows.

The gain G of an antenna is the measure of its radiation directing ability, i.e. the ratio between the power P_1, passing through a unit surface at a distance far enough from the antenna in the main direction of the radiation, and the power density P_0 of the "isotropic antenna":

$$G_t = \frac{P_1}{P_0}. \tag{7.17}$$

The isotropic antenna is an ideal concept without physical reality. It radiates in all directions with the same energy. The power density P_0 of an isotropic antenna at distance r can be obtained by dividing the radiated power P_r by the surface of the sphere with radius r:

$$P_0 = \frac{P_r}{4\pi r^2}. \tag{7.18}$$

The effective area of a receiving antenna is the ratio between the power P_v, transmitted from the antenna into a perfectly matched load, and the power P_a of the wave falling on a unit area of the antenna. If the antenna is in such a position with respect to the plane of the incident wave that maximum energy is received, then:

$$S_T = \frac{P_v}{P_a}. \tag{7.19}$$

The effective surface of the antenna depends on the real area A of the antenna aperture:

$$S = \alpha A, \tag{7.20}$$

where the factor $\alpha<1$ is the effectiveness of the antenna and expresses the fact that the illumination of the antenna aperture is not uniform. In the most advantageous cases $\alpha=0.6$ to 0.7.

The areal aperture β is the angle between the direction of the point at which the gain function is a maximum and the direction of the point at which the power is decreased by a factor of two.

The radiation resistance R_r is the ratio between the emitted power and the square of the effective value of antenna current.

In the case of microwaves strongly bunched beams and high gain can be provided with acceptable geometric dimensions. The most important antenna types used in practice are the horn antenna and the reflector antenna illuminated by a horn.

A horn antenna is made by the continuous widening of a wave guide. The most frequently used reflecting surface is the parabolic mirror *(Fig. 7.28)* which is made by analogy with optics.

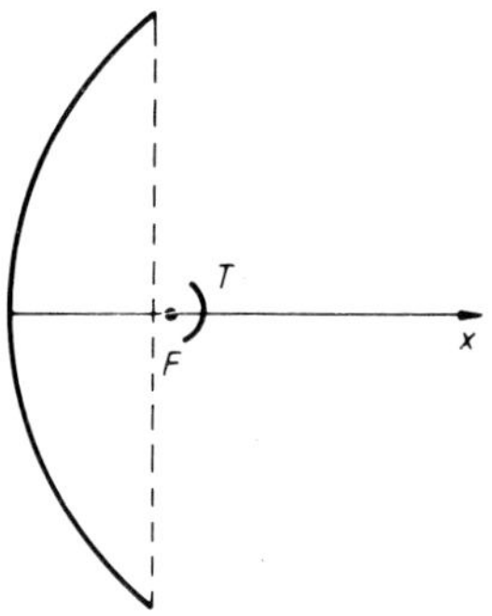

Figure 7.28. Parabolic antenna

7.4 Digital Techniques

Digital techniques (more exactly, binary-digital techniques) means that the signal only distinguishes between two definitely separated states (yes — no, 0 — 1, etc.). In electronic digital circuits a nearly zero level, and a well distinguishable other level, are produced in a voltage or in another electrical quantity. The two positions of the switch therefore produce a digital signal.

Digital signals are less sensitive to noise and distortion than analogue signals. Their application in electronics has increased dramatically since cheap, simple and fast semi-conductor switches have been available.

A simple digital signal (namely that, at a given point and at a given instant, the current, voltage or charge in question is approximately either zero or one) yields only one element of information (1 bit). However, any number can be expressed by means of an (encoded) set in a defined order of sequence. A digital signal containing complicated information is the encoded set of pulses of voltage or current *(Fig. 7.29)*.

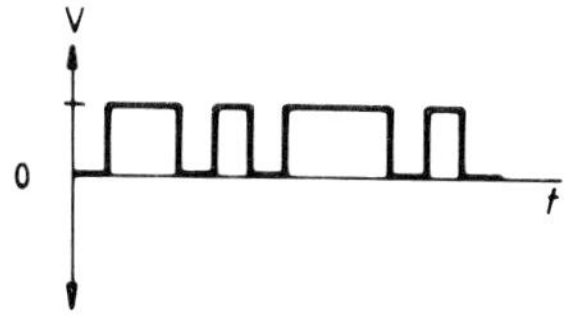

Figure 7.29. Pulse sequence

It is well known, for example, how to express any number by digital signals i.e. by sets of bits on the basis of the binary system (see *Table 7.2*). This number can also be returned to the decimal system and, in *Table 7.2*, four bits are seen to be necessary: this is the BCD code.

Table 7.2. The binary system

0	0000	6	0110	11	1011
1	0001	7	0111	12	1100
2	0010	8	1000	.	
3	0011	9	1001	.	
4	0100	10	1010	.	
5	0101				

There are also other codes using 4 bits for each number which may be more advantageous with respect to current circuit realization or for avoidance of error. There are codes which use five bits for a decimal number. (E.g. there is a fifth, so-called parity, bit added to the four bits which makes the number of ones sum to an even number.) This helps automatic error recognition.

7.4.1 Digital Logic

Digital signals can be combined in different ways, not only by simply putting them into a sequence. There are special rules for these operations, collectively called digital logic, and electronic realizations of the different operations are called logic circuits (gates).

The most simple logic circuit, the inverter, carries out the operation of negation: if the input signal is 0, 1 appears at the output; if the input signal is 1, the output signal is 0. It can be realized by means of an elementary transistor and some resistors: in *Fig. 7.30* the bipolar transistor operates as a controlled switch. The emitter is at earth potential. If the base voltage is high ($U_{in} > 0.8$ V), the sector emitter-collector operates as a short circuit, the output point (the collector) is at a very low potential and it corresponds to the zero level. If the base voltage is low ($U_{in} < 0.6$ V), there is almost no current flow in the transistor, the collector potential is almost identical with the supply voltage and this corresponds to the level 1.

The other logic circuits have more than one input. In the AND circuit, signal 1 appears on the output only if signal 1 is on all inputs. In the OR circuit, the signal 0 appears on the output only if 0 is on each input. The operation of the NAND

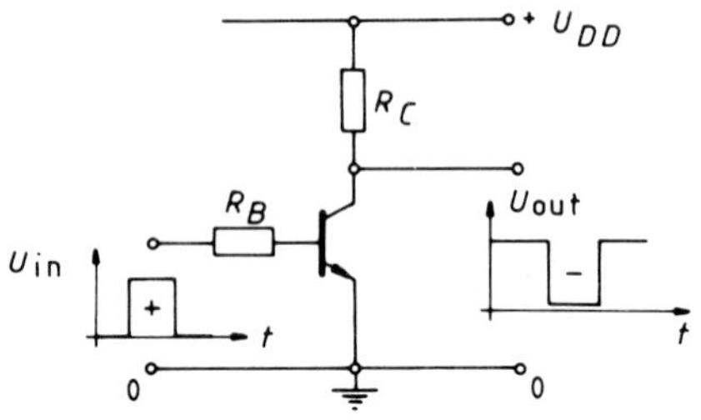

Figure 7.30. Inverter gate using transistors

circuit can be regarded as if an inverter had been connected after an AND circuit. The operation of the NOR circuit can be explained similarly. *Table 7.3* presents these relations and the usual symbols of logic circuits.

Modern techniques also enable NAND and NOR circuits to be developed in a more simple way than transistors *(Fig. 7.31)*. Today, however, more complex logic circuits are available in the form of integrated circuits as elements for more complicated tasks. Such elements can easily be unified in a system if basic elements working on the same principle are used. The most commonly known current circuit systems are:

— the systems called *TTL*, I^2L and very rapid *ECL*, which contain only bipolar transistors and resistors;

— systems containing *MOS* transistors alone, in which are present only elements with an *N*-channel or elements with both *N*- and *P*-channels, the latter having characteristics which mirror each other symmetrically. The latter system is the *CMOS*

Table 7.3. Logic functions

	Input 1	Input 2	Output
AND	0	0	0
	0	1	0
	1	0	0
	1	1	1
OR	0	0	0
	0	1	1
	1	0	1
	1	1	1
NAND	0	0	1
	0	1	1
	1	0	1
	1	1	0
NOR	0	0	1
	0	1	0
	1	0	0
	1	1	0

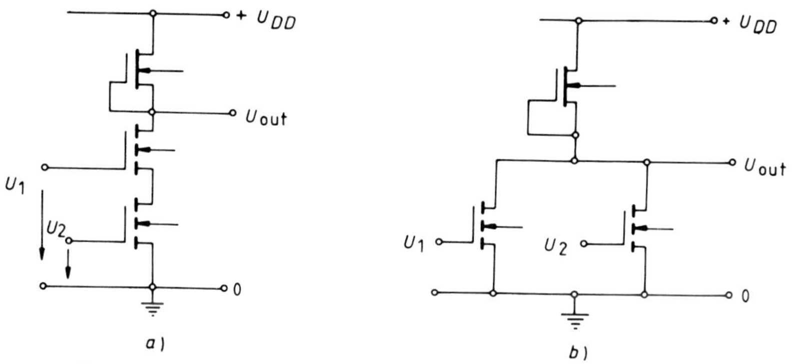

Figure 7.31. Gated NAND *a)* and NOR *b)* using MOS transistors

logic. Newly developed *MOS* circuits are compatible, i.e. they can be combined with *TTL* circuits.

The central processor units of computers consist of a very large number of logic circuits. In big computers a control unit and programs belong naturally to the system together with several different memory units for data storage. The connection to the user and the outward connection are maintained by peripheral output and input units.

Technical development has produced the microprocessor, i.e. the central unit integrated onto one single semi-conductor chip. The microprocessor was at first applied in cases where the computer was used for some special task, e.g. for process control. Later a new generation of microcomputers was developed from the cheap microprocessors by combining them with further similarly cheap units. The micro-computer gives the main services of a complete computer for an insignificant fraction of the price a decade earlier.

7.4.2 Memory Circuits

The flip-flop is a digital circuit with two inputs in which the output can be brought steadily to the level 0 or 1 by short 1 signals or by pulses fed to the inputs. The most simple example is the *R–S* flip-flop which can be imagined as two inverters with positive feed-back to each other. The connections between the setting input *S* and the resetting input *R*, also the operation output *Q* (and the inverting output $\bar{Q}$ working always in the opposite sense to *Q*) are listed in *Table 7.4*.

Table 7.4. ***R–S*** **flip-flop**

R	*S*	*Q*	$\bar{Q}$
0	0	Previous state	
0	1	1	0
1	0	0	1
1	1	not allowed	

Flip-flops can store 1 bit of information and this is just one way of using it.

In semi-conducting memories the multitude of elementary cells are placed side by side. The information can be read out and rewritten several times.

7.4.3 Counting Circuits

The flip-flop can also be used in counting units *(Fig. 7.32)*. It counts the number of pulses in a set. The flip-flop is supplemented by further elements in such a way that the state of the output *Q*, can only change if a change of state of a definite direction (either from 1 to 0 or from 0 to 1) takes place at the input *T*. One output is connected with the input *T* of the next flip-flop, hence this takes notice of every second pulse.

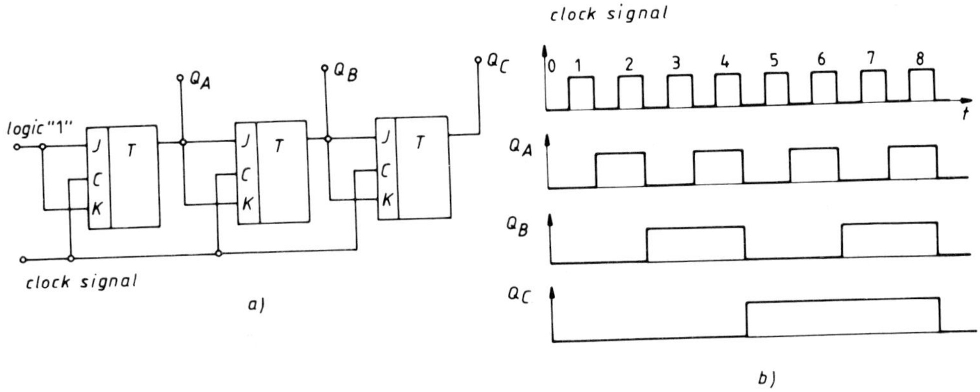

Figure 7.32. Two types of symbols for a counter element

The result obtained, the series of subsequent outputs, gives the number of pulses. As has been shown, four subsequent stages are sufficient for the binary display of the ten figures 0 to 9; for the display of the ten figures of the decimal a further four circuits are needed, and so on. The outputs of a set of four flip-flops always give one decimal in an increasing sense.

The decoding required for the display can be made simply so that, from the four outputs *Q*, those having the value “1” at a certain decimal according to the code transmit this level to the inputs of an AND circuit. If “1” appears on the output of this circuit, a further electronic circuit illuminates the figure in question according to *Table 7.2.* Naturally the logic circuits OR and NAND, or NOR can also be used with corresponding modification.

The connection further enables the frequency of a periodic series of pulses to be divided by a certain power of 2.

The counter can also be supplemented in such a way that each new pulse decreases by one the quantity read, i.e. it counts down. Devices which can be set for counting up or down, by changing the auxiliary voltage level between 1 and 0, have practical importance.

7.4.4 Analogue-digital Conversion

The essence of the converter is the comparison of a measured analogue value with a quantity expressed by a quantized value, i.e. a digitized number.

The circuit for the comparison is called a comparator. It is essentially a high gain amplifier, the voltage to be determined enters the inverting (−) input and the reference voltage the other (+) input. Due to the high gain the amplifier is already overmodulated for small differences: it works as a switch and yields a digital signal. The principle of the parallel-working analogue-digital converter is shown in *Fig. 7.33*. Its disadvantage is that as many as 100 comparator circuits are needed to give an accuracy of 1 percent.

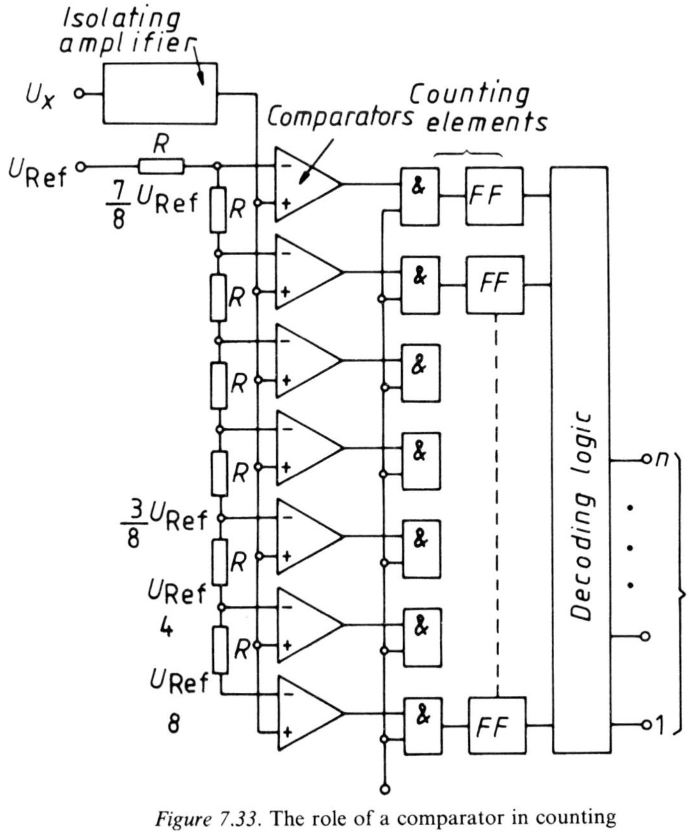

Figure 7.33. The role of a comparator in counting

Figure 7.34 shows a simpler solution. The reference signal set to 127 mV in the example enters a binary decreased attenuator. The output signal of the comparator controls the switches through a controller. They are at first in short circuit. In the first cycle the highest resistance (64 R) becomes free. If the input signal is less than 64 mV, the "0" signal appearing on the output of the comparator brings the switch to its off-position. In the next cycle the switch of the member 32 R comes to its

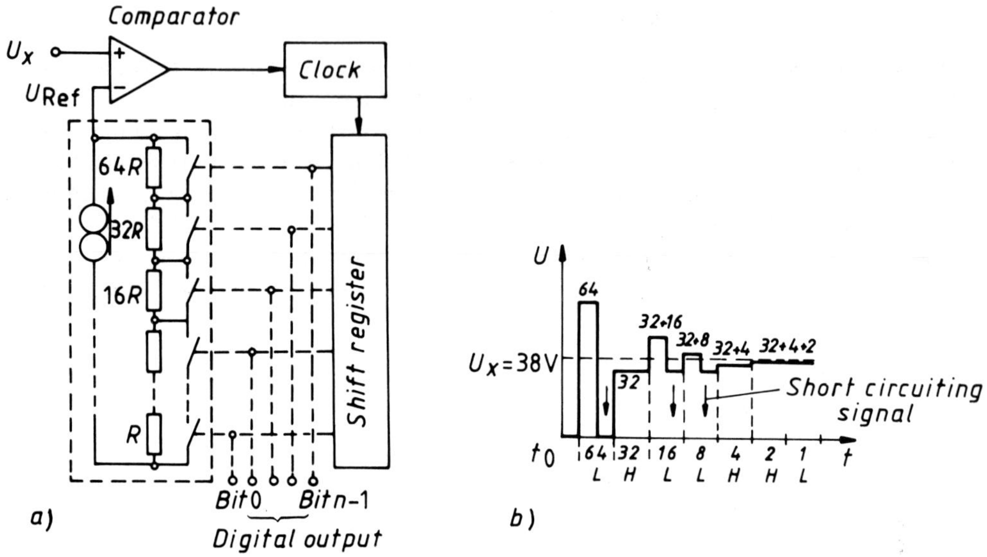

Figure 7.34. Series Analogue/Digital converter with attenuator

on-position. If the input signal is greater than 32 mV, the signal "1" appears on the output of the comparator, which leaves the former switch in its on-position and adds 16 mV to the reference value by bringing the next switch into its on-position. If the input signal is less than 48 mV, the "0" signal of the comparator switches on the 8 mV section instead of the 16 mV, etc.

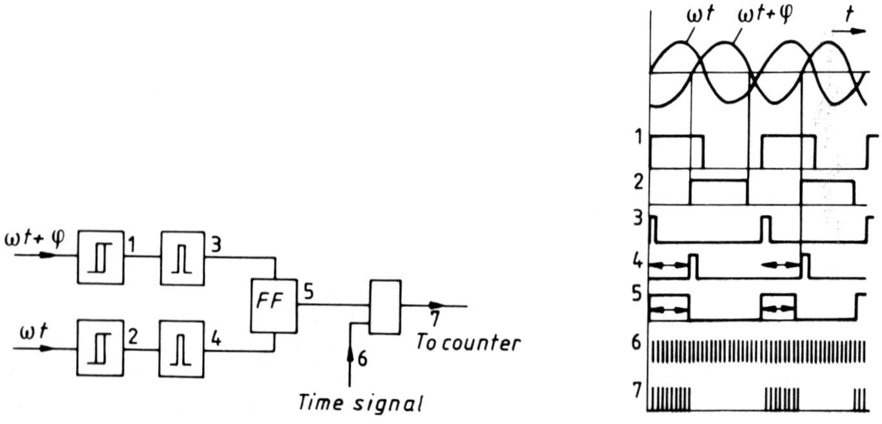

Figure 7.35. Digital phase-meter

7.4.5 Digital Phase-meter Circuits

The operation of a digital phase-meter is explained in *Fig. 7.35*. The signal to be measured and the reference signal each come to a squaring circuit *(1, 2)* and to a pulse-generating circuit *(3, 4)*; they are equal for both signals. The pulse containing the phase information operates a bistable flip-flop circuit *(5)*; the latter produces a gate signal *(7)* with a width equal to the phase-difference. The gate circuit transmits the time-measuring signal *(6)* while the gate is open, and the counter counts the time signals.

7.5 Electro-optics

Light is an electromagnetic radiation of very short wavelength, i.e. very high frequency. The human eye is sensitive to a very narrow frequency band (from about 4 to 8×10^5 GHz) *(Fig. 7.36)*, but the adjacent radiation bands some octaves wide, i.e. the lower frequency infrared and the higher frequency ultraviolet radiations, are also called light (see Section 1.1).

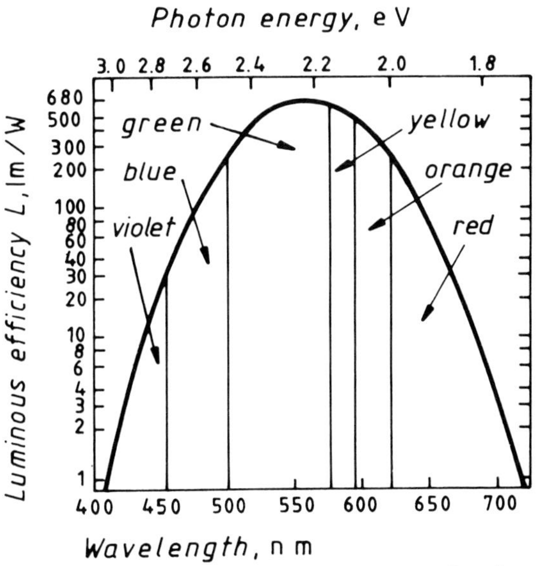

Figure 7.36. Light sensitivity of the human eye as a function of the wavelength

The intensity of radiation is the power falling on a unit of surface, with dimension W/m^2.

Independently of intensity, the sensation of light in the eyes is the basis of the measurement of the flux of visible light. The unit of light flux is the lumen (lm). At the frequency where the eye is most sensitive, a monochromatic radiation of power 1 W corresponds to a light flux of 680 lm. The light intensity is the light flux falling on a unit surface.

7.5.1. Electronic Light Sensors

The photo-electric phenomenon is the fact that during light-absorption individual electrons are able to take up energy quantums $h\nu$ of the light (h is Planck constant and ν the frequency of the light wave). In the interior of semi-conductors the absorbed light quantum (photon) frees a bound electron, i.e. a hole electron pair is produced, if:

$$h\nu \geqq W_g, \tag{7.21}$$

where W_g is the difference between the energy of the bound valence electrons and the free conducting electrons, i.e. the width of the so-called forbidden band.

Materials of high W_g are sensitive only to ultraviolet light or light of a shorter wavelength (higher $\nu = \frac{c}{\lambda}$), but silicon is sensitive to the whole visible spectrum and also the neighboring infrared.

Photoresistance. The new charge carriers produced decrease the measurable resistance between the two metal contacts of a semi-conductor crystal chip. To a first rough approximation, with a given external voltage the current increases linearly with increase of light intensity, but the proportion is a function of frequency. It follows from the equation that this would be so even if all photons produced a hole electron pair, but in reality the so-called quantum efficiency is considerably less than 1 due to reflection generating no absorption, etc.

The semi-conducting photoresistor working in such a way has also, of course, its original conductance; this is the origin of the so-called dark current. The difference between the actual current and the dark current does not exactly follow the variations of light intensity for different reasons. Non-linear phenomena (distortions) also occur, and the operation is slow, i.e. it cannot follow rapidly changing light signals.

Photodiode. The construction of the photodiode enables the entering photon to generate a hole electron-pair in the neighbourhood of the *PN*-junction if Eq. (7.21) holds. Between the two sides of the junction the field generated by the field-charge layers (which is even amplified in the case of off-voltage) separates the two charge carriers. Holes come to the more negative *P*-side and electrons to the more positive *N*-side. The generated photo-current flows therefore in the off-direction. At a given wavelength it follows the light intensity and its variations linearly, almost without inertia. If the electrodes of an illuminated photodiode are closed by a cut-off (or as approximation the electrodes are connected to the poles of a resistor of very high resistance without any external voltage), then holes are accumulated on the *N*-side and electrons on the *P*-side, and a measurable potential difference arises. Here the physical basic connections are similar to the idealized equation of a *PN*-junction. The generated voltage is proportional to the logarithm of light intensity. (Hence there are two possibilities for measurement: current produces a linear scale, voltage a logarithmic scale.) An illuminated photodiode has a short-circuit current and a no-load voltage, therefore it can be used as an electric energy source (solar cell).

Several special arrangements have been developed for increasing light sensitivity. In the PIN diode there is a pure (not doped) layer, which improves the photon efficiency.

Light sensors with internal amplification. The avalanche photodiode uses the amplifying phenomenon that, with a high field strength, some of the charge carriers attain high energies and can free a further hole-electron pair by collision with the atomic lattice. By an avalanche-like increase even 100 times greater currents can be produced, but the transfer of rapid variations is somewhat worse than with the PIN diode.

The phototransistor is also an amplifier element; it is essentially a bipolar transistor, as described previously, in which the base is mainly illuminated by the external light. In case of $h\nu \geqq W_g$ the entering photons increase the number of holes and electrons. The effect is the same as if the base current were increased. Phototransistors are seldom used.

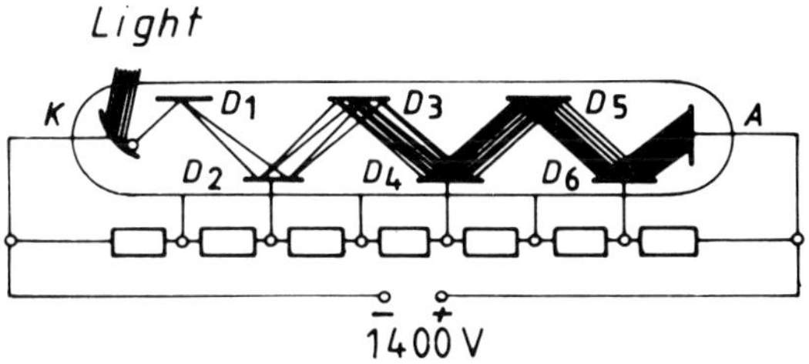

Figure 7.37. Photoelectron multiplier

A vacuum device (special electron tube), the photoelectron-multiplier *(Fig. 7.37)*, is also used for the detection of weak light signals. Its working principle is based on photoemission and secondary emission.

Certain materials are able to emit electrons into free space as a result of being hit by electrons of high enough energy (>2 eV). The photocathode in the multiplier is made from such a material.

Due to the effect of light, emission electrons leave the cathode. With a proper array of electrodes, and the use of screening plates, the accelerated electrons hit the D electrode (dinode) with high velocities and make free secondary electrons. One accelerated electron causes the separation of more electrons. If the number of dinodes is increased and the secondary emission is used several times, an easily measurable current appears at the anode. The value of this anode current can be obtained from the following connection:

$$I_a = I_k \sigma^n, \tag{7.22}$$

where I_k = the primary cathode current,
σ = the secondary emission factor depending on the material and on the potential difference of the adjacent dinode (its usual value is 2 to 4),
n = the number of dinodes (8 to 13 in practice).

Thus a photoelectron multiplier carries out both light detection and an internal amplification of 10^4–10^6.

The photoelectron multiplier is used primarily for the detection of weak light energies (e.g. for distance meters using light) and to carry out special measurement tasks.

7.5.2 Incoherent Light Sources

Light emitting diodes. If on-directed current flows through a *PN*-junction, the vast majority of the charge carriers which passed through the junction are recombined within very short distances. Recombination means in a physical sense that "free" electrons become bound in the semi-conductor. During this process the energy of the electrons decreases by a value corresponding to the width of the forbidden band. The released energy is transformed into heat or electromagnetic radiation. In the latter case the energy leaves in the form of photons; due to the condition (7.21), the wavelength is:

$$\lambda \approx \frac{hc}{W_g}, \tag{7.23}$$

where c is the velocity of light.

It depends on the band structure of the semi-conducting material whether the radiating or the non-radiating mechanism dominates in the recombination process. In silicon, for example, the probability of non-radiating recombination is very high. The most well-known material in which radiation recombination dominates is gallium arsenide (GaAs). Since the width of the forbidden band is 1.4 eV, the wavelength of the emitted light is 0.89 μm, i.e. it is an invisible infrared light. (Human eyes perceive light between 0.4 and 0.75 μm.)

The light of a light emitting diode is incoherent. The light spectrum has a band-width of some 10 nm. The radiated power is some percent of the DC power coming from the on-direction voltage source. The transients of the on and off states of the GaAs diode are very short (of the order of some ns-s), therefore this diode is very suitable for digital signal transfer. Within wide limits the number of emitted photons is proportional to the current, thus analogue signal transfer is also easily realizable.

Light-emitting diodes with visible light. In order to emit visible light the diode must be made from a semi-conductor, the forbidden band of which is at least 1.6 eV wide, but for the emission of the yellowish green light ($\lambda \approx 0.55$ μm) for which the human eye is the most sensitive, a forbidden band at least 2.4 eV wide is necessary.

The former conditions are satisfied by the gallium–arsenide–phosphide diode composed of three materials; this diode gives a well visible red light. Yellowish green light can be obtained from a properly doped gallium–phosphide diode; with different doping this diode also gives red light. (The efficiency of the radiating recombination process, however, is better for red light than for yellowish green. This roughly compensates for the difference in eye sensitivity.)

LED-s with visible light are usually used in the digital displays of instruments.

Digital displays. In illuminated digital displays all of the seven segments are light--emitting diodes (made mostly of gallium–arsenide–phosphide with red light). The array is presented in *Fig. 7.38*. Depending on the different numbers, the output of the counter gives a positive voltage of about 1.5 V to the *P*-side of the diodes forming the corresponding segments. The *N*-sides are at earth potential.

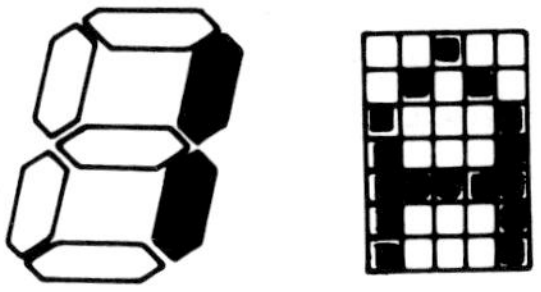

Figure 7.38. Alphanumerical display with 7 segments and 35 picture points

For the display of letters, punctuation and numbers, the more complicated alphanumerical displays are suitable; they always contain seven lighting points in each of five columns (diode matrix).

The display device in some instruments is a gas-discharge lamp. The different numbers are electrodes supplied with negative voltage. The light source is the glow-discharge surrounding the electrode. The discharge requires, however, a voltage of at least about 50 V.

A special cathode-ray tube, which is similar to a TV-screen, can also serve as a display. The deflecting signals of the electron beam come from an encoding device: more data are displayed on the screen at any one time.

The nematic liquid-crystal cell with 7 segments is a very frequently used display. It is not a light source but a passive device. The liquid crystal is an organic liquid whose molecules are ordered in the rest position so that the liquid is transparent. This order is disintegrated by the effect of an electric field and the optical characteristics (reflexion, refraction) change. The liquid crystal cell has an inertia of the order 0.1 s, its power supply is a maximum of some μW.

7.5.3 Coherent Light Sources (Lasers)

In natural light sources photon emission is a spontaneous, statistical process. The phase of oscillation changes rapidly and irregularly in time and the frequency is not quite constant.

As has been discussed in detail in Subsection 1.5.5, there exists an induced emission as well as the spontaneous emission, which means that the rise of one photon brings about the rise of another. In this case the phase and frequency of the emitted oscillations remain constant. Under natural conditions the induced emission is so weak that it cannot be detected. It becomes considerably high when the energy distribution of the electrons is reversed: more electrons are on a higher energy level than on a lower level. (Thermodynamical equilibrium prescribes the opposite of this.)

It is also necessary that the photons already emitted pass through the light source device several times, i.e. that the light waves leave only after having been reflected within it several times. Such a light source is constructed as an optical resonator surrounded by semi-transparent mirror surfaces.

Semi-conducting lasers. Reversed energy distribution can arise at a highly doped *PN*-junction with a high on-direction current. On the one hand the number of electrons at the energy level corresponding to the top of the valency band is low due to the large number of holes, on the other hand a large number of electrons come to the *P*-side to the bottom of the conducting band due to the strong current, and that means an energy higher by W_g *(Fig. 7.39)*.

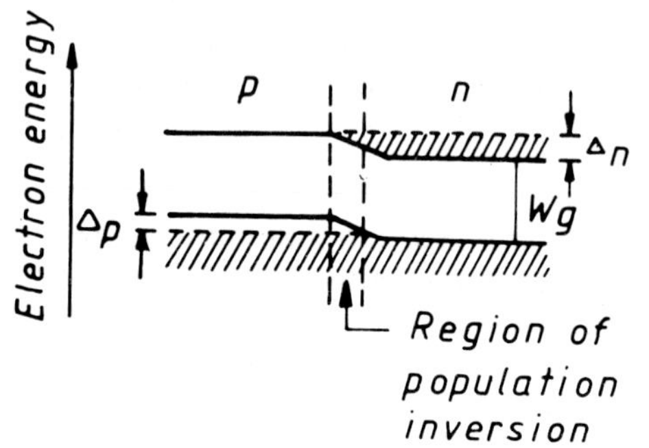

Figure 7.39. Inversion stage of the semi-conductor laser

An optical resonator is also necessary for semi-conductor lasers, but here the resonator is the crystal itself, made by orienting and splitting a larger crystal. The refraction coefficient of semi-conductors is very high (that of gallium arsenide is $n=3.6$), therefore there is such a big jump at the boundary surface air-semi-conductor, and so much light will be reflected, that the positive feed-back and hence the agitation of laser oscillation is ensured. Losses occur due to material faults, therefore much cooling is necessary.

The basic material of the *PN*-junction of a semi-conductor laser is gallium arsenide, which emits infrared light at $\lambda = 0.89$ nm. The construction is much more complicated than that of simple light-emitting diodes. The diode works at first as a light-emitting diode but, by increasing the current beyond a certain limit, there is a sudden increase of light intensity: this is the beginning of laser operation.

Other solid lasers. Some specially doped glasses and insulating crystals can also work as lasers. They mostly emit short light pulses of high intensity. To reverse the energy distribution, photons of higher energy than that of their own emission must be provided from an external source, e.g. from a flash light. With present techniques extremely short ($t < 10^{-8}$ s) "giant pulses" can be produced by such lasers.

Gas-lasers are described in Subsection 1.5.5.

7.5.4 Light Modulation

The emission of light diodes can be modulated in intensity by changing the current flowing. This is valid to a higher degree for semi-conductor lasers. The modulating signal can be not only a continuous time function but also a digital signal, i.e. a pulse or a sequence of pulses.

In continuously operated gas-lasers the modulator must be placed in the path of the light beam. The modulator is an optical cell, whose light transmission can be controlled electronically. In the cell there is a transparent material in which the electro-optical effect occurs. The main point is that the polarized light is split by the electric field into two components. The Pockel effect occurring in certain crystals, e.g. in potassium-dihydro-phosphate (KDP), is proportional to the field; the Kerr effect occurring in nitrobenzene (liquid) is proportional to the square of the field. The modulator consists of a polarized filter (if the light is not polarized originally), or a material with an electro-optical effect placed between two electrodes, and the analyser (analysing filter).

The inertia of the electro-optical modulator is very low. It can be used up to a frequency of several hundred MHz with a suitable electrode system.

A mechanically oscillating crystal, or a liquid with ultrasound standing waves, can also work as a modulator because density and refractive index vary periodically within it.

Sometimes giant light pulses repeated at greater time intervals may be necessary for measurement purposes. They can be generated so that the loss or the Q factor of the optical resonator of the laser is controlled electronically; the energy at the higher level is collected during a longer time interval and the energy drop to the lower level, together with the induced light emission, takes place within a very short time interval.

The demodulation of intensity-modulated light is not discussed here; the change of photocurrent produced in the light sensors, described in Subsection 7.5.1, follows the variation in the number of photons received and thus yields a demodulated signal.

Frequency or phase modulation of light is not used in present measurement techniques.

8 Distance Measuring Instruments Using the Principle of Interference

Interference is the consequence of uniting two wave-fronts (Subsection 1.1.3). In the simplified case of interference, the amplitude and frequency of the two wave-fronts are equal; in this case only the two phase-fronts differ, namely:

$$a_1 = A \cos 2\pi \left(f \cdot t - \frac{x}{\lambda} \right),$$

$$a_2 = A \cos 2\pi \left(f \cdot t - \frac{x - \varphi}{\lambda} \right), \tag{8.1}$$

where a_1 and a_2 are the instantaneous characteristics (magnitude) of the two wave-fronts, A is the common amplitude of the two oscillations, f the common frequency, t the propagation time, x the path corresponding to the instantaneous phase, λ the wavelength and φ the phase difference between the two oscillations.

The resulting oscillation has the form:

$$a = a_1 + a_2 = 2A \cos \frac{\pi\varphi}{\lambda} \cos \left[2\pi \left(f \cdot t - \frac{x}{\lambda} \right) + \frac{\pi\varphi}{\lambda} \right]. \tag{8.2}$$

It can be seen that the frequency of the oscillation produced by the interference is the same as that of the components a_1 and a_2, but the phase and instantaneous magnitude of the resulting oscillation depends on the phase difference of the components. If the phase difference between the component oscillations is zero, the amplitude of the resulting oscillation is twice the amplitude of the components, and if the phase difference between the components is 180°, the two oscillations extinguish each other.

If the two oscillation components have the same amplitude, but differ in frequency (in which case the wavelengths are also different), or in equations:

$$a_1 = A \cos 2\pi \left(f_1 t - \frac{x}{\lambda_1} \right),$$

$$a_2 = A \cos 2\pi \left(f_2 t - \frac{x}{\lambda_2} \right), \tag{8.3}$$

the result of the interference is much more complicated. The resulting frequency is in the general case:

$$a = 2A \cos 2\pi \left(\frac{f_1 + f_2}{2} t - \frac{\lambda_1 + \lambda_2}{2\lambda_1\lambda_2} x \right) \cos 2\pi \left(\frac{f_1 - f_2}{2} t - \frac{\lambda_1 - \lambda_2}{2\lambda_1\lambda_2} x \right). \tag{8.4}$$

If the frequencies of the two components are nearly equal, i.e. $f_1 \sim f_2$ and correspondingly $\lambda_1 \sim \lambda_2$ also, Eq. (8.4) is reduced to:

$$a = 2A \cos 2\pi \left(f_1 t - \frac{x}{\lambda_1} \right) \cos 2\pi \left(\frac{\Delta f}{2} t - \frac{\Delta\lambda}{2\lambda_1^2} x \right). \tag{8.5}$$

The resulting oscillation of Eq. (8.5) is regarded as an oscillation whose frequency is identical with the frequency f_1 of the basic oscillation, but whose amplitude varies between zero and $2A$. The wavelength of the so-called beat between the equal amplitudes of the resultant is:

$$\lambda = \frac{\lambda_1 \lambda_2}{\lambda_1 - \lambda_2} \simeq \frac{\lambda_1^2}{\Delta\lambda}. \tag{8.6}$$

The frequency of the beat is equal to the frequency difference Δf.

8.1 Distance Measurement by means of the Interference of Light Waves of the Same Frequency

The basic principle of this distance measurement method is that distance is measured as the path difference corresponding to the phase difference of two light beams of the same frequency interfering with each other, i.e.:

$$D = \frac{N\lambda_0}{2n}, \tag{8.7}$$

where N = the order of interference,
λ_0 = the wavelength of the light in vacuo,
n = the phase refractivity of the light of wavelength λ_0 (Subsection 9.3.1).

For absolute distance measurement the length unit is the wavelength of light; the distance expressed in units of wavelength is later converted into accepted units of length.

The longest distance measurable by absolute measurement depends on the coherence time of the light source. Coherence time is the time interval within which radiations emitted from two different points of the light source remain able to interfere. The coherence time of non-laser type light sources is about 10^{-8} s, the longest distance measurable is 3 m; the coherence time of lasers (Subsection 1.5.5) is 10^{-3} s, the theoretically measurable longest distance is consequently 300 km. In practice, however, the distance measurable in average conditions is only about 300 m due to the fluctuation of the atmosphere.

The equality of frequency of interfering light waves can be ensured by splitting the light beam into two and re-unifying the two beams.

The international definition of the metre was 1,650,763.73 times the wavelength in vacuo of the radiation emitted by a krypton atom at the transition of its electrons from the path of energy level $5d_5$ to the path of energy level $2p_{10}$. The reliability of this definition is about 10^{-8} to 10^{-9}, the previous definition having had a reliability of 2×10^{-7}; furthermore the so-called secondary unit of length is relatively easily reproducible, since light wavelengths can be compared easily and with a high accuracy using a laser light source.*

The Michelson array has become the most frequently used method of using the interference of light sources of the same frequency. A block diagram of this array is shown in *Fig. 8.1*. The light from a laser *1* is split into two beams by the half-silvered mirror *2*. The reflected beam is reflected by the mirror *3* which is perpendicular to the direction of the displacement to be measured; it passes through the mirror *2* and meets the receiver *4*. The two positions of the mirror *5* represent the two end points of the distance to be measured. The mirror can be moved between its two extreme positions by means of a precision device.

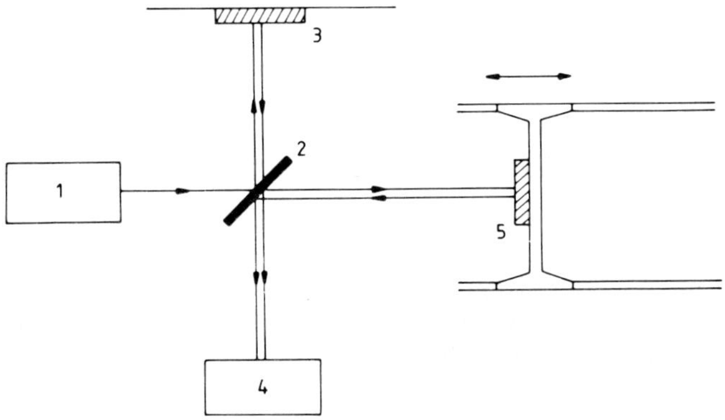

Figure 8.1. Block diagram of the Michelson asymmetric interferometer: *1* laser; *2* half-silvered mirror; *3* fixed mirror; *4* receiver; *5* movable (measuring) mirror

As result of the interference of the separated beams an interference pattern, consisting of bright and dark linear or concentric fringes, is generated in the focal plane of the optical system of the receiver. By moving the mirror *5* in one direction one can observe the continuous shift of the interference fringes, linearly or radially. If by means of a diaphragm only a small part of the interference pattern is made visible, the luminosity of the field of view changes periodically (by the motion of the mirror *5*). The most intense luminosity corresponds to the path difference $N\lambda$, the minimum luminosity to the path difference $(2N+1)\cdot\lambda/2$. Hence if, by some means, the displacement of the mirror is determined between its extreme position and the adjacent position of maximum luminosity, the fractional part of the order of interference is thereby determined. (It is assumed in this case that, when the mirror is at the other end of the distance to be measured, maximum luminosity has already been ensured in some manner.) The integer part of the order is generally determined by counting the number of luminosity changes.

* Following the more accurate determination of the velocity of light in vacuo ($c = 299{,}792.458 \pm 0.001$ km/s), the General Commission of Weights and Measures accepted in a resolution in 1983 a new definition of the metre, according to which 1 metre is the distance travelled by light in vacuo during 1/299,792.458 s. The new definition does not change the length of the metre determined on the basis of the krypton definition.

Distance meters working on the basis of interference (interferometers) require as a basic condition a light source of maximum possible frequency stability. Laser equipment on the market have an unsatisfactory frequency stability and the frequency of laser light must be kept constant by means of an external stabilizer. The principle of stabilization is negative feed-back: the instantaneous difference of the frequency of laser light from the required value is transformed into a proportional error signal. The error signal varies the distance between the resonator mirrors of the laser in the direction and by the amount necessary. The frequency deviation is determined by the use of an external frequency standard (a laser with a stabilized frequency) or by the comparison of certain components of the laser beam, e.g. by comparing the light fluxes of two radiation modes. The error signal compensates for the change of distance between the mirrors caused by a change of temperature or atmospheric pressure, either by means of the inverse piezoelectric effect (one resonator mirror is fixed on a small piezoelectric plate) or by the change of the exciting voltage (in this case the temperature of the active material changes, changing the temperature expansion of the resonator). Permanent frequency stability of 10^{-9} to 10^{-10} can be attained by piezoelectric stabilization, and of 10^{-7} to 10^{-8} by voltage stabilization.

The path of the movable mirror must be such that the reflected light beam remains unchanged in position along the whole length of the path. The construction of a correspondingly uniform path of motion does not cause any difficulty if triplet prisms are used as reflectors (Subsection 1.3.3).

The oscillations of the moving mirror can be several times greater then the wavelength of the light (especially at start and stop). Therefore the determination of the direction of motion is important, as well as the counting of the interference fringes. For the so-called "forward-backward" counting (see Section 10.2) two signals with a phase-shift of 90° must be produced. For this purpose the phase-mirror, the stage mirror or the small plate $\lambda/4$ are used: the first two separate the beam into two parallel beams with a relative phase-shift of 90°, while the latter generates within the beam two components of perpendicular polarization which have a relative phase-shift of 90° and can be separated by a polar filter.

The change of luminosity of the two interference images are converted by a photoelectric converter into electrical sinusoidal signals, which are then squared by a trigger stage. The rising and falling edges (intensity changes) of the square signal always generate one pulse to be counted; hence 4 pulses are generated within one period of luminosity, i.e. the shift can be measured in units corresponding to one eighth of the light wavelength.

The speed of the mirror motion is limited by the inertia of the elements of detection and pulse-generation. Modern electronic elements can detect signals with frequencies of 5 to 10 MHz; this corresponds to a speed of 5 to 10 m/s.

A block diagram of the comparator for levelling staffs, operating on the principle described, is shown in *Fig. 8.2.* The beam-spread of the frequency-stabilized light emitted from the laser source *1* is narrowed and the light then passes through the partially reflecting mirror *2*, also called a half-silvered mirror, which also serves as a phase-mirror. The reflected part of the light is reflected from the rigidly fixed triplet prism *3*, the other part passes through the mirror *2* and is reflected from the reflector *5* (also a triplet prism) fixed on the measuring carriage. This carriage supports the

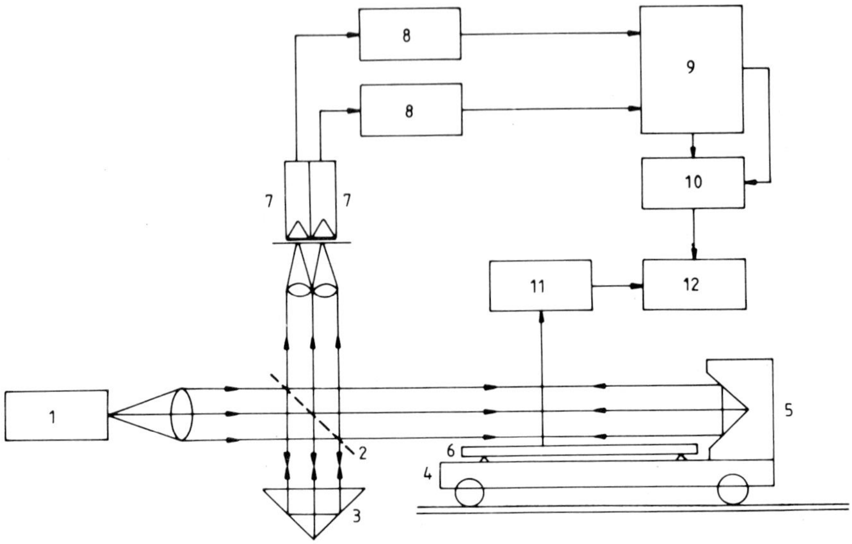

Figure 8.2. Interference comparator for testing levelling staffs: *1* laser; *2* half-silvered mirror; *3* fixed mirror (reflector); *4* measuring carriage; *5* moving mirror; *6* levelling staff; *7* diode pair for counting and detecting the direction; *8* trigger; *9* direction detector and signal converter; *10* counter; *11* photoelectric microscope; *12* data processor

levelling staff to be calibrated. The change of luminosity of the interference fringes is converted into a pair of sinusoidal signals and then the pair of triggers *8* converts them into square signals. The pulses to be counted are produced by the direction detector and signal-converter *9* and are counted by the counter *10*. The photoelectric microscope *11* always produces one pulse when the luminosities of the two halves of the field of view are equal and a minimum (i.e. in each case when the imaginary dividing line between the two halves of the field of view coincides with one of the graduation lines on the staff). The pulses counted between two pulses of the photoelectric microscope yield, after corresponding conversion, the length of a graduation unit. The results are processed and tabulated or plotted graphically by the data processor unit *12* (computer + peripherals).

By means of the levelling staff comparator developed at the Technical University, Karlsruhe, the graduation errors of precise levelling staffs (with an invar tape) can be determined with a standard deviation less than ± 5 μm.

A modified variant of this equipment makes the process of graduation more accurate, replacing a later calibration of the graduated invar tape. The photoelectric microscope is substituted by a pulse-laser of high output capacity. When the number of pulses counted corresponds to a displacement of the levelling staff by one graduation unit, a giant light pulse is generated by the electrical signal coming from the data-processing unit. This giant signal burns off the un-marked coating of the invar tape by a matrix in the form of a graduation line. The advantage of the equipment is that a later calibration, and the necessity for taking it into account when processing precise levelling, is rendered superfluous.

8.2 Distance Measurement Using the Doppler-effect: the Laser Doppler-interferometer

If the distance between an electromagnetic radiation source and the receiver device changes, i.e. the two equipments move with a radial velocity v_r with respect to each other, the frequency of the received radiation changes due to the Doppler-effect; the approximate frequency of the signal received is:

$$f_D = f_1 \pm \Delta f = f_1\left(1 \pm \frac{v_r}{v}\right), \tag{8.8}$$

where f_1 = the frequency of the signal emitted,
Δf = the change of frequency,
v = the propagation velocity of the radiation.

If the receiver and transmitter are united in one unit and the radiation is reflected by a reflector device, then:

$$f_D = f_1\left(1 \pm \frac{2v_r}{v}\right). \tag{8.9}$$

The frequency difference between the two signals is:

$$\Delta f = f_D - f_1 = \pm \frac{2v_r}{\frac{v}{f}} = \frac{v_r}{\frac{\lambda}{2}}, \tag{8.10}$$

from which:

$$v_r = \pm \frac{\lambda}{2}\Delta f. \tag{8.11}$$

From the measured frequency difference Δf the radial velocity v_r can be computed, and if the velocity v_r is integrated over the time interval during which the radial displacement ΔS takes place:

$$\Delta S = \int_{t_1}^{t_2} v_r \, dt = \frac{\lambda}{2}\int_{t_1}^{t_2} \Delta f \, dt. \tag{8.12}$$

Thus a relative displacement can be determined by the Doppler-effect, but not absolute distance.

The block diagram of the Laser–Doppler–interferometer *HP-5526 A* (Hewlett–Packard, USA) is shown in *Fig. 8.3*. The optical resonator of the laser light source *1* with a stabilized frequency is placed in a constant magnetic field: due to the consequent Zeeman-effect, the emitted radiation is separated into two parts with the slightly differing frequencies f_1 and f_2 and with circular polarization of opposite senses. The frequency difference $f_1 - f_2$ is 1.8 MHz.

The light with the compound frequency f_1 and f_2 is divided at the half-silvered mirror *2*: one component goes to the reference photodiode *7*, the other component to the external light-separating prism *3* which also serves as an optical filter; it passes the light component with frequency f_1, and reflects the light component with frequency

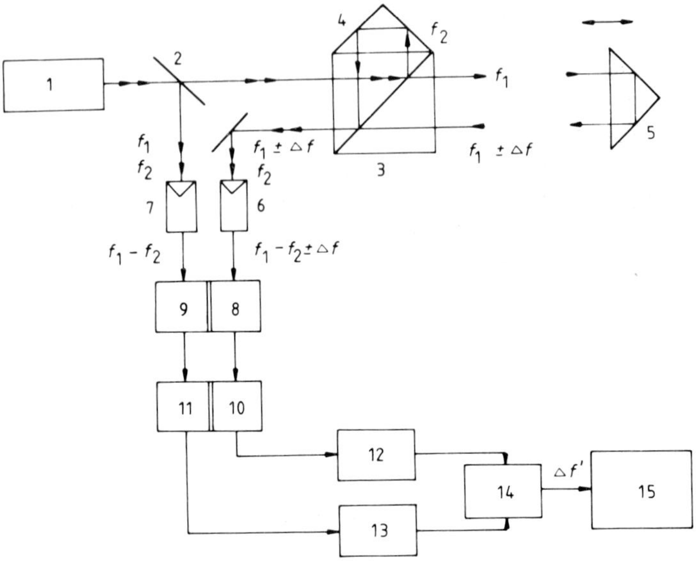

Figure 8.3. Block diagram of the Doppler-interferometer *HP-5526 A* (Hewlett–Packard, USA): *1* Zeeman-laser; *2* half-silvered mirror; *3* light-separating prism and optical filter; *4* fixed reflector (mirror); *5* movable reflector; *6* measuring photodiode; *7* reference diode; *8* and *9* amplifier and trigger stage; *10* and *11* frequency doublers; *12* and *13* counters; *14* difference counter; *15* data processor

f_2 to the fixed reflector *4* mounted with the prism *3* in one unit. The light component with frequency f_1 passes through the light-separating prism and is then reflected from the movable reflector *5* back to this prism, but in the mean time its frequency is changed by $\pm\Delta f$ due to the Doppler-effect. The lights reflected from the moving and the fixed reflectors are again united at the separating plane of the light separating prism *3* and meet the measuring photodiode *6* as a light with the compound frequency $(f_1 \pm \Delta f)$, f_2. The diodes *6* and *7* also mix the difference: the frequency of the electrical output signal of the measuring channel is $f_1 - f_2 \pm \Delta f$, the frequency of the output signal of the reference channel is $f_1 - f_2$. Amplifier *8* and trigger *9* produce a square signal from the detected and from the mixed signal respectively. Frequency doublers *10* and *11* increase the resolving power of the measurement. Now each of the two signals goes to a counter *12, 13*, and the difference between the two counts is computed in the difference computer *14*: the frequency difference $\Delta f'$ obtained in this way is the basis of the distance measurement. The data processor *15* computes the instantaneous value $v_r = \pm f' \frac{\lambda}{4}$ of the radial velocity of the moving reflector device ($\Delta f' = 2\Delta f$, where Δf is the Doppler shift, therefore the multiplication is by $\lambda/4$ instead of $\lambda/2$); the integral of the velocities yields the radial displacement ΔS of the moving reflector with respect to the original position.

Since the wavelength used (the instantaneous wavelength λ of the He–Ne Laser radiation) differs from the laser wavelength in vacuo, which is $\lambda_0 = 0.6328$ µm, the refractive index n of the air along the measured distance must be determined. The data processor computes the average refractive index n from the data given by the meteorological sensors, and makes the correction $\lambda_0/n=\lambda$ before computing the instantaneous velocity. The displacement obtained as a final result of the processing appears on the display of the data processor.

The effective range of the displacement measurement is about 60 m under normal conditions, and the accuracy of the display is 0.01 μm. The highest measurable velocity is 0.3 m/s, the accuracy of display is 0.002 mm/s.

Accessories developed for the interferometer make possible precise determination of both flatness (tilt) and straightness. The accessory for tilt measurement and its beam path is shown in *Fig. 8.4*. If the reflector is fixed then $\Delta f_1=\Delta f_2=0$. If the reflector moves, but there is no angular rotation (its tilt remains constant), the frequency differences Δf_1 and Δf_2 are equal but not zero, hence the difference of the differences remains zero. By the proper choice of the separation of the two measuring beams (i.e. of the geometric dimensions of the reflector) the frequency difference $\Delta f_1-\Delta f_2$ can be made to express the tilt angle in seconds of arc.

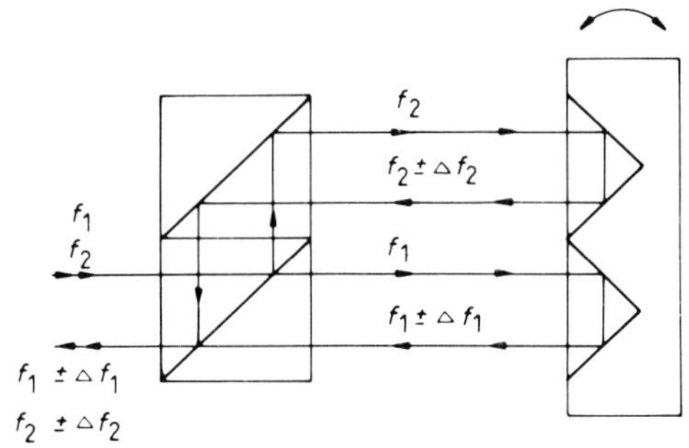

Figure 8.4. Tilt measuring accessory for the interferometer *HP-5526 A*

The reliability of the determination of the change of tilt is

$$\pm 0.1'' \text{ within a tilt of } \pm 100'',$$

$$\pm 1'' \text{ within a tilt of } \pm 1000'',$$

$$\pm 4'' \text{ within a tilt of } \pm 10^\circ.$$

The accessory for the straightness determination and its beam path is shown in *Fig. 8.5*. The two light components are separated by a Wollaston prism (see *Fig. 1.19*) and they are directed to the reflectors in two directions which make equal angles with the entrance beam, but of opposite sense. The reflector device consists in this case of two plane mirrors. If the mirror system does not move, then $\Delta f_1=\Delta f_2=0$. If the mirror system moves (but has no cross divergence) the frequency differences Δf_1 and Δf_2 are not zero but they are equal, hence the difference of the differences remains zero. Finally, if the moving path of the mirror system diverges from the straight line set out by the direction of the entrance beam, the frequency differences

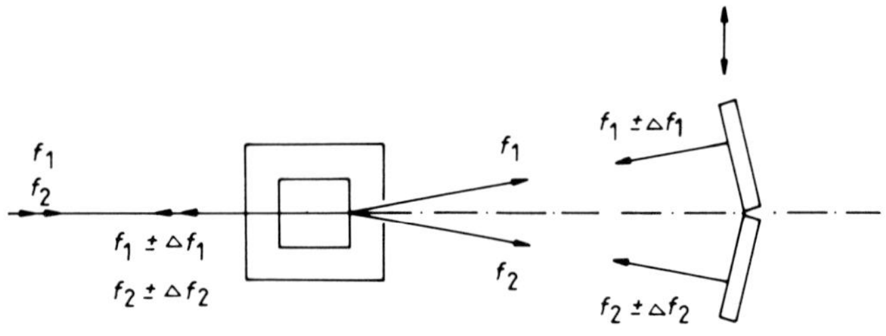

Figure 8.5. Accessory for the interferometer *HP-5526 A* for straightness determination

Δf_1 and Δf_2 differ, hence the difference of the differences is no longer zero: the linear difference (deviation) can be determined from the frequency difference $\Delta f_1 - \Delta f_2$.

The reliability of the reduction to the straight line is ± 0.01 μm over a range of 3 m, ± 0.1 μm over a range of 30 m.

8.3 Distance Determination by Optical Multiplication

In the case of absolute distance measurement, described in Sections 8.1 and 8.2, the unit of distance is the wavelength of the interfering beams. In relative distance measurement the phenomenon of interference is used to determine the position or instant at which the distance to be measured is an integer multiple of a standard length arbitrarily chosen (the usual name of the method is optical multiplication).

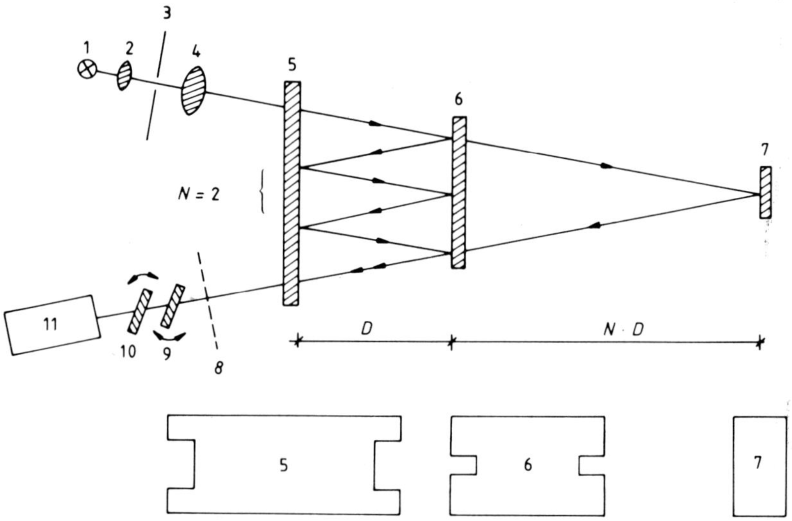

Figure 8.6. Diagram of the Väisälä interference comparator: *1* light source; *2* condenser lens; *3* diaphragm; *4* collimator lens; *5, 6* and *7* mirrors; *8* optical grating; *9* and *10* compensators for the path difference; *11* spectroscopic telescope

The best known array for relative measurement is the Väisälä interference comparator *(Fig. 8.6)*. The point image of the light source *1* is generated by the lens *2* in the plane of the diaphragm *3*; this plane is at the same time the focal plane of the collimator lens *4*. From this lens a coherent light beam goes to the mirror *5*, passing through the slit at the edge of the mirror without hindrance. The upper and lower part of the beam is reflected from the mirror *6*, while the central part goes to the mirror *7*. The angle of incidence is set so that the beam component reflected from mirror *7* meets the beam component multiplied by reflection between the mirrors *6* and *5*. The two beams are diffracted at the optical grating *8*, and the interference pattern generated can be observed in the spectroscopic telescope *11*. If the path lengths of the two beams are equal, the interference pattern in the observing telescope is symmetrical, but with a small path difference the symmetry is distorted. The path difference can be eliminated by the plane parallel plate compensators *9* and *10* and thus a symmetrical interference pattern can be produced. The correction applied can be read on the compensators.

By means of the device i.e. by a slight shift of the mirror *7* the distance between mirrors *6* and *7* can be changed to become an integer multiple of the distance between the mirrors *5* and *6*. The measure of the optical multiplication is identical with the number of reflections from mirror *5*. In order to express the distance to be measured (between mirrors *6* and *7*) in units of length, the basic distance between mirrors *5* and *6* must be determined in units of length. This can be done by means of a quartz standard, using the method of interference with Newton's rings.

When measuring longer distances (several hundred metres) the fluctuations of the atmosphere strongly decrease the visibility of the fringes. In spite of this it was possible to measure the 864 m long base line at Nummela (Finland) with a standard deviation of some hundredths of mm.

9 Electronic Distance Measuring Instruments

9.1 Theory of Electronic Distance Measurement

Electromagnetic waves are transmitted from a device at one end of the distance to be measured towards the other end. The waves are reflected from another device at the other end back to the transmitter. The time between emission and rearrival of a specially marked part of the electromagnetic waves is the so-called travel time; if the travel time is measured the slope distance D between the two end points can be determined as

$$D = \frac{v\tau}{2}, \tag{9.1}$$

where v = the average propagation velocity of the transmitted waves, and
τ = the travel time.

The electromagnetic waves are modulated in the transmitter, i.e. some characteristics of them (e.g. frequency, phase, amplitude, or position of the polarization plane) are altered. Direct measurement of the travel time is usually substituted by a comparison of the modulating phase of the transmitted and the reflected signal.

Due to the double transit of the distance to be measured, the following difference occurs between the two signals:

$$\Delta\varphi = 2\pi f\tau, \tag{9.2}$$

where f is the modulation frequency.

From Eqs (9.1) and (9.2) one obtains

$$D = \frac{v}{4\pi f}\Delta\varphi,$$

or by substituting $v = f\lambda$:

$$D = \frac{\Delta\varphi}{2\pi}\frac{\lambda}{2}. \tag{9.3}$$

In order to determine the distance, the phase difference $\Delta\varphi$ obtained from phase comparison is either zeroed by changing the modulation frequency or it is measured: in the latter case the modulation frequency must not be changed.

The method of changeable modulation frequency (measuring frequency) has been used until now only in few instruments. Its essential point is that, by a continuous change of measuring frequency, $\Delta\varphi$ is zeroed and the corresponding modulation frequency f_1 measured. The following zero position is then determined by changing the measuring frequency, and in this position the modulation frequency f_2 is measured again. If $f_2 > f_1$, then:

$$D = n\frac{v}{2f_1} = (n+1)\frac{v}{2f_2}, \tag{9.4}$$

and substituting from this the integer $n = \dfrac{f_1}{f_2 - f_1}$ gives:

$$D = \frac{v}{2(f_2 - f_1)}. \tag{9.5}$$

In practice f_1 and f_2 are not modulation frequencies belonging to adjacent zero positions but, in order to increase accuracy, there are ν zero positions between them. Hence:

$$D = \frac{v\nu}{2(f_{\nu+1} - f_1)}. \tag{9.6}$$

The value of ν must be determined by counting in the course of the frequency change.

In the method of constant measuring frequency, $\Delta\varphi$ is measured. Since the phase difference can in general be measured only within one period, therefore from Eq. (9.3) it is only possible to compute that portion of the distance to be measured (given by half the travel time) which is less than the modulation half wavelength,* as determined by the half period. The integer half wavelengths included in the distance (i.e. the whole periods in the travel time) must be determined as a separate task, which is called the resolving of ambiguity or the ensurance of unambiguity.

Unambiguity can be ensured by the use of several measuring frequencies (modulation frequencies). With two measuring frequencies:

$$D = \left(n_1 + \frac{\Delta\varphi_1}{2\pi}\right)\frac{v}{2f_1} = \left(n_2 + \frac{\Delta\varphi_2}{2\pi}\right)\frac{v}{2f_2}, \tag{9.7}$$

where n_1 and n_2 are integers unknown for the moment,

$$n_1 = \frac{(n_2 - n_1)f_1 + \dfrac{\Delta\varphi_2}{2\pi}f_1 - \dfrac{\Delta\varphi_1}{2\pi}f_2}{f_2 - f_1}. \tag{9.8}$$

For determining $n_2 - n_1$ the approximate value of the measured distance must be known in advance with an accuracy of at least:

$$\frac{\lambda_1\lambda_2}{4(\lambda_1 - \lambda_2)},$$

where λ_1 and λ_2 are the wavelengths of the modulation frequencies f_1 and f_2.

The use of a third measuring frequency gives a control for the value of n_1:

$$n_1 = \frac{(n_3 - n_1)f_1 + \dfrac{\Delta\varphi_3}{2\pi}f_1 - \dfrac{\Delta\varphi_1}{2\pi}f_3}{f_3 - f_1}. \tag{9.9}$$

* The modulating half wavelength is usually known as scale.

The advantage of the method is that, by a proper choice of the ratio of measuring frequencies, the subsequent numbers or number groups of the total measurement results (distance or travel time) can be obtained by production of the sets of distance- or time-scales in decades. The distance- or time-scales can also be produced indirectly: in this case frequencies necessary for the scales are given by the differences between the basic frequency and the other frequencies, and the measured result can be obtained from the differences between the difference measured at the basic frequency and the phase differences measured at the other frequencies.

9.2 Microwave Distance Measurement

9.2.1 Propagation of the Carrier Wave, the Microwave Relationship

For microwave distance measurement it is necessary to ensure that the transmitted energy propagates within a spatial angle as narrow as possible. The band of the electromagnetic carrier waves should minimize the energy-reducing effect of the atmosphere. In order to achieve low beam-spread and low attenuation, a carrier wavelength of some cms is used in microwave distance meters (carrier wavelength in earlier types is 10 cm, in recent types 3 cm, exceptionally 0.8 cm).

The transmitted microwave energy decreases during propagation; the energy decrease can be characterized by the transmission loss. The transmission loss is the ratio between transmitted and received energy; it consists of attenuation in free space (in vacuo) and additional attenuation.

Attenuation in free space expresses the fact that only a part of the transmitted microwave energy is utilized by the receiver antenna, even if propagation takes place in vacuo (in a homogeneous medium in which EM waves are not absorbed, reflected or spread). The cause of attenuation in free space is the imperfect directioning of the transmitted energy (low degree of bundling) and the finite size of the receiver antenna surface. The attenuation in vacuo at distance D is:

$$A_v = \frac{4\pi D^2}{G_t S_r}, \tag{9.10}$$

where G_t = the transmitter antenna gain,
S_r = the effective surface of the receiver antenna (see Chapter 7).

The additional attenuation includes attenuation caused by the atmosphere, by reflection and by wave deflection and attenuation inside the instrument, all of which are considerable.

The attenuation due to the atmosphere manifests itself in such a way that part of the transmitted microwave energy generates a polarization current in the water droplets always present in air. This is accompanied by heating, hence the energy of the electromagnetic field is partially transformed into heat. Water molecules act as elementary dipoles and radiate electromagnetic energy in all directions, the consequence of which is wave spread. The attenuation caused by the atmosphere is high only if the carrier wave is shorter than 4 cm. It can be determined by means of a nomogram as a function of wavelength and of the quantity of precipitation present in the air (mist, rain).

In the millimetre band of microwaves another absorption phenomenon can be observed. Certain air molecules have a constant electric moment, while others have a constant magnetic moment. If the frequency of the electromagnetic radiation is equal to the eigenfrequency of a molecule then resonance occurs, and this forced oscillation is maintained by a part of the energy of the EM field. In the millimetre microwave band the resonance effect of oxygen and of water vapour molecules is considerable.

Under the conditions of distance measurement in surveying, part of the microwave beam transmitted by the transmitter antenna reaches the receiver antenna after being reflected from the ground. The attenuation caused by a ground reflection can be computed theoretically if the height of the transmitter and receiver antennas, their distance apart and the wavelength of the radiation are known. Due to the large number of factors to be taken into account, however, the loss caused by the reflection can be determined in practice only from experimental observations.

Ground reflection not only causes attenuation, but also changes the result of the distance measurement, because the signal received is the signal resulting from the interference of the signal "propagating in a straight line" and the signal reflected from the ground. Due to the path difference of the two signals the result is burdened with a phase error (distance error).

If it is assumed that (1) a single reflection takes place during the propagation and the reflecting surface is plane; (2) the waves propagate along a straight line, i.e. the path curvature is negligibly small; (3) the propagation takes place in one direction, in other words the transmitter is at one end and the receiver at the other, then the distance error e can be expressed by a mathematical formula. For practical use this formula requires a knowledge of the difference of the path lengths Δ of the signal reflected from the ground and of the signal propagating directly, the amplitude ratio of the reflected and direct signals, and the wavelengths of the carrier wave λ_c and of the modulation wave λ_m.

A good approximation for the path difference Δ is $\frac{2h^2}{D}$ if the transmitter and receiver are at the same height h above the ground and this height is considerably less than the distance D to be measured.

The amplitude ratio a depends on the shape of the direction diagram of the transmitter antenna (Chapter 7) and on the unevenness of the reflecting surface. The effect of the reflection can be neglected if:

$$\tan\left(\frac{\Delta\varphi}{2} + 0.5^\circ\right) \leq \frac{2h}{D},$$

where $\Delta\varphi$ is the beam width of the antenna (the angle made by the rays through the points $0.7 \times P_{\max}$ of the direction diagram). By increasing the antenna diameter and decreasing the wavelength of the carrier the beam width can be decreased.

In the majority of the cases occurring in practice $a<0.1$, in which case the following is a good approximation:

$$e=\frac{\lambda_m}{2\pi}a\cos\frac{2\pi\Delta}{\lambda_c}\sin\frac{2\pi\Delta}{\lambda_m}, \tag{9.11}$$

where e is the distance error caused by ground reflection; λ_m the wavelength of the modulation wave, λ_c the wavelength of the carrier wave; Δ the path difference; a the amplitude ratio.

From Eq. (9.11) it may be seen that, by changing the carrier wavelength (the frequency of the carrier), the distance error changes periodically. If the results measured at several carrier frequencies are plotted on a diagram, one obtains the so-called swing curve (Subsection 9.2.3). The mean value of measurements over one whole period of the curve is not burdened by the error resulting from ground reflection. The condition for the development of at least one whole period is:

$$\Delta \geqq \frac{c}{\Delta f_c},$$

where c is the propagation velocity (3×10^8 m/s), Δf_c the band over which it is possible to change the carrier frequency.

It can also be concluded from Eq. (9.11) that the modulation wavelength is directly proportional to the distance error, i.e. a decrease of the modulation wavelength (an increase of the modulation frequency) strongly decreases the resultant distance error, thus making superfluous repeated measurement at several carrier frequencies.

Attenuation caused by wave deflection can be calculated in a relatively simple way if the wavelength λ of the transmitted waves is considerably less than the height h of the obstacle causing the wave deflection, and if the latter is considerably less than the distance D_1 or D_2 between the obstacle and the nearer antenna. The amplitude ratio (the reciprocal of attenuation) A/A_0 corresponding to the value of the parameter

$$v=h\sqrt{\frac{2}{\lambda}\left(\frac{1}{D_1}+\frac{1}{D_2}\right)}$$

can be taken from tables or determined by nomograms.

Attenuation within the instrument is the result of bad matching between its high frequency elements and the wave guides transmitting the energy (see Chapter 7). This attenuation can be minimized by proper dimensioning.

Besides its energy decreasing effect, the atmosphere also influences the propagation velocity and the propagation direction of the transmitted microwaves.

The propagation velocity of a defined phase-front of the EM field is:

$$v_f=\frac{c}{\sqrt{\varepsilon'\mu'}}, \tag{9.12}$$

where c = the propagation velocity of EM waves in vacuo (in electronic distance measurement the value $c = 299{,}792.5 \pm 0.4$ km/s is used, as accepted by the International Union of Geodesy and Geophysics in 1957),*
ε' = the dielectric constant of the wave-propagation medium,
μ' = the magnetic permeability of the wave-propagation medium.

The term $\sqrt{\varepsilon'\mu'}$ is called the phase refractive index of phase, its notation is n_f. Since μ' can be regarded to a good approximation as 1 for gases, $n_f \sim \sqrt{\varepsilon'}$.

If the medium is an ideal insulator for EM waves, ε' is independent of wavelength, in other words the components of any complex oscillation propagate with the same velocity. The equal propagation velocities of the different phase-fronts are in this case equal to the propagation velocity of the energy of the EM field. Air can be regarded as an ideal insulator for microwaves, therefore the propagation velocity of microwaves in the atmosphere is independent of wavelength and the index referring to the phase-front is superfluous in formulae for propagation velocity (9.12) and refractive index (see Subsection 9.3.1).

The refractive index of air with respect to microwaves is determined by supposing the air to be an ideal gas mixture consisting of dry air and water vapour. The refractive index depends through density of the medium on the thermodynamic state-characteristics (temperature, pressure) of the components, hence it can be computed from temperature, pressure and partial water vapour pressure of the air. In 1960 the International Union of Geodesy and Geophysics proposed that the following formula, determined by *Essen* and *Froome*, should be used:

$$N = (n-1) \times 10^6 = \frac{103.49}{T}(p-e) + \frac{86.26e}{T}\left(1 + \frac{5748}{T}\right), \tag{9.13}$$

where T = temperature (K),
p = atmospheric pressure (Hgmm),
e = partial vapour pressure (Hgmm).

Temperature and atmospheric pressure can be measured directly, partial water vapour pressure can be computed from the results of psychrometric measurements. The most frequently used instrument is the Assmann aspiration psychrometer; it consists of two mercury thermometers in one common frame. A ventilator connected to the frame is operated by a clockwork drive and maintains a constant air circulation around the mercury bulbs of the thermometers. One of the bulbs is covered with a textile sock, which is moistened with distilled water before measurement. Due to evaporation the moistened thermometer shows a lower temperature. The evaporation depends on the water vapour content and the temperature of the surrounding air, hence the partial vapour pressure can be calculated from the "dry" and "wet" temperatures. For the calculation the simplified Sprung formula is usually used:

$$e = e' - A(t - t')p, \tag{9.14}$$

* A more precise value for the propagation velocity in vacuo was accepted in 1983; $c = 299{,}792.458 \pm 0.001$ km/s.

where

t and t' = temperatures read on the dry and the wet thermometers respectively (°C),
p = atmospheric pressure (Hgmm),
A = the psychrometric constant (if the speed of the air circulation is not less than 2.5 m/s and the wet bulb is surrounded by water, then $A = 1/1510$),
e' = the saturated water vapour pressure corresponding to air of temperature t' and a water vapour content of 100 percent (it can be taken from tables).

Aneroid barometers are almost always used for measuring atmospheric pressure (Subsection 3.3.3).

The effect of errors in the different meteorological elements (temperature, atmospheric pressure, partial water vapour pressure) on the error of the refractive index gives information about the accuracy required in the determination of these elements. The values of the partial derivatives of Eq. (9.13) in average meteorological conditions ($t = +15\,°C$, $p = 755$ Hgmm $= 1007$ mbar, $e = 10$ Hgmm $= 13.3$ mbar) are:

$$\frac{\partial n}{\partial t} \times 10^6 = -1.4; \qquad \frac{\partial n}{\partial p} \times 10^6 = +0.4; \qquad \frac{\partial n}{\partial e} \times 10^6 = +5.9.$$

The accuracy of the determination of the refractive index (and consequently the accuracy of the computed propagation velocity and distance) depends decisively on the accuracy of the determination of the partial water vapour pressure. It must be mentioned, however, that the refractive index of air changes in space and time. The propagation velocity of the EM waves should theoretically be computed from the mean refractive index, integrated both over the path between the two ends and over the duration of the measurement. In practice the changes of meteorological data are regarded as linear both in space and time; the required meteorological data are measured at both ends at the beginning and the end of the measurement, and the refractive index is computed from the corresponding mean values. Since the validity of such an approximation is somewhat dubious, especially with respect to the spatial variation of the refractive index, increased accuracy of the meteorological measurements at the ends contributes only slightly to the improvement of the accuracy of distance measurement.

Another effect of the atmosphere is to bend the path of the EM waves as a consequence of the fact that the refractive index of air varies along the path. Except for a 1 to 2 m thick near-surface air layer, the refractive index decreases in the lowest 100 to 200 m of the atmosphere in proportion to height and the vertical gradient of the refractive index is thus constant. The curvature of the ray path is therefore also constant and, when viewed from above, the path is a convex circular arc (EM beams are inclined towards the Earth) whence:

$$-\frac{\mathrm{d}n}{\mathrm{d}h} = \text{const} = \frac{1}{\rho}, \tag{9.15}$$

where ρ is the radius of curvature of the path of the EM radiation.

In the so-called standard atmosphere of the temperate zones, this radius of curvature is four times the Earth radius, i.e. $\rho = 4R$.

The propagation conditions of microwaves can be more easily reviewed if both the Earth curvature and the curvature of the beam are reduced by $1/\rho$. In this case the path curvature is zero and, instead of the curvature of the Earth, one obtains the so-called virtual curvature of the Earth:

$$\frac{1}{R_{\text{virt}}}=\frac{1}{R}-\frac{1}{\rho}. \tag{9.16}$$

By substituting $\rho=4R$ into Eq. (9.16) the virtual Earth radius is $R_{\text{virt}}=4R/3$. If the Earth is represented by a sphere of radius R_{virt}, the paths of the microwaves are straight.

By introducing the so-called modified refractive index:

$$m=n+\frac{h}{R} \tag{9.17}$$

the virtual Earth curvature is obtained as the gradient of m, namely:

$$\frac{\mathrm{d}m}{\mathrm{d}h}=\frac{1}{R}-\frac{1}{\rho}=\frac{1}{R_{\text{virt}}}. \tag{9.18}$$

Exceptional propagation conditions may occur in a non-standard atmosphere. In the case of substandard propagation the gradient of the modified refractive index is greater than it is in the case of the standard atmosphere. In such a case the radio horizon (the effective range of microwave distance measurement) decreases. In the case of super-standard propagation the gradient of the modified refractive index is less than the standard value and the radius of the radio horizon (the effective range) increases. In the case of a zero gradient the radio horizon becomes infinite, i.e. the propagation of microwaves is not restrained by Earth curvature (ray path curvature and real Earth curvature are equal). Under certain conditions in a moderately thick layer it is even possible for an inversion of the modified refractive index to develop; in such a case one speaks of super-refraction: the inversion layer becomes a duct (atmospheric wave-guide).

9.2.2 Main Constructional Elements of Microwave Distance Meters

The carrier wave of the microwave distance meters used in geodesy and surveying is produced by a reflex klystron, or in recent types by a Gunn diode. The frequency of the carrier signal is modulated by the measuring signal of some 10 MHz generated by a stabilized crystal oscillator. The modulation signal controls the voltage of the repeller of the reflex klystron. In the case of a Gunn diode oscillator an extra phase modulator must be used. This is a phase-shifter operating with a so-called varicap diode, the capacity of which can be changed by the modulation voltage. The frequency-modulated carrier signal is transmitted by an antenna towards the measuring station set up at the other end of the distance to be measured; the same antenna receives the signal transmitted from the other station. The polarization planes of the signals emitted by the measuring stations are usually perpendicular to each other. The signal of intermediate frequency, produced by mixing the transmitted and received signals, contains the information corresponding to the measured distance in the form of frequency modulation and the reference basis in the form of amplitude modulation. The modulation signals are separated from the modulated carrier signal by means of corresponding demodulators. The phase difference between the demodulated low frequency signals is determined by phase-meter circuits.

The description of the different construction elements and circuits can be found in Chapter 7.

9.2.3 Principle of the Tellurometer

The working principle of microwave distance meters used nowadays, i.e. the principle of the tellurometer, was developed by *T. L. Wadley* in the second part of the nineteen-fifties. The first microwave distance meter operating on this principle was the model *MRA-1*. It consisted of two measuring units, a master unit and a remote unit *(Fig. 9.1)*, which were set up at the ends of the line to be measured.

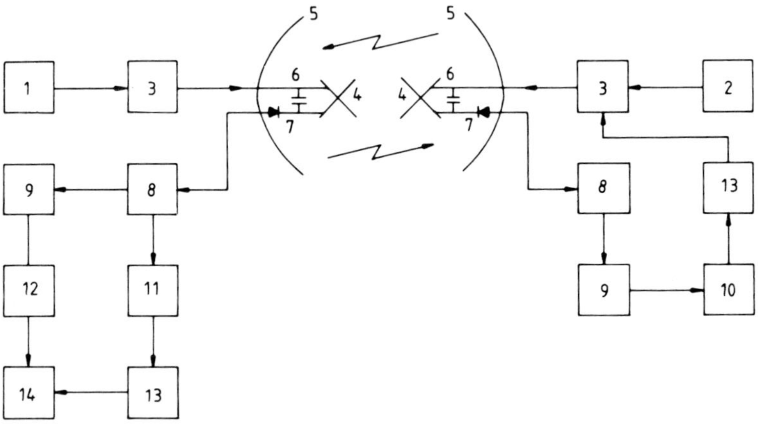

Figure 9.1. Blok diagram of the distance meter *MRA-1* (Tellurometer, UK): *1* measuring signal generator; *2* subsignal generator; *3* klystron; *4* primary transmitter; *5* secondary transmitter; *6* connected condenser; *7* mixer diode; *8* IF-amplifier; *9* AM-detector; *10* pulse-former; *11* FM-detector; *12* time circle former; *13* pulse-amplifier; *14* cathode-ray tube

The EM microwave carrier is generated by the klystrons *3* of the master and the remote unit. The frequency of the carrier is modulated in the master unit by the pattern signal of the frequency oscillator *1* and in the remote unit by the subsignal of the auxiliary pattern frequency oscillator *2*. The frequency modulated signals are transmitted towards each other by the antennas of the units: the signals reach the receiver antenna after travelling the distance to be measured. Two frequency modulated signals reach the mixer crystals *7* of both units: one is the signal received, the other the signal of the unit itself. The complex oscillation produced by the mixing of the two signals goes to the intermediate-frequency amplifier *8*, the band-centre and band-width of which is set in such a way that only one component can pass through it: the frequency of this signal, which is amplified, is equal to the difference between the two carrier frequencies (33 MHz).

The IF-signal of the remote unit is amplitude-modulated by the frequency difference between the measuring signal and the subsignal (1 kHz). The modulation of 1 kHz is separated from the basic oscillation of 33 MHz by the AM-detector (demodulator) *9* of the remote unit. The 1 kHz sine signal goes into the pulse-generator *10* and is converted there at its positive-going zero transits into a series of pulses with a repetition frequency of 1 kHz. This series modulates the frequency of the remote carrier signal similarly to the subsignal.

Hence the EM signal transmitted at the remote station is twice frequency-modulated: continuously by the subsignal and intermittently by the pulses.

The frequency of the signal, which is amplified by the IF-amplifier of the master station, is also equal to the difference of the two carrier frequencies (33 MHz), but the signal is modulated twice: on one hand it is amplitude-modulated by the 1 kHz frequency difference of the measuring signal and the subsignal (similarly to the signal at the remote station), on the other it is frequency-modulated by the 1 kHz repetition frequency of the pulse series. The phase-difference between the two modulations contains the information about the distance to be measured.

The AM-detector (demodulator) of the master station separates the sine amplitude-modulation of 1 kHz. A circuit which generates a circular diagram *12* generates from the demodulated sine signal the so-called basic time circle, which is displayed on the screen of the cathode-ray tube. The IF detector (phase-discriminator) of the master separates the pulse frequency-modulation of a repetition frequency of 1 kHz. The pulses are lead after amplification to the Wehnelt electrode of the cathode-ray tube. At the instant of the generation of a pulse the cathode-ray tube cuts off so that a gap, of length equal to that of the pulse, appears in the circular diagram. The phase position of the pulse, which contains the information about the distance to be measured, can be read on the screen of the cathode-ray tube.

The basic measuring frequency of the distance-meter is 10 MHz, hence the period is 10^{-7} s = 100 ns. The screen of the cathode-ray tube is radially graduated into 100 divisions; the phase position of the pulse can be read directly in nanoseconds (tenths of ns can be estimated).

For the determination of distance, decimals of the travel time higher than 100 ns are also necessary. For their determination several measuring frequencies and several subfrequencies are used. By a proper choice of a series of measuring frequencies, the higher decimals of the travel time can also be obtained.

The use of the tellurometer has the following advantages in comparison with a direct distance measurement.

— The low frequency signal of 1 kHz carries the distance information in the instrument. No broadband amplifier is necessary to transmit this signal with a low phase error, hence the circuitry of the instrument is considerably simplified.

— The external effective phase delay depending on the travel time is proportional to the 10 MHz measuring frequency, and the internal harmful phase delay is proportional to the 1 kHz indicating differential frequency, hence the relative phase error is 1/10,000 of the relative phase error of the direct distance measurement.

— If the additional phase errors of the indicating frequencies of the master and remote units are ψ_1 and ψ_2, then the resultant phase error of an instrument operating on the basis of the tellurometer principle is $\psi_1 - \psi_2$ but, in direct distance measurement, the resulting phase error is $\psi_1 + \psi_2$. As the IF amplifiers in master and remote have the same structure, $\psi_1 \sim \psi_2$ so that the effect of the additional phase errors on travel time is compensated effectively.

— In order to increase accuracy two sub-frequencies are mixed in the remote unit with the basic measuring frequency of the master, one 1 kHz higher and the other 1 kHz lower than the master frequency. If both sub-frequencies are used during measurement ("lower" and "upper" mixing), the result obtained can be regarded as if the phase measurement had been carried out at twice the measuring frequency (the circle graduation unit of the cathode-ray tube is now 0.5 ns instead of the former 1 ns). In this case the result is also free of phase error due to the difference between the geometrical and electrical zero points of the cathode-ray tube.

— An increase of accuracy is again the purpose of the so-called reversal, i.e. the sudden reversal of the phase of the pulses of the remote unit by 180°. This method eliminates error caused by possible eccentricity of the circle graduation of the cathode-ray tube, as well as phase error occurring at the generation of the series of pulses.

— The carrier frequency of the distance meter is about 3 GHz, hence the wavelength of the carrier is about 10 cm. By mechanically tuning the klystron cavity in both units the carrier frequency can be continuously changed on both stations between 2.75 GHz and 3.15 GHz. If there is a large enough path difference between the microwave beam propagating directly between the two stations and the beams reflected from the ground, then the effect of the ground reflection can be decreased by changing the carrier frequency within a proper band from one fine measurement to another (made on the basic measuring frequency), maintaining thereby the necessary difference of 33 MHz between the carriers of the two units. The results of such measurement sets give the curve of the ground swing and the mean of this curve gives a more accurate final result.

Radio telephone connection can be established between the two stations (this is not illustrated in *Fig. 9.1*). In the operation mode "speak" the carrier frequency is modulated by the acoustic signal; demodulation takes place in the FM detector. In the operation mode "measure" a sound corresponding to the measuring signal of 1 kHz can be heard in the headphone or loudspeaker.

9.2.4 Microwave Distance Meters in General Use

9.2.4.1 Distance Meters of the Firm Tellurometer

The operating principles of the *MRA-1* distance meter of the British firm Tellurometer have been partially described in the previous section.

The instrument *MRA-2* is an improved version of the *MRA-1*. The more important alterations are: the two units are the same and they can be interchanged as master or remote; transistors are used instead of electron tubes, the mass and power requirement of the instrument being consequently considerably lower; the transmitter can be separated from the measuring unit.

In order to decrease the effect of ground reflection it became necessary to narrow the beam by using a shorter carrier wavelength. In the distance meter *MRA-3* a carrier wavelength of 3 cm (carrier frequency about 10 GHz) was used instead of 10 cm (carrier frequency about 3 GHz). A danger of the increased carrier frequency is that, due to its unavoidable fluctuations, the carrier frequency difference between the two stations may deviate from the prescribed value to a degree higher than the band-width of the IF amplifier. Therefore a continuous connection between the two units is ensured by an automatic frequency control system (AFC). Further characteristics of the instrument are: power supply is built in, the front panel is formed as the antenna, the electronic units are mounted on plug-in printed boards, transistors replace all electron tubes except the klystron. The *MRA-3* is manufactured in two variants. The variant *MK-3* has a cathode-ray tube phase-meter, the reference base is displayed as a circle, the measuring signal as a straight line. The variant *MK-2* operates with a resolver. The latter was developed to meet a military requirement; its civil counterpart is the type *MRA-101*.

The *MRA-4* is the first distance meter in which a carrier wavelength of 8 mm has been used. As well as the high carrier frequency (about 35 GHz), the pattern frequency is also increased (the pattern frequency for fine measurement is 75 MHz, this is equivalent to the use of 150 MHz, i.e. a scale of 1 m, if the measurement is made at "lower" and "upper" mixing). Due to these modifications the effect of ground reflection has been greatly diminished.

The *MRA-5* and *MRA-6* were again developed to meet military requirements. The carrier wavelength is 3 cm, the basic pattern frequency is 75 MHz. The measurement is partly automated, phase measurement and the display of the results are digitized. In the operation mode "fine reading" the display accuracy is 1 cm, in the mode "coarse reading" it is 10 cm; a coarse measurement can be carried out even between slowly moving units. The transmitter of the *MRA-5* can be fixed on the measuring unit, or can be set up at a maximum distance of 25 m from it.

The *CA-1000* is a variant of the *MRA-5* for civil use. Its construction is aimed at having a distance meter of small size, small mass and low power requirement. The electrical part of the instrument mostly contains integrated circuits as elements, even the reflex klystron being substituted by a semi-conductor element (Gunn-diode). The pattern frequency series is generated by a frequency synthesizer from the signal of a single quartz oscillator. It has a quasi-digital phase-meter: the phase difference between the two signals can be decreased in steps corresponding to 10 percent of the total period, the remainder can be read by the deflection of the pointer of the phase-meter. The modulated signals are transmitted towards each other by the simple, small horn-antennas of the units, and the duplex connection maintains verbal connection even during the measurement between master and remote. When continuously changing the carrier frequency, the pointer of the phase-meter deflects proportionally to the error due to ground reflection. From this deflection the observer obtains information about the error, or he can choose that carrier frequency which gives the average of the ground swing period, repeated measurements at different carrier frequencies thus becoming superfluous.

The variant *CA-1000 D* is capable of dynamic measurement, i.e. of following a moving station; results can be stored or processed in units matched by the interface *RS 232*.

Table 9.1. Technical data and characteristics

Type	Manufacturing firm	Carrier frequency generator Carrier frequency (GHz)	Frequency of fine measurement (MHz)	Number of frequency patterns	Antenna		Power requirement (W)
					diameter	half value width (°)	
Electrotape DM 20	Cubic Corp. (US)	klystron 10.0 to 10.5	7.5	5	33	6	
Distameter 3	Ertel–Grundig (FRG)	klystron 10.0 to 10.5	7.5	5	35	6	30
GET-B1	Finommechanikai Vállalat (Hungary)	klystron 2.7 to 3.2	10	4	44	10	50
GET-A2		klystron 9.2 to 9.6	10	4	49	6	30
Microfix 100C	Global Surveys Ltd (UK)	16.2	40	5	normal 20 narrowed 36	6 3.5	6
PEM 2	(GDR)	klystron 9.2 to 9.6	10	5	38	6	60
RG 10	RAWAR (Poland)	klystron 8.8 to 9.3	7.5	4	29	7	32
Distomat DI 50	Wild–Albis (Switzerland)	klystron 10.2 to 10.5	15	5	36	6	50
SIAL MD-60 C(S)	Siemens–Albis (Switzerland)	klystron 10.3	150	5	35	6	38
RDG (VRD)	(Soviet Union)	klystron 2.7 to 3.2	10	4	43	10	67
Luch		klystron 8.6 to 8.9	10	4	38	6	80

of some microwave distance meters

Phase measurement A = analogue K = cathode ray tube R = resolver D = digital	Effective range (km)	Mass (kg)	Dimension (cm)	Standard error of distance measurement $\pm(A+B\times10^{-6}D)$		Remarks
				A (mm)	B (mm/km)	
$A(R)$	50	11.5	33 × 36 × 29	10	3	
$A(R)$	50	10.8	32 × 28 × 19	20	3	Orientation antenna rod, filter circuit decreasing ground reflextion
$A(K)$	50	15.0		50	3	Master and remote operation not interchangeable
$A(R)$	50	14.0 5.0	40 × 36 × 14 40 × 15 × 21	20	3	Antenna can be removed
D	30 50	3.4	27 × 27 × 17	15	3	Tracking with 45° horn-type antenna, controller-display unit can be removed
$A(K)$	50	14.0		30	3	
$A(R)$	40	10.0	33 × 18 × 38	30	3	
D	50	15.0 + 8.0	47 × 24 × 24 44 × 25 × 20	20	3	Antenna can be removed
D	150	15.0 12.0 10.0	36 × 36 × 18 36 × 36 × 22	10	3	Only one carrier frequency; MD-60S: antenna can be removed
$A(K)$	30	16.0		50	3	Master and remote units cannot be interchanged
$A(R)$	40	14.0 + 8.0	40 × 40 × 20	30	3	Antenna can be removed

Table 9.1. (continued)

Type	Manufacturing firm	Carrier frequency generator Carrier frequency (GHz)	Frequency of fine measurement (MHz)	Number of frequency patterns	Antenna		Power requirement (W)
					diameter	half-value width (°)	
MRA-1	Tellurometer (UK)	klystron 2.7 to 3.2	7.5	5	44	10	70
MRA-2		klystron 2.7 to 3.2	7.5	5	44	10	40
MK-2 MRA 3 MK-3		klystron 10.2 to 10.4	7.5	5	25	9	32
MRA-101		klystron 10.0 to 10.4	7.5	5	33	6	31
MRA-4		klystron 34.5 to 35.1	75	6	33	2	
MRA-5		klystron 10.1 to 10.5	75	6	33	6	42
MRA-6		klystron 10.1 to 10.5	75	6	33	6	40
CA 1000 (D)		Gunn-diode 10.1 to 10.5	25	5		10	4.5
CMW 6	Tellumat	16.0 to 16.5		4		4.5	20
CMW 20	Tellumat	Gunn-diode 34.5 to 34.9			28 × 18	1.2 to 1.5	12

The firm Tellurometer has changed its name in the meantime to *Tellumat* and sells the distance meter type *CMW 6* under this name. A novelty of the instrument is the microprocessor-controlled automatic distance measurement and self-checking, and the computer built into the instrument; meteorological data can be entered on its keyboard and it computes the atmospheric correction, sometimes even the reduced

Table 9.1. (continued)

Phase measurement A = analogue K = cathode ray tube R = resolver D = digital	Effective range (km)	Mass (kg)	Dimension (cm)	Standard error of distance measurement $\pm(A+B\times10^{-6}D)$		Remarks
				A (mm)	B (mm/km)	
$A(K)$	50	25.0		50	3	Master and remote unit cannot be interchanged
$A(K)$	50	18.0		50	3	
$A(R)$ $A(K)$	50	16.0	35 × 33 × 30	15	3	Antenna can be removed
$A(R)$	50	7.0	38 × 37 × 18	15	3	
$A(R)$	50	20.0	37 × 37 × 31	3	3	
D	50	12.0 + 6.0	31 × 35 × 16 31 × 35 × 30	10	3	Antenna can be removed
D	50	7.0	22 × 26 × 38	10	3	
(D)	30	1.6	30 × 15 × 8	15	5	Master and remote units cannot be interchanged
D	50	7.0	26 × 38 × 22	10	3	Data can be recorded
D	25	4	25 × 18 × 15	5	3	Data can be recorded; Speak connection during measurement not possible

distance. The results can be stored and processed in units matched by the interface *RS 232*.

The *CMW 20* already meets the characteristic data of short-range electro-optical distance meters with respect to its mass, dimensions and accuracy: it can be attached to the alidade of a theodolite. The harmful effect of ground reflection is decreased to a minimum by a 8 mm carrier wave corresponding to a 35 GHz carrier frequency. Microphone and loudspeaker needed for finding and communication are built into the computer and display unit. The measurement is automated. Due to the identical construction of the two units, the roles of master and remote can be interchanged by pressing a button. The following commands can be entered into the computer unit: which of the stations should be the master; choice of the unit of distance displayed (metre, English or American length units); measuring unit of the vertical circle reading (degree, grad or artillery angle-unit) which is to be keyed in for the reduction of slope distances and for the calculation of elevation differences. The sequence of data entry is: vertical circle reading, air pressure, temperature read on the "wet" thermometer (instead of the latter relative air humidity can be entered in precent, in which case the psychrometric temperature measurement is superfluous), air temperature, longitudinal eccentricity of the measuring stations. The data entered in the remote station can be transmitted into the computer unit of the master by pushing a button. Distance measurement can also be carried out in the "tracking" mode (repeated quick measurement, computation and display). On the proper command the following results are displayed: slope distance, distance reduced to the horizontal and elevation difference (of course the latter only if the vertical circle reading has been entered). The computer is able to store the results of as many as 50 measurements and the data can also be transmitted into external memories; when full the internal memory can be erased and data recording continued. The data recorded are: longitudinal eccentricity, meteorological readings, vertical circle reading and slope distance. The shortest measurable distance is only 20 m, at which the intensity of the measuring signal is decreased by a damping unit switched on automatically. Measured distances are displayed with mm accuracy up to 10 km, but at longer distances the smallest displayed digit is cm.

The characteristics of the different tellurometers can be found in *Table 9.1*.

9.2.4.2 Other Instruments Operating on the Basis of the Tellurometer Principle

The distance meter *GET-B1* (Finommechanikai Vállalat, Enterprise for Fine Mechanics, Hungary) and the distance meter *RDG* and its improved variant *VRD* (manufactured in the Soviet Union) resemble the Tellurometer *MRA-1* both in appearance and in operating principle: they have the common characteristic that the transistorized converter is built in.

The distance meter *PEM 2* (GDR) has a carrier wavelength of 3 cm, the basic pattern frequency is 10 MHz, the phase-meter is a cathode-ray tube.

The distance meter *RG-10* (RAWAR, Poland) has a carrier wavelength of 3 cm, the pattern frequency of fine measurement is 7.5 MHz, the phase-meter consists of an inductive phase-shifter and a phase detector. Master and remote are the same.

The distance meter *GET-A2* (Finommechanikai Vállalat, Enterprise for Fine Mechanics, Hungary) also consists of two identical units, and the transmitters are separated from them. The pattern frequencies are generated by a frequency synthesizer from the signal of a single quartz oscillator. Only prototypes of this instrument have been manufactured.

The distance meter *Luch* (Soviet Union) consists of three identical units. The transmitter can be removed from the measuring unit to a distance of 25 m.

The *Electrotape DM-20* (Cubic Co. US) is similar to the variant *MK-2* of the Tellurometer *MRA-3*. The continuous link between the two stations during measurement is ensured by an automatic frequency control (AFC) system.

An interesting characteristic of the *Distameter 3* (Ertel–Grundig, FRG) is that the pattern frequencies are generated by the frequency synthesizer of the unit acting as remote. The antenna system can be separated; the ground reflection effect is decreased by a special circuitry which filters out the improperly polarized part of the return microwave beam. Rough pointing is facilitated by an orienting bar antenna. Only a prototype of the *Distameter 8* — using an 8 mm carrier wavelength — has been manufactured.

The units of the *Distomat DI-50* (Wild–Albis, Switzerland) are the same, and the transmitter can be removed from the measuring unit up to a distance of 30 m. Phase measurement and data display is digital, the distance measurement is partly automated, the pattern frequencies must be changed by the observer.

The distance meter *SIAL MD-60* (Siemens–Albis, Switzerland) is an improved variant of the *DI-50*. The carrier wavelength is 3 cm, the modulator frequency is unusually high: 150 MHz. The instrument contains a supplementary phase-modulator as well as the usual reflex klystron. Designers believed that the high modulator basic pattern frequency was sufficient to diminish the effect of ground reflection, therefore they did not use a change of carrier frequency. The single fixed carrier frequency of the remote station (in the remote operating mode) is 10.335 GHz, in the master it is 10.6 MHz lower. As soon as the automatic link has been established between the two stations, the program control unit of the remote station activates the pattern frequency generator. At the master the program controller is started by pushing the start knob. The measurement is automated, phase-measurement and display are digitized. The result is stored in the storage unit with correct digits until the next partial result enters. The identical digits of the partial results are compared by the control unit and the final result computed from the corrected partial results is displayed as an eight-digit number with millimetre resolution on the display. Ten partial results are used for the final result: the instrument always makes two measurements on each of the five pattern frequencies. The distance meter *SIAL MD 60* has two equivalent variants: in the variant *MD 60 C* transmitter and measuring unit are built together, in the variant *MD 60 S* they can be removed from each other up to a distance of 25 m.

The distance meter *Microfix 100 C* (Global Surveys Ltd, UK) has two identical units; the carrier wavelength is 2 cm, thus the beam can be bundled better than the beam of meters with a carrier wavelength of 3 cm. For distance measurement two parabolic antennas with different free apertures are used; for dynamical measurement (i.e. for quick distance measurements between moving stations) horn-type antennas with large apertures are used. The controller-display is connected by cables to the distance meter and they can be separated by a distance of up to 25 m. The measurement is fully automated, the time needed for one measurement being less than 5 s; for a dynamical measurement it is 0.5 s. The result can be displayed on both instruments independently of whether it is used as master or remote. Corrections and reduction of the distance can be computed by the operation-processor of the controller-display.

Microphone and loudspeaker used for communication are built into the controller-display unit.

The characteristics of the instruments outlined in this section are listed in *Table 9.1*.

9.2.5 Accuracy of Microwave Distance Measurement

The standard error m_D (mm) of microwave distance measurement is usually given as the sum of two error components, one independent of the distance and one proportional to the distance D (km):

$$m_D = \pm(A + B \times 10^{-6} D).$$

Typical values for modern instrument types are: $A \sim 10$ to 15 mm, $B \sim 3$ mm/km.

According to a somewhat simplified concept A is the standard deviation of the fine measurement (i.e. A is the reliability of the phase-measurement) and B is the standard deviation of the scale of the fine measurement (or the reliability of the frequency stability). In reality the error component independent of distance also contains the standard error of the additive constant, and the distance-dependent error component depends also on the cyclic phase errors and on the effect of ground reflection.

The temperature-dependent effect of the frequency error can be decreased by putting the quartz-oscillator in a thermostatically controlled oven. The maximum frequency stability is $\pm 1 \times 10^{-6}$. For the small *CA-1000* no oven was used in order to decrease power consumption; an electrical temperature compensation system was used instead, ensuring a relative frequency stability of $\pm 5 \times 10^{-6}$.

Due to the ageing of quartz, the modulation frequency changes with time; this can be decreased by artificially ageing the quartz before using it in the circuit. In order to maintain the required accuracy it is advisable to calibrate the pattern frequencies in the laboratory at least once a year and, if they change, they should be readjusted.

The refractive index error depends somewhat on the accuracy of the refractive index determination at the point (i.e. on the reliability of the meteorological measurements), but it depends mostly on the difference between the measured refractive index and the average refractive index along the line. The refractive index at the ends can be determined by careful meteorological measurements with a relative reliability of ± 1 to $\pm 2 \times 10^{-6}$ but, under poor conditions, this value can deviate from the average along the line by as much as 8 to 10×10^{-6}. This error can be decreased slightly by waiting for more advantageous weather or by distributing the surveying program over different parts of the day.

Cyclic phase errors are caused by electrical cross-talk between the transmitter and receiver systems, or by a reflection within the instrument itself, and by the non-linearity of the phase-meter. The error due to cross-talk can be decreased by electrical screening, the error due to internal reflection by proper electrical matching. The non-linearity of the phase-meter causes a considerable distance error in the case of inductive phase shifters (resolvers). This error effect is decreased by repeating the phase measurement with a 180° phase-reversal of one of the signals. The determination of the systematic error, by repeatedly moving one of the stations and comparing standard distances with measured distances, is rather complicated; hence this method of phase-shifter calibration is not used in practice.

Distance error due to ground reflection occurs occasionally, depending on the local geometrical conditions of the measurement and on the unevenness of the reflecting surface. This error can also be decreased by a better bundled microwave beam, i.e. by using a shorter carrier wavelength and also an antenna with a higher gain, but the most effective method is to decrease the scale of the fine measurements (i.e. to increase the pattern frequency). For a given instrument type the effect of possibly occurring ground reflection can be further decreased by fine measurements made with different carrier frequencies and by averaging the results which vary periodically.

The electrical zero points of the master and the remote do not in general lie on the plumb line through the centring points of the instruments, hence the two stations have a common additive constant. This constant changes over a long period due to the ageing of the electronic elements, but it can be checked by repeated calibration.

In order to determine the additive constant to a high accuracy, a distance is divided internally into several sections and all possible combinations of these sections are measured. The unknowns of the condition equation system are the lengths of the different section combinations and the additive constant. The number of redundant measurements increases with increase in the number of divisions. If the lengths of the section combination are known with the necessary accuracy, then there are only two unknowns in the condition equation system: the additive constant and the multiplying constant (the latter cannot be determined from the measurement of section combinations of unknown length).

The additive constant can also be determined by a simpler method, i.e. by the comparison of a large number of independent distance measurements carried out on a base line of known length. It is an important requirement that the error of the refractive index should be negligible in the measurements and that there should be no ground reflection.

9.3 Electro-optical Distance Measurement

9.3.1 Propagation of the Optical Carrier Wave, Electro-optical Relationship

The carrier wave of electro-optical distance meters is visible light (wavelength: 0.4 to 0.7 μm), or infrared radiation (wavelength: 0.8 to 1.0 μm). Due to the very short wavelength the optical beam can be bundled to a very high degree (the divergence angle is about 10,000 times less than that of the microwave beam).

The energy of the optical radiation decreases during its propagation; this energy decrease can also be expressed by the transmission loss which consists of free space attenuation (in vacuo) and additional attenuation.

When assuming a non-laser light source and a reflector which does not affect the divergence of the light beam, free space attenuation over a distance $2D$ (i.e. transmitter and receiver are built in a single unit and the passive reflector is set up at a distance D from the instrument) is:

$$A_v = \frac{16 S_s D^2}{d_r^2 \pi f_t^2}, \tag{9.19}$$

where S_s = the light emitting surface of the light source,
d_r = the diameter of the objective lens of the receiver,
f_t = the focal length of the objective lens of the transmitter.

The light flux in lumens at the objective of the receiver (i.e. after travelling the distance $2D$):

$$\Phi_v = 15.4B\frac{d_t^2 d_r^2}{D^2}, \tag{9.20}$$

where B = the light density of the light source in cd/m^2,
d_t and d_r = the diameters of the objective lenses of transmitter and receiver, respectively, in m,
D = the distance in km.

The largest additional attenuation is that caused by the atmosphere and attenuation within the instrument. There is no ground reflection due to the small divergence angle.

Attenuation in the atmosphere is the consequence of collision with floating particles and of the resulting wave spread. For the characterization of atmospheric attenuation, the concept daily visibility distance has been introduced: this is the distance at which the human eye can detect an object whose light density differs by 2 percent from that of its environment. Atmospheric attenuation is expressed by the factor $e^{-15.6D/s}$ for a light path of $2D$ at a daily visibility distance s (e is the base of Napierian logarithms).

For the propagation of infrared radiation, the attenuation caused by molecular resonance mentioned in Subsection 9.2.1 is important as well as wave spread due to collision with floating particles. Resonance occurs at some resonant frequencies of water vapour, ozone and carbon dioxide molecules; it can be avoided by the choice of an appropriate wavelength for the radiation used to measure distances.

The different optical elements of an instrument and reflector decrease light energy, the effect being characterized by the attenuation factor. If the attenuation factor of the modulator is a_m, that of the optical system of the transmitter a_t, that of the optical system of the receiver a_r and that of the reflector (prism) a_p, then the total attenuation of the instrument and reflector is characterized by the product of the different factors, or $a_{instr} = a_m a_t a_r a_p$.

When measuring a distance D (i.e. when the path is $2D$), the light flux falling on the photocathode is from Eq. (9.20):

$$\Phi_r = 14.4Ba_{instr}\frac{d_t^2 d_r^2}{D^2}e^{-15.6D/s}. \tag{9.21}$$

The photo-current generated by the light flux Φ_r is:

$$i_{photo} = k'\Phi_r,$$

where k' is the factor which characterizes the emission of the photocathode. A condition of electro-optical distance measurement is that i_{photo} must exceed the existing average noise level.

It should be mentioned that the reflector used in electro-optical distance measurement, i.e. the triplet prism, makes the reflected beam convergent, therefore an increase of the surface of the reflector increases the received light flux Φ_r as long as the light beam illuminates the whole surface of the reflector.

Similarly to microwaves, the atmosphere also has an influence on the propagation velocity of electro-optical radiation and it even changes the direction of the wave propagation. Equation (9.12) is also valid for this case:

$$v_f = \frac{c}{\sqrt{\varepsilon'\mu'}}; \qquad n_f \sim \sqrt{\varepsilon'}.$$

The atmosphere, however, cannot be regarded as an ideal insulator for optical carrier waves. Since the dielectric constant ε' depends on wavelength, the components of compound oscillation propagate with different velocities depending on the wavelengths of the oscillations. Thus the propagation velocity of the different phase fronts differs from the propagation velocity of the energy of the EM field: for the former the phase velocity v_f already mentioned is characteristic, while for latter the group velocity v_{gr}. The following connection is valid between the two velocities:

$$v_{gr} = v_f - \lambda \frac{dv_f}{d\lambda}. \tag{9.22}$$

Analogously to the phase refractive index $n_f = \frac{c}{v_f}$, the concept of the group refractive index, i.e. $n_{gr} = \frac{c}{v_{gr}}$, can be introduced; it is to a good approximation:

$$n_{gr} = n_f - \lambda \frac{dn_f}{d\lambda}. \tag{9.23}$$

The air is a dispersive medium for optical carrier frequencies. For normal dispersion the group-velocity is lower than the phase-velocity, or the group refractive index is higher than the phase refractive index. When calculating the refractive index, the air must also be taken as an ideal gas mixture of dry air and water vapour. As empirical formulae for the refractive index, the International Union of Geodesy and Geophysics recommended in 1960 that the following be used.

The group-refractive index for standard state air (dry, temperature 0 °C, pressure 760 Hgmm and 0.03 volume percent CO_2 content) according to Barrell and Sears is:

$$N_{0,gr} = (n_{0,gr} - 1)10^7 = 2876.04 + 3\frac{16.288}{\lambda^2} + 5\frac{0.136}{\lambda^4}, \tag{9.24}$$

where the length λ of the optical carrier wave is taken in μm.

The connection between the refractive index of air at a temperature t, pressure p and partial water vapour pressure e, and that of standard state air, is according to Barrell and Sears, referring to light of the same wavelength:

$$n_{gr} - 1 = \frac{n_{0,gr} - 1}{1 + \alpha t}\frac{p}{760} - \frac{5.5 \cdot 10^{-8} e}{1 + \alpha t}, \tag{9.25}$$

where t = temperature (°C),
p = atmospheric pressure (Hgmm),
e = partial vapour pressure (Hgmm),
α = 0.00366, the temperature expansion coefficient of gases.

The methods of determination of the different meteorological elements can be found in Subsection 9.2.1. Information about the accuracy required of the measurements can be obtained from the effect on the refractive index of a unit error in the determination of the different meteorological elements. Under average meteorological conditions ($t=+15\,°C$, $p=755$ Hgmm $=1007$ mbar, $e=10$ Hgmm $=13.3$ mbar), the partial derivatives of Eq. (9.25) have the values:

$$\frac{\partial n}{\partial t}10^6=-1.0; \qquad \frac{\partial n}{\partial p}10^6=+0.4; \qquad \frac{\partial n}{\partial e}10^6=-0.05.$$

The very low effect of the partial water vapour pressure can be neglected. As stated in Subsection 9.2.1, an increase of accuracy of meteorological measurement at the ends increases the accuracy of distance measurement only to a small degree.

It is well known that the refractive index determined at different points of the measuring beam does not represent the average physical state of the air along the beam with sufficient accuracy. Hence experiments are made with the aim of determining the average refractive index along the line to be measured. The basis of the dispersion method is the dispersive behaviour of the air against light waves. The basic idea of the method comes from *Prilepin* (1957). If two monochromatic light sources of different colour are used for the measurement of the same distance, the two results obtained are not equal due to the different propagation velocities of the two optical carrier waves. Knowing the wavelength of the optical carrier and the approximate distance, the path difference can be computed for the given distance in air of a known physical state (e.g. of standard state). By comparing the calculated and the measured path difference one can determine the average refractive index characteristic of the physical state of the air along the line during the measurement, and thus the corrected distance.

Neglecting the effect of the partial water vapour pressure, the average refractive index and the refractivity N can be given in the following form:

$$N=(n-1)10^6=A(\lambda)f(p,t), \tag{9.26}$$

where $A(\lambda)$ = a known quantity expressing the wavelength dependency of the refractive index,

$f(p,t)$ = the function of known form expressing the effect of the average of atmospheric pressure and temperature (p, t are unknown).

For the optical carrier wavelengths used, λ_1 and λ_2:

$$N_1=A_1f(p,t)$$

and

$$N_2=A_2f(p,t)$$

whence

$$N_{1,2}=A_{1,2}f(p,t), \tag{9.27}$$

where $N_{1,2}=N_1-N_2$, and $A_{1,2}=A_1-A_2$.

The results of the two distance measurements of the unknown distance D are:

$$D_1=n_1D=(1+N_1\cdot 10^{-6})D$$

and

$$D_2 = n_2 D = (1 + N_2 \cdot 10^{-6}) D \tag{9.28}$$

from which:

$$N_{1,2} = \frac{D_1 - D_2}{D} 10^6 \tag{9.29}$$

and

$$N_1 = \frac{D_1 - D}{D} 10^6. \tag{9.30}$$

Dividing Eq. (9.30) by Eq. (9.29) it is:

$$\frac{N_1}{N_{1,2}} = \frac{D_1 - D}{D_1 - D_2}. \tag{9.31}$$

But as

$$\frac{N_1}{N_{1,2}} = \frac{A_1}{A_{1,2}},$$

the end result is the distance corrected by the determined average refractive index:

$$D = D_1 - A(D_1 - D_2). \tag{9.32}$$

The formula implies two approximations: on one hand it does not take into account the effect of partial water vapour pressure, on the other since $A = \frac{n_1 - 1}{n_1 - n_2}$ and the exact values of n_1 and n_2 are unknown (as very task is to determine one of these values), only an approximate value can be calculated for A. If the wavelengths of the two optical carriers are $\lambda_1 = 0.458$ μm (blue) and $\lambda_2 = 0.514$ μm (green), then:

$$A = 55.983 - 19.036 \frac{e}{p} + 3.84 \left(\frac{e}{p} \right)^2, \tag{9.33}$$

where for e and p it is sufficient to substitute the partial water vapour pressure and the atmospheric pressure determined at one end of the line.

By increasing the difference of the two carrier frequencies chosen, and by decreasing at least one of the optical carrier wavelengths, the accuracy can be improved; this depends mainly on the development of a corresponding laser light source.

The promising experimental results in the determination of the dispersive refractive index did not lead to the construction of a portable instrument for a long time. The *Georan-1* constructed in the National Physical Laboratory (UK) is one of the electro-optical distance meters in which the atmospheric correction to the distance measured can be determined by the dispersion method. In the block diagram of the instrument *(Fig. 9.2)* the light source *1* is an argon-ion laser which emits pulses. The light source radiates at several discrete wavelengths between the 0.458 μm (blue) and 0.514 μm (green): for measurement these two carrier frequencies are used. The modulating signal of about 500 MHz, generated by the measuring signal generator *2*, changes the optical characteristics of the potassium-di-deuterium-phosphate (KD*P) crystals *4* and *10* enclosed in the cavity resonator *3*: due to its effect the polarization plane of the polar light emitted by the light source is changed at the

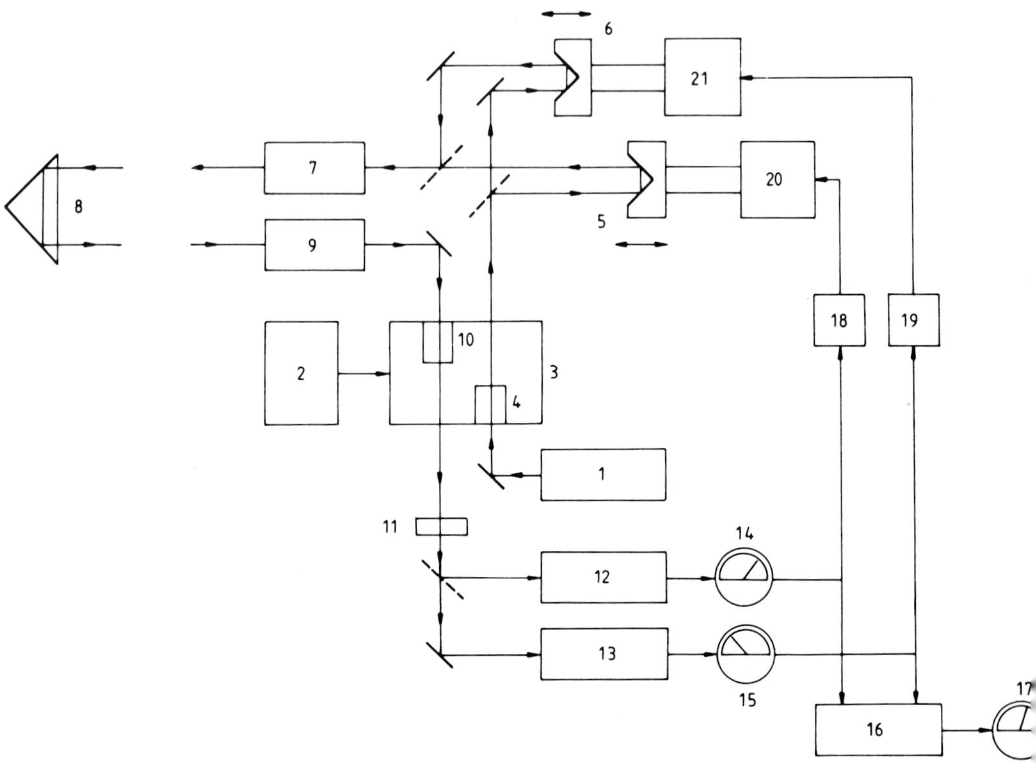

Figure 9.2. Blok diagram of *Georan-1* (National Physical Laboratory, UK): *1* argon-ion laser; *2* measuring-signal generator; *3* modulator cavity resonator; *4* electro-optical crystal (modulator); *5* and *6* internal optical line; *7* transmitter optical system; *8* reflector device; *9* receiver optical system; *10* electro-optical crystal (phase-detector); *11* analyzer; *12* and *13* photomultiplier; *14* and *15* micro-ammeter (zero-indicator); *16* difference amplifier; *17* zero difference indicator; *18* and *19* phase follower device; *20* and *21* servo-motor

rate of the modulating frequency. The two polarization-modulated light components are separated by selective mirrors and directed to one of the optical lines *5* and *6* in the instrument, where the components are forced to pass along a definite and changeable supplementary path. After that the two components are unified and leave the instrument through the optical transmitter system *7*. The light reflected from the reflector *8* re-enters the instrument through the electro-optical crystal *10* set in the cavity resonator; this plays the role of a phase-detector. Now the light passes through the analyser *11* and the two components are selected by selective mirrors and each is directed to a photomultiplier *12* and *13*.

By changing the length of the optical line inside the instrument the total path travelled by the light must be an integer (odd) multiple of the modulating half-wavelength. In this case the polarization plane of the light leaving the phase-detector crystal, and that of the light entering the modulator crystal, have the same orientation. Since the transmitting plane of the analyzer is perpendicular to the polarization plane of the laser light, no light passes through the analyzer if the condition mentioned is fulfilled, and therefore the anode currents of the photomultipliers are zero. The photo-currents generated are indicated by the micro-ammeters *14* and *15*; the photo-current difference is indicated after amplification in *16* by the micro-ammeter *17*. The anode currents of the photomultipliers are reduced to zero by the phase-followers *18* and *19* and by the servo motors *20* and *21* by changing the supplementary path. When the servo systems are at rest (i.e. there is no difference current) the path difference between the two measuring channels is just the optical path difference necessary for the determination of the average refractive index. (The length of the supplementary path can be read with an accuracy of 0.01 mm.) By continuously recording the signal proportional to the anode current flowing in the different channels, the variations of the path difference can also be determined.

Unambiguous distance measurement is made possible by the indirect distance scale series (3 m, 30 m, 300 m, 3000 m) obtained by the division by decades of the basic measuring frequency of 500 MHz.

The experimental model of the *Georan-I* weighs 25 kg, the built-in rechargeable 12 V battery 10 kg, the power requirement of the equipment is 45 W. From measurements carried out over a 4.4 km test line the assumed effective range is 30 km, and the relative standard deviation of distance measurement is less than $\pm 0.5 \times 10^{-6}$ for distances longer than 10 km.

The accuracy of the determination of the average refractive index can be improved by taking into account the effect of partial water vapour pressure. Since the refractive index of light in air is, however, practically independent of partial water vapour pressure, the instrument using two optical carrier waves must be supplemented by a third channel used microwaves to determine the water vapour pressure along the line.

In the distance meter constructed at the University of Washington *(Fig. 9.3)* the two optical carrier frequencies are transmitted by a helium-neon laser of 5 mW power (0.6328 μm, red) and by a helium-cadmium laser (0.4416 μm, blue), respectively. The electro-optical crystal modulator is controlled by a modulating signal of frequency 3 GHz. The condition that the modulated light should be an integer (odd) multiple of the modulating half-wavelength is ensured by changing the measuring frequency (i.e. by the modulation wavelength), instead of by a supplementary optical line, consequently the result of the distance measurement is obtained from a frequency measurement. The carrier frequency of the third microwave channel is 9.6 GHz (the carrier wavelength is about 3 cm). The measurements carried out in the three channels are automatically sampled and recorded by the instrument every 10 s. The distance corrected by the average refractive index along the line and its standard deviation is computed by the built-in computer. From measurements carried out along 2 to 10 km test lines, the relative standard deviation of distance measurement is about 1×10^{-7} (0.1 mm/km), but due to its dimensions the instrument in its original form is not suitable for high precision distance measurement (e.g. the study of horizontal crustal movements) where it is necessary to move the instrument frequently.

The above mentioned idea was realized in the *Terrameter* which is a portable instrument used for high precision distance measurement (e.g. setting-out in the nuclear accelerator *CERN*, Switzerland).

The refracting effect of the atmosphere also affects the propagation of the optical carrier waves, the curvature $1/\rho$ of the beam path being uniform in the lower 100 to 200 m layer of the atmosphere; from above the beam path is a convex circular arc.

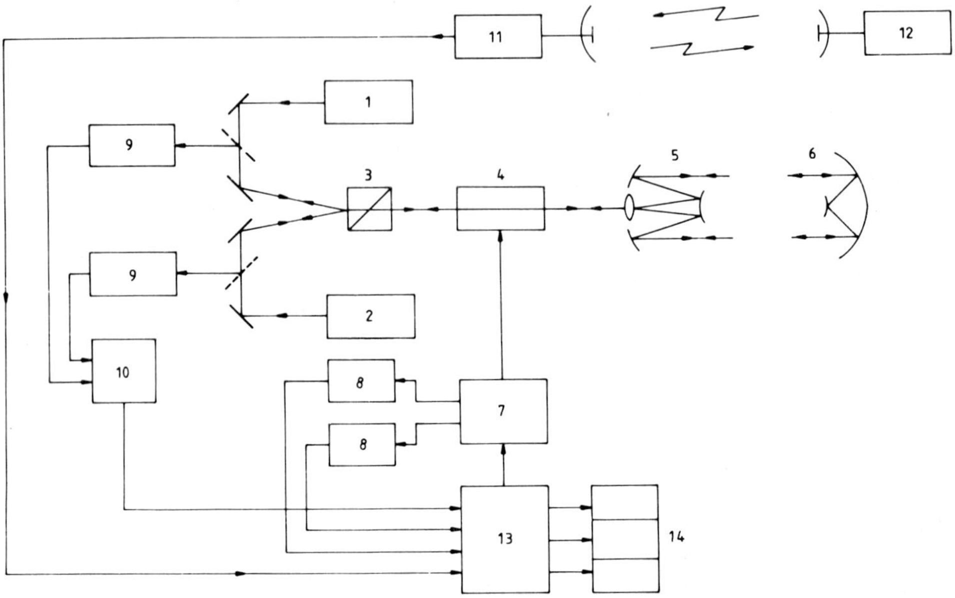

Figure 9.3. Block diagram of a three channed dispersion distance meter (University of Washington); *1* He–Ne laser; *2* He–Cd laser; *3* Wollaston prism; *4* electro-optical crystal modulator; *5* transmitter-receiver optical system; *6* reflector device; *7* measuring-signal generator; *8* frequency counter; *9* photomultiplier; *10* AD-converter; *11* and *12* master and remote stations of the microwave channel; *13* data processor unit; *14* display

The bending effect of the atmosphere on optical carrier waves is usually characterized by the so-called refraction coefficient, i.e. by:

$$k = \frac{R}{\rho}, \tag{9.34}$$

where R is the Earth radius and ρ the radius of curvature of the light path.

The extreme values of the refraction coefficient in temperate climate zones are:

— in summer, around noon, in clear weather (during highest irradiation): $k \sim 0.13$,

— in summer, at night, in clear weather (during the highest emission): $k \sim 0.30$.

Variations of the radius of curvature do not affect the effective range. The result of the measurement, however, is not the length of the straight line connecting the two ends, but the arc length of the refraction curve between them; the difference between the arc and the chord needs to be taken into account only when measuring distances longer than 20 km.

9.3.2 Important Constructional Elements of Electro-optical Distance Meters

The carrier signals of the electro-optical distance meters used in geodesy and surveying are either visible light or infrared radiation. The carrier signal is generated by the light source (radiation source). The amplitude (sometimes the orientation of the polarization plane) of the carrier signal is modulated in general by a measuring signal of a few times 10 MHz, generated by a stabilized quartz oscillator. Usually an external modulator is necessary for modulation, but the emission of the light emitting diodes (LED) can be simply modulated by a change of the generating current. The amplitude-modulated carrier signal is transmitted by the sending optical system of the transmitter toward the passive reflector set up at the other end of the line to be measured. The reflected signal is detected by the optical system of the receiver. The modulating signal is separated from the amplitude-modulated carrier signal by demodulating circuits (AM detectors). As regards demodulators the photomultiplier acts as an amplifier and sometimes also as a mixer; with photodiodes separate mixer and amplifier circuits are used. The phase difference between the demodulated and mixed (low frequency) measuring signal, and the also mixed (low frequency) reference signal, can be determined by means of phase-meter circuits.

As regards light sources, the average power demand of a filament lamp — now almost obsolete — is 20 to 30 W (12 V, DC), its average light density is 8×10^6 cd/m^2. In good visibility the effective maximum range is 2 to 3 km by day, 10 to 12 km by night. Its disadvantage is that white light is emitted, hence the disturbing effect of background light in day-time (white noise) is very high.

Longer effective ranges can be achieved by using gas discharge tubes; that most frequently used is the mercury-vapour lamp. It consists of three tungsten electrodes, two of them the working electrodes and the third the igniter. At the instant of ignition the mercury in the tube vaporizes at a temperature of 700 to 900°C and at an internal pressure of 70 to 80 times atmospheric pressure; the mercury vapour produces a narrow arc of light between the working electrodes. A special device generates the ignition voltage of 50 to 60 V and controls the current (the conductivity of the tube increases to a high degree after ignition). The average power requirement of mercury-vapour lamps used in electro-optical distance meter is 100 to 150 W (at 12 V DC). Due to this high requirement a motor generator is used as a power supply. The light density of the mercury-vapour lamp is about 4 to 5×10^8 cd/m^2. With good visibility an effective range of 10 km (by day) and of 25 km (by night) can be achieved with a mercury-vapour lamp as a light source. Its disadvantage is that its discontinuous spectrum is composed of many components with different wavelengths, therefore the disturbing effect of background light cannot be eliminated in day-time.

The construction of gas-lasers, LED-diodes and laser diodes are described in Chapter 7.

The transmitter and receiver optical systems must ensure the intensity of the modulated light signal necessary for distance measurement. With a conventional (non-laser) light source the light flux emitted from the optical system of the transmitter is proportional to the square of the diameter of the objective lens. The effective range strongly depends not only on the light flux emitted, but also on the ability of the transmitter to shape the emitted light into a parallel (non-divergent) beam (thus concentrating it on a reflector with a not very large reflecting surface). The divergence angle of the beam decreases as the focal length of the objective lens increases. Hence the objective lens of the transmitter has a large diameter and a long focal length. The *Cassegrain* optical system with mirrors or mirrors and lenses (Subsection 1.3.5) is very frequently used; it has a small mass with respect to its illuminating power, its dimensions are not large with respect to its focal length and its mass distribution is favourable. It is also very advantageous to use this system as an optical receiver: the large diameter is needed to increase the light flux obtained, and the long focal length to filter out the disturbing background light.

The diameter of the light beam emitted by a laser light source is about 1 mm, its divergence angle is about 6 min of arc. The task of the transmitter optical system is to decrease the divergence angle in this case also but, since the original divergence angle is already relatively small and the telescope decreases it, the laser beam can be held parallel by the telescope to such a degree that the divergence of the beam is mainly determined by diffraction alone. A simple Galilean telescope, or sometimes a telescope with mirror, is used to decrease the divergence angle. The divergence of the beam with a decreased aperture is 1/15 to 1/20 of the original, whence the diameter of the beam decreases 15 to 20 times.

An important element of the receiver optical system of distance meters with a laser light source is the interference filter (Subsection 1.1.3), by means of which the monochromatic property of the laser light can be advantageously utilized. The Fabry-Perot interference filter consists of a layer of low refractive index evaporated between two layers of high reflectivity; the three layers together are supported by a colour filter glass plate. As result of the interference effect arising after reflection between the layers of high reflectivity, only a light of defined wavelength can pass through the filter; this wavelength depends on the incidence angle of the light and on the optical thickness of the filter. The colour filter filters out the interferences of higher order occurring due to multiple reflection. Although the interference filter also damps the efficient signal, the disturbing background day-light can be almost wholly filtered out: the signal-to-noise ratio can be increased 10,000 times.

A very important requirement of reflectors used in electro-optical distance measurement is simple pointing and the highest possible reflectivity. The most frequently used reflector is the triplet prism. This is the vertex of a glass cube, cut along a surface which is normal to the diagonal through the vertex. A light beam entering through the cut surface is totally reflected from each of the three sides and emerges parallel to its entering direction and symmetrical to the central ray falling on the vertex. The condition for an exactly parallel reflection is that all three plane angles of the prism must be exactly 90°; that can be ensured with present manufacturing technology to an accuracy of ± 1 sec of arc. The effect on the light beam of error in a plane angle is discussed in detail in Subsection 1.3.3. The error of pointing the axis of symmetry of the prism towards the distance meter has no effect if it is less than 20°.

Besides the passive reflectors generally used for electro-optical distance meters in geodesy and surveying, experiments have also been carried out with active reflectors (secondary stations or remote units). The concept of active reflection was first applied in the electro-optical distance meter *Geoponder* constructed at the National Physical Laboratory (UK).

The instrument operates on the principle of microwave distance measurement *(Fig. 9.4)*. At one end of the line the master station is set up, at the other end the remote station. Even the role of the remote unit is the same as for microwave distance meters: it demodulates the signal arriving from the master, then it modulates the phase-information referring to the distance to be measured using the light of its own light source, and transmits it back to the master; there the measuring signal and the reference signal are established by detecting the double modulation, and the phases of the two signals are compared. Similarly to the principle of the *Tellurometer*

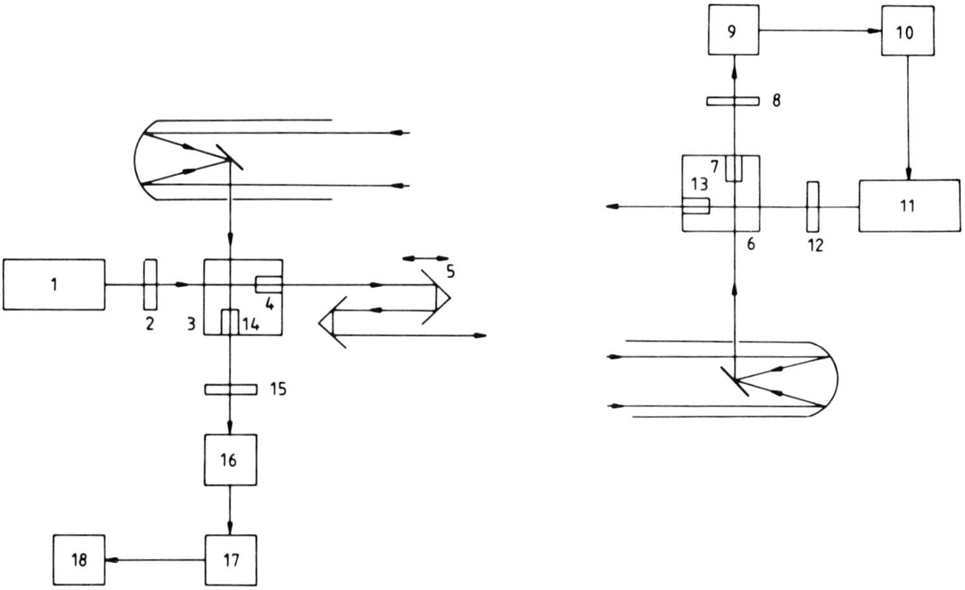

Figure 9.4. Block diagram of the *Geoponder* distance measuring system (National Physical Laboratory, UK): *1* and *11* light source (He–Ne laser); *2* an *12* polarizer; *3* and *6* modulator cavity resonator; *4* and *13* electro-optical crystal (modulator); *5* internal optical line; *7* and *14* electro-optical crystal (phase-detector); *8* and *15* analyzer; *9* and *16* photomultiplier; *10* pulse former; *17* amplifier; *18* cathode-ray tube

(Subsection 9.2.3), unambiguous distance measurement is ensured by producing several measuring and auxiliary frequency pairs.

Advantages of instruments operating with active reflection are:

— the measuring beam travels without amplification only the distance to be measured (compared with twice the distance with a passive reflector), and therefore the range is at least twice that of conventional electro-optical distance meters,

— the inaccuracy of determination of the meteorological factors decreases the accuracy of electro-optical distance measurement considerably less than that of microwave distance measurement, thus the increase of the range hardly decreases the accuracy of the distance measurement,

— the two units have the same construction (they are compatible as master and remote stations), consequently they can also be used as conventional electro-optical distance meters with a passive reflector.

Among phase-measuring methods, optical phase-shifting can be only realized in electro-optical distance meters — due to its characteristics. The optical phase-shifting device is a movable mirror- and prism-system, by means of which the modulated beam can be forced to travel a supplementary, exactly measurable path. The range of path length variation must not be less than the half-wavelength of the lowest modulation frequency (the largest scale) and therefore optical phase shifting can only be used in distance meters which have a high enough basic modulation frequency (its order of magnitude is 100 MHz). In addition, the establishment of a series of indirect scales is necessary in order to ensure the unambiguity of the distance measurement (in this case the frequencies of the coarse measurements do not differ considerably from the basic measuring frequency). After the optical phase-shifting the modulated light goes to a reference photomultiplier, which detects the minimum (sometimes the maximum) of the light signal. With light minimum the anode current of the photomultiplier is minimum, and it can be seen on the micro-ammeter connected to the anode circuit. In this case the path travelled by the light is an integer multiple of the modulating half-wavelength, hence the value read on the optical phase-shifter is the distance remaining within the scale (the fine reading). The advantage of this phase-shifter is that it is simpler and more accurate than electrical phase shifting, and furthermore that it gives distances expressed directly in units of length.

9.3.3 Some More Frequently Used Electro-optical Distance Meters

9.3.3.1 Distance Meters with Long and Medium Range

The *Geodimeter 2A* (AGA, Sweden) was the first electro-optical distance meter to achieve a relatively wide use. In its phase-meter the phases of high frequency electrical signals were compared but, due to the lack of an electrical phase-shifter, this was done by means of optical phase-shifting with a large optical calibrating device, constructed to correspond to the 15 m scale of the fine measurement. Consequently setting up the instrument was difficult and the measurement itself lengthy. The *Geodimeter 4B* and *Geodimeter 6* also had high frequency phase-meters, but the phase compensation was facilitated by an electrical phase-shifter. Due to the non-linear characteristic of the high-frequency phase-shifter, distances read on it had to be computed from tables, and the phase-shifters had to be calibrated from time to time. The phase-meters of the later types of distance meter compared the phase of low-frequency signals.

The long-range distance meter *Geodimeter 8* (AGA, Sweden) has a He–Ne laser as a light source and a series electro-optical crystal modulator. The long range is ensured not only by the light power of 5 mW, but also by the 15 × decrease of the divergence angle of the beam and by an interference filter which filters out the disturbing background light.

Geodimeter 600 (AGA, Sweden) is an improved variant of the medium-range distance meter *Geodimeter 6BL*. The role of the optical systems in the transmitter and receiver of the *Geodimeter 6BL* has been interchanged; in the *Geodimeter 600* the Kerr cell has been substituted by an electro-optical crystal modulator, thus increasing the 25 km maximum range of the *6BL* to 40 km.

The *Geodolite 3G* (Spectra Physics, US) has a He–Ne laser as a light source, and an electro-optical crystal modulator. The measuring unit consists of a transmitter-receiver optical unit, of a controlling-measuring unit and of a display unit. Distance measurement is automated, the data display digitized, and the display resolution is 10^{-3} foot (0.3 mm).

The light source of the distance meter *Quartz* (Soviet Union) is a He–Ne laser, its modulator a Kerr cell.

In the *Rangemaster* (Laser System and Electronics, US) there is a He–Ne laser light source and an electro-optical crystal modulator. The distance measurement is automatic, phase-measurement and data display are digitized. The atmospheric correction reduced to units of length and the additive constant can be entered beforehand.

In the medium-range distance meter *Geodimeter 6A* (AGA, Sweden) the filament-lamp light source can be substituted by a mercury-vapour lamp, and its modulator is a Kerr cell. The electrical power needed for the mercury vapour lamp is supplied by a motor generator.

The *Geodimeter 6B* is an improved version of the *6A*. It maintains the accuracy of the *6A* by using a proper measurement and calculation method. If the approximate length of the distance to be measured is known in advance, however, then the end-part of the distance up to 200 m can be measured by a faster method with an accuracy of ± 1 cm, as the distance can be read directly on the display unit which has been somewhat modified with respect to that of the *6A* and is connected to the phase-meter. Its electronic construction differs from that of the type *6A* in that the difference of the corresponding measuring and auxiliary frequencies are automatically kept constant by an automatic frequency control (AFC) system. The *Geodimeter 6BL* is again an improved variant of the *6B*, having a He–Ne laser of 1 mW power as the light source.

The light source of the distance meter *EOS* (Jenoptik, GDR) is an electric filament lamp; it has an ultra-sound modulator, therefore the modulation frequency of the measuring light is twice the frequency of the modulating electrical signal. As a consequence of the loose connection between the modulated light and the modulating electrical signal, a reference is established by separating 2 percent of the measuring light and demodulating it.

All variants of the *Ranger series* (Laser Systems and Electronics, US) have the same main features as those of the Rangemaster, mentioned earlier. The range of Ranger I is 4 km, of Ranger II 6 to 7 km, of Ranger III and IV 12 km and of Ranger V 25 km.

The characteristic technical data of long- and medium-range distance meters are listed in *Table 9.2*.

Table 9.2. Technical data and characteristics of some long-range

Type	Manufacturing firm	Light source	Modulator	Scale of fine measurement (m)	Number of frequency patterns for coarse measurement
Geodimeter 2A	AGA (Sweden)	Filament lamp (mercury vapour lamp)	Kerr cell	15	2
Geodimeter 4B		Filament lamp (mercury vapour lamp)	Kerr cell	5	2
Geodimeter 6		Filament lamp (mercury vapour lamp)	Kerr cell	5	2
Geodimeter 6A (B)		Filament lamp (mercury vapour lamp)	Kerr cell	5	3
Geodimeter 6BL		He–Ne laser 1 mW	Kerr cell	5	3
Geodimeter 8		He–Ne laser 5 mW	KDP	5	3
Geodimeter 600		He–Ne laser 1 mW	KDP	5	3
Ranger IV V	Laser System & Electronics (USA)	He–Ne laser 2.2 mW 3.0 mW	KDP	10	3
Rangemaster		He–Ne laser 3 mW	KDP	10	3
Geodolite 3G	Spectra-Physics	He–Ne laser 5 mW	KDP	3	4
Quartz	(Soviet Union)	He–Ne laser 1 mW	Kerr cell	5	3
EOS	Jenoptik (GDR)	Filament lamp	Ultra-sound	2.5	3

9.3.3.2 Short-range Distance Meters

The demands of everyday routine surveying on distance measuring instruments can be stated as follows.

— The minimal range of a distance meter should be 500 m, but measuring distances longer than 2 to 3 km are usually not necessary.

— The instrument should be of small size, small mass and low energy requirement, and the small size energy source should provide continuous operation for at least four hours.

— The measuring beam of the instrument should be movable simultaneously with the line of sight of the theodolite in order to carry out angle measurement and distance

and medium-range distance meters

Power requirement (W)	Phase-measurement A = analogue D = digital	Range (km)		Mass (kg)	Dimension (cm)	Standard error of distance measurement $\pm(A+B\times10^{-6}D)$	
		by day	by night			A (mm)	B (mm/km)
140 (400)	A		15 (30)	94	88 × 43 × 35 87 × 46 × 53	20	1
60 (350)	A	2 (5)	8 (25)	17	36 × 30 × 36	10	2
30 (300)	A	3 (6)	15 (25)	16	22 × 36 × 53	10	2
30 (300)	A	5 (10)	15 (15)	16	22 × 36 × 53	5	2
26	A	25		15	22 × 36 × 53	5	1
75	A	60		23	55 × 31 × 26	5	1
26	A	40		15	22 × 36 × 53	5	1
57	D	12 25		17 16	31 × 22 × 43	5 10	2 2
120	D	60		32	28 × 28 × 66	5	2
400	D	70		49	86 × 51 × 40	1	1
140	A	30	50	25	45 × 34 × 28	10	2
70	A	10	15	37	30 × 40 × 36	5	2

measurement with one single pointing. It is desirable that pointing and measuring in both telescope positions should not be hindered by the attached distance meter.

— The process of distance measurement should be automated as much as possible, data display should be digitized. It is desirable that data can be recorded in a field data collector unit.

— Besides measuring the slope distance it should be possible to determine the distance reduced to the horizontal and also the elevation difference.

— In order to permit quick setting out, it is desirable that "tracking" (i.e. repeated quick determinations the slope distance, and perhaps also the reduced distance and elevation difference) be possible.

— The instruments should consist of a minimum number of units, connecting cables should not disturb the observer and the instrument should be easy to handle.

The *Distomat DI10* (Wild, Switzerland) was the first short-range distance meter on the market. Its transmitter unit fits on the telescopes of Wild theodolites, and it is connected to the measuring unit by two cables. Phase measurement is automated; the output signal of the phase detector activates a DC motor which turns the rotatable part of the inductive phase-shifter in the direction of decreasing phase difference. During the first (coarse) phase of the measurement the measuring frequency pattern is continuously changed from 13.5 MHz to 15 MHz. At the end of this fast frequency change the number of integer 10 meters of the distance can be read on the data display by means of a circle graduated into 100 units, on which the result of the coarse measurement is indicated. This is automatically followed by the fine measurement during which the pattern frequency remains constant. At the end of the fine measurement a three digit number appears on another, numbered display disc which is divided into 1000 units. The three digits give the fine reading (a proportion of the integer 10 metres) in metres, 10 centimetres and centimetres (mm can be estimated). The results of coarse and fine measurements are projected side by side, as a five digit number, on the opaque glass screen of the display. The additive constant can be entered before measuring.

The measuring unit of the *Distomat DI 3* can be force-centred, attached to the tribrach of the theodolites *Wild T1*, *T16* or *T2*, and the theodolite itself is fitted on top of the measuring unit. The transmitter unit is fitted on the telescope; it contains the transmitter and receiver optical systems, the light source, the demodulator, the high-frequency unit and the elements of the internal optical path. The measuring unit contains the phase-meter and data display unit, also a pre-programmed computer. The distance measurement is automated and the data display digitized. By entering the vertical circle reading in the computer, the distance reduced to the horizontal and the elevation difference can also be displayed. The atmospheric corrections reduced to length units can also be entered.

The *Distomat DI 3S* is an improved variant of the type *DI 3*. The theodolite telescope can be transited into the second telescope position with the transmitter attached. The computer is mounted below the cover of the measuring unit. The reading on the vertical circle can be entered on its keyboard as well as the oriented direction of the line, thus the coordinate differences ΔY and ΔX can also be displayed. In the repeated measurement mode (tracking) the measured slope distance appears on the data display for one second every three seconds. The measurement can be checked in the mode "normal measurement": after carrying out the distance measurement, depression of the "start" button on the keyboard permits the pointer of a control instrument to deflect proportionally to the standard deviation of the result obtained.

The *Distomat DI 4 (Figs 9.5* and *9.6)* consists of a single unit and it can be fitted to the telescope of the Wild theodolites. The telescopes of the theodolites *T1* and *T16* can also be transited with the distance meter attached. The distance measurement is automated, the data display digitized. The additive constants corresponding to two types of Wild prism are pre-programmed, the atmospheric correction and the possible reductions reduced to length units can be entered. Characteristics of the state of the power source and of the strength of the received signal can be displayed, as well as the slope distance. By means of the keyboard linked to the meter the vertical circle reading can be entered into the microcomputer, and the distance reduced to the

Figure 9.5. The distance meter *Distomat DI 4* (Wild, Switzerland)

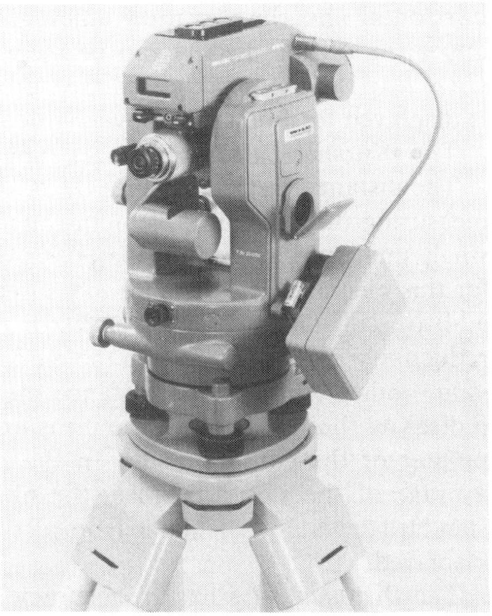

Figure 9.6. The distance meter *Distomat DI 4* attached to the theodolite *T2* (Wild, Switzerland)

Figure 9.7. The distance meter *Distomat DI 20* attached to the theodolite *T2* (Wild, Switzerland)

horizontal and the elevation difference can therefore also be computed. The data appearing on the display can be recorded by the transistorized field data recorder *GRE 2*. In the mode "repeated measurement" (tracking) the measured slope distance is displayed for one second every three seconds.

The *Distomat DI 4L* is a longer range variant of the distance meter *DI 4*.

The distance meter *Distomat DI 20 (Fig. 9.7)* can be classified with respect to the measuring distance as a medium-range instrument. The instrument consisting of a single unit can be fitted on the telescope of Wild theodolites modified in the factory for this purpose. The distance measurement is automated, the display digitized. The additive constant as well as a factor (i.e. the atmospheric correction and possible reductions) can be entered. In the mode "normal measurement" the displayed digital value indicates the slope distance measured. In this case the standard deviation indicating the reliability of the measurement is proportional to the deflection of the pointer of the control meter. In the mode "extended distance measurement" the average of the slope distance, the number of measurements carried out and the standard deviation calculated from all distance measurement results are automatically displayed.

The *Distomat DI 5* has an accuracy higher than that of the *DI 4L*, and it can be connected to the electronic theodolite *Theomat T2000* without a cable. By means of the keyboard *GTS 3*, connected to the distance meter, the reading on the vertical circle can be entered into the computing unit and both the reduced distance and the elevation difference can be displayed. Having entered the oriented direction of the measured distance, the coordinate differences between the instrument and reflector station can also be displayed. When setting-out the required distance is entered and the distance measured, and thus the length and direction of the correction to the point set out, can be displayed.

The *Distomat DI 1000* is a more modest (and consequently cheaper) variant of the new Wild distance meters. Due to its relatively short measuring range it is primarily used in engineering survey, cadastral survey and tacheometric detail survey. The instrument can be attached to the telescope of the theodolite types *Wild T1*, *T16* and *T2*. The keyboard *GTS 5* can be connected to the distance meter, by means of which the reading on the vertical circle can be entered, and thus the reduced distance and the elevation difference can be displayed. When setting-out the required horizontal distance is entered on the keyboard and, after completing the distance measurement, the correction (the difference between the required distance and the actual one) appears on the display.

The *Distomat DI 3000* is similar to the type *DI 20*. It is a time-pulse instrument with a laser diode radiation source. The pulse ratio is 15 MHz, the measuring time is 3.5 s (standard mode), 0.8 s (rapid mode) or 0.3 s (tracking mode). The *DI OR 3002* is a model of the *DI 3000* that needs no reflectors for determining distances up to 250 m with an accuracy of about 5 mm to 10 mm.

The distance meter *MA 100* (Tellurometer, UK) is an independent instrument which can be interchanged with a theodolite for forced centring. The frequency of fine measurement is 75 MHz, this corresponds to 150 MHz (to a scale of 1 m) for a measurement in "upper" and "lower" mixing. The measurement is partly automated, but the switching on of the different measuring frequencies and the recording of the partial results is the task of the surveyor. The phase-measurement and the data display is digitized, the resolution of the display is 0.1 mm. The measured distance data can be recorded.

The distance meter *CD-6*, also manufactured by Tellurometer, can be fitted on the upper part of a theodolite and the telescope can be transited under the attachment. The distance measurement is automated, the data display digitized. In the operation modes "coarse" and "fine" measurement, the result of the measurement on the corresponding measuring frequency appears every 2 seconds on the display with a resolution of 10 mm and mm, respectively.

An interesting feature of the distance meter *HP 3800* (Hewlett–Packard, US) is that the measurement on the internal calibration path is carried out almost simultaneously with the external measurement: the two modulated light signals, of frequency 30 Hz, are alternate brought to the demodulator by means of an electronically controlled diaphragm. The phase measuring is analogue and is reduced to voltage measurement: the phase detector transforms the phase-difference of the compared signals into voltage. The voltage necessary for phase measuring (compensation voltage) can be set by constant and variable resistors. After compensating the voltage, the digital value of the slope distance can be read directly on the position switches of the resistors; the distance is obtained in type *3800A* in feet, and in type *3800B* in metres.

The distance meter *HP 3805 A* is an improved variant of the type *3800*. The distance measurement is automated, the data display digitized. The processor unit of the instrument calculates the results of the measurement and indicates if their standard deviation exceeds the pre-programmed value. The atmospheric correction reduced to length units can be entered. In the mode "tracking" the slope distance is displayed every 2 seconds.

With respect to its range the distance meter *HP 3808* can be classified as a medium-range distance meter. The distance measurement is automated, the data display digitized. In addition to the slope distance the value of the standard deviation can also be displayed, and the displayed data can be recorded with a transistorized data recorder of type *HP 3851*. The atmospheric correction can be entered. In the mode "tracking" the slope distance is displayed every 2 seconds.

An interesting feature of the distance meter *HP 3810A* is that the elevation angle is also measured to the accuracy required for the horizontal reduction of the distance. The following method is used for its automatic measurement: three insulated sectors of the vertical circle (circular ring) are immersed in an electrolyte and the change of the electrical resistance between them is measured. The distance measurement is automated, the data display digitized, and the atmospheric correction can be entered. In addition to the slope distance, the elevation angle, the horizontal distance and the elevation difference can be displayed, even in the tracking mode: in this case the calculated distance components appear at 3 second intervals, and the measured angle in half second intervals.

Each of the types *3800 A/B*, *3805 A*, *3808 A* and *3810 A* are independent distance meters; they can be interchanged with a theodolite with forced centring.

The distance meter *DM 2000* (Kern, Switzerland) is an independent instrument. Its interesting feature is that the reference base is produced by demodulation of the modulated light. Distance measurement is automated, data display digitized. An external data recorder can be linked to the meter.

The distance meters *DM 500, DM 501, DM 502* and *DM 503,* also made by Kern, can be fitted on the telescope of the Kern theodolites *K0-S, K1-M* and *DKM-2AE,* and the telescope can be transited together with the attached distance meter. In order to decrease the distance error caused by cross-talk, the transmitter optical system is below the telescope and the receiver optical system above; the frame of the electronic unit connects the two optical systems. To link the distance between the objective lenses of the transmitter and the receiver, the roof-edge variant of the usual triplet prism is used for the reflector. The distance measurement is automatic after setting in the received signal level and the data display is digital. The meter is also connected electrically to the theodolite, the power source cable being plugged into the illumination-connector of the theodolite.

The *DM 501* operates as an improved variant of the *DM 500* in that it also operates in the tracking mode, and the slope distance is displayed every 2 seconds. The *DM 502 (Fig. 9.8)* is of higher precision than the former two; in the "tracking" mode the value of the measured distance can be transmitted to the prism position by additional modulation of the measuring beam. The receiver unit *RD 10* can be attached to the reflector device, on which the distance can be displayed. An external data recorder of the type *R48* (fully transistorized) can be linked to the distance meter. This instrument can be linked not only to the optical Kern theodolites *K0-S, 1-M* and *DKM 2-AE*, but also to the electronic Kern theodolites *E 1* and *E 2*. In the latter slope

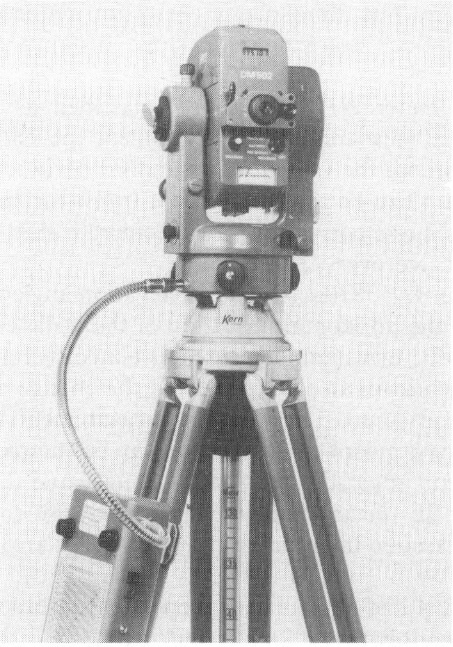

Figure 9.8. The distance meter *DM 502* attached to theodolite *DKM-2AE* (Kern, Switzerland)

Figure 9.9. The distance meter *DM 503* attached to theodolite *DKM-2AE* (Kern, Switzerland)

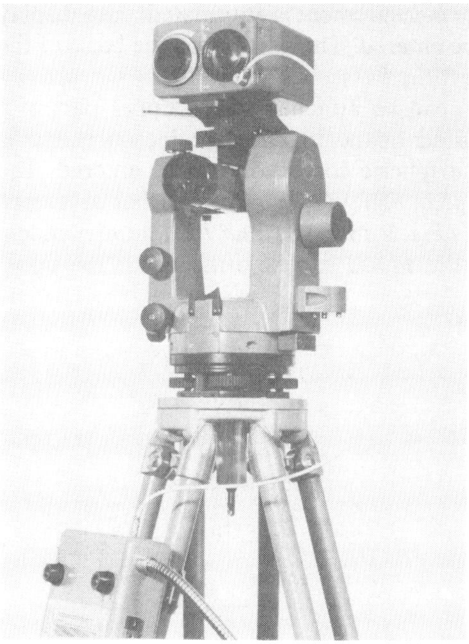

Figure 9.10. The distance meter *DM 102* attached to theodolite *KO-S* (Kern, Switzerland)

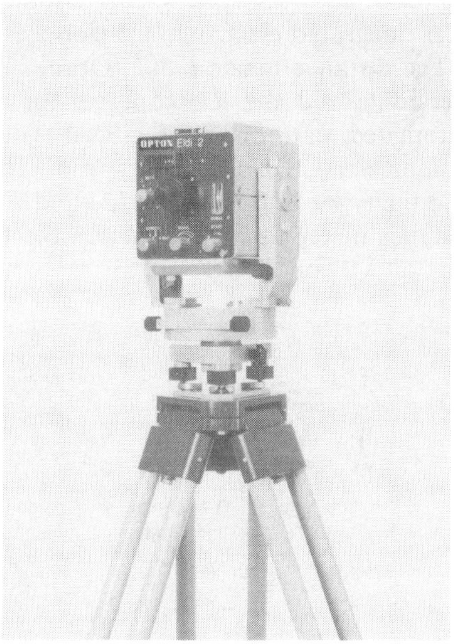

Figure 9.11. The distance meter *Eldi 2* (Opton, FRG)

distance, horizontal distance and elevation difference can be transmitted to the display of the reflector device. The distance meter linked to the electronic theodolite, together with the data recorder *R 48* and the pocket computer *HP-41 C(V)* which is interfaced by *DIF41*, forms an electronic tacheometer of modular structure (see Chapter 10).

The *DM 503 (Fig. 9.9)* has a longer range and higher accuracy than the *DM 502*, and the additive constant, the atmospheric correction and a distance reduction if necessary, can all be entered. The entered data appear automatically on the display when the device is switched on. The normal measuring time is 8 s, but additional measurement series can be observed and the displayed end result is the arithmetic mean of the whole series. From the start of measurement the display continuously shows the percentage of the time during which a signal of a strength suitable for measurement is received in the receiver, thus the observer is given information about the frequency and time duration of the interruption of the measuring beam.

The distance meters *Kern DM 102 (Fig. 9.10)* and *DM 103* are variants of the *DM 502* and *503*, but they can be attached to the telescopes of both Kern and other instruments. The meter is fixed on top of the telescope and consequently the telescope cannot be transited, but the distance between the transmitter and the receiver optical systems has been minimized within the instrument. A simple triplet prism can be used as a reflector.

The distance meter *SM 11* (Opton, FRG) is an instrument similar to a larger theodolite, and the transmitter and receiver optical systems are coaxial and identical with the telescope of the theodolite. Distance measurement is automated, data display digitized. The atmospheric correction can be entered. This meter was the basis of the later integrated electronic tacheometer *Reg Elta 14* (see Chapter 10).

The distance meter *Eldi 2* (Opton, FRG) can be attached to the upper part of a theodolite and the telescope can be transited below it. Distance measurement is automated, data display digitized. The atmospheric correction can be entered. The variant *Eldi 2 Mining* is damp-proof. The *Eldi 2 (Fig. 9.11)* has been further developed into the types *Eldi 1 (Fig. 9.12)* and *Eldi 3 (Fig. 9.13)*. The *Eldi 1* is an independent distance meter fitted with a teleobjective; the *Eldi 3* can be attached to the upper part of a theodolite.

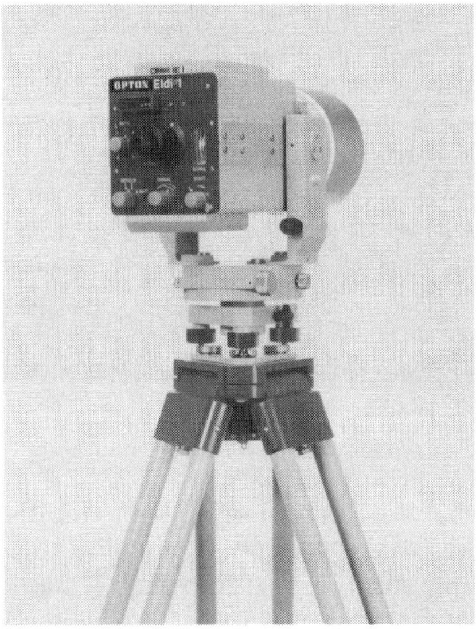

Figure 9.12. The distance meter *Eldi 1* (Opton, FRG)

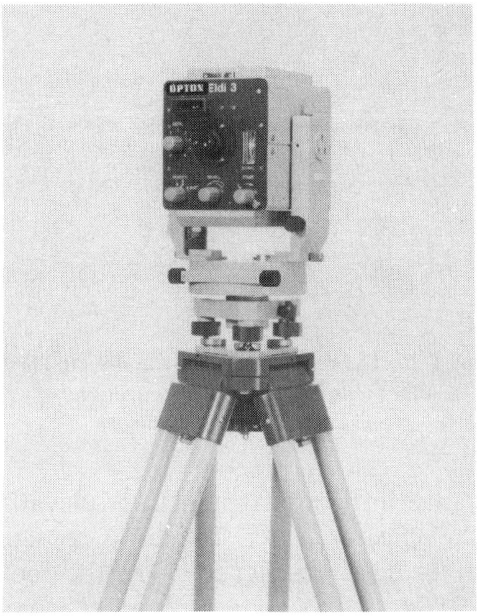

Figure 9.13. The distance meter *Eldi 3* (Opton, FRG)

The distance meter *SM 4* is also manufactured by Opton *(Fig. 9.14)*: the light-source and demodulator are built into the telescope of its theodolite. The telescope can be transited beneath the electronic unit attached to the top of the theodolite. Distance measurement is automated, data display digitized. The atmospheric correction can be entered and the meter can also be used in the tracking mode. This was the basis of the construction of the electronic tacheometers *Elta 4* and *Elta 46(R)* (see Chapter 10).

The distance meters *RSM 3* and *RSM 4 (Fig. 9.15)* are improved variants of the *SM 4*; the horizontal and vertical circles of the *RSM 3* are graduated in the 360° system, those of the *RSM 4* in the new 400^g system. Distance measurement as well as the readings of the vertical circle are automatic, consequently there is an automatic digital display of both the horizontal distance and the elevation difference. Elevation difference and slope distance, or elevation difference and horizontal distance, can be displayed at the same time on the double display. In the tracking mode the displayed data appear at 1 second intervals.

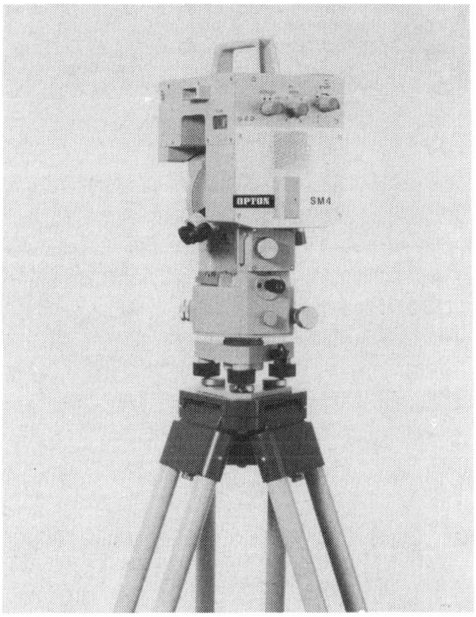

Figure 9.14. The distance meter *SM 4* (Opton, FRG)

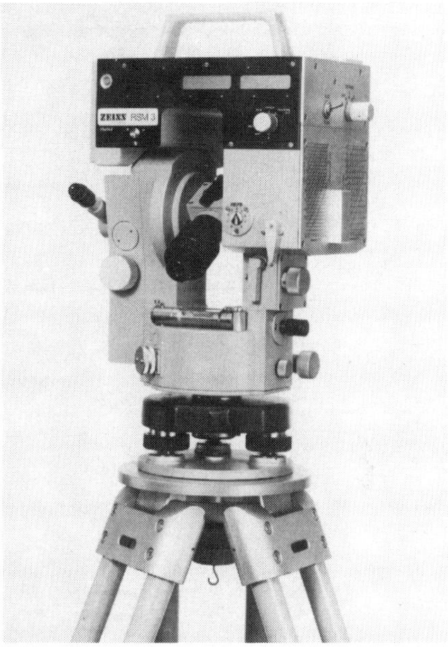

Figure 9.15. The distance meter *RSM 3* (Opton, FRG)

The distance meter *EOK 2000* (Jenoptik, GDR) is an independent instrument, being interchangeable with Jenoptik theodolites with forced centring. It has an analogue phase-meter, the value of the distance is obtained from the partial results of measurements carried out on the three measuring frequencies, and the necessary subtractions can be avoided by zeroing the display. The distance measurement is unambiguous up to a range of 500 m.

The theodolite unit of the distance meter *EOT 2000* (Jenoptik, GDR) is a modified Jenoptik theodolite *Theo 010 A*. The theodolite telescope is at the same time the observing telescope, the optical system for transmitting the light, and the transmitter and receiver optical units for the measuring beam. The light source, demodulator, internal optical path and the whole electronic system are enclosed in an electronic unit placed beneath the theodolite unit. The distance measurement is automated, the data display digitized. The following data can be entered via an external keyboard: the value of the atmospheric correction, the reading of the zenith angle in one or both telescope positions, the absolute height of the station and the instrument height. After measuring the distance and entering the data, the horizontal distance, elevation difference and absolute height of the observed point, can be displayed (the latter when using the prism supporting pole belonging to the instrument outfit). The displayed data can be recorded in an external field data recorder.

The *Geodimeter 12 A* (AGA, Sweden) can be attached to the telescopes of different types of theodolite. Distance measurement is automatic, data display digital, and the atmospheric correction can be entered. Electronic pointing is facilitated by an acoustic

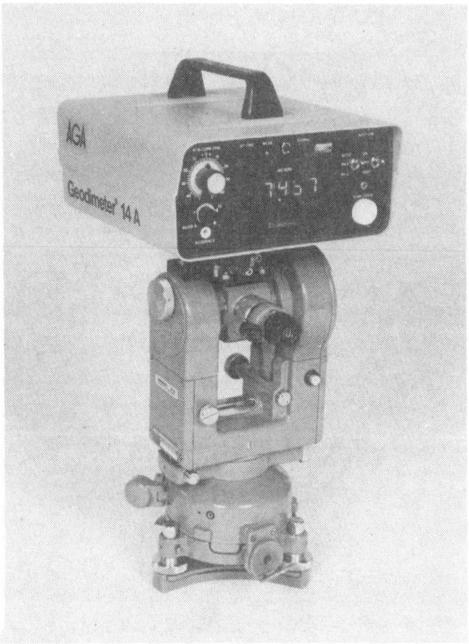

Figure 9.16. The distance meter *Geodimeter 14 A* attached to the theodolite *Wild T2* (AGA, Sweden)

Figure 9.17. The distance meter *Geodimeter 112* attached to the theodolite *Wild T2* (AGA, Sweden)

indicator. The *Geodimeter 10* is a somewhat simplified variant of the *12 A*: it cannot be used in the tracking mode and its range is shorter than that of the *12 A*. The range of the *Geodimeter 14 A (Fig. 9.16)* has been considerably increased by increasing the power of light source and by modifying the optical system. Besides the normal operation mode the distance meter can also be used in another mode; in this the result is more reliable, due partly to the repeated determination of standard deviation and partly to the automatic "weighting" of the results (the weight is proportional to the intensity of the received signal).

The *Geodimeter 120* has been brought to the market by AGA as the first member of a new family of distance meters. A quite new feature of the instrument is that the zenith angle of the measuring beam is measured by an electrical pendulum, hence the reading of the vertical circle is automated. The distance meter can be attached to the telescopes of different types of theodolite; the distance measuring beam must first be made parallel to the line of sight of the telescope and then, after the line of sight has been set horizontal, the displayed reading of the vertical circle must be set to 90° (100^g) by means of an adjusting potentiometer. Now the slope distance as well as the horizontal distance, the elevation difference and the angle between the measuring beam and the plumb line can be displayed. The displayed results can be recorded automatically by the transistorized data recorder *Geodat 120*. The meter can also be used in the tracking mode, when the display is repeated twice a second.

The *Geodimeter 110* corresponds to the earlier *Geodimeter 10*. The decimals of a kilometre of the distance also appear on the display in spite of the fact that the maximum length of an unambiguous measurement is 1 km. When the distance measurement is completed, two additional automatic measurements follow with the extreme values (+50 and −50 mm/km) of the atmospheric correction, which can be entered. The difference between the two results multiplied by 10,000 yields, after rounding up, the decimals of a km.

The *Geodimeter 112 (Fig. 9.17)* operates in the mode "standard deviation-tracking" as well as in the normal mode: if the standard deviation of the results exceeds the programmed value, the fine measurement program is automatically continued. The distance measurements can be automatically recorded on the transistorized data recorder *Geodat 122* or *Geodat 124-Alpha* (alphanumeric). The *Geodimeter 112* can also be used in the tracking mode.

The *Geodimeter 116 (Fig. 9.18)* can be used in both the normal mode and the tracking mode, and the angle between the measuring beam and the vertical is measured by an electrical pendulum with an accuracy of ±20 mgon (as a supplementary attachment a tilt-meter of the accuracy of ±10 mgon can also be built in). Slope distance, horizontal distance, elevation difference and the elevation angle of the measuring beam can be displayed in both modes, and in the normal mode the results can be recorded by the devices *Geodat 122* or *Geodat 124-Alpha*. By means of the separate *R.O.E.* unit (*Remote Object Elevation* determination) the elevation of near-inaccessible points can be measured indirectly. In such cases the prism must be set up at an accessible height in the vertical of the point to be measured. After the

Figure 9.18. The distance meter *Geodimeter 116* attached to the theodolite *Wild T2* (AGA, Sweden)

measurement has been carried out by observing the prism, the computer continuously computes the intersection of the line of sight and the plumb line of the prism by means of the elevation angle between the horizontal and the measuring beam. By pointing the line of sight to the point to be measured, using the vertical slow motion screw, the elevation of the point can be read.

Two extra attachments can be purchased which facilitate setting out. With the *Unicom* the observer at the instrument can transmit commands to the person holding the reflector prism by means of a one-way light-telephone. The infrared measuring beam of the distance meter is modulated by the audio-signal. The audio-signal, demodulated by the demodulator beside the prism, can be heard from the loudspeaker fixed on the prism-supporting pole. The *Tracklight* is a light-projector device which can be mounted on the handle of the instrument. The person holding the prism sees a red or green flashing light at the instrument, depending on the direction in which he is out of its line of sight. The red and green light sectors are separated by a white light giving frequent double flashes. The divergence of the white light beam is equal to the divergence of the measuring beam, so that setting out with Unicom is possible within the divergence angle of the white light-beam.

In the *Geodimeter 122 (Fig. 9.19)* all the additionally purchasable accessories of the *Geodimeter 116* — except *R.O.E.* — are built in, namely a more precise tilt-meter, *Unicom* and *Tracklight*.

Figure 9.19. The distance meter *Geodimeter 122* attached to the theodolite *Wild T2* (AGA, Sweden)

The *Geodimeter 220* unites the useful characteristics of all the former types. The *R.O.E.* unit and *Unicom* are built into the instrument itself, while the *Tracklight* unit can be adapted to be attached to the bottom of the instrument. Before measurement both the multiplying constant (meteorological constant) and the additive constant can be entered into the computing unit. In the "tracking" operation mode the new measured result appears on the liquid-crystal display every 0.4 second. The reliability of the automatic tiltmeter, needed for the reduction of the slope distance to the horizontal, is ± 20 sec of arc if the elevation angle does not exceed 20°.

Figure 9.20. The distance meter *Fen 2000* (Fennel, FRG)

The *Geodimeter 210* is a long range variant of the *220.*

It is worth mentioning that the novelty of the distance meters *Fen 2000 (Fig. 9.20)* and *Fen 4000* (Fennel, FRG) is that the distance measurement is based on the measurement of the travel time of pulses of infrared radiation.

The light source is a laser-diode, and the computer unit statistically processes the transmitted and reflected series of pulses and then calculates the travel time of the radiated energy, and from that the distance measured. The advantage of this solution is that, for distances shorter than about 300 m, very simple reflectors pressed from plastic, such as a cycle rear reflector, can be used for distance measurement and fine pointing within the divergence cone of the beam of aperture 6 min of arc is unnecessary. The time needed for distance measurement with normal distance meters is 1 s, with the "rapid" types it is only 0.01 s, and the standard deviation of the measurement in the latter case is only ± 2 cm. The distance meter can be attached to the upper part of a theodolite, but it can also be attached to a ranging rod and used as an "electronic measuring tape". It has only three controlling elements: the main switch, the keyboard for entering the atmospheric correction and the start button. The data display is digital and the smallest displayed quantity is 1 mm.

In the variant *Fen 2000 S* the shortest measurable distance is only a few centimeters (practically zero) and the standard deviation of measurement over the whole measuring range is ± 5 mm. The type *Fen 10,000 Marine* has been made for hydrographic purposes. The reliability of measurement up to 20 m is ± 20 mm, over longer distances (up to 10 km) ± 10 cm. The time needed for a measurement is only 0.01 sec, hence the instrument is primarily suitable for determining the distance to a moving object.

The characteristic of the type *Fen 101* is that, up to a certain distance (which is 50 m for the variant "A", 100 m for the variant "B" and 250 m for the variant "C"), the

Table 9.3. Technical data and characteristics of so

Type	Manufacturing firm	Optics Transmitter (I) Receiver (II) Coaxial (III) Biaxial (IV) Ø (mm)	Scale of fine measurement (m)	Phase-measurement A = analogue, D = digital	Power requirement (W)	Range (km) Number of prisms		Mass (kg)
1	2	3	4	5	6	7		8
Geodimeter 10	AGA (Sweden)	IV 60	10	*D*	17	0.7 1	1.2 3	2.5
Geodimeter 12 A		IV	10	*D*	17	1.2 1	3.0 8	2.5
Geodimeter 14 A (16)		IV 60	10	*D*	17	6.0 1	15.0 (10.0) 16	2.5
Geodimeter 76 (78)		IV	10	*D*	60	1.5 (5.0) 1	5.0 (10.0) 3 (9)	8.2
Geodimeter 110 (112)		IV 60	10	*D*	17	1.4 (2.5) 1	3.1 (5.5) 8	2.6
Geodimeter 116 120 122		IV 60	10	*D*	17	1.0/1 1.2/1 2.5/1	2.3/8 3.0/8 5.5/8	2.8
Geodimeter 220 210		IV	10	*D*	4	1.6/1 2.3/1	3.2/8 4.5/8	1.3
DM 60	Cubic Corp. (US)	IV 50	2	*D*	15	1.0 1	2.0 3	6.8
HDM 70 Minitape	Cubic Western	IV		*D*			2.0 3	

-range electro-optical distance meters

mensions (cm)	Standard error of distance measurement $\pm(A+B\times10^{-6}D)$		Degree of automation	Additive/multiplicative constant can be entered	Construction of the instrument	Pointing	Computation of D_v and D_H	Tracking	Remarks
	A (mm)	B (mm/km)							
9	10		11	12	13	14	15	16	17
× 19 × 9	10		3	N/Y	3	3			
× 19 × 9	5	5	3	N/Y	3	3			
× 19 × 9	normal 5 stand. dev.-checking	 10	3	N/Y	3	3			see Fig. 9.16
× 14 × 19	10	1	3	N/Y	1	1			light source: He–Ne laser modulator; Kerr cell
× 19 × 9	normal 5 stand. dev.-checking 5	 5 3	3	N/Y	3	3			no stand. dev.-checking with type 110; see Figs 9.18, 9.19
× 19 × 9	normal 5 stand. dev.-checking 5	 5 3	3	N/Y	3	3	2	1 + 2 + 3	no stand. dev.-checking with type 116
5 × 11 × 9	normal 5 stand. dev.-checking 5	 5 3	3	Y/Y	3	3	2	1 + 2 + 3	
× 18 × 36	5	10	2	N/Y	3	2 + 4			
	5	5	3	N/Y	3	3 + 4	2	1 + 2 + 3	

Table 9.3. (continued)

Type	Manufacturing firm	Optics Transmitter (I) Receiver (II) Coaxial (III) Biaxial (IV) Ø (mm)	Scale of fine measurement (m)	Phase-measurement A = analogue, D = digital	Power requirement (W)	Range (km)		Ma (k
						Number of prisms		
1	2	3	4	5	6	7		8
Fen 2000 (4000)	Fennel (FRG)	III	measures travel time	display D		2.0 (2.5) 1	2.5 (4.5) 2 (6)	2.
Fen 10,000 Marine		III	measures travel time	display D		2.5/1	10/6	2.
Fen 101 A B C		III	measures travel time			without reflector device 0.05 0.10 0.25		3. 3. 3.
HP 2800 A B	Hewlett–Packard (US)	III I = 50 II = 110	10	A	12	3.0 3		7.
HP 3085 A		III I = 50 II = 110	10	D	18	0.5 1	1.6 3	10.
HP 3808 A		III I = 50 II = 110	10	D	18	3.0 1	10.0 6	9.
HP 3810 A		III I = 50 II = 110	10	D	18	0.5 1	1.6 3	11.
EOK 2000	Jenoptik (GDR)	IV I = 32 II = 35	5	A	8	2.5 9		12.
EOT 2000		III 40	10	D	13	0.4 1	2.0 21	11.
DM 1000 (2000)	Kern (Switzerland)	III 110	10	D	11	1.5 1	3.0 7	10.
DM 500 (501)		III 28	10	D	11 (6)	0.3 (1.0) 1	0.5 (1.6) 3	1.

Table 9.3. (continued)

imensions (cm)	Standard error of distance measurement $\pm(A + B \times 10^{-6}D)$ A (mm)	B (mm/km)	Degree of automation	Additive/multiplicative constant can be entered	Construction of the instrument	Pointing	Computation of D_v and D_H	Tracking	Remarks
9	10		11	12	13	14	15	16	17
× 12 × 10	5	5	3	N/Y	3	2+4			light source: infrared laser diode in pulse mode; see Fig. 9.20
× 12 × 10	100		3	Y/Y	3	2+4		(1)	light source: infrared laser diode in pulse mode
× 12 × 10 × 12 × 10 × 12 × 10	10 10 100		3	Y/Y	3	2+4		(1)	light source: infrared laser diode in pulse mode
× 31 × 15	5	7	1	N/Y	1	1			distance measurement with type A in feet
× 26 × 15	5	10	3	N/Y	1	1			
× 32 × 28	5	1	2	N/Y	1	1			
× 26 × 15	5	10	3	N/Y	1	1	2	1+2+3	
× 35 × 29	10		1	Y/N	1	1			
	10		2	N/Y	4	3+4	1		
× 24 × 26	10		2	N/Y	1	1			
× 7 × 17	10 (5)	(5)	2 (3)	N/N	3	3+4		(1)	

Table 9.3. (continued)

Type	Manufacturing firm	Optics Transmitter (I) Receiver (II) Coaxial (III) Biaxial (IV) Ø (mm)	Scale of fine measurement (m)	Phase-measurement *A* = analogue, *D* = digital	Power requirement (*W*)	Range (km)		M (…
						Number of prisms		
1	2	3	4	5	6	7		
DM 502 (503)	Kern (Switzerland)	III 28	10	*D*	4	1.2 (2.5) 1	2.0 (5.0) 3	1
DM 102		III 28	10	*D*	4	1.2 1	1.7 3	1
Autoranger	Keuffel & Esser (US)	III 40	10	*D*	15	1.0 1	3.0 2	2
Autoranger A		III	10	*D*		1.6 1		
Autoranger III X		III	10	*D*		2.0 1		3
Microranger	Laser System & Electronics (US)	III	10	*D*		1.0 1	1.5 3	3
ND 160 (250)	Nikon (Japan)	III	10	*D*	9	1.6 (2.5) 3		3.
PM-81	Pentax (Japan)	III	10	*D*	2.5	1.4 1	2.0 3	2.
PX 10D		III 54	10	*D*		1.4 1	1.7 3	6.
Beetle 500S 1000S 1600S	Precision International Inc. (US)	III	10	*D*	18	0.5/1 0.7/1 1.0/1	1.0/3 1.4/3 1.6/3	5.
Citation CI-410 (450)		III	10	*D*	9	1.0 (1.6) 1	2.2 (3.0) 11	2. (2.
SDM 1C (3C)	Sokkisha (Japan)	III	10	*D*	10	1.0 1	1.6 3	3. (17.
SDM 3EG		III	10	*D*	9	2.5		7.
RED 2		III	10	*D*		2.0 1	5.0 9	2.
RED MINI		III 40	10	*D*	3	0.3 1	0.5 3	0.

Table 9.3. (continued)

Dimensions (cm)	Standard error of distance measurement $\pm(A+B\times 10^{-6}D)$		Degree of automation	Additive/multiplicative constant can be entered	Construction of the instrument	Pointing	Computation of D_v and D_H	Tracking	Remarks
	A (mm)	B (mm/km)							
9	10		11	12	13	14	15	16	17
18 × 7 × 17	3	5 (2)	3	(Y)/(Y)	3	3 + 4		1	see Figs 9.8, 9.9
18 × 7 × 17	5		3	N/N	3	3		1	see Fig. 9.10
22 × 14 × 11	5	6	3	N/Y	3	3		1	
	1...2	3	3	N/Y	3	3		1	
	5	5	3	N/Y	3	2 + 4	1	1	
	5	2	3	N/Y	3	3		1	
	5	5	3	Y/Y	3	2 + 4	2	1 + 2 + 3	
18 × 21 × 19	5	5	3	Y/Y	3	2	1	1	see Figs 9.22, 9.23
16 × 19 × 35	5	5	3	Y/Y	4	3 + 4	1	1	
	10		3	N/Y	3	2	1	1	
20 × 15 × 12	5	5	3	Y/Y	3	3	1	1	see Fig. 9.21
	5	5	2		3 (2)	3 (3 + 4)			
	5	5	3	Y/Y	4	3 + 4	1	1	
14 × 18 × 9	5	5	3	Y/Y	3	2	1	1	
17 × 8 × 8	5		3	Y/Y	3	3	1	1	

Table 9.3. (continued)

Type	Manufacturing firm	Optics Transmitter (I) Receiver (II) Coaxial (III) Biaxial (IV) Ø (mm)	Scale of fine mea-sure-ment (m)	Phase-mea-sure-ment *A* = ana-logue, *D* = dig-ital	Power re-quire-ment (*W*)	Range (km) Number of prisms		Mass (kg)
1	2	3	4	5	6	7		8
SM 3	(Soviet Union)	III	5	*A*	5	2.0		10.0
MSD-1M		IV	1	*A*	2.5	0.3 1	0.5 3	10.0
MA 100	Tellurometer (UK)	III 175	2	*D*	15	1.0 1	2.0 4	17.0
CD-6		IV 60	10	*D*	18	0.8 1	2.0 7	3.7
DM A1 (C3)	Topcon (Japan)	III	10	*D*	8 (10)	2.5 (3.2) 3		2.6 (2.8)
GTS 2		III 40	10	*D*		1.4 1	2.0 3	6.0
Distomat DI 10	Wild (Switzerland)	IV 70	10	*A*	15	0.6 3	2.0 9	3.0 + 15.0
Distomat DI 3 (S)		IV 35	20	*D*	14 (17)	0.4 (1.0) 3	0.9 (2.0) 9	2.0 + 5.0 (4.0)
Distomat DI 4 (L)		IV 35	30	*D*	5	1.0 (2.5) 1	2.0 (5.0) 11	2.0
Distomat DI 20		IV 50	33	*D*	5	6.0 1	9.0 11	3.7
Distomat DI 5		IV 35	30	*D*	5	2.5/1	5.0/11	2.0
Distomat DI 1000		IV 50	33	*D*	4	0.5/1	0.8/3	1.1
Distomat DI 3000		IV 50	meas. travel time	*D*	5	6.0 1	9.0 11	1.7
SM 11	Opton (FRG)	III 120	10	*D*	12	0.5 1	2.0 19	14
Eldi 1		III 110	10	*D*	5	I.2.0 II.2.5 1	4.0 5.0 19	8

Table 9.3. (continued)

imensions (cm)	Standard error of distance measurement $\pm(A+B\times10^{-6}D)$		Degree of automation	Additive/multiplicative constant can be entered	Construction of the instrument	Pointing	Computation of D_v and D_H	Tracking	Remarks
	A (mm)	B (mm/km)							
9	10		11	12	13	14	15	16	17
	10		1	Y/N	1	1			
	2		1	N/N	1	1			damp-proof
× 37 × 28	1.5	1	2	N/Y	1	1			
× 14 × 8	5	5	2	N/Y	3	2 + 4		1	
	5	5	3	Y/Y	3	2 + 4	1	(1)	
× 20 × 17	5	5	3	Y/Y	4	3 + 4	1	1	
× 15 × 10 + × 19 × 36	10		2	Y/N	2 + 3	3			
× 11 × 7 + × 28 × 9 (7)	5	5	3	Y/Y	2 + 3	3	1	1	
	5	5	3	Y/Y	3	3 + 4	1	1	see Figs 9.5, 9.6
× 19 × 9	5	1	3	Y/Y	3	3	1	1	see Fig. 9.7
	5	5	3	Y/Y	3	3 + (4)	1	1	
	5	5	3	Y/Y	3	3 + (4)	1	1	
	5	1	3	Y/Y	3	3	1	1	light source: infrared laser diode in pulse mode
× 18 × 46	10		2	N/Y	4	3 + 4			
× 12 × 33	I.5 . . . 10 II.10 20	2 2	2	N/Y	1	1			three-fold electrical amplification in measuring range 2 (Fig. 9.12)

Table 9.3. (continued)

Type	Manufacturing firm	Optics Transmitter (I) Receiver (II) Coaxial (III) Biaxial (IV) Ø (mm)	Scale of fine measurement (m)	Phase-measurement A=analogue, D=digital	Power requirement (W)	Range (km) / Number of prisms		Ma (k
1	2	3	4	5	6	7		8
Eldi 2	Opton (FRG)	III 60	10	*D*	5	I.0.7 II.1.0 1	2.0 3.0 19	4.
Eldi 3		III 40	10	*D*	5	I.0.4 II.0.7 1	I.1.0 1.5 7	3.
SM 4		III 40	10	*D*	5	I.0.7 II.1.0 1	1.5 2.0 7	7
SM 41		III 40	10	*D*		I.1.0 II.1.4 3	1.5 1.8 9	6
RSM 3 (4)		III 40	10	*D*		I.1.0 II.1.5 1	1.6 2.0 3	7

Legend of symbols used in the Table:
In column 11: 1=non-automated; 2=partly automated; 3=wholly automated
In column 12: N=no; Y=yes
In column 13: 1=independent; 2=transmitter and receiver units seperated; 3=attachable to a theod
4=built-in one instrument.
In column 14: 1=distance meter and theodolite interchangeable; 2=connected but separate poin
3=connected and common pointing; 4=telescope can be transited
In column 15: 1=semi-automatic; 2=automatic
In column 16: 1=slope distance; 2=horizontal distance; 3=elevation difference
See also *Figures 9.21, 9.22* and *9.23.*

energy of the beam reflected from a relatively smooth surface at the end of the distance to be measured (e.g. a concrete wall) is sufficient to carry out the measurement. The accuracy of the measurement is ±1 cm with variants “A” and “B”, and ±10 cm with variant “C”. In addition to the sighting telescope it is advisable to use a small He–Ne laser for marking the end point of the distance to be measured.

In *Table 9.3* the technical data and characteristics of the short-range electro-optical distance meters are listed. Some comments are necessary for the correct interpretation.

— Distance measurement has been taken as semi-automated if the signal level received before measurement has to be regulated by the the observer, even if the measurement takes place automatically after pressing the start button.

Table 9.3. (continued)

Dimensions (cm)	Standard error of distance measurement $\pm(A+B\times10^{-6}D)$		Degree of automation	Additive/multiplicative constant can be entered	Construction of the instrument	Pointing	Computation of D_v and D_H	Tracking	Remarks
	A (mm)	B (mm/km)							
9	10		11	12	13	14	15	16	17
4 × 12 × 16	I.5 II.10. 20	2 2	2	N/Y	3	2+4			as Eldi 1 type Eldi 2 Mining is damp-proof (Fig. 9.11)
4 × 12 × 16	I.5. . .10 II.10. . .	2	2	N/Y	3	2+4			as Eldi 1 see Fig. 9.13
× 34 × 13	I.5. . . II.10. 20	2 2	2	N/Y	4	3+4			as Eldi 1 see Fig. 9.14
× 23 × 15	I.10 II.20	2 2	3	N/Y	4	3+4			as Eldi 1
× 37 × 15	5. . .10	2	3	N/Y	4	3+4	2	1+2+3	circle graduation: RSM 3 360 RSM 4 400^{g} (Fig. 9.15)

— By entering the additive constant different prism-constants can be taken into account. In some types of instrument the constants of two types of prism are pre-programmed. The multiplying constant is usually the atmospheric correction, but it may also be a combined correction: the scale factor expressing the reduction to sea level and the map projection reduction.

— A great number of types of instrument can be used independently as well as combined with a theodolite. From the point of view of practical application, the combination with a theodolite is the more valuable. In the case of attachable distance meters no distinction has been made between those that can be used with the

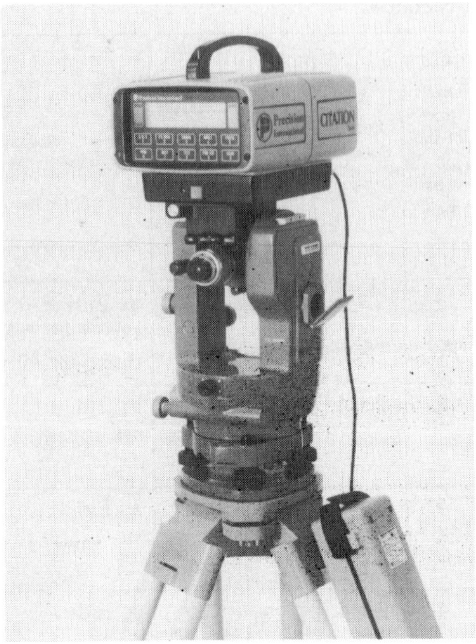

Figure 9.21. The distance meter *Citation CI-450* attached to the theodolite *Wild T2* (Precision International, US)

Figure 9.22. The distance meter *PM-81* (Pentax, Japan)

Figure 9.23. The distance meter *PM-81* attached to the theodolite *TH-10 D* (Pentax, Japan)

theodeolites of only one manufacturer and those that can be used with the theodolites of several.

— Pointing has been taken as simultaneous if the distance meter is attached to the theodolite telescope and it has been assumed that the line of sight and the measuring beam are aligned.

— The calculation of horizontal distance and elevation difference has been taken as semi-automated if the observer has to enter the reading of the vertical circle into the computer unit for the reduction.

9.3.3.3 High Precision Distance Meters

The usual accuracy of ±5 to 10 mm of distance meters constructed for short-range measurement is not always sufficient. High precision setting out deformation analysis some geodynamic tasks and, last but not least, the calibration of the above-mentioned meters require a distance meter with an inner precision of some tenths of a millimetre.

The high precision electro-optical distance meter *Mekometer ME-3000* (Kern, Switzerland) has been developed in the National Physical Laboratory (UK) and some of its electronic units are manufactured by COM-RAD Electronic Equipment Ltd (UK). The instrument itself is shown in *Fig. 9.24*, its block-diagram in *Fig. 9.25*. The light source *1* (xenon flash-lamp) produces light pulses of duration 1.5 μs and of frequency 100 Hz, and the effective wavelength of the radiation (i.e. that belonging to the centre of gravity of the area below the curve of energy distribution) is 0.485 μm. The light falls on the modulator-cavity resonator *3* after passing through the

Figure 9.24. The distance meter *Mekometer ME-3000* (Kern, Switzerland)

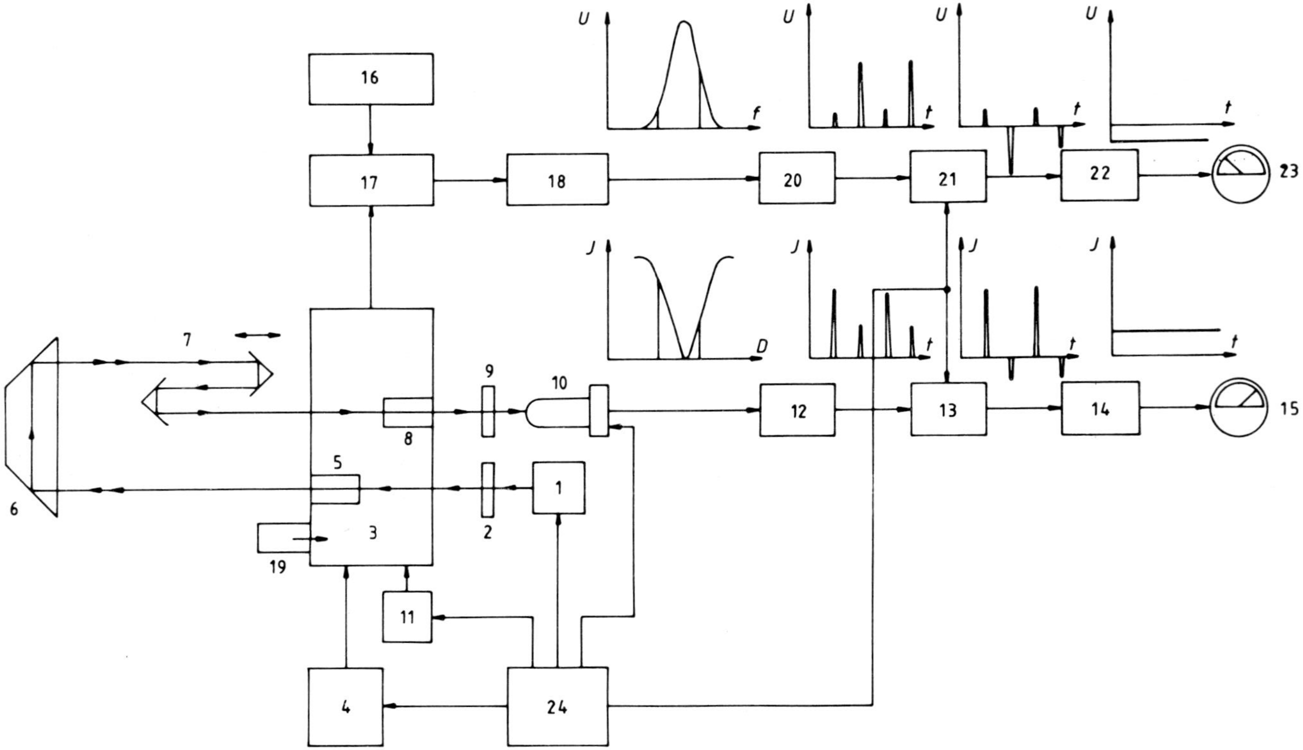

Figure 9.25. Block diagram of the electro-optical distance meter *Mekometer ME-3000* (Kern, Switzerland): *1* light source (xenon flash-lamp); *2* polarizer; *3* modulator cavity resonator; *4* modulating signal generator; *5* electro-optical crystal (modulator); *6* reflector device; *7* optical phase shifter; *8* electro-optical crystal (phase-detector); *9* analyzer; *10* photomultiplier; *11* varactor diode; *12* detector; *13* synchronizer detector; *14* mtegrator; *15* micro-ammeter (zero-indicator); *16* subsignal generator; *17* mixer and nine-fold frequency multiplier; *18* standard cavity resonator; *19* tuner piston of the modulator cavity resonator; *20* detector; *21* synchronizer detector; *22* integrator; *23* milli-voltmeter (resonance-indicator); *24* control unit

polarizer *2*. The basic modulating frequency of about 500 MHz is produced from the 15 MHz signal of the quartz oscillator *4* by a multi-stage frequency multiplier. The frequency of the modulating pulses is also 100 Hz, their duration 40 μs, and the phase is such that the modulating pulse covers the whole light pulse. The optical properties of the potassium dihydrogen-phosphate (KDP) crystal *5* in the modulator cavity are changed in the rhythm of the modulating signal, in consequence of which the light emitted from the modulator is elliptically polarized and the oblateness changes periodically. The light is reflected from the reflector device *6* set up at the other end of the distance to be measured, passes through the optical phase-shifter *7*, and then passes through the second KDP crystal *8* in the modulator cavity; this crystal acts as the phase-detector. Now the light passes through the analyzer *9* and meets the cathode of the photomultiplier *10*. If the oscillation transmitting planes of the analyzer and of the polarizer are perpendicular to each other, the photo-current is:

$$I=\frac{1}{2}\left\{1-\cos eJ_0\left[\frac{2\pi U}{U_{\max}}\cos\frac{2\pi D}{\lambda}\right]\right\}, \qquad (9.35)$$

where e = the oblateness expressing the degree of elliptical polarization,
J_0 = the zero order Bessel function,
U = the modulating voltage,
$U_{\max}$ = the voltage necessary to achieve maximum electro-optical effect,
D = the distance to be measured (more exactly half the distance travelled by the modulated light).

The light current is minimum (almost zero) if $D=N\frac{\lambda}{2}+\frac{\lambda}{4}$, or $2D=(2N+1)\frac{\lambda}{2}$, i.e. if the path travelled by the modulated light is an odd multiple of the modulating half wavelength. Light minimum can be obtained by changing the optical phase-shifting; a micro-ammeter connected to the anode-current circuit of the photomultiplier acts as an indicator.

The photo-current changes slowly in the neighbourhood of its minimum, which makes it difficult to adjust the optical phase-shifter correctly. In order to increase accuracy, the modulating frequency produced by the oscillator and the frequency multiplier is rhythmically changed with a frequency of 100 Hz between the values $f+\Delta f$ and $f-\Delta f$. The abrupt frequency change Δf is produced by the varactor diode *11* changing the capacity of the modulator cavity. The output signal of the detector *12* is thus two series of pulses, the elements of which belong alternately firstly to the frequency $f+\Delta f$ and then to the frequency $f-\Delta f$ (i.e. to the respective modulating wavelength). The synchronizing detector *13* changes the polarity of one of the series, and the integrator *14* produces a resultant signal level proportional to the intensity difference between the two series; this signal level is indicated by the micro-ammeter *15*.

If the optical phase-shifter is correctly adjusted, the two pulse series consist of elements with equal absolute values, thus the resultant signal is zero. In the opposite case the pointer of the micro-ammeter is deflected due to the effect of the resultant signal (see *Fig. 9.25* which shows this position). With zero resultant signal one can read out from the optical phase-shifter the fine reading of the distance measured, which is less than the 30 cm scale of the fine measurement.

Besides the exact determination of the fine reading, the other condition of high precision distance measurement is a properly stabilized measuring frequency pattern. A so-called $\lambda/4$ standard cavity resonator is used for the stabilization of the 500 MHz basic frequency pattern of the distance meter. In order to decrease the dimensions of the cavity resonator, the resonance frequency is 4.5 GHz, consequently the frequency pattern must be multiplied nine times. The frequency pattern series, which differ from each other by a small amount (the distance meter establishes a series of indirect distance scales), are produced from the basic frequency by adding the signal of the subsignal generator *16*, and the mixing and the nine-fold frequency multiplication takes place in the mixer and frequency multiplier stages *17*. The nine times higher measuring frequency pattern goes to the standard cavity resonator *18*. The tuning plunge *19* of the measuring cavity resonator is used to set the resonance exactly. In the resonant state the output voltage of the standard cavity resonator is maximum, thus the tuning can be checked by the milli-voltmeter connected to the output circuit. Due to the pulsing of the modulating frequency, however, the output signal of the detector *20* is a series of pulses the elements of which correspond alternately firstly to the point with abscissa $f+\Delta f$ and then to the point with abscissa $f-\Delta f$ of the resonance curve. The synchronizing detector *21* reverses the polarity of one of the pulse series, and the integrator *22* produces a resultant signal level proportional to the intensity difference between the two pulse series, which can be read on the milli-voltmeter *23*.

If the measuring frequency is exactly tuned, i.e. $9f = f_r$, the two pulse series consist of equal elements, and the resultant signal is thus zero. If not (the situation shown in *Fig. 9.25*), the pointer of the milli-voltmeter deflects. The distance measurement is precise only if the measuring frequency is adjusted to give resonance, i.e. the pointer of the milli-voltmeter is central.

If there is a vacuum in the standard cavity resonator, the resonant frequency is $f_r = \dfrac{c}{F(L)}$, where c is the propagation velocity of electromagnetic waves in vacuo and $F(L)$ is a so-called shape-factor depending on the shape and dimensions of the resonator. Thus, in order to ensure the stability of the pattern frequency, it is necessary that the resonator dimensions should not change. In the *Mekometer* the resonator is connected to the inner air-space of the instrument by a very small aperture, therefore the resonant frequency is $f_r = \dfrac{c}{n_m F(L)}$, where n_m is the microwave refractive index of the air filling the cavity. Although in this case the resonant frequency changes from moment to moment, due to the variation of the refractive index of the air, the wavelength corresponding to the resonance remains constant. In this way the instrument produces a length unit independent of the condition of the surrounding air, i.e. scale stabilization is established instead of frequency stabilization.

The wavelength of the light propagating in air, corresponding to the resonance frequency f_r, is $\lambda = \dfrac{c}{n_{gr} \cdot f_r}$, where n_{gr} is the group-refractive index of the light at the effective wavelength. Substituting the resonant frequency gives:

$$\lambda = \frac{n_m}{n_{gr}} \cdot F(L). \tag{9.36}$$

Hence one of two conditions of scale stabilization is that the dimensions of the standard cavity resonator remain unchanged. By properly shaping the resonator (a quartz-resonator body fixed in a copper or aluminium housing) the frequency change caused by thermal expansion can be kept to a minimum, and this small variation (which can also be caused by the ageing of the elements) can be determined from a temperature-frequency connection defined by repeated calibration of the instrument.

Another condition of scale stabilization is that the ratio n_m/n_{gr} must be constant. Since the microwave refractive index of air strongly depends on partial water vapour pressure, and at the same time the light refractive index is practically independent of it, this condition can only be fulfilled if the air filling the standard cavity resonator is dry. Therefore the air enters the cavity resonator through a drying cartridge fixed in front of the intercommunicating aperture of the resonator.

For noise protection, the photomultiplier amplifies only during light pulses. The properly timed pulses of the control unit *24* ensure a synchronized operation of the units. This controller controls the light source, the measuring signal generator, the varactor diode producing the frequency pulsing, the photomultiplier and the synchronizing detector.

The polarity of the pulse series produced by the synchronizing detectors can be reversed during fine measurement, thus the electrical zero point error and the error of the mechanical zero point of the indicator instruments do not affect the result.

In addition to the 30 cm scale of fine measurement there are further scales of: 3 m, 30 m, 300 m and 3000 m, thus distance measurement is unambiguous within the 3000 m range of the instrument. The position of the optical phase-shifter is read out by an electronic device; the calculation of differences needed for the interpretation of the results of the coarse measurements, as well as their reduction, are calculated by a computer unit. The display units have seven segments formed by hot wires (heating filaments), the smallest unit displayed being 0.1 mm. The inner accuracy of distance measurement is $\pm(0.2\ \text{mm}+1\ \text{mm/km})$, the instrument mass is 16 kg, its power requirement is 16 W, and the power source is a rechargeable 12 V battery built into an external current converter.

The electro-optical distance meter *Geomensor 204* is manufactured by COM-RAD Electronic Equipment Ltd (UK); the same firm produced some electronic units for the *Mekometer ME-3000*. An interesting feature of the distance measuring method is that the measuring (pattern) frequency can be changed exactly, within certain limits, thus the path travelled by the light can be set to be an integer multiple of the modulating wavelength. The changeable measuring frequency makes the use of an optical phase-shifter unnecessary.

The light source of the meter is a xenon flash lamp as used already in the *Mekometer*, and the methods of modulation and light detection are also the same. The minimum of the photo-current, proportional to the detected light intensity, is set by changing the modulating frequency and is indicated on the micro-ammeter. With minimum photo-current:

$$2D=N_i\lambda_i=N_i\frac{c}{n_i f_i}, \tag{9.37}$$

where D is the distance to be measured, N_i an integer yet unknown, c is the propagation velocity of electromagnetic radiation in vacuo, n_i is the average of the group refractive index of air taken along the line, f_i is the modulating frequency which must be measured.

For the measurement of the distance (for determining the integer N with sufficient accuracy) the modulating frequency must be measured at several minima. The following hold for any two minima:

$$n_0 f_0 - n_M f_M = M \frac{c}{2D}, \tag{9.38}$$

and

$$2D = M \frac{n_i f_i}{n_0 f_0 - n_M f_M} \cdot \frac{c}{n_i f_i}, \tag{9.39}$$

where f_i is an arbitrary measuring frequency, but belonging to a light minimum between the values f_0 and f_M, the term

$$M \frac{n_i f_i}{n_0 f_0 - n_M f_M}$$

is an integer from Eq. (9.37), and M is also an integer;

$$M = M' \frac{n_0 f_0 - n_M f_M}{n_0 f_0 - n_{M'} f_{M'}}, \tag{9.40}$$

where M' is also an integer:

$$M' = \frac{n_0 f_0 - n_{M'} f_{M'}}{n_0 f_0 - n_1 f_1}. \tag{9.41}$$

In Eq. (9.41) f_0 and f_1 are modulating frequencies corresponding to two adjacent minima.

It should be mentioned that the subscripts 0, 1, M' and M of the refractive index n indicate that the average of the refractive index along the line cannot be regarded as constant; the averages of the refractive index along the line must be determined during all the four (f_0, f_1, f'_M, f_M) frequency measurements.

The limit of unambiguous distance measurement depends on the range of frequency change. With the distance meter *Geomensor 204* the distance can be unambiguously determined up to 10 km from the values measured at the four minima f_0, f_1, f'_M, f_M.

Measurement and calculation are carried out as follows.

a) After setting one minimum of the photo-current, n_0 is determined and f_0 measured (the method of determining the refractive index is described later).

b) After setting the adjacent minimum, n_1 is determined and f_1 measured.

ca. *If $D < 100$ m:*

caa. Putting $M = 1$ the two approximate distances D_i are computed, where $i = 0, 1$.

cab. The mean value of the distance is computed and displayed.

cac. After setting an arbitrary minimum, n_j is determined and f_j measured.

cad. By means of the approximate distance the precise distance D_j is computed.
cae. The previous steps are repeated, and the mean value and the standard deviation displayed, until the measurement is regarded as sufficiently precise.

cb. *If 100 m* $< D < 750$ *m:*
cba. The modulating frequency is increased by about 10 MHz and, after setting the first minimum, n_M is determined and f_M measured.
cbb. M is calculated with $M' = 1$.
cbc. Using this value of M, the approximate distance D_i is computed, where $i = 0$, 1, M.
cbd. The mean values of the distance are computed and displayed.
cbe. After setting an arbitrary minimum, n_j is determined and f_j measured.
cbf. Using the approximate distance the exact distance D_j is computed.
cbg. The previous steps are repeated, and the mean value and the standard deviation displayed until the measurement is regarded as sufficiently precise.

cc. *If* $D > 750$ *m:*
cca. The modulating frequency is increased by about 500 kHz, the first minimum is set, $n_{M'}$ is determined, $f_{M'}$ measured.
ccb. The modulating frequency is increased by about 10 MHz, the first minimum is set and n_M is determined, f_M measured.
ccc. M' and M are computed from the results of the two measurements.
ccd. From the computed values of M' and M the approximate distances D_i are computed, where $i = 0$, 1, M', M.
cce. The mean value and the standard deviation of the distance is computed and displayed.
ccf. The arbitrary minimum is set, n_j is determined and f_j measured.
ccg. The exact distance D_j is computed by means of the approximate distance.
cch. The previous two steps are repeated, and the mean value and standard deviation displayed until the measurement is regarded as sufficiently precise.

The reliability of frequency measurement is better than 10^{-7}, therefore the internal accuracy of the instrument is $\pm(0.1\ \text{mm} + 0.1\ \text{mm/km})$. Due to the high internal precision the reliability of distance measurement depends on the accuracy of the determination of the average refractive index along the line.

The instrument has a built-in module which is a sensor of air temperature and pressure and which computes from the atmospheric data the refractive index correction and the atmospheric correction, respectively, with a reliability of $\pm 5 \times 10^{-7}$. The computed value of the atmospheric correction can be displayed and, at the command of the observer, both the measured and the corrected distance can be displayed. Further external module sensor units are commercially available; if they are set up along the line to be measured they compute automatically the air temperature and pressure, also the atmospheric correction, and transmit them together with their distance from the instrument by radio. The selective receiver, which can be built into the instrument, can receive information from a maximum of four external sensors, and the approximate mean refractive index along the line is computed by a microcomputer.

For the fine adjustment of the minima required for the measurement, the photo-current in the anode current circuit of the photomultiplier can be zeroed by fine tuning the modulating frequency. The instantaneous photo-current is indicated by the micro-ammeter. Due to the rapid variations of the refractive index along the line, the wavelength of the modulating frequency also changes, as may be seen by the oscillation of the pointer of the instrument around its central position. To increase the reliability of zero determination, the modulating frequency is controlled by feed-back of the photo-current. A particular intensity and direction of the photocurrent activates a servo device which changes the modulating frequency in such a way that the photo-current is zeroed. Thus the measuring (pattern) frequency does not remain constant after fine tuning has been completed, but it oscillates in accordance with the frequency measurement which is a 10 second average of its fluctuations.

On the basis of the oscillation amplitude and the possible tendency to change, the microcomputer in the instrument calculates the amount by which a 10 second average can deviate from the average that would be obtained by measurements carried out over a properly long time interval. This probable deviation of the 10 s mean value can be displayed. If necessary, the measurement program can be extended by a further 10 s measurement. By displaying the previous average and probable deviation, the observer can judge the increasing effect of the repeated measurement on the accuracy.

The main technical data of the instrument are: mean square error of the instrument constant: ± 0.1 mm; the internal error component depending on the distance: 0.1 mm/km; the mean square error of the atmospheric correction: $\pm 5 \times 10^{-7}$. The liquid crystal display has eight digits, the smallest displayed value is 0.1 mm. The effective range with 30 km visibility is 5 km with one prism, 8 km with three prisms and 10 km with seven prisms. The time to heat up is 3 s, the time necessary for a single distance measurement is 1 min. The mass of the instrument is 26 kg, its power requirement 24 W, and a 12 V accumulator may be used as an external power source.

9.3.4 The Accuracy of Electro-optical Distance Measurement

The standard error of electro-optical distance measurement is usually given as the sum of a component independent of the distance and a component proportional to the distance:

$$m_D = \pm(A + B \times 10^{-6} D).$$

The components characterizing modern distance meters are $A \sim 5$ to 10 mm, $B \sim 1$ to 2 mm/km. The accuracy of long- and medium-range distance meters is mainly determined by the component B, the accuracy of short-range meters is mainly determined by the component A.

According to a usual but somewhat simplified concept, A is the standard error of the fine measurement, i.e. the characteristic of the reliability of phase measurement, while B is the standard error of the scale of the fine measurement, i.e. the characteristic of the reliability of frequency stability. In reality the distance-independent error component A contains also the standard error of the additive constant, and the distance-dependent error component B also depends on the standard error of the determination of the refractive index, on cyclic phase errors, and also on the phase inhomogeneity occurring in modulation and demodulation.

Frequency error is temperature dependent; its effect is considerable mainly in long- and medium-range distance meters. In these meters the quartz oscillators are put into a thermostatically controlled oven, therefore a relative frequency stability of $\pm 1 \times 10^{-6}$ can be achieved. The effect of frequency error is not so important in short-range distance meters, because the resulting distance error is proportional to the distance measured. In such meters simpler electrical temperature-compensation solutions are used instead of an oven, or the frequency pattern is not stabilized at all. By means of partial temperature compensation a relative frequency stability of about $\pm 3 \times 10^{-6}$ can be achieved, the frequency fluctuation of unstabilized quartz oscillators being $\pm 1 \times 10^{-5}$.

Due to the ageing of the quartz, the modulating frequency changes in time. The change can be decreased by suitable technology (by artificial ageing before use). In order to ensure proper accuracy in distance measurement, the frequency patterns must be calibrated at least once a year in a laboratory and, if there is a change, they should be readjusted to the correct values.

The error of the refractive index depends partly on the accuracy of its determination at the station (i.e. on the reliability of the atmospheric measurements), but it depends to a greater extent on the deviation of the derived refractive index from its average along the line. Although by careful meteorological measurement the refractive index at a station can be determined with a relative standard error of $\pm 1 \times 10^{-6}$, in unfavourable conditions this value can deviate from the average by 4 to 5×10^{-6}. Distance error caused by refractive index error is distance-dependent, therefore its influence is considerable on long- and medium-range distance meters. The effect of the error can be decreased if the measurements are carried out in suitable weather (cloudy and slight breeze) when a strong layering of the refractive index cannot develop, or if the distance is measured in different parts of the day.

Cyclic phase error can be caused by different factors. In the instrument itself the electrical or optical cross-talk between measuring signal and reference signal may cause a cyclic phase error of amount $\frac{A_1}{A_2} \sin \varphi$, where A_1 is the amplitude of the part due to the cross-talk of the reference signal, A_2 is the amplitude of the measuring signal ($A_1 \ll A_2$), and φ is the phase-difference carrying the distance information. The error is independent of the method of measuring phase and it occurs more or less in every type of distance meter.

Another characteristic cyclic phase error of short-range distance meters is caused by the phenomenon that part of the measuring signal travels some multiple of the proper path of the signal due to reflection between the optical system of the instrument and the prism. The effects of the two types of cyclic phase error can be considerably different at different distances, therefore they cannot be taken into account by computation.

A third group of cyclic phase errors is caused by the non-linearity of the phase-meter. The effect of the error is considerable when a resolver (inductive phase-shifter) is used. This error is systematic and has a period of 180°. Its effect can be decreased if the phase measurement is carried out over different sections of the phase-shifter, by changing the phase of one signal by 90° and 180°. Due to its systematic character, this error is easy to determine: the reflector is repeatedly transferred within the half-wavelength of the fine measurement by suitable equal steps (distances), and the standard distances are compared to the measured distances.

In a modulated light beam emerging from an LED or from a light modulator, the modulation phase varies across the cross section normal to the propagation direction, in other words the radiation is phase-inhomogeneous. The phase of the demodulated electrical signal consequently depends on the point on the cathode of the demodulator (photomultiplier or photodetector) on which the reflected beam falls. If only part of the emitted radiation is reflected and if the projected beam does not illuminate the whole surface of the cathode (or if the light spot falls on different parts of the cathode surface), the phase inhomogeneity gives rise to a distance error depending on distance and on pointing. This error can be decreased if a received signal of maximum intensity is established when pointing. In short range distance meters the signal level is usually automatically controlled by a control device corresponding to the requirements of automated and rapid measurement, hence there is no means of checking the pointing.

The electrical and optical zero points of the distance meter and the reflector usually do not lie in the plumb line of the station point, consequently the instrument and the reflector used have a compound additive constant. The prism constant is easy to determine by measuring the geometrical data and knowing the refractive index of the prism material. The instrument constant also depends on the state of its electronic units. There is a relatively rapid change in these units after switching on, and a slow change due to ageing can also be observed. The effect of the rapid change can be considerably decreased if the internal and external measurements (i.e. measurements to the reflector and of the internal path of known length) follow each other quickly. The remaining error arising as a consequence of phase inhomogeneity, cross-talk and non-linearity of the phase-meter, causes a fluctuation of the instrument constant. Due to slow changes such as ageing, the instrument constant changes in time and the amount of the change can be determined by repeated calibration.

The easiest way to determine the instrument constant of short-range electro-optical distance meters is to measure an unknown distance and then to measure the same distance in two parts from an internal point. The difference between the two results is the instrument constant. The reliability of the constant is ensured in practice by carrying out the necessary number of superfluous measurements: a base line of nearly 1 km is divided by five intermediate points into six sections and all possible combinations of the sections are measured. The condition equations of the 21 measured distances contain 7 unknowns (i.e. the six section distances and the instrument constant), hence the number of superfluous measurements is 14. If the non-linearity of the phase-measuring system is tested independently, the lengths of the sections must be an integer multiple of the scale of the fine measurement. If the section lengths are known with proper accuracy, the condition equations contain only two unknowns: the additive constant and the multiplying constant of the instrument. (It should be mentioned that the multiplying constant cannot be determined by the measurement of sections of unknown length.)

The instrument constants of medium- and long-range distance meters can also be determined by the method described, but usually it is sufficient to compare the known length of one base line to the results of a large number of independent distance measurements. It is an important requirement that the effect of refractive index error should be negligible during the measurements.

10 Electronic Theodolites and Tacheometers

More and more complete automation of surveying and mapping work has been required for surveying and especially for cadastral surveying, for decades. In such automation the graphical or numerical end product, i.e. the map or the data for a surveying data bank, must be processed from the field measurement data with a minimum of human intervention. A very important stage of such an automated process is the so-called recording tacheometer which automatically records the surveying data.

Experiments for automatically recording distance measurements had been carried out already before short-range electro-optical distance meters came into general use. From all optical distance measurement methods, measurement with a subtense bar was chosen for this purpose; correspondingly a reliability of the order of ± 1 sec of arc was to be ensured in automatic circle reading.

The first prototype of an electronic theodolite, the *Digigon* (Breithaupt, FRG), was made at the beginning of the sixties. Although this instrument had all the basic properties of modern electronic theodolites, it did not come into production.

The *Fennel* (FRG) recording theodolite (Subsection 2.3.2) and a similar *Kern* theodolite (Switzerland) were also constructed in the sixties. A common feature of both instruments is that, instead of reading the graduated circles, they photograph them with a camera attached to the instrument; the numbered graduation lines are substituted by code strips, and there is always an un-numbered division line in each graduation unit. The photos can be evaluated in the office by the *Zuse Z84* processor device (FRG). Automatic photoelectric evaluation consists of "reading out" the encoded graduation values and of measuring the distance between the images of two diametrically opposite graduation lines which are projected side by side; the latter is the so-called "truncated (fine) reading". This reading can be determined by a photoelectric interpolator (micrometer) with an accuracy of 0.1 mgon. Supplementary information (e.g. serial number of the station) can be recorded in encoded form during measurement by an external number-setting device. The total information recorded on film is transferred to punched tape without any human intervention.

At the end of the sixties a recording theodolite was also constructed by *MOM* (Hungarian Optical Works) (Subsection 2.3.2). The encoded angle unit of the graduated circles of the theodolite *Ko-B1* is 10 min of arc. The images of diametrically opposite graduation lines, projected side by side, are brought into coincidence five times a second by an optical micrometer rotating with constant angular velocity. Coincidence is detected photoelectrically, and the time necessary for establishing a coincidence, which is proportional to the value of the fine reading, is measured by pulse counting. The accuracy of the display is 1 sec of arc. The reading of the graduated circle appears on the display of the electronic unit connected to the theodolite. Independently of the display, the encoded information entered by the number-setting device of the instrument, as well as the readings of the horizontal and vertical circles (i.e. the mean value of three or five readings on each according to choice), are recorded on punched tape by a tape-puncher operated from an accumulator. A small run of this instrument was produced.

The development of recording theodolites was given an impetus with the introduction of distance meters constructed for the measurement of short distances: as a result, recording electronic tacheometers were developed. Two methods of automatic circle reading became widely used: the method of decoding and the incremental or counting method. The methods used in the different types of instrument differ from each other essentially only in the method of refining (interpolation) the graduation unit.

10.1 Circle Reading by Decoding

The method of decoding the graduated circles is also called the absolute method, because the reference base (zero line) is physically marked on the graduated circle. The principle of operation is illustrated in *Fig. 10.1.*

Light from the light source *1* passes through the encoded glass circle *2* on which the number of the graduation is marked in a binary code: each corresponding digit

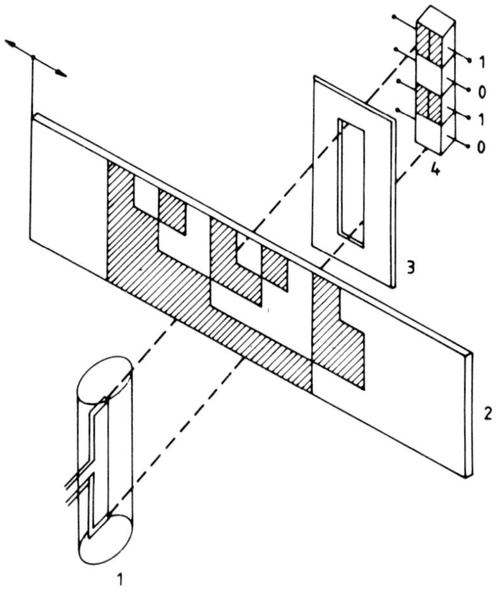

Figure 10.1. The principle of reading by decoding (optical scanning): *1* light source; *2* carrier of the encoded graduations; *3* diaphragm; *4* series of photodiodes

of the two figures of the binary code are represented by a transparent or an opaque element. The smallest division depends on the number n of the code-rings, being 360° (400^g)/2^n. Now the light passes through the diaphragm *3*, the width of which corresponds to the width of the code-strip, and falls on the series of photodiodes *4* placed behind it; the number of diodes is identical with the number of code rings. The output signal of the photodiodes is brought into the proper form of a decimal number sequence by a parallel-to-serial converter; the serial signal can be displayed by means of the decoder and display, or it can be recorded independently of the display.

Besides the optical scanning mentioned, magnetic scanning can also be realized; similarly to the optical method, this is free of physical contacts. In some types of instrument encoding is achieved by means of contacts sliding on the code-tracks. The advantage of this method is doubtless its simplicity, but its disadvantage is the unavoidable wearing and dirtying of the sliding contact, and the resolving power achievable is less than that of optical or magnetic scanning.

10.2 Circle Reading by the Incremental (Counting) Method

This method of reading the graduated circle is called relative because the reference base (zero line) is not marked on the circle, thus the instrument in itself is only capable of measuring the angle of rotation. The principle of operation is illustrated in *Fig. 10.2*.

Light from the source *1* passes through the graduated glass circle *2* on which the graduation units are represented by transparent and opaque strips of equal width. The smallest graduation unit depends on the width of one strip. The light now passes

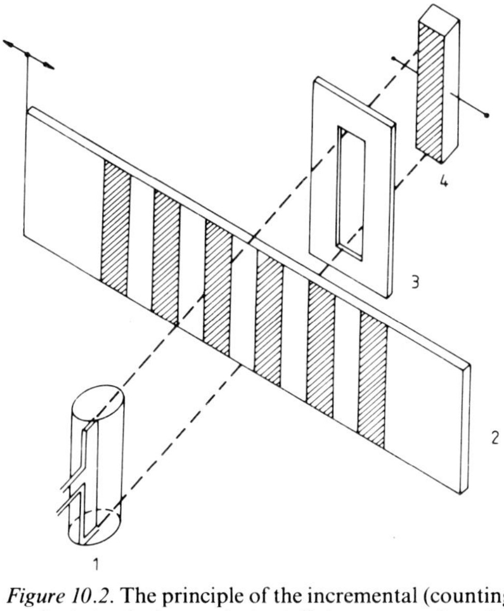

Figure 10.2. The principle of the incremental (counting) method (optical scanning): *1* light source; *2* optical graduations; *3* diaphragm; *4* photodiode

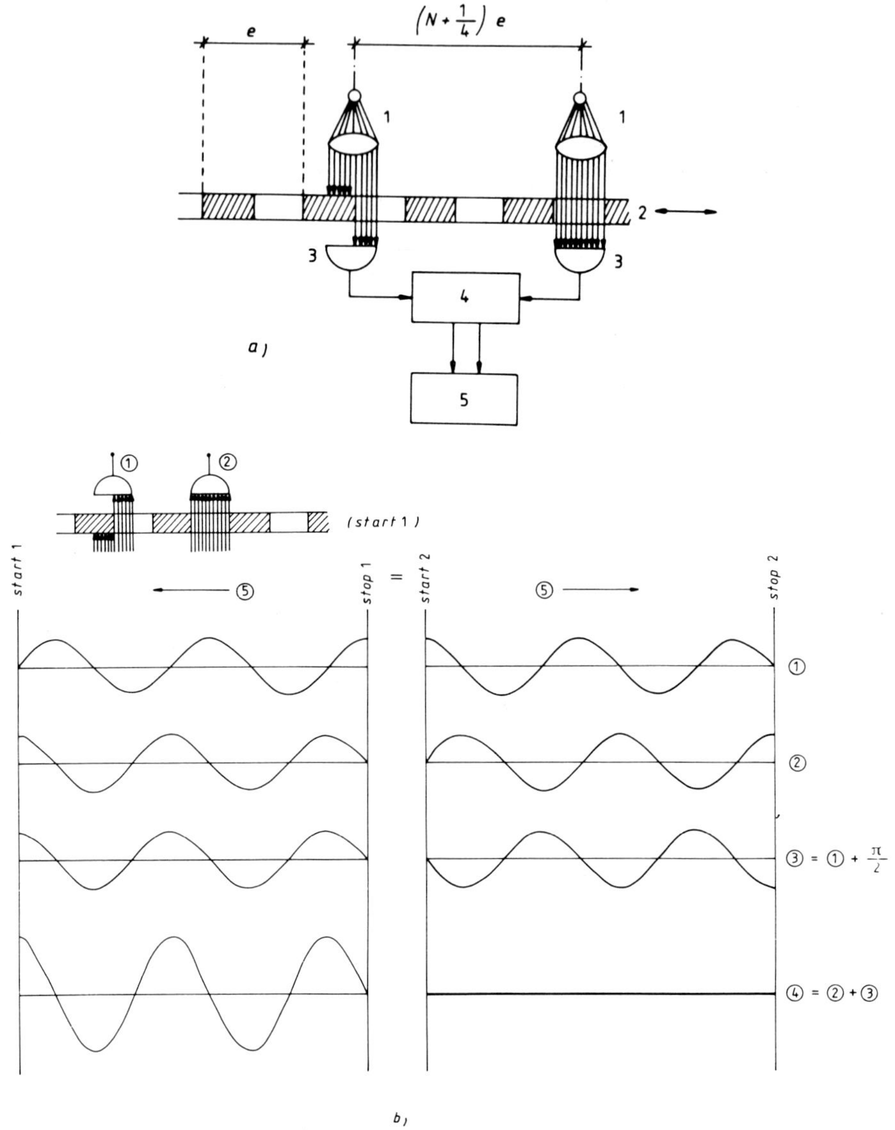

Figure 10.3. The principle of the direction detection: *a) 1* light source; *2* moving optical graduations; *3* photodiode; *4* direction detector; *5* forward-backward counter; *b) 1* counting signal; *2* direction detecting signal; *3* modified counting signal; *4* resultant signal; *5* direction of motion of the optical grid

through the diaphragm *3*, whose width is equal to the band-width of the graduated circle, and then falls on the photodiode behind the diaphragm. If the graduated circle rotates the quantity of light falling on the photodiode changes nearly sinusoidal and, by counting the periods of the sinusoidal photo-current, the amount of rotation can be determined. The result of the counting, however, only corresponds to the angle of rotation if the rotation is uni-directional from the beginning to the end. In practice such a movement cannot be ensured, therefore the instrument is fitted with a second, so-called direction-determining channel. This channel is of the same construction as the counting one, but the two channels are relatively displaced by $\left(N + \frac{1}{4}\right)$ graduation units. The principle of direction detection is shown in *Fig. 10.3, a, b.* Having changed the phase position of the counting signal by 90°, the resultant of the modified counting signal and the direction-detecting signal depends on the direction of the rotation. The resultant signal controls the control device: when rotating in one direction the number of periods counted is added to the instantaneous content of the counter, but when rotating in the opposite direction it is subtracted.

This solution can also be realized with both magnetic and electrical scanning. The advantage of optical scanning lies in its high resolving power: an optical grid can contain as many as 200 lines within 1 mm.

The resolving power of an optical grid can be further increased by the utilization of the so-called Moiré effect. If two optical grids are placed one above the other in such a way that their graduation lines make a small angle, the so-called Moiré fringes appear, consisting of light and dark elements. A longitudinal movement of one of the optical grids (i.e. perpendicular or nearly perpendicular to the grid lines), makes the fringes shift in the cross-direction, the latter shift being several times greater than the longitudinal movement. The shift of the fringes can be detected by photoelectric devices, similarly to the movement of the optical grid.

The processing of the detected electrical signals can easily be followed on the example of the earlier mentioned theodolite *Digigon.* The block diagram of the electronic reading device of the instrument is shown in *Fig. 10.4.* The projector prism

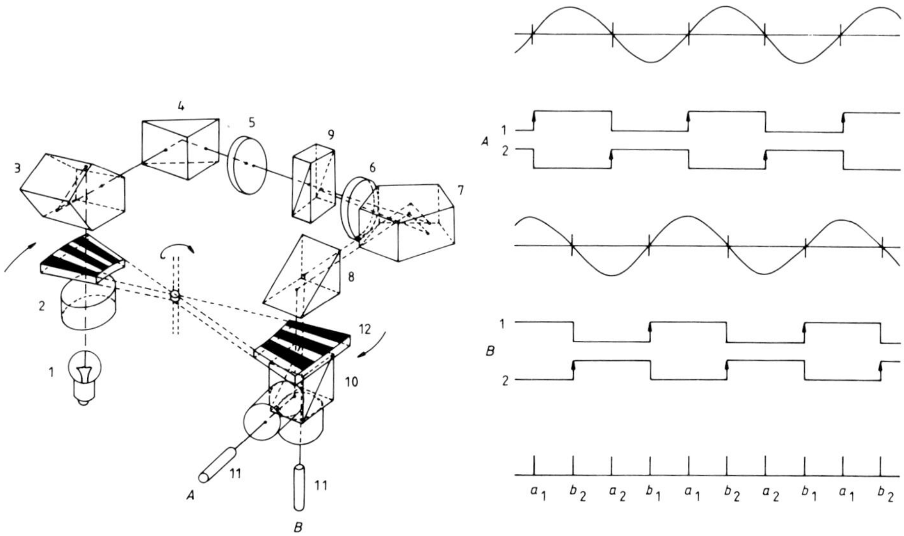

Figure 10.4 The circle reading device (horizontal circle) of the *Digigon* (Breithaupt, FRG): *1* light source; *2* to *8* projecting prisms and lenses; *9* light splitting and phase-delaying Wollaston prism; *10* light-splitting prism (polar filter); *11* photodiodes of the measuring channels *A* and *B; 12* optical grid

and lens system projects the images of diametrically opposite graduation lines onto each other. The circle is graduated into 5000 units. Due to the opposite motion of these diametrically opposite images, 10,000 light periods (fringes) fall on the photodiode during one whole revolution of the circle, which is rigidly fixed to the alidade; the length of one period thus corresponds to an angular value of 0.04 gon $= 4^c$. The square signal *A 1* is produced from the sinusoidal photo-current by a trigger, then the logic circuits produce square signal *A 2*, which is the inverse (i.e. the mirror image with respect to the time axis) of the signal *A 1*. In both signals the rising edges are displaced at 0.04 gon. At points on these rising edges (of square signals) a differentiator stage produces a counting pulse, the separation between adjacent elements of both series of pulses a_1 and a_2 corresponding to an angular value of 0.02 gon. In order to determine the direction of rotation the projected signal is doubled and its phase is shifted by 90° by a Wollaston prism. From the shifted signal (cosine signal) two further series of pulses are produced by means of the same conversion as for the sine signal; the separation of adjacent elements of both series b_1 and b_2 also correspond to 0.02 gon. The corresponding elements of the series b halve the separation of the corresponding elements of the series a, consequently the separation between adjacent elements of the two united series give an angular value of 0.01 gon, and this is further interpolated to 0.001 gon by an electronic interpolator.

As mentioned previously only the angle of rotation can be measured by the incremental method, therefore a zero line must be marked on the graduated circle before angle measurement, in other words the circles must be initiated. In this context initiation does not mean a physical mark, but only that the counter determining the angle of rotation is zeroed at the instant of initiation. The initiation of the horizontal circle does not require any extra operation, because the position of the zero line is not constrained. For the vertical circle, however, the position of the zero line is defined by the horizontal or by the plumb line (this can be achieved by an electronic pendulum or by a spirit level), and the content of the counter must be erased when the line of sight of the telescope points in the direction of the zero line.

10.3 General Characteristics of Recording Electronic Tacheometers

Electronic tacheometers can be classified according to their construction into two groups:

— tacheometers with a modular system, where the distance meter unit, the processor unit, the keyboard and the data recorder unit can be attached as modules to an electronic theodolite;

— total station tacheometers, where the units of angle measurement, distance measurement, microprocessor, keyboard and sometimes even the recorder unit are united in the instrument.

A general feature of electronic tacheometers is that the readings of the horizontal and vertical circles as well as the slope distance determination are all automatic. The task of the observer is to set up at the point, to set the vertical axis vertical and to point at the reflector device (with some instruments also to set in the signal level) and then to start the automated measurement. The phases of the measurement process controlled by the microprocessor are as follows.

— Determination of the slope distance and its correction, the additive and multiplying factors having been entered previously. Entering the additive constant beforehand is necessary if a non-standard or different reflector is used, or if there is a longitudinal eccentricity. The multiplying constant is usually the atmospheric correction.

— Reading the horizontal and vertical circles and correcting them for collimation error and index error; the correction values are entered beforehand or determined by special measuring programs.

— Display of the slope distance and of the readings of the two graduated circles in units chosen in advance (metric or foot system, and degrees or gons, respectively). Recording of the three results in the data collector with corresponding identity codes on extra command.

— Computation of the distance reduced to the horizontal, and of the elevation difference between the horizontal axes of the instrument and of the reflector device, taking into account Earth curvature and average ray curvature (refraction coefficient $k=0.13$). The distance reduced to the horizontal is also multiplied by a factor either entered beforehand or determined by a special measuring program, in this case the result being reduced to sea level or to the map projection plane.

— Display of the horizontal circle reading, of the reduced distance (reduced to a sea level or projection surface) and of the elevation difference in the chosen units of length. On extra command the two computed items are recorded in the data collector together with their corresponding identity codes.

Some types of electronic tacheometers are fitted with attached or built-in pre-programmed microprocessors which are able to compute the basic data of further point determination or setting-out from the results of the processor-controlled measurement. To distinguish them, such tacheometers are called computer-tacheometers.The most frequently used programs for program-controlled measurements are:

a) The choice of a free station point, i.e. the determination of the unknown horizontal coordinates of the station from direction and/or distance measurements to points with known horizontal coordinates. If there are redundant measurements, the microcomputer computes and adjusts the coordinates of the station (Helmert transformation) and displays them together with the standard error of the determination (position error) and the adjusted value of the scale factor, and finally it orients the horizontal circle electronically (i.e. it “sets” the zero line into the direction of the coordinate axis $+x$).

b) The unknown absolute height of the station is determined from trigonometric levelling to points of known height and distances measured or computed from the coordinates. The microprocessor computes the height of the instrument station, adjusting it if there are redundant measurements, and displays it together with the standard error of the determination.

c) The orientation of the horizontal circle from the results of measuring from a station with known coordinates to a station with known coordinates, with adjustment if there are redundant measurements.

Table 10.1. Technical data and characteristics

Type	Manufacturing firm	Distance meter unit: scale of fine measurement	Distance meter unit: range (km) / number of prisms		Distance meter unit: standard error of distance measurement $\pm(A+B\times10^{-6}D)$: A (mm)	B (mm/km)	Angle measuring unit: method of circle reading; accuracy (mgon) (Hz / V)	Angle measuring unit: telescope magnification; field of view (m) minimum focussing distance (m)
1	2	3	4		5		6	7
Geodimeter 700 (710)	AGA (Sweden)	5	3.0 3	(5.0) 6	5	11	counting 0.5 1.0	30 × 2.8/100 1.7
Geodimeter 140 (136)		10	2.5 (1.0) 1	5.5 (2.5) 8	normal 5 stand. dev.- -checking 5	 5 3	counting 0.6 (1.0) 0.6 (1.0)	30 × 2.6/100 1.3
Geodimeter 142		10	2.5 1	5.5 8	normal 5 stand. dev.- -checking 2	 5 3	counting 0.6 0.6	30 × 2.6/100 1.3
HP 3820A	Hewlett– –Packard (US)	10	1.0 1	4.0 6	5	2	decoding 0.6 0.6	30 × 2.6/100 5.0
RETA	Jenoptik (GDR)	10	1.0 1	3.0 21	5	2	decoding 0.1 0.1	30 × 2.5/100 1.5
RECOTA		10	1.0 1	3.0 21	5	2	decoding 0.1 0.1	30 × 2.5/100 1.5
E2 + DM 503	Kern (Switzer- land)	10	2.5 1	5.0 7	3	2	counting 0.1 0.1	32 × 2.7/100 1.5
Vectron + + Autoran- ger	Keuffel & Esser (US)	10	1.0 1	2.0 3	5	6	counting 1.0 1.0	30 × 2.4/100 1.8

…t some electronic tacheometers

Mass (kg)	Data carrier	Construction of instrument	Grade of automation	Pointing	Additive/multiplicative factor can be entered	Tracking	Number of simultaneously displayed data	Program-controlled measurement	Applicable as industrial measuring system (IMS)	Remarks
8	9	10	11	12	13	14	15	16	17	18
31.0	1 (2)	2	1	2	N/Y	1+2+3	2	0	N	light source He–Ne laser modulator Kerr-cell
8.5	3	3	2	2	Y/Y	1+2+3	1	2	Y	no standard deviation-cheking with model 136
9.5	3	3	2	2	(Y)/Y	1+2+3	1	2	Y	
11.0	3	3	2	2	Y/Y	1+2+3	1	0	Y	light source infrared laser diode
10.0	3	3	2	2	Y/Y	1+2+3	I.3 II.2	0	N	
12.8	3	3	2	2	Y/Y	1+2+3	I.3 II.2	2	N	
8.7 +	3	1	2	2	Y/Y	1+2+3	I.5 II.2	2	Y	
10.0	3	1	2	1	N/Y		2	2	Y	one displayed result is the slope distance

Table 10.1. (continued)

Type	Manufacturing firm	Distance meter unit: scale of fine measurement	Distance meter unit: range (km) / number of prisms		Distance meter unit: standard error of distance measurement $\pm(A+B\times 10^{-6}D)$: A (mm)	B (mm/km)	Angle measuring unit: method of circle reading; accuracy (mgon) (Hz / V)	Angle measuring unit: telescope magnification; field of view (m) minimum focussing distance (m)
1	2	3	4		5		6	7
Tachymat TC 1 (L)	Wild (Switzerland)	30	1.0 (2.5) 1	2.0 (5.0) 11	5	5	counting 0.6 1.0	26 × 3.2/100 2.0
Theomat T 2000 + DI 4 (L)		33	1.0 (2.5) 1	2.0 (5.0) 11	5	5	counting 0.01 0.01	32 × 2.7/100 1.7
TC 2000		30	2.5 1	5.0 11	3	2	counting 0.01 0.01	32 × 2.7/100 1.7
RegElta 14	Opton (FRG)	10	0.5 1	1.2 7	5...10		decoding 1.0 1.5	25 × 3.0
Elta 2		10	I.2.0/3 II.2.5/3	3.0/9 4.0/9	I.5 II.10	2 2	decoding 0.1 0.1	30 × 2.5
Elta 20 (3)		10	I.1.6 (1.2) II.2.5 (2.0) 3	2.5 (2.0) 4.0 (3.0) 9	I.5 II.10	2 2	decoding 0.2 (0.3) 0.2 (0.3)	30 × 2.5
Elta 4 (41)		10	I.1.2/3 II.1.6/3	 2.2/9	I.5 II.10	2 2	counting 1.0 1.0	25 × 3.0
Elta 46 (R)		10	I.1.6/3 II.2.0/3	 3.0/9	I.5 II.10	2 2	counting 1.0 1.0	25 × 3.0
Elta 40		10	I.0.5/1 II.0.7/1	0.7/3 1.5/3	I.5 II.10	2 2	counting 2 2	2.8 × 1.3

Legend for Table 10.1

In column 9: 1 = punched tape; 2 = magnetic tape; 3 = semi-conductor memory
In column 10: 1 = modular system; 2 = processor unit separate; 3 = unified (total station)
In column 11: 1 = partly automated; 2 = totally automated

Table 10.1. (continued)

Mass (kg)	Data carrier	Construction of instrument	Grade of automation	Pointing	Additive/multiplicative factor can be entered	Tracking	Number of simultaneously displayed data	Program-controlled measurement	Applicable as industrial measuring system (IMS)	Remarks
8	9	10	11	12	13	14	15	16	17	18
13.0	2	3	2	2	Y/Y	1+2	2	0	N	
10.0	3	1	2	2	Y/Y	1+2+3	I.3 II.2	2	Y	data collector: magnetic bubble memory
10.0	3	3	2	2	Y/Y	1+2+3	I.3 II.2	2	Y	
31.0	1	2	1	2	N/Y		1	1	N	
13.5	3	3	2	2	Y/Y	1+2+3	1	2	Y	three-fold electrical amplification in the second measurement range
13.5	3	3	2	2	Y/Y	1+2+3	1	1	N	as Elta 2
6.5	—	3	2	2	N/Y	1+2+3	3	0	N	as Elta 2 data cannot be recorded
6.5	(3)	3	2	2	N/Y	1+2+3	3	0	N	as Elta 2 data can be recorded with model 46 R
8.5	3	3	2	2	N/Y	1+2+3	3	0	N	as Elta 2

In column 12: 1 = separate pointing; 2 = common pointing
In columns 13 and 17: N = no; Y = yes
In column 14: 1 = slope distance; 2 = horizontal distance; 3 = elevation difference
In column 16: 0 = impossible; 1 = limited to basic programs; 2 = possible

d) Calculation of the rectangular coordinates of points measured by polar method.
e) Computation of polar data for setting out points whose rectangular coordinates have been determined. In the first step the microcomputer computes and displays the required polar values for setting out (horizontal circle reading and horizontal distance) from the coordinates of the station point and the point to be set out. Having set out the point, it should be regarded as preliminary and the microcomputer computes and displays its rectangular and polar corrections.
f) Indirect distance determination when the microcomputer computes the unknown distance between two points from the results of polar measurements made to them without orientation of the horizontal circle.

The light source of electronic tacheometers is usually a light emitting diode, in some cases a laser diode, or even a He–Ne laser; the effective range of distance measurement is 1 to 5 km depending on the light source. In the well-known formula for the standard error $m = \pm(A + B \times 10^{-6} D)$, the value of A is 5 to 10 mm and that of B is 1 to 5 mm/km. The method of circle reading is either absolute (reading by decoding) or relative (counting); interpolation for the fine reading usually differs from one instrument type to another. The accuracy of the horizontal circle reading is 0.1 to 1 mgon (0.3 to 3 sec of arc), that of the vertical circle 0.1 to 2 mgon (0.3 to 6 sec of arc). The data collector in earlier types is a punched tape or a magnetic tape cassette, and in modern types a semi-conductor solid-state memory.

Technical data for better known electronic tacheometers are listed in *Table 10.1.* The table also contains data for types of instrument in which circle reading and distance measurement is automated, but the measured data cannot be recorded.

A comparison of the accuracy of distance measurement and angle measurement shows that the reliability of angle measurement is considerably higher over short distances. This is the basis of instrument assemblies known collectively under the name "industrial measuring system" (IMS). The measuring system consists of two electronic theodolites which are linked on-line via proper interfaces to a desktop-computer. The instruments are set up at the ends of a base line which either is known or can be determined with an extra measurement program. The necessary orientation measurements are carried out and the three coordinates of the points to be measured are determined by spatial intersection. The advantage of the method is that no reflector needs to be used on the points to be measured (even their marking is usually unnecessary). The reliability of the determination of points depends not only on the accuracy of the electronic theodolites but also on the intersection angle and on the reliability of pointing. In a favourable case the standard error can be decreased to some tenths of a mm. A further advantage of the method is that the desktop computer can provide a quick coordinate transformation (e.g. the coordinates of points which are measured from several base-lines can be transformed into one coordinate system which is eventually connected to the object to be measured). Peripherals of the computer may provide a simultaneous graphical display. In *Table 10.1* reference can be found as to whether the given instrument can be used as an element of an industrial measuring system.

10.4 Electronic Tacheometers in General Use

One of the electronic tacheometers of modular construction is the instrument assembly Vectron/Autoranger (Keuffel & Esser, US). Here the distance meter unit can be attached to the alidade of the electronic theodolite and connected to it electrically. The processor unit and transistorized data recorder are constructed together and can be connected to the theodolite. A magnetic casette recorder for data storage and a printer can also be attached to the measuring system. After measuring a slope distance, the following quantities can be displayed and stored: slope distance, readings on the horizontal and vertical circles, distance reduced to the horizontal, elevation difference, all three coordinates of the observed point. The processor unit permits complicated surveying tasks to be computed (such as e.g. traversing) besides calculating the mean value and the standard deviation; programs of limited length made by the user can also be stored in the program memory.

The variant *E2 (Fig. 10.5)* of the electronic theodolite *E1* (Kern, Switzerland) with the electro-optical distance meter *DM 501, DM 502* or *DM 503* (also manufactured by Kern), the transistorized data recorder *R48* or *Alphacord 128,* and the pocket computer *HP 41 C/CV*(Hewlett–Packard, US), together form an electronic tacheometer of modular construction *(Fig. 10.6)*. The circle reading is incremental and the optical graduation of the circles contains 40,000 elements (20,000 graduation lines). Part of the graduated circle, which contains about 200 graduation lines, is projected by an optical system of prisms and lenses onto the diametrically

Figures 10.5. The electronic theodolite *E2* (Kern, Switzerland)

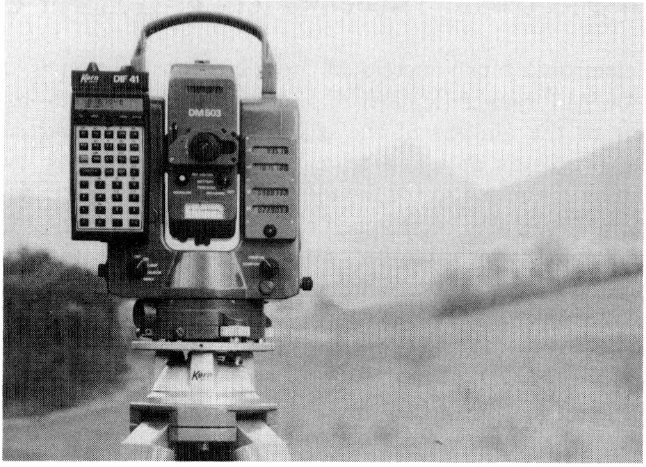

Figure 10.6. The electronic tacheometer of modular construction of Kern, Switzerland (electronic theodolite *E2*, distance meter *DM 503*, interface *DIF 41*, pocket computer *HP 41 CV*)

opposite part. The magnification of the projected image is 1.005 times, due to which the graduation and its image of different magnification produce a system of Moiré fringes. The fringes are parallel to the graduation lines and the length of a whole period is 2.2 mm. Within one period of the fringe system, magnified to 4 mm, four photodetectors detect the intensity change of the incoming infrared radiation, and these detectors are always shifted by 1 mm (corresponding to a phase shift of 90°). Due to the diametrical projection the images of the two graduated parts move in opposite directions when the instrument is rotated, thus one period of the Moiré fringe system corresponds to an angular rotation of 0.01 gon.

The coarse reading is produced by a reading device by counting the periods of the photo-current of the first photodiode; counting is possible up to a maximum angular velocity of 2.5 revolution/s. The smallest coarse reading unit is 0.01 gon. During fine reading the circles are fixed with respect to the reading devices. The voltages $u_1 \ldots u_4$ occur at the outputs of the four photodiodes:

$$u_1 = U_1 \sin \varphi$$

$$u_2 = U_2 \sin(\varphi + 90°) = U_2 \cos \varphi$$

$$u_3 = U_3 \sin(\varphi + 180°) = -U_3 \sin \varphi$$

$$u_4 = U_4 \sin(\varphi + 270°) = -U_4 \cos \varphi,$$

where φ is the phase of the first photodiode (determining the coarse reading) in the whole period being the unit of coarse reading.

The signals are transformed and the voltages are subtracted in pairs:

$$u_1 - u_3 = (U_1 + U_3) \sin \varphi,$$

$$u_2 - u_4 = (U_2 + U_4) \cos \varphi.$$

The condition $U_1 = U_2 = U_3 = U_4$ is ensured when the diodes are adjusted in the factory, hence:

$$\frac{u_1 - u_3}{u_2 - u_4} = \frac{\sin\varphi}{\cos\varphi} = \tan\varphi .$$

After the AD conversion, $\tan\varphi$ and φ can be computed. Since the computed value of φ is an angle with sign in the range $-90^\circ < \varphi < +90^\circ$, the processor must also check the sign of $\cos\varphi$ in order to give the correct phase angle:

if $\cos\varphi$ is positive, the phase angle is $360^\circ + \varphi$, subtracting 360° if this is greater than 360° (i.e. if φ is positive);

if $\cos\varphi$ is negative, the phase angle is $180^\circ + \varphi$.

The fine reading is $\varphi^\circ/360^\circ$ times the 0.01 gon unit of the coarse reading.

It is sufficient to make the interpolation with an accuracy of 1 percent, consequently it is sufficient to position the diodes and adjust their voltages with an accuracy of 10 percent.

The advantage of the fine measurement method used is that the Moiré fringes are produced by 200 pairs of graduation lines, thus the effect of individual graduation errors is decreased.

In order to mark the zero line of the vertical circle the telescope must be turned through the zenith with its objective lens upwards. At the instant when the line of sight and the vertical axis are aligned, a fixed diaphragm coincides with a diaphragm which rotates with the vertical circle. This produces a pulse from a photodiode which initiates the counting of the pulses of the coarse measurement on the vertical circle and, at the same time, it also initiates the horizontal circle.

Fine readings are repeated at intervals of 0.3 s and the accuracy of the display is 0.1 mgon or 1 sec of arc.

The compensator is not inserted into the optical train of the vertical circle reading. Although the processor unit corrects the readings on the horizontal and vertical circles for the dislevelment of the vertical axis, this correction can be separated from the reading and displayed separately. The compensator has two axes, the horizontal reference surface being represented by the surface of silicone oil contained in a small chamber. The infrared radiation of a light-emitting diode is reflected from this surface when horizontal to a photodiode with a large surface divided into four segments. If the position of the vertical axis changes, the point in which the reflected beam meets the surface of the photodetector also changes, and the illumination of the segments of the detector changes correspondingly. The components of the dislevelment of the vertical axis, in the direction of the tilting axis and perpendicular to it, can be determined from the four photo-currents by subtraction and after AD conversion.

When using the instrument in the tacheometer mode, the horizontal and vertical circle readings, the horizontal distance and the elevation difference all appear on the four displays of the theodolite unit, while the slope distance is displayed on the distance-meter unit. The measured data (horizontal and vertical circle readings and slope distance) are stored in the data recorder *R48*. The program memory of the electronic pocket computer *HP 41 C/CV* can be used freely by the user but, by means of the program system compiled by the producing firm, a measurement controlled by the fixed program can also be carried out; this can be done in the mode "dialogue" by using the keyboard of the computer. The data needed for the program-controlled measurement must be entered into the appointed data storage registers of the computer; the program memory can also store computed results to a limited amount. The information needed at the prism point (such as e.g. the instantaneous corrections to the setting out) are transmitted to the prism in modulated form by the measuring beam of the distance-meter unit, and the digital data appear on the two displays of the receiver *RD 10* (see also Point 9.3.3.2) coupled to the prism. The measured and computed results stored in the program memory of the computer *HP 41 C/CV* can be entered directly into another computer by means of the data converter *DL 40* and a correct interface.

The electronic theodolite *Theomat T2000* (Wild, Switzerland) *(Fig. 10.7)*, with the electro-optical distance meter *DI 4*, *DI 4L*, *DI 5* or *DI 20* (also made by Wild) and the magnetic bubble memory *GRE 3*, together form an electronic tacheometer of modular construction. The electronic method of circle reading was developed by the French firm Sercel in cooperation with the firm Wild. Both the vertical and the horizontal circles are graduated into 1024 units. In spite of this the circle reading is absolute because the reading device measures the angular rotation with respect to a fixed index (photodetector); the direction of the other leg of the angle is represented by an index (photodiode) rotating with the telescope around the horizontal and the vertical axis, respectively. The coarse reading could be made by counting the periods of photo-current occurring during a rotation through the angle φ, but for that it would be necessary to turn the moving photodetectors into the direction of the fixed

Figure 10.7. The electronic theodolite *Theomat T2000* (Wild, Switzerland)

photodetectors before measuring: in other words it would be necessary to make in some way a coarse zero reading on both circles. This can be avoided, however, if the reading devices start to operate only when the pointing has been completed, i.e. the indices of both circles have stopped moving. On the command "read" each graduated circle is rotated by an electromotor. From the instant at which the marked graduation line passes the index representing the zero line, the counter starts to count the periods of the photo-current until the marked graduation line passes the second index line.

The fine measurement (interpolation) gives the part of the angle between the two index lines which remains within one period of coarse measurement. The measuring method is the same as the digital phase measuring method used for distance measurement, and it is reduced to time measurement (pulse counting). Counting starts when the preceding graduation line crosses the zero index line, and stops when the graduation line which precedes the second index line (which is fixed to the alidade) crosses the second index line, resulting in n_i pulses. A second counter continues to count until the next graduation line passes the zero index; the result is the pulse number N_i which is proportional to the unit of smallest division. Hence the fine measurement gives n_i/N_i-times the unit of smallest division.

After each measurement of a fine reading, i.e. after having carried out a pulse count, the processor unit generates and stores in the summing memory the computed value of the fine reading. This needs time, thus there is no measurement during the next period, and then the measuring and computing process is repeated. The end result of the fine measurement is the arithmetic mean of the $1024/2=512$ results determined during one revolution of the circle. From the measuring principle the end result is the arithmetic mean of values measured in 512 circle positions, therefore the effect of graduation error is small and negligible. In order to eliminate the error due to the eccentricity of the graduated circles, two diametrically opposite reading devices are used on both circles (i.e. two fixed index lines and two index lines moving with the telescope). The horizontality of the index line of the vertical circle is ensured by a liquid-compensator inserted into the beam path; consequently the optical grid of the vertical circle consists of transparent and opaque elements, while the optical grid of the horizontal circle consists of reflecting and non-reflecting elements.

The theodolite has one processor unit, therefore horizontal and vertical circle readings follow each other in time. On the command "read" the two electromotors accelerate the two graduated circles to the required velocity within 0.3 s. Next the horizontal coarse and fine circle readings are made simultaneously, firstly on one of the index pairs and then on the other. The whole measuring program only takes one second, including the acceleration, and then the electrical and electronic units are switched off automatically. The accuracy of circle reading (the smallest displayed unit) is 0.01 mgon or 0.1 sec of arc. On two of the three displays of the theodolite unit the following data pairs can be displayed: horizontal and vertical circle readings; the vertical circle reading and the slope distance; the horizontal circle reading and the horizontal distance; the elevation difference and the height of the observed point; the coordinates x and y of the observed point. Each of these data pairs can be recorded. The third display is used for the mode "dialogue" between the operator and the central control unit by using the keyboard of the theodolite unit. The processor and data collector unit *GRE 3* with a magnetic bubble memory stores the data continuously — without power supply for a long time — until their deliberate erasure, and the store is very rapidly accessible. The program memory can be freely used by the user and, in addition, a measurement controlled by a fixed program can also be made by means of a module that can be built in. The data recorder can be directly coupled to a computer through a proper interface, and it is capable of both sending and accepting data.

The *Theomat T 2000 S* has the same construction as the *Theomat T 2000*, but its optical system has been improved. It is used primarily for tasks in engineering survey, but other uses include being an element of an industrial measuring system (IMS) (when connected to a computer).

The electronic theodolite *T 1000* is a simpler (and consequently cheaper) product. If combined with an electro-optical distance meter of the type *DI 4*, *DI 4L*, *DI 5* or *DI 20*, it forms a modular electronic tacheometer.

Of the integrated measuring systems the *RegElta 14* (Opton, FRG) was the first electronic tacheometer produced in quantity. The instrument is similar to a theodolite and consists of a distance measuring unit, an angle measuring unit, an attachable data recorder unit (tape punch), and of a later developed computer unit *Eltac*. The distance meter unit is identical with the electro-optical distance meter *SM 11* (Point 9.3.3.2). The method of circle reading is decoding and micrometer interpolation. The code discs are graduated into 400 units and geared, the disc pairs are coupled or released by a screw. After an approximate pointing the disc pairs of the horizontal and vertical circles are connected mechanically; at this instant the electrical code reader starts to operate, producing the coarse reading whose smallest unit is 1 gon. When exactly pointed, by the use of slow motion screws, the instrument can be rotated around a second horizontal and vertical axis, respectively, within an angle of 1.6 gon. This rotation is converted by means of a precise mechanism into a linear shift of a measuring wedge, and the encoded information giving the position of the wedge, i.e. the fine reading, is generated by an electronic decoder. For reading the vertical circle the fine reading is projected to the photodiodes by a lens suspended as a pendulum, thus acting as an index compensator. The program boxes of the computer unit *Eltac* are changeable and thus permit the carrying out of some measurements controlled by simpler programs.

The *Geodimeter 700* and its later variant *710* (AGA, Sweden) consist of angle measuring and distance measuring units, of an attachable processor unit and a data recorder (in model *700* it is a tape punch, in model *710* a magnetic tape cassette). The light source of the distance meter is a He–Ne laser, the modulator is a Kerr cell. The method of phase measurement is analogue, but the inductive phase shifter is activated by a stepping motor coupled with a counter device, thus the phase measurement is automated and the data display digitized. The circle reading is incremental; circle readings are always established by means of two concentric circles divided into 10,000 units. The graduation lines of the two circles are shifted relatively by 1/4 graduation unit; the diametrically opposite graduation lines are projected onto each other by an optical system consisting of prisms and lenses, so that 20,000 light periods correspond to one whole revolution of the alidade. The projected images of the graduation lines are not parallel to the diametrically opposite ones, therefore Moiré fringes appear in the projection, and their motion is detected by a photoelectric device with a resolving power of 2.5 mgon. The measurement of the rotation angle of the horizontal circle is further refined by another electronic interpolator with a resolving power of 0.5 mgon. The index compensator of the vertical circle is an electrical pendulum. The two results which can be displayed together on the display of the processor unit are: the readings of the horizontal and vertical circle; the horizontal circle reading and the elevation difference; the horizontal circle reading and the horizontal distance; the vertical circle reading and the slope distance. All displayed data can be recorded.

In the electronic tacheometer *Tachymat TC-1* (Wild, Switzerland) *(Fig. 10.8)* the measuring, processor and recorder units are united into one instrument. The circle reading is incremental and both horizontal and vertical circles are graduated into 12,500 units. The reading device consists of two diametrically opposite elements, using diffraction for position detection. The optical train is shown in *Fig. 10.9*; for the sake of clarity the beams are drawn as passing through the circle instead of being reflected from it, and the mirror image of the receiver side is shown in the figure. The beam emitted by the infrared ray source *1* of the two position sensor elements is made parallel by a collimator lens. Having passed through the analyzer *2* the parallel beam is diffracted; reflection and diffraction occur again at the reflecting cover strips of the graduated circle *3*. The reflected beam again passes through the analyzer *4* and, due to the effect of the third diffraction, two parallel beams are produced which fall on the receiver detector. After the first diffraction the beam of rays is split into components of + and − signs. Now both components are diffracted again on the

Figure 10.8. The electronic tacheometer *Tachymat TC-1* (Wild, Switzerland)

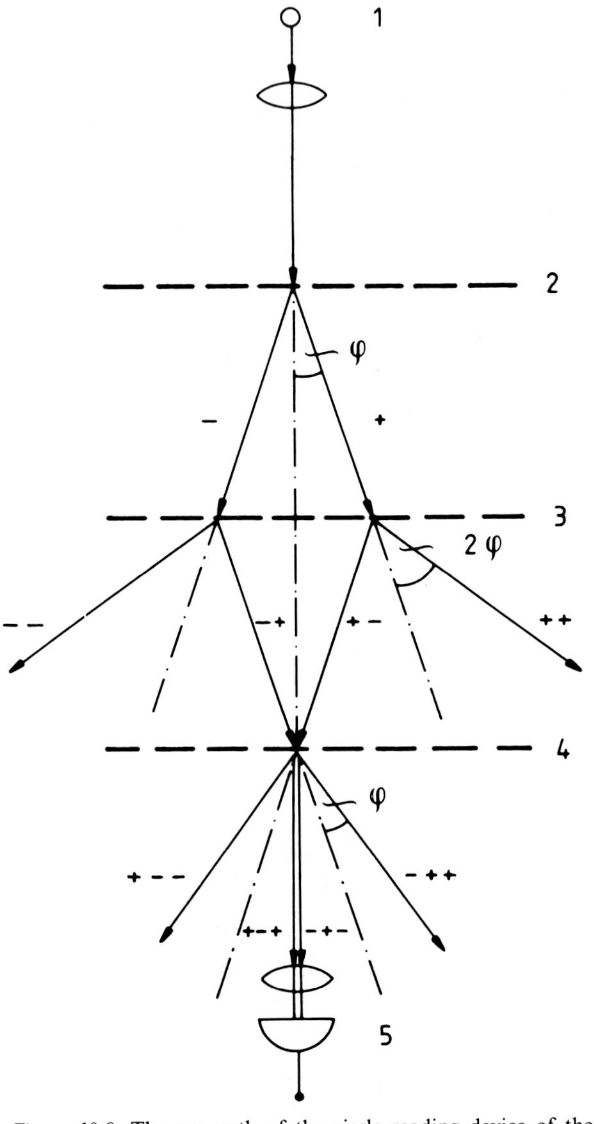

Figure 10.9. The ray path of the circle reading device of the *Tachymat TC-1* (Wild, Switzerland): *1* infrared radiation source; *2* analyzer; *3* graduated circle; *4* analyzer; *5* photodetector

graduation of the circle, the second diffraction angle being twice the first, and consequently four components are produced (their signs being: + +, + −, − +, − −). The components + − and − + are united on the analyzer after reflection, while due to the third diffraction the components + − + and − + − pass onto the photodetector *5*. Due to the effect of the rotation of the circle — as a result of the different angles of incidence — a phase difference arises between the two components falling onto the detector. As a result of interference, amplitude modulation arises with a period which corresponds to half the value of the grid (double interpolation). At the same time Moiré fringes are also produced by the interference of the two components; the shift of the fringes is detected by four photodetectors, and the phase difference between the output of adjacent detectors is 90°. From the sums and differences of the signals four further electrical signals can be produced, making altogether eight signals which are always shifted by 45°. Including the double interpolation already mentioned each reading device makes possible a sixteen times interpolation, i.e. the two diametrically opposite devices together permit a thirty two times interpolation, giving a circle reading accuracy of 1 mgon. The index compensator of the vertical circle has a pendulum-like suspension; the compensator can be disengaged from the measurement and then the zero line of the vertical circle is determined by the vertical axis of the instrument. The following data pairs can be recorded: horizontal and vertical circle reading; vertical circle reading and slope distance; horizontal circle reading and horizontal distance; elevation difference and height of the point observed; coordinates x and y of the point observed.

The type *TC 2000* (Wild, Switzerland) unites the electronic theodolite *Theomat T 2000* and the electro-optical distance meter *Distomat D15* into an integrated measuring system.

The electronic tacheometer *HP 3820 A* (Hewlett–Packard, US) has as a light source a GaAs laser diode emitting infrared radiation. The identically patterned graduated circles consist of three (concentric) parts. The central part is encoded in eight zones, i.e. the circle for the coarse reading is divided into $2^8=256$ units, the reading being made by optical decoding. The reading is refined by double interpolation. Beside the innermost code ring there is a transparent ring with a band width changing sinusoidally with a period of $2^7=128$; two groups of photodiodes (always four) in diametrically opposite positions detect the luminosity of one single period at points of 90° phase difference. The interpolation carried out with the four photodiodes divides the period into 1000 parts, therefore the accuracy of the interpolation is about 3 mgon (128,000th part of 400 gons). A finer interpolation is made possible by the optical grid beside the outermost code ring. The grid has $2^{12}=4096$ units, thus with 1000 times interpolation the accuracy of fine reading is 0.1 mgon.

The interpolation takes place in both cases as follows: due to the position of the diodes the photo-current of the four diodes are in turn:

$$i_1=I_0+i\sin\varphi$$

$$i_2=I_0+i\sin(\varphi+90°)=I_0+i\cos\varphi$$

$$i_3=I_0+i\sin(\varphi+180°)=I_0-i\sin\varphi$$

$$i_4=I_0+i\sin(\varphi+270°)=I_0-i\cos\varphi,$$

where φ is the phase angle characterizing the position of the graduated circle.

The difference of pairs of photo-currents is:

$$i_1-i_3=2i\sin\varphi$$

$$i_2-i_4=2i\cos\varphi.$$

The radiation of the infrared LED illuminating the optical grid is modulated sinusoidally by a frequency of 375 Hz, and the form of the modulating signal is:

$$i = I \sin \omega t.$$

This gives the two differential photo-currents as:

$$i_1 - i_3 = 2I \sin \varphi \sin \omega t$$

$$i_2 - i_4 = 2I \cos \varphi \sin \omega t.$$

After changing the phase of the differential photo-current $(i_1 - i_3)$ by 90°, the two signals are summed by an operational amplifier. The form of the resulting signal containing the phase-information is:

$$i_\varphi = 2I(\sin \varphi \cos \omega t + \cos \varphi \sin \omega t) = 2I \sin(\omega t + \varphi).$$

By comparing the phases of the resulting signal and the modulating signal the angle φ giving the instantaneous position of the circle can be determined. The method of determination is the same as that of dynamic distance measurement carried out with moving reflectors (tracking), and interpolation is made by the same phase measuring unit which compares the phases of the two signals resulting from the distance measurement. The interpolation is the same on the internal and the external path. Since for a line-grid the change of luminosity is not exactly sinusoidal, a diaphragm with a sinusoidally changing aperture is placed in front of the photodiode

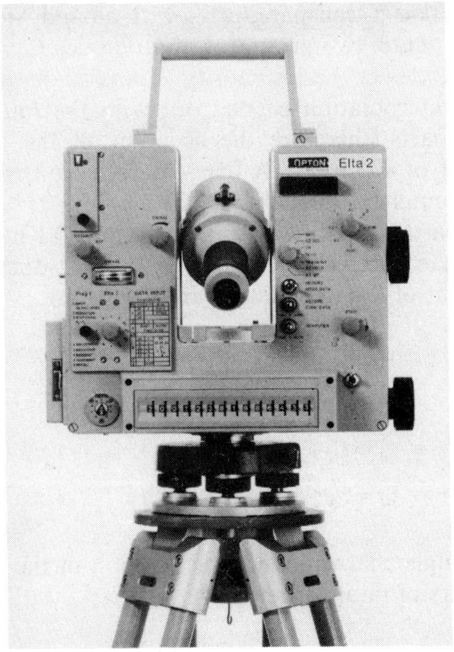

Figure 10.10. The electronic tacheometer *Elta 2* (Opton, FRG)

group of the outer zone. In order to eliminate eccentricity two reading devices in diametrically opposite positions are used on both circles.

The tilt of the vertical axis is determined automatically. Its sensor is a double axis autocollimator, hence the mirror plane is always horizontal. If the position of the vertical axis changes, the direction of the infrared beam reflected from the mercury surface also changes, and so does the spot on the screen of the receiver. The two and perpendicular components of the displacement are determined by a reading device with the same construction as the interpolating device for circle reading. The processor unit corrects the circle readings corresponding to the amount and direction of the tilt of the vertical axis. The following results can be displayed and recorded: slope distance, horizontal distance, horizontal circle reading, vertical circle reading, magnitude and direction of the tilt of the vertical axis.

On the horizontal and vertical circle of the electronic tacheometer *Elta 2* (Opton, FRG) *(Fig. 10.10)* the coarse reading is encoded, and the central lines of the different code bands are marked by division lines without numbering. These central lines are repeated on an inner ring which is concentric with the encoded part and has the same width, but the encoded bands are not repeated *(Fig. 10.11)*. An optical system consisting of prisms and lenses projects the diametrically opposite part of the graduated circle onto the reading device in such a way that the images of the code bands fall on blank places on the inner ring, and the images of the division lines fall on the division lines of the inner ring. The coarse reading is generated by the photodetector block both from the code at the reading device and from the projected

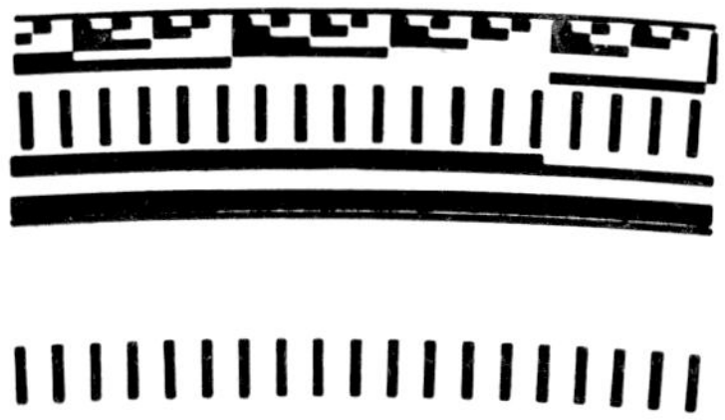

Figure 10.11. Part of the graduated circle of the *Elta 2* (Opton, FRG)

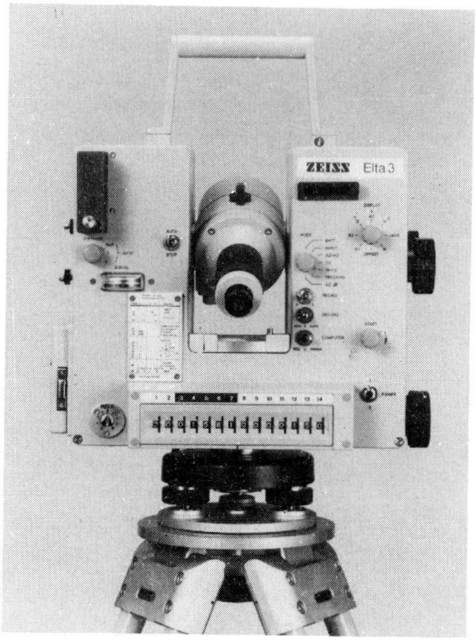

Figure 10.12. The electronic tacheometer *Elta 3* (Opton, FRG)

code band, thus giving a check. For the fine readings the graduation lines are brought into coincidence by means of the plane parallel glass plate of an optical micrometer. Coincidence is detected by a photoelectric microscope, and the fine reading is established by determining the position of the plane parallel glass plate with an electronic interpolator. The beam path of the vertical circle passes through the index compensator which has a pendulum-like suspension. The accuracy of the fine reading is 0.1 mgon on both circles.

By means of the data input program of the plug-in-program-box *PROG 12*, the instrument height, the scale factor and corrections for the collimation and index error, determined from the circle readings on the same point in both telescope positions, can be entered into the data storage of the processor unit. The circle readings of the first telescope position are later automatically corrected by the processor unit. By means of the program-box *PROG 12* a program-controlled measurement can be made. The measured and calculated results can be displayed one by one and they can be recorded individually or in groups corresponding to the program in the transistorized data memory unit *MEM 200* or *MEM 400*, which can be plugged into the instrument. The recorded data can be entered into a computer through the data converter *DAC 100*; *MEM* type memories are able not only to send data but also to accept them. Data and additional information can be entered into the data memory or processor by a keyboard attached to the instrument.

The electronic tacheometers *Elta 20* and *Elta 3 (Fig. 10.12)*, also produced by Opton, are simplified variants of the type *Elta 2*. The principle of the circle reading

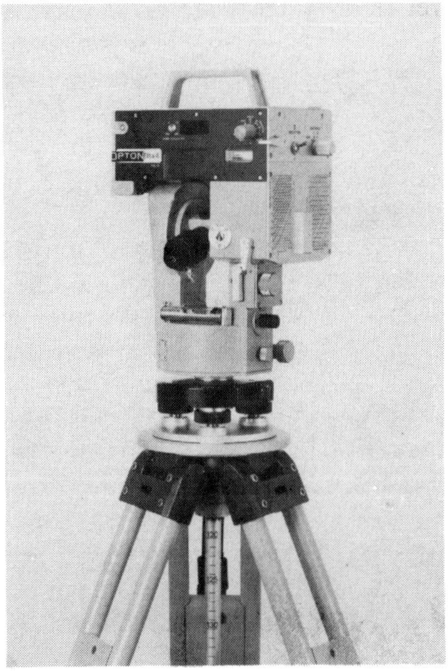

Figure 10.13. The electronic tacheometer *Elta 4* (Opton, FRG)

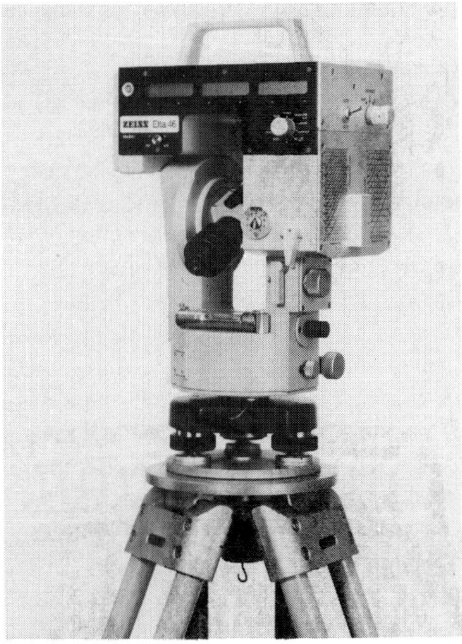

Figure 10.14. The electronic tacheometer *Elta 46* (Opton, FRG)

is the same in all three types, but the interpolation sensitivity, i.e. the accuracy of fine reading, is 0.2 mgon for the *Elta 20* and the *Elta 3*, but in the reading device of the *Elta 3* there is no optical system consisting of prisms and lenses (the reading device has only "one index"). Instead of the program-box *PROG 12* an internal program memory controls the running of some programs (orientation of the horizontal circle; computation of rectangular coordinates from the results of polar point determination, computation of polar data for setting out points determined by their rectangular coordinates). The initial data (the three coordinates of the instrument station, the instrument height, the additive constant, the scale factor, the corrections due to collimation and index error) are entered and established, respectively, in the same way as for the *Elta 2*; the ability to display, record and transfer data is also the same.

The *Elta 4 (Fig. 10.13)* is an electronic tacheometer of simpler construction manufactured by Opton. The circle reading is incremental and the accuracy of reading is 1 mgon. The two circle readings, the slope, horizontal distance and elevation difference can be displayed one at a time, but the displayed results cannot be recorded automatically.

The *Elta 46 (Fig. 10.14)* is an improved variant of the *Elta 4*. Circle reading and its accuracy are the same but, on the liquid-crystal display consisting of three parts, the horizontal and vertical circle readings and the slope distance, or the horizontal circle reading, the horizontal distance and the elevation difference, can be displayed simultaneously. The displayed data can be automatically recorded in the variant *Elta 46 R*.

The *Elta 40* is an electronic tacheometer constructed primarily for the tasks of engineering survey and topography. The measured results and/or the calculated results can be stored in the data store *Rec 200* which can be connected to the instrument. From this store the data are entered into the computer via the data converter *DAC 100*. A direct connection between the instrument and the computer is also possible if the computer has an input interface of the type RS 232C. The tacheometer can be directly connected to the personal computers *Epson HX 20* and *Epson PX 8*, and the corresponding software is also available.

In 1985 Opton started the production of a new series of electronic tacheometers. The first two models of the series "*E*" are called *Elta 3* and *Elta 4*, in spite of the fact that the firm had previously manufactured electronic tacheometers with the same names. The two models have the same construction, the only difference being that the error caused by the tilt of the vertical axis of the *Elta 3* is eliminated from the circle readings. Program-controlled measurement is possible for both models. It is of further interest that both instruments are also manufactured without a distance meter unit (i.e. as electronic theodolites) under the names *Eth 3* and *Eth 4*.

In the electronic tacheometer *Geodimeter 140* (AGA, Sweden) *(Fig. 10.15)* the circle reading device does not contain optical elements and the graduated circles are made of metal. The circle readings are generated by a dynamical method by means of a high frequency electric field, by the contact-free electrical scanning of each graduation line of the graduated circle; the detailed principles of the operation of the reading device are not published by the manufacturer. The accuracy of fine reading is 0.2 mgon. The tilt of the vertical axis is measured by an electronic pendulum and

Figure 10.15. The electronic tacheometer *Geodimeter 140* (AGA, Sweden)

corrections for its instantaneous magnitude and direction are made by the processor. The collimation error is also calculated and taken into account by the processor; it is computed from measurements made to the same point in both telescope positions.

The horizontal and vertical circle readings, the slope distance, the horizontal distance and the elevation difference can be displayed one at a time. The displayed data can be recorded in the transistorized data recorders *Geodat 122*, *Geodat 124* or *Geodat 126*. There are four permanent programs in the program store of the *Geodat 124*, and eight in the *Geodat 126*, by means of which program-controlled measurements can be made. There is free capacity in both program stores in which user programs can be stored. Information recorded in the data recorder can be entered into a computer via the interface RS 232/V24, or by means of a telephone-line (modem), or it can be transferred to a secondary data holder (magnetic tape). It is possible to obtain hard copy of the contents of the data recorders of both the *124* and the *126*; the devices can be connected directly to a printer.

Setting out is facilitated by the observer at the instrument being able to give one-way commands to the person holding the prism by means of a light-telephone connection (*UNICOM*, see also Point 9.3.3.2). The audio-signal modulates the infrared measuring beam of the distance meter unit. The audio-signal is demodulated by the detector beside the prism and can be heard in the loudspeaker on the pole supporting the prism.

The height of points, to which access with the prism is difficult, can also be determined indirectly (*R.O.E.*=*Remote Object Elevation* determination, see also Point 9.3.3.2). In such a case the prism must be placed at an accessible point of arbitrary height in the plumb line of the point whose height is to be determined. After measuring to the prism and setting the mode "indirect height measurement", the processor continuously computes the height of the intersection between the line of sight and the plumb line of the prism by means of the vertical circle reading. By setting the line of sight with the slow motion screw to the point to be measured, the height of the point can be read.

The *Geodimeter 136* is of the same construction as the *Geodimeter 140*, but its effective range and accuracy of circle reading are lower.

The *Geodimeter 142* is an improved variant of the model *140*. Its novelty is that, if the instrument is displaced during measurement, the computing unit corrects not only the circle readings but also the numerical value of the distance measured, i.e. the measured results refer to the initial instrument stand-point. The corrections derive from the electrical pendulum with two axes used in the model *140*.

The electronic tacheometer *RECOTA* (Jenoptik, GDR) *(Fig. 10.16)* has a horizontal and a vertical circle, in both of which the outermost ring contains 4000 graduation lines, the distance between two adjacent lines thus corresponding to 0.1 gon. In the next concentric zone the number of graduation lines is 400 (1 gon), while the third zone contains the encoded numbering of the units of gons. The inner three zones are empty and the projector system, consisting of prisms and lenses, projects in them the images of the diametrically opposite parts of the circle. The double image of the field of view of the reading microscope is projected to a rotating optical deflecting element, which alternately projects to a photodetector block the parts of the field of view necessary to establish the horizontal and vertical circle readings. The fine circle readings are carried out by a common electronic interpolator. The two circle readings

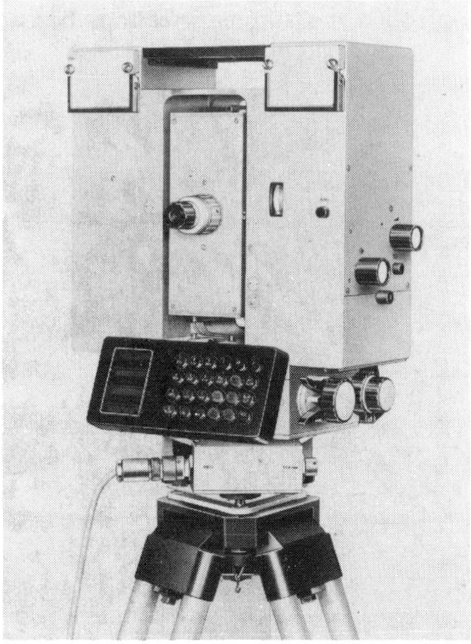

Figure 10.16. The electronic tacheometer *RECOTA* (Jenoptik, GDR)

appear simultaneously on the display with an accuracy of 0.1 mgon; each displayed value is the mean of eight readings. The optical train for the vertical reading passes through the index compensator. The component of the tilt of the vertical axis in the direction of the horizontal axis is determined by a single axis electronic tilt-meter, and the corresponding correction is taken into account by the processor unit when establishing the horizontal circle reading. The corrections due to the collimation error and the index error are obtained from pointings made to the same point in both telescope positions. The computed corrections can be entered into the processor unit via the keyboard.

The program boxes *PR 4* and *PR 5* can be plugged into the instrument. They are supplied by the manufacturer with permanent programs by means of which the corresponding measuring tasks can be carried out. The measured and computed data appear on the display according to the program used, the code numbers identifying the data appearing in a separate display window. The displayed values can be recorded in the transistorized data recorder *Micronic 445* (made in Sweden). A control hard copy of the contents of the data recorder can be obtained, and the recorder can be connected to the printer *Micronic M 20* without an interface. The information stored in the recorder can be entered into a computer either directly via interface RS 232/V24 and a telephone line (modem), respectively, or it can be transferred to a secondary data holder (magnetic tape) without an interface.

Jenoptik developed the transistorized data recorder *FSI 1016* which can be plugged into the instrument. The data stored in the recorder can be entered into a computer via the data converter *DTK-1*. The recorder can also accept data through the data converter.

The electronic tacheometer *RETA (Fig. 10.17)* is a simplified variant of the *RECOTA*. A program-controlled measurement cannot be carried out with it. The

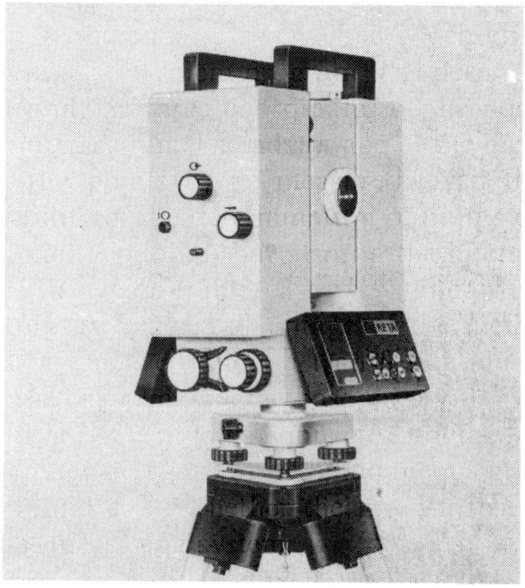

Figure 10.17. The electronic tacheometer *RETA* (Jenoptik, GDR)

measured and computed data (horizontal and vertical circle readings, slope distance, horizontal distance, elevation difference) can be recorded by means of the data recorder used in the *RECOTA*.

As well as the more important technical data of electronic tacheometers some qualifying points of view about their use are also listed in *Table 10.1*. For the correct interpretation of the data some comments are necessary.

— The separation of the instruments into groups according to their construction does not mean a classification according to quality. A modular construction is preferable to a unified construction only in that the instrument can be purchased in separate units, and the units can also be used separately for different tasks at the same time. On the other hand, the use of a measuring system in which the parts are unified into one unit (sometimes including the power source) is more comfortable and needs less adjustment (the alignment of the line of sight of the theodolite and the measuring beam of the distance meter must be checked from time to time in instruments of a modular system).

— Distance measurement has been taken as partially automatized if the level of the received signal must be adjusted by the observer before measurement.

— With electronic tacheometers of modular construction, pointing is considered as simultaneous if the distance meter is attached to the telescope of the theodolite unit, assuming that the line of sight and the measuring beam are in alignment.

— By entering the additive constant the different prism constants can be taken into account. In some instruments the additive constants for two types of prism are pre-programmed. The multiplying constant is usually equal to the atmospheric correction, but in a wider sense it can also be a compound correction: the scale factor expressing the reduction to sea level, reduction to a map projection, and the systematic scale error of the geodetic network.

— In the case of a simultaneous display of several results, the number of displays in the first and in the second telescope positions are also given if necessary.

— In the column "program-controlled measurement" a distinction is made when only some basic programs can be carried out and user programs cannot be stored in the program memory.

Part III

Office Instruments and Equipment for Processing

The processing of measurements is carried out in offices using computers, mapping and map measuring equipments. Only equipments for mapping and measuring on maps will be introduced here.

Maps represent on a plane the data of horizontal and vertical position determinations reduced to a given scale.

Mapping is the set of operations which result in the possibility of drawing the form of the surface of the Earth at a given scale from angles and distances measured for the purpose of position determination. Mapping consists of two main operations: a) plotting (scaling) the points to be mapped; control points and detail points, b) connection of the mapped points, drawing of key elements and lettering. For precise mapping practice demands an accuracy of 0.1 to 0.2 mm which is at the limit of perceptibility by the naked eye. To reach such an accuracy special aids for drawing and plotting have been developed: from simple diagonal scales and other types of scales and manual mapping devices up to automatic precise plotting machines. Mapping equipment, and operations such as plotting and drawing with it, were independent procedures until the appearance of automatic plotters. The connection between plotting and drawing operations has been made possible by the automation of map drawing by uniting them in a single operation carried out with the same apparatus.

Mapping devices include all the manual equipment and partly or completely automatic plotting machines which carry out map editing and drawing (Chapters 12 and 13).

The map as a scaled reduced image can be used to measure certain real dimensions of the features plotted on them with some given standard deviation. This usually means measurement of distances and areas. A special case of distance measurement on maps is the measurement of the coordinates of characteristic mapped points. Digitizers used for measuring the coordinates of mapped points are treated separately from other map measuring devices (Section 13.1). Similarly, instruments for measuring areas from the survey plot are described separately (Chapter 14).

11 Grid Square and Coordinate Plotting Devices

The plotting (scaling) of map points consists of the accurate drawing of angles and short distances relative to specified points and directions on the drawing. One set of specified points is fictitious: these are the intersections of rectangular lines which are, in most cases, parallel to the coordinate system used. These rectangular lines form a grid on the map with a suitable grid density (most frequently $\Delta = 5$–10 cm). The intersections (grid points) are separated by $n \cdot \Delta$ metres, where n depends on the scale of the map. The lengths of distances measurable depend on the equipment used.

Manual equipment allows only shorter distances to be plotted, while plotting machines allow the plotting of the longest distances which can be contained in a map sheet. Therefore the method of constructing consecutive points depends on the instruments used.

With manual equipment horizontal control points must be plotted from their coordinates relative to the grid points as additional specified points. Detail points are then plotted from these specified points using local — orthogonal or polar — coordinates, i.e. they are plotted directly from the measurements. Plotting by angles with the necessary accuracy is more difficult than plotting by distances only, therefore plotting is usually carried out in an orthogonal system, i.e. using only distances.

Coordinate plotting machines plot every point in the coordinate system of the map sheet — i.e. from the value of the coordinates reduced to the sheet grid — which is why these instruments are also called coordinatographs.

Distinction must be made between grid square plotting devices, manual point plotting instruments and coordinated plotting machines. Devices for redrawing maps at a different scale, termed pantographs, belong in a wider sense to the family of mapping instruments.

11.1 Grid Square Plotting Devices

The simplest instrument for drawing a grid is the primary plotting scale, a straight copper rule about 45 mm wide and 5 mm thick bevelled on its upper surface. Its length is greater than the diagonal of the boundary line of the network by about 10 cm. The graduation of the scale corresponds to the actual scale of the map, e.g. when metric units are used, it is graduated in mm-s *(Fig. 11.1)*. A rectangular isosceles triangular rule, of the same thickness and with sides of 8 to 10 cm, is used with the primary scale. One side of the triangle has a vernier and is placed against the graduation of the straight rule. The other side has a bevelled drawing edge with a graduation corresponding to that of the straight rule. The primary scale can be used

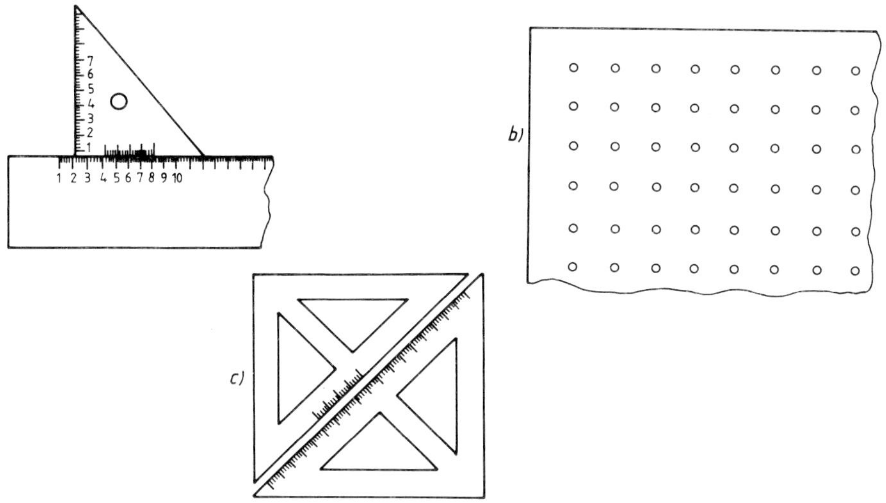

Figure 11.1. a) detail of a major plotting scale; *b)* detail of a grid plotting plate; *c)* the Majzik triangle pair

in general to plot any distance less than 1 m; when scaling, the triangle is moved along the straight rule and the appropriate distance is set by the vernier. The primary rule should be checked for straightness of the drawing and graduation accuracy, and the triangle for the accuracy of its right angle. The maximum deviation of the drawing edge from a straight line and the perpendicularity error of the rule must not exceed 0.03 mm. The graduation error over the whole length of 1000 mm may be ± 0.1 mm. Calibration of the primary scale is carried out by comparing it with a standard metre, taking into account the thermal coefficient of expansion of the rule.

An instrument commonly used due to its simplicity is the grid templet. It is often a 2 mm thick plane plate, made of an alloy with a low thermal coefficient of expansion, which has a highly finished surface and holes of 3 mm diameter at the corners of a 5×5 cm rectangular grid *(Fig. 11.1, b)*. A sprung pin in a metal casing, which fits centrally into these holes, is used to plot the points of the network. The plotting accuracy of the grid points is ± 0.1 mm. The distances between the holes should be checked from time to time by means of a precise glass scale as the holes wear during use. Defective or worn grid templets must not be used.

11.2 Equipment and Devices for Manual Point Plotting

11.2.1 Equipment for Plotting by Distance and Angle

Points can be plotted with manual equipment from either orthogonal or polar coordinates.

Simple equipment for the former is the plotting scale. Plotting from polar coordinates should be carried out with the type of protractor which can also be used for plotting distances.

Plotting scales are usually metal rule systems which consist of two or more parts with precisely machined surfaces, supplied with a graduation corresponding to the actual map scale and with a vernier reading system.

A well-known item of plotting equipment is a set-square, termed the "Majzik triangle pair" (see *Fig. 11.1, c*), which consists of two isosceles triangles made of metal plate about 2 mm thick. The hypotenuse of one of the triangles is graduated along its full length, and the other triangle has its two right-angled sides bevelled and a vernier graduation on its hypotenuse. The advantage of the Majzik triangle pair is that, with a single layout of the pair, two perpendicular distances can be plotted and drawn. As the graduation unit on the hypotenuse is $\sqrt{2}$ times longer than that on the corresponding perpendicular sides, setting and reading is more accurate.

Protractors are used for plotting polar coordinates. Protractors used for precise mapping have a horizontal circle, graduated in grades, which consists of a fixed and a rotating part. On a metal scale, radially attached to the rotating part, a vernier and a pricking needle are mounted in order to set and mark the distance. The metal scale is rigidly connected to the rotating circle, and it has a vernier graduation which is used together with the graduation of the fixed circle to set angles.

Figure 11.2 shows a diagrammatic sketch of the protractor (polar coordinatograph) type *5100* made by the Swiss firm Haag-Streit. The rotating circle has bearings. The effective radius of the rule is 200 mm. The circle is graduated to 2 minutes of arc, with estimation to 1 minute, and the accuracy of reading the metal scale is 0.05 mm. The rule can be purchased for scales 1 : 500/1000/2000. The accuracy of plotting is ±0.02 mm.

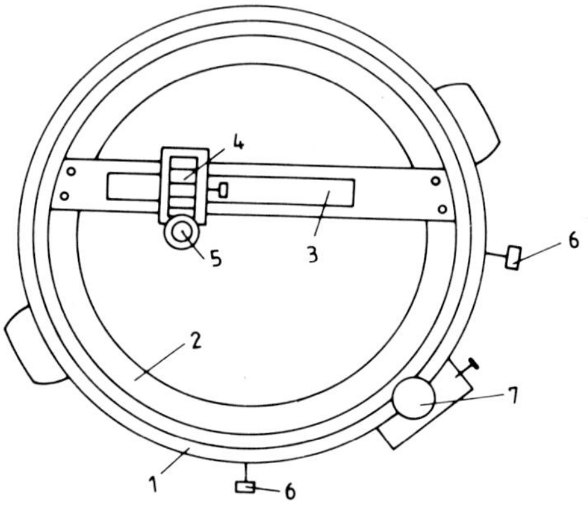

Figure 11.2 Sketch of the Haag–Streit *5100* polar coordinatograph: *1* fixed circle; *2* rotatable circle; *3* rule graduation; *4* vernier; *5* pricking microscope; *6* centring screw; *7* circle division reader

11.2.2 Pantographs

Pantographs are described in the present chapter, included with the equipment for point plotting. They are used to redraw maps at a smaller or larger scale. Its main parts are: the pole which is fixed after the pantograph is set up, the guided pin which is to be guided along the lines on the map to be copied, and lastly the plotting pin which plots the map at the new scale. These parts are connected by a hinged rule system. The operating principle is based on the equality of the ratio of the sides of similar triangles. There are several ways in which its construction is realized. Here the so-called *Milanese pantograph* is mentioned, its operating principle being as follows *(Fig. 11.3)*: the pole P is at one corner of the parallelogram $PEVD$. The triangles PBR and PVD are similar. The rules forming the sides of the parallelogram are connected by hinges at the points P, B, C, D, ensuring the similarity of the triangles even if the angle α changes. The distances x and y can be changed. The ratio of the old and new scales is expressed by the ratio of the distances m and n, i.e.

$$x=\frac{b}{\frac{m}{n}}=\frac{nb}{m} \quad \text{and} \quad y=\frac{a}{\frac{m}{n}}=\frac{na}{m}.$$

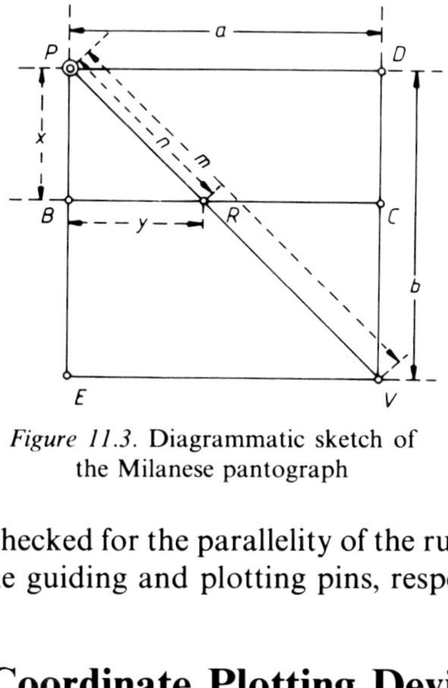

Figure 11.3. Diagrammatic sketch of the Milanese pantograph

Pantographs should be checked for the parallelity of the rules, for their straightness and for the centrality of the guiding and plotting pins, respectively.

11.3 Coordinate Plotting Devices

Coordinate plotting devices with manual setting are mechanical coordinatographs. They use a rectangular axis system to span quite long distance, usually a drawing area of 1200 × 1500 mm, thus they can be used to plot both the grid points of the network and coordinates computed from measurements; sometimes they are even used for measuring these coordinates from the map.

11.3.1 Coordinatographs with Manual Setting

Mechanical coordinatographs appeared at the end of the last century *(Fig. 11.4)*, being used initially for plotting grid squares as well as horizontal control points for maps. The equipment can be mounted on a table or it may consist of a stable metal construction built on the drawing table, with the coordinate axes represented by two mutually perpendicular guiding axles fitted with gear racks. One of the axles is fixed with respect to the table, the other can be moved parallel to itself. The moving axle is fitted with a carriage which is movable by hand. The carriage can be set to any point of the area flanked by the axles. Coordinatographs cannot be used directly for the measurement of distance in an arbitrary direction. Both plotting and measurement of distances on maps are carried out by the measurement of distances parallel to the axes y and x, i.e. independently from each other. A counting mechanism is used to set and read coordinates. The point to be plotted is marked by the pricking needle on the carriage or by some other drawing instrument.

The mechanical coordinatograph *Coradograph* with manual setting, produced by the Swiss firm *Coradi*, is a precision tool. The accuracy of this equipment is: the maximum error of perpendicularity is 0.02 mm/m, the error in setting distance along both axes is ± 0.04 mm/m.

The Swiss firm *Haag-Streit* produces two kinds of coordinatograph. The types *580*, and *6000* of greater dimensions, utilize the constructional principle shown in *Fig. 11.4*. In the type *6000* the axle system is made of stainless steel with guaranteed high stability. The measuring system consists of a precise gear rack and a measuring wheel running on it. The movement of the carriage is determined by an index using a vernier division. Moving parts run on stainless, precision bearings. There are several types of this model *6000* with different table sizes (820–1600 × 1220 mm). The plotting accuracy is 0.04 mm, the reading accuracy on the vernier 0.01 mm. The reading apparatus may have a numerical display. Coordinates are displayed with an accuracy of 0.01 mm.

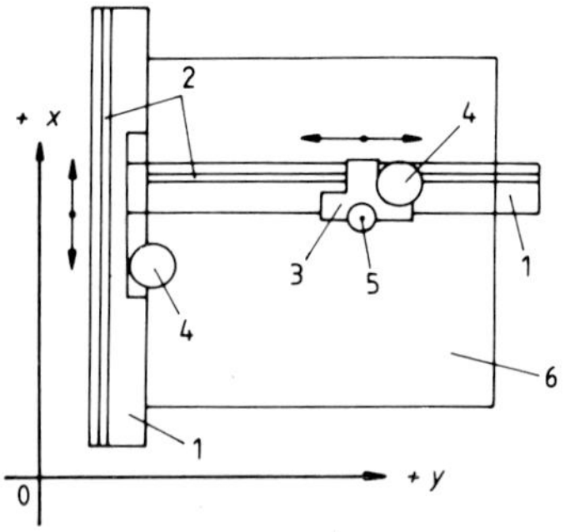

Figure 11.4. Diagrammatic sketch of a coordinatograph with manual setting: *1* control axles; *2* main graduations; *3* moving carriage; *4* counting mechanism; *5* instrument holder; *6* drawing board (maximum 1200 × 1200 mm)

Figure 11.5. Detail of the coordinatograph *ARISTO 4438*, FRG

The smaller models of these coordinatographs, portable types *5200* and *5300*, are used mainly to plot detail points. These types do not have their own table surface. The axles are mounted on a portable metal frame which can be set on any drawing surface. The usual dimensions of the *5200* is 400 × 240 mm, that of the *5300*, 550 × 500 mm. Plotting accuracy is ±0.02 mm.

The firm *ARISTO* in the German Federal Republic manufactures instruments similar to the above, from which the coordinatograph *4438* should be mentioned: it is illustrated in *Fig. 11.5*. It has a rotatable board so that it can be used as both an orthogonal and a polar coordinatograph.

The Austrian firm *Rost* makes the coordinatograph *Unicord* which uses the same principle as shown in *Fig. 11.4* and can be connected to the *Teleconta* data collecting system. Together they form the *Digicart* digitizing equipment. The determination of the coordinates x and y is made by photoelectric scanning.

11.3.2 Automatic Coordinatographs

Automatic coordinatographs *(Fig. 11.6)* provide a transition between mechanical coordinatographs with manual setting and partly or completely automatic plotting machines. They have the same principle of operation as mechanical coordinatographs with manual setting in so far that the movement of the drawing instrument is independent along the two axes. The significant difference is that the movement of the drawing instrument is directed by an electronic control unit through electromechanical transmission. The coordinates to be plotted are transferred in the form of electrical signals from the data medium to the control unit. Automatic coordinatographs consist of the following three main parts: input unit, drafting table with plotting unit, and control unit.

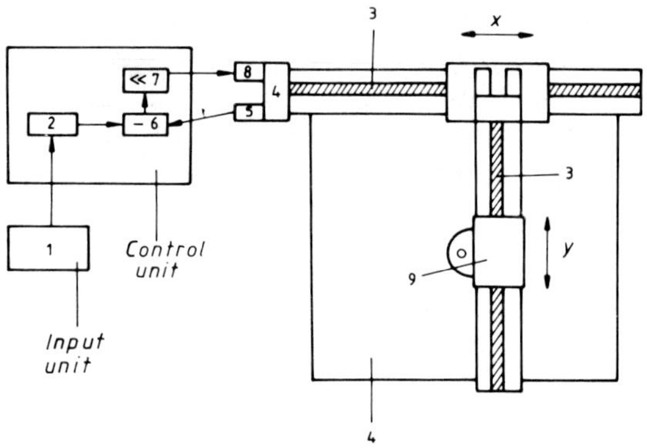

Figure 11.6. Sketch of an automatic coordinatograph

In *Fig. 11.6*, the input unit *1* transmits to the control unit the coordinates y and x of the point to be plotted and stored in a computer. The control unit has mechanical and electrical connections to the drawing table. Both control shafts of the drawing table are fitted with a screw spindle *3*, driven by the servomotor *8* through the transmission *4*. The moving shaft carries the carriage *9* holding the drawing instrument. The driving mechanisms move the drawing instrument to the point on the table where the position of the plotter needle coincides with the coordinates of the point transmitted by the input unit. The task of the control unit is to set the position of the screw spindles to the coordinates corresponding to the pulses transmitted from the reader. The screw spindle is connected through transmission *5* to the counting mechanism. This mechanism corresponds to the immediate reading of mechanical coordinatographs and gives electrical signals which determine the actual position of the spindle screw unambiguously. The counting mechanism signals the actual position of the drawing head. The pulses coming from the input unit give the required value in the decoder *2* of the control unit. The difference of the two signals produced in *6* is amplified by *7* and then controls the servomotor. The sense of rotation of the servomotor depends on the sign of the difference, and if both signals are identical, i.e. when they coincide, the rotation is stopped and control of the drawing instrument is automatically established in order to plot the position of the point. As accessories for the drawing instruments are supplied which enable the plotted points to be pricked and circled.

Coradi developed its mechanical coordinatograph into an automatic plotter under the name *Coradomat*. Its table weighs is 700 kg and is of metal construction with a working area of 1300 × 1600 mm. The control unit is supplied with a keyboard from which the equipment can also be manually operated with the keyboard as the input unit. The decoder is built together with an electronic counting device which makes possible both a change of scale and coordinate transformation. The control circuit checks 70 differences in a second. The velocity of the screw spindles is also determined by the control unit in four stages. When coincidence is reached the point is automatically marked and the new coordinates are entered on an electrical control signal. The maximum speed of the carriage is 55 mm/s. The maximum error of plotting is 0.05 mm. The equipment can also be used with corresponding supplementation, in an inverse mode of operation, as a digitizer (Chapter 13).

Opton developed an electrical control device for the drawing table of the mechanical coordinatograph of the stereoplanigraph *C8*, which resulted in the automatic coordinatograph *Kordimat (Fig. 11.7)*. The input unit here is a punched card or punched tape reader.

Figure 11.7. Kordimat automatic coordinatograph (Opton, FRG)

The following data are stored after reading from the data medium: x and y coordinates with a maximum of 6 digits, sign, number of the points and coded information for printing symbols if necessary. The control shaft and the carriage are moved by a screw spindle, and the movement corresponding to the coordinates stored in the control unit is ensured by a mechanical rotation counter fitted with electrical contacts. The rotation speed of the coincidence device can be modified in relation to the rotation of the spindles by means of a gear train. This is the solution for a change of scale. When coincidence is obtained the drawing head operates automatically: the rotating pricking needle marks the position of the point. The main source of error during the operation of automatic coordinate plotters is curvature of the shafts and error in their perpendicularity.

Some important technical data for automatic coordinatographs are presented in *Table 11.1.*

Table 11.1. Main technical data for some coordinatographs

Producing firm, model name	Country	Type	Operation mode	Working area	Accuracy	Drawing facility	Reading device	Scale
ARISTO 4422	FRG	portable	manual setting	200 × 200 mm	± 0.1 mm	pricking needle	index, magnifying glass	1 : 1000/2000/2500
4416		portable	manual setting	650 × 100 mm	± 0.1 mm	pricking needle	vernier, magnifying glass	
4438–8738		coordinato-graph with rotat-able board	manual setting	800 mm ∅	± 0.02 mm	pricking needle	vernier, magnifying glass, digital display	
ARISTOMAT 105 M		flatbed coordinato-graph	manual setting, automatic plotting	1200 × 1500 mm	± 0.02 mm	pricking needle	vernier, magnifying glass, digital display	
Rost Unicord	Austria	flatbed coordinato-graph	manual setting	1200–800 mm				
Haag-Streit 5300	Switzerland	portable	manual setting	550 × 500 mm	± 0.02 mm	pricking needle	vernier, magnifying glass	1 : 250/500/1000/2000
6241		flatbed coordinato-graph	manual setting	820 × 1220 mm	± 0.04 mm	pricking needle	vernier, magnifying glass, digital display	1 : 250/500/1000/2000

12 Equipment for Automatic Mapping and Plotting

The use of electronic computers in the processing of surveying measurements in the office has led to the numerical display of map information.

Different kinds of instruments have been developed for the graphical perception of large quantities of numerical information, to display the results. A significant stage in the development of computers has been the advent of computer-controlled graphic output units. They are able to display the results of processing in a graphical form without human intervention. The main types of graphic output units are: screen displays, automatic digital and analogue-digital mechanical plotting machines.

The function of the screen display differs significantly from the present needs of map drawing. It enables one to study certain parts of the map projected on a screen. Its use in mapping at present is either as an interactive display connected to a data bank or as a supplement to automatic plotting.

There are many types of plotting machines. Digitally controlled flatbed plotters are very widely used in mapping today. Their mechanism is similar to that of the automatic coordinatographs described in Subsection 11.3.2, but the movements are controlled here by an electronic control unit. Electronic plotting machines can be considered as special printing peripherals.

For automatic processing it should be presumed that the measurement data, or the numerical values computed from them, are recorded in some data recording media. Data recording for processing can occur either in the field, simultaneously with obtaining the measurement data, or from field notes as a first step in the course of the office work.

12.1 Data Recorders and Data Recording Instruments

The following types of data recording are distinguished:
- making records by hand or manual marking,
- passive data recording,
- active data recording,
- automatic data recording.

Conventional, hand-written, but properly transformed field records are worth mentioning here only because, due to their format, they can be used directly as data documents from which the data input is made directly onto punched tape or some other data carrier.

With passive field data recording, the record of the data is made by keyboarding. The data entered via a keyboard appears at first on a display, where it can be checked and corrected if necessary.

An active data recorder, together with the corresponding peripheral and storage units, can be regarded as a microcomputer. Its advantage against a passive data recorder is its interchangeable program storage and Random Access Memory (RAM) of high capacity.

Automatic field data recording is made e.g. by electronic tacheometers (see Chapter 10). Here, after the different automation levels of the measurement process, measured data are digitally displayed and recorded on data carriers in an encoded form.

12.1.1 Use of Special Punched Cards

The method of mark sensing can be classified as making records by manual marking; punched cards of special form, with a maximum number of columns of 27, are used for this purpose. The usual form of the card is shown in *Fig. 12.1.* The content of the document is marked on the card with a pencil of high graphite content. The pencil traces are detected and automatically punched by a copy-punching device developed especially for this purpose. To ensure the electrical conductivity of the pencil-line it is very important that the graphite particles should contact each other, consequently a soft pencil (2B, 3B) must be used. The marked cards are put into the duplicating punch; data are punched in the proper columns after the necessary preparation. This means that the numbers detected on the card are punched into a column of the same card.

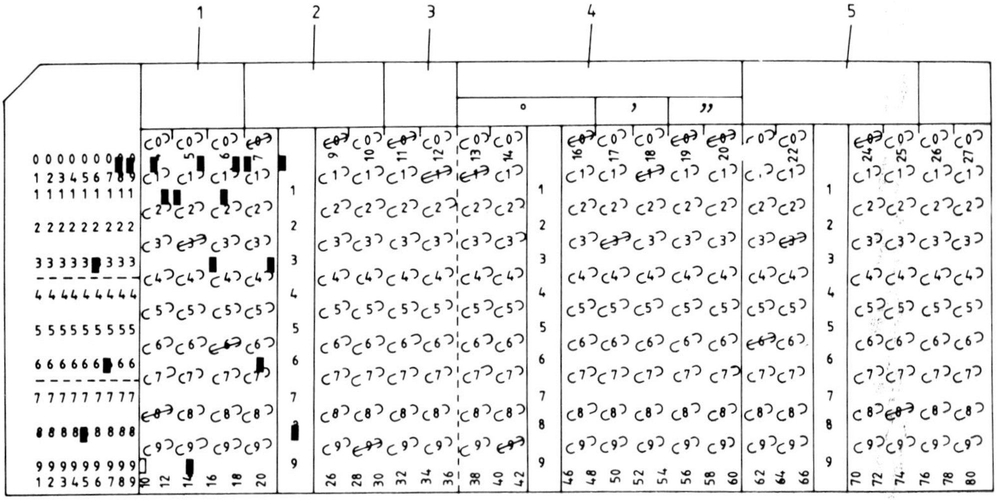

Figure 12.1. Punched card for the mark sensing method: *1* block number; *2* serial number of the observed point; *3* station number; *4* direction angle; *5* distance

A practical solution of data recording is the use of 80 column cards made for the "PORT-A-PUNCH" method, but here only the even-numbered columns can be used, i.e. 40 columns. For punching and over-punching the numbers, a rectangular punched card (normal size), perforated in advance, is used. These cards are put in a plastic card holder with a rubber underlay. This underlay is divided into 12 rows according to the punching requirements. There is a transparent plastic ꜔'ate above the card in which there are small holes at points corresponding to the columns of the card.

The device includes a point pricker of the size of a ball-pen, by means of which the required digit can be punched through the holes of the covering plastic plate. When punching, the perforated rectangle falls or is pricked into the holes of the robber underlay.

12.1.2 Field Cassette Recorder System (KATA)

Portable devices such as the above, also developed in Hungary and elsewhere, belong to the group of passive data recorders. It is a data recording device applicable to field conditions, independent of a main electrical supply, it operates from a battery and it is constructed to use minimum power.

The data recorder consists of the following parts as shown in *Fig. 12.2*: *1* keyboard, *2* data recorder, *3* interface, *4* TV control, *5* field instrument which can be

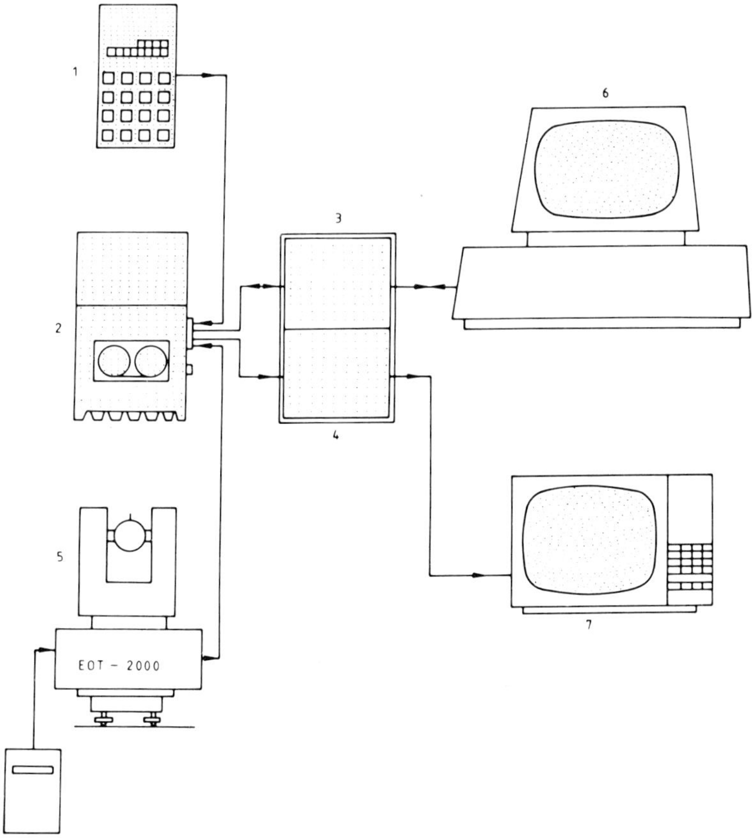

Figure 12.2. Field data-recording system with cassette

connected directly (such as e.g. the electronic tacheometer *EOT-2000*), *6* computer, *7* TV receiver.

The data-recording keyboard *(Fig. 12.3)* has a display with a microprocessor and an independent memory. The information stored in the memory can be modified and compiled according to the microprogram. Numerical information (a word) of a maximum of 10 digits can be displayed on the liquid crystal display. A further character gives the serial number of the information and the operation mode of the keyboard, respectively.

The display unit shows the information, consisting of a maximum of 10 characters, its serial number and the operation mode of the keyboard. The information displayed can be modified if necessary.

The data recording keyboard is connected by a multicore cable to the field cassette recorder. Current is supplied to the keyboard from the data recorder. The keyboard is contained in a metal box, it is light and easy to handle.

The functional parts of the keyboard unit are as follows:

16 keys in the arrangement shown in *Fig. 12.3*: numbers from 0 to 9; "–" as a sign only at the beginning of the word, but the last character typed can be erased by this key; *E* means "forward" in the memory; *H* means "backward" in the memory; *P* means erase the memory; *BL* means enter the characters on the display into the memory; *L* means transfer the data to the cassette recorder.

Letters *P*, *H*, *E* have two functions: they can also be used as information characters. They can occur anywhere except at the beginning of a word.

The buffer-store of the keyboard has a capacity of 256 bytes; it is used for the intermediate storage of the information entered by the keyboard. The words stored

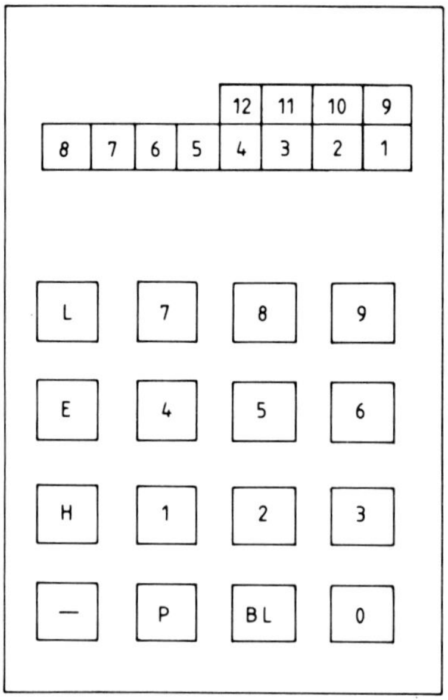

Figure 12.3. Keyboard of the field data recorder

in the memory are divided by an information-separating character. Information stored in the memory can be recalled by the functions "backward" and "forward" respectively, and displayed. With the function "backward" the word counter decreases by one; the function "forward" can be used only if "backward" was used in the previous step. Saturation of the memory is indicated by the character showing the operation mode. After the contents of the memory have been transferred into the data recorder unit, the memory can be refilled. The memory is erased when the power supply fails or is disconnected; this is indicated by the appearance of the operation mode character on the display of the keyboard unit.

The manipulation of the keyboard. When switching on the supply voltage, number *1* appears in the 11th digit. If in the course of entering data a word contains erroneous information, then the words following the corrected word are erased from the memory. To make a correction, firstly the required correct information is generated on the display (by erasing the characters present and entering new characters), then the key *BL* is pressed. Now the word counter increases by *1* and the other positions (characters) on the display are erased. Code-part ASCII (7 + 1 bit).

If the word last entered comes near the end of the memory but there is still room for one word of 10 characters, then a number "–" appears in the display position with *L* in the first position. This shows the length of the word which can still be entered into the memory. If the number is 8 then a word of 8 characters can be entered, after which the number changes to 0. If there is insufficient room in the memory for the word in the display and the key *BL* is pressed, then the word counter remains unchanged and letters *L* appear in the positions 1–10.

A block diagram of the keyboard unit is shown in *Fig. 12.4*. The signal connections TI 0-3 and TO 0-3 contain a 4 × 4 switch matrix, at the crossing points of which the

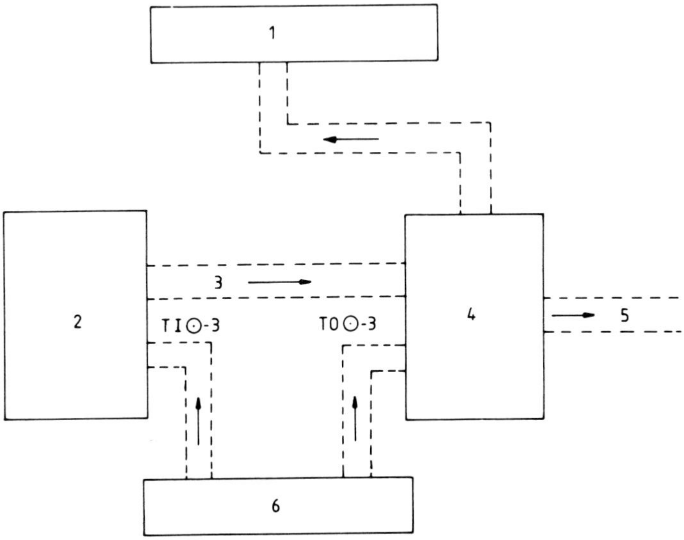

Figure 12.4. Blok diagram of the keyboard of the field data recorder: *1* display unit; *2* CPU; *3* bus; *4 I/O* interface and RAM memory; *5* serial output; *6* keyboard

keys are placed. The microprocessor continuously scans the keyboard. When it finds a depressed key it generates the corresponding code and commands the display unit.

The magnetic tape recorder is an independent unit *(Fig. 12.5)*. It is a universal type data recorder and it can be directly connected to the electronic distance meter *EOT-2000* (Chapter 10) as well as to an independent keyboard unit. The recorder is based on the conventional magnetophone technique; it has its own power source (dry battery). The digital signal transferred from the previously mentioned devices are automatically recorded on a cassette.

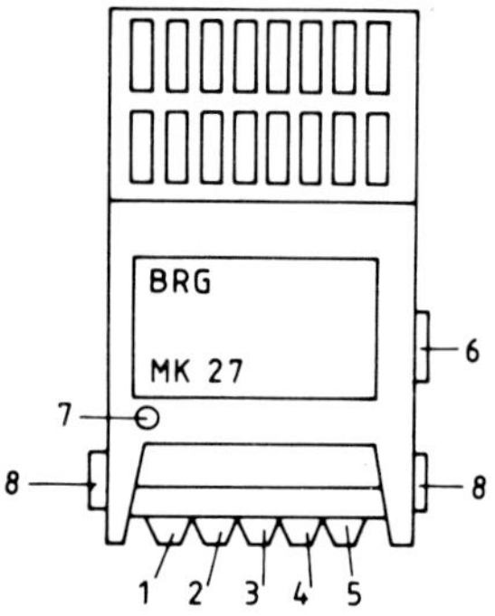

Figure 12.5. Magnetic tape recorder unit for the field data recorder: *1* rapid rewind; *2* loading of tape; *3* stop; *4* read-write; *5* rapid wind forward; *6* connection; *7* battery check; *8* handle

The data store is interchangeable and can be easily interfaced to any instrument. The electrical construction is characterized by the use of CMOS circuits of high reliability and low power consumption. The tape recorder can be operated in both reading and writing operational modes. Its functions are: to record data on magnetic tape, and to read and transfer recorded data and to provide protection against error.

The device is started by connecting it to an accumulator, then the magnetic tape recorder is connected by cable to the distance meter or to the keyboard unit. By pressing knobs 4 and 5 (see *Fig. 12.5*) the data recorder winds the magnetic tape to the sensor head, then it stops. Blocks of 256 bytes are accepted by the device. If the length of the information block is shorter than 256 bytes the missing characters are automatically filled with $x-$s. When data is input (writing) the different information blocks are recorded on the tape twice, one after the other, in order to be able to correct when re-reading. Synchronization bytes are also recorded in order to distinguish between the doubled blocks. If, when reading, the tape recorder finds an error in one of the two identical blocks, the Interface gives a sound signal. If the number of errors reaches 8, the ERROR lamp (red LED) lights. If error is found in both parts of a block pair, the Interface gives a continuous sound signal and the data recorder stops. The tape can be restarted by pressing the knob RESET. In the position WRITE of the READ/WRITE switch of the interface, it is possible to record on the tape information arriving from the computer.

The field magnetic tape data-recorder can be connected to the desktop computer *EMG 666* or to a conventional television receiver.

Before inputting the data into a computer care should be taken that the first 256 bytes be transferred from the tape into the memory of the recorder. After the input of these first 256 bytes into the operative memory of the computer, the tape recorder automatically transfers the next 256 bytes into the buffer.

12.1.3 Field Data Recorder GEOMEMO

The field data recorder GEOMEMO *(Fig. 12.6)* is jointly manufactured by Geotechnica (FRG) and the Society of Geodetic and Electronic Systems.

The GEOMEMO is an active data recording unit for use both in the field and in the office. It has an effective data ordering program, by means of which directly accessible data are stored and the possibility of error correction is ensured. Furthermore data acquisition at the point of data generation may be manual, semi-automatic or automatic. The instrument operates from a battery and is light (1100 g). Besides recording it can also be used for preparation, checking and data delivery. It is fitted with a unit which transfers the stored data directly from the connection point to the memory.

Figure 12.6. Field data recorders GEOMEMO and GEODÜM

The intelligence of the GEOMEMO is characterized by: easy handling, the possibility of correcting the stored data, addressing the files, search routines, the automation of entering the numbers, the use of optical and acoustic checking and of attention-calling signals, and the possibility of altering the schedule.

The modifiable BCD interfaces provide:

— semi-automatic data transfer from a measuring station,

— continuous data acquisition in different fields of application.

For entering data, either the keyboard or the modifiable BCD interfaces can be used. The functions of the keys are enter, correction erase, search, store, transfer. Codes and measured data can be well distinguished.

The CMOS microprocessor provides the control in each operation mode and a controlled data-flow; stored data are retrieved and displayed within a fraction of a second. The modifiable BCD interfaces ensure data transfer without a supplementary interface. The format can be established for display on either a printer or a screen. The display is given by two liquid crystal displays (LCD), each having six characters.

The acoustic control functions operate unambiguously at the end of operations, for error correction and when storage is completed.

The data memory of the GEOMEMO is a semi-conductor capable of storing 12K/24K characters. One file consists of 5 segments, with 6 digits in each segment. The memory makes an automatic battery check, giving an acoustic signal if the battery is exhausted; it provides correction from the keyboard, and erasure of data, and it ensures the priority of the data required.

A rechargeable NiCd accumulator can be used as a power source for the GEOMEMO, which requires 450 mA/h, but if the data transfer device GEODÜM is used, then a charger is necessary *(Fig. 12.6)*. The speed of data transfer is 110–2400 baud. The size of the GEOMEMO is 45 × 105 × 230 mm.

12.1.4 Field Data Recorder MEMOBOARD

The MEMOBOARD is the active field data recorder of Hartel Datentechnik (FRG) *(Fig. 12.7)*. It looks like a plane table which can be slung over the shoulder. Its size is 320 × 320 × 25 mm and it weighs 1.6 kg.

The keyboard, embedded in rubber, is in the upper right hand corner of the MEMOBOARD; it contains 12 input keys and 13 function keys. To the left of the keyboard there are two liquid crystal displays (LCD) with 8 characters each, and in the upper left hand corner there is a place for a pocket calculator (e.g. *HP 41*), which can be attached through a special interface. The central part of the table is used for attaching a drawing. The electronic elements and the rechargeable accumulators are within the instrument behind moisture-protecting insulation. The semi-conductor memory has a capacity of 12–72 Kbytes which can be increased in 12 Kbyte units. The working capacity of the accumulator is 48 hours without recharge. The memory holds data for some weeks.

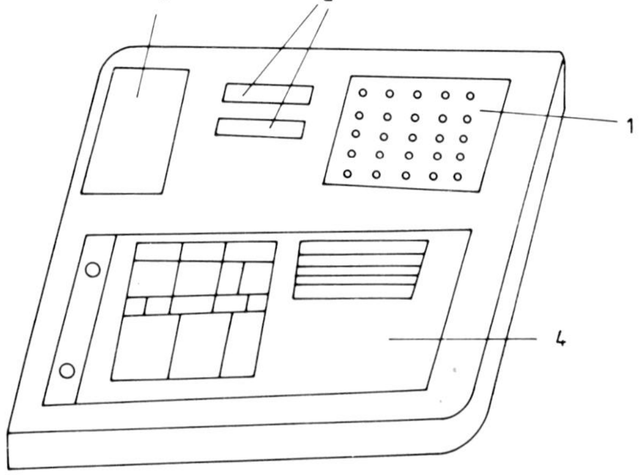

Figure 12.7. MEMOBOARD: *1* keyboard; *2* LCD with 8 characters; *3* pocket calculator; *4* sketch of measured detail

The microprograms in the ROM memory permit a free choice of record format — thus the recorder can be used for many kinds of tasks — and they also permit the retrieval of a record in the RAM memory from the serial number of the station. A printer can be connected to the recorder.

12.1.5 Field Data Recorder REC 100

The electronic recorder *REC 100* (Opton, Oberkochen, FRG) *(Fig. 12.8)* permits measured and computed data to be entered into the data memory, including the corresponding 14 decimal code numbers, and stored data to be recalled.

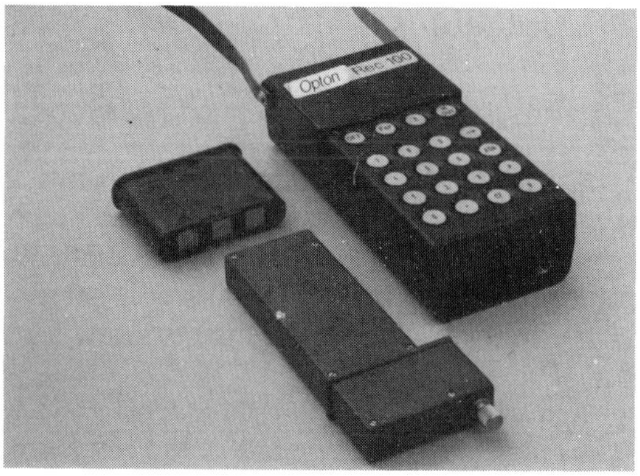

Figure 12.8. Electronic recorder *REC 100* (Opton, FRG)

The operation of the recorder can be summed up as follows. The data format can be changed:

— within a data series by fixing the number of blocks;

— within a block by floating the decimal places.

Data recall: by entering the address it is possible to display a series of stored data.

When entering data:

— the data index increases automatically by 1 after each entry into store;

— it is possible by floating the point to choose any decimal position within the 14 decimal index;

— the device indicates by a flashing accepted the positions in which entered data will be index;

— the identifying number of the measurement, and the horizontal and vertical angles, are entered in a cyclic sequence;

— the device displays the data entered;

— the data can be corrected before storage in the memory.

During data recording:

— having finished entering the data, the complete data series is stored in the next free position of the data carrier memory;

— in order to distinguish reduced values from unreduced, the data series can be automatically marked.

In order to protect the data:

— in the stored data series individual values can be neither erased nor corrected, in order to ensure the documentary character of the record;

— the memory can be erased only totally and only by pressing a definite key;

— the error made is indicated by an error display.

Some special operations can be carried out by the device:

a) The brightness of the display can be varied, thus changing daylight conditions can be compensated.

b) For the sake of a better overview, the display of the index number, which consists of 14 digits, is divided into groups by points.

c) In order to conserve the battery, the device is automatically switched off after 3.5 minutes if no key is activated.

12.1.6 Field Data Recorder AGA Geodat 120

The *AGA Geodat 120* (AGA Geotronics, Stockholm) is a field data recorder of the size of a pocket calculator, which can be connected by an interface directly to different data processors. The interface can be used with modems and acoustic devices. It can be directly connected to the electronic distance meters *AGA Geodimeter 120* and *112* (see Chapter 10). In the transistorized memory data for 600 to 1000 measured stations can be stored. The maximum capacity is 32 Kbytes.

The data are displayed on a LED display with 15 characters *(Fig. 12.9)*. All the usual information to be recorded can be entered by the keyboard. Many abbreviations can be used which can also be displayed.

Figure 12.9. Data recorder *AGA Geodat 120*

The *AGA Geodat 120* can be used equally with a desktop calculator or with a micro-, mini- or main-frame computer, with which the data can also be printed.

In order to prevent data loss, the on and off switches are protected. The other key functions are protected against erroneous or accidental entry by a so-called time-delay.

The instrument can be used at temperatures between −20 °C and +50 °C. It is water-proof and works reliably under unfavourable environmental conditions. Its size is 88 × 172 × 36 mm, its weight: 0.6 kg.

The *AGA Geodat 122* and *124* are new types of this data recorder. They can be connected primarily to the *Geodimeter 136*, with which the horizontal circle reading, the zenith angle and the slope distance can be recorded automatically by pressing a key.

12.1.7 Electronic Data Recorder MICROBOOK

The field data recorder MICROBOOK (Micronic, Sweden) has been constructed for recording special measured surveying data. It includes the capability of debugging. Data measured in the field can be recorded in a format proper for computer analysis.

Recorded data can be transferred equally to a computer or to a plotter. The capacity of the memory enables data for 800 measured stations to be stored, the memory holding the data for about 100 days. Alphanumeric data can be entered into the memory through the keyboard which has 20 keys. It is water-proof and fitted with a transparent plastic cover for use in the field in rain.

The dimensions of the recorder are: 170 × 85 × 35 mm and its weight is 0.5 kg.

The routines provided with the MICROBOOK include those for distance measurement, traversing, levelling, triangulation, computation of coordinates, etc.

12.1.8 Pocket Calculators as Data Recorders

Large firms dealing with instrument production use pocket calculators more and more for data recording purposes. Such calculators used for these purposes usually have the following capabilities:

— acceptance of different data, the ability to check and store measured data on the spot;

— display for the visual control of the data to be recorded, the ability to recall the recorded data;

— transfer of the data in the data store to a data carrier;

— establishment of data handling routines;

— built-in power source;

— hand portable.

The pocket-calculator *HP 41 CV* can also be used as a data recorder, since:

— it contains 319 data registers (and it can be extended by a further 600 registers);

— the capacity of the micro-cassette is 127 Kbytes;

— recorded data can be checked by the supplementary printer device;

— the dimensions of the calculator and the micro-cassette are 30 × 20 cm.

The computer *EPSON HX-20* (Japan) is contained in a portable brief-case and has a ROM memory of 32 to 64 Kbytes and a RAM memory of 16 to 32 Kbytes. The speed of the dot matrix printer with 24 columns is 42 line/min; connections are ensured by the standard and serial interfaces of the LCD screen (20 columns and 4 lines). Micro-cassettes and ROM memories are available. A NiCd accumulator of 1200 mA/h capacity is used and the memory holds the data for 40 hours after switching off the device.

These and many other pocket calculators capable of data recording are already used as a standby for surveying instruments.

The *Wild GRE 3 (Fig. 12.10)* is a versatile instrument for recording data, carrying out computations and processing them.

Figure 12.10. Data recorder *Wild GRE 3*

Four model variants can be assembled from the elements; they have the following different capacities and temperature ranges:

— 32 Kbytes or about 1000 blocks of data from 0 °C to +5 °C;

— 32 Kbytes or about 1000 blocks of data from −20 °C to +50 °C;

— 128 Kbytes or about 4000 blocks of data from 0 °C to +50 °C;

— 128 Kbytes or about 4000 blocks of data from −20 °C to +50 °C.

The data are displayed on three LCD-s. One of the displays contains 4 alphanumeric symbols, each of the other two have 8 symbols, also the decimal point and the sign. There are 18 push-buttons on the instrument which produce and record the necessary identifiers. The instrument includes a microprocessor of the type *NSC 800 CMOS* with a 16 Kbyte program storage unit.

The dimensions of the instrument are 245 × 85 × 57 mm and its weight is 630 g, with a built-in accumulator of 170 g. The 0.2 Ah NiCd accumulator ensures operation for the whole of a working day.

The magnetic bubble memory stores 32 Kbyte (4000 words or 1000 blocks), or 128 Kbyte (16,000 words or 4000 blocks);

1 word = 16 characters, 1 character = 4 bits,

1 byte = 2 characters, or 8 bits.

The *GRE* data terminal ensures automatic data transfer and recording from the distance meter *Theomat Wild T-2000*, or, with a special connection, from the *Distomat* and from the instruments *Wild DI4*, *DI4L*, *DI5* and *DI20*.

The manual data input by push-button enables the user to record data measured by optical instruments (theodolites, levelling instruments and distance meters). The *GRE 3* can be connected to computer terminals, to printers, to modems, etc. This enables data to be transferred to and from them and, further, programs to be read from the computer into the *GRE 3*.

It is also possible to record data from the *GRE 3* onto cassettes.
The *GRE 3* ensures a connection with surveying instruments in the field, with the computer and with interactive graphical systems in the office.

12.2 Automatic Plotting Machines

Computer controlled plotting machines can be used not only to plot map points from their coordinates, but also to draw connections between points, and to draw automatically map symbols and inscriptions. Plotting machines belong according to their operation to the drawing category or to the matrix printing category. Plotters can be further subdivided into analogue and digitally controlled systems. The majority of devices used at present for drawing belong to the plotting category. Some machines of the matrix printing category are also used for this purpose in screen systems connected to displays to record map detail projected on the screen.

12.2.1 Plotters

The mode of operation of drawing machines can be both analogue and digital depending on the method of positioning in the x–y plane. Here only digitally controlled drawing machines will be dealt with; for brevity they are called plotters.
Positioning is the motion of a writing tool along a prescribed path. The path is described by digital information generated or transmitted by a computer to the control system of the plotter. The writing tool is perpendicular to the x, y plane, which is identical with the drawing medium, and it can also be moved in the z direction. This motion in the z direction has two positions: mov and draw. The movements in the x and y directions are made by servomotors. Most plotters have an incremental mode of operation. The basic idea of incremental operation is as follows: any curve can be infinitely closely approximated in theory by a series of at least four sufficiently short straight lines selected from directions which are perpendicular to each other. Such short lines are called increments. The elementary shifts in the x and y directions

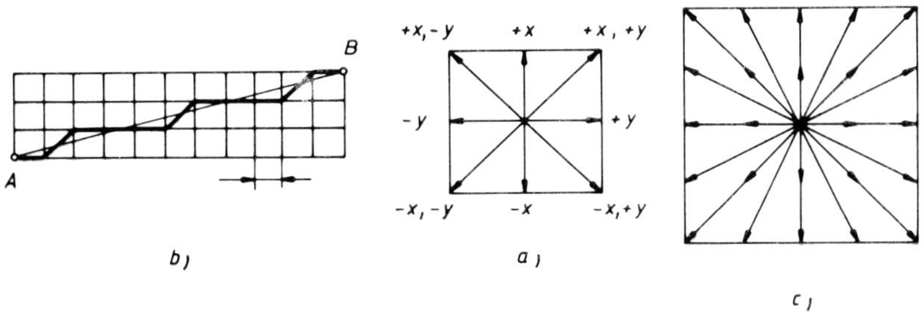

Figure 12.11. Elementary steps of plotters with incremental operation: *a)* elementary staps in 8 directions; *b)* approximation to the line with movements in 8 directions; *c)* 24 elementary movements in 16 directions

are called the basic steps of the plotter. As their length cannot be decreased below a certain limit, the accuracy of the approximation to the curve is enhanced by increasing the possible directions of the elementary movements. This can be of several kinds depending on the solution of the step-control. A simultaneous step in both directions x and y results in a movement at 45° of distance equal to $\sqrt{2}$-times the basic *(Figs 12.11, a* and *b)*. Thus 8 elementary directions and 8 elementary steps are generated. The approximation to the curve can be further improved by using 16 elementary directions and 24 elementary movements constructed from two elementary step-lengths, one of which is twice the other *(Fig. 12.11, c)*.

Plotters consist of three main functional units whose technical solutions influence the operational characteristics of the plotters, which are:

— the mechanical construction,
— the electrical drive,
— the writing tool.

The mechanical construction consists of a mechanical drive in each of the directions x and y, and the frame carrying it. On the basis of their mechanical construction there are two types of plotter:

— the flatbed plotter and
— the drum plotter.

Flatbed plotters draw on a flat surface which is horizontal or tiltable. The highest accuracy reached at present is given by flatbed plotters. The frame is absolutely vibration-free. This frame carries the drawing surface which, for precise plotters, is a highly finished surface. The drawing medium can be fixed on the flatbed mechanically, electrostatically or by vacuum. In the latter case there are holes of diameter a few tenths of a mm — the so-called sucking holes — in the working area of the flatbed. Vacuum pumps can be operated in those parts of the flatbed which correspond to standard sheet sizes.

The movement of the writing tool — similarly to mechanical coordinatographs — is made along the x axis by a moving axle, in the y direction by a carriage mounted on the axle. The axle is held in precise guides at both ends. The transmission of the mechanical drive can be any of a great number of solutions. The drive of the moving axle is made by screw spindles, ball spindles or by a gear rack. The whole construction is thermally compensated. In some plotters movements in the x direction are made by the flatbed moving with respect to the fixed lateral axle. Transmission is often by steel band. Due to the great mass of the moving axle, making its movement deformation-free is a considerable mechanical problem. Details of the practical realizations can be found in the technical documentation of the different products, together with the procedures to test and regulate the mechanical system.

Drum-type plotters are less complicated and less expensive than flatbed *(Fig. 12.12)*. The movement in the x direction is generated by a cylinder which carries the drawing base and is rotatable about its axis in both directions. The movement in the y direction is generated by the writing tool which is movable over the surface of the cylinder. The drawing medium lies exactly on the precisely finished cylinder surface which is made of metal, rubber or plastic. There are several methods of gripping the drawing medium. A frequent solution is to use pulling claws at both ends of the drawing medium in the longitudinal direction, which engage a series of holes.

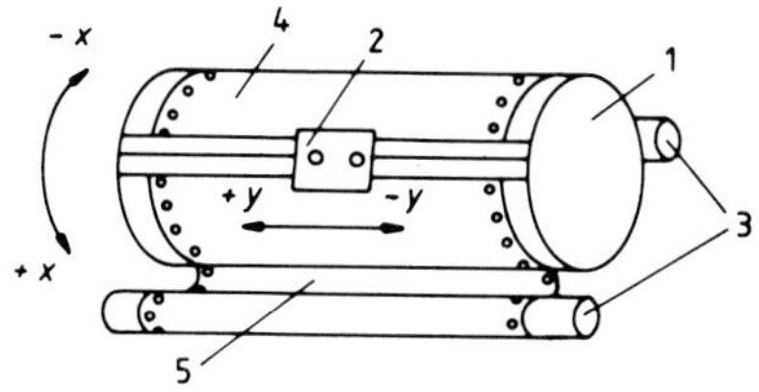

Figure 12.12. Construction of a drum plotter: *1* drum; *2* writing mechanism; *3* spindles; *4* drawing base; *5* paper

The drawing base rolls up and down, to and from two cylinders with parallel axes. Transmission is ensured in one direction by the drum, and the carriage is usually moved by a steel band. The dimension of the drawing base is theoretically infinite in the x direction, in the y direction it is equal to the width of the drum. In some plotters the length of the drawing base is also limited in the x direction.

The electric drive consists of servomotors and their control. They generate the force to move the writing tool. Independently of whether the plotter is a flatbad or drum type, two kinds of electric drive can be used:

— stepping motor

— analogue servomotor.

The stepping motor is generally simpler, and therefore also less precise, and it is used in fast plotters. In precise plotters both solutions are used.

The stepping motor is a digital electromotor. At the receipt of a control pulse its rotor rotates by a step angle α, then it stops again. Thus the rotation of the rotor is intermittent and only discrete angular positions may occur. Consequently this kind of electric drive can be considered as a finite integrator, and it is applicable for incremental control. The stepping motor can also be constructed so that the control pulse brings it into a synchronous position without missing a step, and it remains synchronised. After the vibrations have been damped the writing tool stops without overrun. In some stepping motors the step angle is small, the step speed is high, and a scanning speed of several ten thousands of steps/second can be attained when the steps are continuous.

The drive by analogue servomotors is realized either by AC or by DC motors. The necessary reliability can be achieved by using motors with high resolution, low capacity and high acceleration, and by braking with multiple feedback and closed loop control.

The writing tool carries the clamping head of the drawing tool. The latter is moved in the x and y directions, respectively, by the different mechanical and electrical solutions. Clamping heads are usually constructed to hold 2, 4, or even more drawing tools. The clamping hole has a precision finish, and ensures a centered, clearance-free clamp. The motion of the drawing tool in the z direction is made by an electromagnet or an electromotor initiated by its control pulse.

The drawing tool is generally a ball-point pen making a line of width 0.1 mm, or a ruling pen making a line width of 0.2 to 0.3 mm. There are models in which the liquid refill is controlled by the speed of the writing tool. Thus the drawing of uniform and solid lines is ensured equally for both thin and bold lines, which is important when copying. Special drawing tools also used in drawing maps are the engraver and the light spot projector. Both can be used only in connection with a drawing table.

The engraver is a pre-cut chisel-shaped engraving tool. The width of the chisel edge is equal to the nominal width of the line drawn. The edge is always set by electromotor to be perpendicular to the direction of drawing. With curves that means a direction perpendicular to the tangent, and this is consequently called the tangent-controlled method of plotting. Engraving ensures the nominal width of lines to an accuracy of some hundredths mm: due to the tangent control, the line width is uniform and eventual wear of the tool does not influence it. Tangent controlled engravers are hold by a special clamping device.

A light spot projector draws by using a light beam projected on a light sensitive drawing base. The highest accuracy can be obtained at present by this drawing tool and, for drawings containing many key symbols and texts, the highest speed can also be obtained by it. The illumination is provided by short flashes when the projector stops or by continuous light during movement. The short flash photographs a key symbol on the light sensitive drawing base. The key symbol is selected automatically by a corresponding control pulse from a key set on a glass disc, the so-called symbol-disc.

Continuous light is used to draw lines. The symbols on the glass disc are selected by rotating the disc. A schematic view of the main part of the projector is seen in *Fig. 12.13*. The symbols on the disc can be photographed either at original size or magnified. The precision light spot projector allows the line width to be controlled to an accuracy of a hundredth mm. To achieve this value the unevennesses of the board surface should be kept below ± 0.1 mm. The adjustment of the projector and of the drawing table should be checked from time to time.

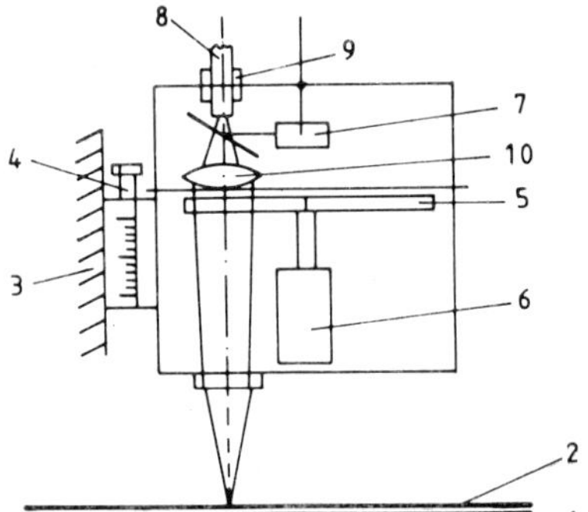

Figure 12.13. Main part of the light spot projector: *1* drawing board; *2* light sensitive drawing base; *3* carriage; *4* fine motor; *7* light meter: *8* light conductor with fibre optics: *9* condenser; *10* objective lens

12.2.2 Matrix Printers

In mapping, matrix printers are less important than plotters. The mechanical construction is similar to that of drum plotters in so far as the drawing base is given a translational movement instead of the drawing tool, and there is a series of pins plotting a series of dots or a dot matrix. The drawing consists of a multitude of dots, its operation thus being identical with certain matrix printers. These drawing instruments can also be used as printers. They are used in cartography for connection to screen displays. The drawing by the semi-graphic raster-system on the screen can be copied into paper by control pulses transmitted to pins in positions corresponding to the luminous dots. The permanent hardcopy provides a schematic sketch for orientation purposes.

12.2.3 Main Technical Parameters of Plotters

Below are enumerated the physical parameters of discrete area drawing machines which influence their use in cartography.

Working area: a working area corresponding at least to a map sheet is necessary. The usual dimensions correspond to standard sizes according to DIN A3-A0. In the case of drum plotters, the width of the paper is significant: it is usually between 27 and 94 cm.

Basic step: this is the theoretical lower limit of position accuracy (Subsection 12.2.1). For the best known equipment the basic step is between 0.25 and 0.005 mm.

Resolution: this is the maximum small distance within which two neighbouring points cannot be resolved.

Reproducibility: this is the small distance within which the writing tool is able to regain the plotted position of any point.

Accuracy: means the distance between the theoretical and the actual position of the drawing tool. It is usual to express accuracy as a percentage of the maximum possible movement of the tool. The accuracy of the best known equipment is between 0.25 and 0.01 mm.

Mechanical construction: flatbed, drum.

Maximum speed of drawing: usual values are 5 to 100 cm/s.

Maximum acceleration: usual values 1 to 4 g.

Writing tool: the majority of plotters have a carriage for ball-point pens and ruling pens with a line width of 0.2 mm; the necessary quality of drawing can be provided by them. For mapping, a ruling pen adjustable in steps of 0.1 mm line width is the correct tool, but it gives an acceptable quality of drawing only if the refill is controlled. The number of drawing tools on a single carriage is between 1 and 8 and they are automatically changed when drawing. Mapping quality can be best provided by the tangent controlled engraver and by the light spot projector.

Drawing base: a wide variety of drawing bases is necessary for map bases, therefore tools should be used which do not necessitate the use of specially shaped drawing bases (e.g. with a perforated edge).

12.2.4 Programming

The programming of discrete area plotters entails the use of the basic software supplied by the manufacturer. This section will consider the independent use of plotters as an exclusive graphical output device in some technical computation system. There are also applications when the plotter is a part of an interactive system including a screen display. The independent basic software consists of two parts:

— interpolation programs,
— graphics subroutines.

The interpolation programs mainly compute approximations to straight lines, circles, and sections of higher order curve from elementary steps, and they thus take immediate control of the writing tool. These programs are mainly fixed programs (firmware) i.e. they are burnt into ROM-s (read-only memories), and therefore they have the properties of hardware.

Graphics subroutines are the collections of programs for elementary partial drawing tasks which enable arbitrary drawing to proceed. If the subroutines are written in a high-level program language, then the user of the plotter can build them into his own programs. Basic graphics subroutines are as follows:

— zero setting
— choice of tools
— drawing a section of a straight line
— plotting an arc of a circle
— drawing alphanumerical characters
— lettering
— drawing of map symbols
— transformations (rotation, translation, scale change)
— mirroring
— preparation of windowing parallel to the coordinate axes
— interpolation of higher order curves.

The basic software is a part of the softwares used in interactive graphics systems, therefore its parameters strongly influence the general use of the plotter.

12.2.5 Modes of Operation

The plotter may be operated:

— on-line
— off-line.

For on-line operation *(Fig. 12.14)* the plotter is linked to a channel of the host computer and it operates as a peripheral of the computer. In the off-line operation *(Fig. 12.15)* the plotter and its control unit are independent systems which contain as an input peripheral a magnetic tape unit which contains, together with the plotter software control, commands to the plotted as generated by the computer from coordinates.

Independently of the mode of operation the plotter-controller may be under microprocessor control when the interpolation programs and other basic graphics subroutines are built into the firmware. Plotters differ with respect to which operations are solved by software and which by firmware.

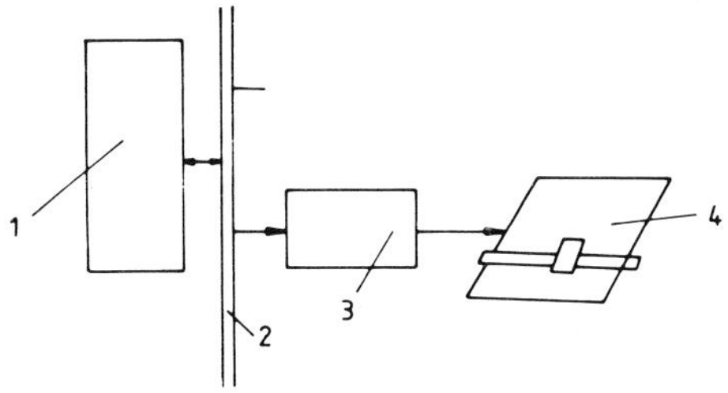

Figure 12.14. On-line operation of a plotter: *1* computer; *2* bus; *3* control unit; *4* plotter

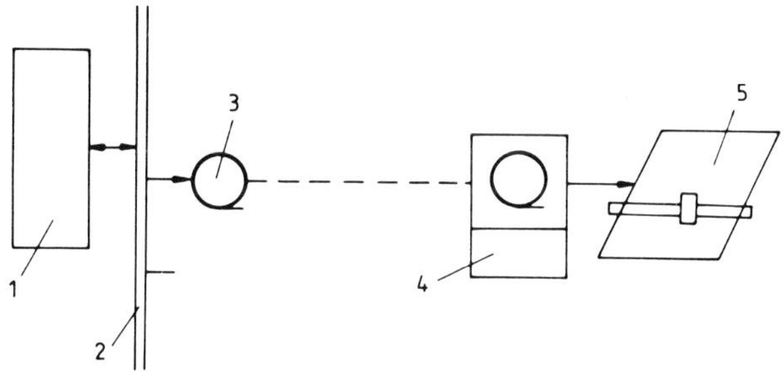

Figure 12.15. Off-line operation of a plotter: *1* computer; *2* bus; *3* magnetic tape; *4* magnetic tape and control unit; *5* plotter

12.2.6 Some Well-known Plotters

California Computer Products (US) has a very wide selection of plotters. The plotters *Calcomp 945* and *Calcomp 965* are especially suitable for mapping *(Fig. 12.16)*. They have a mechanical construction similar to drum plotters, but the working area is limited in the directions of both axes. The working area of the model 945 is of the size A1, that of the model *965* is A0 (ISO standard). The electrical drive consists of feed-back controlled servomotors. The carriage can hold 4 tools, the maximum drawing speed in the x-direction is 762 mm/s. The acceleration along the axis is optionally 1g, 2g, 3g or 4g (g is the acceleration of gravity). The mode of operation can be both on-line and off-line. In an on-line system it operates with the controller *Calcomp 907* or *951*, with an asynchronous interface type RS 232C having a transfer speed of 19,200 baud. In off-line mode of operation the controllers *Calcomp 91X* or *92X* can be used.

The *Calcomp 7000 (Fig. 12.17)* is a plotting system with off-line operation; it consists of a *Calcomp 749* precision drawing table, a *Calcomp 925* controller and a magnetic tape unit. An on-line system can also be built through the standard interface RS 232C, when no magnetic tape unit is necessary. The carriage can hold 4 drawing tools and these ensure very high quality drawing (Four-pen pressurized inking system). The precisely plane surface of the drawing board is in a rigid frame. The carriage runs on a light but rigid supporting frame. The controller enters data into the plotter by digital increments and with a high velocity. After decoding, the digital data control

Table 12.1. Some characteristic data of plotters

Manufacturer model	Mechanical construction* working area	Basic step (mm)	Resolution (mm)	Accuracy (mm)	Maximum acceleration in axial direction velocity (mm/s)	Writing tool**	Control interface
1	2	3	4	5	6	7	8
Calcomp 945, 965	PD A1, A0				1/2/3/4g 762	PT	907/958 91 × /92 × , RS 232C
Calcomp 7000	PT	0.005			767	4PT TC E	925 RS 232C
Kuhlmann FR	PT A1, A0	0.025		0.1	17–100	4PT	
Wild TA 10	PT 100 × 120 cm			0.015	2 m/s^2 250	2/4 PT TC E	RS 232C
Wild TA 2	PT 120 × 120 cm			0.020	5 m/s^2 300		RS 232C
Opton DZ 7	PT 932 × 1192 mm		0.01		100–280	3 PT E	HP 1000–HR IB bus
Ferranti Cetec EP 230/231	PT 1220 × 915 mm	0.0125	0.025	0.04	100	8 PT LP	PDP 11
Benson 2532	PT 1220 × 1680 mm	0.006	0.050	0.100	0.2 g 300	4 PT TC E	to magnetic tape RS 232C

Benson 1232	PD 730 × 500 mm		0.1		310	3 PT	
DCS F 201	PT A0	0.025		0.25	250	1/3 PT	DCS M IP V24
DCS F 100/1000	PT A0	0.001		0.15	133	1/3 PT TC E	V24
Gerber 6233 B	PT 25 × 30 inch		0.0001 inch	0.0003 inch	$105 \frac{\text{inch}}{\text{min}}$	E LP	to magnetic tape
Aristo	PT		0.01	0.04	5 m/s^2	4 PT	to magnetic tape
Aristomat 305 M	1200 × 1800 mm				$60 \frac{\text{m}}{\text{min}}$	TC E LP	

Explanation:
* In this column: PT plotter with table; PD plotter with disc
** In this column: PT plotting tool; TC tangent control; E engraver; LP light spot projector

Figure 12.16. Plotters *Calcomp 945* and *Calcomp 965* (Calcomp, US)

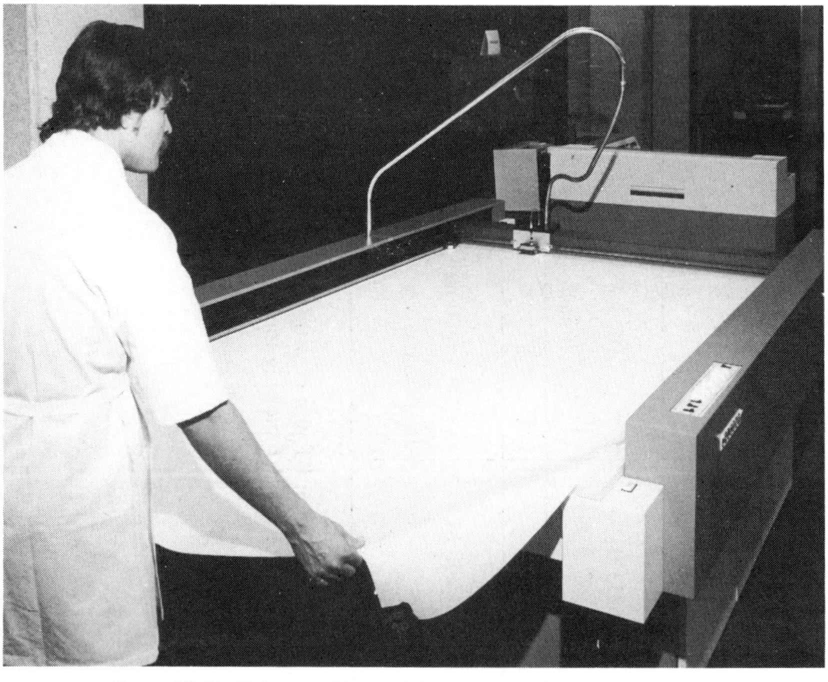

Figure 12.17. *Calcomp 7000* precision drawing table (Calcomp, US)

the high capacity servomotors. Optical encoders mounted on the axes generate position and speed data by feed-back. Accelerometers on the carriage generate a feed-back signal to optimize the line quality. Both speed and acceleration are set in accordance with the necessary drawing quality by the following parameters:

— acceleration;

— scale 1 : 1, 1 : 2, 1 : 0.5, 1 : 0.25;

— time required to raise and lower the drawing tool: it can be set to between 8 and 80 ms in 10 steps, depending on the drawing tool used;

— maximum axial speed also in 10 steps up to 76.7 cm/s.

The basic step is 0.005 mm, ensuring a very high drawing quality. In addition to special pens, tangent controlled engravers can also be used. The controller *Calcomp 925* contains a 16K–32K memory and can be supplemented by teletype and by the CALEDIT software which contains different graphics operations. All *Calcomp* models are supplied with basic software containing graphics subroutines in a high level program language (FORTRAN).

The *Kuhlmann FR* plotter (FRG) is controlled by a microprocessor, the working area is of the sizes DIN A1 or DIN A0, the electrical drive is derived from scanning motors, the basic step is 0.025 mm, and the drawing speed can be chosen by the program to be between 1 and 6 m/min; the carriage can hold 4 tools, and the repetition accuracy is given by the manufacturer as ± 0.1 mm.

The Opton *Model DZ7* is mechanically constructed as a tiltable plotter. The moving axis is guided precisely at both ends. The electronic control is provided by a microprocessor with a 32kB memory. The working area is 932 × 1192 mm, the resolution 0.01 mm, the speed between 100 mm/s and 280 mm/s. The carriage can hold one or three drawing tools.

The program system DZ7-AS can be used in an interactive graphics system, and it contains the graphics program package GRAPH-T1 for use as independent basic software. The variant DZ7C can be used with an interface IEC in an on-line system, while the variant DZ7A is used as a mapping tool connected to an analogue interpretation device.

Ferranti Cetec Graphics Limited (Scotland) manufactures four variations of its precision plotter, which include the high precision EP23/231 and the EP330/331 which is a special plotter for cartographic purposes. Its mechanical construction differs from usual solutions in so far that there is no moving axis: instead the whole working area moves below the standing axis holding the carriage. The basic step is 0.0125 mm and the working area is 1220 × 915 mm or 1520 × 1220 mm, respectively. The highest speed is 100 mm/s. The carriage offers an automatic choice between 8 drawing tools, and the line width can be changed between 0.2 and 1.0 mm. The carriage can be supplied with line spot projectors. The plotter is delivered with the program package Ferranti Cetec Modular Plotting Software (MPS). The MPS is a collection of programs written in the assemblar language DEC MARCO, and it contains all the drawing tasks necessary to run in the operation system RSX-11M. Due to its modular construction, sets of programs for any task can be composed. The plotter can be operated on-line through an interface DMA or off-line with a paper tape or magnetic tape unit. Some of the main technical data for several plotters are presented in *Table 12.1*.

13 Automatic Equipment for Map Production

13.1 Digitizers

Digitizing is used to acquire data for map construction. Numerical data (e.g. coordinates) are produced by digitization from graphic or photographic data on any kind of data carrier.

There are two basic methods of digitizing data:

— measuring graphic data one by one, so-called linear digitization;

— automatic monitoring of all points of the complete data set, i.e. scanning, which produces the result of digitization in the form of a raster-matrix.

13.1.1 Devices for Linear Digitization

Linear digitizing equipment contains the following major components independently of its operational principle: measuring device, controlling electronics, input/output interface and data store output.

For equipment operating in two dimensions, the base for the graphic material is fixed on a plane flatbed or on a cylindrical surface (a drum). In both cases a plane system of coordinates is used. This system can be, however, either orthogonal or polar.

The polar digitizer detects rotation and linear shift. It can be used on smaller areas than the orthogonal system and it has a variable accuracy. Angle measurements are made electro-optically from a disc with incremental graduation. The cursor is mounted at the end of a shaft, the length of which can be changed.

Equipment using orthogonal plane coordinates can be separated into two main types: manually controlled instruments and automatic line-tracking equipment.

With a manually controlled digitizer the operator follows the line to be digitizer with a cursor. The advantage of this type of instrument is its relatively low price, and that the basic material to be digitized does not have to be prepared; it has about the same speed as automatic line tracking. The digitizing has the following three variants.

a) With recording by single points, the operator sets the cursor on the point to be measured and starts the recorder with a switch.

b) With incremental recording a constant path interval between 0.1 and 9.9 mm is set on the control panel; recording is made automatically at the end of each interval.

c) In equipment with a constant time difference, a time interval between 0.01 and 9.9 s can be set and recording is made automatically at corresponding time intervals. Its advantage is that sections of lines which can be tracked easily are recorded by comparatively fewer points than sections of great irregularity.

Most manually controlled digitizers operate on electronic principles. The base plate is usually mounted on a tiltable frame, it contains the measuring system with the cursor connected by a cable and it is freely movable. The cursor is usually a glass plate at the electronic centre of a coil which induces an electromagnetic field. A cross-hair or a small circle on the plate is used for point setting. The field of view of the cursor is often magnified. The cursor weighs a few dozen grams and is built into an easily movable housing on which buttons for the different modes of operation are located. Supplementary data can be entered using the keyboard.

The control unit contains a microprocessor, supplemented in certain types by a PROM (Programmable Read Only Memory). By pressing a button on the control unit, two kinds of coordinate systems can be used; thus both absolute coordinates and incremental coordinates can be obtained. In the former case the origin can be freely allocated, in the latter the origin is always the point at which the last coordinates were measured.

The interface enables data to be transferred to the computer or other peripherals to be connected. The display unit can be either a fast printer, a tape punch or a magnetic tape unit, or a graphics display.

With the majority of map digitizers the measuring system operates on the inductive principle. Its basic principle is that there is on the working plate a grid which consists of wires in the directions of the x and y coordinates which detect the magnetic field induced by the coil in the cursor. This grid produces amplitude modulated AC signals according to the movement of the cursor. The signals are transformed by the analogue electronics firstly into phase-modulated signals, then into square waves. Analysis of these waves determines how many times the cursor crossed a grid wire during its movement. The remaining distance can be determined by the ratio of the voltages induced in the neighbouring grid lines. The output of the digitizing electronics consists of pulses which correspond to the incremental changes when the cursor crosses the grid lines. The system collects the pulses in a digital counter which determines the x and y coordinates of the cursor.

The internal accuracy of precise digitizing equipment is between 0.025 and 0.1 mm and depends on the resolution of the instrument: this is the shortest distance within which coordinate differences between two neighbouring points are perceptible by the measuring system. The measurement error can be as much as 4 to 5 times the resolution of the cursor due to the eccentricity of the cursor and to rounding-off error in the transformation between different scales.

Orthogonal digitizers may also operate with automatic line tracking. A series of pulses, produced photoelectrically by a high sensitivity photodiode, controls the cursor. When using such an instrument, the cursor is put on the chosen line and it is controlled by servomotors to move along this line so that its centre lies equi-distant from both edges of the line. The position data of the cursor is displayed at given distances or time-intervals. Difficulties arise if the distances between corners and along strongly curved lines are too short. In such cases the operator must guide the cursor manually.

Digitization tasks are greatly facilitated by the possibility of selecting drawing codes, operational codes, key figures and symbols from the keyboard or by a menu of logic solution (from the field of the menu, designated by a square grid, table coordinates can be allocated as identifiers or operational signals). The keyboard is often mounted on the cursor itself. The menu also contains mnemonic codes and graphs. The menu itself is fixed on the table outside the sheet. A table coordinate can be allocated to an identifier point of the menu. The digitizing program system automatically allocates the appropriate sign, key figure or code number to the coordinate in question.

Digitizer RA-01. The instrument *RA-01* manufactured by the Fine Mechanics and Electronic Instrument Producing Cooperative FOK-GYEM (Hungary) has a resolution of 0.25 mm, i.e. it can be used only for work of lower accuracy.

The working table with the dimension A0 contains the electric wires in the directions x and y which detect the alternating magnetic field induced by the cursor. The working table is mounted on tiltable and height-adjustable legs which provide arbitrary positions. It contains a bridge which is easily movable in the x and y directions and on which the cursor can be fixed. Independent use of the cursor head is also possible.

The control and display units, containing control switches and digital displays of the different coordinate positions and the equipment status displays, are mounted on the working table.

The cursor contains a coil which induces an alternating magnetic field due to the inducing current from the control unit. The centre of the cross-hair at the centre of the coil should be used for tracking the line to be digitized.

The logic system includes the control circuitry, the coordinate registers and the circuits which process the signals arriving from the cursor.

The output unit contains the circuits of the interface units supplied as options.

During digitization the actual position of the cursor with correct sign is visible on the numerical display.

In the continuous mode of operation the logic system automatically produces a series of characters which corresponds to the distance from the origin. In the single

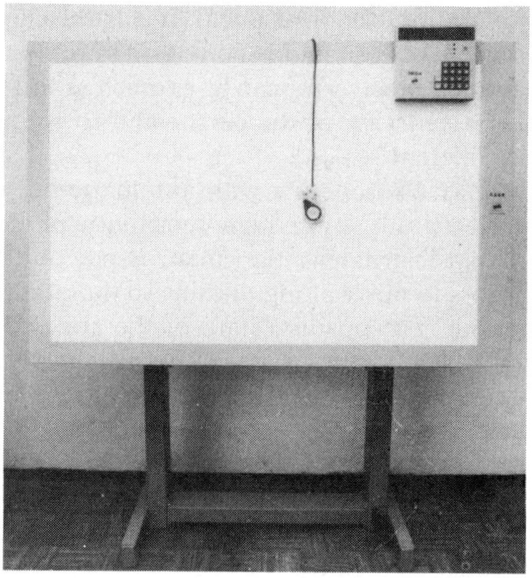

Figure 13.1. The digitizer *RA-06* (FOK-GYEM, Hungary)

mode of operation the operator initiates the realization of the character series by the pulse transmitter. This character enters through the output unit into the data storing device. The rate of digitization is determined by the output peripheral.

The activation of the STOP switch of the operating unit conserves the content of the coordinate registers, thus enabling the cursor to be moved freely without influencing the content of the registers.

The output is of standard parallel signals according to the British Standard Interface (of 8 bits). The codes are identical with the formats given in the code table of ASCII.

Digitizer RA-06. A more developed variant of the digitizer *RA-01* is the digitizer *RA-06* consisting of the following units *(Fig. 13.1)*:

1. The dimensions of the working area of the table can be changed from 475 × 750 mm to 675 × 1050 mm. The resolution is 0.1 ± 0.02 mm. Numerical display is ±9999.9.
2. The maximum speed of tracking with the cursor is 1 m/s; in other details its construction is identical with that of the *RA-01*.
3. The keyboard of the display enables one to program the parameters controlling the operation of the digitizer and to change the operational mode between ADATBEVITEL (DATA ENTRY), STOP and MÉRÉS (MEASUREMENT). Signal lamps indicate the actual mode of operation.

In the mode ADATBEVITEL (DATA ENTRY) the following parameters controlling the operation can be programmed:

— coordinates in mm or in inches
— resolution in the x direction from 0.1 to 2.5 mm
— resolution in the y direction from 0.1 to 2.5 mm
— alignment of the direction of the x axis; both the axes $+x$ and $+y$ can be aligned in several directions by rotating the axes x and y around the origin by 90, 180 or 270 grades. DATA ENTRY modes of operation are as follows:

STEP: the coordinates x and y can be transmitted to the peripherals by the switch STEP on the positioner.

AUTO ORIGIN: by pressing the switch STEP the measured increment is added to the coordinates previously transmitted to the peripherals.

CONTXY: the first digit of the two-digit numbers shows how many x grid lines, and the second digit how many y grid lines, are to be crossed at the instant of the transmission of the coordinates x and y to the peripherals.

TIME: depending on the increment set, coordinates are transmitted to the peripherals.

REMOTE F: coordinates are transmitted at certain instants independently of the serial or parallel input peripherals.

The format of the output blocks may be: binary, BCD or ASCII;
block length x: max. 8 digits and sign;
block length y: max. 8 digits and sign;
magnification: max. 5:1;
reduction: max. 1:5;
output peripherals: max. 2 parallel (TTL-level BSI) and 1 serial asynchronous (CCITT V.24).

In the mode of operative STOP the cursor can be moved freely without changing the last measured coordinate. This mode of operation is important if a map greater than the dimension of the working area is being digitized.

In the mode of operation MÉRÉS (MEASUREMENT) the line can be tracked with the cursor. The display shows the x and y coordinates of the cursor as measured from the origin, in a format corresponding to the chosen parameters.

4. Using the keyboard, characters corresponding to ASCII codes can be transmitted to the output of the digitizer in the operation mode MÉRÉS (MEASUREMENT), thus the coordinates can be annotated with arbitrary supplementary information.

5. The control electronics are controlled by a microprocessor. It processes the data measured in an *A/D* (analogue/digital) manner, it controls the display and through the intervace cards the peripherals, and it receives the characters entered by the keyboard. The electronics have a parallel input-output interface. The block scheme of the *RA-06* is shown in *Fig. 13.2.*

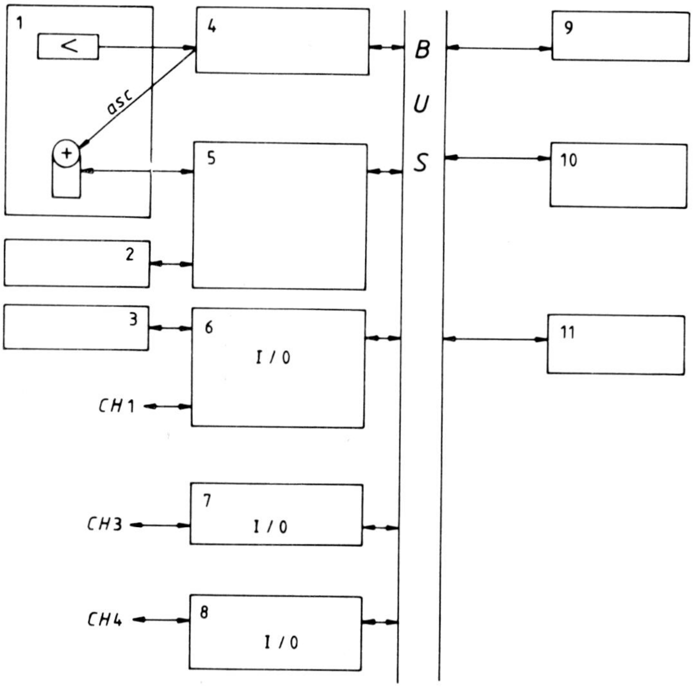

Figure 13.2. Block scheme of the digitizer *RA-06: 1* working table; *2* display; *3* keyboard; *4* measuring unit; *5* drive; *6* parallel unit; *7* parallel unit; *8* series unit; *9* CPU; *10* APU; *11* memory

Digitizer RDR-1. The digitizer system *RDR-1* with microprocessor intelligence of the Telecommunication Company (Hungary) provides computerized processing and repeated display of the data.

The basic module necessary for digitization consists of:

— a digitizing table (active surface 800 × 600 mm),

— digitizing electronics,

— display,
— cursor and keyboard,
— tape punch unit EP-35 to store the data.

The basic module can be extended by the microprocessor *Z-80,* by the display *ADP 2000 (2052),* by a tape reader and a tape punch, and by a cassette recorder. A floppy disc, line printer and plotter can also be interfaced to the system.

The resolution of the system is 0.1 mm. The origin is freely relocatable within the working area by software or by hardware according to the user's choice.

In a continuous mode of operation the shortest unit of digitization is, for digitization of equally spaced points, $x_{min}=y_{min}=0.1$ mm, and for equal times: $t_{min}=$ $=300$ ms.

Single items of data are transferred by pressing the button STEP on the cursor.

The movement of the cursor induces sinusoidally changing voltages and the analogue signals thus produced are transformed by the digitizer electronics into a digitally processable series, into binary numbers. The absolute x and y coordinates produced are transformed by the electronics into the coordinate system of the table and they are displayed on the two by four decade display, then the coordinates are punched. The microprocessor, interfaced with the basic module by software, carries out the following functions:
by software
— data management,
— rounding-off,
— preparation for the data bank,
— control,
— different kinds of computations and corrections,
— management of storate elements,
— selection of the operational mode of digitization.

The block scheme of the digitizer system is presented in *Fig. 13.3.*

Cartometer KAR-A2 (Polish product). The cartometer *KAR-A2* is an instrument to transform graphics information into numerical information; it can be used mainly on large scale maps, drawings, serial-photos, etc., for measuring and recording the coordinates of points. The numerical information corresponds to the x and y coordinates of points selected with the help of the optical system. The coordinates appear in a numerical form on the displays on the front plate of the electronic block. They can be transmitted to the following peripherals:
— desktop computer,
— tape punch,
— printer.

The cartometer combined with a programmable computer enables the following operations to be carried out:
— determination of an area within lines given by a series of points,
— transformation of point coordinates from a local system into a geodetic system,
— determination of earth masses by digitizing topographic maps for construction works, for planning the lines of public utilities,
— plotting the map position of a few points given by coordinates.

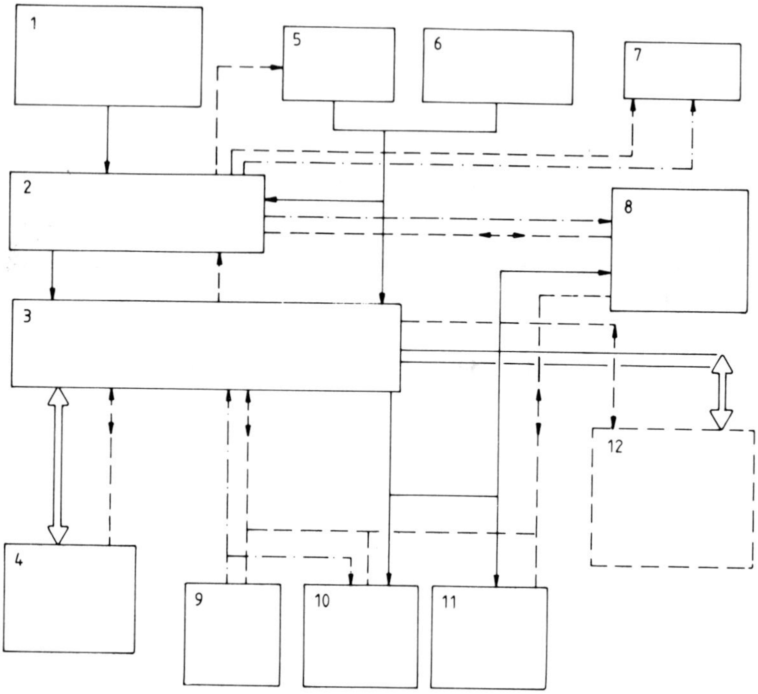

Figure 13.3 Block scheme of the digitizing system *RDR-1*: *1* digitizing table; *2* digitizing electronics; *3* microprocessor *Z-80 PDU 36*; *4* display *ADP 2000*; *5* cursor; *6* keyboard; *7* indicator; *8* punch unit *EP-35*; *9* reading device; *10* magnetic store, cassette or tape; *11* line printer unit *DZM 180*; *12* floppy disc 900/1800 MOM (Hungarian Optical Works)

Coordinates can be stored as a list on the line printer or on paper tape. The special programs used in this equipment enable the user to undertake the following tasks:
— construction of a numerical terrain model from topographic maps,
— establishment of numerical data bases for planimetric maps,
— establishment of numerical bases for the unification of property lots,
— collection of surveying data for soil quality recording.
Its technical data and characteristics are:
— active area 594 × 420 mm,
— resolution 0.05 mm.
Capability of measuring coordinates:
— in millimetres,
— in pulses (one pulse corresponding to 0.05 mm).
The standard error of the determination of point coordinates is: ± 0.1 mm.
Digital displays for changes of the x, y coordinate axes. Accessories:
— point pricker for the cursor,
— cover for the working table.

Pointing is made by an illuminated optical pointing system. The magnification of the reading system is 2 ×, field of view 30 mm diameter, current supply 220 V ± 10%, power consumption 75 VA, temperature limits +5 °C to +40 °C.

The dimensions of the graphics pulse converter in its operational position are: 590 × 545 × 160 mm, the dimensions of the cover are: 780 × 300 × 200 mm. The shipping weight of the graphics pulse converter is about 20 kg. The dimensions of the electronic block are: 350 × 217 × 172 mm; its weight is about 7 kg, shipping weight about 10 kg.

The system can be operated:

— with a desktop computer such as the *Hewlett–Packard HP 25*, or any similar model,

— with standard interface BS 4421 and ISO-7 encoder (e.g. MERA DT 105S),

— with a standard character printer (16 characters/line).

The cursor system can also be supplied with a point pricker which enables points to be marked. A Hewlett–Packard *HP 25* or *HP 67* desktop computer can be adapted for the cartometer.

Digicord. Geopolar, Helsinki, produces smaller geodetic and automation instruments. In 1979 it developed the system *Digicord,* a system of microcomputer intelligence for manual map plotting and digitization.

It is used to digitize maps rapidly, to carry out certain changes on maps, or to supplement a data base in data banks. The working area of the table is 600 × 1000 mm. A digitization accuracy of ±0.1 mm can be obtained with the reading microscope.

Measured digital data can be labelled by code numbers and then stored, similarly distances and areas can also be stored. Area determination can be carried out by determining the corner points of a plane figure or by precisely following the boundary lines of the figure.

Aristogrid (Aristo, Federal Republic of Germany). This manually controlled electronic two-dimensional coordinate meter works on an inductive principle *(Fig.*

Figure 13.4. The digitizer *Aristogrid*

13.4) and it does not contain mechanical moving parts. The resolution is 0.025 mm, the absolute accuracy ±0.125 mm. The speed of digitization may reach 762 cm/s. The three main parts of the *Aristogrid* are the working table with the cursor, the electronic control unit and the output device. The working area is 76 × 91 or 107 × 183 cm, and it is covered by white polyesther plastic with built in metal wires at a density of 0.5 inches in two perpendicular directions, in inductive connection with the cursor. Several kinds of cursors can be used with the equipment. The standard cursor has a diameter of 3 to 5 cm and is made of a transparent plastic material *(Fig. 13.5)* with engraved concentric circles and cross-hairs in order to ensure high accuracy. It has three data storing buttons, a fourth button to store the coordinate of the actual position of the cursor, and a switch to relocate the origin of the table system of coordinates.

Figure 13.5. Cursor of the *Aristogrid*

The electronic unit includes a keyboard consisting of 28 characters (digits 0 to 9, 14 letters or symbols and four keys for special functions), which has a locking system which impedes incidental data entry. Four different scales can be set on the instrument by the factors 1.0; 0.5; 0.25 and 0.1. The instrument enables the user to select dimensions with different grid layouts for printed circuits and applications to microelectronics. It automatically rounds off the coordinates of the points being digitized to the x, y values of the next grid point of a given density. The necessary base grid interval can be selected between 0.025 and 2.5 cm by a switching frame on the control box. The firm Aristo has also developed a digitizing system for general mapping tasks, in which the switches on the cursor are extended by operational and numerical symbols.

The origin of the x, y coordinate system can be located at any point on the working area of the digitizer, thus graphic material can be placed on the most appropriate part of the table. The origin can be located by pressing the button on the cursor or by a switch on the keyboard. The x and y axes can be arbitrarily rotated.

Digitization of continuous lines is facilitated by continuous recording which has two modes of operation: with a time-base method (the time-base lies between 1 s and 10 ms) and with a distance-base method ($\Delta x + \Delta y = 0.1$ to 25 mm). The actual distance between recordings is limited to the digitizing accuracy. The system contains "operator sensor logic" to override recording in the time method if the operator stops the cursor movement.

The following output peripherals can be connected to the *Aristogrid:*

a) Teletype ASR 33 consisting of an 8-channel paper tape punch, of a typewriter-style set of letters with full alphanumerical keyboard, and of a tape reader.

b) High speed, 50 characters/s paper tape punch permitting very rapid point digitizing and keyboard entry, which generates an 8-channel paper tape using a code system to be selected on the control unit.

c) Communications Terminal Bendix 4380 providing a high speed on-line data transfer to the computer in an asynchronous mode of operation.

d) A seven or nine-channel magnetic tape recorder, providing data recording with a rate of 700 steps/second, controlled by logic in the Aristogrid. Magnetic tapes are IBM compatible.

There are five switches as follows: RECORD-1, RECORD-2, RECORD-3, SET and HOLD. The first three switches enable three different formats to be recorded on the output. The format determines how many coordinate digits should be recorded, and it also stores the necessary constant data, carriage return, linefeed, etc. Pressing the SET button zeros the system or sets the value set by the operator on the rotary switch on the control panel. The switch HOLD has two functions: the first stores the values of the measured coordinates, the second is used to shift the drawing on the working table if required. In the latter case the switch is to be set correspondingly, when the drawing can be shifted arbitrarily and then the switch is activated again. In that case the coordinates of the selected point remain fixed irrespective of the new position of the drawing.

The keyboard consists of a maximum of 20 rotary switches which can be used to enter fixed numerical data for identification of the output.

The buffer interface units PDP-8 and PDP-11 enable coordinate data to be entered into the computer. Data transfer is initiated by the operator or by the devices which carry out mass digitization.

The bidirectional computer interface unit ensures a connection between the Aristogrid and the user's computer by switching the digital devices of the computer to the operational channel. This connection transforms the Aristogrid into a digital peripheral. The interface unit carries out all the tasks of communication with the computer and enables 8-bit information to be transferred in both directions.

By a proper digitization, a data base can be established which enables drawings to be made with the drawing programs developed for the *Aristomat* without any further change. For this purpose the firm Aristo developed a special cartography-oriented menu which contains inter alia: surface symbols, point symbols, line symbols, hatching styles, further interpolation supplemented by line symbols. These record on the punched tape the command codes, which are required to start the plotter, and point coordinates.

13.1.2 Digitization Method and Accuracy of Graphically Represented Data and Maps

The method of digitization is used to determine the geometry and the structure of the contents of maps, and additionally to produce numerical data from information which is only available in graphical form. The structure is determined by the corresponding thematic and geometrical form of the planimetry on the maps. Elements which are identical at different logic levels form a chain.

The structure measured by the digitizer is complete at all levels, i.e. all its boundary points are measured several times. The accuracy-increasing effect of repeated measurements is provided by weighted averaging.

The uniformity of the map can be ensured by measuring identical points near the sheet edges, also at the intersections of the sheet edges and the 50 m or 100 m grid lines. The coefficients so obtained require multiple transformation.

The reliability of the digitization can be checked from the points of view of the completeness of structural elements and of the accuracy of the digitized coordinates. The completeness of the structures can be checked by testing property boundaries, as in the structure each segment of a partial lot boundary must appear twice.

It is important to know the positional reliability of digitized coordinates. Their accuracy can be checked as follows:

— by an analysis of the scatter of the coordinates of the points obtained from repeated measurements without averaging;

— by comparing the recorded average coordinates of these points with their numerically determined coordinates determined before mapping.

The mean square error m of digitized map coordinates is given as the mean square error of the differences between the numerical and the digitized coordinates of the points in the following form:

$$m^2 = m_D^2 + m_T^2 + m_E^2,$$

where m_D = is the mean square error of the digitizer,

m_T = is the mean square error of mapping (scale conservation by the medium, reliability of map drawing),

m_E = is the mean square error characterizing the reliability of the digitizing.

On an Astralon-base map of scale 1:1000, the linear mean square error with 4 repetitions is:

$$m = \pm 0.21 \text{ mm}.$$

13.2 Screen Displays

13.2.1 The Role of Screen Displays in Computation Technique

The use of computer techniques has become necessary today in many fields of human activity, including surveying and geodesy, in order to provide a better and quicker solution of problems.

One of the main problems in computer techniques is information exchange between man and computer. The most natural methods would be the use of acoustic and pattern recognition techniques, which are at present only in an initial stage of development, as they correspond to communication forms used among people. The computer-man connection using interactive displays does, at least partially, realize this requirement. The text or graphics information displayed on the screen is perceived by man by those of his perceptors which have the widest range of frequencies, and the response can be given quite simply by using entry tools (keyboard, positioning sphere, light pen, graphic tablet, etc.). Such systems which provide human intervention are called interactive systems.

The aim of interactive man-computer communication using a screen display is to obtain rapid and concise information about the status of the computer, which enables simple responses to be made to the program running in the computer. Thus a purpose-oriented dialogue can be realized between computer and man in which the task can be well defined; the part necessitating high calculating capacity is carried out by the computer program, while the complicated decisions with many degrees of freedom are made by the man who has the correct expertise. This system leads to an economic solution for complicated tasks incorporating many computations.

Another task of the systems which include interactive displays is to facilitate the use of methods where both the parameter entry and the dialogue continue in a form corresponding to the needs of special branches, e.g. of surveying, and only a minimum level of expertise in computer technique is demanded from the user.

In surveying, screen display and graphic input have special importance as map-like representation is their basic form of information.

In the following, screen displays will be treated under the headings:
— classification of screen displays,
— their construction,
— different types of screens,
— devices for data entry,
— their programming,
— data transfer.

Finally, two examples from practice will be presented.

13.2.2 Classification of Screen Displays

Screen displays used in computer techniques are classified into three major groups: alphanumerical, semigraphic and graphics displays. In surveying, all three kinds of displays find application.

Alphanumerical displays can be used to represent readible information consisting of letters, numbers and special symbols. This means that different characters can be written on the screen. The choice of the characters to be displayed and their number depend on the specifications of the display unit. The most widely used set of characters is that of the code table ASCII. Modern screens can display simultaneously about 2000 characters.

Semi-graphic displays usually have a raster system. A point grid can be displayed on the screen. The number of points determines the resolution of the image. The range of raster point numbers lies between 256 × 256 and 512 × 512. Such a resolution provides only a coarse display, a schematic representation. This kind of representation is most frequently used in process control. In addition to this raster, alphanumerical characters can also be written on the screen. The raster-style display permits the use of TV techniques, therefore the device can be manufactured at quite a low price.

Graphics displays present a fine structured image. The range of point numbers is between 1024 × 1024 and 4096 × 4096. The structure of drawing consists of short and long vectors. The picture tubes have large dimensions. A drawing is produced continuously. By this device mechanical mapping and architectural drawings can equally be plotted. Intervention tools — positioning sphere, light pen (Section 13.3) — provide an interactive mode of operation.

13.2.3 Construction of Screen Displays

All screen displays consist of three main parts: screen, input device(s), and control unit *(Fig. 13.6)*.

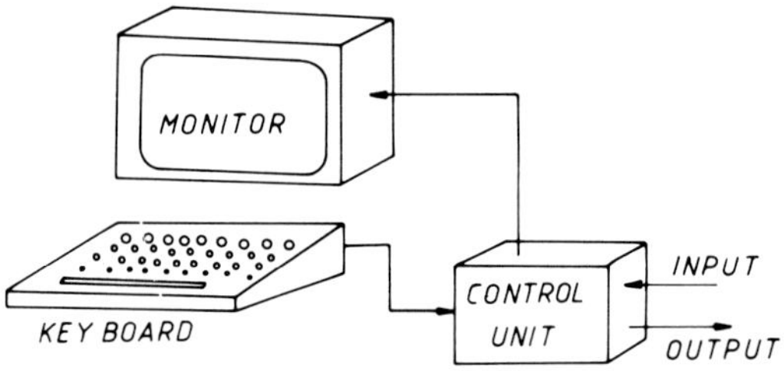

Figure 13.6. Main components of screen displays

The screen is used to display the image required. It can be both vacuum-tube and solid state based and it may emit light or control the transmission or reflection of external light. Both black-and-white and colour images may be generated with it.

The overwhelming majority of presently purchased screens is based on the principle of the cathode-ray tube. The most common data input device is a keyboard, which is unavoidable for all types of screen display. The keyboard is used to enter alphanumerical characters, and it contains keys which initiate different functions. The movable image point to mark a position on the screen, i.e. the cursor, belongs to the category of other data input devices. Marking can be carried out by special rapid tools, such as the positioning sphere and the light-pen; they are necessary for semi-graphic and graphic devices.

The task of the control unit is to ensure the operation of the screen display. The more independent and intelligent the display terminal is, the more complex are the tasks that can be solved by it. The increasing use of highly integrated electronic circuits has resulted in the computer-like construction of modern display units. The controller contains a microprocessor, a store, and further interfaces for the input and output units. The image is generated by special purpose processors. The store contains the program which controls the operation. Input and output information is also stored here together with the graphic information to be displayed.

13.2.4 Types of Screens

Screens are classified according to their operational principle, i.e. whether they emit light or regulate the transmission of external light. An example of the former group is the most commonly used screen display, the so-called cathode-ray tube, while from the latter group the liquid crystal display can be mentioned.

A further classification is based on their construction: vacuum-tube and solid state types. Even another classification is based on whether it is possible to address each point on the screen, or whether only larger contiguous areas can be marked. Or further: can the screen be used for colour display, or only for black-and-white representation (*Table 13.1*).

Table 13.1. Classification of screens

Type of screen	Light is		Addressability		Refreshment		Colour image	
	emitted	regulated	point	matrix	necessary	no	possible	impossible
Cathode-ray tube	+		+		+		+	
Storage tube	+		+			+		+
Plasma panel	+			+		+	+	
Light diodes	+		+	+	+		+	
Liquid crystal		+		+	+			+
Electrophoretic		+		+		+	+	
Electrochromic		+		+		+	+	

Light emitting displays have the feature that they need a less illuminated environment as the diffuse light incident on them strongly decreases the contrast. Their advantage is that they need no extra light source, but high voltages are generally necessary. To this group belong the cathode-ray tube, the storage tube and plasma panel displays. The storage tube display has the smallest luminosity but, in spite of this, it can reproduce the finest drawing and it can thus be used in only slightly illuminated rooms.

Figure 13.7 shows a Tektronix storage tube display.

In the construction represented in the figure, the three above-mentioned units are built into the same box. The screen and the keyboard are visible, and the controller electronics is hidden in the boy below the screen. The screen shows clearly that it is a graphics display as both diagrams and text are visible simultaneously. A display of this type is necessary for surveying purposes.

Displays which control light transmission operate by changing the transmission and reflection, respectively, of light emitted by an external light source. Liquid crystal, electrophoretic and electrochromic displays belong to this group. This group has an important advantage over light emitting displays in that they can be used even in strong environmental illumination. They are driven by low voltages and power in many cases being close to the level usual in integrated circuits.

Plasma panel or solid state displays would ideally allow flat screens to be produced, giving them a further advantage over cathode-ray tube type screens which require a large amount of space.

Figure 13.7. Tektronix storage tube display

In the area of the more and more widely spread colour displays, the leading role is unambiguously played by cathode-ray tubes. For colour display, the colour TV monitor is usually used for alphanumerical and semi-graphic displays, but for high resolution graphics displays the so-called *Penatron-tube* can be used. In the latter two different coloured phosphorus layers are used and the colour effect depends on the penetration depth of the light beam into these layers.

For solid state displays, colour display is much more complicated. With light emitting devices, the generation of colour images is more easily possible. There are light emitting diodes in different colours but it would be very expensive to produce a raster of greater dimensions from them. The situation is even more difficult with displays which regulate the transmission of external light, where the colour effect is determined by the electrolyte or by the material contained in it.

It seems that cathode-ray tubes will continue to play a leading role in screen displays for quite a long time. Their advantages stem from TV techniques (low price, high resolution, colour image, etc.). Taking into account the rapid development of semi-conductor techniques, in the more distant future flat solid state displays fitted to highly integrated elements will have a leading role.

13.2.5 Devices for Data Input

The keyboard. The most basic data input device for screen displays is the keyboard which consists of two main parts: alphanumerical and functional keys. The alphanumerical keyboard is a type-writer-style keyboard for letters, numbers and symbols and for initiating elementary operations. Their services are not only noise-free and fast, but also rich, as they enable special forms and the simpler graphic symbols to be displayed.

The functional keyboard plays only the usual supplementary role when in the normal mode of operation, e.g. it enables numerical data entry to be used as in calculators. In the functional mode of operation, however, pressing a single key activates either the corresponding sequences of the firmware burnt into the store of the screen display, and of the software dumped there from the memory of the controller computer, respectively, or the corresponding sequences of the special program in the controlling computer, it may thus initiate a complicated sequence of operations. *Figure 13.8* shows the keyboard of a widely used display.

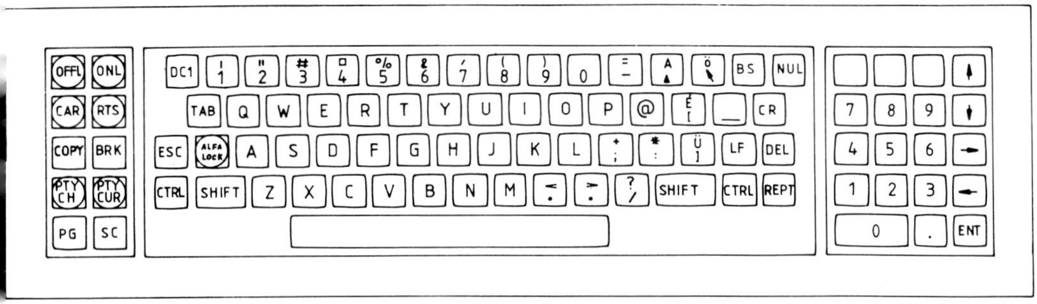

Figure 13.8. Keyboard of a screen display

Tools for position identification and marking. The simplest tools for position marking are the keys which move the *cursor*. They solve the typical text-construction tasks occurring with alphanumerical displays.

With more developed types of graphics representation, the effectiveness of information exchange is based on the substitution of lengthy specifications by a simple reference when operator commands are to be communicated. In such cases, one side of the interactive communication is the program which presents its results in many cases in a complicated graphical form, on the other side is the operator who makes decisions about the future activity of the program on the basis of the graphical information (eventually modified by colour codes and intensity changes): how to continue the operation of a given system, of a complex process. Such a connection needs a data input device which enables any raster-points, or any image position on the screen, to be located and marked.

The most simple tool for position identification and marking is the joy-stick, which may change the position of a *light marker* on the screen in eight directions, similarly to a compass dial. If the light marker has reached the designated position on the screen, the operator accordingly informs the computer system by the action key on the joy-stick. The positioning sphere, which is rotatable in any direction by hand and is a widely used input device for graphics displays, uses a similar principle. In both cases the encoded values of the vertical and horizontal coordinates of the light marker are generated by potentiometers or by a light mask screening photodiodes.

With the images of graphics displays, which contain several millions of raster points localized at the surface of the screen, the most effective and quick tool for identification and position marking is the light pen. The optical system in the tip of the "writing mechanism", and the simultaneous signals of the electron beam which locate any raster point of the screen and cause a flash there, enable the position of the light pen to be identified at any instant. Activation is usually made by the micro-switch in the light pen.

Another kind of input tool for graphics information is the graphics tablet. This can be used to enter map-like drawings. The necessary drawing or its characteristic points can be copied from paper on a flatbed, using an electric pen which emits a visible track. An arbitrarily interpretable area can also be formed on the tablet, this further promotes data entry.

13.2.6 Data Transfer

Data transfer and mutual exchange of information between computer and screen display can be realized with different interface methods.

The interfacing can be parallel or serial.

In case of a *parallel* interface the codes of the actual data are transferred in a parallel mode. This means that, in a given code system, as many wires are necessary as there are bits included in the code. During the connection all the bits are transferred in one elementary transfer step, therefore this method provides a very fast data transfer but it is very expensive. This is because separate wires are required, not only for all the bits but also for certain control signals. Thus parallel data transfer is only used where rapid data transfer is necessary over short distances.

With serial interfacing the codes, which are transferred within the computer and in the terminal in parallel form, are transformed for the duration of the data transfer into serial information. In this case the bits follow each other along a wire and, when they have reached their destination, the receiver restitutes them into a parallel code. The advantage of this method is that it is inexpensive. It needs only a single wire and the parallel-serial and serial-parallel transformations can be solved quite simply. Its disadvantage is that the transfer becomes slow due to the serial transfer form and due to synchronisation errors. This method alone is used over long distances, but even for short distance data transfer it is used more widely than the parallel method.

Serial transfer can be asynchronous or synchronous. In the former case the character codes are situated after the serial transformation, between a *start* and one or two *stop bits*. These provide synchronization at the receiver side, i.e. error-free restoration of the code. With synchronous transfer the transmitter and receiver clocks, which determine the steps of both the receiver and the transmitter, are synchronized by a previously determined exchange of messages, and the synchronization is continued simultaneously with the data transfer.

Asynchronous transfer is mostly used if the usual case is to transfer short messages, dialogues consisting of only a few characters. It is used in so-called *teletype-compatible displays*. Synchronous transfer is adopted for information transfer in larger blocks within computer networks.

Characters are usually transferred in the code system ASCII consisting of seven bits. This system enables 128 different codes to be interpretated (punctuation marks, numbers, letters and control characters) *(Table 13.2)*. There are systems where an eight bit is also used when generating the codes; they are parity bits to check whether the transfer is error-free. In IBM systems the eight-bit code set EBCDIC is used, which enables 256 standard codes to be interpreted.

The connection can be unidirectional (semiduplex) and bidirectional (duplex). In the duplex case simultaneously transmission and reception can be ensured, while they

Table 13.2. ASCII-codes

Bits														
b_7								0	0	0	0	1	1	1	1
b_6								0	0	1	1	0	0	1	1
b_5								0	1	0	1	0	1	0	1
	b_4	b_3	b_2	b_1	Column / Row	0	1	2	3	4	5	6	7
	0	0	0	0	0	NUL	(TC_7)DLE	SP	0	((a))③	P	\ ④	p
	0	0	0	1	1	(TC_1)SOH	DC_1	!	1	A	Q	a	q
	0	0	1	0	2	(TC_2)STX	DC_2	" ⑥	2	B	R	b	r
	0	0	1	1	3	(TC_3)ETX	DC_3	£②⑦	3	C	S	c	s
	0	1	0	0	4	(TC_4)EOT	DC_4	$ ②	4	D	T	d	t
	0	1	0	1	5	(TC_5)ENQ	(TC_8)	%	5	E	U	e	u
	0	1	1	0	6	(TC_6)ACK	(TC_9)SYN	&	6	F	V	f	v
	0	1	1	1	7	BEL	(TC_{10})ETB	' ⑥	7	G	W	g	w
	1	0	0	0	8	FE_0(BS)	CAN	(	8	H	X	h	x
	1	0	0	1	9	FE_1(HT)	EM	)	9	I	Y	i	y
	1	0	1	0	10	FE_2(LF) ①	SUB	*	⑧	J	Z	j	z
	1	0	1	1	11	FE_3(VT)	ESC	+	; ⑧	K	(|) ③	k	③
	1	1	0	0	12	FE_7(FF)	IS_4(FS)	,	<	L	③	l	③
	1	1	0	1	13	FE_5(CR) ①	IS_3(GS)	–	=	M	(|) ③	m	③
	1	1	1	0	14	SQ	IS_2(RS)		>	N	^④⑥	n	‾④⑤
	1	1	1	1	15	SI	IS_1(US)	/	?	O	_	o	DEL

are possible only one after the other in the semiduplex case. Today the duplex system is more often used.

The speed of asynchronous transfer is 75 to 9600 bits/s (baud) in the majority of cases, higher speeds being used only exceptionally. With telephone line connections the typical speed is 1200 baud.

Shorter connections are established by cable, longer ones by telephone line. Telephone lines can be autonomous (ordinary) or rented. In both cases transformer units, so-called *modems*, must be used at both ends of the line. It is also possible to rent a private direct telephone line, when the line-matcher GDN must be used.

With asynchronous serial transfer the specifications of the standard CCITT V.24 are usually used — its American counterpart is RS232. In networks, for block-wise transfer the protocol HDLC (High Level Data Link Control) recommended by ISC is used, while IBM has its own specifications, i.e. the protocol SDLC (Synchronous Data Link Control).

13.2.7 Programming of Screen Displays

The exchange of information between the computer and the linked screen display is carried out in a definite manner, according to the specifications of the communication protocol. This is a system of regulations determining the form of characters or series of characters in which information can be transmitted. The information can be either text or a graphic image, but commands to prepare some arbitrary function or to carry out the command itself can also be transferred.

From the point of view of the screen display, the interpretation of the communication protocol is made by an internal program. The interpreting program — as shown in *Fig. 13.9* — consists of two main parts. The task of the program F1 (firmware 1) to generate the text or the graphic image from the incoming message which corresponds to the rules of the communication protocol. The task of the program F2 (firmware 2) is to prepare the output message corresponding to the regulations of the communication protocol. This activity is carried out by interpreting the codes entering via the keyboard. *Figure 13.9* shows that, if the output of the screen display is fed back to the input, then the results of pressing the keys will be immediately visible (*off-line* operation). If the output is interfaced to the corresponding input circuits of the computer (*on-line* operation), then the first task can be accomplished only with the help of the computer. Since modern screen displays also contain elements which carry out basic computer functions, their construction has the character of a computer and it is therefore possible to make their operation more intelligent. This means that, in addition to the mode of operation presented in *Fig. 13.9* the display terminals are capable of performing more simple or more complex tasks independently and autonomously by additional special programs. These programs are either built into a read only memory *(fixed firmware)* or the computer transmits them into a memory which also provides a writing capability (RAM).

To perform the information exchange between screen terminal and computer, an interpreting program is also necessary for the computer. This task is carried out by

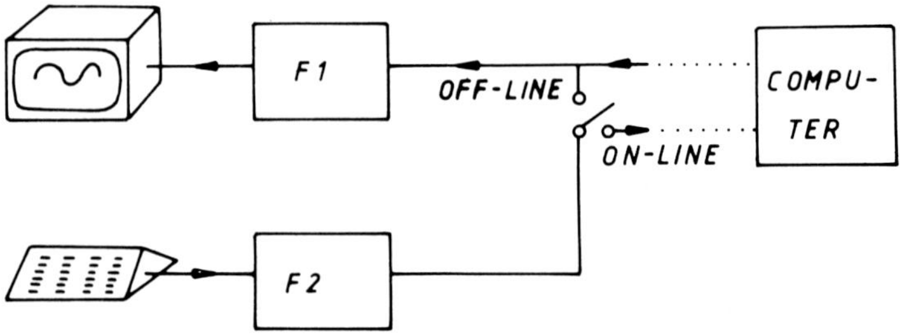

Figure 13.9. Program structure of a screen display

a peripheral handling program *(handler)*, either built-in into the operation system given in the high level programming language used or in a specialized program. This means that arbitrary program and data sets can be transferred between the system peripherals (disc, line printer, plotter, etc.) and the terminal.

In addition to system programs, problem-oriented user programs can be implemented to solve complex problems (e.g. process control, computerized planning, etc.). The programs can be independent special programs in *Assembly* language (special program systems) or programs in a general high level program language. The advantage of the latter is that their operation, the process controlled by them, and their mode of operation can be quickly modified and developed. Due to the advantages of programming in high level languages they are more and more often used in graphics tasks embodying visual man-computer interaction. Graphics operational systems which are extremely effective in solving more complicated tasks, and high level graphics languages which promote the solution of high level graphics problems, have been developed. They simplified the often very complicated tasks of constructing and planning images to the usual forms and definition processes of traditional image-constructing activity.

13.2.8 Fields of Application

Interactive graphic systems containing screen displays have many fields of application in geodesy and surveying, partly in connection with image and map processing and generation, partly in data bank types of problems.

Main applications in geodesy are:

— applications in surveying,
— records of public utilities,
— photogrammetry,
— image processing,
— cartography.

Figure 13.10. The *Aristogrid* drawing processing system

In order to present some applications, firstly a general system for image input and processing will be described, then a photogrammetric system will be presented.

Figure 13.10 shows the geometric data processing system *Aristogrid CD 400*. The interactive screen system ensures quick and reliable input of drawing and map data, their processing and archivation.

The system contains a computer PDP 11/35 and a maximum of six screen stations. The stations are of different structure, they are supplied with digitizer units for data entry and data can be modified by a problem identification tablet. The different types of distribution can be determined by the user.

The programs of the system are written in FORTRAN IV, therefore further implementation is simple.

Technical construction. 1. The central unit consists of the following: the DEC computer PDP 11/35 with an operative memory of 32 K words, expandable up to 128 K words. The system includes fixed and floating point arithmetic, real-time clock, and a printer unit with 132 characters/line. For storage and archiving two discs with capacities of 4.8 Mbytes are available; greater quantities of data are stored in a 9 channel magnetic tape unit with a speed 45 inches/s and a density of 800 bits/inch.

2. Digitizing station: this consists of a digitizer, an alphanumeric input and output unit and a screen unit. The digitizer is completely electronic. Its working area is 900 × 1200 mm, its resolution 0.0254 mm. The unit is interfaced immediately to the central unit.

The alphanumerical input and output unit has a 12 character display and 64 keys to carry out the dialogue between the system and the user.

The screen display is a Tektronix type 4014 storage tube screen with a resolution of 4096 × 4096 points.

Figure 13.11. The *Wildmap* interactive graphics system

3. Modifying station: modifications are entered by a graphics tablet supplied with an electric stylus. The working are is 700 × 700 mm. The tablet is provided with a Tektronix 4014 type display and with the alphanumeric input-output unit described previously.

Up to now only the preparation of maps has been dealt with. Interactive graphics systems (Section 13.3) can, however, also process data other than graphics. It is thus possible to establish an information system. This method enables e.g. a house to be annotated both with available graphical elements, e.g. the coordinates of the corner points and the lines connecting them, and with other graphical symbols, such as e.g. house number, owner, number of stories, etc. The data in the information system are accessible and can be collated and retrieved under arbitrary headings.

Figure 13.11 shows the interactive graphics system *Wildmap*.

13.3 Interactive Graphics Systems

The most modern method of map plotting and application is the use of the computer-aided digital map. It is stored in a computer. The graphics tools of the computer — digitizer, interactive graphics display, plotter — are used for generation and plotting. The general name for computer-aided graphics processing is Computer Aided Design (CAD). The hardware architecture of CAD systems — in systems of different orders of magnitude — includes one or more Interactive Graphics Work stations (IGW) *(Fig. 13.12)*. These consist of one micro- or minicomputer and an environment of graphics peripherals and data handling tools. The IGW can be stand-alone or on-line. The map data is transferred to the processor of the IGW, in the former case by magnetic tape or floppy or cartridge discs, respectively, in the latter case by an on-line computer link.

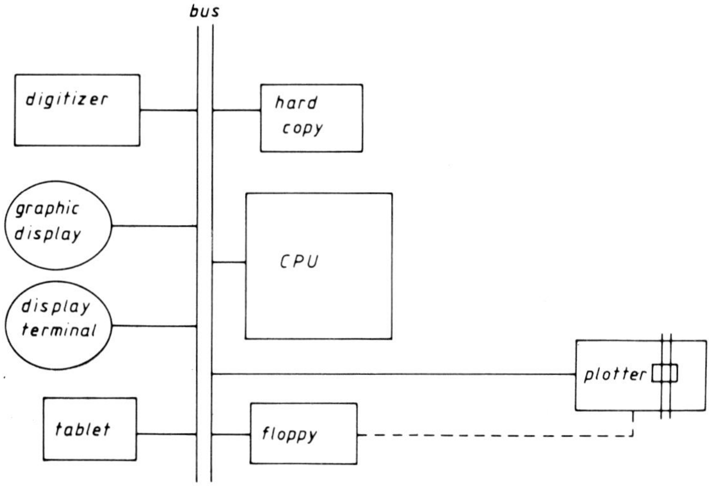

Figure 13.12. General form of an interactive graphics work station

There are general and purpose-oriented interactive graphics systems. The systems for mapping purposes can be classified into three types:
— intergrated systems with a large computer
— data acquisition and editing systems with a minicomputer
— purpose-oriented systems with a microcomputer.

In interactive graphics systems with large computers the central machine has a central memory of a capacity of some Mbytes, and magnetic disc auxiliary memory of some hundred Mbytes. The data bank created there is managed by on-line IGW-s. The operating software has to ensure the quick management of both graphics and text data bases.

Interactive graphics systems with large computers have been developed by the Intergraph Corporation (USA); they are effectively applicable in all fields of surveying and cartography. The Intergraph software integrates the spatial information into two data banks. Graphics data are managed by the Interactive Design Graphics Software (IDGS). The characteristics of the graphics elements are stored in the data base of the Data Management and Retrieval System (DMRS). The two data bases are managed from graphics work stations located on central units DEC VAX 11/730, 11/751 or 11/780. The main parts of the Intergraph system are, as shown in *Fig. 13.13*, the control computer *1*, the data base store *2* and the graphics work-stations *3*. The digital storage of public utility networks, together with the basic map, is carried out by a special graphics and data bank software (FIDS) developed by Intergraph.

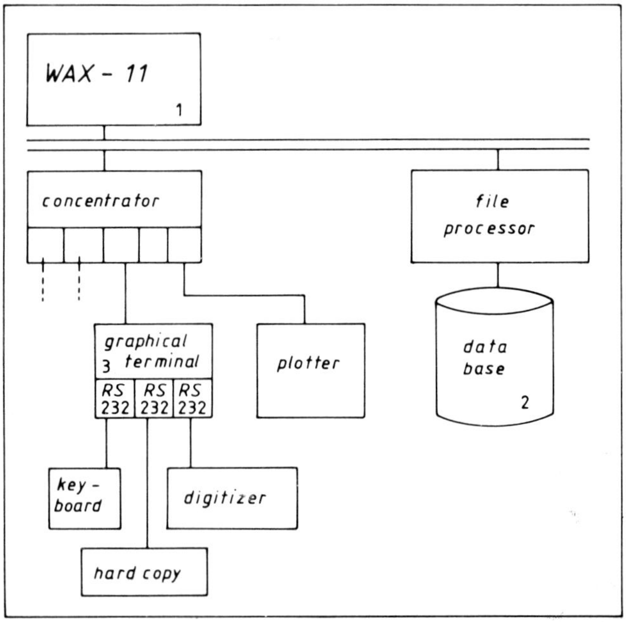

Figure 13.13. Main parts of the interactive cartographic system Intergraph: *1* host computer; *2* data base store; *3* graphics work station

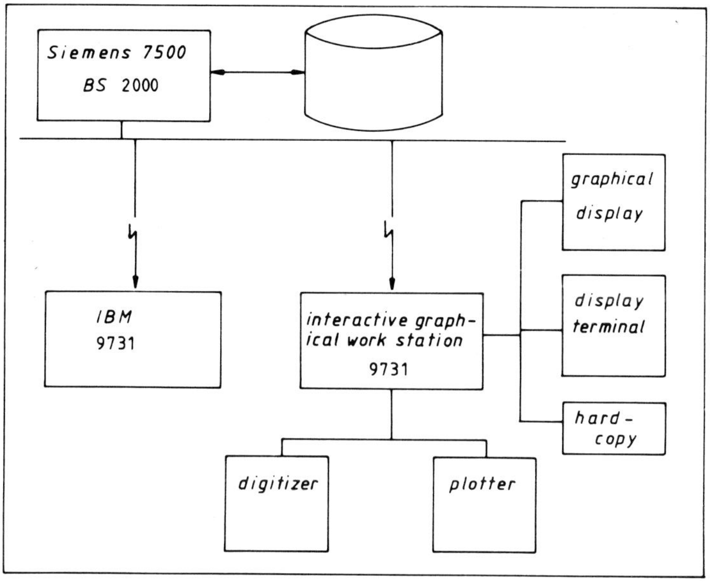

Figure 13.14. Interactive graphics system SICAD (Siemens, FRG)

The interactive graphics system of Siemens Aktiengesellschaft (Federal Republic of Germany) is the Siemens Computer Aided Design (SICAD) based on a Siemens 7500 computer *(Fig. 13.14)*. The graphics data bank, stored on a magnetic disc auxiliary store of 126 Mbyte to 1 Gbyte, is managed from interactive graphics work stations Model 9731. Graphics work stations contain graphics displays, digitizing table, alphanumerical display, hardcopy tool and plotter. In surveying, the system SICAD aids the data acquisition, data storage, selection, analysis and plotting phases. Some minicomputer systems, such as the Contraves *Gradis 2000*, provide on-line digitization and interactive map plotting.

The interactive digitizing and editing graphics system of the *Gradis 2000* has the DEC PDP 11 as the central computer, with the operational system RSX-11M. The data base of the Gradis is divided according to map sheets and it can be split into levels within each division. The central computer serves 4 to 8 graphics work stations. These work stations are controlled by microprocessor and contain a high resolution Tektronix storage tube display, a precise digitizing table and a keyboard *(Figs 13.15, 13.16)*. *Coragraph, Calcomp* and *DZ7* drawing tables can be interfaced to the Gradis system. The software of the graphics work stations facilitates a high level digitization procedure by its subroutines for coordinate geometry and for two dimensional graphics. The *Gradis 3000* is the result of a further development into three dimensional graphics.

The DIGITAL COMPUTER SYSTEM GmbH (DCS)-S 16 or SIRIUS has the VICTOR PC as a central unit. Graphics operations are carried out by a graphics display containing 400 × 800 image points, a digitizer F/201-D and a precision drawing table F/1000. The software operates in the operational systems CP/M or MS-DOS

Figure 13.15. Interactive graphics system *Gradis 3000* (Contraves, Switzerland)

Figure 13.16. Interactive graphics system *Geoplot* (DCS GmbH, FRG)

and it contains the following main surveying tasks, aided by a data bank DCS-PULIT: processing of terrestrial measurements, coordinate geometry tasks, digitization, digital terrain model, and drawing programs.

The firm Opton, Oberkochen, has integrated its equipment for photogrammetric interpretation into a graphics system controlled by a Planicomp computer *(Fig. 13.17)*. The processing instruments C100, C120 and C130 and the analytical orthophoto system Orthocom Z2, with processors of the family HP 1000, offer a

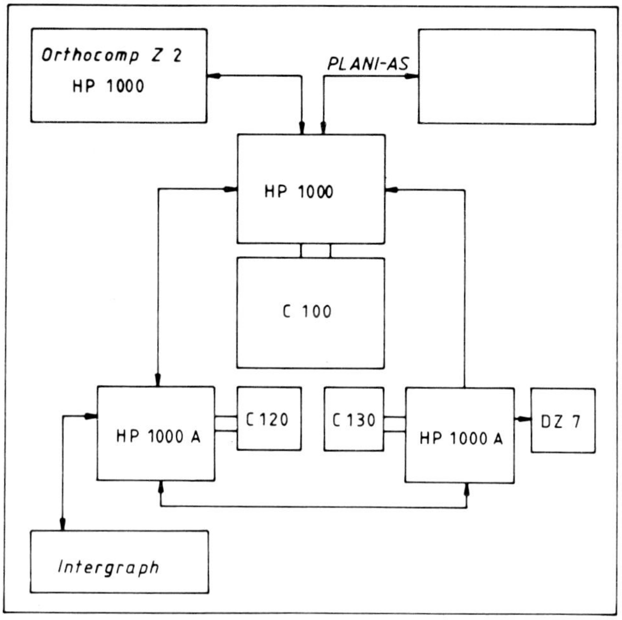

Figure 13.17. Planicomp computer system (Opton, FRG)

wide range of solutions for mapping tasks, either separately or integrated into a single system, as shown in *Fig. 13.17.* The most important element in an effective use of the Planicomp system is the basing of the software on the basic software of the HP 1000; it consists of basic programs and user's programs denoted AL, and it provides off-line processing for both analogue and digital instruments. From the main modules of the program system, Plan-AS includes aero-triangulation and digital terrain model; DZ7-AS is used for automatic mapping. The photogrammetric interpretation system can be operated interfaced with the Intergraph system, creating an effective interactive graphics system.

Microcomputer systems are practically off-line operating IGW-s. As an example the Wild GEOMAP should be mentioned, which contains surveying and drawing software, and on-line digitization and plotting based on a Tektronix 4054 graphics microcomputer.

The main characteristics of some important interactive graphics systems are presented in *Table 13.3.*

Table 13.3. Characteristic data of some interactive graphic systems

Manufacturer type	Control computer	Characteristic peripherals*	Software**	Use***
Intergraph CO Intergraph	DEC VAX-11	Interact IGM Interpro IGM Intermap IGM Gerber P D	IDGS/DMRS DBHS FIDS	S, PH, MPW, LU, SP, RMR, CEP, CA
Siemens AG SICAD	Siemens Sys. 7500/BS 2000	9731 IGM 9750 G P D	SICAD KARTOG-RAPHIE SESAM DBHS	S, RER, MPW, SP, TM
IBM	IBM Sys. 370/158 IBM 3033/N 6 MBy	IBM 3377 IGM	IMS DBHS KIV ARSM	S, RER, CEP
GIXI GRAFIXI 2D/3D CAD/MAPPING	MITRA 115/125	Tektronix IGM Benson P Benson D	CISIGRAPHE	TM
Ferranti Cetec CLUMIS	DEC PDP-11	COMDIG IGM EP 130/2300/330 P, D	CLUMIS MPS	S, MPW
Contraves AG GRADIS 2000 GRADIS 3000	DEC PDP-11 DEC VAX	GRADIS IGM DZ7 P Calcomp 965 D	GRADIS 2000 GRADIS 3000	S, MPW
DIGITAL CUMPUTER SYSTEM GMBH DCS GEOPLOT	DCS-16 SIRIUS/VICTOR	F 100 P F 201 D	DCS PULI DCS GEOPLOT S112, S100, S110, S125, S140	S, CEP, AR
OBERKOCHEN Planicomp	HP 1000	Orthocomp C 100 Planicart Intermap	Plani-AS DZ7-AS GEFIO STRIM PATM, PAT8, HIFI	PH
GEOS	HP 9845	Floppy disc	GEOS	S
ARISTO Graphic SYSTEM GMBH ARISTOGRID 103	PDP 11 LSI	ARISTOGRID D IGM Floppy disc ARISTOMAT P	I/A digitizer SW	S
Summagraphic DATAGRID Sys.	Summadraft 8000 S	IGM P, D	I/A digitizer SW	S

Table 13.3. (continued)

Manufacturer type	Control computer	Characteristic peripherals*	Software**	Use***
Wild GEOMAP	Tektronix 4054 A	Floppy disc TA 10/TA2 P D	GEOMAP	S
Calcomp IGS	Calcomp CC16/40	IGM P, D	Calcomp Engineering Drafting SW (EDS)	General graphical tasks
Hewlett–Packard VERMESSUNGS-SYS 9845	HP 9845	Floppy disc P	RF SW	S

Explanation:

* In this column P plotter; D digitizer.

** In this column DBHS data base handling system.

*** In this column S surveying; RER real estate register; PH photogrammetry; MPW mapping of public works; LU land use; SP settlement planning; RMR raw material register; CEP civil engineering planning; TM thematic mapping; AR architecture; CA cartography.

14 Instruments to Measure Areas

Areas of plotted figures can be determined either indirectly (area determination) or by using a special area measuring device (planimeter).

The basic principle of area determination is to divide the area to be determined into geometrical figures — rectangles, trapezoids, triangles or segments of a parabola — the area of which can easily be computed. The sum of the geometrical figures gives the area of the plotted figure.

14.1 Simple Tools for Area Determination

A simple tool to determine the area of a triangle is the so-called hyperbola graph *(Fig. 14.1)*. It is based on the fact that, for any point on a (rectangular) hyperbola with the axes x and y of a Cartesian system of coordinates as asymptotes, the product $x \cdot y$ is constant and can thus be found from the plotted set of hyperbolae. Such graphs can be used for the determination of the area of triangles and quadrangles.

To determine the area of a triangle, the graph should be placed on the figure so that one of its vertices B lies at the origin of the hyperbolae 0, and the vertex C lies on the y axis. The graph should then be shifted parallel to itself until the third vertex A lies on the y axis, when the area can be immediately read at C from the set of hyperbolae.

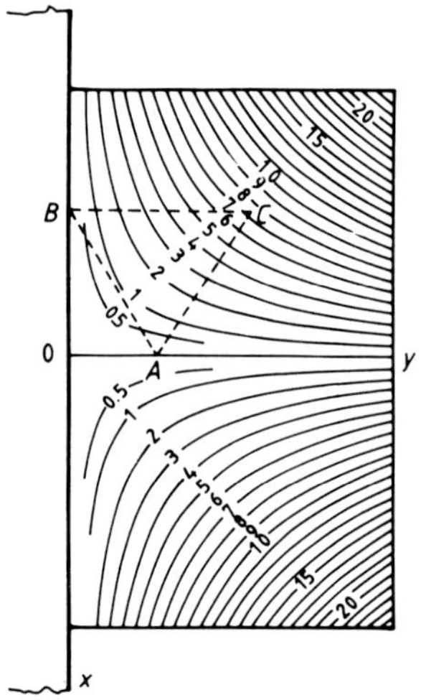

Figure 14.1. Hyperbola graph

For a quadrangle the graph should be positioned so that the diagonal *BC* is on the *y* axis, and one of the vertices is at the origin; then the graph is shifted parallel to itself until the vertex *A* lies on the *y* axis; the position of the vertex *C* is now read. For the next step the quadrangle is shifted parallel to itself again to bring the point *B* on the *y* axis. The position of the vertex *C* is then read again. The sum of the two values obtained is the area of the quadrangle. *Figure 14.1* shows the Kloth set of hyperbolae.

The area of figures with curved boundaries can be determined from a measure of parallel strips or squares. With parallel strips the widths of the strips are generally the same, and they are small enough for the bounding curve of the figure to be considered as a straight line between adjacent strip divisions. Thus the area is divided into trapezoids and two elementary areas which are bounded partly by straight, partly by curved lines. In the parallel strips it is sufficient to measure the length of the central line of the trapezoid. For this purpose a pair of compasses is useful *(Fig. 14.2)*. The compasses have jointed legs and their opening is measured by a counting mechanism and can be previously set. If the central length of the strips are summed with the compasses, then the number of its openings plus the last opening gives the sum of the central lengths which, multiplied by the width of the strips, gives the area. The area of the elementary end areas, with straight and curved bounds is computed as the product of the length of the straight line multiplied by the maximum departure of the curve from it, multiplied by two thirds.

The area measuring harp can also be used for area determination by parallel strips *(Fig. 14.3)*. This is a metal frame between whose legs thin fibres are attached. The harp is placed on the area to be determined and the fibres then divide the area into

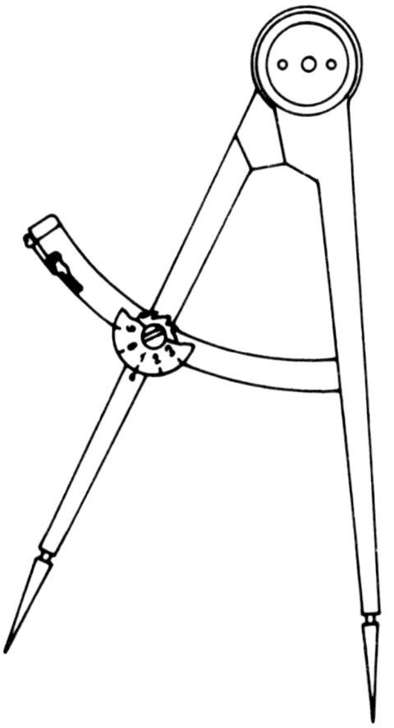

Figure 14.2. A pair of compasses

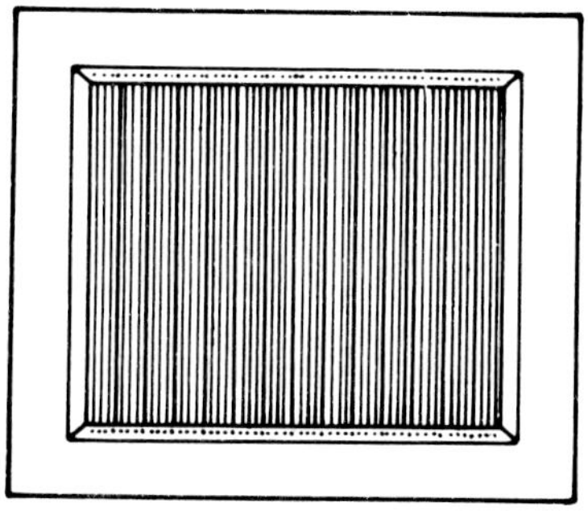

Figure 14.3. The Adler harp

trapezoids. In this case the central lines of the trapezoids are also measured and their sum multiplied by the width of the strips. An example is the Alder harp *(Fig. 14.3)*.

For area determination by division into squares, a square grid engraved on a glass plate and generally having an area unit of 1 mm^2 (1×1 mm) is placed on the area. Area determination consists of counting the number of complete squares within the boundary of the area, and the area of all the partial squares (determined by estimation) is added to this value.

14.2 Planimeters

The planimeter is an instrument for measuring the area of plotted plane figures using a mechanical tool which gives a value proportional to the area. The planimeter is an integrator for the numerical determination of the value of integrals of the form

$$y = \int f(x)\mathrm{d}x. \qquad (14.1)$$

In practice the planimeters used are usually those which require to be guided around the perimeter of the area to be determined. There are many varieties of traditional planimeters, but they have in common a measuring point (stylus) touching the paper which can be freely moved in any direction, and a measuring wheel which rotates when the measuring point is moved in certain directions. The wheel has a counting mechanism which displays the number of rotations made during the motion of the point.

Mechanical integration can be performed either in polar or in Cartesian coordinates, correspondingly distinction is made between polar and orthogonal planimeters.

In a Cartesian system of coordinates one has

$$T = \int y\mathrm{d}x, \qquad (14.2)$$

where the integration is to be made around the complete perimeter of the plane figure. In a polar system of coordinates:

$$T = \int \frac{1}{2}\rho^2 \mathrm{d}\rho, \qquad (14.3)$$

where the integration is again to be made around the complete perimeter.

Planimeters can also be classified according to another aspect, depending on whether the distance between the measuring point and the measuring wheel is constant or not. Planimeters with a variable arm length are called rotational planimeters, those with a constant arm length are called bar planimeters.

14.2.1 Orthogonal Bar Planimeters

In the basic type of orthogonal planimeter the measuring point is at one end of a thin rigid bar, at the other end of which the measuring wheel is mounted *(Fig. 14.4)*, with its axis parallel to the bar. The standing axis of the bar is supported in a corresponding sleeve. The sleeve fits into the longitudinal slot of a rule fixed on the paper and can be moved only along the slot, i.e. on a straight, forced path. Thus the

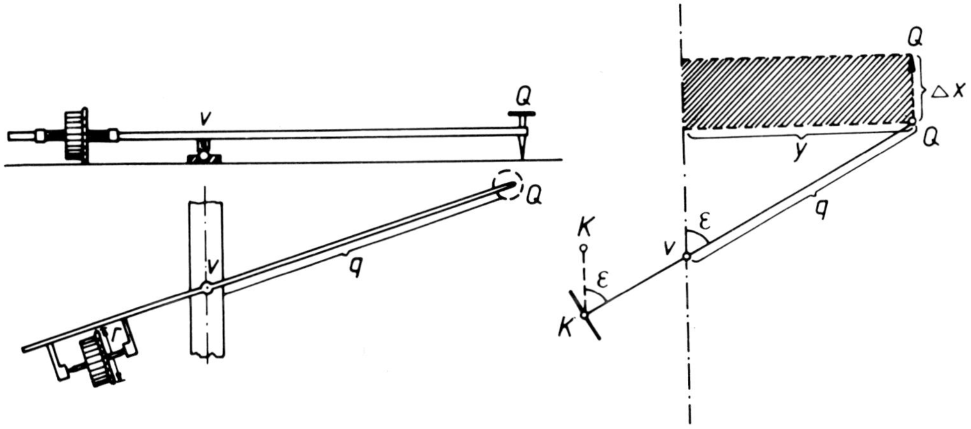

Figure 14.4. Orthogonal bar planimeter

measuring point can be rotated about the standing axis of the bar (the sleeve), and it can be moved along a straight line parallel to the forced path, together with the bar and the sleeve. Therefore it can be moved in any direction on the paper, and its movement causes rotation of the measuring wheel which is in contact with the paper. The amount of rotation is read on the revolution counter of the wheel.

The wheel does not rotate if the measuring point is moved along the straight line of the forced path, as in this case the wheel is moving in a direction perpendicular to its own plane, i.e. it slips. The line, movement along which does not cause the measuring wheel to rotate, is called the base line of the planimeter. In the present case the base line is the straight line of the forced path.

It can be shown that, if the measuring point is moved by an amount Δx along a straight line parallel to the base line, then the rotation of the measuring wheel $\Delta\beta$ is proportional to the area between the base line, the line of the movement, and two perpendicular lines from the end points of the movement to the base line. Thus, if the measuring point is guided around the perimeter of a plane figure, then the area is:

$$T = rq\beta,$$

where β is the angular rotation, q the bar length between measuring point and axis, r the radius of the measuring wheel. In practice the number of revolutions is measured, not the angular rotation. If the number of revolutions equivalent to the angular rotation is denoted by n, then $\beta = 2\pi n$, and the area is

$$T = (2\pi qr)n = T_1 n, \tag{14.4}$$

where T_1 is the constant of the planimeter.

The *Hohmann–Coradi* orthogonal bar planimeter *(Fig. 14.5)* contains a carriage structure revolving on cylinders. Both heavy cylinders *1* are rigidly connected to a common axle and they contact the paper on a wide, finely grooved surface: i.e. if it is pulled or pushed forwards or backwards by the measuring bar *2* used as a carriage-pole, it moves always along the same straight path. The area to be measured is unlimited in one direction.

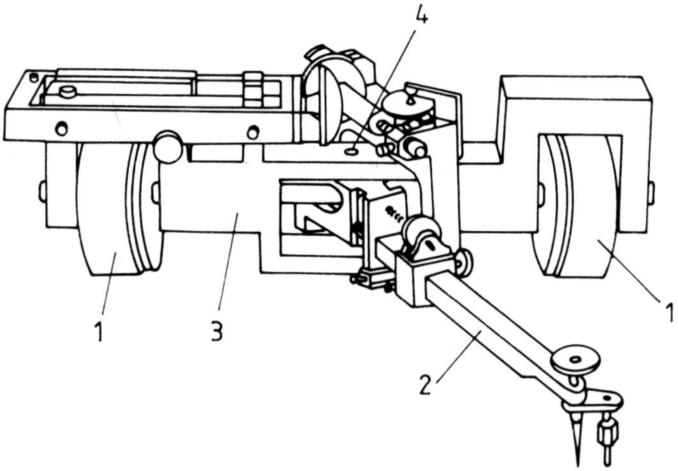

Figure 14.5. The Hohmann–Coradi orthogonal bar planimeter

The solid frame *3* is fastened to the cylinder and at its centre the measuring arm is rotatable around the standing axis *4*. At the end of the arm there is a measuring point, which can be rotated around the standing axis *4* of the arm, and it can be moved longitudinally together with the complete carriage structure (parallel to the direction of the forced movement), i.e. the measuring point can be moved in any direction.

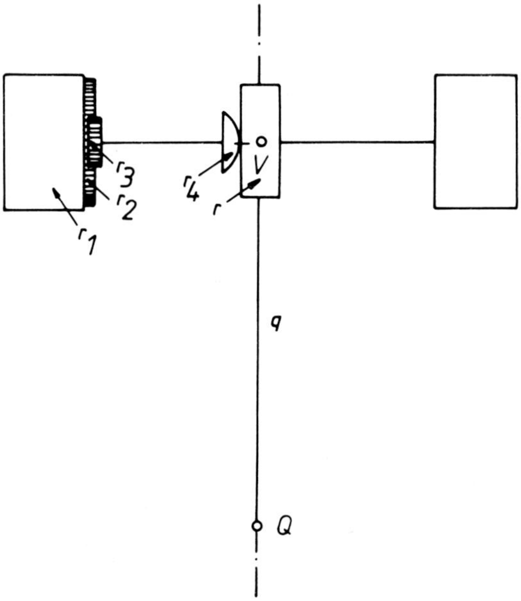

Figure 14.6. Diagrammatic sketch of a Hohmann–Coradi orthogonal bar planimeter in its basic position

On one of the axles there is a gear wheel with very fine teeth *(Fig. 14.6)*. Its radius r_2 is somewhat less then the radius r_1 of the cylinder. When the carriage moves, this gear wheel rotates through an angle which corresponds exactly to the rotation of the cylinder. It engages another gear wheel of a smaller radius r_3, which rotates above it around a parallel axis. The axle of the small gear wheel ends in a spherical cap (with radius r_4).

The measuring cylinder of radius r is supported in bearings on the metal frame mounted to the measuring bar. Its axis is parallel to the axis of the measuring arm, and the cylinder fits the spherical cap. The cylinder moves together with the measuring arm, thus the direction of its axis may change. It is continuously connected to the spherical cap by properly seated springs. The measuring cylinder is supplied with a revolution counter.

The measuring arm is moved by the spherical cap. It may rotate only if the planimeter is pushed forwards or backwards. It does not rotate if the cylinder touches the apex of the cap.

The base line of this planimeter is a straight line parallel to the forced path described by the measuring point if the measuring cylinder touches the apex of the spherical cap *(Fig. 14.6)*.

The rotation of the measuring cylinder $\Delta\beta_x$, which corresponds to the movement of the point Δx parallel to the base line, is proportional to the shaded area ΔT in *Fig. 14.7*, i.e.

$$\Delta\beta_x = Cy\Delta x. \qquad (14.5)$$

In other words, the cylinder with radius r_1, and the gear wheel rigidly connected to it with radius r_2, rotate through an angle $\Delta\alpha$, due to the apex movement Δx. Assuming perfect rolling, $\Delta x = r_1 \Delta\alpha$, i.e.

$$\Delta\alpha = \frac{1}{r_1}\Delta x. \qquad (14.6)$$

The larger gear wheel of radius r_2 rotates the smaller gear wheel of radius r_3 through an angle $\Delta\alpha'$. Both move by equal amounts, thus $r_2\Delta\alpha = r_3\Delta\alpha'$; using Eq. (14.6), this gives:

$$\Delta\alpha' = \frac{r_2}{r_3}\Delta\alpha = \frac{r_2}{r_1 r_3}\Delta x. \qquad (14.7)$$

The spherical cap rotates through the same angle $\Delta\alpha'$, thus the point of contact between measuring cylinder and spherical cap moves through an arc $\rho\Delta\alpha'$. This movement rotates the measuring cylinder through an angle $\Delta\beta_x$, and it is necessary that

$$r\Delta\beta_x = \rho\Delta\alpha' \qquad (14.8)$$

if free rolling is again assumed. *Figure 14.8* shows separately the spherical cap and the measuring cylinder. From this figure: $\rho = r_4 \sin\varphi$, and from *Fig. 14.7* $\sin\varphi = y/q$, therefore

$$\Delta\beta_x = \frac{1}{r}\rho\Delta\alpha' = \frac{r_4}{rq}y\Delta\alpha' = \frac{r_2 r_4}{r r_1 r_3}y\Delta x = Cy\Delta x, \qquad (14.9)$$

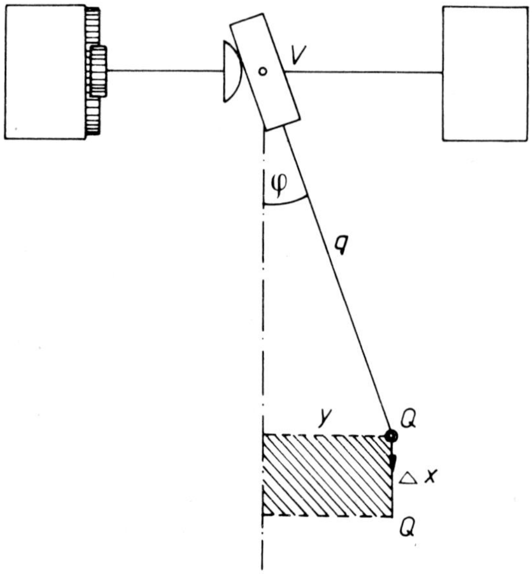

Figure 14.7. Diagrammatic sketch of a Hohmann–Coradi orthogonal bar planimeter in a general position

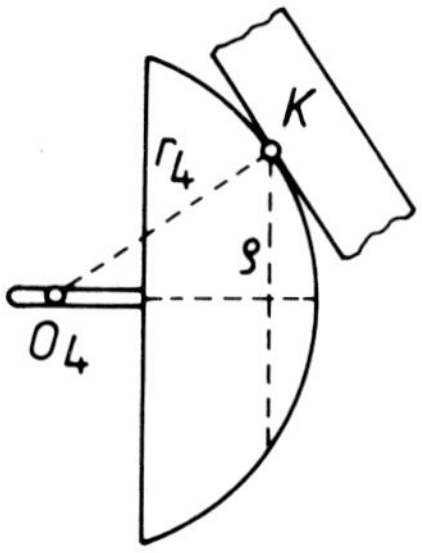

Figure 14.8. The sperical cap and the measuring wheel

which demonstrates the law of Eq. (14.5). If the area T is perambulated by the measuring point, then the number of revolutions is proportional to the area itself:

$$\beta = C \int y \mathrm{d}x = \frac{r_2 r_4}{r r_1 r_3} \frac{1}{q} T,$$

i.e.

$$T = \frac{r r_1 r_3}{r_2 r_4} q\beta,$$

and if the number of revolutions n is introduced instead of β,

$$T = T_1 n, \tag{14.10}$$

where

$$T_1 = 2\pi \frac{r r_1 r_3}{r_2 r_4} q. \tag{14.11}$$

The length q of the arm of the planimeter can be changed, and therefore the planimeter can be used with different constants. In practice the length is chosen so that the constant is an integer.

14.2.2 Polar Bar Planimeters

The polar planimeter *(Fig. 14.9)* is a device to measure areas which enables the area of the plotted figure to be determined by mechanical integration in a polar system of coordinates.

The device has a point which remains fixed after setting the instrument and is called the pole. Constructionally the pole *2* is the point extruding from the heavy

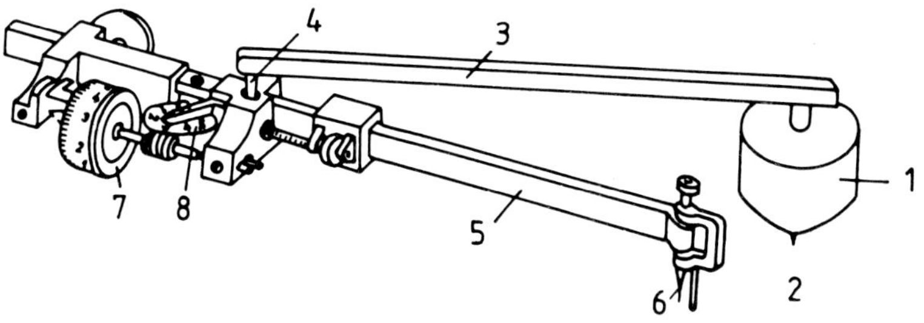

Figure 14.9. The *Amsler–Coradi* polar planimeter

weight (pole body) *1*. The polar arm *3* is rigidly mounted on the pole body and it may rotate around the pole. A hinge *4* extrudes downwards at the end of the polar arm and the measuring arm *5* can rotate around it. The measuring point *6*, is at one of its ends, the measuring wheel *7* at the other. The axis of the measuring wheel is parallel to the axis of the measuring bar and has a rotation counter *8*. The polar arm is usually high enough to allow the measuring arm to be guided below it.

The *Amsler–Coradi* polar planimeter has a circle as its base line, namely the base circle which is described by the measuring arm when the plane of the measuring wheel contains the pole. In this case the wheel does not rotate as it is always perpendicular to the direction of movement.

A movement of the point parallel to the base circle corresponds to an angular rotation $\Delta\beta$, which is proportional to the area between the line traced by the point, the base circle, and lines drawn from the two end points of the movement perpendicular to the base circle.

If an arbitrary figure is perambulated by the measuring point, and if the pole of the planimeter is outside the area, then the angular rotation of the measuring wheel (β) is proportional to the area T and:

$$T = rq\beta = (2\pi rq)n = T_1 n. \tag{14.12}$$

If the pole is inside the area to be measured, then the area of the base circle (T_0) is to be added. In that case the area of the figure is:

$$T = T_1 n + T_0. \tag{14.13}$$

The measurement method with all planimeters is the same, namely the measuring point must be guided around the perimeter of the figure. The planimeter must be located so that the measuring point can be guided around the complete perimeter. The point must be set on an arbitrary point of the perimeter and the revolution counter read. The point is then guided around the perimeter back to the starting point and a second reading is made. The difference of the two readings is the number of revolutions n.

The mean square error of a measurement with the Hohmann–Coradi planimeter is $m = \pm 0.01\sqrt{T}$, with the Amsler–Coradi planimeter $m = \pm 0.03\sqrt{T}$.

14.2.3 Electronic Digital Planimeters

There are today several types of instruments within this family, which include instruments with an electronic display of the area and planimeters consisting of electronic elements. The precise electronic planimeter OTTPLAN 700/710 (A. OTT, Kempten, Federal Republic of Germany) is described — it operates on the principle of the polar planimeter *(Fig. 14.10)*.

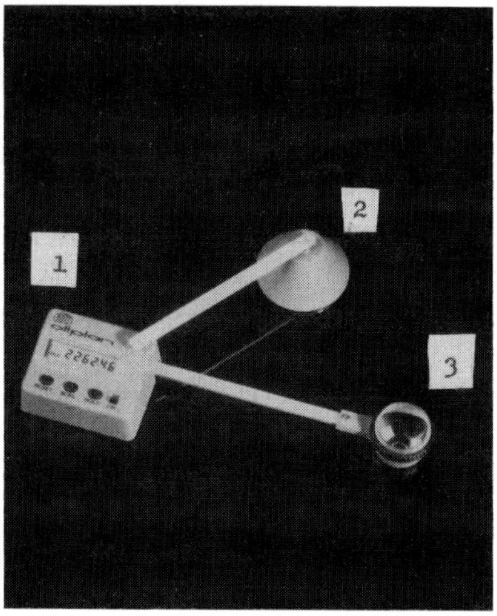

Figure 14.10. The OTTPLAN 700/710 electronic digital planimeter

The digital planimeter consists of the following:
— a measuring wheel *1* supplied with a measuring unit and operational buttons,
— a pole fixed by a weight *2*,
— a measuring arm with a movable measuring mark *3*,
— a charging set or a 24 V battery.

The cover of the measuring contains a liquid crystal display with 6 digits. The decimal point can be anywhere and values less than zero are indicated by a negative sign.

The switch must be switched to "ON" when the value 0 appears on the display, and the scale factor is 1.00000. By pressing the "HOLD" button rotation of the measuring wheel is not detected by the planimeter.

By pressing the button "SCAL" the decimal point moves one decimal place to the left or right if the "HOLD" button is not pressed. When pressing this button in the mode "HOLD", the scale factor is displayed. If the "RESET" button is pressed, the display is zeroed and, if there were a minus sign, it is also erased but the decimal position set by the button "SCAL" remains unaltered. In addition, the "RESET" button erases any ERROR message.

The pole may be located either outside or inside the area to be measured. An external position is preferable. In such a case the planimeter should be located so that, when the measuring mark is placed at about the centre of gravity of the area to be measured, the pole arm and the measuring arm should be approximately perpendicular to each other. The measuring mark should be placed at a starting point on the perimeter of the area to be measured, then it must be guided along the perimeter in a clockwise sense. The value on the display multiplied by 10 gives the area in mm^2. If the scale of the map differs from 1 : 1, then a scale factor between 1 and 9.99999 must be set and should be used.

By pressing the "HOLD" button, the displayed value can be stored. This enables partial areas to be summed.

As an option the OTTPLAN planimeter can be interfaced with a 24 V data interface, enabling it to be connected to a computer or to a printer.

To calibrate and test, the digital planimeter, the equipment includes a control rule. If a circle with a radius selected on this rule is perambulated with the measuring mark with the aid of the rule, then this value can be compared to the precise circle areas given in a table attached to the planimeter. The accuracy of the instrument is ± 0.6 percent, its dimensions $425 \times 230 \times 110$ mm, its weight 2.8 kg.

Appendix
Some Important Firms Producing Surveying Instruments

Austria — Swarovski Optik K. G., Absam (earlier separately Gebr. Miller GmbH Innsbruck)
China — several State Instrument Factories
England — Vickers Ltd, Vickers Instruments York (earlier name: Cooke, Thrughton & Simms)*
Hall & Watts Ltd, St. Albans (earlier Rank Precision Industries Ltd, Leicester)
Hilger & Watts (earlier name: E. R. Watts & Sohn)
W. F. Stanley & Co., London*
Tellumat Ltd (earlier Tellurometer Ltd) — electronic distance meters (also outside England like South Africa, the USA, Australia and Canada)
France — Société D'Optique, Précision, Electronique et Mécanique (SLOM), which was formed as the union of ESSEL and SOPELEM
Société de Recherches et de Perfectionnements Industriels des Etablissements H. Morin-Secretan (SRPL), Puteaux (union of Secretan, H. Morin and S.R.P.I.)
Chasselon, Cachan/Seine
Coppin, Paris
German Democratic Republic** — Kombinat VEB Carl Zeiss JENA (JENOPTIK, JENA GmbH)
VEB Freiberger Präzisionsmechanik (Division of Kombinat VEB Carl Zeiss JENA), Dresden
Hungary — Hungarian Optical Works (MOM), Budapest
India — National Instruments and Ophthalmic Glass, Calcutta
Italy — Filotecnica Salmoiraghi, Milano
Officine Galileo, Firenze
Japan — Fuji Surveying Instr. Co. Ltd, Tokyo
Sokkisha Co. Ltd, Tokyo
Tokyo Optical Co. (TOPCON)
Nippon Kogaku (NIKON), Tokyo
Asahi Precision Co. (Pentax), Tokyo
Poland — Polskie zaklady optyczne (PZO)
Soviet Union — Geofizika
EOMS
Charkov Geodetical Instrument Factory

* Vickers Ltd and W. F. Stanley & Co. recently stopped producing surveying instruments.
** Feinmass has stopped producing surveying instruments.

Sweden — AGA Geotronics AB Lidingő, Stockholm — electronic surveying instruments

Switzerland — Kern & Co. AG, Aarau
Wild Heerbrugg AG, Heerbrugg

USA — manufacturer/distributor addresses

ABR Engineering & Surveying Instruments, Inc.
Burbank, California

AGL Corporation
Jacksonville, Arizona

ALLEN Precision Equipment, Inc.
Atlanta, Georgia

Berger Instrument Company
Boston, Massachusetts

Clausen Instrument Co., Inc./Jena
Raleigh, North Carolina

Cubic Precision, Survey Systems Division
Tullahoma, Tennessee

Geodimeter–Geotronics of North America Inc.
Novato, California

CLS Industries
Hawthorne, California

Hewlett Packard Company (computers for surveying computations)
Palo Alto, California

IR Incorporated
Orlando, Florida

Kern Instruments, Inc.
Brewster, New York

Laser Alignment Incorporated
Grand Rapids, Michigan

Lewis & Lewis
Ventura, California

The Lietz Company
Overland Park, Kansas

Nikon, Inc., Surveying Division
Torrance, California

Pentax Corporation
Englewood, Colorado

Schonstedt Instrument Company
Reston, Virginia

SECO Manufacturing Co., Inc.
Redding, California

SMI Surveyors' Module Inc.
Kingsport, Tennessee

Spectra-Physics, Inc., Construction Division
Dayton, Ohio

Topcon Instrument Corporation of America
Paramus, New Jersey
Warren–Knight Instrument Company
Philadelphia, Pennsylvania
Wessco
Denver, Colorado
Western Atlas International, Aero Service
Houston, Texas
David White Instruments
Germantown, Wisconsin
Wild Leitz USA, Inc.
Rockleigh, New Jersey
Carl Zeiss, Inc.
Thornwood, New York

West Germany* — F. W. Breithaupt & Sohn, Kassel
GEO Fennel Führer & Co., Baunatal (earlier Otto Fennel, Kassel) — marketing
Opton Feintechnik GmbH, Oberkochen (Württemberg) (Kombinat Carl Zeiss, Oberkochen)
Theis Co., Breidenbach
Wallace Tiernan GmbH, Günzburg/Donau — barometers
Hewlett–Packard GmbH, Frankfurt/Main } electronic computers for surveying computations
Aristo-Werke, Hamburg-Altona } electronic computers for surveying computations

* Ertel Werke (München) and Franke Co. (Giessen) have been closed down.

Selected Bibliography*

Ackerl, F.: Geodäsie und Photogrammetrie, Teil I: Instrumente und Verfahren der Vermessung. Wien: Verlag Georg Fromme & Co., *1950*.

Almássy, Gy.: Mikrohullámú kézikönyv (A manual of microwaves). Műszaki Könyvkiadó, Budapest, *1973*.

Apenko, M. I.—Dubovik, A. S.: Prikladnaja optika. Moscow, *1971*.

Aristo-Werke: Digitalisierungsgerät Aristogrid 8970-10.

Bárány, N.—Mitnyán, L.: Optimechanikai műszerek (Opto-mechanical instruments). Műszaki Könyvkiadó, Budapest, *1971*.

Bárány, N.: Finommechanikai kézikönyv (A manual of precision instruments). Műszaki Könyvkiadó, Budapest, *1974*.

Berg, J.: Utsättnigs- och kontrollmätninsmetoden vid byggnads produktion. Statens institut för byggnadsforsking, Stockholm, *1969*.

Berghaus, W.: Untersuchungen über Fokussiergang und Ziellinie geodätischer Fernrohre. Zeitschrift für Instrumentenkunde, *1939*.

Cimerman, V.: Atlas Geodetskih Instrumenata. Technicka Knjiga, Zagreb, *1960*.

Chebotarev, A. S.: Geodeziya. Izdatyelstvo Geodezicheskoy Lityeratury, Moscow, *1962*.

Cooper, M. A. R.: Modern Theodolites and Levels. (2nd ed.) Granada/London/Toronto/Sidney/New York, *1982*.

Deumlich, F.: Koordinatentheodolit. VEB Vermessungstechnik, Dresden, *1965*.

Deumlich, F.: Instrumentenkunde der Vermessungstechnik. (8. Auflage.) VEB-Verlag Bauwesen, (East) Berlin, *1988*; In English: Walter de Gruyter, Berlin/New York, *1982* (after the 7th ed.)

Dummer, G. W. A.—Granville, J. W.: Miniature and microminiature electronics. Sir Isaac Pitman and Sons Ltd., London, *1964*.

Eichholtz, K.: MW 77 — Ein Vermessungskreisel hoher Genauigkeit, kurzer Messzeit und kleiner Abmessungen. FIG Congress, Stockholm, *1977*.

Fialovszky, L.: Teodolit dlya automaticheskogo opredeleniya koordinat. Geodeziya i Kartografiya, Moscow, *1965*.

Fialovszky, L.: Eine Entwicklungsvariante des Kreisuntersuchungsverfahrens nach Heuvelink. Zeitschrift für Vermessungswesen, *1968*.

Fialovszky, L.: An Improvement of the Heuvelink Method of Circle Division Control Measurement. MOM Review 2, Budapest, *1971*.

Girsberg, M. A.: Geodeziya I. Izdatyelstvo Nedra, Moscow, *1967*.

Glissman, T.: Eine Methode zur experimentellen Untersuchung der Ziellinie von Nivelliergeräten mit Planplatte. Zeitschrift für Vermessungswesen, *1969*.

Grafarend, E.: Der klassische nordsüd-suchende Kreisel. Vermessungstechnische Rundschau, *1967*.

* Supplementary references are given in the footnotes.

Gregerson, L. F.: An investigation of the MOM's Gi-B2 gyroscopic Theodelite. The Canadian Surveyor, *1970.*

Grossmann, W., Kahmen, H.: Vermessungskunde Bd. I, 16. Aufl. *1985.* Bd. II, 14. Aufl. *1986* and Bd. III, 11. Aufl. *1979.* Walter de Gruyter & Co., Berlin/New York.

Gruber, O.: Optische Streckenmessung und Polygonierung. Herbert Wichmann Verlag, Berlin, *1955.*

Guide Book of Japanese optical precision instruments. Hosokawa Printing Co. Ltd, Tokyo, *1967.*

Gushev, N. A.: Markseidersko-geodezicheskie pribori i instrumentovedenie. Nedra, Moscow, *1968.* (2nd ed.)

Gushev, N. A.: Ob opitnom obraztse avtoredukcionnogo nivelira s dvustoronney liniey vizirovaniya. Geodeziya i Kartografiya, *1971.*

Haupt, W.: "Forschungsarbeiten Vermessungskreisel" des Institutes für Markscheidewesen der TH Clausthal 1947–69. Rellesman-Festschrift, *1970.*

Hazay, J.: Geodéziai kézikönyv (A manual of geodesy). Közgazdasági és Jogi Kiadó, Budapest, *1956.*

Hildebrand, S.: Feinmechanische Bauelemente. VEB Verlag Technik, Berlin, *1967.*

Híradástechnikai kislexikon (Handbook of telecommunication). Műszaki Könyvkiadó, Budapest *1976.*

Hodam, F.: Optik in der Längenmesstechnik. VEB Verlag Technik, Berlin, *1961.*

Hodam, F.: Technische Optik. VEB Verlag Technik, Berlin, *1966.*

Hofmann, O.: Ein neues astasiertes Pendel für genaue Neigungsmessungen und automatische Stabilisierung von Ziellinien. Jenaer Jahrbuch *1958.* Verlag VEB Gustav Fischer, Jena.

D'Hollander, R.: Topographie générale. Tome I–II. Editions Eyrolles, Paris, *1970–1971.*

Jordan—Eggert—Kneissl: Handbuch der Vermessungskunde. 10. Aufl. Bd. II. *1963*, Bd. III. *1958*, Bd. VI. *1966.* Metzlersche Verlagsbuchhandlung, Stuttgart.

Jacobs, D. H.: Fundamentals of Optical Engineering. New York, *1943.*

Kahmen, H.: Mikroprozessoren und elektronische Tachymeter. Zeitschrift für Vermessungswesen, *1979.*

Kahmen, H.: Elektronische Messverfahren in der Geodäsie. (2nd ed.) Verlag Wichmann, Karlsruhe, *1978.*

Karovic, K.: Eine neue Methode zur experimentellen Untersuchung der Lage der Ziellinie bei optischen Fernrohrsystemen. Optik, *1963.*

Klopocinski, J.: Teodolit do pomiaru wspotrzadnych. Pzglad Geodenija, *1964.*

Kochetov, F. G.: Koordinatniy teodolit Tk-G. Geodeziya i Kartografiya, No. 5, *1973.*

Kochetov, F. G.: Novoye v tyehnyike i metogyike inzhenyernogeodezicheskih rabot na stroityelnoy ploschadke. Rotaprint obchestva Znanyiye MSSR, *1969.*

Kondraschkow, A. W.: Elektrooptische Entfernungsmessung. VEB Verlag für Bauwesen, Berlin, *1961.*

König, A.—Köhler, H.: Die Fernrohre und Entfernungsmesser. Springer-Verlag, Berlin–Göttingen–Heidelberg, *1959.* (3rd ed.)

Kruger, M.: Spravochnyik konstruktora optikomechanyicheskih priborov. Izd. Mashinostroenie, Leningrad, *1968.*

Laurila, H. S.: Electronic Surveying and Navigation. J. Wiley and Sons, New York, *1976.*

Malcev, M. D.—Karakulina, G. A.: Prikladnaya optika i opticheskiye izmerenyiya. Moscow, *1968.*

Meisenheimer, D.: Geodätische Instrumente und Geräte. Verlag Chmielorz, Wiesbaden, *1985.*

Meissner, O.: Sehr genaue Schlauchwaagemessungen und praktische Anwendungen. Soproni Bánya- és Földm. Karok Közl., *1956.*

Müller, P.—Löbel, G.—Schmid, H.: Lexikon der Datenverarbeitung. Verlag Moderne Industrie Wolfgang Dummer & Co., München, *n.d.*
Ochsenhirt, H.: Zur Untersuchung von Feinnivellierlatten mit Invarband. Zeitschrift für Vermessungswesen, *1956.*
Oltay, K.—Rédey, I.: Geodézia (Geodesy). Tankönyvkiadó, Budapest, *1964.*
Prihoda, A. G.—Finkelstein, E. I.: Noviy opticheskiy mikrobarometr OMB-3P i perspektivi yevo primeneniya. Geodeziya i Kartografiya, *1971.*
Prihoda, A. G.—Finkelstein, E. I.: Barometricheskoe nivelirovanie. Moscow, *1972.*
Riekher, R.: Fernrohre und ihre Meister. VEB Verlag Technik, Berlin, *1957.*
Rola, F.: Sposoby eliminovania wpiwu temperatury w nivelacji hydrostatycznej. Geodezja i Kartografia, Warsaw, *1965.*
Rola, F.: Zastosowanie reperow przenosnych w nivelacji hydrostatycznej. Geodezja i Kartografia, *1962.*
Rusinov, M. M.: Technicheskaya optika. Maschgiz., Leningrad, *1961.*
Scheel, G.: Systematische Fehler des hydrostatischen Nivellements und Verfahren zu ihrer Ausschaltung. Veröffentlichungen der Deutschen Geodätischen Komission. Reihe B, Heft 27, Frankfurt, *1956.*
Schlemmer, H.: Aktuelle Instrumentenkunde. Verlag Wichmann, Karlsruhe, *1982.*
Seeber, G.: Inertiale Vermessungssysteme und ihre Anwendungsmöglichkeiten in der Geodäsie. Zeitschrift für Vermessungswesen, *1979.*
Silar, J.: Souradnicovy theodolit MOM Te-K1. Geodeticky a Kartograficky obzor, *1965.*
Smith, W. J.: Modern optical engineering. New York, *1966.*
Sneddon, I.: Hydrostatic Levelling. A Review, The Australian Surveyor, *1972.*
Soltau, G.: Ein Beitrag zur Untersuchung von geodätischen Instrumenten mit automatisch einspielenden Höhenableseeinrichtungen. Wiss. Arb. d. Ist. für Verm.-Wesen an der TH. Hannover, No. 28, *1968.*
Stolitzka, G.: Untersuchung der Ziellinienänderung eines Nivelliers beim Umfokussieren. Österr. Zeitschrift für Verm., *1968.*
Szymonski, J.: Instrumentoznawstwo geodezyjne. (2nd ed.) Warsaw, I—III., *1968—1972.*
Thierbach, H.—Barth, W.: Die automatische Präzisionsschlauchwaage "ELWAAG". Zeitschrift für Vermessungswesen, *1979.*
Torochkov, V. I.: Giroteodoliti. Nedra, Moscow, *1969.*
Voronkov, N. V.: Giroskopicheskoe orientirovanie. Moscow, Voenizdat, *1968.*
Wittke, H.: Einführung in die Vermessungstechnik. (4th ed.) Ferd. Dümmler's Verlag, Bonn, *1971.*
Yeliseyev, S. V.: Geodezicheskiye instrumenti i pribori. Geodezizdat, Moscow, *1973.* (3rd ed.)
Zavrov, V. N.—Zhitonürskij, I. B.: Noviye markseyderskie girocompasi MV 2 i MT 1. Ugolj, *1965.*
Zetsche, H.: Elektronische Entfernungsmessung. Verlag K. Wittwer, Stuttgart, *1979.*
Zwickert, E.: Konzepte registrierender Vermessungsinstrumente (1959–1979). Zeitschrift für Vermessungswesen, *1980.*

Name and subject index